Che[illegible] Analysis and Design Fundamentals

2nd Edition

Chemical Reactor Analysis and Design Fundamentals

2nd Edition

James B. Rawlings
Department of Chemical Engineering
University of California, Santa Barbara
Santa Barbara, California

John G. Ekerdt
Department of Chemical Engineering
The University of Texas
Austin, Texas

Santa Barbara, California

This book was set in Lucida using LaTeX, and printed and bound by Worzalla.

Cover design by Cheryl M. and James B. Rawlings, and John W. Eaton.

Nob Hill Publishing, LLC
Cheryl M. Rawlings, publisher
Santa Barbara, CA 93101
orders@nobhillpublishing.com
http://www.nobhillpublishing.com

Library of Congress Control Number: 2020940698

Printed in the United States of America.

First **Edition**
  First Printing   March 2002
  Second Printing   December 2004
Second **Edition**
  First Printing   August 2012
  Second Printing   November 2013
Paperback **Edition**
  Third Printing   July 2020

To Cheryl, Melanie and Blake,

and

Carol, Alice and Barbara,

for their love and encouragement.

Preface to the Second Edition

During the decade that elapsed since the first printing of the text, several trends have become clear. First, the decision to emphasize fundamental concepts and rely on convenient, high-level computational languages to deduce consequences of these fundamentals has been broadly validated. If anything, this trend has only accelerated since the printing of the first edition. The authors therefore decided to make the computational appendix available on the web and removed it from the printed text, enabling more rapid updating of the material as computational languages such as Octave and MATLAB evolve. This reduction in total printed pages allowed addition of new material while maintaining a reasonable length. See www.nobhillpublishing.com for the computational appendix.

A second trend that has continued unabated during the last decade is the increasingly prominent role of manufacturing solid or particulate products in the chemical process industries. No general chemical engineering reactor design textbook provides students with the tools to describe, analyze, or design reactors for this important class of materials. To address this need, we have added a Chapter 10, covering particulate reactors and the population balances required to describe particulate products having a particle size distribution. Since the manufacture of value-added products using biological cells is a prime example of this class of systems, this new chapter is also a logical place to discuss the balances required to describe and analyze bioreactors.

Finally, the trend to use discrete, stochastic models and stochastic simulation to augment the core continuous, deterministic models of classical chemical reaction engineering has also only increased during the last decade. In some fields, such as systems biology, the use of stochastic simulation has become a dominant method for system analysis and design. The fundamental concepts of stochastic kinetics were already introduced in Chapter 4 of the first edition as a way to model and understand reaction kinetics at the small scale. Stochastic simulation is again found highly relevant in describing particulate reactors, which often operate in regimes of small particle number where understanding and quantifying sources of variability become important.

In addition to their widely accepted place in graduate education, we look to a time in the near future when familiarity with the fundamental principles of particulate systems and population balances will also be considered an indispensable component of the education of *undergraduate* chemical and biological engineers.

JBR
Madison, Wisconsin

JGE
Austin, Texas

Added for the second edition, third printing

The second edition, third printing, is available as a paperback to reduce the cost to the students.

Preface

Chemical reactors are at the core of the chemical engineering discipline, and chemical reactor analysis and design is one of the distinguishing courses that clearly separates the chemical engineers from the other engineering professionals. Given that chemical reactor analysis and design is a mature and stable topic in the curriculum of chemical engineering, however, it is natural to ask what is the motivation for a new text on this topic.

We offer our motivations here. This book grew out of the combined experience of the two authors teaching this subject to undergraduates for more than 30 years. Given the rapidly changing landscape of scale and type of reactors of interest to practicing chemical engineers (chemical vapor deposition reactors, pharmaceutical fermentors, micro-reactors, as well as traditional catalytic crackers, bulk polymerization reactors, etc.), it seems unwise to emphasize one industrial sector and treat its reactor types in detail. Practicing chemical engineers work in a broad array of industrial sectors, and many will change sectors during their careers. If chemical engineering has any important distinguishing characteristic, it is a set of fundamentals that apply to all scales and all types of reaction and transport processes involving chemical change.

This book is all about reactor fundamentals. Rather than presenting many facts about reactors, we focus on the framework for how to think about reactors—a framework for thinking that enables one, with some experience, to establish any of these facts for oneself, and discover new facts given new situations. All engineering and science textbooks do this to some extent; in this text, we will do it to a rather large extent.

Computations matter in this subject. Reactor fundamentals, like the fundamentals in any subject, are few in number. But the diversity of the consequences of these fundamentals is enormous. Computational approaches provide a powerful and general approach to systematically investigating these consequences without making unrealistic simplifying assumptions. We attempt to exploit the significant advances in computing algorithms, software, and hardware in order to revise and streamline the presentation of reactor fundamentals. We focus on two high-level languages intended for numerical computation, Octave and

MATLAB. Octave is freely available for a variety of hardware platforms and can be downloaded from www.octave.org. MATLAB is commercially available from The MathWorks, Inc., and is becoming a commonly available tool of industrial engineering practice. These languages allow us to focus on essentials and ignore programming details, which is the goal of any "high-level" language.

Students should not feel compelled to recall the detailed information in the figures, but recall only the concepts, principles and main results. Students will have the computational tools to recreate the figures in this or any other textbook on this subject. For example, all calculations required for the figures in this text were performed with Octave. Students should not feel compelled to memorize design equations for reactor types. The goal is to develop sufficient expertise so that students can set up appropriate models from the basic principles for each new problem they encounter. That is the time-tested way to instill confidence that one can analyze a new situation, which we fully expect to be the experience of practicing engineers.

Newly practicing engineers certainly will need to learn the economics of manufacturing their main products, main reactor configurations, detailed energy-recovery schemes, contacting patterns, new catalysts, and reactor monitoring and control systems. Rather than make inadequate efforts to include all of this information as a survey during the reactor design course, we feel this material deserves separate coverage, or can be learned during the early years of engineering practice. We do not regard it as poor preparation that some of this information is not included in university curricula; rather, we regard learning this material on the job as part of the maturation process of the practicing professional chemical engineer.

Finally, we hope the text conveys some of the excitement that chemical engineers feel for chemical reactions and chemical reactors. Understanding the fundamentals prepares you to investigate new situations. Grounding in the fundamentals enables you to explore with confidence and be creative. Enjoy.

JBR
Madison, Wisconsin

JGE
Austin, Texas

Acknowledgments

We benefited greatly from the help and advice of many colleagues during the preparation of this text. Taking Warren Stewart's graduate reactor modeling course greatly influenced the first author's way of thinking about this subject. More recently, Warren also provided many helpful comments on computational issues involving reaction-diffusion, and parameter estimation. Harmon Ray contributed many insightful and helpful comments on a variety of topics, especially on issues of reactor mixing and industrial polymerization reactors. Bob Bird provided many helpful comments on presenting the material related to transport phenomena. He also provided detailed feedback on a review copy of the text.

Marty Feinberg read an early version of the notes, and provided detailed feedback on many issues including reaction rates and presenting computational material. Yannis Kevrekidis used a preliminary version of the notes as part of a graduate reactor modeling course at Princeton. His feedback and encouragement are very much appreciated. John Falconer also provided many helpful comments.

Tom Edgar provided much-needed departmental support and encouragement when the authors began writing. Roger Bonnecaze, Buddie Mullins and C. D. Rao used early versions of the notes and Octave software in teaching the undergraduate course at The University of Texas. Their feedback was very helpful.

Several graduate students contributed to the software, which is an essential part of this text. Ken Muske developed some examples for the chapter on mixing. John Campbell worked on programs for managing thermochemical data, Sankash Venkatash contributed programs on parameter estimation and reaction-diffusion, and Dominik Nagel contributed programs to calculate potential-energy surfaces. Chris Rao provided helpful feedback on mixing issues, and Matt Tenny and Eric Haseltine provided helpful input on the cover design. Eric also helped prepare the solution manual. We have benefited from the feedback of many classes of undergraduates both at the University of Wisconsin and The University of Texas. We appreciate the enthusiasm with which they received some very early and incomplete drafts of the notes.

Our publisher deserves special recognition for providing enough flexibility so that we could realize our original vision, while still insisting that we actually finish the project.

Finally, we would like to acknowledge the essential contributions of John Eaton. When the two authors first outlined the goals of this writing project, we envisioned creating in parallel a convenient computational environment. The original goal of the computational environment was to enable students to quickly and conveniently define and solve new reactor design problems as a means to gain deeper understanding and appreciation of chemical reactor fundamentals. Starting from this simple idea, John Eaton created the Octave software project. We could not have written this text in the way we planned without Octave. Every figure in the text requiring calculation was prepared with Octave. John contributed in a myriad of other ways. He designed the typesetting macros and the text version control system. We maintained more than 10 versions of the source files on more than 7 different computing systems at various stages of production. With John's expert guidance, we easily managed this complexity; in fact, the computing tools worked so well, that the computational aspects of the book production were enjoyable.

Added for the second edition. The authors would like to acknowledge Dave Anderson of the UW Department of Mathematics for helpful discussion of stochastic particle growth models, and John Yin of the UW Department of Chemical and Biological Engineering for providing pictures of viral infections of cells.

Contents

1 Setting the Stage **1**

1.1 Introduction 1
1.2 Some Classifications and Terminology 2
1.3 Scale 7
1.4 Some Examples 10
1.4.1 Chemical Vapor Deposition 10
1.4.2 Hydrodesulfurization 12
1.4.3 Olefin Polymerization 15
1.4.4 Hepatitis B Virus Modeling 18
1.5 An Overview of the Text 20

2 The Stoichiometry of Reactions **27**

2.1 Introduction 27
2.2 Chemical Reactions and Stoichiometry 28
2.3 Independent Reactions 35
2.4 Reaction Rates and Production Rates 41
2.5 Computational Aspects of Stoichiometry 46
2.5.1 Computing Production Rates from Reaction Rates 47
2.5.2 Computing Reaction Rates from Production Rates 48
2.5.3 Measurement Errors and Least-Squares Estimation 49
2.6 Summary and Notation 53
2.7 Exercises 55

3 Review of Chemical Equilibrium **63**

3.1 Introduction 63
3.2 Condition for Reaction Equilibrium 64
3.2.1 Evaluation of the Gibbs Energy Change of Reaction 75
3.2.2 Temperature Dependence of the Gibbs Energy . 76
3.3 Condition for Phase Equilibrium 79
3.3.1 Ideal Mixtures 80
3.3.2 Nonideal Mixtures 82
3.4 Heterogeneous Reactions 82
3.5 Multiple Reactions 88

3.5.1 Algebraic Approach 88
3.5.2 Optimization Approach 92
3.6 Summary and Notation 94
3.7 Exercises 98

4 The Material Balance for Chemical Reactors 109
4.1 General Mole Balance 109
4.2 The Batch Reactor 110
4.2.1 Analytical Solutions for Simple Rate Laws 111
4.3 The Continuous-Stirred-Tank Reactor (CSTR) 127
4.4 The Semi-Batch Reactor 131
4.5 Volume Change Upon Reaction 132
4.6 The Plug-Flow Reactor (PFR) 144
4.6.1 Thin-Disk Volume Element 144
4.6.2 Steady-State Operation 145
4.6.3 Volumetric Flowrate for Gas-Phase Reactions . . 146
4.6.4 Volumetric Flowrate for Liquid-Phase Reactions 146
4.6.5 Single Reaction Systems 147
4.6.6 Multiple-Reaction Systems 150
4.7 Some PFR-CSTR Comparisons 156
4.8 Stochastic Simulation of Chemical Reactions 162
4.9 Summary and Notation 171
4.10 Exercises 174

5 Chemical Kinetics 189
5.1 Introduction 189
5.2 Elementary Reaction 190
5.3 Elementary Reaction Kinetics 195
5.4 Fast and Slow Time Scales 207
5.4.1 The Reaction Equilibrium Assumption 209
5.4.2 The Quasi-Steady-State Assumption 218
5.5 Rate Expressions 226
5.6 Reactions at Surfaces 237
5.7 Summary and Notation 255
5.8 Exercises 258

6 The Energy Balance for Chemical Reactors 273
6.1 General Energy Balance 273
6.1.1 Work Term 274
6.1.2 Energy Terms 275
6.2 The Batch Reactor 275

6.3 The CSTR 284
6.3.1 Steady-State Multiplicity 288
6.3.2 Stability of the Steady State 293
6.3.3 Sustained Oscillations, Limit Cycles 303
6.4 The Semi-Batch Reactor 314
6.5 The Plug-Flow Reactor 315
6.5.1 Plug-Flow Reactor Hot Spot and Runaway 320
6.5.2 The Autothermal Plug-Flow Reactor 324
6.6 Summary and Notation 330
6.7 Exercises 336

7 Fixed-Bed Catalytic Reactors **351**
7.1 Introduction 351
7.2 Catalyst Properties 353
7.2.1 Effective Diffusivity 356
7.3 The General Balances in the Catalyst Particle 359
7.4 Single Reaction in an Isothermal Particle 361
7.4.1 First-Order Reaction in a Spherical Particle . . . 362
7.4.2 Other Catalyst Shapes: Cylinders and Slabs . . . 370
7.4.3 Other Reaction Orders 373
7.4.4 Hougen-Watson Kinetics 376
7.4.5 External Mass Transfer 381
7.4.6 Observed versus Intrinsic Kinetic Parameters . . 386
7.5 Nonisothermal Particle Considerations 388
7.6 Multiple Reactions 391
7.7 Fixed-Bed Reactor Design 396
7.7.1 Coupling the Catalyst and Fluid 396
7.7.2 Logarithmic Transformation 412
7.8 Summary and Notation 413
7.9 Exercises 416

8 Mixing in Chemical Reactors **429**
8.1 Introduction 429
8.2 Residence-Time Distribution 430
8.2.1 Definition 430
8.2.2 Measuring the RTD 431
8.2.3 Continuous-Stirred-Tank Reactor (CSTR) 435
8.2.4 Plug-Flow Reactor (PFR) and Batch Reactor 438
8.2.5 CSTRs in Series 440
8.2.6 Dispersed Plug Flow 442
8.3 Limits of Reactor Mixing 449

8.3.1 Complete Segregation 451
8.3.2 Maximum Mixedness 454
8.3.3 Mass Transfer and Limits of Reactor Mixing . . . 460
8.4 Limits of Reactor Performance 463
8.4.1 A Single Convex (Concave) Reaction Rate 463
8.4.2 The General Case 469
8.5 Examples in Which Mixing is Critical 474
8.6 Summary and Notation 482
8.7 Exercises 486

9 Parameter Estimation for Reactor Models **497**
9.1 Experimental Methods 497
9.1.1 Analytical Probes for Concentration 497
9.1.2 Experimental Reactors for Kinetics Studies . . . 502
9.1.3 Characterizing Catalysts and Surfaces 506
9.2 Data Modeling and Analysis 509
9.2.1 Review of the Normal Distribution 510
9.2.2 Eigenvalues and Eigenvectors 511
9.2.3 Least-Squares Estimation 514
9.2.4 Least Squares with Unknown Variance 520
9.2.5 Nonlinear Least Squares 523
9.2.6 Design of Experiments 528
9.2.7 Parameter Estimation with Differential-Equation Models 534
9.3 An Industrial Case Study 547
9.4 Summary and Notation 557
9.5 Exercises 561

10 Particulate Reactors **573**
10.1 Particle Size Distributions 573
10.2 Applications 574
10.2.1 Crystallization 574
10.2.2 Emulsion Polymerization 577
10.2.3 Biological Cells 579
10.3 Population Balance 580
10.3.1 Single Size Coordinate 581
10.3.2 Boundary Conditions 582
10.3.3 Multiple Internal Coordinates 586
10.4 Multiphase Mass and Energy Balances 588
10.5 Nonsegregated Fermentation Model 594
10.6 Stochastic Models of Nucleation and Growth 599

10.6.1 Modeling Particle Growth and Dissolution 599
10.7 Stochastic Simulation and Deterministic Population Balances 602
10.7.1 Linear rate processes. 603
10.7.2 Thermodynamic limit. 606
10.8 Summary and Notation 608
10.9 Exercises 612

Index **622**

A Computational Methods **634**
A.1 Linear Algebra and Least Squares 634
A.2 Nonlinear Algebraic Equations and Optimization 637
A.2.1 Functions (`function`) 637
A.2.2 Nonlinear Algebraic Equations (`fsolve`) 638
A.2.3 Nonlinear Optimization (`fmincon`) 639
A.3 Differential Equations 641
A.3.1 Ordinary Differential Equations (`lsode`) 642
A.3.2 Octave-MATLAB Compatibility 643
A.3.3 Differential-Algebraic Equations (`daspk`) 645
A.3.4 Automatic Stopping Times (`dasrt`) 647
A.4 Sensitivities of Differential Equations (`paresto`) 648
A.5 Boundary-Value Problems and Collocation (`colloc`) . . 652
A.6 Parameter Estimation with ODE Models (`paresto`) . . . 657
A.7 Exercises 661

List of Figures

1.1 Schematic diagram of a batch reactor. 4
1.2 Expanded view of the internals of a batch reactor. 4
1.3 Schematic diagram of a CSTR. 5
1.4 Schematic diagram of five CSTRs in series. 6
1.5 Schematic diagram of a plug-flow reactor. 6
1.6 Cross-sectional view of a 1 μl combinatorial screening reactor. 7
1.7 Assembled view of 256 1-μL reactors built into a 3-in by 3-in support wafer. 8
1.8 Top view of a combinatorial reactor gas-distribution network. 9
1.9 Polyethylene reactor. 9
1.10 Top view of a single-wafer CVD reactor. 11
1.11 Representative organosulfur compounds. 13
1.12 Hydrodesulfurization process. 14
1.13 Cross-sectional view of a commercial HDS reactor. . . . 15
1.14 Expanded view of a multibed HDS reactor. 16
1.15 Simplified polymerization process. 17
1.16 The chemical events comprising the reproduction cycle of the hepatitis B virus. 19

2.1 Defining the reaction rate, r, for the SiH_2 reaction. . . . 42
2.2 Estimated reaction rates from six production-rate measurements subject to measurement noise. 52
2.3 Estimated reaction rates from 500 production-rate measurements subject to measurement noise. 52

3.1 Thermodynamic system. 63
3.2 Gibbs energy versus reaction extent at constant T and P. 65
3.3 Gibbs energy $\widetilde{G}$ versus reaction extent ε'. 75
3.4 Partial pressures of components A and C versus liquid-phase composition in a nonideal solution. 84
3.5 Gibbs energy versus two reaction extents at constant T and P. 89

3.6 Gibbs energy contours for the pentane reactions as a function of the two reaction extents. 95
3.7 Two systems with identical intensive properties and different extensive properties. 100
3.8 Vessel with semipermeable membrane and chemical reaction $2A \rightleftharpoons B$. 102

4.1 Reactor volume element. 109
4.2 Batch reactor volume element. 111
4.3 First-order, irreversible kinetics in a batch reactor. . . . 113
4.4 First-order, irreversible kinetics in a batch reactor, log scale. 113
4.5 First-order, reversible kinetics in a batch reactor, $k_1 = 1$, $k_{-1} = 0.5$, $c_{A0} = 1$, $c_{B0} = 0$. 116
4.6 Second-order and first-order kinetics in a batch reactor; for second-order, $kc_{A0} = 1$, and for first order, $k = 1$, so the rates are equal initially. 118
4.7 Reaction rate versus concentration for nth-order kinetics, $r = kc_A^n$, $n \geq 0$, $k = 1$ for all orders. 120
4.8 Batch reactor with nth-order kinetics, $r = kc_A^n$, $k_0 = kc_{A0}^{n-1} = 1$, $n \geq 1$. 122
4.9 Batch reactor with nth-order kinetics, $r = kc_A^n$, $k_0 = kc_{A0}^{n-1} = 1$, $n < 1$. 122
4.10 Reaction rate versus concentration for nth-order kinetics, $r = kc_A^n$, $n \leq 0$, $k = 1$ for all orders. 123
4.11 Two first-order reactions in series in a batch reactor. . . 125
4.12 Two first-order reactions in parallel in a batch reactor. . 126
4.13 CSTR volume element. 127
4.14 Reaching steady state in a CSTR. 130
4.15 Semi-batch reactor volume for different monomer addition policies. 142
4.16 Semi-batch reactor feed flowrate for different monomer addition policies. 142
4.17 Semi-batch reactor monomer content for different monomer addition policies. 143
4.18 Semi-batch reactor polymer content for different monomer addition policies. 143
4.19 Plug-flow reactor volume element. 145
4.20 Benzene conversion versus reactor volume. 152
4.21 Component mole fractions versus reactor volume. . . . 152

4.22 Molar flowrate of ethane, ethylene and NO versus reactor volume for ethane pyrolysis example. 157
4.23 Molar flowrate of ethane versus reactor volume for inlet temperatures of 1000, 1050 and 1100 K. 157
4.24 The PFR is smaller than the CSTR for irreversible, nth-order kinetics, positive order, $n > 0$. 158
4.25 The CSTR is smaller than the PFR for irreversible, nth-order kinetics, negative order, $n < 0$. 159
4.26 PFR versus CSTR with recycle and separation. 160
4.27 Overall and per-pass conversion of A as a function of fractional recycle, α. 162
4.28 Choosing a reaction with appropriate probability. 165
4.29 Stochastic simulation of the first-order reactions $A \longrightarrow B \longrightarrow C$ starting with 100 A molecules. 166
4.30 Stochastic simulation of the first-order reactions $A \longrightarrow B \longrightarrow C$ starting with 1000 A molecules. 167
4.31 Stochastic simulation of the first-order reactions $A \longrightarrow B \longrightarrow C$ starting with 4000 A molecules. 167
4.32 Species cccDNA versus time for hepatitis B virus model; deterministic and average stochastic models. 169
4.33 Species rcDNA versus time for hepatitis B virus model; deterministic and average stochastic models. 169
4.34 Envelope versus time for hepatitis B virus model; deterministic and average stochastic models. 169
4.35 Species cccDNA versus time for hepatitis B virus model; two representative stochastic trajectories. 170
4.36 An alternative CSTR-separator design. 180
4.37 Deterministic simulation of reaction $A + B \rightleftharpoons C$ compared to stochastic simulation. 182
4.38 Two 2-CSTR reactor configurations with bypass of reactant A in the second configuration. 184

5.1 Morse potential for H_2 and HF. 196
5.2 Potential-energy surface for the F, H_2 reaction. 197
5.3 Contour representation of the potential-energy surface. 198
5.4 Reaction-coordinate diagram. 199
5.5 Comparison of measured and calculated rate constant versus temperature for trioxane decomposition. 208
5.6 Full model solution for $k_1 = 1, k_{-1} = 0.5, k_2 = k_{-2} = 1$. . 210
5.7 Full model solution for $k_1 = 1, k_{-1} = 0.5, k_2 = k_{-2} = 10$. 210

5.8 Concentrations of B and C versus time for full model with increasing k_2 with $K_2 = k_2/k_{-2} = 1$. 211
5.9 Comparison of equilibrium assumption to full model for $k_1 = 1, k_{-1} = 0.5, k_2 = k_{-2} = 10$. 214
5.10 Normalized concentration of C versus dimensionless time for the series reaction $A \to B \to C$. 221
5.11 Fractional error in the QSSA concentration of C for the series reaction $A \to B \to C$. 224
5.12 Fractional error in the QSSA concentration of C versus dimensionless time for the series-parallel reaction, $A \rightleftharpoons B \to C$. 225
5.13 Molar flowrates of C_2H_6, C_2H_4 and CH_4 corresponding to the exact solution. 234
5.14 Fractional error in the QSSA molar flowrate of C_2H_4 versus reactor volume. 236
5.15 Comparison of the molar flowrates of C_2H_6 and C_2H_4 for the exact solution (solid lines) and the simplified kinetic scheme (dashed lines). 237
5.16 Surface and second-layer (dashed) atom arrangements for several low-index surfaces. 238
5.17 Schematic representation of the adsorption/desorption process. 241
5.18 Fractional coverage versus adsorbate concentration for different values of the adsorption constant, K. 243
5.19 Langmuir isotherm for CO uptake on Ru. 245
5.20 Linear form of Langmuir isotherm for CO uptake on Ru. 245
5.21 Two packing arrangements for a coverage of $\theta = 1/2$. . 247
5.22 Five possible arrangements around a single site. 248
5.23 Dimensionless CO_2 production rate versus dimensionless gas-phase CO and O_2 concentrations. 252
5.24 Dimensionless CO_2 production rate versus a single dimensionless gas-phase concentration. 253
5.25 Adsorbed oxygen concentration versus gas-phase oxygen concentration. 263

6.1 Reactor volume element. 273
6.2 Flow streams entering and leaving the volume element. 275
6.3 CSTR and heat exchangers. 286
6.4 Conversion of A versus reactor temperature. 288

6.5 Steady-state conversion versus residence time for different values of the heat of reaction. 291
6.6 Steady-state temperature versus residence time for different values of the heat of reaction. 291
6.7 Steady-state conversion versus residence time for $\Delta H_R = -3 \times 10^5$ kJ/kmol; ignition and extinction points. 292
6.8 Steady-state temperature versus residence time for $\Delta H_R = -3 \times 10^5$ kJ/kmol; ignition and extinction points. 292
6.9 Steady-state temperature versus residence time for $\Delta H_R = -3 \times 10^5$ kJ/kmol. 294
6.10 Rates of heat generation and removal for $\tau = 1.79$ min. 296
6.11 Rates of heat generation and removal for $\tau = 15$ min. . 297
6.12 Rates of heat generation and removal for $\tau = 30.9$ min. 297
6.13 Eigenvalues of Jacobian matrix vs. reactor temperature. 301
6.14 Eigenvalues of Jacobian matrix vs. reactor conversion. . 301
6.15 Marble on a track in a gravitational field; point A is the unique, stable steady state. 302
6.16 Marble on a track with three steady states; points A and C are stable, and point B is unstable. 302
6.17 Marble on a track with an ignition point (A) and a stable steady state (C). 302
6.18 Steady-state conversion versus residence time. 304
6.19 Steady-state temperature versus residence time. 304
6.20 Steady-state conversion vs. residence time — log scale. 305
6.21 Steady-state temperature vs. residence time — log scale. 305
6.22 Eigenvalues of the Jacobian matrix versus reactor conversion in the region of steady-state multiplicity. 307
6.23 Eigenvalues of the Jacobian matrix versus reactor conversion in the region of steady-state multiplicity. 307
6.24 Real and imaginary parts of the eigenvalues of the Jacobian matrix near points C–F. 308
6.25 Conversion and temperature vs. time for $\tau = 35$ min. . 309
6.26 Phase portrait of conversion versus temperature for feed initial condition; $\tau = 35$ min. 310
6.27 Phase portrait of conversion versus temperature for several initial conditions; $\tau = 35$ min. 310
6.28 Conversion and temperature vs. time for $\tau = 30$ min. . 311
6.29 Conversion and temperature vs. time for $\tau = 72.3$ min. 311
6.30 Phase portrait of conversion versus temperature at showing stable and unstable limit cycles. 312

6.31 Marble on a track with three steady states; point C is unstable and is surrounded by limit cycles. 313
6.32 Plug-flow reactor volume element. 315
6.33 Tubular reactors with interstage cooling. 318
6.34 Temperatures and molar flows for tubular reactors with interstage cooling. 320
6.35 Molar flow of o-xylene versus reactor length for different feed temperatures. 323
6.36 Reactor temperature versus length for different feed temperatures. 323
6.37 Autothermal plug-flow reactor; the heat released by the exothermic reaction is used to preheat the feed. 324
6.38 Coolant temperature at reactor outlet versus temperature at reactor inlet; three steady-state solutions. 328
6.39 Reactor and coolant temperature profiles versus reactor length. 329
6.40 Ammonia mole fraction versus reactor length. 329
6.41 Coolant temperature at reactor outlet versus temperature at reactor inlet. 346
6.42 Volume element moving with the PFR fluid velocity. . . 346

7.1 Expanded views of a fixed-bed reactor. 353
7.2 Volume element in a stationary solid containing energy density E and molar concentrations c_j, with energy flux $\boldsymbol{e}$ and mass fluxes $\boldsymbol{N}_j$. 360
7.3 Hyperbolic trigonometric functions sinh, cosh, tanh. . . 365
7.4 Dimensionless concentration versus dimensionless radial position for different values of the Thiele modulus. 365
7.5 Effectiveness factor versus Thiele modulus for a first-order reaction in a sphere. 368
7.6 Effectiveness factor versus Thiele modulus for a first-order reaction in a sphere (log-log scale). 368
7.7 Characteristic length a for sphere, semi-infinite cylinder and semi-infinite slab. 370
7.8 Effectiveness factor versus Thiele modulus for the sphere, cylinder and slab. 372
7.9 Effectiveness factor versus Thiele modulus in a spherical pellet; reaction orders greater than unity. 374
7.10 Effectiveness factor versus Thiele modulus in a spherical pellet; reaction orders less than unity. 374

7.11 Dimensionless concentration versus radius for zero-order reaction in a spherical pellet. 375
7.12 Dimensionless concentration $\overline{c}$ versus z for large $\tilde{\Phi}$; consider the function $z(\overline{c})$ in place of $\overline{c}(z)$. 378
7.13 Effectiveness factor versus an inappropriate Thiele modulus in a slab; Hougen-Watson kinetics. 380
7.14 Effectiveness factor versus appropriate Thiele modulus in a slab; Hougen-Watson kinetics. 380
7.15 Effect of external mass transfer on pellet surface concentration. 381
7.16 Dimensionless concentration versus radius for different values of the Biot number; first-order reaction in a spherical pellet with $\Phi = 1$. 383
7.17 Effectiveness factor versus Thiele modulus for different values of the Biot number; first-order reaction in a spherical pellet. 384
7.18 Asymptotic behavior of the effectiveness factor versus Thiele modulus; first-order reaction in spherical pellet. 385
7.19 Effectiveness factor versus normalized Thiele modulus for a first-order reaction in nonisothermal spherical pellet. 390
7.20 Dimensionless concentration versus radius for the nonisothermal spherical pellet: lower (A), unstable middle (B), and upper (C) steady states. 392
7.21 Dimensionless temperature versus radius for the nonisothermal spherical pellet: lower (A), unstable middle (B), and upper (C) steady states. 392
7.22 Concentration profiles of reactants; fluid concentration of O_2 (×), CO (+), C_3H_6 (∗). 395
7.23 Concentration profiles of reactants (log scale); fluid concentration of O_2 (×), CO (+), C_3H_6 (∗). 395
7.24 Concentration profiles of products. 397
7.25 Fixed-bed reactor volume element containing fluid and catalyst particles; the equations show the coupling between the catalyst particle and the fluid. 400
7.26 Molar flow of A versus reactor volume for second-order, isothermal reaction in a fixed-bed reactor. 404
7.27 Molar concentrations versus reactor volume. 407
7.28 Dimensionless equilibrium constant and Thiele modulus versus reactor volume. 407

7.29 Fluid molar concentrations versus reactor volume. . . . 410
7.30 Fluid temperature and pressure versus reactor volume. 410
7.31 Reactor positions for pellet profiles. 411
7.32 Pellet CO profiles at several reactor positions. 411
7.33 Catalyst pellet with slab geometry. 417
7.34 Effectiveness factor versus Thiele modulus for different values of the Biot number; second-order reaction in a cylindrical pellet. 424

8.1 Arbitrary reactor with steady flow profile. 430
8.2 Family of pulses and the impulse or delta function. . . . 433
8.3 CSTR and tracer experiment. 435
8.4 CSTR and volume elements in a game of chance. 436
8.5 CSTR residence-time distribution. 437
8.6 Plug-flow reactor with moving front of tracer. 439
8.7 Series of n equal-sized CSTRs with residence time τ/n. 440
8.8 RTD $p(\theta)$ versus θ for n CSTRs in series, $\tau = 2$. 441
8.9 $P(\theta)$ versus θ for n CSTRs in series, $\tau = 2$. 442
8.10 $P(\theta)$ versus θ for plug flow with dispersion number D, $\tau = 2$. 445
8.11 Residence-time distribution $p(\theta)$ versus θ for plug flow with dispersion number D, $\tau = 2$. 445
8.12 Start-up of the tubular reactor; $c_A(t, z)$ versus z for various times, $0 \leq t \leq 2.5$ min, $\Delta t = 0.25$ min. 447
8.13 PFR followed by CSTR (A), and CSTR followed by PFR (B). 448
8.14 Comparison of the effluent concentrations for the two cases shown in Figure 8.13. 450
8.15 Completely segregated flow as a plug-flow reactor with side exits; outlet flows adjusted to achieve given RTD. . 451
8.16 Alternate representation of completely segregated flow (A), maximum mixed flow (B), and an intermediate mixing pattern (C). 452
8.17 Maximum mixed flow as a plug-flow reactor with side entrances; inlet flows adjusted to achieve a given RTD. . 454
8.18 Volume element in the state of maximum mixedness. . 455
8.19 Dimensionless effluent concentration versus dimensionless rate constant for second-order reaction. 459
8.20 CSTR followed by PFR (A) and PFR followed by CSTR (B) as examples of complete and partial mixing. 460
8.21 Adding two liquid-phase feed streams to a stirred tank. 461

8.22 Total concentration of A in the reactor effluent versus particle size. 464
8.23 Particle concentrations of A and B versus particle age for three different-sized particles. 464
8.24 Differentiable convex and concave functions. 465
8.25 Two volume elements before and after mixing. 465
8.26 Convex rate expression and the effect of mixing; rate of the mean (r_m) is less than the mean of the rate (r_s). . . . 466
8.27 Two tubes before and after mixing the entering feed. . . 467
8.28 Mean segregated and mixed concentrations versus θ; curves crossing at θ_1 is a contradiction. 468
8.29 Reaction rate versus concentration of limiting reactant; rate expression is neither convex nor concave. 470
8.30 Inverse of reaction rate versus concentration; optimal sequence to achieve 95% conversion is PFR–CSTR–PFR. . 471
8.31 RTD for the optimal reactor configuration. 472
8.32 Imperfect mixing (top reactor) leads to formation of an A-rich zone, which is modeled as a small CSTR feeding a second CSTR (bottom two reactors). 475
8.33 Conversion of reactant for single, ideal CSTR, and as a function of internal flowrate in a 2-CSTR mixing model. 479
8.34 Yield of desired product C for single, ideal CSTR, and as a function of internal flowrate, $\rho = Q_r/Q_2$, in a 2-CSTR mixing model. 479
8.35 Step response for single, ideal CSTR, and 2-CSTR mixing model with $\rho = 0, 1$. 480
8.36 Conversion of reactant A versus reactor length for different dispersion numbers. 483
8.37 Yield of desired product B versus reactor length for different dispersion numbers. 483
8.38 Reactor configurations subjected to a step test in tracer concentration. 486
8.39 Tracer concentrations in the feed and effluent streams versus time. 487
8.40 Reactor before and after stirrer is started. 489
8.41 Effluent concentration versus time after unit step change in the first reactor. 491
8.42 Some convex functions; differentiability is not required. 492
8.43 Value of $\int_0^\infty f(\theta)p(\theta)d\theta$ as the function $p(\theta)$ varies. . . 493

9.1 Gas chromatograph schematic. 500
9.2 Mass spectrometer schematic. 501
9.3 Volumetric chemisorption apparatus. 507
9.4 Univariate normal with zero mean and unit variance. . . 511
9.5 Multivariate normal for $n_p = 2$. 512
9.6 The geometry of quadratic form $\boldsymbol{x}^T \boldsymbol{A}\boldsymbol{x} = b$. 513
9.7 Measured rate constant at several temperatures. 516
9.8 Transformed data set, $\ln k$ versus $1/T$. 517
9.9 Several replicate data sets, $\ln k$ versus $1/T$. 517
9.10 Distribution of estimated parameters. 518
9.11 Reducing parameter correlation by centering the data. . 519
9.12 Values of χ^2 and F versus the number of data points when estimating 2 and 5 parameters. 521
9.13 Parameter estimates with only 10 data points. 522
9.14 Confidence intervals with known (solid line) and unknown (dashed line) error variance. 523
9.15 Model fit to a single adsorption experiment. 526
9.16 Model fit to all adsorption experiments. 527
9.17 Drawing a line through two points under measurement error: points far apart (A); points close together (B). . . 529
9.18 Effect of next measurement temperature on parameter confidence intervals. 530
9.19 Uncertainty in activation energy E and rate constant $\ln k_m$ versus next measurement temperature. 532
9.20 Uncertainty in activation energy E and rate constant $\ln k_m$ versus number of replicated experiments. 533
9.21 Experimental measurement and best parameter fit for nth-order kinetic model, $r = kc_A^n$. 538
9.22 Monte Carlo evaluation of confidence intervals. 539
9.23 Species cccDNA versus time for hepatitis B virus model; initial guess and estimated parameters fit to data. 542
9.24 Species rcDNA versus time for hepatitis B virus model; initial guess and estimated parameters fit to data. 542
9.25 Envelope versus time for hepatitis B virus model; initial guess and estimated parameters fit to data. 542
9.26 Species cccDNA versus time for hepatitis B virus model. 546
9.27 Species rcDNA versus time for hepatitis B virus model. 546
9.28 Envelope versus time for hepatitis B virus model. 546
9.29 Semi-batch reactor addition of component B to starting material A. 549

9.30 Depiction of an LC curve for determining the concentration of intermediate C and product D. 550
9.31 Base addition rate and LC measurement versus time. . . 551
9.32 Comparison of data to model with optimal parameters. 552
9.33 Total amount of species A, C and D versus time. 553
9.34 Total amount of species B versus time. 553
9.35 Predictions of LC measurement for reduced model. . . . 556
9.36 Fit of LC measurement versus time for reduced model; early time measurements have been added. 558
9.37 Confidence intervals for reduced model without (dashed) and with (solid) redesigned experiment. 558
9.38 Batch-reactor data for Exercise 9.3. 562
9.39 Batch-reactor data for Exercise 9.4; 3 runs with different measurement error variance. 563
9.40 Batch-reactor data for Exercise 9.11. 566
9.41 A second experiment for Exercise 9.11. 566
9.42 Batch-reactor data for Exercise 9.12. 568
9.43 A second experiment for Exercise 9.12. 568

10.1 Growing glycine crystals displaying a distribution of sizes and shapes. 575
10.2 Laboratory crystallizer equipped with real-time imaging of growing crystals. 576
10.3 Growing pharmaceutical drug crystals with a needle-like particle shape. 577
10.4 Well-stirred emulsion polymerization reactor. 578
10.5 Cell infection by a virus. 579
10.6 Basic fermentation system. 580
10.7 Particle size distribution showing particles entering and leaving size class L to $L + \Delta L$. 581
10.8 Particle size distribution near $L = 0$. 583
10.9 Integrating over a small volume element near $L = 0$. . . 584
10.10 Volume element. 586
10.11 Reactor containing two well-mixed phases of matter. . . 588
10.12 The two steady-state biomass and substrate concentrations versus dilution rate; stable (solid), unstable (dashed). 598
10.13 Typical growth rate and dissolution rate as a function of solute concentration in the solution phase. 600

10.14 Tracking 10 particles undergoing different growth-rate dispersion mechanisms; intrinsic growth-rate dispersion (left), and growth-dependent dispersion (right). 601
10.15 Probability density for single particle stochastic growth model at different times. 605

A.1 Estimated reaction rates from 2000 production-rate measurements subject to measurement noise. 636
A.2 Gibbs energy contours for the pentane reactions as a function of the two reaction extents. 640
A.3 Solution to first-order differential equation $dc_A/dt = -kc_A$, and sensitivities $S_1 = \partial c_A/\partial k$ and $S_2 = \partial c_A/\partial c_{A0}$. 650
A.4 Function $c(r)$ and its values at five collocation points. . 652
A.5 Dimensionless concentration versus dimensionless radial position for different numbers of collocation points. 654
A.6 Relative error in the effectiveness factor versus number of collocation points. 655
A.7 Molar flow of A versus reactor volume for second-order, isothermal reaction in a fixed-bed reactor; two approximations and exact solution. 656
A.8 Magnified view of Figure A.7. 656
A.9 Measurements of species concentrations in Reactions A.9 versus time. 658
A.10 Fit of model to measurements using estimated parameters. 658
A.11 Dimensionless concentration versus radius for the non-isothermal spherical pellet: lower (A), unstable middle (B), and upper (C) steady states. 662

List of Tables

4.1 Reactor balances for constant-density and ideal-mixture assumptions. 137
4.2 Reactor balances for general equation of state. 138
4.3 Summary of mole balances for several ideal reactors. . . 172

5.1 Molecular partition function terms. 201
5.2 Parameters for the trioxane reaction. 206
5.3 Fast and slow time-scale models in extents of reactions. 213
5.4 Fast and slow time-scale models in species concentrations. 215
5.5 Exact and QSSA solutions for kinetic Schemes I and II. . 222
5.6 Ethane pyrolysis kinetics. 232
5.7 Chemisorption and physisorption properties. 239
5.8 Gas-phase oxygen concentration and adsorbed oxygen concentration. 263

6.1 Parameter values for multiple steady states. 289
6.2 Selected values of steady-state temperatures and residence times. 294
6.3 Parameter values for limit cycles. 303
6.4 Steady state and eigenvalue with largest real part at selected points in Figures 6.20 and 6.21. 306
6.5 PFR operating conditions and parameters for o-xylene example. 322
6.6 Parameter values for Example 6.6. 327
6.7 Energy balances for the batch reactor. 330
6.8 Energy balances for the CSTR. 331
6.9 Energy balances for the semi-batch reactor. 332
6.10 Energy balances for the plug-flow reactor. 333
6.11 Thermodynamic data for allyl chloride example. 337

7.1 Industrial reactions over heterogeneous catalysts. 355
7.2 Porosity and tortuosity factors for diffusion in catalysts. 359

7.3 Closed-form solution for the effectiveness factor versus Thiele modulus for the sphere, semi-infinite cylinder, and semi-infinite slab. 372
7.4 Controlling mechanisms for pellet reaction rate given finite rates of internal diffusion and external mass transfer. 385
7.5 Kinetic and mass-transfer parameters for the catalytic converter example. 394
7.6 Feed flowrate and heat-transfer parameters for the fixed-bed catalytic converter. 408

8.1 Mass-transfer and kinetic parameters for Example 8.1. . 446
8.2 Mass-transfer and kinetic parameters for micromixing problem. 461
8.3 Reactor and kinetic parameters for feed-mixing example. 476
8.4 Parameters for the dispersed PFR example. 481

A.1 Octave compatibility wrapper functions. 645

List of Examples

2.1 Stoichiometric matrix for a single reaction 30
2.2 Columns of $\boldsymbol{\nu}$ 30
2.3 Rows of $\boldsymbol{\nu}$ 31
2.4 Stoichiometric matrix for CVD chemistry 33
2.5 Conservation of mass 34
2.6 More species than reactions 37
2.7 Maximal set of reactions for methane oxidation 39
2.8 Production rate for SiH_2 45

3.1 Ideal-gas equilibrium 66
3.2 Ideal-gas equilibrium, revisited 68
3.3 Minimum in G for an ideal gas 73
3.4 Chemical and phase equilibrium for a nonideal mixture 82
3.5 Equilibrium composition for multiple reactions 90
3.6 Multiple reactions with optimization 93

4.1 Reaching steady state in a CSTR 128
4.2 Phenol production in a CSTR 130
4.3 Semi-batch polymerization 139
4.4 Changing flowrate in a PFR 147
4.5 Benzene pyrolysis in a PFR 150
4.6 Benzene pyrolysis, revisited 153
4.7 Ethane pyrolysis in the presence of NO 154
4.8 The PFR versus CSTR with separation 159
4.9 Stochastic vs. deterministic simulation of a virus model 168

5.1 Computing a rate constant 205
5.2 Production rate of acetone 226
5.3 Production rate of oxygen 228
5.4 Free-radical polymerization kinetics 230
5.5 Ethane pyrolysis 232
5.6 Fitting Langmuir adsorption constants to CO data 244
5.7 Equilibrium CO and O surface concentrations 249
5.8 Production rate of CO_2 250
5.9 Production rate of methane 252

6.1 Constant-pressure versus constant-volume reactors . . 279
6.2 Liquid-phase batch reactor 281
6.3 Temperature control in a CSTR 285
6.4 PFR and interstage cooling 317
6.5 Oxidation of o-xylene to phthalic anhydride 321
6.6 Ammonia synthesis 325

7.1 Using the Thiele modulus and effectiveness factor . . . 367
7.2 Catalytic converter 393
7.3 First-order, isothermal fixed-bed reactor 399
7.4 Mass-transfer limitations in a fixed-bed reactor 401
7.5 Second-order, isothermal fixed-bed reactor 402
7.6 Hougen-Watson kinetics in a fixed-bed reactor 405
7.7 Multiple-reaction, nonisothermal fixed-bed reactor . . . 408

8.1 Transient start-up of a PFR 446
8.2 Order matters 448
8.3 Two CSTRs in series 457
8.4 Optimal is neither segregated nor maximally mixed . . . 470
8.5 Mixing two liquid-phase streams in a stirred tank 475
8.6 Maximizing yield in dispersed plug flow 480

9.1 Estimating the rate constant and activation energy . . . 515
9.2 Unknown measurement variance and few data points . 521
9.3 Fitting single and multiple adsorption experiments . . . 525
9.4 Fitting reaction-rate constant and order 537
9.5 Fitting rate constants in hepatitis B virus model 541

10.1 Mass and energy balances with multiple phases 588
10.2 Crystallization model 591
10.3 Emulsion polymerization model 592
10.4 Fermentation model 593

A.1 Estimating reaction rates 636
A.2 Two simple sensitivity calculations 649
A.3 Single-pellet profile 653
A.4 A simple fixed-bed reactor problem 655
A.5 Estimating two rate constants in reaction $A \rightarrow B \rightarrow C$ . . . 657

1

Setting the Stage

1.1 Introduction

The chemical reactor lies at the heart of most chemical processes. The design and operation of the reactor often determine the success or failure of the entire process. In the overall process, feedstocks are delivered to the chemical reactor at the appropriate temperature, pressure, and concentrations of species. The chemical reactor is that essential component in which the feed is converted into the desired products. The chemical reactor is the place in the process where the most value is added: lower-value feeds are converted into higher-value products.

The reactor is normally followed by separation processes that separate the products from the unreacted feed and the reactor byproducts. The modern chemical processing plant, whether processing fine or commodity chemicals, manufacturing pharmaceuticals, refining petroleum, or fabricating microelectronic devices, is a highly integrated operation. In addition to meeting production and purity goals, the process design and operation also are influenced by the sometimes conflicting goals of minimizing energy consumption, and minimizing the amount of feed or product that must be kept in storage.

Many aspects of reactor analysis and design are treated in this text, including predicting performance, specifying initial and feed conditions, sizing the reactor, specifying operating conditions, and selecting the reactor type. To accomplish these goals, we construct and illustrate a set of reactor analysis and design principles. As we will see, the principles are small in number, but the consequences of the principles are rich, diverse and complex. These principles can be applied at many size scales to many different types of chemically reacting systems. While the reactor of interest is often a man-made vessel, sometimes the reactor is a living organism. We will illustrate throughout the text, using many different examples, the diversity of systems and issues addressed by chemical reactor analysis and design.

Batch, continuous-stirred-tank, and plug-flow reactors. We model many chemical reactors using three main reactor archetypes: batch, continuous-stirred-tank, and plug-flow reactors. By virtue of their design and the typical operating conditions, many complex chemical reactors can be well approximated by these three simple reactor types. The material and energy balances of these three reactors are sets of first-order, nonlinear ordinary differential equations (ODEs) or nonlinear algebraic equations, or in some situations, differential algebraic equations (DAEs). The great simplification that has become standard practice in introductory reactor design texts is to neglect the momentum balance and a careful treatment of the fluid flow pattern within the reactor.

Concentration, temperature and pressure are therefore the usual dependent variables that are solved as functions of time or distance along the reactor as the independent variable. Sometimes the reactor model is quite simple, such as describing a single reaction in an isothermal reactor, and a single ODE or algebraic equation describes a single species concentration or extent of a single reaction. More often the design involves many reactions and nonisothermal operation, and coupled sets of ODEs or algebraic equations are needed to describe the temperature, pressure and species concentrations. Regardless of the complexity, the design problem is approached in the same manner using the same set of principles. In all but the simplest cases, numerical methods are required to solve the models. Fortunately, high-level programming languages are readily available and easily can be used to solve the complex models. We make extensive use of numerical methods and high-level programming languages in this text to solve complex models.

1.2 Some Classifications and Terminology

Ideal mixing and plug flow. The batch, continuous-stirred-tank, and plug-flow reactors are defined by certain idealized assumptions on the fluid flow. The batch and continuous-stirred-tank reactors are assumed to be ideally well mixed, which means that the temperature, pressure and species concentrations are independent of spatial position within the reactor. The plug-flow reactor describes a special type of flow in a tube in which the fluid is well mixed in the radial direction and varies only in the axial or tube length direction. Plug flow often describes well the limit of fully developed turbulent flow, i.e., flow in the limit of large Reynolds number. If the flow in a tubular reactor is not turbulent (Reynolds Number less than 10^3–10^4), then the flow may not be well modeled as ideal plug flow, and other models are needed to describe the reactor.

Homogeneous and heterogeneous reactions. Another important reactor classification pertains to the phase in which the reaction occurs. The reactants and products of homogeneous chemical reactions are in a single phase. Examples include: steam cracking of ethane to ethylene, and the photochemical reactions of chlorocarbons in the troposphere that lead to ozone destruction for gas-phase reactions; enzymatic isomerization of glucose to fructose, and esterification of an acid and alcohol to produce an ester for liquid-phase reactions; and, the fusing of limestone and charcoal to produce calcium carbide, and the interfacial reaction between strontium oxide and silicon dioxide to form strontium silicate for solid-phase reactions. Sometimes the reactants and products are transported through the reactor in one phase, but the reaction occurs in a different phase, often a solid phase. The heterogeneously catalyzed reactions such as zeolite-catalyzed cracking of high-boiling crude oil and iron-catalyzed ammonia synthesis from hydrogen and nitrogen involve two phases. Sometimes three phases are present simultaneously during the reaction, such as synthesis gas ($CO + H_2$) reactions over iron-based catalysts that are suspended in a high molecular weight alkane to help moderate the reaction temperature and dissolve the reaction product. Multiphase reactions need not involve heterogeneous catalysts. The reaction of liquid *p*-xylene and gaseous O_2 to produce liquid terephthalic acid, occurs via free-radical intermediates in the liquid phase of the reactor. Another two-phase, noncatalytic reaction is the low-pressure epitaxial growth of Si(100) films from gas-phase disilane (Si_2H_6).

Batch, semi-batch and continuous operation. The operation of the reactor can be classified as batch, semi-batch and continuous. In batch operation, the reactor is charged with reactants, the reaction takes place, and after some processing time the contents of the reactor are removed as product. Batch reactors, depicted in Figure 1.1, are often used for liquid-phase reactions, and the manufacture of low-volume, high value-added products, such as specialty fine chemicals, pharmaceuticals and fermentation products. Batch reactors also are used in

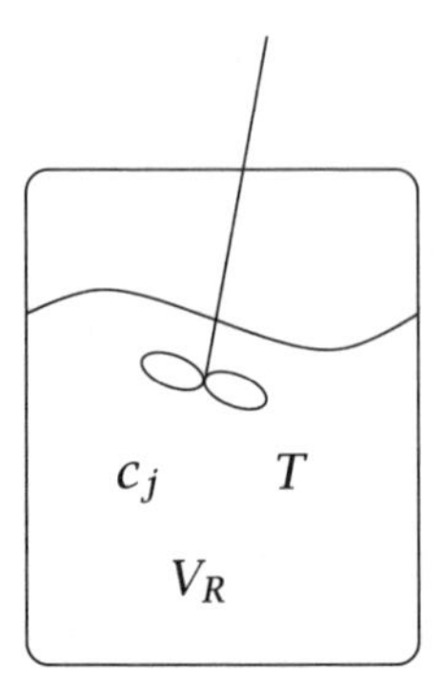

Figure 1.1: Schematic diagram of a batch reactor.

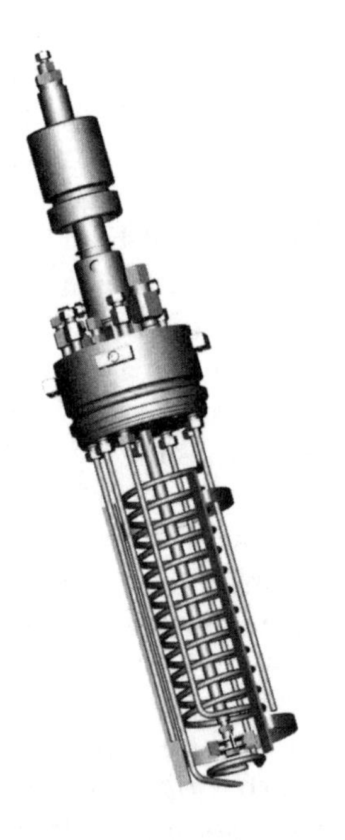

Figure 1.2: Expanded view of the internals of a batch reactor. Courtesy of Autoclave Engineers, Division of Snap-tite, Inc.

situations where it is not practical to implement a continuous process. The great flexibility of batch processing allows the reactor to be used for the manufacture of many different products. Figure 1.2 provides an expanded view of the internals used in a batch reactor showing the cooling coils, tubes for adding (sparge tube) and removing (sample tube) fluid, and baffles to ensure complete mixing. This unit is placed in a cylindrical vessel that comprises the exterior of the batch reactor.

The batch cycle begins by charging the reactants to the vessel, and often heating the reactants to the reaction temperature. The cycle often ends by bringing the contents to a discharge temperature, emptying the vessel, and cleaning the vessel before the next charge of reactants. Product may or may not form during these preparation steps and they often involve manual labor, so manufacturing costs can be considerably higher than a corresponding continuous process.

The semi-batch process is similar to the batch process except feed addition occurs during the batch cycle. Products may also be removed during the semi-batch process. The addition/removal policy allows one to control the reaction rate or heat release during reaction. The semi-batch reactor also may provide more complete

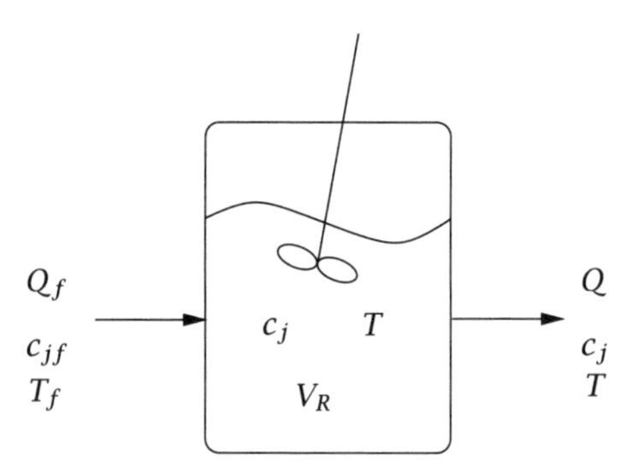

Figure 1.3: Schematic diagram of a CSTR; the effluent composition of the CSTR is identical to the conditions that exist in the reactor.

use of the reactor volume in reactions such as polymerizations that convert lower-density reactants to higher-density products during the course of the reaction.

The continuous-stirred-tank reactor (CSTR) is shown in Figure 1.3. Reactants and products flow into and out of the reactor continuously, and the contents of the reactor are assumed to be well mixed. The well-mixed assumption can be realized more easily for liquids than gases, so CSTRs are often used for liquid-phase reactions. The fluid composition and temperature undergo a step change when passing from the feed stream into the interior of the reactor; the composition and temperature of the effluent stream are identical to those of the reactor.

The CSTR is used extensively in situations where intense agitation is required, such as the addition of a gaseous reactant to a liquid by transfer between the bubbles and the continuous liquid, and the suspension of a solid or second liquid within a continuous liquid phase. Polymerization reactions are sometimes conducted in CSTRs. It is common to employ a cascade or series of CSTRs in which the effluent from the first reactor is used as feed to the second and so forth down the cascade (Figure 1.4). The cascade permits one to realize high conversion of reactant, while minimizing total reactor volume.

The plug-flow reactor (PFR) is a constant cross-section, tubular reactor as depicted in Figure 1.5. Under turbulent flow conditions, the velocity profile becomes plug-like, which greatly simplifies the material and energy balances. The velocity, composition and temperature are functions of only the axial position. In this text, we are usually

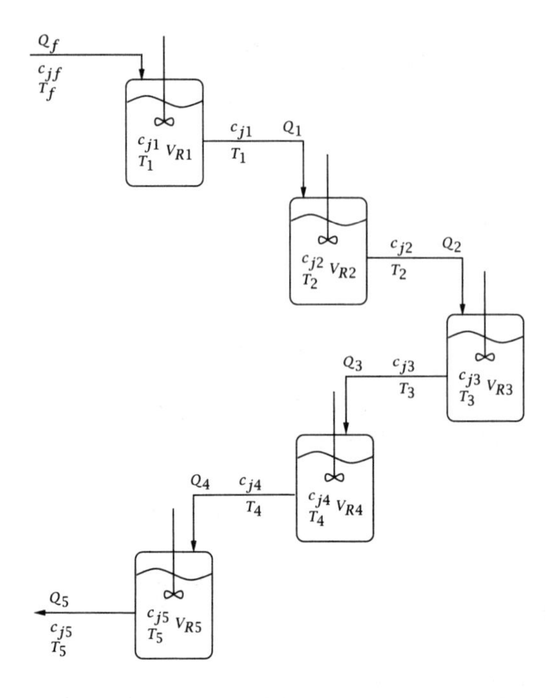

Figure 1.4: Schematic diagram of five CSTRs in series; the effluent of each reactor becomes the feed to the next.

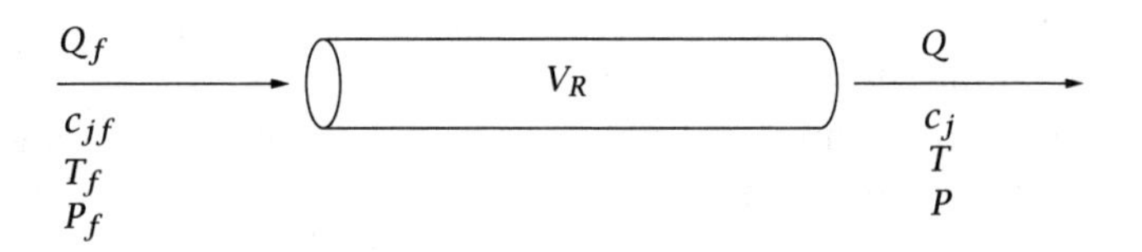

Figure 1.5: Schematic diagram of a plug-flow reactor.

interested in only the steady-state profile in the tube and neglect the dynamics.

Plug-flow reactors are used for gas-phase and liquid-phase reactions. If the PFR is filled with a porous catalyst and the fluid flowing in the void space is turbulent, the reactor is referred to as a fixed-bed reactor. We will see that the isothermal PFR usually leads to higher conversion of reactant per unit volume than the CSTR, i.e., there is more efficient use of volume in a PFR. For this reason, PFRs are employed in situations that require high capacity and high conversion. PFRs also are used in situations involving highly exothermic or endothermic re-

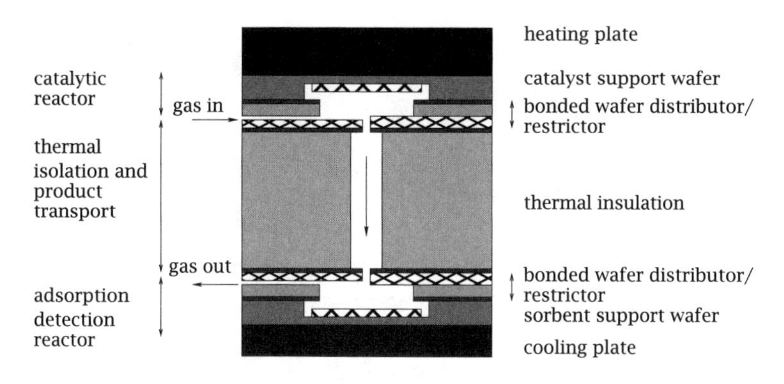

Figure 1.6: Cross-sectional view of a 1 μl combinatorial screening reactor. Copyright 2001, Symyx Technologies, Inc., used with permission.[1]

actions; a bundle of small-diameter tubes can be placed in a furnace for endothermic reactions, or surrounded by a high-temperature boiling fluid for exothermic reactions. PFRs are the reactor of choice for gas-phase reactions due to the problems of mixing gases in a CSTR.

1.3 Scale

The size or volume of chemical reactors varies widely. Reactor volumes can range from hundreds of nanoliters for combinatorial, lab-on-a-chip reactor systems, to several hundred thousand liters for certain petroleum refining operations. In the combinatorial reactors, one is interested in determining if a reaction proceeds and in minimizing the scale of the experiment so many combinations or conditions can be screened rapidly. Figure 1.6 presents a schematic view of 1 μl test reactors that are used for combinatorial screening of heterogeneous catalysts.

Figure 1.7 presents an assembled view of these reactors. As many as 256 different catalyst formulations can be deposited into shallow wells in a 3-in by 3-in support wafer. By changing the gas flowrates, the residence time can be be varied from 10 ms to 1 s. Uniform flow distribution to each of the reactors is but one of the design challenges for this small-scale, massively parallel device. The B-tree flow distribution network to accomplish this is shown in Figure 1.8.

[1]All rights reserved, U.S. Patent No. 5,985,356, 6,004,617. Additional U.S. and foreign patents pending.

Figure 1.7: Assembled view of 256 1-μL reactors built into a 3-in by 3-in support wafer. Copyright 2001, Symyx Technologies, Inc., used with permission.[1]

At the opposite end of the spectrum, the size of petroleum refining reactors is dictated by a desire to maximize production given a fixed reaction rate. These reactors are therefore large. For example, Figure 1.9 displays a 950-L, low-density polyethylene polymerization reactor. This process is discussed further in Section 1.4.3.

The time required for chemical reaction also varies widely. In the batch reactor, the batch time determines the production rate for the reactor. For flow systems we define residence time, which is the reactor volume divided by the feed volumetric flowrate; this quantity has the units of time, and can be thought of as the time required to displace the equivalent of one reactor volume. Batch times and residence times can be as short as milliseconds for high-temperature, gas-phase reactions, such as ammonia oxidation and HCN synthesis in flow reactors, or as

Figure 1.8: Top view of a combinatorial reactor gas-distribution network. Copyright 2001, Symyx Technologies, Inc., used with permission.[1]

Figure 1.9: Polyethylene reactor; this 16-in inner-diameter reactor is designed to operate at 35,000 psi and 600°F; in operation, this reactor is in a vertical configuration. Courtesy of Autoclave Engineers, Division of Snap-tite, Inc.

long as days for liquid-phase fermentation reactions in batch reactors.

The rate of chemical reaction has units of moles/time-volume and is a function of temperature and composition. If the rate is high, as we would anticipate for a high-temperature oxidation or pyrolysis reaction, the time needed for a desired production rate is small. Similarly, if the rate is low, a long time is needed. The reactor volume is determined by the reaction rate and amount of product to be manufactured.

1.4 Some Examples

We live in a chemical world and most of the items you use in daily life result from chemical processes, including the ink on this page and the paper itself. Chances are one or more of the articles of clothing you are wearing are made from synthetic fibers. We are surrounded by and regularly use items that are made from plastics and polymers. The automobiles we drive and the fuels we use to power them depend on products made in chemical reactors. Electronic devices and computer chips require a myriad of chemical reaction steps in their manufacture. Chemical reaction engineering is the core discipline that designs, analyzes and creates the processes to convert natural resources into other intermediate chemicals and final products.

1.4.1 Chemical Vapor Deposition of Silicon-Germanium Alloy Films

The demand for faster microelectronic devices with higher reliability, lower power requirements, and lower cost is driving the current silicon-based devices close to their physical limits. One solution that permits the traditional silicon-based circuit design to be retained is to use silicon-germanium alloys in place of silicon in the active region of the device. Silicon-germanium alloys have higher mobilities than silicon and a lower bandgap enabling faster speeds and lower power. Devices such as heterojunction bipolar transistors and field-effect transistors are fabricated by a series of steps that involve masking the semiconductor wafer with a coating, selectively removing regions of the mask with lithography, implanting dopant ions such as boron or phosphorous, and adding layers of metal and dielectrics to build up the device from the wafer surface [19, 7, 11]. For silicon-germanium alloy-based devices, a thin film (tens of nanometers to microns) is grown onto Si(001) wafers. Typically the reactants are hydride gases, such as disilane (Si_2H_6) and germane (GeH_4), at pressures on the order of 0.1 to several Torr in a process known as chemical vapor deposition (CVD). Device speed and performance are related to the amount of Ge in the alloy, and control of the film composition is critical.

Figure 1.10: Top view of a single-wafer CVD reactor; single-wafer processing generally employs cluster tools that permit a wafer to be shuttled between the different reaction/process chambers without exposing it to contamination or ambient conditions; the top view of one reactor shows the pins on which a wafer rests during the CVD reaction. Courtesy of Applied Materials.

As microelectronics manufacturing moves to ever larger wafers (currently 300-mm diameter wafers are routine), single-wafer reactors are used in the manufacturing processing steps (Figure 1.10). In a single-wafer reactor, the wafer is supported in the horizontal position, a gas distribution system is used to ensure uniform flow of reactants over the wafer surface, and the wafer is heated radiatively to maintain the entire cross section at a constant temperature. When the pressure is low enough, the diffusion lengths of the gas-phase components are long enough that one can assume the gas phase to be well mixed, uniform and independent of position. This means the gas phase can be modeled as a CSTR in which gas enters and leaves the reactor in the flow streams and undergoes reaction at the surface of the wafer.

Much is known about the chemistry that takes place during film growth [5, 20, 15, 2, 12, 13]. This information can be compactly summarized in a fairly small, reduced-order kinetic model that accounts for surface and gas-phase components, and can be used to predict the film growth rate [9]. One such kinetic model consists of eight reactions among nine species. The reactions are

$$
\begin{aligned}
Si_2H_6(g) + 6Si(s) &\xrightarrow{k_1} 6SiH(s) + 2Si(b) \\
Si_2H_6(g) + 4Si(s) + 2Ge(s) &\xrightarrow{k_2} 6SiH(s) + 2Ge(b) \\
Si_2H_6(g) + 5Si(s) + Ge(s) &\xrightarrow{k_3} 6SiH(s) + Si(b) + Ge(b) \\
GeH_4(g) + 4Ge(s) &\xrightarrow{k_4} 4GeH(s) + Ge(b) \\
GeH_4(g) + 3Si(s) + Ge(s) &\xrightarrow{k_5} 2GeH(s) + 2SiH(s) + Si(b) \\
2SiH(s) &\xrightarrow{k_6} 2Si(s) + H_2(g) \\
2GeH(s) &\xrightarrow{k_7} 2Ge(s) + H_2(g) \\
SiH(s) + Ge(s) &\underset{k_{-8}}{\overset{k_8}{\rightleftharpoons}} Si(s) + GeH(s)
\end{aligned}
$$

where the symbols (s), (b) and (g) refer to the wafer surface, the wafer bulk, and the gas phase, respectively. The production rates of the nine species follow from the stoichiometry of the eight reactions and the reaction rates as we will see in Chapter 2, and are given by

$$
\begin{aligned}
R_{Si_2H_6} &= -r_1 - r_2 - r_3 \\
R_{GeH_4} &= -r_4 - r_5 \\
R_{Si(s)} &= -6r_1 - 4r_2 - 5r_3 - 3r_5 + 2r_6 + r_8 \\
R_{Ge(s)} &= -2r_2 - r_3 - 4r_4 - r_5 + 2r_7 - r_8 \\
R_{SiH(s)} &= 6r_1 + 6r_2 + 6r_3 + 2r_5 - 2r_6 - r_8 \\
R_{GeH(s)} &= 4r_4 + 2r_5 - 2r_7 + r_8 \\
R_{Si(b)} &= 2r_1 + r_3 + r_5 \\
R_{Ge(b)} &= 2r_2 + r_3 + r_4 \\
R_{H_2} &= r_6 + r_7
\end{aligned}
$$

The production rates enable us to follow the changing composition of the gas phase, the surface, the film growth rate, and the Si/Ge film alloy composition.

1.4.2 Hydrodesulfurization

Crude oil contains a number of organosulfur compounds (RS), including thiols, sulfides, thiophenes and alkyl-substituted thiophenes [8],

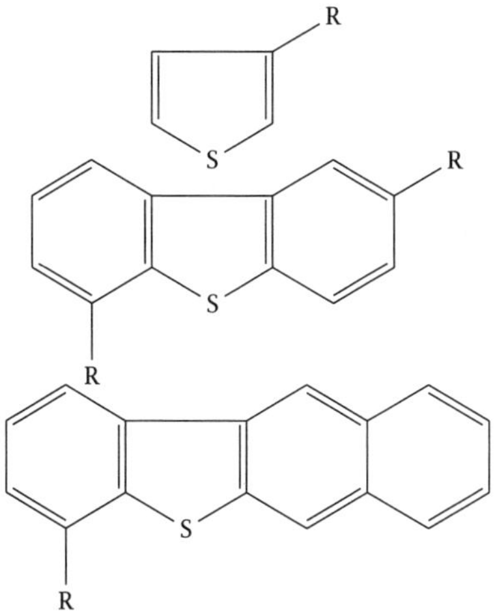

alkyl-substituted thiophene

alkyl-substituted dibenzothiophene

benzonaphthothiophene

Figure 1.11: Representative organosulfur compounds illustrating the aromatic character of the molecules; when the alkyl substitution position is near the S atom, it is difficult to desulfurize the molecule.

some of which are shown in Figure 1.11. The sulfur must be removed from the petroleum feeds to prevent it from poisoning precious metal catalysts that are used in many refinery processes and to meet the restrictions for sulfur content in fuels. The total sulfur content of crude oil can easily be several percent and it must be reduced to several parts per million (ppm). This sulfur is removed by treating the petroleum feed with hydrogen at high pressures in a fixed-bed or trickle-bed catalytic reactor; the process is referred to as hydrodesulfurization (HDS) [3]. During HDS the carbon-sulfur bonds are broken and the hydrogen replaces the sulfur and H_2S is formed.

$$RS + 2H_2 \longrightarrow RH_2 + H_2S$$

The fixed-bed reactors typically operate in the temperature range 525–750 K, and hydrogen pressure range 35–100 atm.

Figure 1.12 presents a simplified process diagram. The feed is heated and mixed with hydrogen in the inlet to a fixed-bed catalytic reactor. HDS is an exothermic reaction and if the reactor temperature becomes too high, undesired side reactions occur, such as the hydrogenation of the unsaturated bonds and hydrogenolysis of the C—C bonds. Multiple beds operating in series and at different temperatures, with inter-

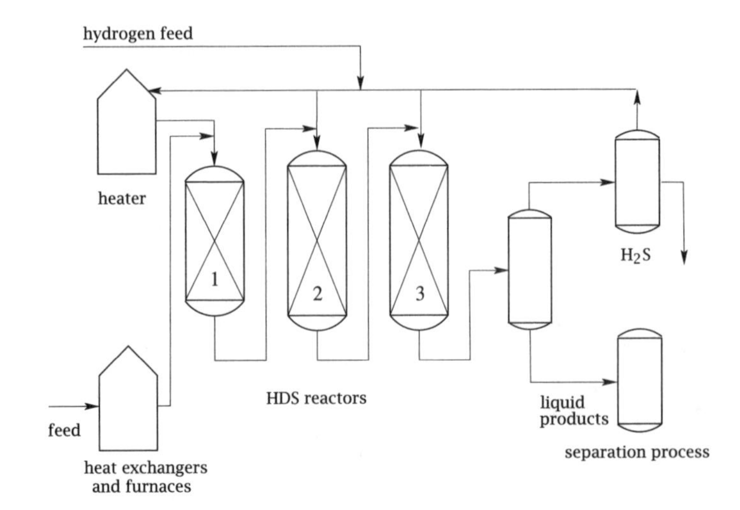

Figure 1.12: Simplified hydrodesulfurization process diagram; three separate fixed-bed catalytic reactors in series; reactor 1 is used as a guard bed in which the metals are removed to not poison the HDS catalyst; reactors 2 and 3 permit control over temperature and sulfur removal. Adapted from McCulloch [8].

bed cooling, is one way to control the temperature and to ensure the complete removal of the sulfur. Separate reactors in series permit the reactor sequence to be changed as catalyst activity degrades with time. Figures 1.13 and 1.14 present two views of a multibed fixed-bed reactor in which the beds are built into one vessel. Within each section containing catalyst, the reactor is modeled as a fixed bed. After cooling the outflow of one of the beds, the reaction is allowed to proceed in the next section. Multibed reactors are generally employed with exothermic reactions, and the fixed-bed sections are operated adiabatically. In this way, the reactor temperature can be controlled by the heat removal. These concepts are discussed in Chapter 6. Figure 1.14 shows an expanded view of the internal construction of a fixed-bed reactor; trays to support the catalyst particles and distribute the flows uniformly across the reactor cross section are shown. Commercial HDS reactors such as shown in Figure 1.14 can hold up to 60,000 kg of catalyst.

Optimal reactor operation and design are two methods to maximize sulfur removal. In addition, considerable research effort is focused on finding new catalysts to improve activity, selectivity and durability, and to adapt to the ever more stringent emission standards [10]. The catalytic reaction chemistry is quite complex. As can be seen from Figure 1.11, the organosulfur molecules are highly unsaturated, and in addition to hydrogenation of the C—S bonds, one wants to avoid saturation of all the C=C double bonds. The catalysts have different types of sites, which complicates the kinetic model [10]. Hydrogen must adsorb dissociatively. The organosulfur molecule must adsorb and orient itself so that the sulfur atom is over an active site, such as an exposed molybdenum atom on a molybdenum disulfide catalyst. The reaction rate is influenced by the relative amounts of reactants (adsorbed hydrogen and organosulfur molecule) and the reactivity of the catalyst. Describing the rate of these kinds of complex reactions on catalyst surfaces is covered in Chapter 5, where we show the rate of the HDS reaction can be expressed by

$$r = \frac{-kc_{H_2}^{\alpha}c_{RS}^{\beta}}{(1 + K_{H_2}c_{H_2} + K_{RS}c_{RS})^{\gamma}}$$

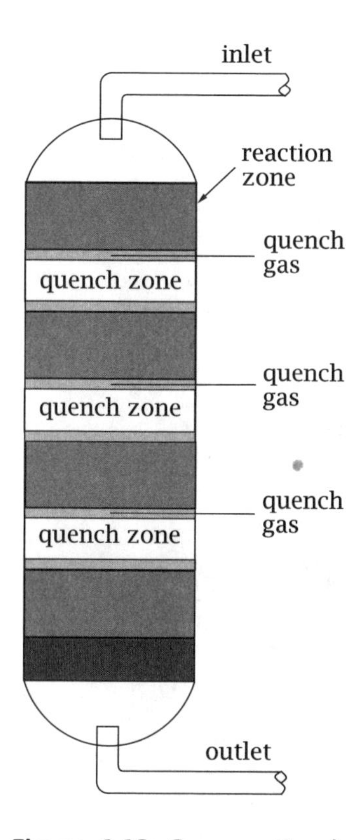

Figure 1.13: Cross-sectional view of a commercial HDS reactor containing four fixed-bed reactors with interstage cooling/quenching; quenching is necessary to limit the temperature increase for this exothermic reaction. Adapted from UOP LLC.

1.4.3 Olefin Polymerization

Polyethylene is a major commodity plastic, with more than 33 billion pounds of the resin produced in the United States in 2000 [1]. Polyethylene encompasses a family of semicrystalline polymers with ethylene as the major building block [6]. The resins are loosely grouped into three classes: low-density polyethylene (LDPE), high-density polyethylene (HDPE), and linear-low-density polyethylene (LLDPE). LDPE is a homopolymer of ethylene with side-chain branching at a frequency

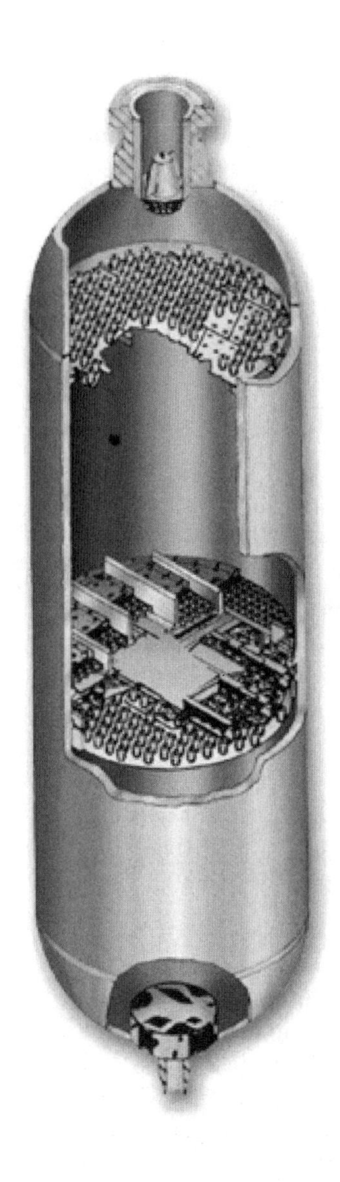

Figure 1.14: Expanded view of a multibed HDS reactor showing the gas distribution tray at the top and the catalyst support tray in the middle; these reactors can range in diameter from 1.5–5 m. Courtesy of UOP LLC.

of 2–50 per 1000 carbons in the chain. Its density is in the range 0.915–0.940 g/cm^3. HDPE can be a homopolymer or, more commonly, a copolymer with butene or hexene, and has a density greater than 0.940 g/cm^3, and a much lower density of side branches than LDPE, 0.5–10 branches per 1000 carbons in the chain. LLDPE is a copolymer of ethylene and small amounts of α-olefins. By changing the density, molecular weight, and branching-group frequency and type, the chemical resistance and mechanical properties can be tailored to a variety of uses. Polyethylene is used in films for bags and packaging, ranging from garbage and trash bags to whole blood storage; coatings for paper, metal and electrical wires; containers; and piping and tubing.

Low-density polyethylene is produced in a free-radical initiated, high-pressure process. Free-radical polymerization mechanisms and kinetics are discussed in Chapter 5. The polymerization process begins with an initiation reaction between ethylene and an initiator to form a primary radical. The primary radical undergoes step-wise chain growth (propagation) until it is terminated by a variety of mechanisms that stop chain growth. The LDPE reactors are either stirred autoclaves or tubular reactors. Figure 1.9 displays a commercial LDPE tubular reactor. Typical tubular reactors have diameters of 15–30 in and lengths of 17–25 ft, and operate in the pressure range of 30,000–45,000 psi and temperature range of 200-300°C. The process temperatures are high enough that the polyethylene is in the melt state, and the pressures are high enough to form a supercritical mixture of ethylene and polyethylene, which acts as the polymerization medium. The polymerization reaction itself is exothermic and the process is operated to prevent excessive temperature increases. The temperature control is accomplished by limiting the amount of ethylene converted within the reactor and/or removing heat through the reactor walls.

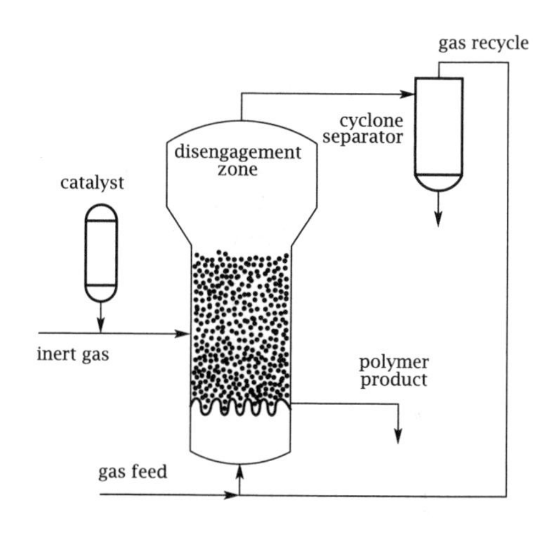

Figure 1.15: Simplified polymerization process. This material is used by permission of John Wiley & Sons, Inc., Copyright ©1996 [6].

High-density polyethylene is produced at lower pressures, from 10–30 atm. HDPE requires a catalyst [23, 18, 6]. Various catalysts have been used including: chromium oxide supported on an inert porous substrate such as silica; a Ziegler catalyst consisting of titanium or vanadium compounds ($TiCl_4$, $TiCl_3$, $VOCl_3$) on a support (silica, $MgCl_2$, graphite, carbon black) and an organoaluminum cocatalyst ($Al(C_2H_5)_3$); and, a metallocene (a complex of zirconium, titanium or hafnium containing cyclopentadienyl rings) and an organoaluminum cocatalyst. HDPE processes can employ a slurry of catalyst, polymer and diluent. One such slurry process involves circulating the reaction mixture at 5–12 m/s around a loop that can range from 0.5–1.0 m in diameter and 200 m in total length. The heat of reaction is removed by a cooling jacket that surrounds the reactor, and the high flowrate ensures turbulent flow within the loop [6]. Ethylene, catalyst and diluent are fed continuously. Polymer particles form and are removed continuously; these particles reside in the reactor loop for 0.5–2.5 hr, and are removed after they settle in a side leg of the loop. Ethylene conversion can reach 95–98%.

Gas-phase polymerization also is used for HDPE in the process represented schematically in Figure 1.15. The cylindrical reactor can be

25–30 ft in diameter and have a five-to-one length-to-diameter ratio. High linear velocities serve to fluidize the catalyst particles that form; fluidization facilitates removing the heat of reaction. Catalyst is added continuously and polymer particles are removed once they reach a particular size (about 500–1000 μm). The reactor diameter is larger at the top, which lowers the linear velocity and acts to disengage the polymer particles from the unreacted gases. These same reactors are used for LLDPE. The choice of catalyst and the presence of inert gases and comonomers are used to regulate the polymer resin properties and molecular weight. The reactor typically operates at 220–370 psig and 160–205°F.

1.4.4 Hepatitis B Virus Modeling

As a final example, we wish to display the wide scope of chemical reaction modeling principles. Consider Figure 1.16, which shows some of the biochemical reaction events that occur in a single cell during the reproduction cycle of the hepatitis B virus [4, p.767]. The understanding and modeling of these biochemical events is an area of current research activity [14, 21, 22, 16, 17]. The following is a simplified but useful model of part of this reproductive system

$$\text{nucleotides} \xrightarrow{\text{cccDNA}} \text{rcDNA} \tag{1.1}$$

$$\text{nucleotides} + \text{rcDNA} \longrightarrow \text{cccDNA} \tag{1.2}$$

$$\text{amino acids} \xrightarrow{\text{cccDNA}} \text{envelope} \tag{1.3}$$

$$\text{cccDNA} \longrightarrow \text{degraded} \tag{1.4}$$

$$\text{envelope} \longrightarrow \text{secreted or degraded} \tag{1.5}$$

$$\text{rcDNA} + \text{envelope} \longrightarrow \text{secreted virus} \tag{1.6}$$

These reactions correspond to the following steps in Figure 1.16:

1. Reaction 1.1 accounts for Steps 5, 9–11.
2. Reaction 1.2 accounts for Step 12.
3. Reaction 1.3 accounts for Steps 5–7.
4. Reactions 1.4 and 1.5 are not present in Figure 1.16, but may prove useful to explain potential loss of active cccDNA.
5. Reaction 1.6 accounts for Steps 13–15.

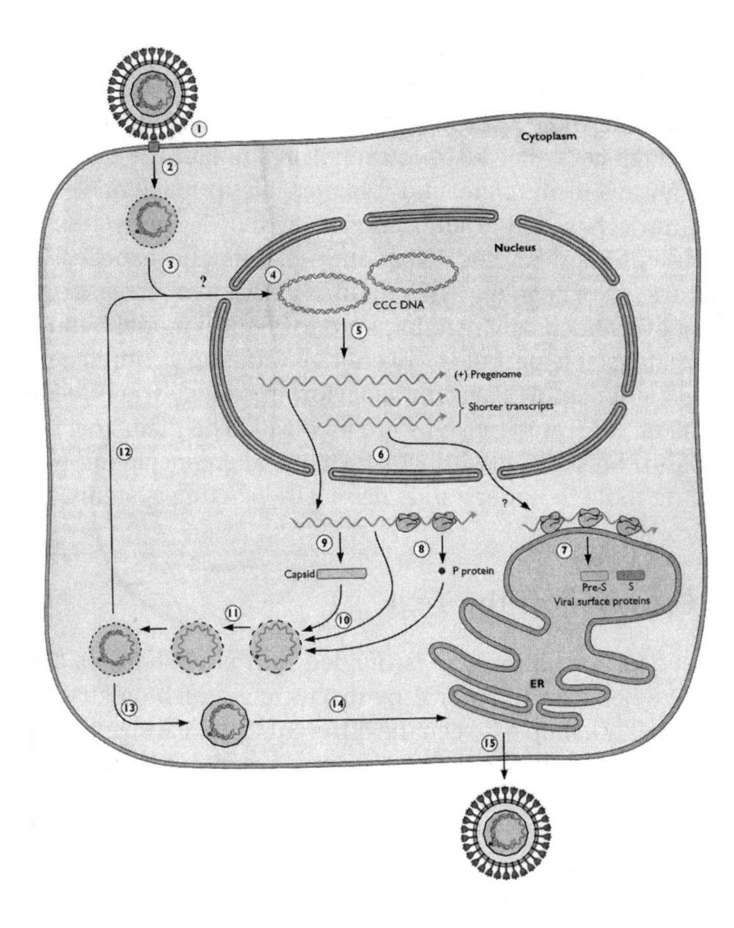

Figure 1.16: The chemical events comprising the reproduction cycle of the hepatitis B virus. Courtesy of ASM Press [4].

When we change to this context, the "chemical reactor" of interest becomes the living cell or, if we also model the cell population, the human liver. The chemical reaction modeling principles remain valid. In Chapter 4 we use this simple model to make quantitative predictions about the evolution of the viral species concentrations. We also show how to model systems that have small concentrations of species, down to less than a few hundred molecules. In Chapter 9 we explore estimating the rate constants that appear in this virus model given the kinds of laboratory measurements that are available.

Tailoring a model to make successful quantitative predictions of a system of interest is still more of an art than a science. We should not underestimate the complexity of some of the systems of interest to chemical engineers. Because we cannot include all the details, our models are always incomplete, and it is possible to make naive use of modeling approaches, and produce models with little connection to reality and little predictive value. This caution is perhaps especially true for biological systems. On the other hand, if we wish to increase our understanding of a chemically reacting system, skillful model building is often an indispensable part of the overall investigation. Simple models often can explain complex system behavior, especially when feedback mechanisms or autocatalytic steps are involved. The main goal of this text is to build the skill set with which chemical engineers apply reaction modeling tools to understand chemically reacting systems.

1.5 An Overview of the Text

Chapter 2. The remaining text is divided into nine chapters and an appendix. We begin in Chapter 2 by discussing stoichiometry or the quantitative relationship between the different chemical species undergoing chemical reaction. We define chemical reaction rate and species production rate, and develop the accounting system for tracking the change in the reaction extent and the species concentration. Since most processes involve multiple chemical reactions, we make free use of matrices and linear algebra to summarize compactly the reaction stoichiometry.

Chapter 3. Next, in Chapter 3, we briefly review the important facts concerning the equilibrium state of a system undergoing chemical reaction. Most chemical processes do not reach equilibrium, but knowledge of the equilibrium state of the system allows one to define limits of reactor performance and identify operating conditions to realize desired production rates. The conditions for equilibrium are developed using the Gibbs energy and the chemical potential or species activity. We also briefly review phase equilibrium so that we are prepared for multiphase reactions. The condition for chemical and phase equilibrium is generally stated as the minimization of an appropriate energy function or maximization of entropy. The use of numerical optimization methods is illustrated for solving complex reaction equilibrium problems involving many reactions.

Chapter 4. In Chapter 4 we develop the material balances for the three reactor types: batch (and semi-batch), continuous-stirred-tank, and plug-flow reactors. We consider homogeneous, single-phase, isothermal reactions at this point. We derive and illustrate the use of the basic material balances for single and multiple reactions using molar concentration or molar flow as the only dependent variable. We also consider the volume change upon reaction and show how to employ the equation of state to complement the species material balances in situations in which the fluid density is not constant. We also introduce the use of stochastic simulation to simulate reaction kinetics in well-mixed reactors.

Chapter 5. We feel it works best to introduce the chemical reactor as soon as possible in Chapter 4, so we delay a comprehensive study of reaction rates and reaction rate expressions until Chapter 5. In Chapter 5 we provide a simple, theoretical framework for predicting the rate of a chemical reaction and for relating detailed statements of the reaction chemistry to reaction rates. We treat both homogeneous and heterogeneous reactions. Because chemical reactions occur at extremely different rates, and species are present in extremely different concentrations, we can often reduce the complexity of the reaction mechanism without changing the main features of the model. We develop the two main procedures for reducing the reaction mechanism complexity: the reaction equilibrium assumption, and the quasi-steady-state assumption. We illustrate the use of these assumptions in developing kinetic expressions, such as free-radical polymerization kinetics. The chapter concludes with a discussion of mechanisms for reactions occurring at the surfaces of solid catalysts.

Chapter 6. In Chapter 6 we develop the energy balance for the chemical reactor. The combined material and energy balances provide us with a rich description of many chemically reacting systems. The coupling of the material and energy balances also provides some surprises. The reactor behavior can become complex and interesting, and we explore some of the complex behavior such as multiple steady states and sustained oscillations. We explore such issues as how to remove or add heat to the system to overcome equilibrium limitations on conversion, and how to integrate a process to use the heat of reaction to preheat the feed entering the reactor.

Chapter 7. Chapter 7 considers the industrially important case of heterogeneous reactions taking place in solid catalyst particles. In the catalyst particle, we must consider the combined reaction-diffusion problem to be able to evaluate the temperature and the species concentrations. The material and energy balances for reaction-diffusion in catalyst particles result in fairly challenging boundary-value problems, and we discuss numerical methods for solving them. Packing the solid catalyst particles in a tube and passing a fluid stream over them produces the fixed-bed reactor. We show how to couple the fluid balances to the catalyst particle balances in order to predict the overall fixed-bed reactor behavior.

Chapter 8. The tradition in introductory reactor design courses is to neglect a careful treatment of the fluid flow and use the simplified, ideal reactors for modeling. In Chapter 8 we explore what to do when the reactor flow pattern is not well represented by these ideal mixing assumptions. We introduce the reactor residence-time distribution and describe the general issues of mixing in chemical reactors. We describe the limitations of these approximate mixing models. In some cases, accurate modeling requires one to solve for the complete velocity profile in the reactor. Although that topic is beyond the scope of this text, computational fluid dynamics software is evolving to the point that this approach is becoming tractable for many problems of interest.

Chapter 9. The fundamental reactor modeling principles covered in Chapters 2–8 provide the framework in which we think about chemical reactors. We understand which phenomena cause which observed reactor behaviors, and which design variables should be changed if we wish to alter the reactor performance. But when we want to make *quantitative* predictions of reactor performance, we require values for the model parameters. It is a simple fact that most of the parameters needed for the chemistries and reactor configurations of interest are not available in the literature. To make these models useful in standard industrial practice, therefore, we must be able to conveniently determine or estimate these parameters from experimental data collected on the system of interest. Chapter 9 covers this important topic of parameter estimation, which is not usually addressed in a systematic manner in introductory treatments of reactor analysis and design.

Chapter 10. Chapter 10 treats more complex situations in which the reactor of interest contains one phase of matter dispersed in a second phase. This dispersion is often a solid or particulate phase dispersed in a liquid. The crystallization and purification of solid crystalline products, such as pharmaceuticals, from a solvent mixture is an important example of this kind of dispersion. The dispersion may also be one liquid phase in small domains that are encapsulated by a separating membrane or stabilization layer and dispersed in a second, continuous liquid phase. Biological cells and emulsion polymers are important examples of this type of dispersion.

The chapter develops the evolution equation for the particle phase's particle size distribution, known as the population balance. In addition, multiphase mass and energy balances are derived for treating the continuous phase. To complement the deterministic, continuous population balance, discrete stochstic models are also introduced to model particle nucleation and growth. The connections and differences between the deterministic and stochastic descriptions are then developed to conclude the chapter.

Appendix A: Computational Methods. Finally, in Appendix A we summarize the numerical methods that have been necessary to solve the reactor models presented in the text. Linear algebra and matrices are introduced to handle reaction stoichiometry involving multiple reactions. Methods for solving ordinary differential equations and differential-algebraic equations are the real workhorses of reaction engineering; they are required in solving the material and energy balances. We also make extensive use of methods to compute the sensitivities of solutions of differential equations to the parameters appearing in the model. These sensitivities are needed for two purposes. First, they help in the problem of estimating model parameters from data. Second, they provide one means for solving the challenging boundary-value problems that arise with simultaneous reaction and diffusion in catalyst pellets. Optimization methods are used extensively in science and engineering. In this text, we used optimization for three main purposes: finding the equilibrium state for complex situations involving multiple reactions and multiple phases; estimating model parameters from data; and solving reactor design problems.

See `www.nobhillpublishing.com` for the computational appendix.

Bibliography

[1] Facts and figures for the chemical industry. *Chem. Eng. News*, 79(26): 43–51, 2001.

[2] T. Bramblett, Q. Lu, T. Karasawa, M. Hasan, S. Jo, and J. Greene. Si(001)2×1 gas-source molecular-beam epitaxy from Si_2H_6: Growth kinetics and boron doping. *J. Appl. Phys.*, 76(3):1884–1888, 1994.

[3] R. J. Farrauto and C. H. Bartholomew. *Fundamentals of Industrial Catalytic Processes.* Blackie Academic and Professional, London, UK, 1997.

[4] S. J. Flint, L. W. Enquist, R. M. Krug, V. R. Racaniello, and A. M. Skalka. *Principles of Virology: Molecular Biology, Pathogenesis, and Control.* American Society for Microbiology, ASM Press, Washington, D.C., 2000.

[5] S. Gates, C. Greenlief, D. Beach, and P. Holbert. Deposition of silane on Si(111)-(7×7) and Si(100)-(2×1) surfaces below 500 °C. *J. Chem. Phys.*, 92 (5):3144–3153, 1990.

[6] Y. V. Kissin and L. W. Pebsworth. Polyethylene. In J. I. Kroschwitz, editor, *Kirk-Othmer Encyclopedia of Chemical Technology*, volume 17, pages 702–784. John Wiley & Sons, New York, fourth edition, 1996.

[7] H. H. Lee. *Fundamental of Microelectronics Processing.* McGraw-Hill Publishing Company, New York, 1990.

[8] D. C. McCulloch. Catalytic hydrotreating in petroleum refining. In B. E. Leach, editor, *Applied Industrial Catalysis*, volume one, pages 69–121. Academic Press, New York, 1983.

[9] S. A. Middlebrooks. *Modelling and Control of Silicon and Germanium Thin Film Chemical Vapor Deposition.* PhD Dissertation, University of Wisconsin, Madison, Wisconsin, 2001.

[10] M. L. Occelli and R. Chianelli. *Hydrotreating Technology for Pollution Control Catalysts, Catalysis, and Processes.* Marcell Dekker, Inc., New York, 1996.

[11] J. D. Plummer, M. Deal, and P. B. Griffin. *Silicon VLSI Technology Fundamentals, Practice and Modeling.* Prentice Hall, Upper Saddle River, New Jersey, 2000.

[12] N. M. Russell and W. Breiland. A surface kinetics model for the growth of $Si_{1-x}Ge_x$ films from hydrides. *J. Appl. Phys.*, 73(7):3525–3530, 1993.

[13] N. M. Russell and J. G. Ekerdt. Kinetics of hydrogen desorption from germanium-covered Si(100). *Surface Science*, 369:51–68, 1996.

[14] C. Seeger and W. S. Mason. Hepatitis B virus biology. *Microbiol. Mol. Biol. Rev.*, 64(1):51–68, March 2000.

[15] K. Sinniah, M. Sherman, L. Lewis, W. Weinberg, J. Yates, and K. Janda. New mechanisms for hydrogen desorption from covalent surfaces: The monohydride phase on Si(100). *Phys. Rev. Lett.*, 62(5):567–570, 1989.

[16] R. Srivastava, D. Loeb, and J. Yin. Computational and experimental analysis of duck hepatitis B virus intracellular kinetics. Annual AIChE Meeting, Reno, Nevada, November 2001.

[17] R. Srivastava, L. You, J. Summers, and J. Yin. Stochastic vs. deterministic modeling of intracellular viral kinetics. *J. Theor. Biol.*, 218:309–321, 2002.

[18] M. P. Stevens. *Polymer Chemistry, an Introduction.* Oxford University Press, New York, 1999.

[19] B. G. Streetman and S. Banerjee. *Solid State Electronics.* Prentice Hall, Upper Saddle River, New Jersey, fifth edition, 2000.

[20] Y. Suda, D. Lubben, T. Motooka, and J. Greene. Adsorption and thermal dissociation of disilane on Si(100)2×1. *J. Vac. Sci. Technol.A.*, 8(1):61–67, 1990.

[21] J. Summers, P. M. Smith, and A. L. Horwich. Hepadnavirus envelope proteins regulate covalently closed circular DNA amplification. *J. Virol.*, 64 (6):2819–2824, 1990.

[22] J. Summers, P. M. Smith, M. Huang, and M. Yu. Morphogenetic and regulatory effects of mutations in the envelope proteins of an avian hepadnavirus. *J. Virol.*, 65(3):1310–1317, 1991.

[23] T. R. Younkin, E. F. Connor, J. I. Henderson, S. K. Friedrich, R. H. Grubbs, and D. A. Bansleben. Neutral, single-component nickel (II) polyolefin catalysts that tolerate heteroatoms. *Science*, 287:460–462, 2000.

2

The Stoichiometry of Reactions

2.1 Introduction

Stoichiometry is defined as the determination of the proportions in which chemical elements combine or are produced and the weight relations in any chemical reaction.[1] In this chapter we explore and develop these quantitative relations between the different chemical species undergoing chemical reaction.

The next section establishes the accounting procedure for tracking chemical change and introduces the stoichiometric matrix. Section 2.3 introduces the concept of linearly independent reactions and discusses the implications of mass conservation on the stoichiometric matrix. Section 2.4 defines the rates of reactions and rates of production of chemical species due to the reactions. Section 2.5 explores the issues involved in calculating production rates given reaction rates and vice versa. We also formulate and solve the least-squares problem of extracting best estimates of reaction rates given production rate measurements containing errors.

Section 2.6 provides a summary of the important concepts and relationships that may prove useful as a study guide or quick reference. References for further study are provided at the end of the chapter. Exercises are provided in Section 2.7 for reinforcing the concepts and to further develop one's understanding. The last several exercises introduce new material and show how the stoichiometry fundamentals presented in this chapter lead into other interesting topics. The reactor analysis book by Aris [1] influenced several sections of this chapter.

[1] Webster's New World College Dictionary, fifth edition, 2004.

2.2 Chemical Reactions and Stoichiometry

We shall consider three primary examples as a means of illustrating the main concepts in this chapter. The first example is one of the reactions responsible for smog formation in the atmosphere. It consists of a single reaction among three species. The second example is the water gas shift reaction, and it illustrates the case of multiple reactions; it consists of three reactions among six chemical species. The third example illustrates the complexity of common industrial reactions of interest, consisting of 20 reactions among 14 species. These reactions have been proposed to describe a silicon chemical vapor deposition process, which is an important step in the production of microelectronic materials.

For the first example, consider two molecules of nitric oxide and one molecule of oxygen reacting to form two molecules of nitrogen dioxide. The stoichiometry of this reaction is

$$2NO + O_2 \rightleftharpoons 2NO_2 \tag{2.1}$$

The convention that we follow is reactants appear on the left-hand side of the chemical reaction symbol, $\rightleftharpoons$, and products appear on the right-hand side. In this example, there is a single chemical reaction and three different chemical species taking part in the reaction, NO, O_2, and NO_2.

The second example is known as the water gas shift reaction. The overall stoichiometry of this reaction is

$$H_2O + CO \rightleftharpoons CO_2 + H_2 \tag{2.2}$$

The rate of this reaction is important in determining the CO/CO_2 ratio in exhaust gases from internal combustion engines, and in determining the H_2 content in the feed for fuel cells. It also is known that the following two reactions are needed to describe what is happening at the molecular level,

$$H_2O + H \rightleftharpoons H_2 + OH \tag{2.3}$$

$$OH + CO \rightleftharpoons CO_2 + H \tag{2.4}$$

Reactions 2.2–2.4 comprise a simple reaction network. There are three chemical reactions and six different chemical species taking part in the three reactions, H, H_2, OH, H_2O, CO, and CO_2.

In order to organize the way we discuss chemical reactions, the following notation is convenient. Let the symbol, A_j, represent the jth species taking part in a reaction. In the first example, we can choose $A_1 = \mathrm{NO}$, $A_2 = \mathrm{O_2}$, and $A_3 = \mathrm{NO_2}$. In the water gas shift example, we can choose $A_1 = \mathrm{H}$, $A_2 = \mathrm{H_2}$, $A_3 = \mathrm{OH}$, $A_4 = \mathrm{H_2O}$, $A_5 = \mathrm{CO}$, and $A_6 = \mathrm{CO_2}$. Using the A_j notation, we can express the water gas shift reactions as

$$\begin{aligned} A_4 \;+\; A_5 &\rightleftharpoons A_6 \;+\; A_2 \\ A_4 \;+\; A_1 &\rightleftharpoons A_2 \;+\; A_3 \\ A_3 \;+\; A_5 &\rightleftharpoons A_6 \;+\; A_1 \end{aligned} \tag{2.5}$$

Reactions 2.5 suppress the identities of the species for compactness. We can further compress the description by moving all of the *variables* to the right-hand side of the chemical reaction symbol and replacing it with an equality sign,

$$\begin{aligned} -A_4 - A_5 + A_6 + A_2 &= 0 \\ -A_4 - A_1 + A_2 + A_3 &= 0 \\ -A_3 - A_5 + A_6 + A_1 &= 0 \end{aligned} \tag{2.6}$$

Again notice the sign convention that **p***roducts* have **p***ositive* coefficients and *reactants* have *negative* coefficients in Equations 2.6.[2] Equations 2.6 now resemble a set of three linear algebraic equations and motivates the use of matrices. Using the rules of matrix multiplication, one can express Equations 2.6 as

$$\begin{bmatrix} 0 & 1 & 0 & -1 & -1 & 1 \\ -1 & 1 & 1 & -1 & 0 & 0 \\ 1 & 0 & -1 & 0 & -1 & 1 \end{bmatrix} \begin{bmatrix} A_1 \\ A_2 \\ A_3 \\ A_4 \\ A_5 \\ A_6 \end{bmatrix} = \begin{bmatrix} 0 \\ 0 \\ 0 \end{bmatrix} \tag{2.7}$$

The matrix appearing in Equation 2.7 provides an efficient description of the stoichiometry for the reaction network, and is appropriately known as the **stoichiometric matrix**. Giving the stoichiometric matrix the symbol $\boldsymbol{\nu}$, and writing $\boldsymbol{A}$ to denote the column vector of the $A_j, j = 1, \ldots, 6$, our final summary of the water gas shift reaction appears as

$$\boldsymbol{\nu A} = \boldsymbol{0}$$

[2]Boldface letters provide a mnemonic device.

The element ν_{ij} in the stoichiometric matrix is the stoichiometric coefficient for the jth species in the ith reaction. The index i runs from 1 to n_r, the total number of reactions in the network, and the index j runs from 1 to n_s, the total number of species in the network. We say that $\boldsymbol{\nu}$ is an $n_r \times n_s$ matrix. After piling up this much abstraction to describe what started out as a simple set of three reactions, let us work a few examples to reinforce the concept of the stoichiometric matrix.

Example 2.1: Stoichiometric matrix for a single reaction

Find the stoichiometric matrix for the nitric oxide example,

$$2NO + O_2 \rightleftharpoons 2NO_2$$

Solution

The nitric oxide example consists of one reaction and three species. We can assign the species to the $\boldsymbol{A}$ as follows: $A_1 = NO$, $A_2 = O_2$, $A_3 = NO_2$. The reaction can then be written as

$$-2A_1 - A_2 + 2A_3 = \begin{bmatrix} -2 & -1 & 2 \end{bmatrix} \begin{bmatrix} A_1 \\ A_2 \\ A_3 \end{bmatrix} = 0$$

The stoichiometric matrix for a single reaction is a row vector, in this case,

$$\boldsymbol{\nu} = \begin{bmatrix} -2 & -1 & 2 \end{bmatrix}$$

□

Example 2.2: Columns of $\boldsymbol{\nu}$

Since we are free to assign chemical species to the A_j in any order we choose, consider what happens if we change the order of the species in the water gas shift example. Instead of using $A_1 = H$ and $A_6 = CO_2$, what is the stoichiometric matrix if A_1 is chosen to be CO_2 and A_6 is chosen to be H?

Solution

Switching the identities of the first and sixth species in the $\boldsymbol{A}$ vector gives us the following modified vector, $A_1' = CO_2$, $A_2' = H_2$, $A_3' = OH$,

$A'_4 = H_2O$, $A'_5 = CO$, and $A'_6 = H$. Reactions 2.5 are then modified to give

$$\begin{array}{ccccccc} A'_4 & + & A'_5 & \rightleftharpoons & A'_1 & + & A'_2 \\ A'_4 & + & A'_6 & \rightleftharpoons & A'_2 & + & A'_3 \\ A'_3 & + & A'_5 & \rightleftharpoons & A'_1 & + & A'_6 \end{array} \tag{2.8}$$

Extracting the modified $\boldsymbol{\nu}$ matrix from Equations 2.8 gives

$$\boldsymbol{\nu}' = \begin{bmatrix} \overset{\downarrow}{1} & 1 & 0 & -1 & -1 & \overset{\downarrow}{0} \\ 0 & 1 & 1 & -1 & 0 & -1 \\ 1 & 0 & -1 & 0 & -1 & 1 \end{bmatrix} \tag{2.9}$$

It is clear from examining Equations 2.9 and 2.7 that switching the identities of species one and six in the $\boldsymbol{A}$ vector has necessitated switching the first and sixth *columns* in the stoichiometric matrix. Therefore one can make the connection between the *columns* of $\boldsymbol{\nu}$ and the *species* taking part in the reactions. More precisely, the jth column of the $\boldsymbol{\nu}$ matrix supplies the stoichiometric numbers of the jth species in all of the reactions. □

Example 2.3: Rows of $\boldsymbol{\nu}$

Just as we are free to assign the species to the $\boldsymbol{A}$ vector in any order we choose, we are also free to express the reactions in any order we choose. We now explore what happens if instead of expressing the water gas shift reaction as it appears in Reactions 2.2–2.4, we express it as

$$OH + CO \rightleftharpoons CO_2 + H \tag{2.10}$$

$$H_2O + H \rightleftharpoons H_2 + OH \tag{2.11}$$

$$H_2O + CO \rightleftharpoons CO_2 + H_2 \tag{2.12}$$

Notice we have written the original third reaction first and the original first reaction third. What is the impact of this change on the $\boldsymbol{\nu}$ matrix?

Solution

Using the original ordering of the $\boldsymbol{A}$ vector, we can ascribe the following elements of a third stoichiometric matrix,

$$\begin{array}{ccccccc} A_3 & + & A_5 & \rightleftharpoons & A_6 & + & A_1 \\ A_4 & + & A_1 & \rightleftharpoons & A_2 & + & A_3 \\ A_4 & + & A_5 & \rightleftharpoons & A_6 & + & A_2 \end{array}$$

$$\boldsymbol{\nu}'' = \begin{matrix} \longrightarrow \\ \\ \longrightarrow \end{matrix} \begin{bmatrix} 1 & 0 & -1 & 0 & -1 & 1 \\ -1 & 1 & 1 & -1 & 0 & 0 \\ 0 & 1 & 0 & -1 & -1 & 1 \end{bmatrix} \tag{2.13}$$

As one might expect, exchanging the orders of the first and third reactions causes us to exchange the first and third *rows* in the $\boldsymbol{\nu}$ matrix as evidenced by comparing the matrices in Equations 2.13 and 2.7. We can therefore make the connection between the **r***ows* of $\boldsymbol{\nu}$ and the **r***eactions*.[3] The ith row of the stoichiometric matrix contains the stoichiometric numbers of all species in the ith reaction. □

From the previous two examples it is clear that one could develop a large number of stoichiometric matrices to describe the same set of chemical reactions. Since there is no reason to prefer one ordering of species and reactions over another, one may permute the columns and rows into any order and maintain a valid stoichiometric matrix.

We now introduce the third example, which is a more complicated reaction network. The following chemistry has been proposed to describe a silicon chemical vapor deposition (CVD) reaction, which is an important process in the production of microelectronic materials.

$$\begin{array}{ll|ll}
SiH_4 \rightleftharpoons SiH_2 + H_2 & & SiH_2 + SiH \rightleftharpoons Si_2H_3 & \\
SiH_4 \rightleftharpoons SiH_3 + H & & SiH_2 + Si \rightleftharpoons Si_2H_2 & \\
SiH_4 + SiH_2 \rightleftharpoons Si_2H_6 & & SiH_2 + Si_3 \rightleftharpoons Si_2H_2 + Si_2 & \\
Si_2H_4 + H_2 \rightleftharpoons SiH_4 + SiH_2 & & H_2 + Si_2H_2 \rightleftharpoons Si_2H_4 & \\
SiH_4 + H \rightleftharpoons SiH_3 + H_2 & & H_2 + Si_2H_4 \rightleftharpoons Si_2H_6 & \\
SiH_4 + SiH_3 \rightleftharpoons Si_2H_5 + H_2 & & H_2 + SiH \rightleftharpoons SiH_3 & \\
SiH_4 + SiH \rightleftharpoons SiH_3 + SiH_2 & & H_2 + Si_2 \rightleftharpoons Si_2H_2 & \\
SiH_4 + SiH \rightleftharpoons Si_2H_5 & & H_2 + Si_2H_3 \rightleftharpoons Si_2H_5 & \\
SiH_4 + Si \rightleftharpoons 2SiH_2 & & Si_2H_2 + H \rightleftharpoons Si_2H_3 & \\
Si + H_2 \rightleftharpoons SiH_2 & & Si + Si_3 \rightleftharpoons 2Si_2 &
\end{array} \tag{2.14}$$

The student should not be dismayed by the complexity of this reaction network. Indeed the principles for analyzing a CVD reactor with this chemistry are exactly the same as the principles for analyzing the

[3] Another mnemonic.

simpler nitric oxide and water gas shift chemistries presented previously. The only difference is that Reactions 2.14 are complex enough that we use a computer to keep track of the algebra for us. In fact, Reactions 2.14 are a simplified version of 120 reactions that were originally postulated for this reaction network [3].

Chemical engineers should also bear in mind that Reactions 2.14 are quite simple compared to many mechanisms that have been proposed for combustion problems in which it is not uncommon to have several hundred reactions. Polymerizations and long-chain-producing reactions consist of thousands of species and associated reactions. Obviously the stoichiometry of these complex problems is intractable if we do not develop a systematic, automated procedure. Developing and understanding that procedure is the topic of the next several sections.

Example 2.4: Stoichiometric matrix for CVD chemistry

Determine the stoichiometric matrix corresponding to Reactions 2.14.

Solution

The first thing we notice is that there are 20 reactions or $n_r = 20$. We then look through all of the reactions and identify the different species taking part. After writing this out we notice that there are 14 different species, $n_s = 14$. A possible assignment to the $\boldsymbol{A}$ vector is: H, H_2, Si, SiH, SiH_2, SiH_3, SiH_4, Si_2, Si_2H_2, Si_2H_3, Si_2H_4, Si_2H_5, Si_2H_6, Si_3. With an $\boldsymbol{A}$ chosen, it is a simple matter to look through Reactions 2.14 and find the stoichiometric coefficients of each species in each reaction. Do not forget the convention that species appearing as products in a given reaction have positive coefficients and those appearing as reactants have negative coefficients. Practice filling out a few rows of the $\boldsymbol{\nu}$ matrix and check it with the values given in Equation 2.15.

Notice that for this example $\boldsymbol{\nu}$ is a 20×14 matrix, and it contains many zero entries. A matrix with many zero entries is called sparse. The large number of zeros simply reflects the physical fact that very few molecules can take part in a particular reaction. All of the reactions in the CVD chemistry, for example, are unimolecular or bimolecular. More will be said about this issue in the discussion of mechanisms in Chapter 5.

$$
\boldsymbol{\nu} = \begin{bmatrix}
0 & 1 & 0 & 0 & 1 & 0 & -1 & 0 & 0 & 0 & 0 & 0 & 0 & 0 \\
1 & 0 & 0 & 0 & 0 & 1 & -1 & 0 & 0 & 0 & 0 & 0 & 0 & 0 \\
0 & 0 & 0 & 0 & -1 & 0 & -1 & 0 & 0 & 0 & 0 & 0 & 1 & 0 \\
0 & -1 & 0 & 0 & 1 & 0 & 1 & 0 & 0 & 0 & -1 & 0 & 0 & 0 \\
-1 & 1 & 0 & 0 & 0 & 1 & -1 & 0 & 0 & 0 & 0 & 0 & 0 & 0 \\
0 & 1 & 0 & 0 & 0 & -1 & -1 & 0 & 0 & 0 & 0 & 1 & 0 & 0 \\
0 & 0 & 0 & -1 & 1 & 1 & -1 & 0 & 0 & 0 & 0 & 0 & 0 & 0 \\
0 & 0 & 0 & -1 & 0 & 0 & -1 & 0 & 0 & 0 & 0 & 1 & 0 & 0 \\
0 & 0 & -1 & 0 & 2 & 0 & -1 & 0 & 0 & 0 & 0 & 0 & 0 & 0 \\
0 & -1 & -1 & 0 & 1 & 0 & 0 & 0 & 0 & 0 & 0 & 0 & 0 & 0 \\
0 & 0 & 0 & -1 & -1 & 0 & 0 & 0 & 0 & 1 & 0 & 0 & 0 & 0 \\
0 & 0 & -1 & 0 & -1 & 0 & 0 & 0 & 1 & 0 & 0 & 0 & 0 & 0 \\
0 & 0 & 0 & 0 & -1 & 0 & 0 & 1 & 1 & 0 & 0 & 0 & 0 & -1 \\
0 & -1 & 0 & 0 & 0 & 0 & 0 & 0 & -1 & 0 & 1 & 0 & 0 & 0 \\
0 & -1 & 0 & 0 & 0 & 0 & 0 & 0 & 0 & 0 & -1 & 0 & 1 & 0 \\
0 & -1 & 0 & -1 & 0 & 1 & 0 & 0 & 0 & 0 & 0 & 0 & 0 & 0 \\
0 & -1 & 0 & 0 & 0 & 0 & 0 & -1 & 1 & 0 & 0 & 0 & 0 & 0 \\
0 & -1 & 0 & 0 & 0 & 0 & 0 & 0 & 0 & -1 & 0 & 1 & 0 & 0 \\
-1 & 0 & 0 & 0 & 0 & 0 & 0 & 0 & -1 & 1 & 0 & 0 & 0 & 0 \\
0 & 0 & -1 & 0 & 0 & 0 & 0 & 2 & 0 & 0 & 0 & 0 & 0 & -1
\end{bmatrix} \tag{2.15}
$$

□

Example 2.5: Conservation of mass

Show that conservation of mass in a chemical reaction can be stated as

$$\boldsymbol{\nu M} = \mathbf{0}$$

in which M_j is the molecular weight of species j.

Solution

In a chemical reaction, the number of molecules is not conserved in general. For example in the nitric oxide reaction, $2NO + O_2 \rightleftharpoons 2NO_2$, *three* reactant molecules react to form *two* product molecules. The mass, however, is conserved in chemical (i.e., not nuclear) reactions. It is clear in the above example that the atoms (N and O) are conserved, so the mass is conserved. Another way to state conservation of mass involves molecular weights of the species. In the nitric oxide reaction, the molecular weights of reactants and products are related by $2M_{NO} +$

$M_{O_2} = 2M_{NO_2}$. Equivalently

$$-2M_{NO} - M_{O_2} + 2M_{NO_2} = \begin{bmatrix} -2 & -1 & 2 \end{bmatrix} \begin{bmatrix} M_{NO} \\ M_{O_2} \\ M_{NO_2} \end{bmatrix} = 0$$

If we put the molecular weights of the species in a vector, $\boldsymbol{M}$, conservation of mass for this single reaction can be written as,

$$\boldsymbol{\nu M} = 0$$

For the water gas shift reaction, using the ordering of the species, $A_1 =$ H, $A_2 = H_2$, $A_3 =$ OH, $A_4 = H_2O$, $A_5 =$ CO, and $A_6 = CO_2$,

$$\boldsymbol{M} = \begin{bmatrix} M_H & M_{H_2} & M_{OH} & M_{H_2O} & M_{CO} & M_{CO_2} \end{bmatrix}^T$$

in which the superscript T means the transpose of the matrix. The transpose of the matrix means to exchange the rows for columns and vice versa. For the first reaction, $H_2O + CO \rightleftharpoons CO_2 + H_2$, we know

$$M_{CO_2} + M_{H_2} - M_{H_2O} - M_{CO} = 0$$

or

$$\begin{bmatrix} 0 & 1 & 0 & -1 & -1 & 1 \end{bmatrix} \begin{bmatrix} M_H \\ M_{H_2} \\ M_{OH} \\ M_{H_2O} \\ M_{CO} \\ M_{CO_2} \end{bmatrix} = 0$$

which is the first row of $\boldsymbol{\nu}$ in Equation 2.7 multiplied by $\boldsymbol{M}$. The second and third reactions simply fill out the second and third rows of $\boldsymbol{\nu}$ so that again, for multiple reactions

$$\boldsymbol{\nu M} = \boldsymbol{0}$$

□

2.3 Independent Reactions

To motivate the discussion of independence of chemical reactions, let us again consider the water gas shift reaction

$$\begin{aligned} H_2O + CO &\rightleftharpoons CO_2 + H_2 \\ H_2O + H &\rightleftharpoons H_2 + OH \\ OH + CO &\rightleftharpoons CO_2 + H \end{aligned} \tag{2.16}$$

The issue of independence centers on the question of whether or not we can express any reaction in the network as a linear combination of the other reactions. If we can, then the set of reactions is not independent. It is not necessary to eliminate extra reactions and work with the smallest set, but it is sometimes preferable. In any case, the concept is important and is examined further. Before making any of these statements precise, we explore the question of whether or not the three reactions listed in Reactions 2.16 are independent. Can we express the first reaction as a linear combination of the second and third reactions? By *linear* combination we mean multiplying a reaction by a number and adding it to the other reactions. It is clear from inspection that the first reaction is the sum of the second and third reactions, so the set of three reactions is not independent.

$$
\begin{array}{rcccccc}
 & H_2O & + & \boxed{H} & \rightleftharpoons & H_2 & + & \boxed{OH} \\
+ & \boxed{OH} & + & CO & \rightleftharpoons & CO_2 & + & \boxed{H} \\
\hline
 & H_2O & + & CO & \rightleftharpoons & CO_2 & + & H_2
\end{array}
$$

If we deleted the first reaction from the network, would the remaining two reactions be independent?

$$H_2O + H \rightleftharpoons H_2 + OH \tag{2.17}$$

$$OH + CO \rightleftharpoons CO_2 + H \tag{2.18}$$

The answer is now yes, because no multiple of Reaction 2.17 can equal Reaction 2.18. There is no way to produce CO or CO_2 from only Reaction 2.17. Likewise there is no way to produce H_2 or H_2O from only Reaction 2.18.

This discussion is *not* meant to imply that there is something *wrong* with the first reaction in Reactions 2.16. Indeed if we focus attention on the second reaction, we can again ask the question whether or not it can be written as a linear combination of the first and third reactions. The answer is yes because the second reaction is the first reaction minus the third reaction.

$$
\begin{array}{rcccccc}
 & H_2O & + & \boxed{CO} & \rightleftharpoons & \boxed{CO_2} & + & H_2 \\
- & \{\, OH & + & \boxed{CO} & \rightleftharpoons & \boxed{CO_2} & + & H \,\} \\
\hline
 & H_2O & + & H & \rightleftharpoons & H_2 & + & OH
\end{array}
$$

So the first and third reactions could be chosen as the independent set of two reactions. Finally, the third reaction in Reaction 2.16 is equal to the first reaction minus the second reaction, so the first and second reactions could be chosen as an independent set. For this example then, any two of the reactions comprise an independent set. The situation is not always this simple as we will see from the chemical vapor deposition chemistry.

Before making the problem more complicated, we explore how to automate the preceding analysis by exploiting the stoichiometric matrix. If you are familiar with linear algebra, the issue of independence of reactions is obviously related to the rank of the stoichiometric matrix. Familiarity with these concepts, although helpful, is not required to follow the subsequent development. We now consider the stoichiometric matrix for the water gas shift reaction presented in Equation 2.7

$$\boldsymbol{\nu} = \begin{bmatrix} 0 & 1 & 0 & -1 & -1 & 1 \\ -1 & 1 & 1 & -1 & 0 & 0 \\ 1 & 0 & -1 & 0 & -1 & 1 \end{bmatrix} \tag{2.19}$$

We can make an important mathematical connection to the preceding physical arguments. The question of whether or not the ith *reaction* can be written as a linear combination of the other reactions is the same as the question of whether or not the ith *row* of the $\boldsymbol{\nu}$ matrix can be written as a linear combination of the other rows. The linear independence of the reactions in a reaction network is equivalent to the linear independence of the rows in the corresponding stoichiometric matrix.

The **rank** of a matrix is defined as the number of linearly independent rows (or equivalently, columns) in the matrix. Therefore, the number of linearly independent reactions in a network, n_i, is equal to the rank of $\boldsymbol{\nu}$. There are efficient numerical algorithms available for finding the rank of a matrix and a set of linearly independent rows. The focus of our attention is not on the algorithm, but on how we can exploit the results of the algorithm to analyze sets of chemical reactions. You should consult Strang [7] or another linear algebra text for a lucid explanation of the algorithm, Gaussian elimination with partial pivoting.

Example 2.6: More species than reactions

Show that mass conservation implies that any *independent* set of reactions has more species than reactions.

Solution

From Example 2.5, we know that mass conservation is equivalent to

$$\boldsymbol{\nu} \boldsymbol{M} = \boldsymbol{0}$$

Consider the columns of the $\boldsymbol{\nu}$ matrix as column vectors. This matrix-vector multiplication can be expressed as a linear combination of the columns of $\boldsymbol{\nu}$ with the elements of the $\boldsymbol{M}$ vector as the coefficients in the linear combination

$$\begin{aligned}
\boldsymbol{\nu M} &= \begin{bmatrix} \nu_{11} & \nu_{12} & \cdots & \nu_{1n_s} \\ \vdots & \vdots & \ddots & \vdots \\ \nu_{n_i1} & \nu_{n_i2} & \cdots & \nu_{n_in_s} \end{bmatrix} \begin{bmatrix} M_1 \\ M_2 \\ \vdots \\ M_{n_s} \end{bmatrix} \\
&= \begin{bmatrix} \nu_{11} \\ \vdots \\ \nu_{n_i1} \end{bmatrix} M_1 + \begin{bmatrix} \nu_{12} \\ \vdots \\ \nu_{n_i2} \end{bmatrix} M_2 + \cdots + \begin{bmatrix} \nu_{1n_s} \\ \vdots \\ \nu_{n_in_s} \end{bmatrix} M_{n_s} \\
&= \begin{bmatrix} 0 \\ \vdots \\ 0 \end{bmatrix}
\end{aligned} \tag{2.20}$$

The last equation implies the columns of $\boldsymbol{\nu}$ are linearly *dependent* because the molecular weights are nonzero.[4] Because the rows are linearly independent, we conclude there are more columns (species) than rows (independent reactions), $n_s > n_i$ and $\boldsymbol{\nu}$ is a *wide* matrix (i.e., not a square or *tall* matrix).

Notice that one must consider linearly independent reactions for the statement in the example to be true. If we considered an arbitrary network for example,

$$\begin{aligned}
2\text{NO} + \text{O}_2 &\rightleftharpoons 2\text{NO}_2 \\
4\text{NO} + 2\text{O}_2 &\rightleftharpoons 4\text{NO}_2 \\
6\text{NO} + 3\text{O}_2 &\rightleftharpoons 6\text{NO}_2 \\
8\text{NO} + 4\text{O}_2 &\rightleftharpoons 8\text{NO}_2
\end{aligned}$$

[4]One could solve for column j of $\boldsymbol{\nu}$ by moving the remaining terms to the other side of the equality and dividing by M_j, which is nonzero. That is possible as long as one of the M_j multipliers in Equation 2.20 is nonzero. In our case, all of the multipliers are nonzero.

These reactions are obviously not independent because they are the same reaction written four times. In this case

$$\boldsymbol{\nu} = \begin{bmatrix} -2 & -1 & 2 \\ -4 & -2 & 4 \\ -6 & -3 & 6 \\ -8 & -4 & 8 \end{bmatrix}$$

Since we have not used independent reactions, $\boldsymbol{\nu}$ is *tall* and not *wide*. Recall $\boldsymbol{\nu} = \begin{bmatrix} -2 & -1 & 2 \end{bmatrix}$ for the single independent reaction, which is in agreement with the example. □

Maximal sets of linearly independent reactions. Up to this point, we have started with a set of reactions and investigated constructing subsets of these reactions that are linearly independent. Now consider the reverse problem. We start with a set of species, and we would like to know the largest number of linearly independent valid chemical reactions among these species. In other words, a given set of reactions may be linearly independent, but we want to be sure we have not left out some valid reactions. By valid chemical reactions we mean *element conserving*, which is a sufficient condition implying the weaker condition of *mass conserving*.

The following describes a systematic approach to this problem. First, list formation reactions for every species in the list from its elements. We may use molecules of pure elements (O_2) rather than atoms (O) to save work as long as the atoms themselves do not appear in the species list. This set is guaranteed by construction to be a maximal linearly independent set for an *enlarged* species list, which includes the original species plus any new elements introduced in the formation reactions.

Then eliminate through linear combinations of reactions any new elements that were introduced in the formation reactions and that do not appear in the original species list. The remaining set is a maximal linearly independent set of reactions for the original species list.

This procedure is perhaps best illustrated by example.

Example 2.7: Maximal set of reactions for methane oxidation

Consider the oxidation of methane in which the observed species are:

$$\boldsymbol{A} = \begin{bmatrix} \mathrm{CO_2} & \mathrm{H_2O} & \mathrm{CH_4} & \mathrm{CO} & \mathrm{H_2} & \mathrm{O_2} \end{bmatrix}^T$$

We first write formation reactions for all species from the elements. It is necessary to write these reactions in the order in which the species appear in the species list. We can use O_2 and H_2 because O and H are not in the species list

$$\mathrm{C} + \mathrm{O_2} \rightleftharpoons \mathrm{CO_2} \tag{2.21}$$

$$2\mathrm{H_2} + \mathrm{O_2} \rightleftharpoons 2\mathrm{H_2O} \tag{2.22}$$

$$\mathrm{C} + 2\mathrm{H_2} \rightleftharpoons \mathrm{CH_4} \tag{2.23}$$

$$2\mathrm{C} + \mathrm{O_2} \rightleftharpoons 2\mathrm{CO} \tag{2.24}$$

Notice we have introduced the element carbon, which is not in the original species list, to express the formation reactions. We can add C to the list and the enlarged species list is

$$\tilde{\boldsymbol{A}} = \begin{bmatrix} \mathrm{CO_2} & \mathrm{H_2O} & \mathrm{CH_4} & \mathrm{CO} & \mathrm{H_2} & \mathrm{O_2} & \mathrm{C} \end{bmatrix}^T$$

The stoichiometry of the formation reactions are then summarized by

$$\tilde{\boldsymbol{\nu}}\tilde{\boldsymbol{A}} = \boldsymbol{0}$$

in which

$$\tilde{\boldsymbol{\nu}} = \begin{bmatrix} 1 & 0 & 0 & 0 & 0 & -1 & -1 \\ 0 & 2 & 0 & 0 & -2 & -1 & 0 \\ 0 & 0 & 1 & 0 & -2 & 0 & -1 \\ 0 & 0 & 0 & 2 & 0 & -1 & -2 \end{bmatrix} \tag{2.25}$$

Notice we have zeros below the diagonal of the first four rows and columns. We know *by inspection* that these reactions are linearly independent, which is why we wrote them in the first place.[5] So the rank of $\tilde{\boldsymbol{\nu}}$ is four. See Exercise 2.9 for a proof that there are no other valid linearly independent reactions among these species and elements.

We now wish to eliminate carbon from the species list. The approach is to replace formation reactions involving C with independent linear combinations of the four reactions that eliminate C from the set. For example we could replace Reaction 2.21 with the sum of Reaction 2.21 and the negative of Reaction 2.23. Equivalently we replace row 1 in Equation 2.25 with the sum of row 1 and negative of row 3. We always add linear combinations of rows *below* the row on which we are making the zero in order not to disturb the pattern of zeros below the diagonal in the first columns of the matrix. We leave the second row unchanged

[5] Examine the locations of the zeros in the first four rows and columns of $\tilde{\boldsymbol{\nu}}$. Because these portions of the first four rows are independent, so are the entire rows, and, therefore, the reactions.

because the $\nu_{2,7}$ is zero already; i.e., C does not take place in the second reaction. Proceeding down the rows we replace row 3 by twice row 3 minus row 4. When we reach the last row, either a zero already exists in the last column or we remove this last row because we have no rows below the last with which to zero that element. That reduces $\tilde{\boldsymbol{\nu}}$ to

$$\tilde{\boldsymbol{\nu}}' = \begin{bmatrix} 1 & 0 & -1 & 0 & 2 & -1 & 0 \\ 0 & 2 & 0 & 0 & -2 & -1 & 0 \\ 0 & 0 & 2 & -2 & -4 & 1 & 0 \end{bmatrix}$$

Inspection of the zeros in the first three columns of $\tilde{\boldsymbol{\nu}}'$ tells us that the rows are independent, and, therefore, the matrix has full rank. If we now multiply out these equations, the last column of zeros removes the C from the species list and we have

$$\boldsymbol{\nu A} = \mathbf{0}$$

in which

$$\boldsymbol{\nu} = \begin{bmatrix} 1 & 0 & -1 & 0 & 2 & -1 \\ 0 & 2 & 0 & 0 & -2 & -1 \\ 0 & 0 & 2 & -2 & -4 & 1 \end{bmatrix}$$

Therefore the maximal linearly independent set for the original species contains *three* reactions. The stoichiometric matrix above corresponds to the following choice of reactions

$$\begin{aligned} CH_4 + O_2 &\rightleftharpoons CO_2 + 2H_2 \\ 2H_2 + O_2 &\rightleftharpoons 2H_2O \\ 2CO + 4H_2 &\rightleftharpoons 2CH_4 + O_2 \end{aligned}$$

□

If we were going to remove other elements besides C, we would repeat this procedure starting with $\boldsymbol{\nu}$. See also Exercise 2.6.

2.4 Reaction Rates and Production Rates

In order to describe the change in composition in a reactor, one has to know the reaction rates. As an example, we consider the third reaction in the CVD chemistry, Reactions 2.14

$$SiH_4 + SiH_2 \rightleftharpoons Si_2H_6 \qquad (2.26)$$

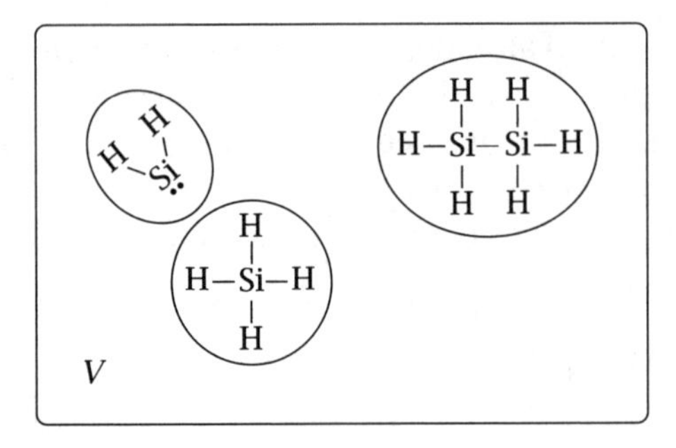

Figure 2.1: Defining the reaction rate, r, for the reaction $SiH_2 + SiH_4 \rightleftharpoons Si_2H_6$.

The **reaction rate**, r, is defined as the number of times this reaction event takes place per time per volume. One can imagine turning SiH_4, SiH_2 and Si_2H_6 molecules loose in a box of some fixed volume V as depicted in Figure 2.1. We define the **reaction extent**, ε, to keep track of the number of times this reaction event occurs. Imagine that we could somehow count up the net number of times an SiH_4 molecule hit an SiH_2 molecule and turned into an Si_2H_6 molecule during a short period of time. The change in the reaction extent, $\Delta\varepsilon$, is the net number of reaction events that occur in the time interval Δt. The reaction rate is then

$$r = \frac{\Delta\varepsilon}{\Delta t V} \tag{2.27}$$

If the forward event (an SiH_4 molecule and an SiH_2 molecule turning into an Si_2H_6 molecule) occurs more often than the reverse event (an Si_2H_6 molecule decomposing into an SiH_4 molecule and an SiH_2 molecule), then the change in ε is positive and the reaction rate is positive. If the reverse event occurs more often than the forward event, then the change in ε and reaction rate are negative. If the system is at equilibrium, then the change in ε is zero and the forward and reverse events occur in equal numbers. The extent ε is a number of molecular change events and therefore the units of r in Equation 2.27 are #/(time·volume). If one divides by Avogadro's number, the units of extent are moles and the units of reaction rate are moles/(time·volume), which are the usual units for extent and reaction rate in this text. Finally, we often deal with physical situations in which we assume the material behaves as a continuum and we can ignore the discrete nature of the molecules. This means we can take the volume V large enough to average the random fluctuations of the molecules, but small enough that there is negligible spatial variation in the average concentrations of the components or the reaction rate within V. Under this continuum assumption, we can speak of the reaction rate as defined at a point in space within some larger reacting system or physical reactor equipment.

Notice in the definition of reaction rate, we are taking the reaction stoichiometry *literally*. We are postulating that these collision and transformation events are taking place at the molecular level. These literal reactions are known as **elementary reactions**. We delay a more complete discussion of elementary reactions and reaction mechanisms until Chapter 5. We will also see that for complex reacting systems, it may be difficult to know whether or not a reaction is an elementary reaction. But that is a separate issue, which we take up later, and that issue does not prevent us from defining the reaction rate.

It is difficult to measure reaction rates directly, because we do not directly sense molecular transformation events. We can measure concentrations, however. It is important to connect the reaction rate to the rate of change of the concentrations of the various species in the reactor, which are the quantities we usually care about in a commercial reactor. We define **production rate**, R_j, as the rate at which the jth species is produced (moles/(time·volume)) due to the chemical reactions taking place. It is clear looking at the stoichiometry in Reaction 2.26 that each time the forward reaction event occurs, an Si_2H_6 molecule is produced. Each time the reverse reaction occurs, an Si_2H_6 molecule is consumed. The production rate of Si_2H_6, $R_{Si_2H_6}$, is therefore directly related to the reaction rate,

$$R_{\mathrm{Si_2H_6}} = r$$

Notice that if r is positive $R_{Si_2H_6}$ is positive as we expect because Si_2H_6 is being produced. Similar arguments lead to relating the other production rates to the reaction rate,

$$\begin{aligned} R_{\mathrm{SiH_4}} &= -r \\ R_{\mathrm{SiH_2}} &= -r \end{aligned}$$

Notice that we have three production rates, one for each species, but only one reaction rate, because there is only a single reaction. If we now introduce the production rate vector, $\boldsymbol{R}$,

$$\boldsymbol{R} = \begin{bmatrix} R_{\mathrm{SiH_4}} \\ R_{\mathrm{SiH_2}} \\ R_{\mathrm{Si_2H_6}} \end{bmatrix}$$

we can summarize the connection between the three production rates and the single reaction rate by

$$\boldsymbol{R} = \begin{bmatrix} -1 \\ -1 \\ 1 \end{bmatrix} r \tag{2.28}$$

Notice that the column vector in Equation 2.28 is just the transpose of the row vector that comprises $\boldsymbol{\nu} = [-1 \ -1 \ 1]$, which follows from Reaction 2.26

Consider what happens to the relationship between the production and reaction rates if there is more than one reaction. Recall the water gas shift reaction,

$$\begin{aligned} \mathrm{H_2O} + \mathrm{CO} &\rightleftharpoons \mathrm{CO_2} + \mathrm{H_2} \\ \mathrm{H_2O} + \mathrm{H} &\rightleftharpoons \mathrm{H_2} + \mathrm{OH} \\ \mathrm{OH} + \mathrm{CO} &\rightleftharpoons \mathrm{CO_2} + \mathrm{H} \end{aligned}$$

Three reaction rates are required to track all three reactions. Let r_i denote the reaction rate for the ith reaction. What production rate of atomic hydrogen, H, results from these three reactions? We notice that H does not take part in the first reaction, is consumed in the second reaction, and is produced in the third reaction. We therefore write

$$R_{\mathrm{H}} = (0)\, r_1 + (-1)\, r_2 + (1)\, r_3 = -r_2 + r_3$$

Consider the second species, H_2. It is produced in the first and second reactions and does not take part in the third reaction. Its production rate can therefore be expressed as

$$R_{H_2} = (1)\,r_1 + (1)\,r_2 + (0)\,r_3 = r_1 + r_2$$

You should examine the remaining four species and produce the following matrix equation,

$$\begin{bmatrix} R_H \\ R_{H_2} \\ R_{OH} \\ R_{H_2O} \\ R_{CO} \\ R_{CO_2} \end{bmatrix} = \begin{bmatrix} 0 & -1 & 1 \\ 1 & 1 & 0 \\ 0 & 1 & -1 \\ -1 & -1 & 0 \\ -1 & 0 & -1 \\ 1 & 0 & 1 \end{bmatrix} \begin{bmatrix} r_1 \\ r_2 \\ r_3 \end{bmatrix} \tag{2.29}$$

The fundamental relationship between the reaction rates and the production rates now emerges. Compare the matrices in Equations 2.7 and 2.29. Notice that the first *row* of the matrix in Equation 2.7 is the same as the first *column* of the matrix in Equation 2.29. Moreover, *each* row of the matrix in Equation 2.7 is the same as the corresponding column of the matrix in Equation 2.29. In other words, the two matrices are transposes of each other. We can therefore summarize Equation 2.29 as

$$\boxed{\boldsymbol{R} = \boldsymbol{\nu}^T\,\boldsymbol{r}} \tag{2.30}$$

in which $\boldsymbol{\nu}^T$ denotes the transpose of the stoichiometric matrix. Equation 2.30 implies that one can always compute the production rates from the reaction rates. That computation is a simple matter of matrix multiplication. The reverse problem, deducing the reaction rates from the production rates, is not so simple as it involves solving a set of equations. We will see in the next section under what conditions that solution can be found.

Example 2.8: Production rate for SiH_2

What is the production rate of SiH_2 in terms of the reaction rates for the CVD example?

Solution

$$\begin{aligned}
&SiH_4 \rightleftharpoons SiH_2 + H_2 && (2.31) \\
&SiH_4 \rightleftharpoons SiH_3 + H && (2.32) \\
&SiH_4 + SiH_2 \rightleftharpoons Si_2H_6 && (2.33) \\
&Si_2H_4 + H_2 \rightleftharpoons SiH_4 + SiH_2 && (2.34) \\
&SiH_4 + H \rightleftharpoons SiH_3 + H_2 && (2.35) \\
&SiH_4 + SiH_3 \rightleftharpoons Si_2H_5 + H_2 && (2.36) \\
&SiH_4 + SiH \rightleftharpoons SiH_3 + SiH_2 && (2.37) \\
&SiH_4 + SiH \rightleftharpoons Si_2H_5 && (2.38) \\
&SiH_4 + Si \rightleftharpoons 2SiH_2 && (2.39) \\
&Si + H_2 \rightleftharpoons SiH_2 && (2.40)
\end{aligned}$$

$$\begin{aligned}
&SiH_2 + SiH \rightleftharpoons Si_2H_3 && (2.41) \\
&SiH_2 + Si \rightleftharpoons Si_2H_2 && (2.42) \\
&SiH_2 + Si_3 \rightleftharpoons Si_2H_2 + Si_2 && (2.43) \\
&H_2 + Si_2H_2 \rightleftharpoons Si_2H_4 && (2.44) \\
&H_2 + Si_2H_4 \rightleftharpoons Si_2H_6 && (2.45) \\
&H_2 + SiH \rightleftharpoons SiH_3 && (2.46) \\
&H_2 + Si_2 \rightleftharpoons Si_2H_2 && (2.47) \\
&H_2 + Si_2H_3 \rightleftharpoons Si_2H_5 && (2.48) \\
&Si_2H_2 + H \rightleftharpoons Si_2H_3 && (2.49) \\
&Si + Si_3 \rightleftharpoons 2Si_2 && (2.50)
\end{aligned}$$

Looking at Reactions 2.31–2.50, we note that SiH_2 takes part in Reactions 2.31, 2.33, 2.34, 2.37, and 2.39–2.43. Extracting the stoichiometric numbers of SiH_2 for each of these reactions gives[6]

$$R_{SiH_2} = r_1 - r_3 + r_4 + r_7 + 2r_9 + r_{10} - r_{11} - r_{12} - r_{13}$$

or

$$R_{SiH_2} = \begin{bmatrix} 1 & 0 & -1 & 1 & 0 & 0 & 1 & 0 & 2 & 1 & -1 & -1 & -1 & 0 & 0 & 0 & 0 & 0 & 0 & 0 \end{bmatrix} \boldsymbol{r}$$

SiH_2 was chosen as the fifth species in Exercise 2.4 so the row vector above is indeed the transpose of the fifth column of the $\boldsymbol{\nu}$ matrix in Equation 2.15. You may wish to choose another component such as SiH_4 and check another column of the $\boldsymbol{\nu}$ matrix. □

2.5 Computational Aspects of Stoichiometry

As we have seen in this chapter, problems of realistic and complex reaction stoichiometry involve matrices and linear algebra. After the fundamental concepts are in place, application of the fundamentals requires computational tools. Moreover, if the computing environment is organized properly, the experience of solving nontrivial problems reinforces the understanding of the fundamental concepts and further

[6]We are so sure you will not forget that products have positive stoichiometric numbers and reactants have negative ones that we will not repeat it again.

prepares one to apply the fundamentals in realistic and complex industrial situations. In Appendix A, we briefly summarize Octave and MATLAB as high-level programming languages for numerical solution of reactor analysis and design problems. Octave is freely available for a variety of hardware platforms and can be downloaded from www.octave.org. MATLAB is commercially available from The MathWorks, Inc., and is becoming a commonly available tool of industrial engineering practice.

2.5.1 Computing Production Rates from Reaction Rates

As discussed previously, computing $\boldsymbol{R}$ from $\boldsymbol{r}$ is a simple matter of matrix multiplication. Consider again the water gas shift reaction chemistry,

$$\begin{bmatrix} 0 & 1 & 0 & -1 & -1 & 1 \\ -1 & 1 & 1 & -1 & 0 & 0 \\ 1 & 0 & -1 & 0 & -1 & 1 \end{bmatrix} \begin{bmatrix} \mathrm{H} \\ \mathrm{H_2} \\ \mathrm{OH} \\ \mathrm{H_2O} \\ \mathrm{CO} \\ \mathrm{CO_2} \end{bmatrix} = \begin{bmatrix} 0 \\ 0 \\ 0 \end{bmatrix} \tag{2.51}$$

In Chapter 5 we discuss means for predicting reaction rates given species concentrations, but for now just assume we know the three reaction rates are, in some chosen units of moles/(time·volume),

$$\begin{bmatrix} r_1 \\ r_2 \\ r_3 \end{bmatrix} = \begin{bmatrix} 1 \\ 2 \\ 3 \end{bmatrix}$$

The production rates of the six species due to these reactions are then computed as

$$\begin{bmatrix} R_{\mathrm{H}} \\ R_{\mathrm{H_2}} \\ R_{\mathrm{OH}} \\ R_{\mathrm{H_2O}} \\ R_{\mathrm{CO}} \\ R_{\mathrm{CO_2}} \end{bmatrix} = \begin{bmatrix} 0 & 1 & 0 & -1 & -1 & 1 \\ -1 & 1 & 1 & -1 & 0 & 0 \\ 1 & 0 & -1 & 0 & -1 & 1 \end{bmatrix}^T \begin{bmatrix} 1 \\ 2 \\ 3 \end{bmatrix} = \begin{bmatrix} 1 \\ 3 \\ -1 \\ -3 \\ -4 \\ 4 \end{bmatrix}$$

The effect of the three reactions is to produce H, H_2 and CO_2, and to consume OH, H_2O and CO at the given rates. Please perform this calculation for yourself.

2.5.2 Computing Reaction Rates from Production Rates

Another common task arising in the analysis of reactors and reaction mechanisms is to measure production rates by monitoring changes in species concentrations, to help infer the corresponding reaction rates. To make the concept clear, consider the simple isomerization reactions between species A, B and C,

$$\mathrm{A} \rightleftharpoons \mathrm{B} \tag{2.52}$$

$$\mathrm{B} \rightleftharpoons \mathrm{C} \tag{2.53}$$

$$\mathrm{C} \rightleftharpoons \mathrm{A} \tag{2.54}$$

The stoichiometric matrix for these reactions is

$$\boldsymbol{\nu} = \begin{bmatrix} -1 & 1 & 0 \\ 0 & -1 & 1 \\ 1 & 0 & -1 \end{bmatrix}$$

and the production rates and reaction rates are related by

$$\begin{bmatrix} R_A \\ R_B \\ R_C \end{bmatrix} = \begin{bmatrix} -1 & 0 & 1 \\ 1 & -1 & 0 \\ 0 & 1 & -1 \end{bmatrix} \begin{bmatrix} r_1 \\ r_2 \\ r_3 \end{bmatrix}$$

It might appear at first glance that we can compute $\boldsymbol{r}$ given $\boldsymbol{R}$ because we have three equations and three unknowns, but because the reactions are not linearly independent, such is not the case. It is clear from inspection of Reactions 2.52–2.54 that the third reaction is the sum of the first two. Computing the rank of $\boldsymbol{\nu}$ confirms that only two reactions are independent.

Before we continue with the full set of three reactions, we explore what happens if we use an independent set. If we omit the third reaction, for example,

$$\mathrm{A} \rightleftharpoons \mathrm{B}$$

$$\mathrm{B} \rightleftharpoons \mathrm{C}$$

the production rates and new reaction rates are related by

$$\begin{bmatrix} R_A \\ R_B \\ R_C \end{bmatrix} = \begin{bmatrix} -1 & 0 \\ 1 & -1 \\ 0 & 1 \end{bmatrix} \begin{bmatrix} \hat{r}_1 \\ \hat{r}_2 \end{bmatrix}$$

By inspection, the first equation gives

$$\hat{r}_1 = -R_A$$

and the third equation gives

$$\hat{r}_2 = R_C$$

The second equation tells us $R_B = \hat{r}_1 - \hat{r}_2 = -(R_A + R_C)$. In other words, using the linearly independent reactions, we can compute both reaction rates and we find a restriction on the possible production rates. We explore subsequently what happens when this restriction is violated by the production-rate measurements.

If we now return to the original set of three isomerization reactions, we can deduce that r_3 is not determined by the production rates, and that from the first equation

$$r_1 = -R_A + r_3 \tag{2.55}$$

and from the third equation

$$r_2 = R_C + r_3 \tag{2.56}$$

and r_3 is arbitrary. The second equation places the restriction $R_B = r_1 - r_2$, which upon substitution of Equations 2.55 and 2.56 gives $R_B = -(R_A + R_C)$ as before. Because r_3 is arbitrary, we cannot deduce the reaction rates from production rates, which is characteristic of using sets of reactions that are not linearly independent.

So we conclude that measuring production rates is not enough to tell us reaction rates. We require more information. As we discuss in Chapter 5, the extra information usually is provided by postulating a reaction mechanism and applying the laws of mass action to the elementary reactions. In that case we seek to determine the rate *constants* from the production rate measurements, not the reaction rates themselves. We take up that important problem in Chapter 9.

Finally, we emphasize that using linearly dependent sets of reactions presents no problem at all if one is interested only in computing the production rates from given reaction rates, which is the usual case when working with reaction mechanisms.

2.5.3 Measurement Errors and Least-Squares Estimation

As a final topic in this chapter, we explore what happens when the extra conditions on the production rates are violated by the data. Imagine the following two linearly independent reactions of the water gas shift are taking place

$$H_2O + CO \rightleftharpoons CO_2 + H_2$$
$$H_2O + H \rightleftharpoons H_2 + OH$$

so the production rates are given by

$$\begin{bmatrix} R_H \\ R_{H_2} \\ R_{OH} \\ R_{H_2O} \\ R_{CO} \\ R_{CO_2} \end{bmatrix} = \begin{bmatrix} 0 & -1 \\ 1 & 1 \\ 0 & 1 \\ -1 & -1 \\ -1 & 0 \\ 1 & 0 \end{bmatrix} \begin{bmatrix} r_1 \\ r_2 \end{bmatrix} \tag{2.57}$$

We can compute the production rates when the two reaction rates are

$$\begin{bmatrix} r_1 \\ r_2 \end{bmatrix} = \begin{bmatrix} 1 \\ 2 \end{bmatrix}$$

and we obtain

$$\boldsymbol{R} = \begin{bmatrix} -2 & 3 & 2 & -3 & -1 & 1 \end{bmatrix}^T \tag{2.58}$$

Now if $\boldsymbol{R}$ is determined by measuring species concentrations, unmodeled effects undoubtedly cause discrepancy between predicted and measured values of $\boldsymbol{R}$. For example, let's assume that the production rate of the first species, H, is in error by a small amount so the measured $\boldsymbol{R}$ is $\boldsymbol{R} = \begin{bmatrix} -2.1 & 3 & 2 & -3 & -1 & 1 \end{bmatrix}^T$. The effect of this error is to make the equations inconsistent, so there no longer is an exact solution for the reaction rates, which is not too surprising because we have six equations and only two unknowns in Equation 2.57. A system with more equations than unknowns is called over-determined. Normally we would not expect to find a solution for arbitrary $\boldsymbol{R}$, only those $\boldsymbol{R}$ that are generated by multiplying an $\boldsymbol{r}$ by $\boldsymbol{\nu}^T$. When the equations are inconsistent, one is usually interested in knowing the values of the two reaction rates that come *closest* to satisfying the six equations simultaneously. If we measure how close the equations are to zero by squaring the error in each equation and summing over all equations, then we are using the classic least-squares approach.

Computing least-squares solutions to over-determined equations is a useful computation in linear algebra. If one is given $\boldsymbol{R}$ and trying to *solve* for $\boldsymbol{r}$ from

$$\boldsymbol{R} = \boldsymbol{\nu}^T \boldsymbol{r} \tag{2.59}$$

then the least-squares solution is given by

$$\boldsymbol{r} = (\boldsymbol{\nu}\,\boldsymbol{\nu}^T)^{-1}\boldsymbol{\nu}\boldsymbol{R}$$

in which the superscript -1 indicates a matrix inverse. If the reactions are linearly independent, then the matrix product $\boldsymbol{\nu}\,\boldsymbol{\nu}^T$ has an inverse and the least-squares solution is unique. If the reactions are not linearly independent this inverse does not exist and the least-squares solution is not unique as before.

If we compute the least-squares solution to the inconsistent data given above, we find $\boldsymbol{r} = [\ 0.983 \quad 2.03\ ]^T$ instead of the *correct* value $r = [\ 1 \quad 2\ ]^T$. Notice a small error in the H production rate has translated into small errors in both inferred reaction rates.

As a final example, let's consider the case in which we have repeated measurements of the production rates, all of which are subject to small random errors. If we add small amounts of random noise to the data given in Equation 2.58, we produce the following six measurements.

$$\boldsymbol{R}_{\text{meas}} = \begin{bmatrix} -2.05 & -2.06 & -1.93 & -1.97 & -2.04 & -1.92 \\ 2.94 & 3.02 & 3.04 & 2.93 & 3.06 & 3.04 \\ 2.01 & 1.94 & 2.01 & 1.92 & 2.01 & 2.04 \\ -2.98 & -2.98 & -2.98 & -2.99 & -2.96 & -2.96 \\ -1.03 & -1.03 & -0.98 & -1.07 & -0.95 & -1.08 \\ 0.97 & 1.05 & 1.06 & 1.09 & 1.00 & 1.07 \end{bmatrix}$$

If we take each column of $\boldsymbol{R}_{\text{meas}}$, that is each production-rate measurement, and compute the least-squares estimate of $\boldsymbol{r}$ for that measurement, we obtain six estimates of the reaction rates, one for each measured production rate. We can perform these operations in one matrix equation via

$$\boldsymbol{r}_{\text{est}} = (\boldsymbol{\nu}\,\boldsymbol{\nu}^T)^{-1}\boldsymbol{\nu}\boldsymbol{R}_{\text{meas}}$$

and the result is

$$\boldsymbol{r}_{\text{est}} = \begin{bmatrix} 0.97 & 1.03 & 1.03 & 1.06 & 0.98 & 1.05 \\ 2.01 & 1.99 & 1.98 & 1.92 & 2.03 & 1.96 \end{bmatrix}$$

Figure 2.2 shows the estimated reaction rates for the six production-rate measurements. Notice that we obtain reasonable estimates of the true reaction rates, $\boldsymbol{r} = [\ 1 \quad 2\ ]^T$. Next consider what happens when we have many measurements available. Figure 2.3 displays the estimated reaction rates given 500 production-rate measurements. Notice that the estimated rates are again scattered about the true value, the

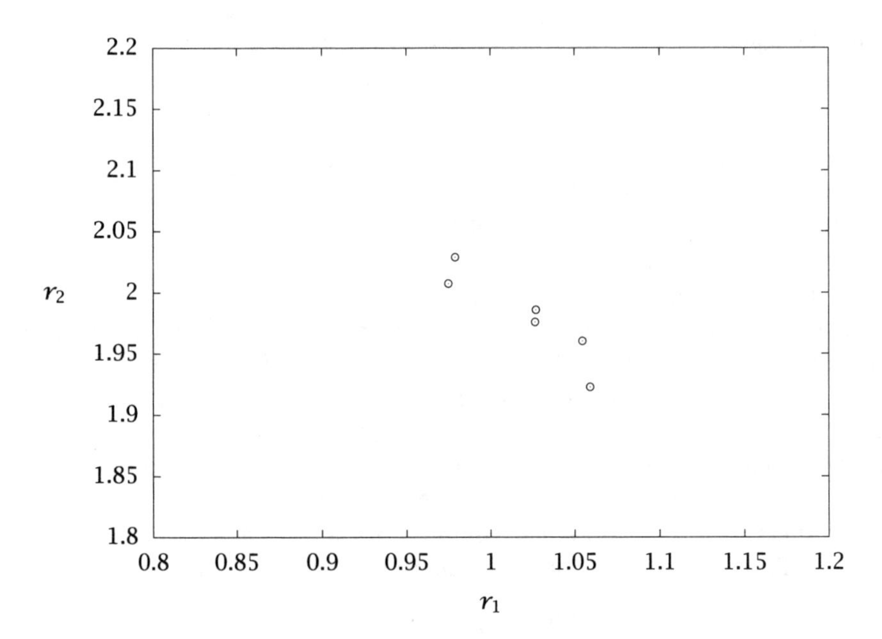

Figure 2.2: Estimated reaction rates from six production-rate measurements subject to measurement noise.

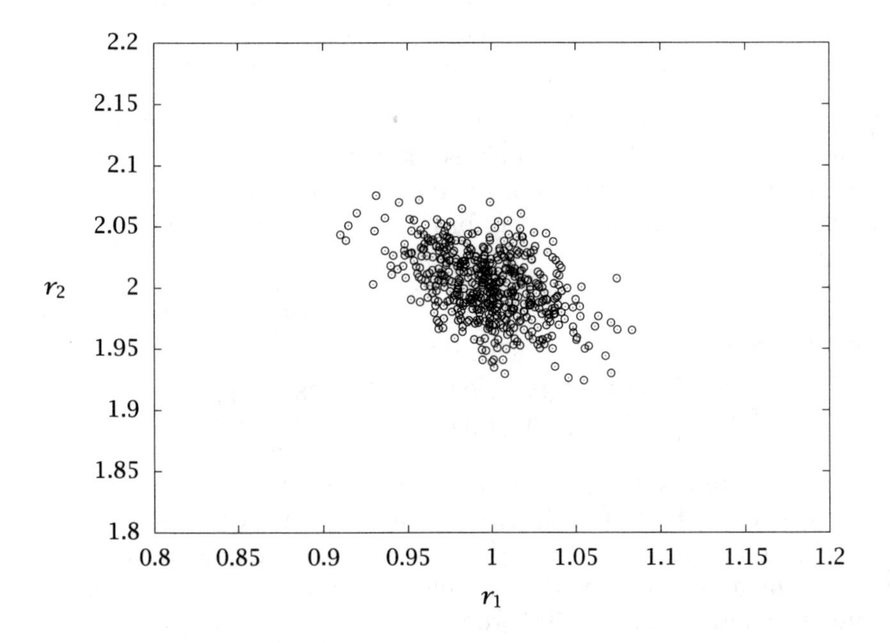

Figure 2.3: Estimated reaction rates from 500 production-rate measurements subject to measurement noise.

mean of these values is close to the true value, and the shape of the frequency distribution for the reaction rate estimates begins to emerge. You might want to generate the same plot with 5000 random measurements and see how much more detail is apparent in the probability distribution of estimates. Further discussion of parameter estimation from data is delayed until Chapter 9, but engineers benefit from thinking about extracting information from data and models at an early stage.

2.6 Summary

In this chapter we have introduced the compact notation for keeping track of the stoichiometry of chemical reactions,

$$\sum_{j=1}^{n_s} \nu_{ij} A_j = 0, \qquad i = 1, 2, \ldots, n_r$$

in which A_j represents chemical species j, $j = 1, \ldots, n_s$ and n_s is the number of species in the reaction network. The stoichiometric coefficients are contained in the stoichiometric matrix $\boldsymbol{\nu}$, in which ν_{ij} is the stoichiometric coefficient for species j in reaction i, $i = 1, \ldots, n_r$ and n_r is the number of reactions in the network. We can summarize the reaction stoichiometry with one vector equation

$$\boldsymbol{\nu A} = \mathbf{0}$$

A set of reactions is linearly independent if no reaction in the set can be written as a linear combination of the other reactions in the set. Linear independence of reactions is equivalent to linear independence of the rows of $\boldsymbol{\nu}$. The rank of a matrix is the number of linearly independent rows (equivalently columns) of the matrix, so the rank of $\boldsymbol{\nu}$ is the number of linearly independent reactions in the network.

The reaction rate is a fundamental concept that allows quantitative prediction of rates of conversions of reactants to products. We define the rate of reaction i, r_i, to be the net number of times a reaction event occurs per time per volume. Given the rates of all reactions, we can calculate directly the production rates of all species,

$$R_j = \sum_{i=1}^{n_r} \nu_{ij} r_i, \qquad j = 1, \ldots n_s$$

or as an equivalent vector equation,

$$R = \nu^T r \tag{2.60}$$

Given the rates of reactions, it is a simple matter to compute the species production rates with Equation 2.60. One cannot solve uniquely the reverse problem, in general. Given observed production rates, computing the corresponding reaction rates requires additional information, such as rate expressions for the elementary reactions in a reaction mechanism. If the set of chemical reactions is linearly independent, then one can uniquely solve the reverse problem. If the observed production rates contain experimental errors, there may not exist an exact solution of reaction rates, $\boldsymbol{r}$, that satisfy Equation 2.60. In this situation, one is normally interested in finding the reaction rates that *most closely* satisfy Equation 2.60. The closest solution in a least-squares sense is easily computed with standard linear algebra software.

Notation

- a_{jl} formula number for element l in species j
- A_j jth species in the reaction network
- E^l lth element comprising the species
- i reaction index, $i = 1, 2, \ldots, n_r$
- j species index, $j = 1, 2, \ldots, n_s$
- M_j molecular weight of the jth species
- n_i number of independent reactions in reaction network
- n_r total number of reactions in reaction network
- n_s total number of species in reaction network
- r_i reaction rate for ith reaction
- R_j production rate for jth species
- ε_i extent of reaction i
- ν_{ij} stoichiometric number for the jth species in the ith reaction

2.7 Exercises

Exercise 2.1: Finding independent sets of reactions

Consider the following set of chemical reactions,

$$2N_2O_5 \rightleftharpoons 2N_2O_4 + O_2 \tag{2.61}$$

$$N_2O_5 \rightleftharpoons NO_2 + NO_3 \tag{2.62}$$

$$NO_2 + NO_3 \rightleftharpoons N_2O_5 \tag{2.63}$$

$$NO_2 + NO_3 \rightleftharpoons NO_2 + O_2 + NO \tag{2.64}$$

$$NO + N_2O_5 \rightleftharpoons 3NO_2 \tag{2.65}$$

$$NO_2 + NO_2 \rightleftharpoons N_2O_4 \tag{2.66}$$

(a) Determine the stoichiometric matrix, $\boldsymbol{\nu}$, and the species list, $\boldsymbol{A}$, for this reaction system so the reaction network is summarized by

$$\boldsymbol{\nu A} = \boldsymbol{0}$$

(b) Use Octave, MATLAB, or your favorite software package to determine the rank of the stoichiometric matrix. How many of the reactions are linearly independent?

(c) Now that you have found the number of independent reactions, n_i, which n_i of the original set of 6 reactions can be chosen as an independent set? Try guessing some set of n_i reactions and determine the rank of the new stoichiometric matrix. Stop when you have determined successfully one or more sets of n_i independent reactions.

Hint: you want to examine the rank of sub-matrices obtained by deleting rows (i.e., reactions) from the original stoichiometric matrix. In Octave, if you assign the original stoichiometric matrix to a name, `stoi`, then you can obtain the rank of the stoichiometric matrix associated with deleting the fifth reaction, for example, by

```
stoi2 = [stoi(1:4,:);stoi(6,:)]
rank(stoi2)
```

Do you see how the indices in forming `stoi2` work out? Notice we do not have to enter any more matrices after we build the original stoichiometric matrix to test the ranks of various reaction networks.

(d) What do you think of a colleague's answer that contains Reactions 2.62 and 2.63 in the final set. Can this be correct? Why or why not?

Exercise 2.2: The stoichiometric matrix

(a) What is the stoichiometric matrix for the following reaction network [2]? By inspection, how many of the reactions are linearly independent? How would you check your answer if you had access to a computer?

$$
\begin{aligned}
H_2 + I_2 &\rightleftharpoons 2HI \\
I_2 &\rightleftharpoons 2I \\
I + H_2 &\rightleftharpoons HI + H \\
H + I_2 &\rightleftharpoons HI + I
\end{aligned}
$$

(b) Given a stoichiometric matrix for a reaction network with n_s species and n_r reactions

$$\sum_{j=1}^{n_s} \nu_{ij} A_j = 0, \qquad i = 1, 2, \ldots, n_r$$

What is the production rate of the jth species, R_j, in terms of the reaction rates for the reactions, r_i?

Exercise 2.3: Finding reaction rates from production rates

Consider again the water gas shift reaction presented in Equation 2.51. Assume the production rates have been measured and are, in some units of moles/(time·volume),

$$
\begin{bmatrix} R_{H} \\ R_{H_2} \\ R_{OH} \\ R_{H_2O} \\ R_{CO} \\ R_{CO_2} \end{bmatrix} = \begin{bmatrix} -1 \\ 3 \\ 1 \\ -3 \\ -2 \\ 2 \end{bmatrix}
$$

(a) If you choose the first two reactions as a linearly independent set, what are the two reaction rates that are consistent with these data. Is this answer unique?

(b) Repeat the calculation if you choose the second and third reactions as the linearly independent set of reactions. Is this answer unique?

(c) How can these reaction rates differ, when the production rates are the same? Can we determine which set of reactions is really causing this measured production rate?

Exercise 2.4: Independent reactions for bromine hydrogenation

Consider the following set of chemical reactions [4, 5],

$$
\begin{aligned}
H_2 + Br_2 &\rightleftharpoons 2HBr \\
Br_2 &\rightleftharpoons 2Br \\
Br + H_2 &\rightleftharpoons HBr + H \\
H + Br_2 &\rightleftharpoons HBr + Br \\
H + HBr &\rightleftharpoons H_2 + Br \\
2Br &\rightleftharpoons Br_2
\end{aligned}
$$

(a) Determine the stoichiometric matrix, $\boldsymbol{\nu}$, and the species list, A, for this reaction system so the reaction network is summarized by

$$\boldsymbol{\nu} A = \mathbf{0}$$

(b) Use Octave or MATLAB to determine the rank of the matrix using the `rank` function. How many reactions are linearly independent?

(c) Now that you have found the number of independent reactions, n_i, which n_i of the original set of six reactions can be chosen as an independent set? Try guessing some set of n_i reactions and determine the rank of the new stoichiometric matrix. Stop when you have determined successfully one or more sets of n_i independent reactions.

Exercise 2.5: Independent reactions for methane oxidation

Consider a mixture of CO, H_2, and CH_4 that is fed into a furnace with O_2 and produces CO, CO_2, and H_2O. The following chemical reactions have been suggested to account for the products that form.

$$\begin{aligned}
CO + \frac{1}{2}O_2 &\rightleftharpoons CO_2 \\
H_2 + \frac{1}{2}O_2 &\rightleftharpoons H_2O \\
CH_4 + 2O_2 &\rightleftharpoons CO_2 + 2H_2O \\
CH_4 + \frac{3}{2}O_2 &\rightleftharpoons CO + 2H_2O
\end{aligned}$$

(a) Are these reactions linearly independent? What is the number of linearly independent reactions, n_i?

(b) List all sets of n_i linearly independent reactions. Which reaction is included in all of the linearly independent sets of reactions? Why?

Exercise 2.6: Methane oxidation and maximal independent sets

(a) List a maximal set of linearly independent reactions if O_2 as well as C are not observed as species in the methane oxidation reactions in Example 2.7.

(b) Repeat if H_2, O_2 and C are not observed as species.

Exercise 2.7: Production rates from reaction rates

(a) Consider the following set of chemical reactions,

$$\begin{aligned}
2N_2O_5 &\rightleftharpoons 2N_2O_4 + O_2 \\
N_2O_5 &\rightleftharpoons NO_2 + NO_3 \\
NO_2 + NO_3 &\rightleftharpoons N_2O_5 \\
NO_2 + NO_3 &\rightleftharpoons NO_2 + O_2 + NO \\
NO + N_2O_5 &\rightleftharpoons 3NO_2 \\
NO_2 + NO_2 &\rightleftharpoons N_2O_4
\end{aligned}$$

Determine the rates of production of each component in terms of the rates of each reaction.

(b) Butene isomerization reactions are shown below.

$$\begin{aligned}
1-\text{butene} &\underset{k_{-1}}{\overset{k_1}{\rightleftharpoons}} \quad \text{cis}-2-\text{butene}\\
1-\text{butene} &\underset{k_{-2}}{\overset{k_2}{\rightleftharpoons}} \text{trans}-2-\text{butene}\\
1-\text{butene} &\underset{k_{-3}}{\overset{k_3}{\rightleftharpoons}} \quad \text{isobutene}\\
\text{cis}-2-\text{butene} &\underset{k_{-4}}{\overset{k_4}{\rightleftharpoons}} \text{trans}-2-\text{butene}\\
\text{cis}-2-\text{butene} &\underset{k_{-5}}{\overset{k_5}{\rightleftharpoons}} \quad \text{isobutene}
\end{aligned}$$

Determine the rates of production of each component in terms of the rates of each reaction.

Exercise 2.8: Restrictions from element balancing

Let the s species, $A_j, j = 1, \ldots, s$ be comprised of the e elements, $E^l, l = 1, \ldots, e$. Writing the chemical formulas for the species in the usual way

$$\begin{aligned}
A_1 &= E^1_{a_{11}} E^2_{a_{12}} \cdots E^e_{a_{1e}}\\
A_2 &= E^1_{a_{21}} E^2_{a_{22}} \cdots E^e_{a_{2e}}\\
&\vdots\\
A_s &= E^1_{a_{s1}} E^2_{a_{s2}} \cdots E^e_{a_{se}}
\end{aligned}$$

in which a_{jl} is the formula number for species j corresponding to element l. Show that any chemical reaction, $\sum_j \nu_j A_j = 0$ satisfies the following e equations to balance the elements

$$\begin{bmatrix} \nu_1 & \nu_2 & \cdots & \nu_s \end{bmatrix} \begin{bmatrix} a_{11} & a_{12} & \cdots & a_{1e} \\ a_{21} & a_{22} & \cdots & a_{2e} \\ \vdots & \vdots & \ddots & \vdots \\ a_{s1} & a_{s2} & \cdots & a_{se} \end{bmatrix} = \begin{bmatrix} 0 & 0 & \cdots & 0 \end{bmatrix}$$

If we define the $s \times e$ matrix $\mathcal{A}$ to hold the formula numbers, we can express the element balance by

$$\boldsymbol{\nu}\mathcal{A} = \mathbf{0} \tag{2.67}$$

Determine $\boldsymbol{A}$, $\boldsymbol{E}$and $\mathcal{A}$ for the species hydrogen (molecular), oxygen (molecular) and water.

Exercise 2.9: Null space and fundamental theorem of linear algebra

Equation 2.67 in Exercise 2.8 is begging to be analyzed by the fundamental theorem of linear algebra [7], so we explore that concept here. Consider an arbitrary $m \times n$ matrix, $\boldsymbol{B}$. The null space of matrix $\boldsymbol{B}$, written $\mathcal{N}(\boldsymbol{B})$, is defined to be the set of *all* vectors $\boldsymbol{x}$ such that $\boldsymbol{B}\,\boldsymbol{x} = \mathbf{0}$. The dimension of $\mathcal{N}(\boldsymbol{B})$ is the number of linearly independent vectors $\boldsymbol{x}$ satisfying $\boldsymbol{B}\,\boldsymbol{x} = \boldsymbol{0}$. One of the remarkable results of the fundamental theorem of linear algebra is the relation

$$\text{rank}(\boldsymbol{B}) + \dim(\mathcal{N}(\boldsymbol{B})) = n$$

The numerical support for computing null spaces is excellent. For example, the Octave command `null(B)` returns a matrix with columns consisting of linearly independent vectors in the null space of matrix $\boldsymbol{B}$.

(a) Armed with this result, consider Equation 2.67 and establish that the number of linearly independent reactions that satisfy the element balances is

$$i = s - \text{rank}(\boldsymbol{\mathcal{A}})$$

(b) Determine $\boldsymbol{A}$, $\boldsymbol{E}$ and $\boldsymbol{\mathcal{A}}$ for the methane oxidation in Example 2.7.

(c) Determine rank($\boldsymbol{\mathcal{A}}$) and i for this example. Do you obtain the same size maximal set as in Example 2.7?

Exercise 2.10: Limits on numbers of independent reactions

(a) Species A, B and C are observed in a reacting system. What is the largest possible number of linearly independent reactions among these species, n_{max}?

(b) What is the smallest possible number, n_{min}?

(c) List a chemical example that has a maximal set with n reactions for each n in $n_{\text{min}} \le n \le n_{\text{max}}$.

Exercise 2.11: Generating possible independent reaction sets

(a) If one has s species comprised of e elements, what are the largest and smallest numbers of linearly independent reactions that can be written among the s species? Give an example with more than one species in which you cannot write any reactions.

(b) If $s > e$, the usual case, how many reactions can be written? Prove you can write at least one valid reaction, or find a counterexample to this statement.

Exercise 2.12: An alchemist's view of stoichiometry

In the early 1700s, the alchemists were unaware of the defining role that electrons played in bond formation and transformation of chemical species. The structure of the nucleus, elucidation of bond formation with electron sharing and the construction of the periodic table were major triumphs of the chemical and physical sciences in the nineteenth and early twentieth centuries [6, pp.113–166]. Before these developments, chemical transformation of the elements themselves, such as lead to gold, was on the table as an early goal of research.

Imagine a more flexible chemical world in which we demand conservation of *only* mass for a valid chemical reaction; any mass-conserving rearrangements of the protons and neutrons in the nuclei are also considered valid chemical reactions.

Assume we observe H_2, O_2 and H_2O as chemical species.

(a) Write a few *mass-conserving* reactions among these species that are not valid chemical reactions.

(b) How many linearly independent mass-conserving reactions can be constructed? You may want to use the result of Example 2.5 and the idea of the null space introduced in Exercise 2.9 to be sure you have accounted for all linearly independent reactions.

(c) If you enforce the stronger condition of element balancing, then, by inspection, how many linearly independent valid chemical reactions are possible?

Exercise 2.13: Generalizing independent reactions

Consider s species comprised of e elements, in which c of the species are chemical compounds, i.e., not elements. Let i be the number of reactions in the maximal independent set. Show

$$\max(0, s - e) \leq i \leq \min(s - 1, c)$$

Hint: use the result $i = s - \text{rank}(\mathcal{A})$ from Exercise 2.9 and determine upper and lower bounds on the rank of $\mathcal{A}$ based on s, e and c.

Exercise 2.14: Eliminating reaction intermediates

Consider the following reaction mechanism with five reactions and eight species, A–H.

$$\text{A} \longrightarrow \text{B} + \text{C} \tag{2.68}$$

$$\text{C} \longrightarrow \text{B} + \text{D} \tag{2.69}$$

$$\text{A} + \text{B} \longrightarrow \text{E} + \text{F} \tag{2.70}$$

$$\text{F} \longrightarrow \text{G} + \text{B} \tag{2.71}$$

$$\text{B} + \text{F} \longrightarrow \text{H} \tag{2.72}$$

and assume that species B, C, and F are highly reactive intermediates.

(a) What is the maximum number of linearly independent linear combinations of these five reactions that do not contain species B, C, and F as reactants or products. Justify your answer.

(b) List one set of these independent reactions that contains only small, integer-valued stoichiometric coefficients.

Exercise 2.15: Reaction rates from production rates

Consider the two reactions

$$\text{A} \rightleftharpoons \text{B}$$

$$2\text{B} \rightleftharpoons \text{C}$$

The following production rates were observed in the laboratory for this mechanism:

$R_A = -4.0$ mol/(time vol) $\quad R_B = 2.2$ mol/(time vol) $\quad R_C = 1.0$ mol/(time vol)

(a) From these measurements, provide a least-squares estimate of the two reaction rates. Recall the least-squares estimate formula is

$$\boldsymbol{r}_{\text{est}} = (\boldsymbol{\nu}\,\boldsymbol{\nu}^T)^{-1}\boldsymbol{\nu}\boldsymbol{R}_{\text{meas}}$$

(b) Write out the production rates for all the species in terms of the two reaction rates.

(c) Calculate the three production rates using the estimated reaction rates. Compare this result to the measured production rates. Comment on why the two sets of production rates are or are not different from each other.

We should point out that these limits on numbers of reactions in Exercises 2.8–2.13 presume that conserving elements in the chemical reactions is the *only* restriction. If other quantities are conserved, the method of analysis remains valid, but the answers change. We will see one such conserved quantity, the surface site, when we study reactions with catalytic surfaces in Chapter 5.

Bibliography

[1] R. Aris. *Elementary Chemical Reactor Analysis.* Butterworths, Stoneham, MA, 1989.

[2] S. W. Benson and R. Srinivasan. Some complexities in the reaction system:$H_2 + I_2 = 2HI$. *J. Chem. Phys.*, 23(1):200–202, 1955.

[3] M. E. Coltrin, R. J. Kee, and J. A. Miller. A mathematical model of the coupled fluid mechanics and chemical kinetics in a chemical vapor deposition reactor. *J. Electrochem. Soc.*, 131(2):425–434, 1984.

[4] K. F. Herzfeld. Zur Theorie der Reaktionsgeschwindigkeiten in Gasen. *Annalen der Physik*, 59(15):635–667, 1919.

[5] M. Polànyi. Reaktionsisochore und Reaktionsgeschwindigkeit vom Standpunkte der Statistik. *Zeitschrift für Electrochemie und angewandte physikalische Chemie*, 26:49–54, 1920.

[6] B. L. Silver. *The Ascent of Science.* Oxford University Press, New York, 1998.

[7] G. Strang. *Linear Algebra and its Applications.* Academic Press, New York, 1980.

3

Review of Chemical Equilibrium

3.1 Introduction

This chapter is intended to remind the reader of some important facts concerning the equilibrium state of a system that can undergo chemical reaction. It is by design a review, and assumes the reader is familiar with the fundamental concepts of classical thermodynamics as discussed in the many introductory textbooks on the subject [15, 6, 10, 17, 5, 4].

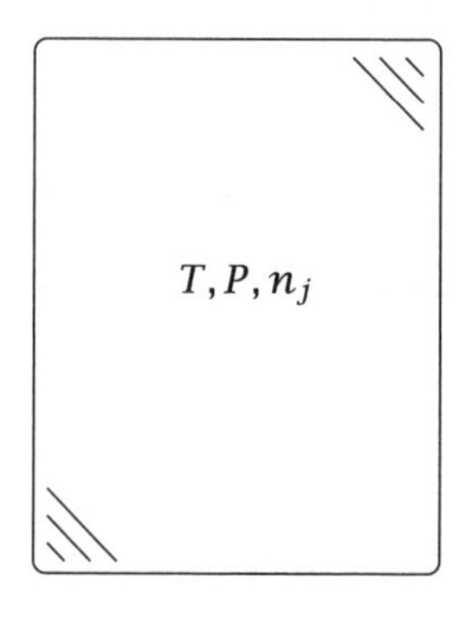

Figure 3.1: Thermodynamic system.

Although essentially all reactors in operation are *not* at chemical equilibrium, it is still important to understand what equilibrium is and how to draw quantitative conclusions from considering the equilibrium state of a system. Chemical reactors are not at equilibrium because the net rates of chemical reactions are seldom allowed to become zero. The calculation of the equilibrium state, however, can allow one to draw conclusions about the limits of reactor performance that would be achievable if this state were reached. In addition, the consideration of the thermodynamic restrictions on reactor performance may motivate one to make operations or design changes that allow these restrictions to be changed and reactor performance improved. For these reasons we begin with a review of chemical equilibrium.

Consider the single-phase, multicomponent system depicted in Figure 3.1. It is convenient for our purposes to describe the state of this system by the variables temperature, T, pressure, P, and the number of moles of each component, n_j. Specifying the temperature, pressure, and number of moles of each component then completely specifies the equilibrium state of the system. The Gibbs energy of the system, G, is the energy function of these state variables that conveniently describes the condition of equilibrium. In particular the equilibrium state has a minimum Gibbs energy with respect to any changes in the number of moles at fixed T and P. The difference in Gibbs energy between two states at slightly different temperatures, pressures, and number of moles of each component is developed in standard textbooks to be

$$dG = -SdT + VdP + \sum_j \mu_j dn_j \tag{3.1}$$

in which S is the system entropy, V is the system volume and μ_j is the chemical potential of component j.

3.2 Condition for Reaction Equilibrium

Consider a closed system in which the n_j can change only by the single chemical reaction,

$$\sum_j \nu_j A_j = 0$$

Given the definition of reaction extent in Chapter 2, for a small change in the extent of this single reaction, the numbers of moles of each component change by

$$dn_j = \nu_j d\varepsilon$$

Substituting this relation into Equation 3.1 gives an expression for the change in the Gibbs energy due to a small change in the reaction extent,

$$dG = -SdT + VdP + \sum_j \left(\nu_j \mu_j\right) d\varepsilon \tag{3.2}$$

For the closed system, G is only a function of T, P and ε. The expansion of a total differential in terms of the partial derivatives allows one to deduce from Equation 3.2

$$S = -\left(\frac{\partial G}{\partial T}\right)_{P,\varepsilon} \tag{3.3}$$

$$V = \left(\frac{\partial G}{\partial P}\right)_{T,\varepsilon} \tag{3.4}$$

$$\sum_j \nu_j \mu_j = \left(\frac{\partial G}{\partial \varepsilon}\right)_{T,P} \tag{3.5}$$

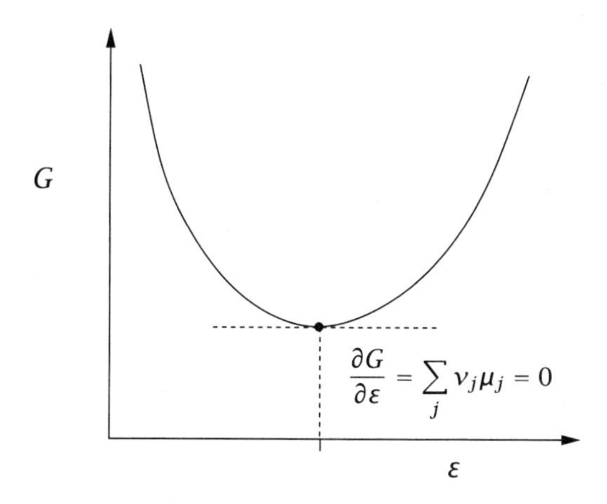

Figure 3.2: Gibbs energy versus reaction extent at constant T and P.

Consider a plot of G at some fixed T and P versus the reaction extent, sketched in Figure 3.2. A necessary condition for the Gibbs energy to be a minimum is that the derivative be zero, or in light of Equation 3.5,

$$\boxed{\sum_j \nu_j \mu_j = 0} \tag{3.6}$$

The direct use of Equation 3.6 to compute equilibrium composition is inconvenient because chemical potential is often expressed in terms of other quantities such as activity or fugacity. One can express μ_j in terms of the activity of component j,

$$\mu_j = G_j^\circ + RT \ln a_j \tag{3.7}$$

in which a_j is the activity of component j in the mixture referenced to some standard state and G_j° is the Gibbs energy of component j in the same standard state. The activity and fugacity of component j are related by

$$a_j = \frac{f_j}{f_j^\circ}$$

in which f_j is the fugacity of component j and f_j° is the fugacity of component j in the standard state.

Standard state. It is common convention in reaction equilibrium calculations to use the following standard state for gaseous, liquid, and solid mixtures: pure component j at 1.0 atm pressure and the system temperature. Sometimes the gaseous mixture standard state is defined to be the pressure corresponding to unit fugacity, but this pressure is close to 1.0 atm for most gases. It is obvious from the definition of the standard state that G_j° and f_j° are not functions of the system pressure or composition, because the standard state is 1.0 atm pressure and pure component j. G_j° and f_j° are, however, strong functions of the system temperature.

Multiplying both sides of Equation 3.7 by ν_j and summing yields

$$\sum_j \nu_j \mu_j = \sum_j \nu_j G_j^\circ + RT \sum_j \nu_j \ln a_j \tag{3.8}$$

The term $\sum_j \nu_j G_j^\circ$ is known as the standard Gibbs energy change for the reaction, ΔG°. Using the equilibrium condition, Equation 3.6, and simple logarithm addition and exponentiation identities yields

$$\Delta G^\circ + RT \ln \prod_j a_j^{\nu_j} = 0 \tag{3.9}$$

Equation 3.9 motivates the definition of the equilibrium constant, K,

$$K = e^{-\Delta G^\circ / RT} \tag{3.10}$$

Rearrangement of Equation 3.9 and exponentiation leads to the following simple condition for chemical equilibrium

$$\boxed{K = \prod_j a_j^{\nu_j}} \tag{3.11}$$

From the definitions of K (Equation 3.10) and the standard state, it is clear that K also is a function of the system temperature, but not a function of the system pressure or composition.

The following example illustrates these concepts and the calculation of the equilibrium composition in the simplest possible setting, an ideal-gas mixture.

Example 3.1: Ideal-gas equilibrium

The reaction of isobutane and linear butenes to branched C_8 hydrocarbons is used to synthesize high octane fuel additives. One such reaction is

$$\begin{array}{ccccc} \text{isobutane} & + & \text{1-butene} & \rightleftharpoons & \text{2,2,3-trimethylpentane} \\ \text{I} & + & \text{B} & \rightleftharpoons & \text{P} \end{array}$$

Determine the equilibrium composition for this system at a pressure of 2.5 atm and temperature of 400 K. You may assume the gases are ideal at this temperature and pressure and this reaction is the only one that occurs. The standard Gibbs energy change for this reaction at 400 K is −3.72 kcal/mol [18].

Solution

We recall that the fugacity of a component in an ideal-gas mixture is equal to its partial pressure,

$$f_j = P_j \tag{3.12}$$

The quantity f_j° is the fugacity of component j in the standard state, which is pure component j at 1.0 atm pressure and the system temperature, 400 K. From Equation 3.12, $f_j^\circ = 1.0$ atm because the partial pressure of a pure component j at 1.0 atm total pressure is 1.0 atm. The activity of component j is then simply

$$a_j = \frac{P_j}{1 \text{ atm}} \tag{3.13}$$

Notice that when we suppress the 1.0 atm in Equation 3.13 in subsequent discussion, the units of P_j have to be atm or a_j is not dimensionless as it should be.

We now attempt to calculate the equilibrium composition from Equation 3.11,

$$K = \frac{a_P}{a_I a_B} \tag{3.14}$$

in which P represents the trimethylpentane, I represents isobutane, and B represents the butene. Since ΔG° is given at the system temperature, we can simply calculate K at 400 K from Equation 3.10, $K = 108$. Substituting the activity of the components in Equation 3.14 then gives

$$K = \frac{P_P}{P_I P_B}$$

If we choose to use the mole fractions of the components, y_j, as the unknown variables rather than the partial pressures, we have

$$K = \frac{y_P}{y_I y_B P} \tag{3.15}$$

in which P is 2.5 atm. The problem is that we have three unknowns, the gas-phase mole fractions at equilibrium, and only one equation,

Equation 3.15. We can of course add the equation that mole fractions sum to one,

$$\sum_j y_j = 1$$

but we still have three unknowns and only two equations. In other words, the equilibrium composition cannot be determined with the information given. All we know at this point is that any set of mole fractions satisfying Equation 3.15 (and summing to one) could be an equilibrium composition. It is important to realize that the mere statement that a system is at equilibrium at a certain temperature and pressure does not necessarily tell one everything about the system. □

The next example illustrates what kind of additional information is required to be able to find the equilibrium composition.

Example 3.2: Ideal-gas equilibrium, revisited

Compute the equilibrium composition of Example 3.1 with the following additional information. The gas is contained in a closed vessel that is initially charged with an equimolar mixture of isobutane and butene.

Solution

We show how the initial composition of the gas in a closed vessel allows us to determine the equilibrium composition. Notice in particular that we do not know the volume of the vessel or the initial number of moles, just that the vessel initially contains an equimolar mixture of the two reactants. We proceed to describe the mole fractions of all components in terms of a single variable, the extent of reaction. Let n_{j0} represent the unknown number of moles of each component initially contained in the vessel. Since the vessel is closed, the number of moles at some other time can be related to the reaction extent from the stoichiometry using $n_j = n_{j0} + \nu_j \varepsilon$,

$$n_I = n_{I0} - \varepsilon, \qquad n_B = n_{B0} - \varepsilon, \qquad n_P = n_{P0} + \varepsilon \tag{3.16}$$

Summing Equations 3.16 produces

$$n_T = n_{T0} - \varepsilon$$

in which n_T is the total number of moles in the vessel. The total number of moles decreases with reaction extent because more moles are consumed than produced by the reaction. Dividing both sides of Equations 3.16 by n_T produces equations for the mole fractions in terms of the reaction extent,

$$y_I = \frac{n_{I0} - \varepsilon}{n_{T0} - \varepsilon} \qquad y_B = \frac{n_{B0} - \varepsilon}{n_{T0} - \varepsilon} \qquad y_P = \frac{n_{P0} + \varepsilon}{n_{T0} - \varepsilon} \tag{3.17}$$

Dividing top and bottom of the right-hand side of the previous equations by n_{T0} yields,

$$y_I = \frac{y_{I0} - \varepsilon'}{1 - \varepsilon'} \qquad y_B = \frac{y_{B0} - \varepsilon'}{1 - \varepsilon'} \qquad y_P = \frac{y_{P0} + \varepsilon'}{1 - \varepsilon'} \tag{3.18}$$

in which $\varepsilon' = \varepsilon / n_{T0}$ is a dimensionless reaction extent that is scaled by the initial total number of moles. Notice that the mole fractions are now expressed in terms of the initial mole fractions, which are known, and a reaction extent, which is the single unknown. Notice also that the mole fractions in Equations 3.18 sum to one for all ε'. Substituting Equations 3.18 into Equation 3.15 then provides the single equation to determine ε',

$$K = \frac{(y_{P0} + \varepsilon')(1 - \varepsilon')}{(y_{B0} - \varepsilon')(y_{I0} - \varepsilon')P}$$

Multiplying through by the denominator and collecting terms gives

$$(y_{B0} - \varepsilon')(y_{I0} - \varepsilon')KP - (y_{P0} + \varepsilon')(1 - \varepsilon') = 0$$

The equilibrium equation is quadratic in ε'. Substituting in the given initial composition, $y_{P0} = 0, y_{B0} = y_{I0} = 1/2$ gives

$$\varepsilon'^2(1 + KP) - \varepsilon'(1 + KP) + (1/4)KP = 0$$

The two solutions are

$$\varepsilon' = \frac{1 \pm \sqrt{\frac{1}{1+KP}}}{2} \tag{3.19}$$

The correct solution is chosen by considering the physical constraints that mole fractions must be positive. If the positive sign is chosen in Equation 3.19, the reaction extent is greater than 1/2, which is physically impossible because it would result in negative numbers of moles of the reactants. The negative sign is therefore chosen, and the solution is $\varepsilon' = 0.469$. The equilibrium mole fractions are then computed from Equation 3.18 giving

$$y_I = 5.73 \times 10^{-2} \qquad y_B = 5.73 \times 10^{-2} \qquad y_P = 0.885$$

The equilibrium at 400 K favors the product trimethylpentane. The reaction has moved from the initial condition of pure reactants to an equilibrium state of relatively pure product. □

Second derivative of G. Before leaving the equilibrium of ideal gases, it is instructive to show that the equilibrium state is in fact a minimum in the Gibbs energy and that this minimum is unique. The procedure in the preceding example determined the value of the reaction extent for which $\partial G/\partial \varepsilon = 0$. To show that this value corresponds to a minimum (and not a maximum or inflection), we would like to show $\partial^2 G/\partial \varepsilon^2 > 0$ at this value of reaction extent as depicted in Figure 3.2. If we can show that $\partial^2 G/\partial \varepsilon^2 > 0$ for every ε, then the minimum is unique.

We begin by performing the same steps we used to determine the mole fractions in Example 3.2, only for an arbitrary number of components. Recall that for the closed system undergoing a single reaction, we can describe the number of moles by

$$n_j = n_{j0} + \nu_j \varepsilon \geq 0 \tag{3.20}$$

Since the reaction extent only has physical significance when it corresponds to a nonnegative number of moles for all components, we have the inequality in Equation 3.20. Summing over all components yields

$$n_T = n_{T0} + \bar{\nu} \varepsilon > 0 \tag{3.21}$$

in which $\bar{\nu} = \sum_j \nu_j$. Notice $\bar{\nu} = 0$ if moles are conserved in the reaction. The total number of moles is strictly positive to preclude the case of no material in the system, an uninteresting case. Dividing Equation 3.20 by Equation 3.21 gives

$$y_j = \frac{n_{j0} + \nu_j \varepsilon}{n_{T0} + \bar{\nu} \varepsilon} \tag{3.22}$$

Defining the dimensionless extent as before,

$$\varepsilon' = \varepsilon / n_{T0} \tag{3.23}$$

Equation 3.22 can be written as

$$y_j = \frac{y_{j0} + \nu_j \varepsilon'}{1 + \bar{\nu} \varepsilon'} \tag{3.24}$$

We remark in passing for later use that

$$\frac{n_T}{n_{T0}} = 1 + \bar{\nu} \varepsilon' > 0 \tag{3.25}$$

We now proceed to calculate $\partial^2 G/\partial \varepsilon^2$. We start with Equation 3.5, repeated here,

$$\left(\frac{\partial G}{\partial \varepsilon}\right)_{T,P} = \sum_j \nu_j \mu_j \tag{3.26}$$

From Equation 3.23, we can express the derivative with respect to the dimensionless reaction extent,

$$\frac{1}{n_{T0}}\left(\frac{\partial G}{\partial \varepsilon'}\right)_{T,P} = \sum_j \nu_j \mu_j \tag{3.27}$$

Differentiation with respect to ε' yields

$$\frac{1}{n_{T0}}\left(\frac{\partial^2 G}{\partial \varepsilon'^2}\right)_{T,P} = \sum_j \nu_j \left(\frac{\partial \mu_j}{\partial \varepsilon'}\right)_{T,P} \tag{3.28}$$

The chemical potential of a component in an ideal gas is

$$\mu_j = G_j^\circ + RT\ln(P_j/1\ \text{atm}) = G_j^\circ + RT\left[\ln y_j + \ln P\right] \tag{3.29}$$

Substituting Equation 3.24 into Equation 3.29 yields

$$\mu_j = G_j^\circ + RT\left[\ln(y_{j0} + \nu_j\varepsilon') - \ln(1 + \bar{\nu}\varepsilon') + \ln P\right]$$

Differentiating this equation with respect to ε' (T and P are constant) gives

$$\left(\frac{\partial \mu_j}{\partial \varepsilon'}\right)_{T,P} = RT\left[\frac{\nu_j}{y_{j0} + \nu_j\varepsilon'} - \frac{\bar{\nu}}{1 + \bar{\nu}\varepsilon'}\right]$$

Substituting into Equation 3.28 gives

$$\frac{1}{n_{T0}RT}\left(\frac{\partial^2 G}{\partial \varepsilon'^2}\right)_{T,P} = \sum_j \nu_j \left[\frac{\nu_j}{y_{j0} + \nu_j\varepsilon'} - \frac{\bar{\nu}}{1 + \bar{\nu}\varepsilon'}\right]$$

Performing the summation on the two terms yields

$$\frac{1}{n_{T0}RT}\left(\frac{\partial^2 G}{\partial \varepsilon'^2}\right)_{T,P} = \sum_j \frac{\nu_j^2}{y_{j0} + \nu_j\varepsilon'} - \frac{\bar{\nu}^2}{1 + \bar{\nu}\varepsilon'}$$

Multiplying through by $1 + \bar{\nu}\varepsilon'$ gives

$$\frac{1 + \bar{\nu}\varepsilon'}{n_{T0}RT}\left(\frac{\partial^2 G}{\partial \varepsilon'^2}\right)_{T,P} = \sum_j \frac{\nu_j^2}{y_{j0} + \nu_j\varepsilon'}(1 + \bar{\nu}\varepsilon') - \bar{\nu}^2 \tag{3.30}$$

Although Equation 3.30 may not appear very useful, if we define two variables,

$$a_j = \frac{\nu_j}{\sqrt{y_{j0} + \nu_j\varepsilon'}} \qquad b_j = \sqrt{y_{j0} + \nu_j\varepsilon'}$$

Equation 3.30 can be written as

$$\frac{1+\bar{\nu}\varepsilon'}{n_{T0}RT}\left(\frac{\partial^2 G}{\partial \varepsilon'^2}\right)_{T,P} = \sum_j a_j^2 \sum_j b_j^2 - \left(\sum_j a_j b_j\right)^2 \tag{3.31}$$

The Cauchy inequality [2], also known as the Schwarz inequality, states that for any two vectors $\boldsymbol{a}$ and $\boldsymbol{b}$

$$\sum_j a_j b_j \leq \left(\sum_j a_j^2\right)^{1/2} \left(\sum_j b_j^2\right)^{1/2} \tag{3.32}$$

The equality is achieved if and only if the two vectors are nonnegatively proportional ($a_j = kb_j$ for every j, with $k \geq 0$). Since $a_j = \frac{\nu_j}{y_{j0}+\nu_j\varepsilon'} b_j$, it is clear that the vectors $\boldsymbol{a}$ and $\boldsymbol{b}$ are not proportional, and strict inequality holds in Equation 3.31. Since $1+\bar{\nu}\varepsilon' > 0$ from Equation 3.25 and n_{T0}, R and T are positive, we can conclude

$$\left(\frac{\partial^2 G}{\partial \varepsilon'^2}\right)_{T,P} > 0 \tag{3.33}$$

Due to the definition of ε', Equation 3.33 is equivalent to

$$\left(\frac{\partial^2 G}{\partial \varepsilon^2}\right)_{T,P} > 0$$

which is what we set out to show. Notice that this inequality is true for every ε corresponding to physically meaningful compositions. Since the G function has positive curvature at every ε, Figure 3.2 is indeed the correct picture for an ideal gas, and the equilibrium state is unique. Aris provides an alternative treatment of this same question [1].

Evaluation of *G*. As the last example on the equilibrium of an ideal gas, we calculate the actual function $G(T,P,\varepsilon')$. As we shall see, the equilibrium extent then can be determined by plotting G versus ε'.

The Gibbs energy of a single phase can be computed from the chemical potentials and mole numbers of the components from (see Exercise 3.6 for a derivation)

$$G = \sum_j \mu_j n_j \tag{3.34}$$

The chemical potential of a component in an ideal gas is given by Equation 3.29, repeated here,

$$\mu_j = G_j^\circ + RT\left[\ln y_j + \ln P\right]$$

Substituting the above into Equation 3.34 gives

$$G = \sum_j n_j G_j^\circ + RT \sum_j n_j \left[\ln y_j + \ln P\right] \tag{3.35}$$

Note that for this single reaction case, $n_j = n_{j0} + \nu_j \varepsilon$, which gives

$$\sum_j n_j G_j^\circ = \sum_j n_{j0} G_j^\circ + \varepsilon \Delta G^\circ \tag{3.36}$$

We now define a modified Gibbs function to measure changes from the initial standard Gibbs energy and to scale by the temperature and the total initial moles as follows,

$$\widetilde{G}(T, P, \varepsilon') = \frac{G - \sum_j n_{j0} G_j^\circ}{n_{T0} RT} \tag{3.37}$$

Substituting Equations 3.35 and 3.36 into Equation 3.37 gives

$$\widetilde{G} = \varepsilon' \frac{\Delta G^\circ}{RT} + \sum_j \frac{n_j}{n_{T0}} \left[\ln y_j + \ln P\right] \tag{3.38}$$

From Equation 3.20, the ratio n_j / n_{T0} is easily seen to be equal to $y_{j0} + \nu_j \varepsilon'$. Substituting this relation, the definition of K and Equation 3.24 into Equation 3.38 yields

$$\widetilde{G} = -\varepsilon' \ln K + \sum_j (y_{j0} + \nu_j \varepsilon') \left[\ln\left(\frac{(y_{j0} + \nu_j \varepsilon')}{1 + \bar{\nu}\varepsilon'}\right) + \ln P\right]$$

which can be rearranged to

$$\widetilde{G} = -\varepsilon' \ln K + (1 + \bar{\nu}\varepsilon') \ln P + \sum_j (y_{j0} + \nu_j \varepsilon') \ln\left(\frac{y_{j0} + \nu_j \varepsilon'}{1 + \bar{\nu}\varepsilon'}\right) \tag{3.39}$$

Notice that because T and P are known values, $\widetilde{G}$ is simply a shift of the G function up or down by a constant and then rescaling by the positive constant $1/(n_{T0} RT)$. In particular, the shape of the function $\widetilde{G}$ is the same as G, and the minimum with respect to ε' is at the same value of ε' for the two functions.

Example 3.3: Minimum in G for an ideal gas

Plot the $\widetilde{G}$ function for Example 3.2 and show that ε' corresponding to the minimum of $\widetilde{G}$ agrees with the result of Example 3.2.

Solution

We apply Equation 3.39 to the chemistry of Example 3.2,

$$\mathrm{I} + \mathrm{B} \rightleftharpoons \mathrm{P}$$

For this stoichiometry, $\sum_j \nu_j = \bar{\nu} = -1$. The reaction started with an equimolar mixture of reactants, so $y_{P0} = 0, y_{I0} = y_{B0} = 0.5$. Substituting these values into Equation 3.39 gives

$$\begin{aligned}\widetilde{G}(T,P,\varepsilon') = -\varepsilon' \ln K(T) + (1-\varepsilon')\ln P + \\ \varepsilon' \ln(\varepsilon') + 2(0.5-\varepsilon')\ln(0.5-\varepsilon') - (1-\varepsilon')\ln(1-\varepsilon')\end{aligned} \tag{3.40}$$

Recall that the range of physically significant ε' values is

$$0 \leq \varepsilon' \leq 0.5$$

The equilibrium constant at 400 K has been computed to be $K = 108$. The pressure is 2.5 atm. Recall we must use atm for the pressure units because of the way we expressed chemical potential in Equation 3.29. Substituting these values into Equation 3.40 and plotting the results gives Figure 3.3. Because the minimum is close to $\varepsilon' = 0.5$ we see again that the equilibrium favors the product. Magnifying this region of the plot in Figure 3.3 shows that $\varepsilon' = 0.47$ at equilibrium, which is in good agreement with the value 0.469 calculated in Example 3.2. Although we prefer the simple calculation in Example 3.2 to obtain a numerical value for ε', the graphical representation of $\widetilde{G}$ is more informative on issues of whether or not the solution is indeed a minimum and whether or not the minimum is unique. It also gives us confidence that the analysis of the previous section is correct because $\widetilde{G}$ has positive curvature for all ε' of physical interest. □

Notice from Equation 3.39 that for an ideal gas, the pressure enters directly in the Gibbs energy with the $\ln P$ term. The effect of pressure on the equilibrium extent depends on the sign of $\bar{\nu}$. If positive, which means there are more product than reactant molecules in the stoichiometry of the reaction, then an increase in the pressure causes a decrease in the equilibrium reaction extent, in agreement with Le Chatelier's principle. This can be verified by examining the term $(1+\varepsilon'\bar{\nu})\ln P$ in Equation 3.39. If $\bar{\nu}$ is positive, increasing the pressure increases this term more at larger values of ε' than at smaller values, thus pushing the minimum in the $\widetilde{G}$ function to smaller values of ε. Hence for $\bar{\nu} > 0$ an

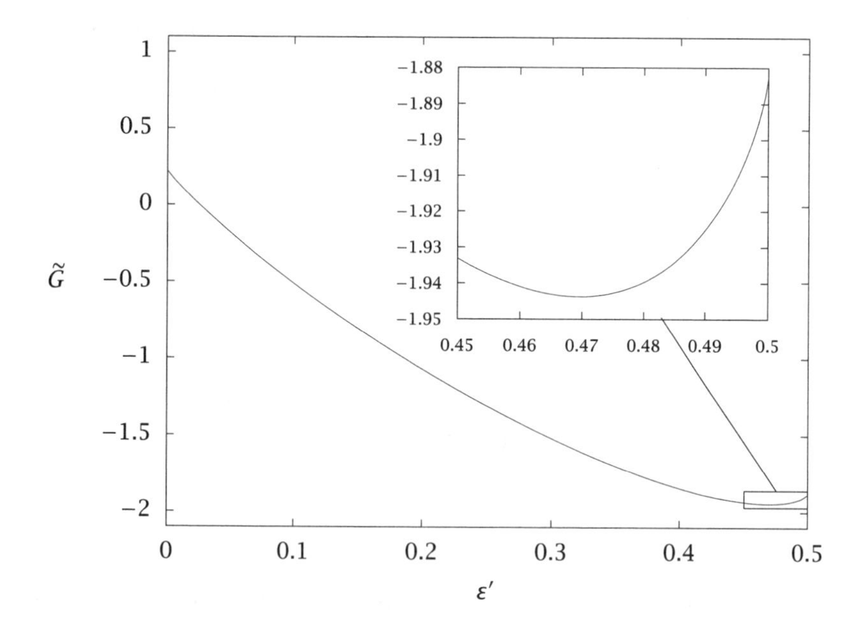

Figure 3.3: Gibbs energy $\widetilde{G}$ versus reaction extent ε'.

equilibrium shift favoring reactants is the effect of a pressure increase. If $\bar{\nu}$ is negative, the opposite situation arises. For $\bar{\nu} = 0$, the pressure does not affect the equilibrium, because the term $(1 + \varepsilon'\bar{\nu}) \ln P$ is then constant with ε' and does not affect the location of the minimum. For single liquid-phase or solid-phase systems, the effect of pressure on equilibrium is usually small, because the chemical potential of a component in a liquid-phase or solid-phase solution is usually a weak function of pressure.

The temperature effect on the Gibbs energy is contained in the $\ln K(T)$ term. This term often gives rise to a large effect of temperature on equilibrium. We turn our attention to the evaluation of this important temperature effect in the next section.

3.2.1 Evaluation of the Gibbs Energy Change of Reaction

We usually calculate the standard Gibbs energy change for the reaction, ΔG°, by using the Gibbs energy of formation of the species, which are commonly available thermochemical data. The standard Gibbs energy of formation of species j is defined as the difference in the Gibbs energy of the species and its constituent elements in some standard state.

$$G^\circ_{jf} = G^\circ_j - \sum_{l=1}^{n_e} a_{jl} G^\circ_{El}$$

The standard state for the elements are usually the pure elements in their common form at 25°C and 1.0 atm. For example, the Gibbs energy of formation of H_2O is

$$G^\circ_{H_2Of} = G^\circ_{H_2O} - G^\circ_{H_2} - \frac{1}{2} G^\circ_{O_2}$$

We then take the appropriate linear combinations of the species Gibbs energy of formation to compute the Gibbs energy change for the reaction at 25°C

$$\boxed{\Delta G^\circ_i = \sum_j \nu_{ij} G^\circ_{jf}} \qquad (3.41)$$

As shown in Exercise 3.11, the Gibbs energy of the elements cancel, leading to Equation 3.41.

Finding appropriate thermochemical data remains a significant challenge for solving realistic, industrial problems. Vendors offer a variety of commercial thermochemical databases to address this need. Many companies also maintain their own private thermochemical databases for compounds of special commercial interest to them.[1]

The standard state temperature 25°C is often not the temperature of the system of interest, so we need to convert from 25°C to the system temperature. To accomplish this conversion, the temperature dependence of ΔG° is discussed in the next section.

3.2.2 Temperature Dependence of the Standard Gibbs Energy

Recall from Equation 3.3 that the change of the Gibbs energy with temperature is the negative of the entropy,

$$\left(\frac{\partial G}{\partial T}\right)_{P,n_j} = -S$$

[1]For educational purposes and to gain some experience with thermochemical databases, you may wish to try the Design Institute for Physical Properties (DIPPR) database. A web-based student version of the database provides students with access to data for 2000 common compounds at no charge: `http://dippr.byu.edu/students/chemsearch.asp`.

This expression can be used to evaluate the change in the Gibbs energy in the standard state, pure component j at a pressure of 1.0 atm,

$$\left(\frac{\partial G_j^\circ}{\partial T}\right)_{P,n_j} = -S_j^\circ \tag{3.42}$$

Notice in this context that the subscripts P and n_j denoting constant pressure and composition are redundant due to the definition of the standard state and are dropped to avoid confusion. Summing Equation 3.42 with the stoichiometric coefficients of a single reaction gives

$$\sum_j \frac{\partial(\nu_j G_j^\circ)}{\partial T} = \sum_j -\nu_j S_j^\circ$$

Defining the term on the right-hand side to be the standard entropy change of reaction, ΔS° gives

$$\frac{\partial \Delta G^\circ}{\partial T} = -\Delta S^\circ \tag{3.43}$$

Let H denote the enthalpy and recall its connection to the Gibbs energy,

$$G = H - TS \tag{3.44}$$

Partial molar properties. Recall the definition of a partial molar property is

$$\overline{X}_j = \left(\frac{\partial X}{\partial n_j}\right)_{T,P,n_k}$$

in which X is any extensive mixture property (U, H, A, G, V, S, etc.). In other words, the partial molar property of component j measures the change in an extensive property as a small amount of component j is added to the mixture at constant T, P and other mole numbers.

Equation 3.44 is valid also for the corresponding partial molar quantities

$$\overline{G}_j = \overline{H}_j - T\overline{S}_j \tag{3.45}$$

Writing Equation 3.45 for a component in its standard state gives $G_j^\circ = H_j^\circ - TS_j^\circ$, and summing with the stoichiometric coefficient yields

$$\Delta G^\circ = \Delta H^\circ - T\Delta S^\circ \tag{3.46}$$

Solving Equation 3.46 for the entropy change and substituting into Equation 3.43 gives

$$\frac{\partial \Delta G^\circ}{\partial T} = \frac{\Delta G^\circ - \Delta H^\circ}{T}$$

Rearranging this equation and division by RT gives

$$\frac{1}{RT}\frac{\partial \Delta G^\circ}{\partial T} - \frac{\Delta G^\circ}{RT^2} = -\frac{\Delta H^\circ}{RT^2}$$

Using differentiation formulas, the left-hand side can be rewritten as

$$\frac{\partial\left(\frac{\Delta G^\circ}{RT}\right)}{\partial T} = -\frac{\Delta H^\circ}{RT^2}$$

which finally can be expressed in terms of the equilibrium constant

$$\boxed{\frac{\partial \ln K}{\partial T} = \frac{\Delta H^\circ}{RT^2}} \tag{3.47}$$

Equation 3.47, known as the van 't Hoff equation, expresses how the equilibrium constant varies with temperature. Equation 3.47 can be integrated between two temperatures to give

$$\int_{T_1}^{T_2} \frac{\partial \ln K}{\partial T} dT = \int_{T_1}^{T_2} \frac{\Delta H^\circ}{RT^2} dT$$

To evaluate the integral one needs to have the standard enthalpy change as a function of temperature. This information can be computed from the partial molar heat capacities. The partial molar heat capacity for component j is defined by

$$\overline{C}_{Pj} = \left(\frac{\partial \overline{H}_j}{\partial T}\right)_{P,n_k} \tag{3.48}$$

Evaluating this expression at the standard state, multiplying by the stoichiometric numbers and summing produces

$$\Delta C_P^\circ = \frac{\partial \Delta H^\circ}{\partial T}$$

in which ΔC_P° is the heat capacity change upon reaction

$$\Delta C_P^\circ = \sum_j \nu_j C_{Pj}^\circ$$

The standard enthalpy change between a reference temperature T_0 and an arbitrary temperature T can be evaluated from

$$\Delta H^\circ(T) = \Delta H^\circ(T_0) + \int_{T_0}^{T} \Delta C_P^\circ dT$$

If the standard heat capacity change of reaction is zero over the temperature range, then ΔH° is constant and Equation 3.47 can be integrated directly to give

$$\ln\left(\frac{K_2}{K_1}\right) = -\frac{\Delta H^\circ}{R}\left(\frac{1}{T_2} - \frac{1}{T_1}\right) \tag{3.49}$$

Notice the H_j° themselves may be strong functions of temperature, but if $\Delta C_P^\circ = 0$, then ΔH° is still independent of temperature and Equation 3.49 is correct. Equation 3.49 also may be a useful approximation, but the error in assuming constant standard change in enthalpy should be examined if heat capacity data are available.

3.3 Condition for Phase Equilibrium

Consider a multicomponent, multiphase system that is at equilibrium and denote two of the phases as α and β. Let T^k and P^k represent the temperature and pressure of phase k and n_j^k represent the number of moles of component j in phase k. The chemical potential of component j in phase k is denoted by $\hat{\mu}_j^k$. The hat is to emphasize that this is the chemical potential of component j in the multicomponent *mixture*. The conditions for equilibrium when more than one phase are present are developed in standard texts to be

$$T^\alpha = T^\beta$$

$$P^\alpha = P^\beta$$

$$\hat{\mu}_j^\alpha = \hat{\mu}_j^\beta, \qquad j = 1, 2, \ldots, n_s \tag{3.50}$$

Since these equations apply to any two phases, the temperatures of all phases are equal and the distinction of temperatures for each phase is unnecessary and one can speak of the system temperature T. Similarly for the pressure. Most of our attention then is focused on the condition that the chemical potential of component j must be equal in all phases.

A loose but useful physical intuition is connected to the equilibrium conditions. If the temperatures of the phases of a two-phase mixture are not equal, one expects to see a transfer of heat from the hotter to the colder phase until the equilibrium condition is satisfied. Similarly if the pressures are not equal, one expects a momentum transfer from the system at higher pressure to the one at lower pressure. Finally, if the chemical potential of a component is higher in one phase than another phase, one expects a mass transfer of the component from the phase with higher chemical potential to the phase with lower chemical potential until the chemical potentials become equal. In reality, of course, the gradients and fluxes do not separate so ideally and we observe the simultaneous transfer of heat, momentum and mass of all components while the system approaches equilibrium from some arbitrary initial condition [3, pp. 767–768].

Again, the chemical potential is often expressed in terms of the fugacity, which is defined by

$$\hat{\mu}_j = \mu_j^\circ + RT \ln \frac{\hat{f}_j}{f_j^\circ} \tag{3.51}$$

in which μ_j° and f_j° denote the chemical potential and fugacity of component j in a standard state. The standard state is again pure component j at the system temperature and 1.0 atm pressure. If we express Equation 3.51 for two phases α and β and equate their chemical potentials we deduce

$$\hat{f}_j^\alpha = \hat{f}_j^\beta \qquad j = 1, 2 \ldots, n_s$$

One can therefore use either the equality of chemical potentials or fugacities as the condition for equilibrium.

3.3.1 Ideal Mixtures

In this and the next section we briefly review the methods for expressing the fugacity of components in a mixture and calculating the phase equilibrium. The simplest rule for evaluating the fugacity of a component in a mixture is the ideal mixture assumption. We review this case first.

Gaseous Solutions

Consider a gas-phase mixture of interest at some temperature and pressure. Let f_j^G denote the fugacity of pure component j in the gas phase at the temperature and pressure of the mixture. The simplest mixing rule for calculating the fugacity of component j in the mixture containing component j at some mole fraction y_j is the following linear mixing rule

$$\hat{f}_j^G = f_j^G y_j \qquad \text{(ideal mixture)}$$

An ideal gas, for example, obeys this mixing rule and furthermore the fugacity of pure j at the mixture T and P is simply the system pressure, $f_j^G = P$. The fugacity of component j in an ideal gas is therefore the partial pressure of component j, P_j, defined as

$$\hat{f}_j^G = P_j := P y_j \qquad \text{(ideal gas)}$$

Liquid (and Solid) Solutions

Again the simplest mixing rule for liquid (and solid) mixtures is that the fugacity of component j in the mixture is the fugacity of pure j at the temperature and pressure of the mixture times the mole fraction of j in the mixture. We let x_j denote the mole fraction to distinguish liquid-phase and solid-phase from gas-phase compositions.

$$\hat{f}_j^L = f_j^L x_j$$

This approximation is usually valid when the mole fraction of a component is near one. In a two-component mixture, one can establish from the Gibbs-Duhem relations that if the first component obeys the ideal mixture, then the second component follows Henry's law:

$$\hat{f}_2^L = k_2 x_2$$

in which k_2 is the Henry's law constant for the second component. Note that it is not generally equal to f_2^L (unless the mixture is ideal over the entire composition range).

Fugacity pressure dependence. Sometimes we know the fugacity of a component at one pressure and we need to evaluate it at another pressure. We start with the standard relation

$$\left(\frac{\partial \ln \hat{f}_j}{\partial P}\right)_{T,x_k} = \frac{\overline{V}_j}{RT}$$

in which $\overline{V}_j$ is the partial molar volume. We integrate between the two pressures of interest to obtain

$$\hat{f}_j\Big|_{P_2} = \hat{f}_j\Big|_{P_1} \exp\left[\frac{1}{RT}\int_{P_1}^{P_2} \overline{V}_j dP\right] \tag{3.52}$$

For condensed phases, the partial molar volume is usually constant over reasonable pressure ranges giving

$$\hat{f}_j\Big|_{P_2} = \hat{f}_j\Big|_{P_1} \exp\left[\frac{\overline{V}_j(P_2 - P_1)}{RT}\right] \tag{3.53}$$

The exponential term in Equations 3.52 and 3.53 is called the Poynting correction factor. For liquids and solids, the partial molar volume is generally small, and the Poynting correction may be neglected if the pressure does not vary by a large amount.

3.3.2 Nonideal Mixtures

For mixtures that display significant deviations from ideality, correcting factors from the ideal mixture assumption are used. For gaseous mixtures, one defines the fugacity coefficient, $\hat{\phi}_j$, as follows

$$\hat{f}_j^G = P y_j \hat{\phi}_j$$

The analogous correcting factor for the liquid phase is the activity coefficient, γ_j.

$$\hat{f}_j^L = f_j^L x_j \gamma_j$$

The introduction of these coefficients obviously has shifted the burden from evaluating the fugacities to evaluating the activity and fugacity coefficients. These coefficients may be available in several forms. Correlations may exist for systems of interest or phase equilibrium data may be available from which the coefficients can be calculated. The following list is a representative sample of the numerous texts devoted to this important topic [12, 13, 19, 14, 6].

3.4 Equilibrium Composition for Heterogeneous Reactions

In this section we illustrate the calculation of chemical equilibrium when there are multiple phases as well as a chemical reaction taking place. The following example illustrates the important issues.

Example 3.4: Chemical and phase equilibrium for a nonideal mixture

Consider the liquid-phase reaction

$$\mathrm{A(l) + B(l) \rightleftharpoons C(l)}$$

that occurs in the following three-phase system.
Phase I: nonideal liquid mixture of A and C only. For illustration purposes, assume the activity coefficients are given by the simple Margules equation,

$$\ln \gamma_A = x_C^2 \left[A_{AC} + 2(A_{CA} - A_{AC}) x_A \right]$$
$$\ln \gamma_C = x_A^2 \left[A_{CA} + 2(A_{AC} - A_{CA}) x_C \right]$$

Phase II: pure liquid B.
Phase III: ideal-gas mixture of A, B and C.

All three phases are in intimate contact and we have the following data:

$$A_{AC} = 1.4 \qquad P_A^\circ = 0.65 \text{ atm}$$
$$A_{CA} = 2.0 \qquad P_B^\circ = 0.50 \text{ atm}$$
$$P_C^\circ = 0.50 \text{ atm}$$

in which P_j° is the vapor pressure of component j at the system temperature.

1. Plot the partial pressures of A and C versus x_A for a vapor phase that is in equilibrium with only the A–C liquid phase. Compute the Henry's law constants for A and C from the Margules equation. Sketch the meaning of Henry's law on the plot and verify your calculation from the plot.

2. Use Henry's law to calculate the composition of all three phases for $K = 4.7$. What is the equilibrium pressure?

3. Repeat for $K = 0.23$.

4. Assume $K = 1$. Use the Margules equation to calculate the composition of all three phases.

5. Repeat 4 with an ideal mixture assumption and compare the results.

Solution

Part 1. The calculation of the partial pressures of A and C in a vapor phase in equilibrium with a liquid-phase mixture of A and C follows from equating the chemical potentials of the components in the different phases. Since the gas phase is assumed an ideal-gas mixture,

$$\hat{f}_A^G = P_A \tag{3.54}$$

In the liquid phase, the fugacity of A is computed from the activity coefficient

$$\hat{f}_A^L = f_A^L x_A \gamma_A \tag{3.55}$$

The fugacity of pure liquid A at the system T and P is not known. The fugacity of pure liquid A at the system temperature and the vapor pressure of A at the system temperature is known, however; it is simply the vapor pressure, P_A°. If the system pressure is not greatly different from the vapor pressure of A at the system temperature, then the following

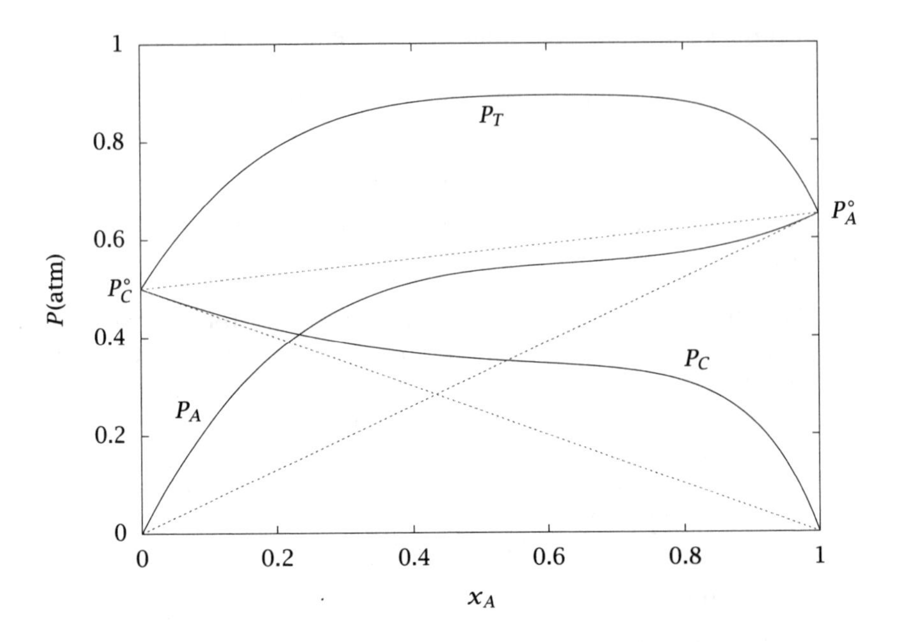

Figure 3.4: Partial pressures of components A and C versus liquid-phase composition in a nonideal solution; solid line: Margules equation; dashed line: ideal mixture.

approximation is valid because, as discussed in Section 3.3.1, fugacities of liquids are only weak functions of pressure

$$f_A^L = P_A^\circ \tag{3.56}$$

Substituting Equation 3.56 into 3.55 and equating to 3.54 gives,

$$P_A = P_A^\circ x_A \gamma_A \tag{3.57}$$

The analogous expression is valid for P_C. Figure 3.4 plots P_A, P_C and $P_T = P_A + P_C$ versus the composition in the liquid phase, x_A, using the Margules equation to compute the activity coefficient in Equation 3.57. Notice that the system exhibits positive deviations from an ideal mixture, which is also plotted in the figure.

Henry's law for component A is

$$\hat{f}_A^L = k_A x_A$$

which is valid for x_A small. Comparing to Equation 3.55 and using Equation 3.56 gives

$$k_A = P_A^\circ \gamma_A$$

which is also valid for small x_A. Computing γ_A from the Margules equation for $x_A = 0$ gives

$$\gamma_A(0) = e^{A_{AC}}$$

So the Henry's law constant for component A is

$$k_A = P_A^\circ e^{A_{AC}}$$

The analogous expression holds for component C. Substituting in the values gives

$$k_A = 2.6, \qquad k_C = 3.7$$

These values can be checked by examining Figure 3.4. The slope of the tangent line to the P_A curve at $x_A = 0$ is equal to k_A. The negative of the slope of the tangent line to the P_C curve at $x_A = 1$ is equal to k_C.

Part 2. For $K = 4.7$, one expects a large value of the equilibrium constant to favor the formation of the product, C. We therefore assume that x_A is small and Henry's law is valid for component A. The validity of this assumption can be checked after the equilibrium composition is computed. The unknowns in the problem are x_A and x_C in the A–C mixture, y_A, y_B and y_C in the gas phase, and the system pressure. There are therefore six unknowns. We require six equations for a well-posed problem. We can equate the fugacities of each component in the gas and liquid phases, and use the fact that the mole fractions sum to one in the gas and A–C liquid phases. Finally, the chemical equilibrium provides the sixth equation and the problem can be solved. In general we would set up six equations in six unknowns and solve the problem numerically, but the structure of this problem allows an analytical solution. Consider first the condition for reaction equilibrium,

$$K = \frac{\hat{a}_C}{\hat{a}_A \hat{a}_B}$$

in which K is evaluated for a liquid-phase standard state. Evaluating the activities first of component A, the component for which we assume Henry's law,

$$\hat{a}_A^L = \frac{\hat{f}_A^L}{f_A^\circ} = \frac{k_A x_A}{f_A^\circ}$$

From the standard state for reactions, f_A° is the fugacity of pure liquid A at the system temperature and 1.0 atm. Again, this value is unknown, but we do know that P_A° is the fugacity of pure liquid A at the system temperature and the vapor pressure of A at this temperature. The difference between 0.65 and 1.0 atm is not large, so we assume $f_A^\circ = P_A^\circ$. Since x_A is assumed small, x_C is assumed near one. The fugacity is then evaluated from

$$\hat{a}_C^L = \frac{\hat{f}_C^L}{f_C^\circ} = \frac{f_C^L x_C}{f_C^\circ}$$

Now f_C^L and f_C° are the fugacities of pure liquid C at the system temperature and the system pressure and 1.0 atm, respectively. If the system pressure turns out to be reasonably small, then it is a good assumption to assume these fugacities are equal giving,

$$\hat{a}_C^L = x_C$$

Since component B is in a pure liquid phase, the same reasoning leads to

$$\hat{a}_B^L = \frac{\hat{f}_B^L}{f_B^\circ} = \frac{f_B^L}{f_B^\circ} = 1$$

Substituting these activities into the reaction equilibrium condition gives

$$K = \frac{x_C}{x_A k_A / P_A^\circ \cdot 1} \tag{3.58}$$

Using $x_C = 1 - x_A$ and solving Equation 3.58 for x_A yields

$$x_A = \left(1 + \frac{k_A K}{P_A^\circ}\right)^{-1} \qquad x_C = \left(1 + \frac{P_A^\circ}{k_A K}\right)^{-1}$$

Substituting in the provided data gives

$$x_A = 0.05, \qquad x_C = 0.95$$

Therefore, from Figure 3.4, the assumption of Henry's law for component A is reasonable.

The vapor compositions now are computed from the phase equilibrium conditions. The fugacities of all components in the ideal-gas mixture are equal to their partial pressures. Therefore, equating these to the fugacities of the liquid phase gives

$$P_A = k_A x_A, \qquad P_B = P_B^\circ, \qquad P_C = P_C^\circ x_C$$

Substituting in the provided data, Henry's law constant and known liquid-phase compositions gives

$$P_A = 0.13 \text{ atm}, \qquad P_B = 0.50 \text{ atm}, \qquad P_C = 0.48 \text{ atm}$$

The system pressure is therefore $P = 1.11$ atm. This low pressure justifies the assumptions we have made about fugacities of pure liquids at the system pressure. Finally, the vapor-phase concentrations can be computed from the ratios of partial pressures to total pressure, which gives

$$y_A = 0.12, \qquad y_B = 0.45, \qquad y_C = 0.43$$

Part 3. For $K = 0.23$ one expects the reactants to be favored so Henry's law is assumed for component C. You are encouraged to work through the preceding development again for this situation. The answers are

$$x_A = 0.97, \qquad x_C = 0.03$$

$$y_A = 0.51, \qquad y_B = 0.40, \qquad y_C = 0.09$$

$$P = 1.24 \text{ atm}$$

Again the assumption of Henry's law is justified and the system pressure is low.

Part 4. For $K = 1$, the equilibrium composition may not allow Henry's law to be assumed for either A or C. In this case we must solve the reaction equilibrium condition using the Margules equation for the activity coefficients,

$$K = \frac{x_C \gamma_C}{x_A \gamma_A}$$

By expressing $x_C = 1 - x_A$, we have one equation in one unknown,

$$K = \frac{(1 - x_A) \exp\left[x_A^2 (A_{CA} + 2(A_{AC} - A_{CA})(1 - x_A))\right]}{x_A \exp\left[(1 - x_A)^2 (A_{AC} + 2(A_{CA} - A_{AC}) x_A)\right]} \tag{3.59}$$

Equation 3.59 can be solved numerically to give $x_A = 0.35$. From this value, the rest of the solution can be computed. Recall that the partial pressures of A and C now must be computed from the fugacity of the liquid phase using the activity coefficients

$$P_j = P_j^{\circ} x_j \gamma_j, \qquad j = A, C$$

The solution is

$$x_A = 0.35, \qquad x_C = 0.65$$

$$y_A = 0.36, \qquad y_B = 0.37, \qquad y_C = 0.28$$
$$P = 1.37 \text{ atm}$$

Using the full Margules equation instead of Henry's law for the other K values gives $x_A = 0.054$ for $K = 4.7$ and $x_A = 0.96$ for $K = 0.23$, in good agreement with the Henry's law solution presented above.

Part 5. Finally, if one assumes that the A–C mixture is ideal, the equilibrium condition becomes

$$K = \frac{x_C}{x_A}$$

which can be solved to give $x_A = 1/(1 + K)$. For $K = 1$, the solution is

$$x_A = 0.5, \qquad x_C = 0.5$$

$$y_A = 0.30, \qquad y_B = 0.47, \qquad y_C = 0.23$$
$$P = 1.08 \text{ atm}$$

As can be seen from comparison with the solution to Part 4, the ideal mixture assumption leads to significant error as might be expected from the strong deviations from ideality shown in Figure 3.4. □

3.5 Multiple Reactions

Determining equilibria for reacting systems with multiple phases and reactions can require significant computational effort. The phase and reaction equilibrium conditions generally lead to mathematical problems of two types: solving nonlinear algebraic equations and minimizing a nonlinear function subject to constraints. In this section it is assumed that the required thermochemical data are available, but finding or measuring these data is often another significant challenge in computing equilibria for systems of industrial interest.

3.5.1 Algebraic Approach

We again consider a single-phase system but allow n_r reactions

$$\sum_j \nu_{ij} A_j = 0, \qquad i = 1, 2, \ldots, n_r$$

Letting ε_i be the reaction extent for the ith reaction, the change in the number of moles of component j for a closed system due to the n_r chemical reactions is

$$n_j = n_{j0} + \sum_i \nu_{ij} \varepsilon_i \tag{3.60}$$

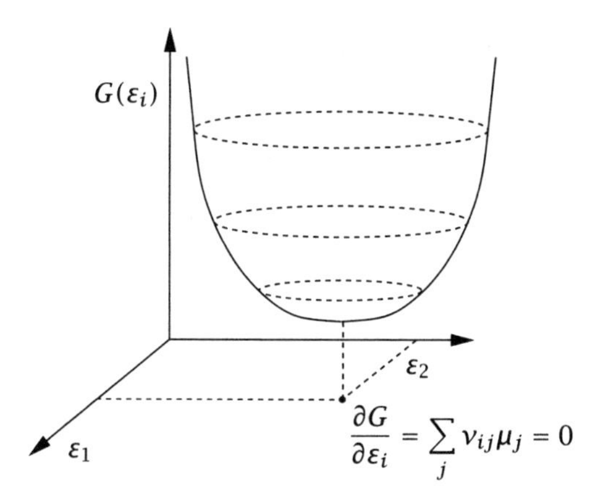

Figure 3.5: Gibbs energy versus two reaction extents at constant T and P.

We can compute the change in Gibbs energy as before

$$dG = -SdT + VdP + \sum_j \mu_j dn_j$$

Substituting in the change in the number of moles with reaction extents, $dn_j = \sum_i \nu_{ij} d\varepsilon_i$, gives

$$\begin{aligned} dG &= -SdT + VdP + \sum_j \mu_j \sum_i \nu_{ij} d\varepsilon_i \\ &= -SdT + VdP + \sum_i \left(\sum_j \nu_{ij}\mu_j \right) d\varepsilon_i \end{aligned} \tag{3.61}$$

At constant T and P, G is a minimum as a function of the n_r reaction extents. Necessary conditions are therefore

$$\left(\frac{\partial G}{\partial \varepsilon_i} \right)_{T,P,\varepsilon_{l \neq i}} = 0, \qquad i = 1, 2, \ldots, n_r$$

as illustrated in Figure 3.5 for the case of two reactions. Evaluating the partial derivatives in Equation 3.61 gives

$$\boxed{\sum_j \nu_{ij}\mu_j = 0, \qquad i = 1, 2, \ldots, n_r} \tag{3.62}$$

In other words, the necessary condition for a single reaction applies for each reaction in the network for the multiple reaction case.

Substituting Equation 3.7 for the chemical potential in terms of the activity into Equation 3.62 gives

$$\sum_j \nu_{ij}\mu_j = \sum_j \nu_{ij}G_j^\circ + RT\sum_j \nu_{ij}\ln a_j$$

Defining the standard Gibbs energy change for reaction i, $\Delta G_i^\circ = \sum_j \nu_{ij}G_j^\circ$ gives

$$\sum_j \nu_{ij}\mu_j = \Delta G_i^\circ + RT\sum_j \nu_{ij}\ln a_j$$

Finally, defining the equilibrium constant for reaction i as

$$K_i = e^{-\Delta G_i^\circ/RT} \tag{3.63}$$

allows one to express the reaction equilibrium condition as

$$\boxed{K_i = \prod_j a_j^{\nu_{ij}}, \qquad i = 1, 2, \ldots, n_r} \tag{3.64}$$

Example 3.5: Equilibrium composition for multiple reactions

In addition to the formation of 2,2,3-trimethylpentane in Example 3.2, it is known that 2,2,4-trimethylpentane may also form with the same stoichiometry,

$$\text{isobutane} + \text{1-butene} \rightleftharpoons \text{2,2,4-trimethylpentane} \tag{3.65}$$

Recalculate the equilibrium composition for this example given that $\Delta G^\circ = -4.49$ kcal/mol for this reaction at 400 K.

Solution

Let reaction 1 be the formation of 2,2,3 trimethylpentane denoted as P_1 and reaction 2 be the formation of 2,2,4-trimethylpentane denoted as P_2. From the Gibbs energy changes, we have

$$K_1 = 108, \qquad K_2 = 284$$

We now set up a table to calculate the four compositions from the extents of the two reactions,

$$n_I = n_{I0} - \varepsilon_1 - \varepsilon_2 \qquad n_{P_1} = n_{P_{10}} + \varepsilon_1$$

$$n_B = n_{B0} - \varepsilon_1 - \varepsilon_2 \qquad n_{P_2} = n_{P_{20}} + \varepsilon_2$$

The total number of moles is then $n_T = n_{T0} - \varepsilon_1 - \varepsilon_2$. Forming the mole fractions yields

$$y_I = \frac{y_{I0} - \varepsilon_1' - \varepsilon_2'}{1 - \varepsilon_1' - \varepsilon_2'} \qquad y_{P_1} = \frac{y_{P_{10}} + \varepsilon_1'}{1 - \varepsilon_1' - \varepsilon_2'}$$

$$y_B = \frac{y_{B0} - \varepsilon_1' - \varepsilon_2'}{1 - \varepsilon_1' - \varepsilon_2'} \qquad y_{P_2} = \frac{y_{P_{20}} + \varepsilon_2'}{1 - \varepsilon_1' - \varepsilon_2'} \tag{3.66}$$

Applying Equation 3.64 to the two reactions gives

$$K_1 = \frac{y_{P_1}}{y_I y_B P} \qquad K_2 = \frac{y_{P_2}}{y_I y_B P}$$

Substituting in the mole fractions gives two equations for the two unknown reaction extents,

$$PK_1(y_{I0} - \varepsilon_1' - \varepsilon_2')(y_{B0} - \varepsilon_1' - \varepsilon_2') - (y_{P_{10}} + \varepsilon_1')(1 - \varepsilon_1' - \varepsilon_2') = 0$$
$$PK_2(y_{I0} - \varepsilon_1' - \varepsilon_2')(y_{B0} - \varepsilon_1' - \varepsilon_2') - (y_{P_{20}} + \varepsilon_2')(1 - \varepsilon_1' - \varepsilon_2') = 0$$

These two equations are solved with the initial condition of an equimolar mixture of reactants, $y_I = y_B = 0.5, y_{P_1} = y_{P_2} = 0$.[2] Using the initial guess from the solution of Example 3.2, $\varepsilon_1' = 0.469, \varepsilon_2' = 0$, gives the solution

$$\varepsilon_1' = 0.133, \qquad \varepsilon_2' = 0.351$$

Other initial guesses result in the same solution, leading us to believe that this solution is the unique, physically meaningful solution.[3] The mole fractions are computed from Equations 3.66 giving

$$y_I = 0.031, \qquad y_B = 0.031, \qquad y_{P_1} = 0.258, \qquad y_{P_2} = 0.680$$

Notice that the existence of the second reaction causes the production of considerably less 2,2,3-trimethylpentane in favor of the 2,2,4 isomer. This is bad news in this example because the production of 2,2,3 is preferable due to its high octane content. The news only gets worse when we consider that a host of other products are also possible, such as the 2,3,3 and 2,3,4 isomers as well as 2,2,3,3-tetramethylbutane

[2] It is best to use a numerical tool to solve these equations when there is more than one reaction.

[3] The mole fractions are real and positive.

and still others. The Gibbs energy of formation of many of these possible products show that they are present in significant amounts at equilibrium. It is clear that one cannot allow the system to reach equilibrium and still hope to obtain a high yield of the desired product. The solution to problems like these is often to develop a catalyst that selectively increases the reaction rate for the desired reaction in comparison to the undesired side reactions. If such a catalyst is available, the reaction is halted after a high yield of the desired product is formed and before the equilibration to the other products can reduce the yield. This use of catalysts to promote the rates of desirable reactions is discussed further in Chapter 5. □

3.5.2 Optimization Approach

The other main approach to finding the reaction equilibrium is to minimize the appropriate energy function, in this case the Gibbs energy [20]. This optimization-based formulation of the problem, as shown in Example 3.3, can be more informative than the algebraic approach discussed above.

We first extend our evaluation of the Gibbs energy for the single reaction case to allow multiple reactions. We start with

$$G = \sum_j \mu_j n_j \tag{3.67}$$

and express the chemical potential in terms of activity

$$\mu_j = G_j^\circ + RT \ln a_j$$

We again use Equation 3.60 to track the change in mole numbers due to multiple reactions,

$$n_j = n_{j0} + \sum_i \nu_{ij} \varepsilon_i$$

Using the two previous equations we have

$$\mu_j n_j = n_{j0} G_j^\circ + G_j^\circ \sum_i \nu_{ij} \varepsilon_i + \left[n_{j0} + \sum_i \nu_{ij} \varepsilon_i \right] RT \ln a_j \tag{3.68}$$

It is convenient to define the same modified Gibbs energy function that we used in Equation 3.37

$$\widetilde{G}(T, P, \varepsilon_i') = \frac{G - \sum_j n_{j0} G_j^\circ}{n_{T0} RT} \tag{3.69}$$

in which $\varepsilon_i' = \varepsilon_i / n_{T0}$. If we sum on j in Equation 3.68 and introduce this expression into Equations 3.67 and 3.69, we obtain

$$\widetilde{G} = \sum_i \varepsilon_i' \frac{\Delta G_i^\circ}{RT} + \sum_j \left[y_{j0} + \sum_i \nu_{ij} \varepsilon_i' \right] \ln a_j$$

Using the definition of the equilibrium constant, Equation 3.63, we can write the energy function as

$$\boxed{\widetilde{G} = -\sum_i \varepsilon_i' \ln K_i + \sum_j \left[y_{j0} + \sum_i \nu_{ij} \varepsilon_i' \right] \ln a_j} \qquad (3.70)$$

To find the equilibrium composition, we minimize this modified Gibbs energy over the physically meaningful values of the n_r extents. The main restriction on these extents is, again, that they produce nonnegative mole numbers, or, if we wish to use intensive variables, nonnegative mole fractions. We can express these constraints as

$$\boxed{-y_{j0} - \sum_i \nu_{ij} \varepsilon_i' \leq 0, \qquad j = 1, \ldots, n_s} \qquad (3.71)$$

Our final statement, therefore, for finding the equilibrium composition for multiple reactions is to solve the optimization problem

$$\boxed{\min_{\varepsilon_i'} \widetilde{G}(\varepsilon_i')} \qquad (3.72)$$

subject to Equation 3.71. The min notation means to minimize the function $\widetilde{G}(\varepsilon_i')$ with respect to the argument ε_i'.

Example 3.6: Multiple reactions with optimization

Revisit the two-reaction trimethylpentane example, and find the equilibrium composition by minimizing the Gibbs energy.

Solution

We start with the modified energy function, Equation 3.70. We recall for an ideal-gas mixture, the activity of component j, is simply its partial pressure divided by 1.0 atm,

$$a_j = \frac{P}{1 \text{ atm}} y_j \qquad \text{(ideal-gas mixture)}$$

Substituting this relation into Equation 3.70 and rearranging gives

$$\begin{aligned}\widetilde{G} &= -\sum_i \varepsilon_i' \ln K_i + \left(1 + \sum_i \bar{\nu}_i \varepsilon_i'\right) \ln P \\ &\quad + \sum_j \left(y_{j0} + \sum_i \nu_{ij}\varepsilon_i'\right) \ln\left[\frac{y_{j0} + \sum_i \nu_{ij}\varepsilon_i'}{1 + \sum_i \bar{\nu}_i \varepsilon_i'}\right]\end{aligned} \tag{3.73}$$

in which

$$\bar{\nu}_i = \sum_j \nu_{ij}$$

The minimization of this function of ε_i' then determines the two equilibrium extents. The constraints on the extents are found from Equation 3.71. For this problem they are

$$\begin{aligned} -y_{I0} + \varepsilon_1' + \varepsilon_2' \leq 0 \qquad & -y_{P_1 0} - \varepsilon_1' \leq 0 \\ -y_{B0} + \varepsilon_1' + \varepsilon_2' \leq 0 \qquad & -y_{P_2 0} - \varepsilon_2' \leq 0 \end{aligned}$$

Substituting in the initial conditions gives the constraints

$$\varepsilon_1' + \varepsilon_2' \leq 0.5, \qquad 0 \leq \varepsilon_1', \qquad 0 \leq \varepsilon_2'$$

Figure 3.6 shows the lines of constant Gibbs energy determined by Equation 3.73 as a function of the two reaction extents. We see immediately that the minimum is unique. Notice in Figure 3.6 that $\widetilde{G}$ is defined only in the region specified by the constraints. The numerical solution of the optimization problem is

$$\varepsilon_1' = 0.133, \qquad \varepsilon_2' = 0.351, \qquad \widetilde{G} = -2.569$$

The solution is in good agreement with the extents computed using the algebraic approach, and the Gibbs energy contours depicted in Figure 3.6. □

Optimization is a powerful tool for solving many types of engineering modeling and design problems. We also rely heavily on optimization tools in Chapter 9 on parameter estimation.

3.6 Summary

In this review of reaction equilibrium, the Gibbs energy was chosen as the convenient function for solving reaction equilibrium problems

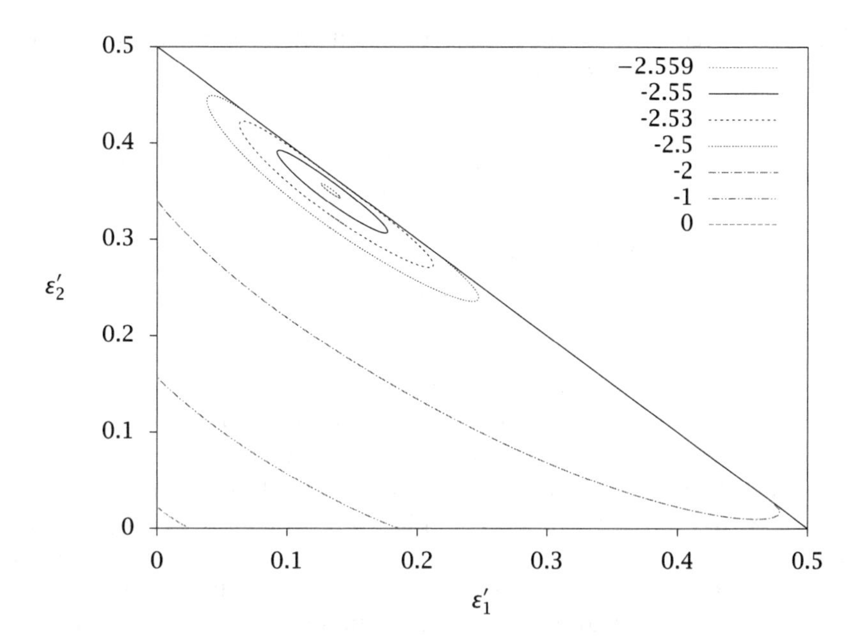

Figure 3.6: Gibbs energy contours for the pentane reactions as a function of the two reaction extents.

when the temperature and pressure are specified. The fundamental equilibrium condition is that the Gibbs energy is minimized. This fundamental condition leads to several conditions for equilibrium such as

$$\sum_j \nu_j \mu_j = 0$$

$$K = \prod_j a_j^{\nu_j}$$

for a single reaction or

$$\sum_j \nu_{ij} \mu_j = 0, \qquad i = 1, \ldots, n_r$$

$$K_i = \prod_j a_j^{\nu_{ij}}, \qquad i = 1, \ldots, n_r$$

for multiple reactions, in which the equilibrium constant is defined to be

$$K_i = e^{-\Delta G_i^\circ / RT}$$

You should feel free to use whichever formulation is most convenient for the problem. The equilibrium "constant" is not so constant, because it depends on temperature via

$$\frac{\partial \ln K}{\partial T} = \frac{\Delta H^\circ}{RT^2}$$

or, if the enthalpy change does not vary with temperature,

$$\ln\left(\frac{K_2}{K_1}\right) = -\frac{\Delta H^\circ}{R}\left(\frac{1}{T_2} - \frac{1}{T_1}\right)$$

The conditions for phase equilibrium were presented: equalities of temperature, pressure and chemical potential of each species in all phases. The evaluation of chemical potentials of mixtures was discussed, and the following methods and approximations were presented: ideal mixture, Henry's law, and simple correlations for activity coefficients.

When more than one reaction is considered, which is the usual situation faced in applications, we require numerical methods to find the equilibrium composition. Two approaches to this problem were presented. We either solve a set of nonlinear algebraic equations or solve a nonlinear optimization problem subject to constraints. If optimization software is available, the optimization approach is more powerful and provides more insight.

Notation

a_j	activity of species j
a_{jl}	formula number for element l in species j
A_j	jth species in the reaction network
$\overline{C}_{Pj}$	partial molar heat capacity of species j
E_l	lth element constituting the species in the reaction network
f_j	fugacity of species j
G	Gibbs energy
$\overline{G}_j$	partial molar Gibbs energy of species j
ΔG_i°	standard Gibbs energy change for reaction i
H	enthalpy
$\overline{H}_j$	partial molar enthalpy of species j
ΔH_i°	standard enthalpy change for reaction i
i	reaction index, $i = 1, 2, \ldots, n_r$
j	species index, $j = 1, 2, \ldots, n_s$

k	phase index, $k = 1, 2, \ldots, n_p$
K	equilibrium constant
K_i	equilibrium constant for reaction i
l	element index, $l = 1, 2, \ldots, n_e$
n_j	moles of species j
n_r	total number of reactions in reaction network
n_s	total number of species in reaction network
P	pressure
P_j	partial pressure of species j
R	gas constant
S	entropy
$\overline{S}_j$	partial molar entropy of species j
T	temperature
V	volume
$\overline{V}_j$	partial molar volume of species j
x_j	mole fraction of liquid-phase species j
y_j	mole fraction of gas-phase species j
z	compressibility factor of the mixture
γ_j	activity coefficient of species j in a mixture
ε	reaction extent
ε_i	reaction extent for reaction i
μ_j	chemical potential of species j
ν_{ij}	stoichiometric number for the jth species in the ith reaction
ν_j	stoichiometric number for the jth species in a single reaction
$\bar{\nu}$	$\sum_j \nu_j$
$\bar{\nu}_i$	$\sum_j \nu_{ij}$
$\hat{\phi}_j$	fugacity coefficient of species j in a mixture

3.7 Exercises

Exercise 3.1: Calculating an equilibrium composition

Consider the following reaction in a closed vessel at a pressure of 1.0 atm and temperature of 500°K

$$\text{isobutane} + \text{1-butene} \rightleftharpoons \text{2,2,3-trimethylpentane}$$
$$\text{I} + \text{B} \rightleftharpoons \text{P}$$

The standard Gibbs energy and enthalpy changes for this reaction at 500 K are $\Delta G^\circ = -4.10$ kcal/mol, $\Delta H^\circ = -20.11$ kcal/mol.

(a) Determine the equilibrium composition for this system for an initial equimolar mixture of isobutane and butene. What assumptions did you make?

(b) What is the equilibrium *conversion* of the reactants in part 3.1a? Assume for economic reasons that you must increase the equilibrium conversion to 97%. At what pressure must you run this reaction at a temperature of 500 K to achieve this conversion?

(c) At what temperature must you run this reaction at a pressure of 1.0 atm to achieve 97% conversion?

(d) What do you know about this system that prevents you from taking these answers too seriously?

Exercise 3.2: Equilibrium and linearly independent reactions

Butene isomerization reactions are shown below.

$$1-\text{butene} \underset{k_{-1}}{\overset{k_1}{\rightleftharpoons}} \text{cis}-2-\text{butene}$$

$$1-\text{butene} \underset{k_{-2}}{\overset{k_2}{\rightleftharpoons}} \text{trans}-2-\text{butene}$$

$$1-\text{butene} \underset{k_{-3}}{\overset{k_3}{\rightleftharpoons}} \text{isobutene}$$

$$\text{cis}-2-\text{butene} \underset{k_{-4}}{\overset{k_4}{\rightleftharpoons}} \text{trans}-2-\text{butene}$$

$$\text{cis}-2-\text{butene} \underset{k_{-5}}{\overset{k_5}{\rightleftharpoons}} \text{isobutene}$$

The Gibbs energy of formation of the components at 400 K are [18]

Components	ΔG_f (kcal/mol)
1-butene	23.10
cis-2-butene	21.94
trans-2-butene	21.33
isobutene	20.23

(a) Determine the minimum number of reactions that are needed to calculate the equilibrium composition of the butenes.

(b) Set up the linear algebra problem to determine the equilibrium extents of the reactions. What assumptions have you made?

(c) Compute the equilibrium composition at 400 K and 1.0 atm pressure starting from pure 1-butene.

(d) Why is isobutene the predominate butene formed?

Exercise 3.3: More than one reaction

Consider the following two gas-phase reactions

$$A + B \rightleftharpoons C$$
$$2C \rightleftharpoons D$$

(a) What are the two equilibrium equations relating the two unknown extents of reaction and the initial gas-phase compositions? You can assume the pressure is low enough that the gases are ideal.

(b) What is the equilibrium gas-phase composition at 2.0 atm total pressure given initial gas-phase composition $y_A = y_B = 0.5$, $y_C = y_D = 0$, and $K_1 = 1.0$, $K_2 = 2.5$? Solve the two nonlinear equations numerically. If a computer is not available, you can use trial and error (partial answer: $y_A = 0.31$).

(c) What is the equilibrium gas-phase composition at 2.0 atm total pressure given initial gas-phase composition $y_A = y_C = 0.5, y_B = y_D = 0$ (partial answer: $y_A = 0.62$)?

Exercise 3.4: Temperature effect on reaction equilibria

Hexane can equilibrate to methyl-substituted pentanes and methylcyclopentane according to

$$\text{hexane} \underset{k_{-1}}{\overset{k_1}{\rightleftharpoons}} \text{methylcyclopentane} + H_2$$

$$\text{hexane} \underset{k-2}{\overset{k_2}{\rightleftharpoons}} 2-\text{methylpentane}$$

Component	$\Delta H_f^{600\text{ K}}$ (kcal/mol)	$\Delta G_f^{600\text{ K}}$ (kcal/mol)
n-hexane	−46.10	43.02
methylcyclopentane	−31.69	45.76
2-methylpentane	−47.63	42.39
hydrogen	0	0

in which $\Delta H_f^{600\text{ K}}$ and $\Delta G_f^{600\text{ K}}$ are the heats of formation and the Gibbs free energy of formation of the compounds at 600 K from the elements, respectively [18]. Find

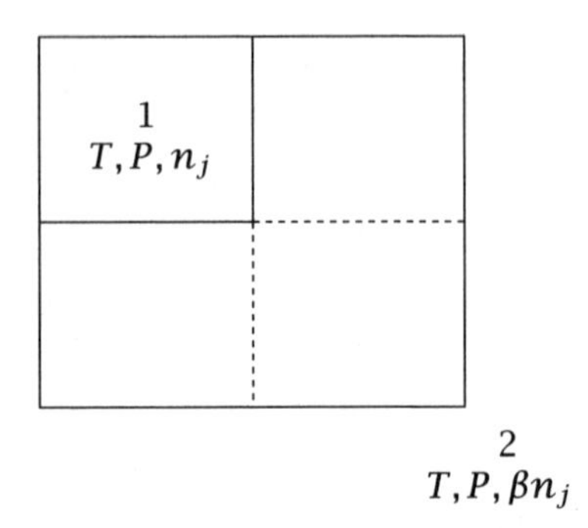

Figure 3.7: Two systems with identical intensive properties and different extensive properties.

the temperature where equal amounts of methylcyclopentane and 2-methylpentane form if pure hexane is allowed to equilibrate at 1.0 atm total pressure. Also find the composition of the mixture at this temperature. At this condition the hydrocarbons are gases.

Exercise 3.5: Other equilibrium conditions

In the text, we focused our attention on minimizing Gibbs energy to determine the chemical equilibrium at fixed temperature and pressure. To be fair to the other energy functions, what system function would be at an extremum (maximum or minimum) if the following were specified. Provide a short derivation verifying your result starting with the second law, $dS_{\text{tot}} \geq 0$.

(a) T and P are fixed.

(b) T and V are fixed.

(c) S and P are fixed.

(d) S and V are fixed.

(e) The system is insulated and V is fixed. In addition to V, what other thermodynamic property is fixed?

(f) The system is insulated and P is fixed. In addition to P what other thermodynamic property is fixed?

Exercise 3.6: Partial molar Gibbs energy and chemical potential

(a) Derive Equation 3.34 in the text. Consider first a system at equilibrium depicted in Figure 3.7 at given $T, P, n_j, j = 1, \ldots, n_s$. Now consider system 2 to be simply β identical copies of system 1. The equilibrium state of system 2 therefore has the same temperature and pressure (intensive properties) as system 1 but the number of moles of each component (extensive property) changes by constant factor β. System 2 is specified by the variables $T, P, \beta n_j, j = 1, \ldots n_s$. Consider a reversible change from system 1 to system 2, imagine slowly removing the partition depicted in Figure 3.7, for example, and integrate Equation 3.1

$$\int_1^2 dG = \int_1^2 \left[-SdT + VdP + \sum_j \mu_j dn_j \right]$$

Because μ_j is an intensive variable, it remains constant during the process, along with T and P. Perform the integration and express G_2 and n_{j2} in terms of G_1 and n_{j1} and β to complete the argument.

(b) Show this same argument produces the following result for any partial molar property

$$X = \sum_j \overline{X}_j n_j$$

in which X is any extensive thermodynamic property.

Exercise 3.7: Creating a second phase with reaction

Consider the gas-phase reaction

$$A + B \rightleftharpoons C$$

Product C has a fairly low vapor pressure, so we are concerned about the formation of a liquid phase in the reactor. The Clausius-Clapeyron equation well represents the vapor pressure of component C as a function of temperature

$$\ln P_C^\circ = c - \frac{\Delta H_{vap}}{RT}$$

The reactor is initially filled with an equimolar mixture of A and B. The equilibrium constant at $T = 298$ K is $K = 8$, the reaction is exothermic with $\Delta H^\circ = -10$ kcal/mol, and the system pressure is $P = 1.0$ atm. Components A and B are not very soluble in liquid C. The heat of vaporization of component C is $\Delta H_{vap} = 5$ kcal/mol, and value of the Clausius-Clapeyron constant is $c = 7.53$.

(a) Over what temperature range does the reactor contain a liquid phase?

(b) If the reaction is endothermic with $\Delta H^\circ = 10$ kcal/mol, over what temperature range does the reactor contain a liquid phase?

Exercise 3.8: The van 't Hoff relation

Methanol can be manufactured by the gas-phase reaction

$$CO + 2H_2 \rightleftharpoons CH_3OH$$

A batch reactor is charged with a H_2 and CO mixture at 1.0 atm and the pressure is maintained at 1.0 atm. The initial mole fraction of H_2 is 0.6 and CO is 0.4. At 400 K, $\Delta G^\circ = -333$ cal/mol and $\Delta H^\circ = -22{,}580$ cal/mol.

Determine the temperature at which the equilibrium mole fraction of H_2 is 0.05.

Exercise 3.9: Semipermeable membrane and chemical reaction

Consider the membrane separation shown in Figure 3.8. An isothermal rigid vessel initially contains pure component B at pressure P_0 and temperature T_0. It is separated by a semipermeable membrane from a source of pure component A at pressure P_e. The membrane is permeable to component A but not component B. The system undergoes chemical reaction

$$2A \rightleftharpoons B$$

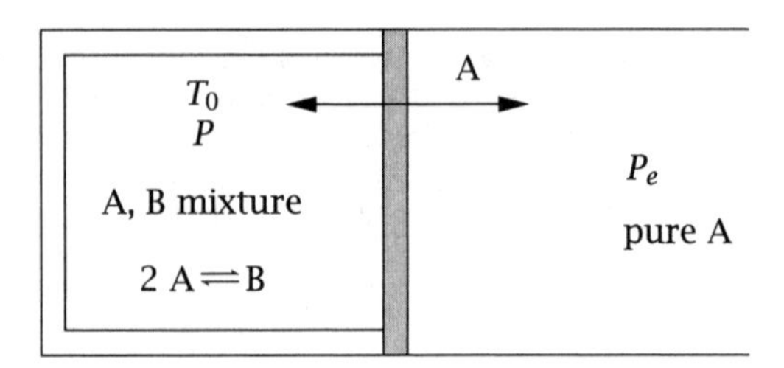

Figure 3.8: Vessel with semipermeable membrane and chemical reaction 2A ⇌ B.

inside the vessel due to the presence of a catalyst. No reaction occurs outside the vessel. The equilibrium constant at temperature T_0 is $K = 0.5$. What is the equilibrium composition and pressure in the vessel if $P_e = 5.0$ atm, $P_0 = 1.0$ atm? An ideal-gas mixture may be assumed.

Exercise 3.10: Reconciling kinetics and thermodynamics

Consider the gas-phase reaction

$$\mathrm{A} \underset{k_r}{\overset{k_f}{\rightleftharpoons}} 2\mathrm{B}$$

Anticipating the discussion of kinetics in Chapter 5, we suppose everyone is familiar with the classical rate expressions for the forward and reverse reactions

$$r_f = k_f c_A, \qquad r_r = k_r c_B^2$$

The forward rate expression follows from assuming that the unimolecular forward reaction is proportional to the concentration of A molecules. The reverse rates expression follows from assuming the bimolecular reverse reaction is proportional to the probability of collisions of two B molecules, which is proportional to c_B^2 given the typically large numbers of molecules. The net reaction rate is given by the difference of forward and reverse rates

$$r = k_f c_A - k_r c_B^2$$

The forward and reverse rate constants,k_f, k_r, quantify the fraction of collisions that result in successful reactions. We expect these rate constants to depend on temperature because the frequency and energy of collisions increases with the mean velocity of the molecules which increases with temperature. A kinetic view of the equilibrium condition is obtained by setting the net reaction rate to zero.

(a) In order for this kinetic view to be in agreement with the results of thermodynamics, show that the rate constants also satisfy the restriction

$$\frac{k_f}{k_r} = \frac{K}{RTz} \frac{\hat{\phi}_A}{\hat{\phi}_B^2}$$

in which z is the compressibility factor of the mixture. What can you conclude about the dependences of rate constants k_f and k_r on the composition?

(b) If A and B form an ideal-gas mixture, what can you conclude about the dependence of k_f, k_r on the composition?

Exercise 3.11: Gibbs energy of formation

Consider the Gibbs energy of formation of species j

$$G_{jf}^{\circ} = G_j^{\circ} - \sum_{l=1}^{n_e} a_{jl} G_{El}^{\circ}$$

in which G_{El}° is the Gibbs energy of element l and a_{jl} is the formula number for element l in species j. Multiply this equation by the stoichiometric coefficient ν_{ij} and sum on j to produce

$$\sum_{j=1}^{n_s} \nu_{ij} G_{jf}^{\circ} = \Delta G_i^{\circ} - \sum_{j=1}^{n_s} \nu_{ij} \sum_{l=1}^{n_e} a_{jl} G_{El}^{\circ} \tag{3.74}$$

Using the result of Exercise 2.8, verify Equation 3.41 in the text for the Gibbs energy of reaction.

Exercise 3.12: Reactions going to completion

In a single-phase system, show the equilibrium condition

$$\sum_j \nu_{ij} \mu_j = 0, \qquad i = 1, \ldots, n_r$$

cannot be satisfied by any reaction going to completion and completely eliminating one of the components. Hint: consider Equations 3.63 and 3.64.

Can reactions go to completion in a multiple-phase system? If no, prove it. If yes, provide an example.

Exercise 3.13: Phase rule

Consider a system at equilibrium with n components and π phases. Let f be the number of independently variable intensive properties of the composite system not allowing chemical reactions. Gibbs's phase rule states

$$f = n + 2 - \pi \tag{3.75}$$

(a) In many textbooks, the following style of argument is used to derive the phase rule. Specify the following *intensive* variables to define the state of the system: temperature, pressure, and $n - 1$ mole fractions in each phase

$$T^k, P^k, x_j^k \qquad j = 1, \ldots, n-1 \qquad k = 1, \ldots, \pi$$

Count these unknowns and show that there are $\pi(n+1)$ unknowns.

Apply the set of equilibrium relations given in Equation 3.50. Count these equations and show there are $(n+2)(\pi - 1)$ equations.

Call f the difference between the number of equations and the number of unknowns, and show that this difference produces Equation 3.75.

In particular, if $f = 0$, we have the same number of equations and unknowns, and are supposed to conclude that the intensive variables of the equilibrium state are therefore uniquely determined. What is *missing* from this argument? Provide sets of two equations with two unknowns that have (i) no solution, (ii) exactly one solution, and (iii) infinitely many solutions.

(b) Rather than reject the phase rule, for which we have abundant experimental evidence, let's work a little harder to derive it, following the discussion by Feinberg [7]. Gibbs himself used the following style of argument [9, pp.87–96]. Start with the usual set of variables

$$dG = -SdT + VdP + \sum_j \mu_j dn_j$$

To obtain a relation among only intensive variables, differentiate Equation 3.67, $G = \sum_j \mu_j n_j$, and subtract to obtain the Gibbs-Duhem relations

$$0 = -SdT + VdP - \sum_j n_j d\mu_j$$

which applies to every phase, so we can write the set of equations

$$\begin{bmatrix} S^1 & -V^1 & n_1^1 & \cdots & n_n^1 \\ S^2 & -V^2 & n_1^2 & \cdots & n_n^2 \\ \vdots & \vdots & \vdots & \ddots & \vdots \\ S^\pi & -V^\pi & n_1^\pi & \cdots & n_n^\pi \end{bmatrix} \begin{bmatrix} dT \\ dP \\ d\mu_1 \\ \vdots \\ d\mu_n \end{bmatrix} = \begin{bmatrix} 0 \\ 0 \\ \vdots \\ 0 \end{bmatrix} \tag{3.76}$$

The phase equilibrium relations tell us that the variables $T, P, \mu_j, j = 1, \ldots, n$ are equal in all phases. Therefore, the number of independently variable intensive properties is simply the number of linearly independent solutions to the set of *linear* equations listed in Equation 3.76.

So, how many linearly independent solutions does Equation 3.76 admit? Use the result of Exercise 2.9 to show

$$f = n + 2 - \text{rank}(\boldsymbol{M})$$

in which $\boldsymbol{M}$ is the $\pi \times (n + 2)$ matrix appearing in Equation 3.76. Because the rank can exceed neither the number of rows (π) nor the number of columns ($n + 2$), we know that f is nonnegative, and that

$$f \geq n + 2 - \pi$$

(c) What assumption about equilibrium states is required before we can conclude $\text{rank}(\boldsymbol{M}) = \pi$ and therefore

$$f = n + 2 - \pi$$

See Noll [11] and Feinberg [7] for further discussion of this issue.

(d) Consider r *linearly independent* chemical reactions and add the reaction equilibrium conditions to Equation 3.76. What are the dimensions of the modified $\boldsymbol{M}$ matrix? Show the modified phase rule with reaction is

$$f = n + 2 - \pi - r$$

Notice we cannot state a phase rule with chemical reaction if we have not defined linearly independent reactions, which is another motivation for defining linearly independent reactions in Chapter 2.

Exercise 3.14: Multiple reactions and ideal-gas equilibrium

We would like to show that, for multiple reactions, the equilibrium state of an ideal gas is unique. The following argument extends the single-reaction discussion starting on page 70 to multiple reactions [20, 16, 8]. In the single reaction case, we showed that the Gibbs energy $\tilde{G}(T,P,\varepsilon)$ has a positive second derivative in the single reaction extent, or

$$\frac{\partial^2 \tilde{G}}{\partial \varepsilon^2} > 0$$

corresponding to Figure 3.2. To reduce notational complexity we consider T and P constant in all partial derivatives of this problem. Assume we have n_r *linearly independent* reactions. Then we must show that the $n_r \times n_r$ *matrix* of second derivatives, $\tilde{\boldsymbol{H}}$

$$\tilde{H}_{ik} = \frac{\partial^2 \tilde{G}}{\partial \varepsilon_i \partial \varepsilon_k}$$

is a positive definite matrix, corresponding to Figure 3.5 for $n_r = 2$.

A matrix $\boldsymbol{H}$ is **positive definite** if

$$\boldsymbol{x}^T \boldsymbol{H} \boldsymbol{x} > 0, \quad \text{for every } \boldsymbol{x} \neq \boldsymbol{0}$$

A matrix $\boldsymbol{H}$ is **positive semidefinite** if

$$\boldsymbol{x}^T \boldsymbol{H} \boldsymbol{x} \geq 0, \quad \text{for every } \boldsymbol{x} \neq \boldsymbol{0}$$

(a) Consider the function $G(T,P,n_j) = \sum_l \mu_l n_l$ and its first derivative $\partial G / \partial n_j = \mu_j$. Taking the second derivative gives a matrix $\boldsymbol{H}$ with elements

$$H_{jl} = \frac{\partial^2 G}{\partial n_j \partial n_l} = \frac{\partial \mu_j}{\partial n_l}$$

Use the ideal-gas chemical potential $\mu_j = G_j^\circ + RT \ln(P y_j)$ and show

$$\frac{\partial \mu_j}{\partial y_l} = \frac{RT}{y_j} \delta_{jl} \qquad \text{in which} \quad \delta_{jl} = \begin{cases} 0, & j \neq l \\ 1, & j = l \end{cases}$$

Using the definition of mole fraction $y_j = n_j / n_T$, show

$$\frac{\partial y_j}{\partial n_l} = \frac{\delta_{jl}}{n_T} - \frac{y_j}{n_T}$$

(b) Using the chain rule

$$\frac{\partial \mu_j}{\partial n_l} = \sum_p \frac{\partial \mu_j}{\partial y_p} \frac{\partial y_p}{\partial n_l}$$

show

$$H_{jl} = \frac{RT}{n_T} \left[\frac{\delta_{jl}}{y_j} - 1 \right] \tag{3.77}$$

(c) To show $\boldsymbol{H}$ is positive semidefinite, we must show that the following quadratic form is nonnegative

$$\boldsymbol{x}^T \boldsymbol{H} \boldsymbol{x} = \sum_j \sum_l x_j H_{jl} x_l \geq 0, \qquad \text{for every } \boldsymbol{x} \neq \boldsymbol{0}$$

Substitute Equation 3.77 into the above equation and show

$$\boldsymbol{x}^T \boldsymbol{H} \boldsymbol{x} = \frac{RT}{n_T} \left[\sum_l \frac{x_l^2}{y_l} - \left(\sum_l x_l \right)^2 \right]$$

Next use the Cauchy inequality, Equation 3.32, to show the right-hand side is non-negative. When is equality achieved in the Cauchy inequality? We have therefore shown $\boldsymbol{H}$ is a positive semidefinite matrix. Note that it is not positive definite, however.

(d) We now find the second derivatives of $\widetilde{G}$ with respect to reaction extent. Given the relation $n_j = n_{j0} + \sum_i \nu_{ij}\varepsilon_i$ use the chain rule to show

$$\frac{\partial \widetilde{G}}{\partial \varepsilon_i} = \sum_j \frac{\partial G}{\partial n_j}\nu_{ij} \qquad \frac{\partial^2 \widetilde{G}}{\partial \varepsilon_i \partial \varepsilon_k} = \widetilde{H}_{ik} = \sum_j \sum_l \nu_{ij} H_{jl} \nu_{kl}$$

Let $\widetilde{\boldsymbol{H}}$ be the matrix of second derivatives in extents. In matrix notation we have shown

$$\widetilde{\boldsymbol{H}} = \boldsymbol{\nu}\boldsymbol{H}\boldsymbol{\nu}^T$$

(e) Finally, use the Cauchy inequality again to show that $\widetilde{\boldsymbol{H}}$ is a positive definite matrix.

Notice the same argument applies if we make the weaker assumption of an ideal mixture rather than an ideal gas.

Exercise 3.15: Ammonia equilibrium

The following gas-phase reaction operates at equilibrium in a constant pressure reactor.

$$N_2 + 3H_2 \rightleftharpoons 2NH_3$$

The feed only contains a 3:1 molar mixture of H_2:N_2. The equilibrium constant at 298 K is $K = 5.27 \times 10^5$ and the heat of reaction, which may be assumed constant, is $\Delta H = -23{,}000$ cal/mol. Determine the temperature to operate the reactor at a pressure of 300 atm that leads to $y_{NH_3} = 0.70$, i.e., mole fraction of NH_3 is 0.7.

Bibliography

[1] R. Aris. *Elementary Chemical Reactor Analysis.* Butterworths, Stoneham, MA, 1989.

[2] E. Beckenbach and R. Bellman. *An Introduction to Inequalities.* The Mathematical Association of America, 1961.

[3] R. B. Bird, W. E. Stewart, and E. N. Lightfoot. *Transport Phenomena.* John Wiley & Sons, New York, second edition, 2002.

[4] H. B. Callen. *Thermodynamics and an Introduction to Thermostatistics.* John Wiley & Sons, New York, 2nd edition, 1985.

[5] K. G. Denbigh. *The Principles of Chemical Equilibrium.* Cambridge University Press, Cambridge, fourth edition, 1981.

[6] J. R. Elliott and C. T. Lira. *Introductory Chemical Engineering Thermodynamics.* Prentice Hall, Upper Saddle River, New Jersey, 1999.

[7] M. Feinberg. On Gibbs's phase rule. *Arch. Rational Mech. Anal.*, 70:219–234, 1979.

[8] M. Feinberg. On the convexity of the ideal gas Gibbs free energy function. Personal communication, 1998.

[9] J. W. Gibbs. On the equilibrium of heterogeneous substances [Trans. Conn. Acad., vol. III, pp. 108–248, 1876; pp. 343–524, 1878.]. In *The Scientific Papers of J. Willard Gibbs*, volume one, pages 55–353. Ox Bow Press, Woodbridge, CT, 1993.

[10] B. G. Kyle. *Chemical and Process Thermodynamics.* Prentice-Hall, Englewood Cliffs, New Jersey, 1984.

[11] W. Noll. On certain convex sets of measures and on phases of reacting mixtures. *Arch. Rational Mech. Anal.*, 38:1–12, 1970.

[12] B. E. Poling, J. M. Prausnitz, and J. P. O'Connell. *Properties of Gases and Liquids.* McGraw-Hill, New York, 2001.

[13] J. M. Prausnitz, R. N. Lichtenthaler, and E. G. de Azevedo. *Molecular Thermodynamics of Fluid-Phase Equilibria.* Prentice Hall, Upper Saddle River, New Jersey, third edition, 1999.

[14] S. I. Sandler, editor. *Models for Thermodynamic and Phase Equilibria Calculations.* Marcel Dekker, New York, 1994.

[15] J. D. Schieber and J. J. de Pablo. *Thermodynamics.* McGraw-Hill, In preparation.

[16] N. Z. Shapiro and L. S. Shapley. Mass action laws and the Gibbs free energy. *J. Soc. Indust. Appl. Math.*, 13(2):353–375, 1965.

[17] J. M. Smith, H. C. Van Ness, and M. M. Abbott. *Introduction to Chemical Engineering Thermodynamics.* McGraw-Hill, Boston, sixth edition, 2001.

[18] D. R. Stull, E. F. Westrum Jr., and G. C. Sinke. *The Chemical Thermodynamics of Organic Compounds.* John Wiley & Sons, New York, 1969.

[19] J. W. Tester and M. Modell. *Thermodynamics and its Applications.* Prentice Hall, Upper Saddle River, New Jersey, third edition, 1997.

[20] W. B. White, S. M. Johnson, and G. B. Dantzig. Chemical equilibrium in complex mixtures. *J. Chem. Phys.*, 28(5):751–755, 1958.

4

The Material Balance for Chemical Reactors

4.1 General Mole Balance

Consider an arbitrary reactor volume element depicted in Figure 4.1, which has inlet and outlet streams with volumetric flowrates Q_0 and Q_1, respectively. The molar concentrations of component j in the two streams are given by c_{j0} and c_{j1} and the production rate of component j due to chemical reactions is R_j. The statement of conservation of mass for this system takes the form,

$$\left\{ \begin{array}{c} \text{rate of} \\ \text{accumulation} \\ \text{of component } j \end{array} \right\} = \left\{ \begin{array}{c} \text{rate of inflow} \\ \text{of component } j \end{array} \right\} - \left\{ \begin{array}{c} \text{rate of outflow} \\ \text{of component } j \end{array} \right\} + \left\{ \begin{array}{c} \text{rate of generation} \\ \text{of component } j \text{ by} \\ \text{chemical reactions} \end{array} \right\} \tag{4.1}$$

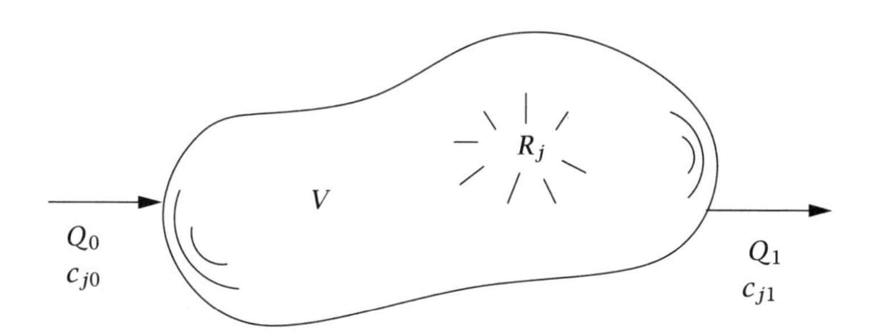

Figure 4.1: Reactor volume element.

In terms of the defined variables, we can write Equation 4.1 as,

$$\frac{d}{dt}\int_V c_j dV = Q_0 c_{j0} - Q_1 c_{j1} + \int_V R_j dV \tag{4.2}$$

Equation 4.2 applies to every chemical component in the system, $j = 1, 2, \ldots, n_s$, including inerts, which do not take place in any reactions. One can, of course, include volume elements with more than two flow streams by summing with the appropriate sign over all streams entering and leaving the reactor. For the balances in this chapter, there will be two or fewer flow streams. Notice also that we are assuming that component j enters and leaves the reactor volume element only by convection with the inflow and outflow streams. In particular, we are neglecting diffusional flux through the boundary of the volume element due to a concentration gradient. The diffusional flux will be considered during the development of the material balance for the packed-bed reactor.

Rate expressions. To solve the reactor material balance, we require an expression for the production rate, R_j, for each component. As shown in Chapter 2, the production rate can be computed directly from the stoichiometry and the reaction rates for all reactions, r_i. Therefore we require an expression for the reaction rates in terms of the concentrations of the species. This topic occupies the majority of Chapter 5. For the purposes of illustrating the material balances in this chapter, we simply use some common reaction-rate expressions without derivation. These rate expressions may be regarded as empirical facts until the next chapter when the theoretical development of the rate expressions is provided.

4.2 The Batch Reactor

The batch reactor is assumed to be well stirred, so there are no concentration gradients anywhere in the reactor volume. In this case it is natural to consider the entire reactor contents to be the reactor volume element as in Figure 4.2, and $V = V_R$. Because the reactor is well stirred, the integrals in Equation 4.2 are simple to evaluate,

$$\int_{V_R} c_j dV = c_j V_R \tag{4.3}$$

$$\int_{V_R} R_j dV = R_j V_R \tag{4.4}$$

Because the reactor is charged with reactants at $t = 0$, and nothing is added or removed from the reactor until the stopping time, the inflow and outflow stream flowrates are zero, $Q_0 = Q_1 = 0$.

Substituting these results into Equation 4.2 gives the general batch reactor design equation,

$$\frac{d\left(c_j V_R\right)}{dt} = R_j V_R \tag{4.5}$$

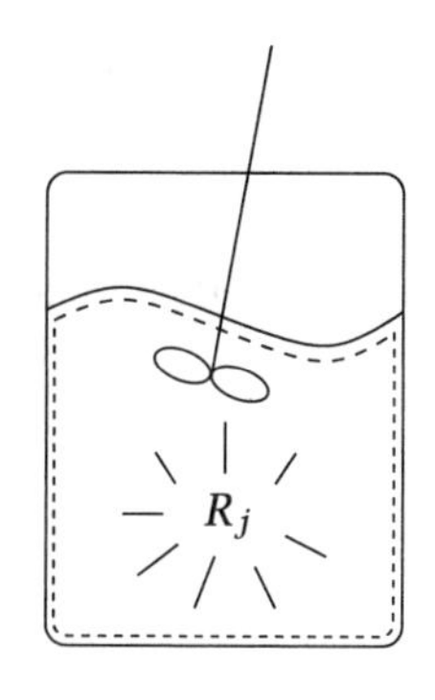

Figure 4.2: Batch reactor volume element.

Equation 4.5 applies whether the reactor volume is constant or changes during the course of the reaction. If the reactor volume is constant, which is sometimes a good approximation for liquid-phase reactions, V_R can be divided out of both sides of Equation 4.5 to give

$$\frac{dc_j}{dt} = R_j \tag{4.6}$$

Be sure to use Equation 4.5 rather than Equation 4.6 if the reactor volume changes significantly during the course of the reaction.

4.2.1 Analytical Solutions for Simple Rate Laws

In complex and realistic situations, the material balance for the batch reactor must be solved numerically. However, if the reactor is isothermal, and the rate laws are assumed to be quite simple, then analytical solutions of the material balance are possible. Analytical solutions are valuable for at least two reasons. First, due to the closed form of the solution, analytical solutions provide insight that is difficult to achieve with numerical solutions. The effect of parameter values on the solution is usually more transparent, and the careful study of analytical solutions can often provide insight that is hard to extract from numerical computations. Secondly, even if one must compute a numerical solution for a problem of interest, the solution procedure should be checked for errors by comparison to known solutions. Comparing a numerical solution procedure to an analytical solution for a simplified problem provides some assurance that the numerical procedure has been constructed correctly. Then the verified numerical procedure can be used with more assurance on the full problem for which no other solution is available.

The next several sections derive analytical solutions for some simple rate laws. Of course, the batch reactor is assumed to be operating at constant temperature in this discussion.

First-order, irreversible. Consider the first-order, irreversible reaction

$$A \xrightarrow{k} B \tag{4.7}$$

in which the reaction rate is given by $r = kc_A$. The units of the first-order rate constant are $(\text{time})^{-1}$. Application of the material balance for a constant-volume reactor gives the following differential equation

$$\frac{dc_A}{dt} = -kc_A \tag{4.8}$$

in which the negative sign arises because the production rate of A is $R_A = -r$ due to the stoichiometry of the reaction. Equation 4.8 requires an initial condition to have a unique solution. We denote the initial concentration of A in the reactor as c_{A0},

$$c_A(t) = c_{A0}, \qquad t = 0$$

The solution to the differential equation with this boundary condition is

$$c_A = c_{A0}e^{-kt} \tag{4.9}$$

which is plotted in Figure 4.3 for several values of the rate constant k. Because the reaction is irreversible, the A concentration decreases exponentially from its initial value to zero with increasing time. The rate constant determines the shape of this exponential decrease. Rearranging Equation 4.9 and taking logarithms gives

$$\ln(c_A/c_{A0}) = -kt$$

which is plotted by using a log scale in Figure 4.4. Notice one can get an approximate value of the rate constant by calculating the slope of the straight line given by $\ln(c_A/c_{A0})$ versus t. This procedure is sometimes recommended as a way to determine rate constants for first-order reactions by plotting experimental concentration data and determining this slope. As will be discussed in more detail in Chapter 9, this procedure is a poor way to determine a rate constant and should be viewed only as a rough approximation.

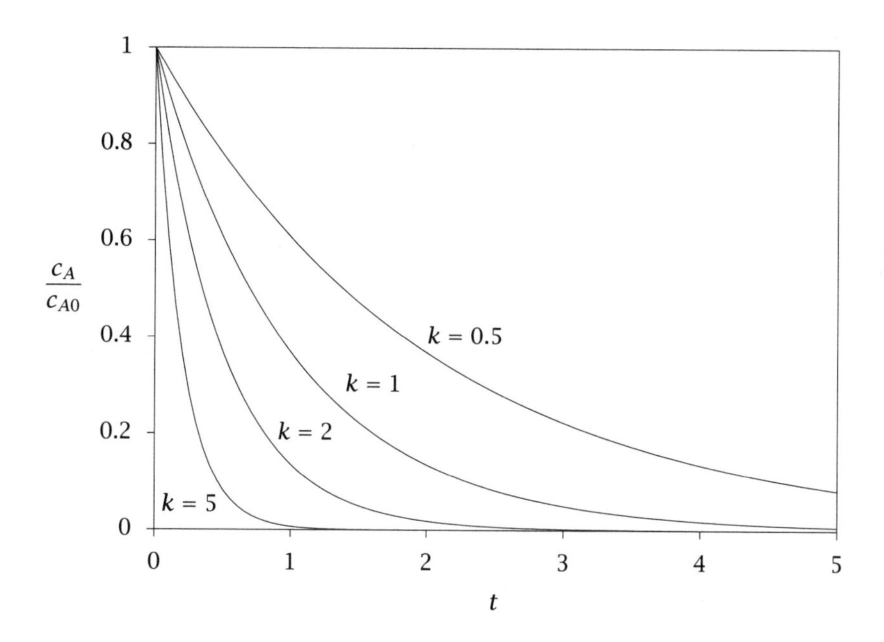

Figure 4.3: First-order, irreversible kinetics in a batch reactor.

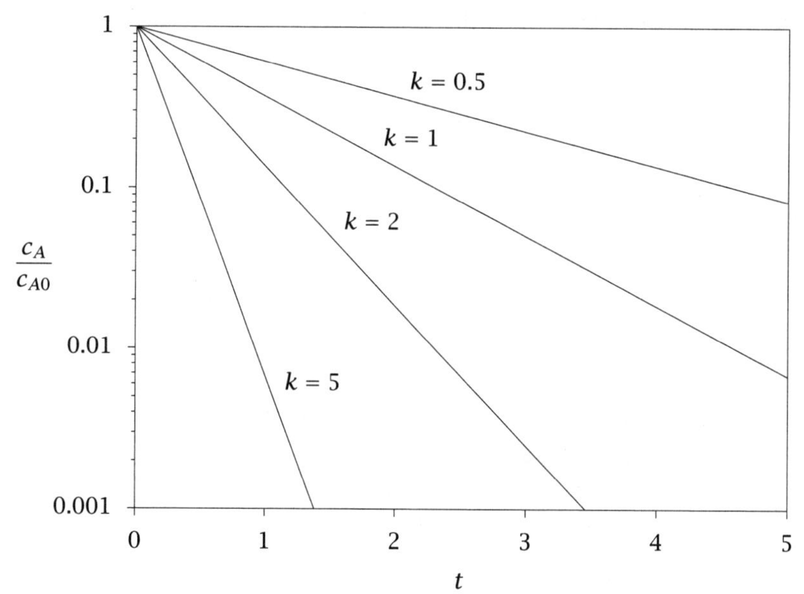

Figure 4.4: First-order, irreversible kinetics in a batch reactor, log scale.

The B concentration is easily determined from the A concentration. One could write down the material balance for component B,

$$\frac{dc_B}{dt} = R_B = kc_A \tag{4.10}$$

and solve this differential equation with the initial condition for B, $c_B(0) = c_{B0}$, after substituting the known solution for $c_A(t)$. It is simpler to note, however, that the sum of concentrations A and B is a constant. Adding Equations 4.8 and 4.10 gives

$$\frac{d(c_A + c_B)}{dt} = R_A + R_B = 0$$

Therefore, $c_A + c_B$ is a constant and independent of time. The value of this constant is known at $t = 0$,

$$c_A + c_B = c_{A0} + c_{B0}$$

which can be rearranged for the B concentration,

$$c_B = c_{A0} + c_{B0} - c_A \tag{4.11}$$

First-order, reversible. Consider now the same first-order reaction, but assume it is reversible

$$\mathrm{A} \underset{k_{-1}}{\overset{k_1}{\rightleftharpoons}} \mathrm{B} \tag{4.12}$$

and the reaction rate is $r = k_1c_A - k_{-1}c_B$. The material balances for A and B are now

$$\begin{aligned} \frac{dc_A}{dt} &= -r = -k_1c_A + k_{-1}c_B \\ \frac{dc_B}{dt} &= r \quad = k_1c_A - k_{-1}c_B \end{aligned}$$

with the same initial condition $c_A(0) = c_{A0}$, $c_B(0) = c_{B0}$. Notice that $c_A + c_B$ remains constant, so c_B can be computed from Equation 4.11. Substituting Equation 4.11 into the material balance for A gives

$$\frac{dc_A}{dt} = -k_1c_A + k_{-1}(c_{A0} + c_{B0} - c_A)$$

which can be rearranged into

$$\frac{dc_A}{dt} + (k_1 + k_{-1})c_A = k_{-1}(c_{A0} + c_{B0}) \tag{4.13}$$

Equation 4.13 is a nonhomogeneous, linear differential equation. The solution can be written as the sum of what is called the particular solution and the solution to the homogeneous equation [2]. One particular solution to the equation is the constant solution

$$c_{Ap} = \frac{k_{-1}}{k_1 + k_{-1}}(c_{A0} + c_{B0})$$

You should substitute this back into Equation 4.13 to check that it is indeed a solution. The homogeneous equation refers to the differential equation with a zero forcing term on the right-hand side,

$$\frac{dc_{Ah}}{dt} + (k_1 + k_{-1})c_{Ah} = 0$$

The solution to this equation already has appeared in the previous section, $c_{Ah} = a\exp(-(k_1 + k_{-1})t)$, in which a is an arbitrary constant to be determined from the initial condition. The full solution to Equation 4.13 is then $c_A = c_{Ah} + c_{Ap}$,

$$c_A = ae^{-(k_1+k_{-1})t} + \frac{k_{-1}}{k_1 + k_{-1}}(c_{A0} + c_{B0}) \tag{4.14}$$

The constant a is now determined from the initial condition. Writing Equation 4.14 for $t = 0$ gives

$$c_{A0} = a + \frac{k_{-1}}{k_1 + k_{-1}}(c_{A0} + c_{B0})$$

Solving this equation for a yields

$$a = c_{A0} - \frac{k_{-1}}{k_1 + k_{-1}}(c_{A0} + c_{B0})$$

Substituting in this value of a into Equation 4.14 and rearranging terms gives the final solution

$$c_A = c_{A0}e^{-(k_1+k_{-1})t} + \frac{k_{-1}}{k_1 + k_{-1}}(c_{A0} + c_{B0})\left[1 - e^{-(k_1+k_{-1})t}\right] \tag{4.15}$$

The B concentration can be determined by substituting Equation 4.11 into 4.15 and rearranging, or more simply, by switching the roles of A and B and k_1 and k_{-1} in Reaction 4.12, yielding

$$c_B = c_{B0}e^{-(k_1+k_{-1})t} + \frac{k_1}{k_1 + k_{-1}}(c_{A0} + c_{B0})\left[1 - e^{-(k_1+k_{-1})t}\right] \tag{4.16}$$

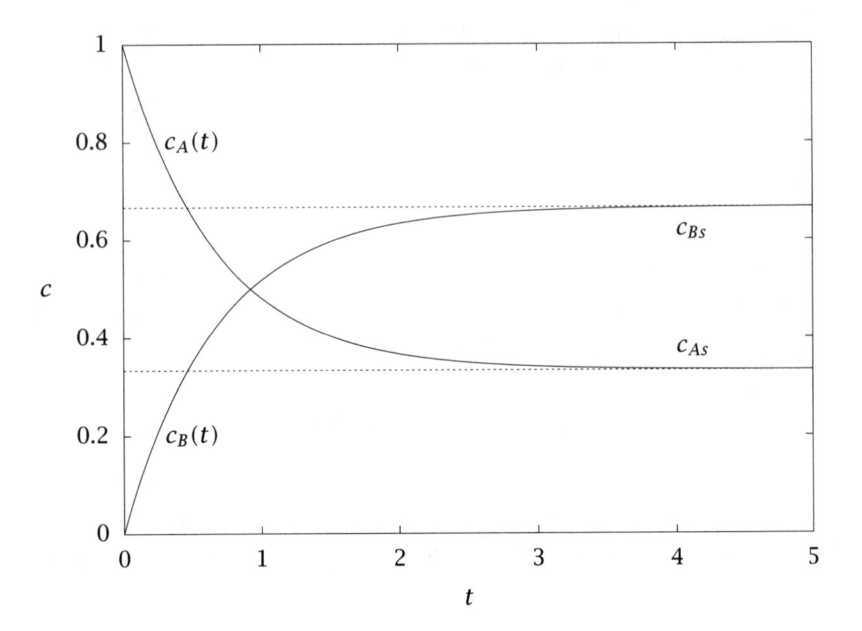

Figure 4.5: First-order, reversible kinetics in a batch reactor, $k_1 = 1$, $k_{-1} = 0.5$, $c_{A0} = 1$, $c_{B0} = 0$.

Equations 4.15 and 4.16 are plotted in Figure 4.5. Notice that with the reversible reaction, the concentration of A does not go to zero as in the irreversible case, but goes to a nonzero steady-state value. We next calculate the values of the steady-state concentrations. Taking the limit $t \longrightarrow \infty$ in Equation 4.15 gives

$$c_{As} = \frac{k_{-1}}{k_1 + k_{-1}}(c_{A0} + c_{B0})$$

in which c_{As} is the steady-state concentration of A. Defining $K_1 = k_1/k_{-1}$ allows us to rewrite this as

$$c_{As} = \frac{1}{1 + K_1}(c_{A0} + c_{B0})$$

Performing the same calculation for c_B gives

$$c_{Bs} = \frac{K_1}{1 + K_1}(c_{A0} + c_{B0})$$

These results are shown in Figure 4.5 for $K_1 = 1/0.5 = 2$ and $c_{A0} = 1$, $c_{B0} = 0$. Notice that because K_1 is larger than 1, the forward reaction is favored and the steady state favors the product B, $c_{Bs} = 2/3$, $c_{As} = 1/3$. For small K_1 values, the steady state would favor the reactant A.

Second-order, irreversible. Consider the irreversible reaction

$$A \xrightarrow{k} B \tag{4.17}$$

in which the rate expression is second order, $r = kc_A^2$. The units of the second-order rate constant are (vol/mol)(time)$^{-1}$. The material balance and initial condition are

$$\frac{dc_A}{dt} = -kc_A^2, \qquad c_A(0) = c_{A0} \tag{4.18}$$

This is our first nonlinear differential equation. Nonlinear differential equations do not have analytical solutions *in general*, but certain ones do. Equation 4.18 can be solved analytically because the equation is separable. Dividing both sides by c_A^2 and putting the time differential on the right-hand side gives

$$\frac{dc_A}{c_A^2} = -kdt$$

in which only a function of c_A appears on the left-hand side and only a function of time appears on the right-hand side. Integrating both sides between the initial condition and a final condition of interest gives

$$\int_{c_{A0}}^{c_A} \frac{dc_A}{c_A^2} = -k \int_0^t dt$$

Performing the integrals gives

$$\frac{1}{c_{A0}} - \frac{1}{c_A} = -kt$$

Finally solving for c_A gives

$$c_A = c_{A0}\left(1 + c_{A0}kt\right)^{-1} \tag{4.19}$$

You should check that this solution does indeed satisfy the differential equation and initial condition.

Figure 4.6 displays this solution and the first-order solution with a rate constant chosen such that the initial rates are equal. Notice that

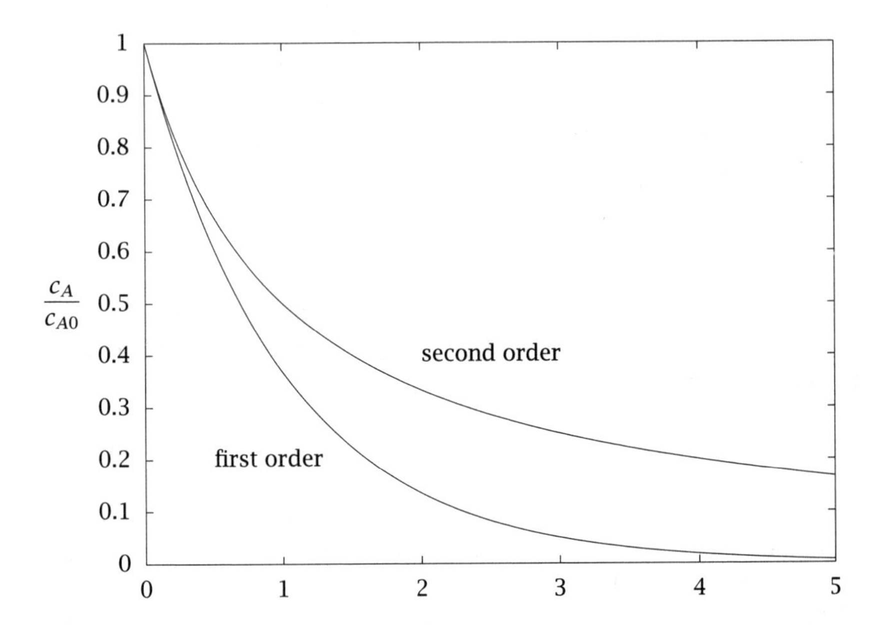

Figure 4.6: Second-order and first-order kinetics in a batch reactor; for second-order, $kc_{A0} = 1$, and for first order, $k = 1$, so the rates are equal initially.

although both solutions have the same qualitative features, the second-order reaction decays more slowly to zero than the first-order reaction at small concentration.

Consider another second-order, irreversible reaction,

$$A + B \xrightarrow{k} C \tag{4.20}$$

in which the rate law is $r = kc_A c_B$. The material balance for components A and B are

$$\frac{dc_A}{dt} = -r = -kc_A c_B \tag{4.21}$$

$$\frac{dc_B}{dt} = -r = -kc_A c_B \tag{4.22}$$

In general we need to solve sets of nonlinear equations numerically, but again the special structure of this problem allows an analytical solution. Notice that we can subtract the material balance for B from the material balance for A to obtain

$$\frac{d(c_A - c_B)}{dt} = 0$$

which implies that in this reaction, $c_A - c_B$ is independent of time. Because this value is known initially, one can solve for the concentration of B in terms of A,

$$c_B = c_A - c_{A0} + c_{B0} \tag{4.23}$$

Substituting this expression into the material balance for A yields

$$\frac{dc_A}{dt} = -kc_A(c_A - c_{A0} + c_{B0})$$

This equation also is separable and can be integrated to give (you should work through these steps),

$$c_A = (c_{A0} - c_{B0}) \left[1 - \frac{c_{B0}}{c_{A0}} e^{(c_{B0} - c_{A0})kt}\right]^{-1}, \qquad c_{A0} \neq c_{B0} \tag{4.24}$$

Notice that if $c_{A0} = c_{B0}$, then from Equation 4.23 $c_A(t) = c_B(t)$ for all t and this case reduces to the solution previously presented in Equation 4.19. One also can obtain that result by taking the limit in Equation 4.24 as discussed in Exercise 4.11. Component B can be computed from Equation 4.23, or by switching the roles of A and B in Reaction 4.20, giving

$$c_B = (c_{B0} - c_{A0}) \left[1 - \frac{c_{A0}}{c_{B0}} e^{(c_{A0} - c_{B0})kt}\right]^{-1}$$

The concentration of component C also can be computed from c_A by noticing that the material balance for C is

$$\frac{dc_C}{dt} = kc_A c_B$$

and therefore, $d(c_A + c_C)/dt = 0$. The concentration of C is given by

$$c_C = c_{A0} - c_A + c_{C0}$$

Notice that if $c_{A0} > c_{B0}$, the steady-state solution is

$$c_{As} = c_{A0} - c_{B0}, \qquad c_{Bs} = 0, \qquad c_{Cs} = c_{B0} + c_{C0}$$

In this case, A starts out in excess and all of the B is depleted. Component A remains at steady state even though the reaction is irreversible.

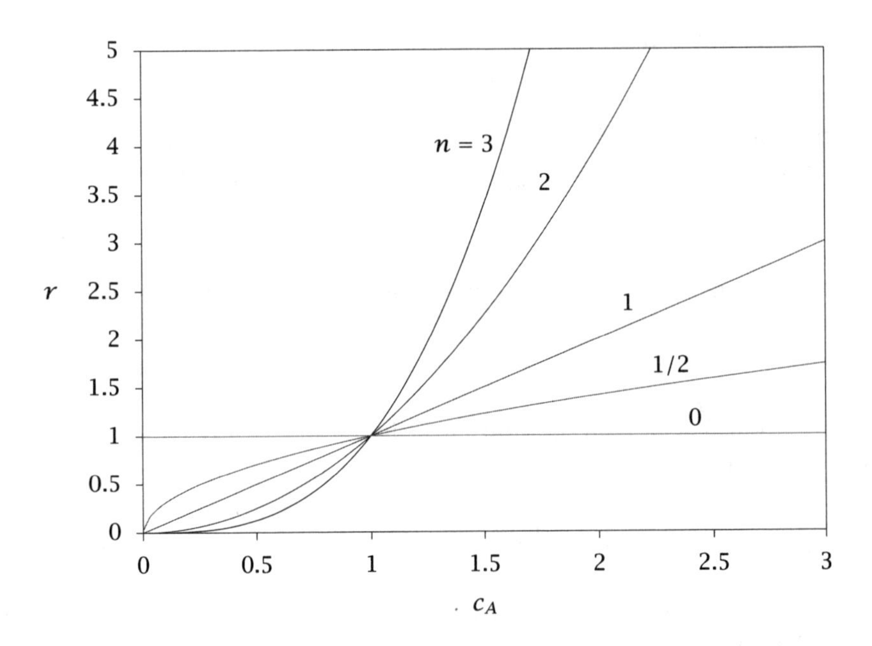

Figure 4.7: Reaction rate versus concentration for nth-order kinetics, $r = kc_A^n$, $n \geq 0$, $k = 1$ for all orders.

The final amount of C is equal to its starting value plus all the B that is present initially. For $c_{B0} > c_{A0}$, the steady-state solution is

$$c_{As} = 0, \qquad c_{Bs} = c_{B0} - c_{A0}, \qquad c_{Cs} = c_{A0} + c_{C0}$$

In this case, B starts out in excess and all of the A is depleted. Component B is present at steady state and the final amount of C is equal to its starting value plus all the A that is present initially.

nth-order, irreversible. The nth-order rate expression $r = kc_A^n$ is displayed in Figure 4.7 for a variety of positive n values. The units of the rate constant are $(\text{vol/mol})^{n-1}(\text{time})^{-1}$. The material balance for Reaction 4.17 with nth-order reaction rate expression is

$$\frac{dc_A}{dt} = -r = -kc_A^n$$

This equation also is separable and can be rearranged to

$$\frac{dc_A}{c_A^n} = -kdt$$

Performing the integration and solving for c_A gives

$$c_A = \left[c_{A0}^{-n+1} + (n-1)kt\right]^{\frac{1}{-n+1}}, \qquad n \neq 1$$

We can divide both sides by c_{A0} to obtain

$$\frac{c_A}{c_{A0}} = [1 + (n-1)k_0 t]^{\frac{1}{-n+1}}, \qquad n \neq 1 \tag{4.25}$$

in which

$$k_0 = k c_{A0}^{n-1}$$

has units of $(\text{time})^{-1}$. The $n = 1$ case already is given in Equation 4.9, which also can be recovered by taking the appropriate limit in Equation 4.25 as discussed in Exercise 4.11. Figure 4.8 shows the behavior of c_A versus t for values of n greater than one. Notice the larger the value of n, the more slowly the A concentration approaches zero at large time. Some care should be exercised when using Equation 4.25 for $n < 1$. First notice in Figure 4.9 that c_A reaches zero in *finite time* for $n < 1$. This time can be obtained by setting the term in brackets to zero in Equation 4.25 yielding

$$t_{\text{zero}} = \begin{cases} 1/k_0(1-n), & n < 1 \\ \infty, & n \geq 1 \end{cases} \tag{4.26}$$

Finally notice in Figure 4.10 that for $n < 0$, the rate decreases with increasing reactant concentration; the reactant *inhibits the reaction.* Inhibition reactions are not uncommon, but care must be exercised in using this kinetic model when the concentrations are small. Notice the rate becomes unbounded as c_A approaches zero, which is not physically realistic. When using an ODE solver to compute solutions that can reach zero in finite time, it is often necessary to modify the right-hand sides of the material balance as follows

$$\frac{dc_A}{dt} = \begin{cases} -kc_A^n, & c_A > 0 \\ 0, & c_A = 0 \end{cases} \tag{4.27}$$

Note that this amendment of the material balance is equivalent to putting a discontinuity in $r(c_A)$ at $c_A = 0$ in Figure 4.10, so, again, models with order less than one, and especially order less than zero, should be examined carefully if the concentration reaches zero during model solution.

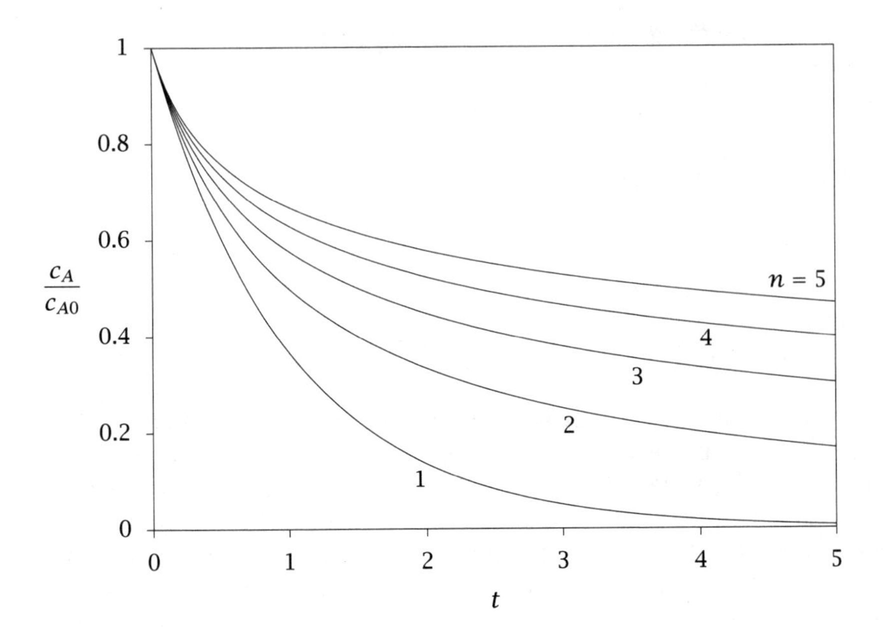

Figure 4.8: Batch reactor with nth-order kinetics, $r = kc_A^n$, $k_0 = kc_{A0}^{n-1} = 1$, $n \geq 1$.

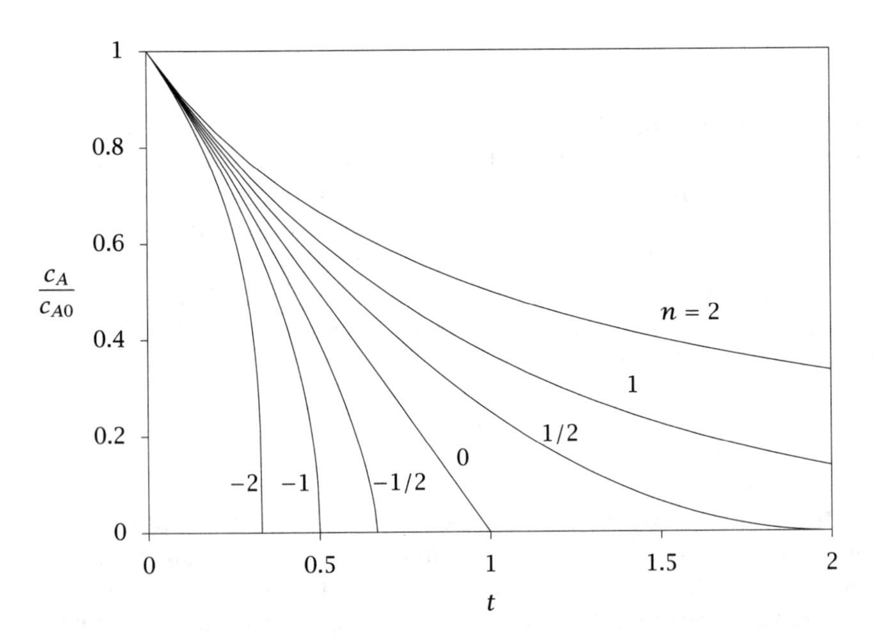

Figure 4.9: Batch reactor with nth-order kinetics, $r = kc_A^n$, $k_0 = kc_{A0}^{n-1} = 1$; note the concentration reaches zero in finite time $t = 1/(k_0(1-n))$ for $n < 1$.

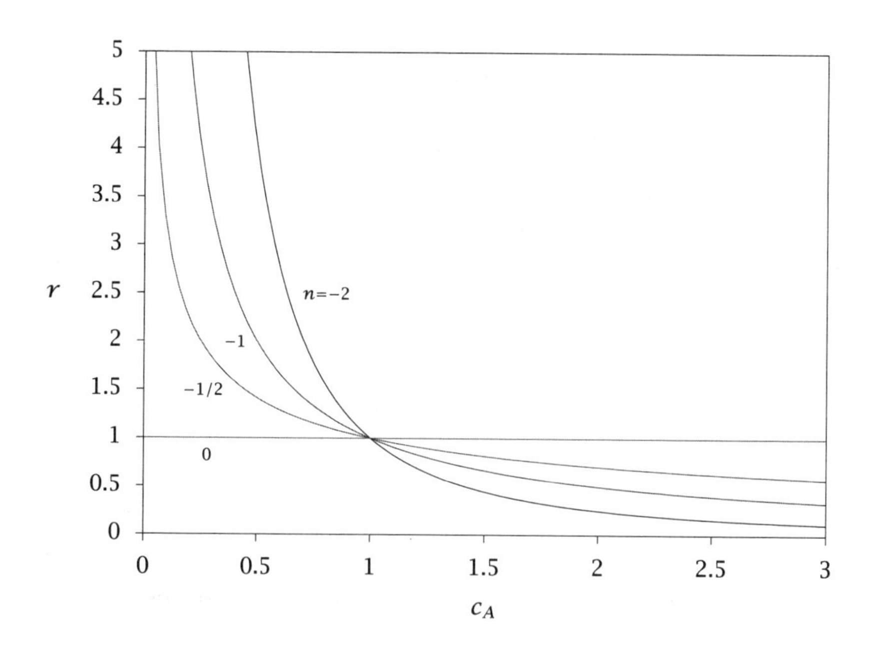

Figure 4.10: Reaction rate versus concentration for nth-order kinetics, $r = kc_A^n$, $n \le 0$, $k = 1$ for all orders.

Two reactions in series. Consider the following two irreversible reactions,

$$\text{A} \xrightarrow{k_1} \text{B} \tag{4.28}$$

$$\text{B} \xrightarrow{k_2} \text{C} \tag{4.29}$$

Reactant A decomposes to form an intermediate B that can further react to form a final product C. Let the reaction rates be given by simple first-order rate expressions in the corresponding reactants,

$$r_1 = k_1 c_A, \qquad r_2 = k_2 c_B$$

The material balances for the three components are

$$\begin{aligned}
\frac{dc_A}{dt} &= R_A = -r_1 &&= -k_1 c_A \\
\frac{dc_B}{dt} &= R_B = r_1 - r_2 &&= k_1 c_A - k_2 c_B \\
\frac{dc_C}{dt} &= R_C = r_2 &&= k_2 c_B
\end{aligned}$$

The material balance for component A can be solved immediately to give $c_A = c_{A0}e^{-k_1 t}$ as before. The material balance for B becomes

$$\frac{dc_B}{dt} + k_2 c_B = k_1 c_{A0} e^{-k_1 t}$$

The homogeneous solution is again $c_{Bh} = a_1 e^{-k_2 t}$. The particular solution can be guessed to be of the form $c_{Bp} = a_2 e^{-k_1 t}$ due to the form of the forcing term on the right-hand side. Substituting the particular solution into the differential equation and solving for a_2 gives

$$a_2 = c_{A0}\frac{k_1}{k_2 - k_1}, \qquad k_1 \neq k_2$$

so the particular solution is known. The initial condition then determines a_1,

$$c_B(0) = c_{B0} = a_1 + a_2$$

Solving for a_1 yields $a_1 = c_{B0} - k_1 c_{A0}/(k_2 - k_1)$. Substituting a_1 and a_2 into the c_B solution gives

$$c_B = c_{B0}e^{-k_2 t} + c_{A0}\frac{k_1}{k_2 - k_1}\left[e^{-k_1 t} - e^{-k_2 t}\right], \qquad k_1 \neq k_2 \tag{4.30}$$

Notice that the case of $k_1 = k_2$ can be handled by taking limits of Equation 4.30. To determine the C concentration, notice from the material balances that $d(c_A + c_B + c_C)/dt = 0$, which implies $c_A + c_B + c_C$ is constant. Therefore c_C can be computed from

$$c_C = c_{A0} + c_{B0} + c_{C0} - c_A - c_B$$

The concentrations of A, B and C are displayed in Figure 4.11 for an initial condition of pure reactant A. Notice that A decays exponentially as before; B forms at intermediate times and then disappears leaving only C as the final product. Exercise 4.10 asks for the optimal time to quench the reaction to achieve the maximum amount of intermediate B.

Two reactions in parallel. Consider next two parallel reactions of A to two different products, B and C,

$$\mathrm{A} \xrightarrow{k_1} \mathrm{B} \tag{4.31}$$

$$\mathrm{A} \xrightarrow{k_2} \mathrm{C} \tag{4.32}$$

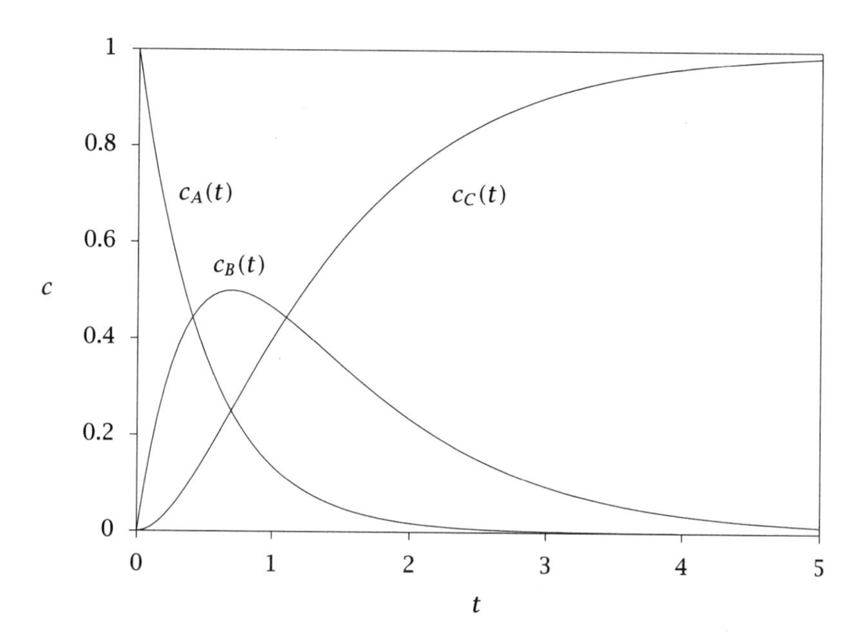

Figure 4.11: Two first-order reactions in series in a batch reactor, $c_{A0} = 1$, $c_{B0} = c_{C0} = 0$, $k_1 = 2$, $k_2 = 1$.

Assume the rates of the two irreversible reactions are given by $r_1 = k_1 c_A$ and $r_2 = k_2 c_A$. The material balances for the components are

$$\begin{aligned}
\frac{dc_A}{dt} &= R_A = -r_1 - r_2 = -k_1 c_A - k_2 c_A \\
\frac{dc_B}{dt} &= R_B = r_1 \qquad\quad = k_1 c_A \\
\frac{dc_C}{dt} &= R_C = r_2 \qquad\quad = k_2 c_A
\end{aligned}$$

The material balance for A can be solved directly to give

$$c_A = c_{A0} e^{-(k_1+k_2)t} \tag{4.33}$$

Substituting $c_A(t)$ into the material balance for B gives

$$\frac{dc_B}{dt} = k_1 c_{A0} e^{-(k_1+k_2)t}$$

This equation is now separable and can be integrated directly to give

$$c_B = c_{B0} + c_{A0} \frac{k_1}{k_1 + k_2} \left(1 - e^{-(k_1+k_2)t}\right) \tag{4.34}$$

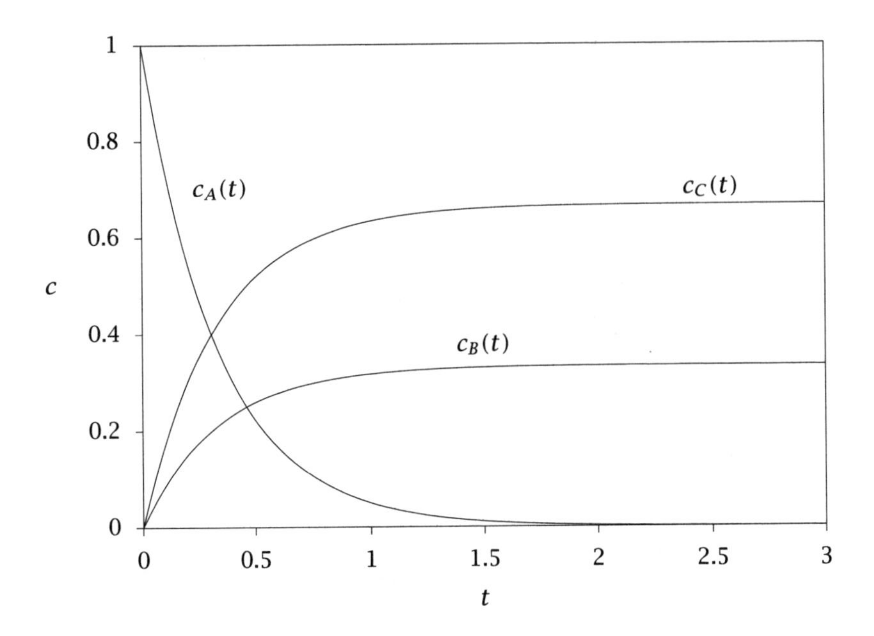

Figure 4.12: Two first-order reactions in parallel in a batch reactor, $c_{A0} = 1$, $c_{B0} = c_{C0} = 0$, $k_1 = 1$, $k_2 = 2$.

Finally, component C can be determined from the condition that $c_A + c_B + c_C$ is constant or by switching the roles of B and C, and k_1 and k_2 in Equation 4.34,

$$c_C = c_{C0} + c_{A0}\frac{k_2}{k_1 + k_2}\left(1 - e^{-(k_1+k_2)t}\right) \tag{4.35}$$

These results are plotted in Figure 4.12. Notice that because the two parallel reactions compete for the same reactant, A, the rate constants determine which product is favored. Large values of k_1/k_2 favor the formation of component B compared to C and vice versa.

When two or more reactions are in this kind of competition, it is convenient to define **selectivity**, **yield**, and **conversion** to quantify the efficiency of the reaction in forming a desired final product. There are several ways to define selectivity, yield and conversion. However, these terms are often invoked without careful definition in the hope that the meaning is clear from context. This practice often confuses the issue and should be avoided.

Point selectivity: The point (or instantaneous) selectivity is the ratio of the production rate of one component to the production rate of another component.

Overall selectivity: The overall selectivity is the ratio of the amount of one component produced to the amount of another component produced.

Yield: The yield of component j is the fraction of a reactant that is converted into component j.

Conversion: Conversion is normally defined to be the fraction of a component that has been converted to products by the reaction network. Conversion has several definitions and conventions. It is best to state the definition in the context of the problem being solved.

4.3 The Continuous-Stirred-Tank Reactor (CSTR)

The continuous-stirred-tank reactor (CSTR) is also a well-stirred reactor so there are no concentration gradients anywhere in the reactor volume. We again consider the entire reactor contents to be the reactor volume element as in Figure 4.13, and $V = V_R$. Since the reactor is well stirred, the relations in Equations 4.3 and 4.4 apply to the CSTR also. The difference between the CSTR and batch reactor is the flow streams shown in Figure 4.13. We denote the feed stream with flowrate Q_f and component j concentration c_{jf}. The outflow stream is flowing out of a well-mixed reactor and is therefore assumed to be at the same concentration as the reactor. Its flowrate is denoted Q. Writing Equation 4.2 for this reactor gives,

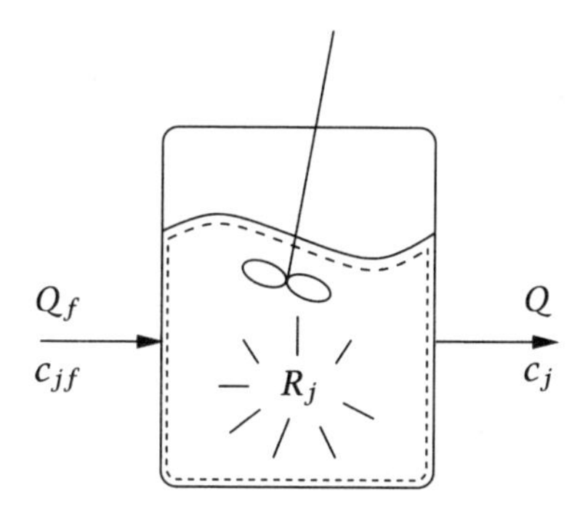

Figure 4.13: CSTR volume element.

$$\frac{d\left(c_j V_R\right)}{dt} = Q_f c_{jf} - Q c_j + R_j V_R, \quad j = 1, \ldots, n_s \tag{4.36}$$

If the reactor volume is constant and the volumetric flowrates of the inflow and outflow streams are the same, Equation 4.36 reduces to

$$\frac{dc_j}{dt} = \frac{1}{\tau}(c_{jf} - c_j) + R_j \tag{4.37}$$

in which

$$\tau = V_R / Q_f$$

is called the **mean residence time** of the CSTR. The residence-time distribution will be discussed further in Chapter 8. We refer to this balance as the constant-density case. It is often a good approximation for liquid-phase reactions.

The steady state of the CSTR is described by setting the time derivative in Equation 4.36 to zero,

$$0 = Q_f c_{jf} - Q c_j + R_j V_R \tag{4.38}$$

Conversion of reactant j is defined for a steady-state CSTR as follows

$$x_j = \frac{Q_f c_{jf} - Q c_j}{Q_f c_{jf}} \qquad \text{(steady state)} \tag{4.39}$$

One can divide Equation 4.38 through by Q_f to obtain for the constant-density case

$$c_j = c_{jf} + R_j \tau \qquad \text{(steady state, constant density)} \tag{4.40}$$

Example 4.1: Reaching steady state in a CSTR

Consider a first-order, liquid-phase reaction in an isothermal CSTR shown in Figure 4.13

$$\text{A} \xrightarrow{k} 2\text{B} \qquad r = kc_A$$

the feed concentration of A is $c_{Af} = 2$ mol/L, the residence time of the reactor is $\tau = 100$ min, and the rate constant is $k = 0.1 \text{ min}^{-1}$.

1. Find the steady-state concentration of A in the effluent for the given feed.

2. Plot the concentration of A versus time for constant feed concentration $c_{Af} = 2$ mol/L if the reactor is initially filled with an inert so $c_{A0} = 0$ mol/L.

3. Plot the concentration of A versus time for constant feed concentration $c_{Af} = 2$ mol/L if the reactor is initially filled with feed so $c_{A0} = 2$ mol/L.

Solution

Part 1. Because the reaction is isothermal and takes place in the liquid phase, we assume the fluid density is constant. Starting with Equation 4.40 for component A

$$c_A = c_{Af} + R_A \tau$$

Substituting the production rate $R_A = -kc_A$ and solving for c_A gives the steady-state concentration

$$c_{As} = \frac{c_{Af}}{1 + k\tau}$$

Substituting in the numerical values gives

$$c_{As} = \frac{2 \text{ mol/L}}{1 + (0.1 \text{ min}^{-1})(100 \text{ min})} = 0.182 \text{ mol/L}$$

Parts 2 and 3. The constant-density mass balance for component A is given by Equation 4.37

$$\frac{dc_A}{dt} = \frac{1}{\tau}\left(c_{Af} - c_A\right) - kc_A \tag{4.41}$$

$$c_A(0) = c_{A0}$$

Equation 4.41 can be arranged in the form

$$\frac{dc_A}{dt} + \left(\frac{1}{\tau} + k\right) c_A = \frac{1}{\tau} c_{Af}$$

a linear differential equation of the same form as Equation 4.13, which was solved earlier. Exercise 4.6 provides another convenient method for solving linear differential equations analytically. Using either of these methods, we find the analytical solution

$$c_A(t) = c_{A0} e^{-((1/\tau)+k)t} + \frac{c_{Af}}{1 + k\tau}\left[1 - e^{-((1/\tau)+k)t}\right] \tag{4.42}$$

Figure 4.14 shows the transient solution for the two initial conditions $c_{A0} = 0$ and $c_{A0} = 2$ mol/L. Notice both solutions converge to the same steady-state solution as time increases even though the starting conditions are quite different. We will see in Chapter 6 that the nonisothermal reactor behavior can be much more complex than that shown in Figure 4.14.

□

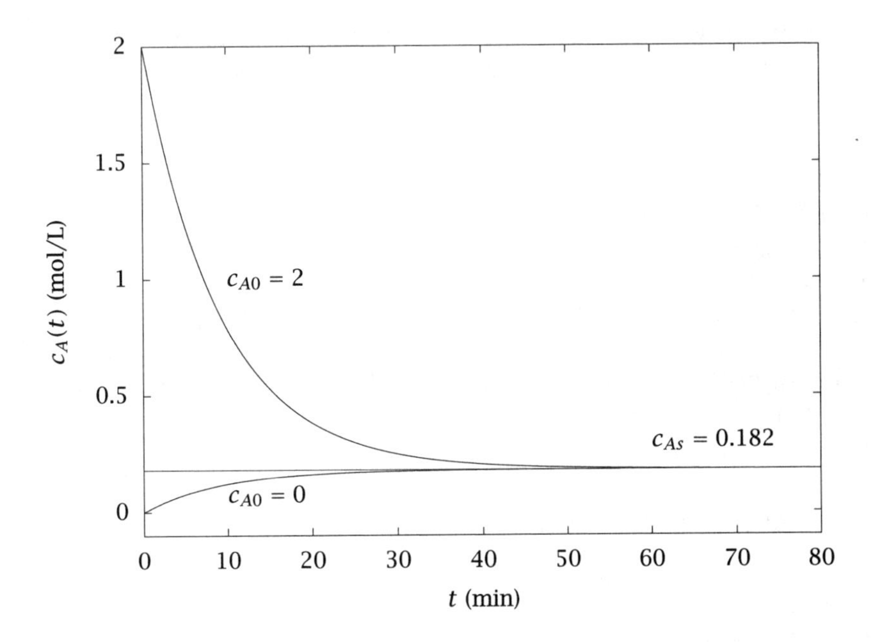

Figure 4.14: Reaching steady state in a CSTR.

Example 4.2: Phenol production in a CSTR

Next we consider a simple reactor-sizing problem. Consider the reaction of cumene hydroperoxide (CHP) to phenol and acetone

$$(C_6H_5)C(CH_3)_2OOH \longrightarrow (C_6H_5)OH + (CH_3)_2CO$$

This liquid-phase reaction is conducted using a small amount of acid as a catalyst [13], and the reaction rate is first order in the concentration of CHP and acid [5]. For the reactor temperature (85°C) and feed conditions, the reaction rate is given by $r = kc_{\text{CHP}}$, which means this is a pseudo-first-order reaction. Find the reactor volume to achieve 85% conversion of CHP at steady state. The flowrate into the reactor is $Q_f = 26.9 \text{ m}^3/\text{hr}$ and $k = 4.12 \text{ hr}^{-1}$.

Solution

The reactants and products are liquids at 85°C and the volume change is neglected. Therefore we assume $Q = Q_f$ and Equation 4.40 applies,

$$c_A = c_{Af} + R_A\tau$$

where A designates CHP. Substituting in for the production rate of CHP, $R_A = -kc_A$, and solving for the CHP concentration gives

$$c_A = \frac{c_{Af}}{1 + k\tau} \tag{4.43}$$

The conversion of CHP (for $Q = Q_f$) is

$$x_A = \frac{c_{Af} - c_A}{c_{Af}} = 1 - \frac{c_A}{c_{Af}}$$

Substituting Equation 4.43 into the above equation gives

$$x_A = \frac{k\tau}{1 + k\tau}$$

Solving for τ in terms of conversion gives

$$\tau = \frac{x_A}{k(1 - x_A)}$$

Substituting the relation $V_R = Q_f\tau$ and solving for V_R gives

$$V_R = \frac{Q_f x_A}{k(1 - x_A)}$$

Substituting in the known values gives the required CSTR volume

$$V_R = \frac{(26.9 \text{ m}^3/\text{hr})(0.85)}{(4.12 \text{ hr}^{-1})(0.15)} = 37 \text{ m}^3$$

□

4.4 The Semi-Batch Reactor

The semi-batch reactor is a cross between the batch reactor and CSTR. The semi-batch reactor is initially charged with reactant, like the batch reactor, but allows a feed addition policy while the reaction takes place, like the CSTR. Normally there is no outflow stream. We then set $Q = 0$ in Equation 4.36 to obtain for the semi-batch reactor

$$\frac{d\left(c_j V_R\right)}{dt} = Q_f c_{jf} + R_j V_R, \quad j = 1, \ldots, n_s \tag{4.44}$$

One may choose to operate a semi-batch reactor to control the reaction rate or heat release during reaction by slowly adding one of the reactants in the feed stream. Compared to the batch reactor, the semi-batch reactor provides more complete use of the reactor volume in reactions such as polymerizations that convert from lower density to higher density during the course of the reaction.

4.5 Volume Change Upon Reaction

The preceding examples in the batch reactor and CSTR sections were solved under the simplifying assumption that the composition of the reaction mixture has no effect on the density of the mixture. This assumption is useful for ideal-liquid mixtures in which the pure components have similar densities, and is a correct statement for ideal-gas mixtures. In this section we establish the appropriate modeling equations for the general case. We start with the material balance for a well-mixed reactor, Equation 4.36,

$$\frac{d\left(c_j V_R\right)}{dt} = Q_f c_{jf} - Q c_j + R_j V_R, \qquad j = 1, 2, \ldots, k, \ldots n_s \tag{4.45}$$

Equation 4.45 covers the batch, CSTR and semi-batch reactors, depending on how we specify Q_f and Q.

If we multiply Equation 4.45 by the molecular weight of species j and sum over all species we obtain,

$$\frac{d(\sum_j c_j M_j V_R)}{dt} = Q_f \sum_j c_{jf} M_j - Q \sum_j c_j M_j + \sum_j R_j M_j V_R \tag{4.46}$$

The term $\sum_j c_j M_j$ is the mass density of the reactor contents, which we denote ρ

$$\rho = \sum_{j=1}^{n_s} c_j M_j \tag{4.47}$$

The term $\sum_j c_{jf} M_j$ is the mass density of the feedstream, ρ_f. We know that conservation of mass in chemical reactions implies $\sum_j R_j M_j = 0$ (see Example 2.5). Substitution into Equation 4.46 leads to

$$\frac{d(\rho V_R)}{dt} = Q_f \rho_f - Q\rho \tag{4.48}$$

Equation 4.48 is clearly a total mass balance, in which the total mass in the reactor changes in time due to the inflow and outflow of mass. Notice that chemical reactions play no role in the total mass balance. The total mass balance is the starting point for the derivation of governing equations for numerous special situations in which the density and the volume of the reactor change, depending on the physical properties of the reaction mixture and the manner in which the reactor is operated.

Notice that because the total mass balance is a linear combination of the mole balances, it is not independent information. We may sometimes prefer to include the mass balance *in place* of one of the mole balances, $j = k$, to ensure that numerical roundoff does not cause small violations of the mass balance. The concentration of the kth species is then obtained by subtraction

$$c_k = \frac{\rho - \sum_{j \neq k} c_j M_j}{M_k}$$

Notice that one should always replace a high-concentration species mole balance with the total mass balance so that catastrophic loss of precision does not occur in the subtraction to obtain c_k.

If we have a single-phase system at equilibrium, the molar concentrations of the components, and the temperature and the pressure completely specify all intensive variables of the system. In this chapter we consider the temperature and the pressure to be known, fixed quantities. Therefore, the density of the reaction mixture, which is an intensive variable, is known if the c_j are known. This relationship is one form of the equation of state for the mixture

$$\rho = \tilde{f}(T, P, c_1, c_2, \ldots, c_{n_s}) \tag{4.49}$$

Substituting the definition of density, Equation 4.47, into Equation 4.49 shows that the equation of state is a single constraining relationship on the set of possible mixture concentrations. Because T and P are fixed constants, it is convenient to express the equation of state as

$$f(c_1, c_2, \ldots, c_{n_s}) = 0 \tag{4.50}$$

For example, we could express the equation of state in terms of the partial molar volumes as

$$\sum_j c_j \overline{V}_j = 1$$

in which $\overline{V}_j$ is the partial molar volume of component j in the mixture. The partial molar volumes are functions of T, P and c_j. If we assume an ideal mixture, this reduces to

$$\sum_j c_j V_j^\circ = 1, \qquad \text{ideal mixture}$$

in which V_j° is the specific volume of pure component j, which is a function of only T and P. We shall assume that a thermodynamic equation of state is valid even when the reactor is not at equilibrium. We briefly review the situation when the density does not vary with composition and then treat the general case.

Constant density. Because the mixture density, ρ, is independent of composition, it does not vary with time either and we can set it to the feed value,

$$\rho = \rho_f$$

The total mass balance then reduces to

$$\frac{dV_R}{dt} = Q_f - Q \tag{4.51}$$

which is sometimes referred to as a "volume balance." This terminology should be avoided because one cannot write a conservation statement for volume. The differential equation for the reactor volume is a consequence of the mass balance and the simplifying assumption of constant-mixture density. The three reactor types we have considered thus far are summarized below.

- Batch reactor. For the batch reactor, $Q = Q_f = 0$. We can therefore conclude from Equation 4.51 that a batch reactor with constant density has constant volume.
- CSTR (dynamic and steady state). If the outflow of the CSTR is regulated so that the CSTR has constant volume, then we can conclude from Equation 4.51 that $Q = Q_f$.
- Semi-batch reactor. In the semi-batch reactor, the reactor is filled during operation so Q_f is specified and positive for some time and $Q = 0$. The solution to Equation 4.51 then determines the change in volume of the reactor during the filling operation.

Nonconstant density. In the general case, we can consider the following variables to fully determine the state of the reactor: T, P, n_j, V_R. We assume throughout that the feed conditions, Q_f, c_{jf}, are specified. In the nonconstant density case, we switch to n_j from c_j because the material balance is fundamentally a balance on total moles in the reactor, not molar concentrations. If we know the n_j and V_R, then the values of the c_j are known, as required in the reaction rate expressions and the equation of state. To specify the right-hand sides of the material balances, Equation 4.45, we also require the value of Q. Because T, P are regarded as known, fixed constants in this chapter, the set of unknowns is n_j, V_R, Q. We therefore have $n_s + 2$ unknowns to determine and require $n_s + 2$ equations. We have the n_s equations from the component mole balances, Equations 4.45. The equation of state, Equation 4.50, provides one additional equation. The final equation is provided by a statement of reactor operation. We consider here three important cases of reactor operations.

1. Constant-volume reactor. The constant-volume reactor can be achieved by allowing overflow of the reactor to determine flowrate out of the reactor. In this situation, V_R is specified as the additional equation.

2. Constant-mass reactor. The constant-mass reactor can be achieved if a differential pressure measurement is used to control the flowrate out of the reactor and the reactor has constant cross-sectional area. The difference in pressure between the top and bottom of the reactor is determined by ρh in which h is the height difference. If the cross-sectional area is independent of height, then the pressure difference is also proportional to ρV_R, the total mass in the reactor. So if the flow controller maintains constant pressure difference, it also maintains constant mass. In this situation ρV_R is specified as the additional equation.

3. Flowrate out of the reactor is specified. This type of operation may be achieved if the flowrate out of the reactor is controlled by a flow controller. In this case $Q(t)$ is specified. A semi-batch reactor is operated in this way with $Q = 0$ until the reactor is filled with the reactants.

The calculations for the nonconstant-density case may be greatly simplified by using a differential-algebraic equation (DAE) solver. All three cases enumerated above can be handled by modifying the residual equations provided to the DAE solver. We do not have to differentiate the equation of state or perform other algebraic manipulations that are required if one uses an ordinary differential equation (ODE) solver.

In the general case, to produce a differential equation for the reactor volume, we begin by taking the differential of the equation of state $f(c_j) = 0$ to obtain

$$\frac{df}{dt} = \sum_j \frac{\partial f}{\partial c_j}\frac{dc_j}{dt} = \sum_j f_j \frac{dc_j}{dt} = 0 \tag{4.52}$$

in which

$$f_j = \frac{\partial f}{\partial c_j}$$

Now differentiating the product term $(c_j V_R)$ gives the relation

$$V_R \frac{dc_j}{dt} = \frac{d(c_j V_R)}{dt} - c_j \frac{dV_R}{dt}$$

Multiplying by f_j and summing gives

$$V_R \sum_j f_j \frac{dc_j}{dt} = \sum_j f_j \frac{d(c_j V_R)}{dt} - \sum_j f_j c_j \frac{dV_R}{dt} = 0$$

in which we have used Equation 4.52. Substituting the material balance into this equation and solving for the rate of change of reactor volume gives

$$\frac{dV_R}{dt} = \frac{\sum_j f_j \left(Q_f c_{jf} - Q c_j + R_j V_R\right)}{\sum_j f_j c_j}$$

Expressing the production rate in terms of the reaction rate yields

$$\frac{dV_R}{dt} = Q_f \frac{\sum_j f_j c_{jf}}{\sum_j f_j c_j} - Q + \frac{\sum_i \Delta f_i r_i V_R}{\sum_j f_j c_j} \tag{4.53}$$

in which Δf_i is defined to be

$$\Delta f_i = \sum_j \nu_{ij} f_j = \sum_j \nu_{ij} \frac{\partial f}{\partial c_j}$$

which is a change in a derivative property upon reaction. For example, if an ideal mixture is assumed for the equation of state, $f_j = V_j^\circ$, the pure component specific volumes, and $\Delta f_i = \Delta V_i^\circ$, the change in specific volume upon reaction i.

The constant-density and ideal-mixture cases for the three reactor operations are summarized in Table 4.1. We have $n_s + 2$ unknowns and the component balances at the top of Table 4.1 provide n_s equations. The remaining two equations for the constant-mass, constant-volume and specified-outflow operations are provided lower in the table. In these cases, these two extra equations are either differential equations or merely specify the values of constants.

The general equation of state case for the three reactor operations is summarized in Table 4.2. The n_s component balances are the same as in Table 4.1. But for the remaining two equations, we have a choice. We can either specify algebraic equations to obtain DAEs for the model, or differentiate the algebraic relations to obtain ODEs. DAEs may be preferred because the ODEs require the partial derivatives of the equation of state, $f_j = \partial f / \partial c_j$, as shown in Equation 4.53. Depending on the form of the equation of state, these derivatives may be complex expressions. In the following example, we set up and solve a nonconstant-density case, in which the products are denser than the reactants.

Unknowns $(n_s + 2)$:		$V_R, Q, n_j, \quad j = 1, \ldots, n_s$		
Component balances:		$\frac{dn_j}{dt} = Q_f c_{jf} - Q c_j + R_j V_R, \quad j = 1, \ldots, n_s$		
Defined quantities:		$n_j = c_j V_R \qquad \rho = \sum_j c_j M_j \qquad \Delta V_i^\circ = \sum_j \nu_{ij} V_j^\circ$		
	(i) constant density: $\rho = \rho_0$		(ii) ideal mixture: $\sum_j c_j V_j^\circ = 1$	
1. vol	$V_R = V_{R0}$	$Q = Q_f$	$V_R = V_{R0}$	$Q = Q_f + \sum_i \Delta V_i^\circ r_i V_R$
2. mass	$V_R = V_{R0}$	$Q = Q_f$	$\frac{dV_R}{dt} = Q_f(1 - \rho_f/\rho) + \sum_i \Delta V_i^\circ r_i V_R$	$Q = Q_f \rho_f / \rho$
3. Q	$\frac{dV_R}{dt} = Q_f - Q$	Q specified	$\frac{dV_R}{dt} = Q_f - Q + \sum_i \Delta V_i^\circ r_i V_R$	Q specified

Table 4.1: Reactor balances for constant-density and ideal-mixture assumptions.

Unknowns ($n_s + 2$):	$V_R, Q, n_j, \quad j = 1, \ldots, n_s$			
Component balances:	$\frac{dn_j}{dt} = Q_f c_{jf} - Q c_j + R_j V_R, \quad j = 1, \ldots, n_s$			
Defined quantities:	$n_j = c_j V_R \qquad \rho = \sum_j c_j M_j \qquad \Delta f_i = \sum_j \nu_{ij} \frac{\partial f}{\partial c_j}$			
Equation of state:	$f(c_1, c_2, \ldots, c_{n_s}) = 0$			
	DAEs		ODEs	
1. vol	$V_R = V_{R0}$	$f(c_j) = 0$	$V_R = V_{R0}$	$Q = Q_f \frac{\sum_j f_j c_{jf}}{\sum_j f_j c_j} + \frac{\sum_i \Delta f_i r_i V_R}{\sum_j f_j c_j}$
2. mass	$\rho V_R = \rho_0 V_{R0}$	$f(c_j) = 0$	$\frac{dV_R}{dt} = Q_f \frac{\sum_j f_j c_{jf}}{\sum_j f_j c_j} - Q + \frac{\sum_i \Delta f_i r_i V_R}{\sum_j f_j c_j}$	$Q = Q_f \rho_f / \rho$
3. Q	Q specified	$f(c_j) = 0$	$\frac{dV_R}{dt} = Q_f \frac{\sum_j f_j c_{jf}}{\sum_j f_j c_j} - Q + \frac{\sum_i \Delta f_i r_i V_R}{\sum_j f_j c_j}$	Q specified

Table 4.2: Reactor balances for general equation of state.

Example 4.3: Semi-batch polymerization

Consider a solution polymerization reaction, which can be modeled as a first-order, irreversible reaction

$$\mathrm{M} \xrightarrow{k} \mathrm{P} \qquad r = kc_M$$

A 20 m^3 semi-batch reactor is initially charged with solvent and initiator to half its total volume. A pure monomer feed is slowly added at flowrate $Q_{f0} = 1\ m^3/min$ to fill the reactor in semi-batch operation to control the heat release.

Consider two cases for the subsequent reactor operation.

1. The monomer feed is shut off and the reaction goes to completion.
2. The monomer feed is adjusted to keep the reactor filled while the reaction goes to completion.

Calculate the total polymer mass production, and the percentage increase in polymer production achieved in the second operation. You may assume an ideal mixture; the densities of monomer and polymer are $\rho_M = 800\ kg/m^3$, $\rho_P = 1100\ kg/m^3$. The monomer molecular weight is $M_M = 100$ kg/kmol, and the rate constant is $k = 0.1\ min^{-1}$.

Solution

While the reactor is filling, the monomer mole balance is

$$\frac{d(c_M V_R)}{dt} = Q_{f0} c_{Mf} - k c_M V_R$$

in which $c_{Mf} = \rho_M / M_M$ is given, and $Q_f = Q_{f0}$ is constant during the filling operation. We denote the total number of moles of monomer by $M = c_M V_R$, and can write the monomer balance as

$$\begin{aligned} \frac{dM}{dt} &= Q_{f0} c_{Mf} - kM \\ M(0) &= 0 \end{aligned} \tag{4.54}$$

For an ideal mixture, the volume is given by

$$\begin{aligned} \frac{dV_R}{dt} &= Q_{f0} + \Delta V k M \\ V_R(0) &= 10\ m^3 \end{aligned} \tag{4.55}$$

in which

$$\Delta V = (1/\rho_P - 1/\rho_M)\, M_M$$

To compute the polymer mass, we note from the stoichiometry that the mass production rate of polymer $\widetilde{R}_P$ is

$$\widetilde{R}_P = -R_M M_M$$

so the mass balance for total polymer $\widetilde{P}$ is given by

$$\frac{d\widetilde{P}}{dt} = \widetilde{R}_p V_R = k c_M M_M V_R = (k M_M) M \tag{4.56}$$

Let t_1 be the time that the reactor fills. Equation 4.54 is of the same form as Equation 4.13 and can be solved in the same manner, yielding

$$M(t) = \frac{Q_{f0} c_{Mf}}{k}\left(1 - e^{-kt}\right), \qquad t \le t_1 \tag{4.57}$$

Substituting this result into Equations 4.55 and 4.56 and integrating gives

$$V_R(t) = V_R(0) + \left(1 + \Delta V c_{Mf}\right) Q_{f0}\, t - \frac{\Delta V Q_{f0} c_{Mf}}{k}\left(1 - e^{-kt}\right), \qquad t \le t_1 \tag{4.58}$$

$$\widetilde{P}(t) = Q_{f0} c_{Mf} M_M \left(t - \frac{1}{k}(1 - e^{-kt})\right), \qquad t \le t_1 \tag{4.59}$$

We can solve Equation 4.58 for the time t_1 when the reactor fills, $V_R(t_1) = 20\ \text{m}^3$. Note we have a single nonlinear equation that must be solved numerically. Substituting in the numerical values and solving for t_1 yields

$$t_1 = 11.2\ \text{min}$$

Note the reactor would have filled in 10 min if the density were constant. The extra time reflects the available volume created by converting some of the monomer to polymer during filling. After t_1 we consider the two operations.

Operation 1. In the first operation, $Q_f = 0$ after t_1. Substituting this value of Q_f into Equations 4.54 and 4.55 gives

$$\frac{dM}{dt} = -kM \tag{4.60}$$

$$\frac{dV_R}{dt} = \Delta V k M \tag{4.61}$$

in which initial condition $M(t_1)$ is given by substituting t_1 in Equation 4.57, and $V_R(t_1) = 20 \text{ m}^3$. We solve Equation 4.60 to obtain

$$M(t) = M(t_1)e^{-k(t-t_1)}, \qquad t_1 \le t$$

and substitution into Equations 4.61 and 4.56 and integration gives

$$V_R(t) = V_R(t_1) + \Delta V M(t_1)\left(1 - e^{-k(t-t_1)}\right), \qquad t_1 \le t \tag{4.62}$$

$$\widetilde{P}(t) = \widetilde{P}(t_1) + M_M M(t_1)\left(1 - e^{-k(t-t_1)}\right), \qquad t_1 \le t \tag{4.63}$$

Notice the reactor volume decreases after t_1 because ΔV is negative. These results are plotted in Figures 4.15–4.18.

Operation 2. Because the reactor volume is constant, we can solve Equation 4.55 for the feed flowrate during the secondary monomer addition

$$Q_f = -\Delta V k M$$

Substitution of this flowrate into Equation 4.54 gives

$$\frac{dM}{dt} = -(1 + \Delta V c_{Mf})kM$$

which can be solved to give

$$M(t) = M(t_1)\exp\left(-(1 + \Delta V c_{Mf})k(t - t_1)\right), \qquad t_1 \le t$$

The polymer mass can be calculated by substituting this result into Equation 4.56 and integrating

$$\widetilde{P}(t) = \widetilde{P}(t_1) + \frac{M_M M(t_1)}{(1 + \Delta V c_{Mf})}\left[1 - \exp\left(-(1 + \Delta V c_{Mf})k(t - t_1)\right)\right], \qquad t_1 \le t$$

These results also are plotted in Figures 4.15–4.18. Notice the final polymer production is larger in Operation 2 because of the extra monomer addition. We can perform an independent, simple calculation of the total polymer in Operation 2. In Operation 2, 10 m^3 of polymer are produced because in an ideal mixture, the volumes are additive. Therefore

$$\widetilde{P}_2 = (V_R - V_{R0})\rho_P = 10 \text{ m}^3 \times 1100 \text{ kg/m}^3 = 11000 \text{ kg}$$

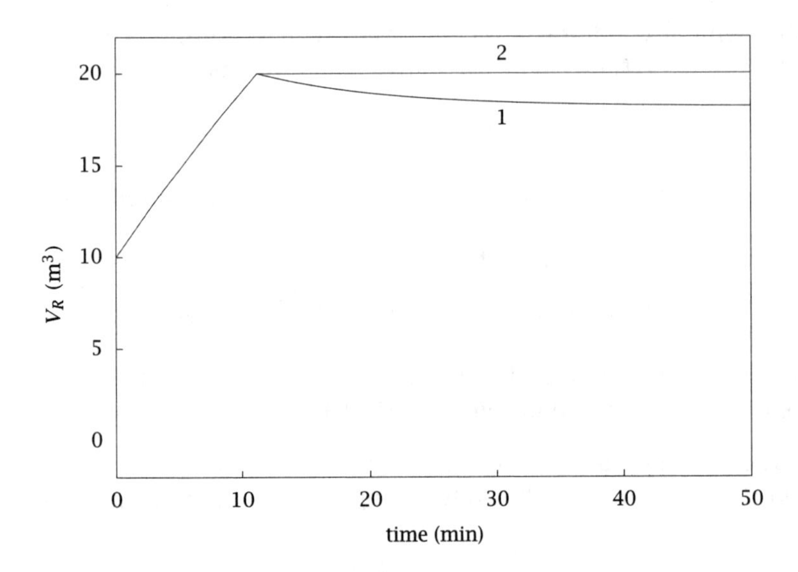

Figure 4.15: Semi-batch reactor volume for primary monomer addition (Operation 1) and primary plus secondary monomer additions (Operation 2).

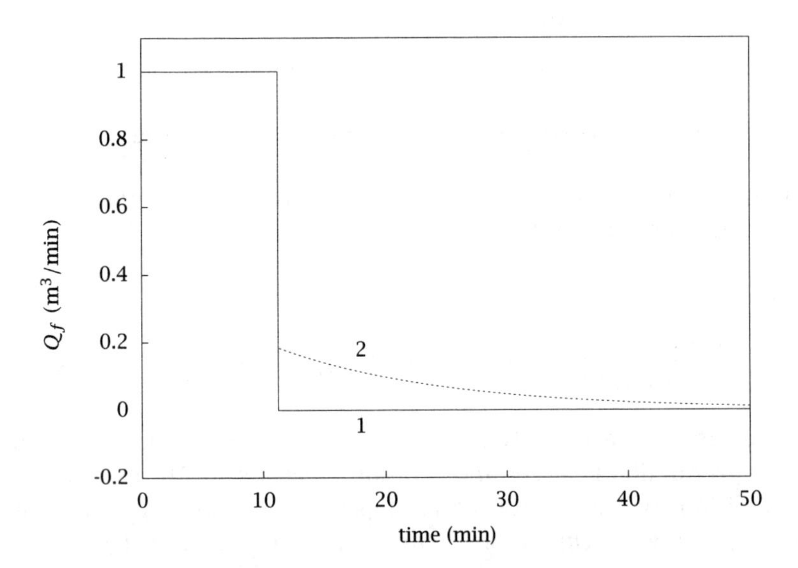

Figure 4.16: Semi-batch reactor feed flowrate for primary monomer addition (Operation 1) and primary plus secondary monomer additions (Operation 2).

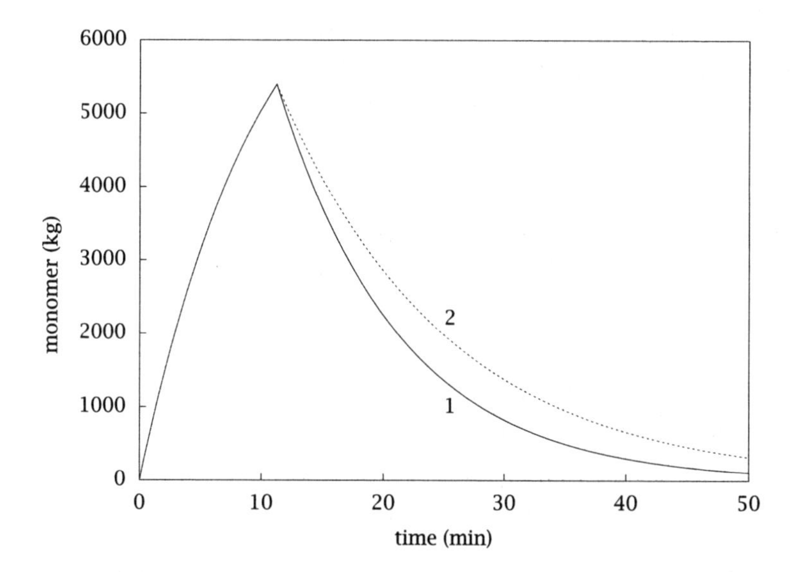

Figure 4.17: Semi-batch reactor monomer content for primary monomer addition (Operation 1) and primary plus secondary monomer additions (Operation 2).

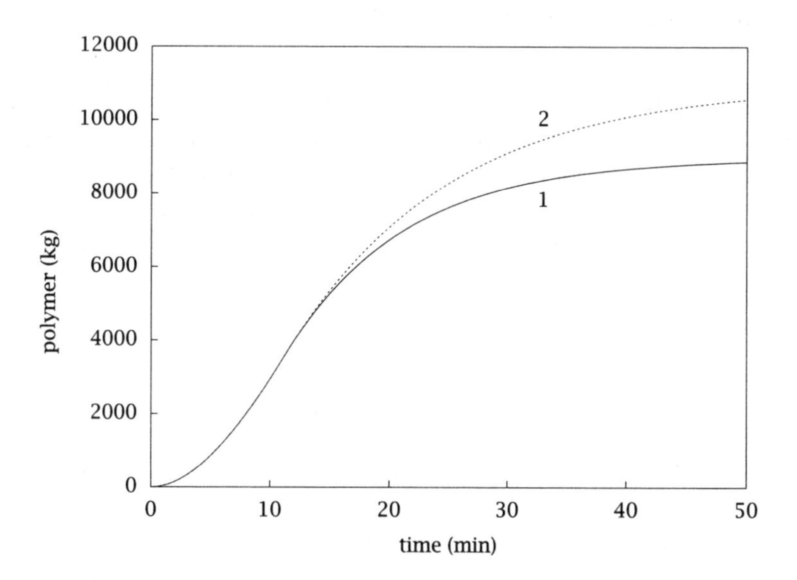

Figure 4.18: Semi-batch reactor polymer content for primary monomer addition (Operation 1) and primary plus secondary monomer additions (Operation 2).

in good agreement with the long-time solution for Operation 2 in Figure 4.18. For Operation 1, we have no simple calculation, so we take the limit as $t \longrightarrow \infty$ in Equation 4.63 to obtain

$$\widetilde{P}_1 = \widetilde{P}(t_1) + M_M M(t_1)$$

Both $\widetilde{P}(t_1)$ and $M(t_1)$ can be evaluated by substituting $t_1 = 11.2$ min in Equations 4.57 and 4.59, respectively. The results are

$$\widetilde{P}(t_1) = 3582 \text{ kg}, \quad M(t_1) = 53.95 \text{ kmol}, \quad \widetilde{P}_1 = 8977 \text{ kg}$$

Note the value $\widetilde{P}_1 = 8977$ kg is in good agreement with the long-time solution for Operation 1 in Figure 4.18. Finally, the increase in production rate is

$$\frac{\widetilde{P}_2 - \widetilde{P}_1}{\widetilde{P}_1} \times 100\% = 22.5\%$$

By using the volume of the reactor more efficiently, the total polymer production increases 22.5%. □

4.6 The Plug-Flow Reactor (PFR)

Plug flow in a tube is an ideal-flow assumption in which the fluid is well mixed in the radial and angular directions. The fluid velocity is assumed to be a function of only the axial position in the tube. Plug flow is often used to approximate fluid flow in tubes at high Reynolds number. The turbulent flow mixes the fluid in the radial and angular directions. Also in turbulent flow, the velocity profile is expected to be reasonably flat in the radial direction except near the tube wall.

4.6.1 Thin-Disk Volume Element

Because of the plug-flow assumption, it is natural to take a thin disk for the reactor volume element as shown in Figure 4.19. The concentration does not change over the volume element because there is complete mixing in the radial and angular directions and because the axial distance Δz is small. The element has volume $\Delta V = A_c \Delta z$ in which A_c is the tube cross-sectional area. Writing Equation 4.2 in this situation gives

$$\frac{\partial\left(c_j \Delta V\right)}{\partial t} = c_j Q\Big|_z - c_j Q\Big|_{z+\Delta z} + R_j \Delta V$$

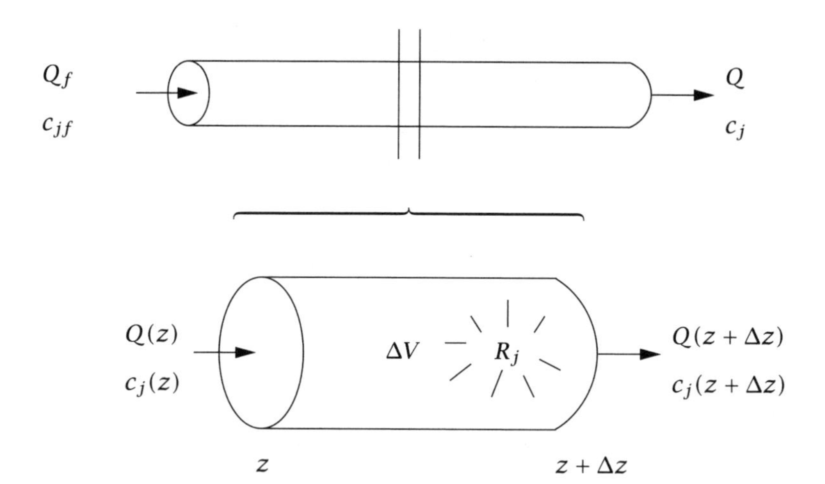

Figure 4.19: Plug-flow reactor volume element.

Dividing the above equation by ΔV and taking the limit as ΔV goes to zero yields,

$$\underbrace{\frac{\partial c_j}{\partial t}}_{\text{accumulation}} = -\underbrace{\frac{\partial \left(c_j Q\right)}{\partial V}}_{\text{convection}} + \underbrace{R_j}_{\text{reaction}} \tag{4.64}$$

If the tube has constant cross section, A_c, then velocity, v, is related to volumetric flowrate by $v = Q/A_c$, and axial length is related to tube volume by $z = V/A_c$, so Equation 4.64 can be rearranged to the familiar form [1, p.584]

$$\frac{\partial c_j}{\partial t} = -\frac{\partial \left(c_j v\right)}{\partial z} + R_j \tag{4.65}$$

4.6.2 Steady-State Operation

One is often interested in the steady-state operation of a plug-flow reactor. Setting the time derivative in Equation 4.64 to zero gives,

$$\frac{d(c_j Q)}{dV} = R_j \tag{4.66}$$

The product $c_j Q = N_j$ is the total molar flow of component j. One also can express the PFR mole balance in terms of the molar flow,

$$\frac{dN_j}{dV} = R_j \tag{4.67}$$

4.6.3 Volumetric Flowrate for Gas-Phase Reactions

To use Equation 4.67 for designing a gas-phase reactor, one has to be able to relate the volumetric flowrate, Q, to the molar flows, $N_j, j = 1, 2, \ldots, n_s$. The important piece of information tying these quantities together is, again, the equation of state for the reaction mixture. The equation of state is a function relating the system temperature, pressure, and all molar concentrations, $f(T, P, c_j) = 0$. Because the molar flow and concentration are directly related,

$$N_j = c_j Q \tag{4.68}$$

the equation of state is also a relation between temperature, pressure, molar flows, and volumetric flowrate. For example, the ideal-gas equation of state, $c = P/RT$, can be stated in terms of molar concentrations, c_j, as

$$\sum_j c_j = \frac{P}{RT}$$

In terms of molar flows, the equation of state is

$$\frac{\sum_j N_j}{Q} = \frac{P}{RT}$$

One can solve the previous equation for the volumetric flowrate,

$$Q = \frac{RT}{P} \sum_j N_j \tag{4.69}$$

Equation 4.69 is used with the mole balance, Equation 4.67, to solve gas-phase PFR problems under the ideal-gas assumption. To evaluate the concentrations for use with the reaction rate expressions, one then rearranges Equation 4.68 to obtain

$$c_j = \frac{N_j}{Q} = \frac{P}{RT} \frac{N_j}{\sum_j N_j} \tag{4.70}$$

4.6.4 Volumetric Flowrate for Liquid-Phase Reactions

Consider the equation of state for a liquid-phase system to be arranged in the form

$$\rho = f(T, P, c_j)$$

The mass density is related to the volumetric flowrate and total mass flow, $M = \sum_j N_j M_j$, via

$$M = \rho Q \tag{4.71}$$

Multiplying Equation 4.67 by M_j and summing on j produces

$$\frac{dM}{dV} = 0, \qquad M(0) = M_f$$

in which M_f is the feed mass flowrate. The total mass balance indicates that the total mass flow is constant with respect to axial position in the PFR because chemical reaction cannot alter the total mass flow. We can solve for the volumetric flowrate by rearranging Equation 4.71

$$Q = \frac{M_f}{\rho} \tag{4.72}$$

If the liquid density is considered constant, $\rho = \rho_f$, then

$$Q = Q_f, \qquad \text{constant density} \tag{4.73}$$

and the volumetric flowrate is constant and equal to the feed value. Equation 4.73 is used often for liquid-phase reactions. If the density variation is significant, however, Equation 4.72 should be used instead. Finally, if we denote the time spent in the tube by $\tau = V/Q$, if Q is constant, we can rewrite Equation 4.66 as

$$\frac{dc_j}{d\tau} = R_j, \qquad \text{constant flowrate} \tag{4.74}$$

which is identical to the constant-volume batch reactor, Equation 4.6. In other words, for the constant-flowrate case, the steady-state profile in a PFR starting from a given feed condition is also the transient profile in a batch reactor starting from the equivalent initial condition.

4.6.5 Single Reaction Systems

Example 4.4: Changing flowrate in a PFR

A pure vapor stream of A is decomposed in a PFR to form B and C

$$\mathrm{A} \xrightarrow{k} \mathrm{B} + \mathrm{C}$$

Determine the length of 2.5 cm inner-diameter tube required to achieve 35% conversion of A. The reactor temperature is 518°C and the pressure is 2.0 atm. Assume the pressure drop is negligible. The reaction rate is first order in A, $k = 0.05\ \mathrm{s}^{-1}$ at the reactor temperature, and the feed flowrate is 35 L/min.

Solution

Writing the mole balance for component A gives

$$\frac{dN_A}{dV} = R_A$$

The production rate of A is $R_A = -r = -kc_A$. Substituting the production rate into the above equation gives,

$$\frac{dN_A}{dV} = -kN_A/Q \tag{4.75}$$

The volumetric flowrate is not constant, so we use Equation 4.69, which assumes an ideal-gas equation of state,

$$Q = \frac{RT}{P}(N_A + N_B + N_C) \tag{4.76}$$

The ideal-gas assumption is reasonable at this reactor temperature and pressure. One can relate the molar flows of B and C to A using the reaction stoichiometry. The mole balances for B and C are

$$\frac{dN_B}{dV} = R_B = r, \qquad \frac{dN_C}{dV} = R_C = r$$

Adding the mole balance for A to those of B and C gives

$$\frac{d(N_A + N_B)}{dV} = 0, \qquad \frac{d(N_A + N_C)}{dV} = 0$$

The stoichiometry does not allow the molar flow $N_A + N_B$ or $N_A + N_C$ to change with position in the tube. Because both of these quantities are known at the tube entrance, one can relate N_B and N_C to N_A,

$$N_A + N_B = N_{Af} + N_{Bf}$$
$$N_A + N_C = N_{Af} + N_{Cf}$$

Rearranging the previous equations gives,

$$N_B = N_{Af} + N_{Bf} - N_A$$
$$N_C = N_{Af} + N_{Cf} - N_A$$

Substituting the relations for N_B and N_C into Equation 4.76 gives

$$Q = \frac{RT}{P}\left(2N_{Af} + N_{Bf} + N_{Cf} - N_A\right)$$

Because the feed stream is pure A, $N_{Bf} = N_{Cf} = 0$, yielding

$$Q = \frac{RT}{P}\left(2N_{Af} - N_A\right)$$

Substituting this expression in Equation 4.75 gives the final mole balance,

$$\frac{dN_A}{dV} = -k\frac{P}{RT}\frac{N_A}{2N_{Af} - N_A}$$

The previous differential equation can be separated and integrated,

$$\int_{N_{Af}}^{N_A} \frac{2N_{Af} - N_A}{N_A} dN_A = \int_0^V -\frac{kP}{RT} dV$$

Performing the integration gives,

$$2N_{Af}\ln\left(N_A/N_{Af}\right) + \left(N_{Af} - N_A\right) = -\frac{kP}{RT}V$$

The conversion of component j for a plug-flow reactor operating at steady state is defined as

$$x_j = \frac{N_{jf} - N_j}{N_{jf}}$$

Because we are interested in the V corresponding to 35% conversion of A, we substitute $N_A = (1 - x_A)N_{Af}$ into the previous equation and solve for V,

$$V = -\frac{RT}{kP}N_{Af}\left[2\ln(1 - x_A) + x_A\right]$$

Because $Q_f = N_{Af}RT/P$ is given in the problem statement and the tube length is desired, it is convenient to rearrange the previous equation to obtain

$$z = -\frac{Q_f}{kA_c}\left[2\ln(1 - x_A) + x_A\right]$$

Substituting in the known values gives

$$z = -\left(\frac{35 \times 10^3\ \text{cm}^3/\text{min}}{0.05\ \text{s}^{-1}\ 60\ \text{s/min}}\right)\left(\frac{4}{\pi(2.5\ \text{cm})^2}\right)\left[2\ln(1 - .35) + .35\right]$$

$$z = 1216\ \text{cm} = 12.2\ \text{m}$$

□

4.6.6 Multiple-Reaction Systems

The modeler has some freedom in setting up the material balances for a plug-flow reactor with several reactions. The most straightforward method is to write the material balance relation for every component,

$$\frac{dN_j}{dV} = R_j, \qquad j = 1, 2, \ldots, n_s$$

Recall the production rate of each species is a simple linear combination of the reaction rates times the appropriate stoichiometric coefficients,

$$R_j = \sum_{i=1}^{n_r} \nu_{ij} r_i, \qquad j = 1, 2, \ldots, n_s$$

The reaction rates are expressed in terms of the species concentrations. The c_j are calculated from the molar flows with Equation 4.68, and Q is calculated from Equation 4.69, if an ideal-gas mixture is assumed.

Example 4.5: Benzene pyrolysis in a PFR

Hougen and Watson [10] analyzed the rate data for the pyrolysis of benzene by the following two reactions. Diphenyl is produced by the dehydrogenation of benzene,

$$2C_6H_6 \underset{k_{-1}}{\overset{k_1}{\rightleftharpoons}} C_{12}H_{10} + H_2 \tag{4.77}$$

Triphenyl is formed by the secondary reaction,

$$C_6H_6 + C_{12}H_{10} \underset{k_{-2}}{\overset{k_2}{\rightleftharpoons}} C_{18}H_{14} + H_2 \tag{4.78}$$

The reactions are assumed to be elementary so that the rate expressions are

$$r_1 = k_1 \left(c_B^2 - \frac{c_D c_H}{K_1} \right) \qquad r_2 = k_2 \left(c_B c_D - \frac{c_T c_H}{K_2} \right) \tag{4.79}$$

in which the subscripts, B, D, T and H represent benzene, diphenyl, triphenyl and hydrogen, respectively. Calculate the tube volume required to reach 50% total conversion of the benzene for a 60 kmol/hr feed stream of pure benzene. The reactor operates at 1033 K and 1.0 atm. Plot the mole fractions of the four components versus reactor volume. The rate and equilibrium constants at $T = 1033$ K and $P = 1.0$ atm are given in Hougen and Watson,

$$k_1 = 7 \times 10^5 \text{ L/mol} \cdot \text{hr} \qquad K_1 = 0.31$$

$$k_2 = 4 \times 10^5 \text{ L/mol} \cdot \text{hr} \qquad K_2 = 0.48$$

Solution

The mole balances for the four components follow from the stoichiometry,

$$\frac{dN_B}{dV} = -2r_1 - r_2 \qquad \frac{dN_H}{dV} = r_1 + r_2$$
$$\frac{dN_D}{dV} = r_1 - r_2 \qquad \frac{dN_T}{dV} = r_2$$

The initial condition for the ODEs are $N_B(0) = N_{Bf}$ and $N_D(0) = N_H(0) = N_T(0) = 0$. Because the number of moles are conserved in both reactions, the total molar flux does not change with reactor volume. The volumetric flowrate can be evaluated from Equation 4.69, which in this case reduces to,

$$Q = \frac{RT}{P} N_{Bf} \tag{4.80}$$

The rate expressions are substituted into the four ODEs and they are solved numerically. The total conversion of benzene, $x_B = (N_{Bf} - N_B)/N_{Bf}$, is plotted versus reactor volume in Figure 4.20. A reactor volume of 403 L is required to reach 50% conversion. The composition of the reactor versus reactor volume is plotted in Figure 4.21. □

In the previous approach, a differential equation is required for each component. Therefore n_s differential equations are solved simultaneously in this approach. One can instead solve the problem by using one differential equation for each independent reaction for a total of n_i differential equations. Because $n_i < n_s$, this second approach produces fewer differential equations. Normally n_i is not much less than n_s, and the computational expenses of the two approaches are similar. Whether we write differential equations for the species or the reaction extents is largely a matter of taste.

In any case, the following procedure is guaranteed to give the minimal set of differential equations. Instead of writing the material balance for each component, define extents for each reaction that satisfy

$$\frac{d\varepsilon_i}{dV} = r_i, \qquad i = 1, 2, \ldots, n_i$$

The initial conditions for these extents can be defined to be zero,

$$\varepsilon_i(0) = 0, \qquad i = 1, 2, \ldots, n_i$$

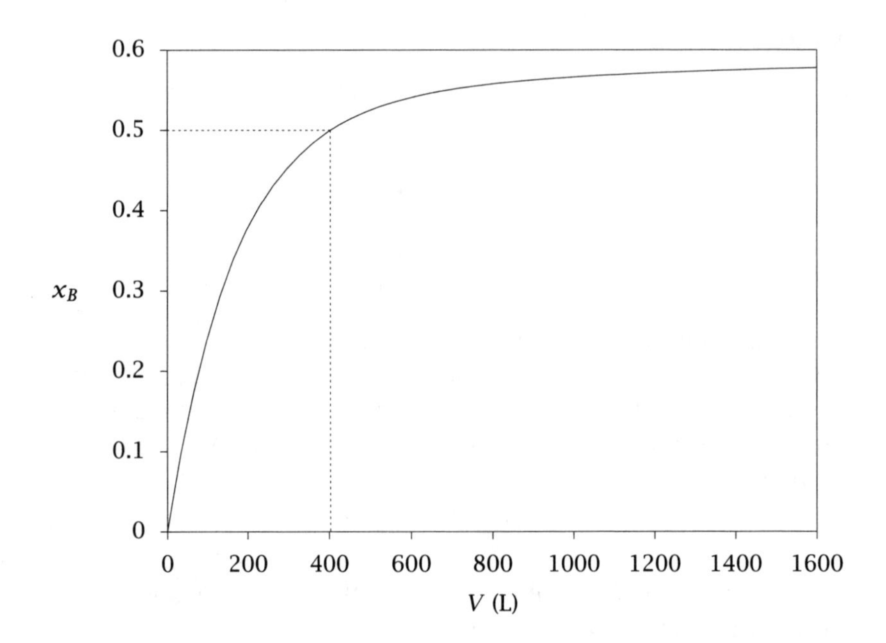

Figure 4.20: Benzene conversion versus reactor volume.

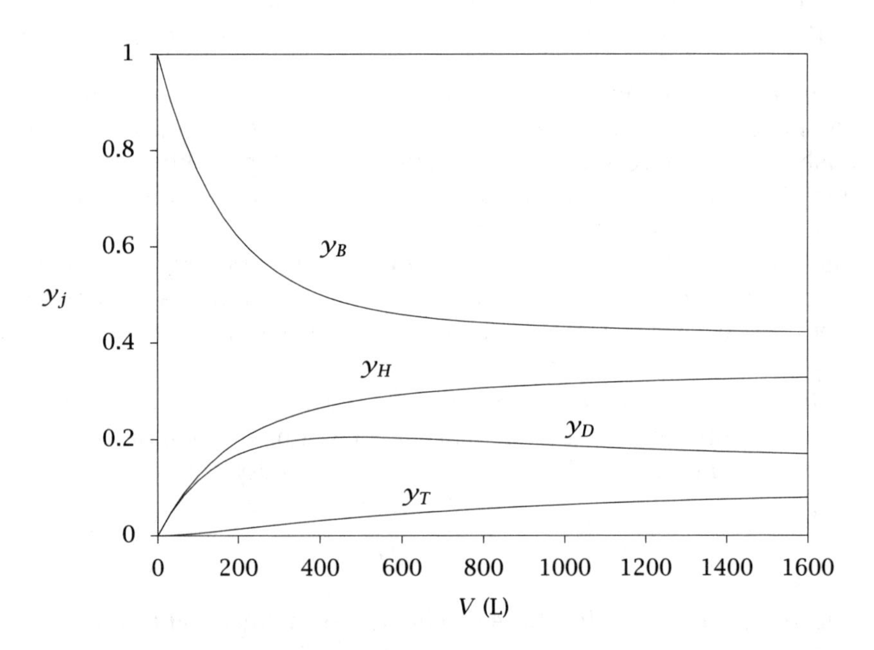

Figure 4.21: Component mole fractions versus reactor volume.

The molar flow for each species can be computed by summing the stoichiometric number times each extent,

$$N_j = N_{jf} + \sum_{i=1}^{n_i} \nu_{ij}\varepsilon_i, \qquad j = 1, 2, \ldots, n_s$$

As with the previous method, the c_j are calculated from the molar flows with Equation 4.68, and Q is calculated from Equation 4.69.

Example 4.6: Benzene pyrolysis, revisited

Rework the previous example using the minimum number of differential equations.

Solution

Defining the extents to track the rates of the two reactions in Equations 4.77 and 4.78 gives

$$\frac{d\varepsilon_1}{dV} = r_1, \qquad \varepsilon_1(0) = 0 \tag{4.81}$$

$$\frac{d\varepsilon_2}{dV} = r_2, \qquad \varepsilon_2(0) = 0 \tag{4.82}$$

The stoichiometry allows all molar flows to be calculated from these two extents,

$$N_B + 2\varepsilon_1 + \varepsilon_2 = N_{Bf} \qquad N_H - \varepsilon_1 - \varepsilon_2 = N_{Hf}$$

$$N_D - \varepsilon_1 + \varepsilon_2 = N_{Df} \qquad N_T - \varepsilon_2 = N_{Tf}$$

Substituting in the feed conditions gives,

$$N_B = N_{Bf} - 2\varepsilon_1 - \varepsilon_2 \qquad N_H = \varepsilon_1 + \varepsilon_2$$

$$N_D = \varepsilon_1 - \varepsilon_2 \qquad N_T = \varepsilon_2$$

These relations can be used along with Equation 4.80 and Equations 4.79 to evaluate the rates r_1 and r_2 in Equations 4.81 and 4.82 in terms of ε_1 and ε_2. In this approach one only has to solve two differential equations rather than the four required in the previous approach. The solutions are the same in the two approaches. □

Example 4.7: Ethane pyrolysis in the presence of NO

Nitric oxide participates in the pyrolysis of ethane to form ethylene by an initiation reaction to form ethyl radicals and a competition reaction for H atoms through the formation of HNO. Laidler and Wojciechowski [11] have suggested the following sequence of elementary reactions to describe pyrolysis

$$
\begin{aligned}
C_2H_6 + NO &\underset{k_{-1}}{\overset{k_1}{\rightleftharpoons}} C_2H_5 + HNO \\
C_2H_5 &\overset{k_2}{\longrightarrow} H + C_2H_4 \\
H + C_2H_6 &\overset{k_3}{\longrightarrow} C_2H_5 + H_2 \\
H + NO &\underset{k_{-4}}{\overset{k_4}{\rightleftharpoons}} HNO
\end{aligned}
$$

The rate expressions and rate constants are given below, in which $k_i = A_i \exp(-E_i/RT)$

$$
\begin{aligned}
r_1 &= k_1 c_{C_2H_6} c_{NO} - k_{-1} c_{C_2H_5} c_{HNO} \\
r_2 &= k_2 c_{C_2H_5} \\
r_3 &= k_3 c_H c_{C_2H_6} \\
r_4 &= k_4 c_H c_{NO} - k_{-4} c_{HNO}
\end{aligned}
$$

i	A_i	E_i (kJ/mol)
1	1.0×10^{14} cm^3/mol s	217.6
−1	1.0×10^{12} cm^3/mol s	0
2	3.0×10^{14} s^{-1}	165.3
3	3.4×10^{12} cm^3/mol s	28.5
4	1.0×10^{12} cm^3/mol s	0
−4	1.0×10^{13} s^{-1}	200.8

Assume the reaction takes place in an isothermal, 1500-cm^3 PFR operating at constant pressure (1.0 atm). The feed to the reactor consists of a mixture of ethane and NO with a molar ratio of 95% ethane and 5% NO. The inlet volumetric flowrate is 600 cm^3/s. Consider a base-case reactor temperature of 1050 K.

1. Develop the necessary design equations to calculate the molar flow of ethane, NO and ethylene, and determine the effluent composition from the reactor. Be sure to specify the initial conditions.

2. Solve the design equations and calculate the molar flowrates of ethane, ethylene and NO as a function of reactor volume for a reactor temperature of 1050 K.

3. Examine the effect of increasing and decreasing the temperature 50 K on the molar flowrate of ethane. Why do you think a small change in temperature creates such a large effect on the amount of ethane reacting?

Solution

The reaction rates are functions of molar concentrations of the components, and the concentrations can be found from the molar flowrates of each component

$$c_j = \frac{N_j}{Q} = \frac{N_j}{\frac{RT}{P}\sum_j N_j} = \frac{P}{RT}\frac{N_j}{\sum_j N_j}$$

We need to know all the molar flows, and they can be found from the PFR mass balance

$$\frac{dN_j}{dV} = R_j = \sum_i \nu_{ij} r_i$$

which produces for this example

$$\begin{aligned}
\frac{dN_{C_2H_6}}{dV} &= -r_1 - r_3 & \frac{dN_{H}}{dV} &= r_2 - r_3 - r_4 \\
\frac{dN_{NO}}{dV} &= -r_1 - r_4 & \frac{dN_{C_2H_4}}{dV} &= r_2 \\
\frac{dN_{C_2H_5}}{dV} &= r_1 - r_2 + r_3 & \frac{dN_{H_2}}{dV} &= r_3 \\
\frac{dN_{HNO}}{dV} &= r_1 + r_4
\end{aligned} \tag{4.83}$$

The rates are calculated from the rate constants and concentrations; for example, the first rate is

$$r_1 = \left(\frac{P}{RT}\right)^2 \left(k_1 \frac{N_{C_2H_6}N_{NO}}{N_{tot}^2} - k_{-1}\frac{N_{C_2H_5}N_{HNO}}{N_{tot}^2}\right)$$

in which

$$N_{tot} = N_{C_2H_6} + N_{NO} + N_{C_2H_5} + N_{HNO} + N_{H} + N_{C_2H_4} + N_{H_2}$$

The feed contains only C_2H_6 and NO at molar flows given by

$$N_{jf} = y_{jf} \frac{P}{RT} Q_f$$

leading to $N_{C_2H_6 f} = 6.62 \times 10^{-3}$ mol/s and $N_{NOf} = 3.48 \times 10^{-4}$ mol/s.

The numerical solution of Equations 4.83 at 1050 K is shown in Figure 4.22. We see the consumption of ethane and the production of ethylene. The ethane molar flows at the three different reactor temperatures are shown in Figure 4.23. The solution shows a small change in ethane molar flow initially, followed by a more rapid change of molar flow with increasing distance in the reactor. The rate is low initially until a sufficient amount of H is formed so the third reaction can begin to contribute to the ethane consumption. The rates are very sensitive to temperature because the activation energies are quite high for the reactions that lead to the loss of ethane. For example, the first reaction that initiates the pyrolysis and the second reaction that produces H, which contributes to the consumption of ethane, have high activation energies. □

4.7 Some PFR-CSTR Comparisons

Given the two continuous reactors in this chapter, the CSTR and the PFR, it is natural to compare their steady-state efficiencies in converting reactants to products. For simplicity, consider a constant-density, liquid-phase reaction with nth-order, irreversible reaction rate

$$\mathrm{A} \xrightarrow{k} \mathrm{B} \qquad r = kc_A^n$$

For this situation, the steady-state PFR material balance is given by Equation 4.74

$$\frac{dc_A}{d\tau} = -r(c_A)$$

We rearrange and solve for the time required to change from the feed condition c_{Af} to some exit concentration c_A

$$\tau = \int_{c_A}^{c_{Af}} \frac{1}{r(c_A')} dc_A'$$

in which the prime is to remind us that this variable is a dummy variable of integration. Figure 4.24 shows the graphical interpretation of this

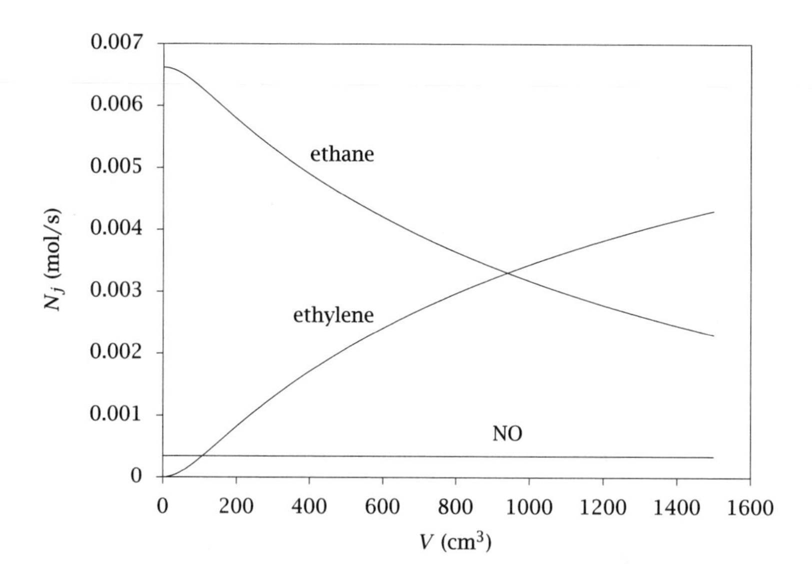

Figure 4.22: Molar flowrate of ethane, ethylene and NO versus reactor volume for ethane pyrolysis example.

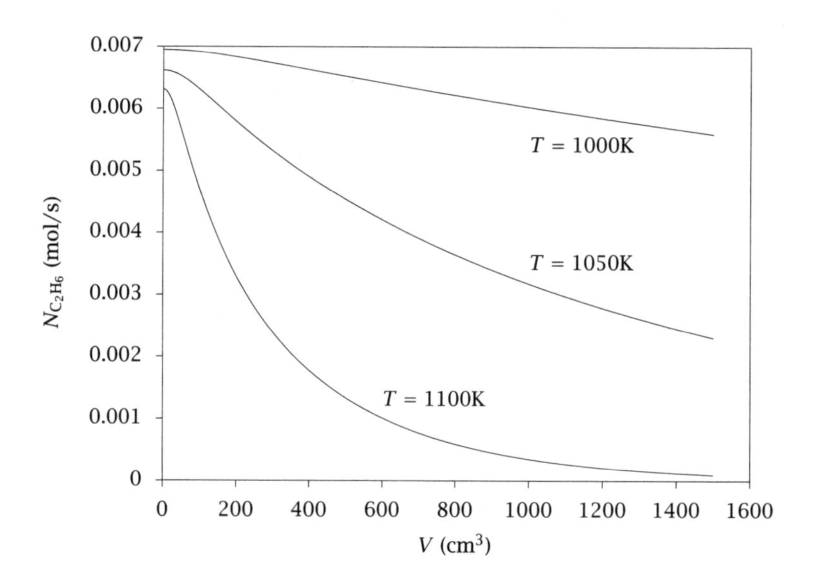

Figure 4.23: Molar flowrate of ethane versus reactor volume for inlet temperatures of 1000, 1050 and 1100 K.

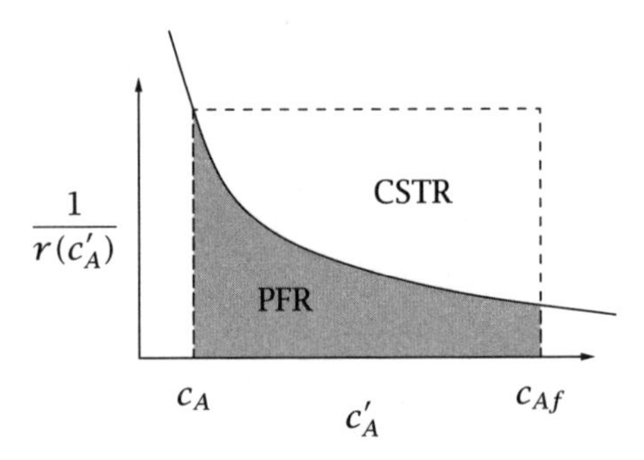

Figure 4.24: Inverse of reaction rate versus concentration; area under curve is the PFR residence time τ; area of the rectangle is the CSTR residence time; PFR is more efficient; irreversible, nth-order kinetics with positive order, $n > 0$.

result. The area under the curve $1/r(c'_A)$ is the total time required to achieve the desired concentration change.

To achieve this same concentration change in the CSTR, we start with Equation 4.40, and solve for τ giving

$$\tau = \frac{c_{Af} - c_A}{r(c_A)}$$

This result also can be interpreted as an area in Figure 4.24. Notice that this area is the height, $1/r(c_A)$, times the width, $c_{Af} - c_A$, of the rectangle shown in Figure 4.24. Obviously if $1/r(c_A)$ is a *decreasing* function of c_A, or, equivalently, $r(c_A)$ is an *increasing* function of c_A, to achieve the same conversion, the PFR time (or volume, $V_R = Q_f\tau$) is less than the CSTR time (volume). We make a final comment to reinforce this point. The reaction rate for the PFR varies with length. The rate is high at the entrance to the tube where the concentration of A is equal to the feed value, and decreases with length as the concentration drops. At the exit of the PFR, the rate is the lowest of any location in the tube. Now considering that the entire volume of the CSTR is reacting at this *lowest* rate of the PFR, it seems intuitively clear that more volume is required for the CSTR to achieve the same conversion as the PFR.

So for this simple, single-reactant case, we only need to know if $r(c_A)$ is an increasing or decreasing function of c_A to know which reactor is more efficient, i.e., requires less volume. If the reaction order is positive (the usual case), the PFR is more efficient, as shown in Figure 4.24. If the reaction order is negative, the CSTR is more efficient,

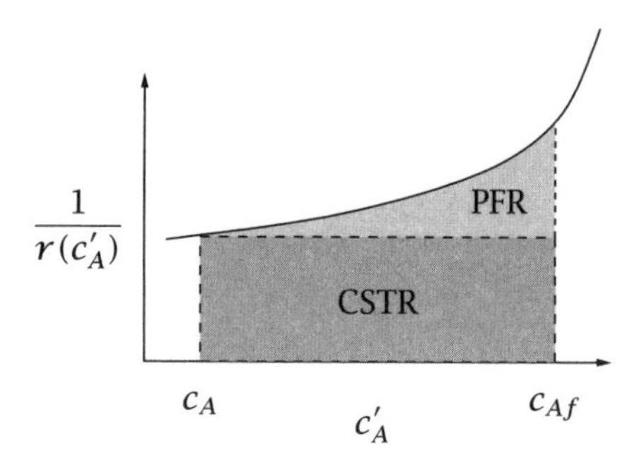

Figure 4.25: To achieve the same conversion, the CSTR is smaller than the PFR for irreversible, nth-order kinetics, negative order, $n < 0$.

as shown in Figure 4.25.

If the reaction rate is neither strictly increasing nor strictly decreasing, the reactor configuration with the smallest volume becomes a series of CSTRs and PFRs. Such an example is shown in Chapter 8, Example 8.4. Levenspiel provides further discussion of these interesting cases [12, pp.128–156].

Example 4.8: The PFR versus CSTR with separation

We have noticed that a PFR achieves higher conversion than an equivalent volume CSTR for the irreversible reaction with first-order kinetics

$$\mathrm{A} \longrightarrow \mathrm{B} \qquad r = kc_A$$

Consider the case in which we add separation. Find a single CSTR and separator combination that achieves the same conversion as the PFR. You may assume a perfect separation of A and B, the feed is a pure A stream, and $k\tau = 1$ for the PFR.

Solution

The PFR achieves a fractional conversion of A

$$x_{\mathrm{PFR}} = 1 - N_A/N_{A0} = 1 - \exp(-k\tau) = 0.632$$

For an equivalent volume CSTR without separation, the conversion of A is

$$x_{\mathrm{CSTR}} = 1 - N_A/N_{A0} = k\tau/(1 + k\tau) = 0.5$$

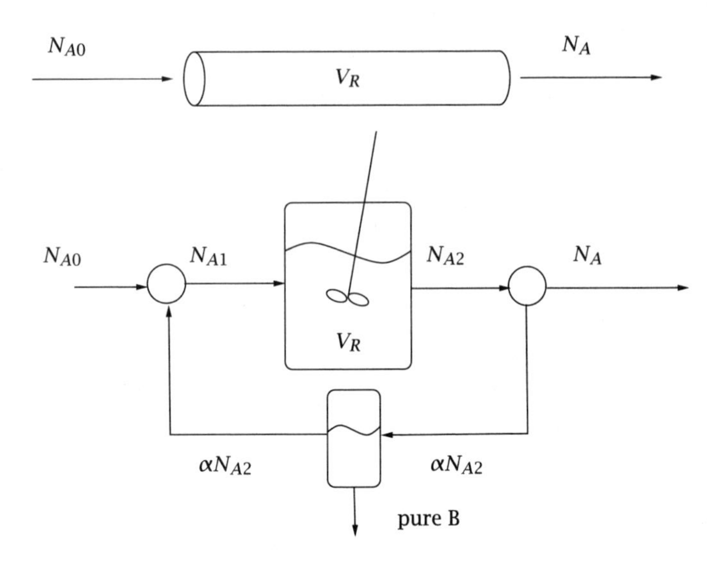

Figure 4.26: PFR versus CSTR with recycle and separation.

The goal is to increase the achievable conversion in the CSTR using separation. Education in chemical engineering principles leads one immediately to consider recycle of the unreacted A as a means to increase this conversion. Consider the flowsheet depicted in Figure 4.26. A fraction of the outflow from the CSTR is recycled, the product B is removed and the unreacted A is combined with the feed as the inflow of the CSTR. Given the assumption of perfect separation, we can achieve essentially complete conversion of A for $k\tau = 1$ with complete recycle, so our goal here is to calculate the fractional recycle, α, that achieves exactly the PFR conversion. For $k\tau < 1$ the achievable conversion is less than one as discussed in Exercise 4.14.

Notice first that the mean residence time of the CSTR, τ', is less than that for the PFR, τ, because the flowrate has increased due to the recycle. Notice that with perfect separation, pure A streams are combined at the mixer, and $Q_0/Q_1 = N_{A0}/N_{A1}$ so

$$\tau' = V_R/Q_1 = (V_R/Q_0)(Q_0/Q_1) = \tau N_{A0}/N_{A1}$$

We may consider four variables to specify the state of the system: $\alpha, N_{A1}, N_{A2}, N_A$; and we can write three component A material balances for the reactor, splitter at reactor exit and mixer at reactor inlet

$$\begin{aligned} \text{reactor:} \quad & N_{A2} = N_{A1}/(1 + k\tau N_{A0}/N_{A1}) \\ \text{splitter:} \quad & N_A = (1-\alpha)N_{A2} \\ \text{mixer:} \quad & N_{A1} = N_{A0} + \alpha N_{A2} \end{aligned}$$

The separator balance is trivial because the separation of A is perfect, and, therefore, the molar flow of A is conserved across the recycle stream. Because the inlet flow of A is not specified, it is convenient to divide the preceding equations by N_{A0}, define dimensionless molar flows, and rearrange to obtain

$$\begin{aligned} \text{reactor:} \quad & \overline{N}_{A2}(1 + k\tau/\overline{N}_{A1}) - \overline{N}_{A1} = 0 \\ \text{splitter:} \quad & \overline{N}_A - (1-\alpha)\overline{N}_{A2} = 0 \\ \text{mixer:} \quad & 1 + \alpha\overline{N}_{A2} - \overline{N}_{A1} = 0 \end{aligned} \tag{4.84}$$

We can specify a single variable as known and solve for the remaining three with the three equations. For example, if we specify the recycle fraction, α, we can solve Equations 4.84 for $\overline{N}_A, \overline{N}_{A1}, \overline{N}_{A2}$, and compute the conversion from $x_A = 1 - \overline{N}_A$. Figure 4.27 shows the resulting conversion of A plotted as a function of α. We can see from Figure 4.27 that the PFR conversion is achieved at about $\alpha = 0.65$. If we want a more accurate answer, we can set $\overline{N}_A = \exp(-k\tau) = 0.3678$ and solve numerically for $\alpha, \overline{N}_{A1}, \overline{N}_{A2}$, with Equations 4.84[1] , and the result is

$$\alpha = 0.6613$$

□

CSTR equivalence principle. Example 4.8 is motivated by a recent result of Feinberg and Ellison called the CSTR equivalence principle of reactor-separator systems [4]. This surprising principle states:

> For a given reaction network with n_i linearly independent reactions, any steady state that is achievable by any reactor-separator design with total reactor volume V is achievable by a design with *not more than $n_i + 1$ CSTRs*, also of total reactor volume V. Moreover the concentrations, temperatures and pressures in the CSTRs are arbitrarily close to those occurring in the reactors of the original design.

[1] Note one can solve this simple problem analytically as well. Eliminate $\alpha\overline{N}_{A2}$ from the second and third equations in Equation 4.84. Substitute the result into the first equation and solve the resulting quadratic equation.

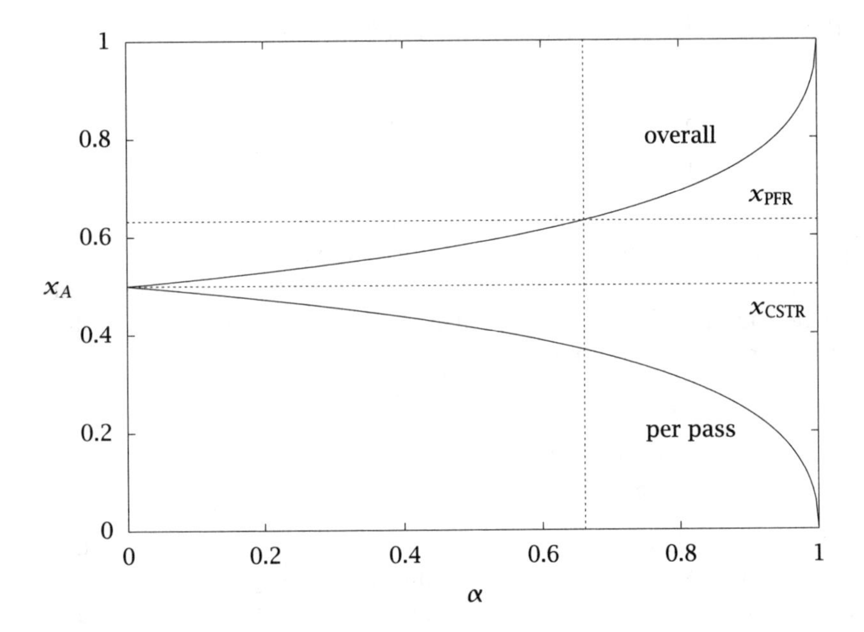

Figure 4.27: Overall and per-pass conversion of A as a function of fractional recycle, α.

Applying this principle to the last example, we know that any achievable concentration of the PFR for a single reaction is achievable with a CSTR and separation. Note the number of CSTRs can be reduced from $n_i + 1$ to n_i in certain situations, such as the one considered in Example 4.8. And we know the concentration in the CSTR will be achieved somewhere in the PFR.

4.8 Stochastic Simulation of Chemical Reactions

We wish to introduce next a topic of increasing importance to chemical engineers, stochastic (random) simulation. In stochastic models we simulate quite directly the random nature of the molecules. We will see that the deterministic rate laws and material balances presented in the previous sections can be captured in the stochastic approach by allowing the numbers of molecules in the simulation to become large. From this viewpoint, deterministic and stochastic approaches are complementary. Deterministic models and solution methods are quite efficient when the numbers of molecules are large and the random behavior is not important. The numerical methods for solution of the nonlinear differential equations of the deterministic models are also highly developed. The stochastic modeling approach is appropriate if the random nature of the system is one of the important features to be captured in the model. These situations are becoming increasingly important to chemical engineers as we explore reactions at smaller and smaller length scales. For example, if we are modeling the chemical transformation by reaction of only a few hundreds or thousands of molecules at an interface, we may want to examine explicitly the random fluctuations taking place. In biological problems, we often consider the interactions of only several hundred or several thousand protein molecules and cells. In sterilization problems, we may wish to model the transient behavior until every last organism is eliminated.

It is perhaps best to illustrate features of the stochastic approach with a simple example. Instead of the common case in which we have on the order of Avogadro's number of reacting molecules, assume we have only a hundred molecules moving randomly in the gas phase and we wish to follow the reaction

$$\mathrm{A} \xrightarrow{k_1} \mathrm{B}$$

$$\mathrm{B} \xrightarrow{k_2} \mathrm{C}$$

in a constant-volume batch reactor. In this section we take reaction statements quite literally. We assume these reactions are not merely observed stoichiometries, but actual molecular events.

The *probability* of a reaction is assumed proportional to the number of combinations of the molecules that can be taken with the reaction stoichiometry. For a uni-molecular reaction, the number of combinations is simply the number of molecules, so the probabilities for the two reactions are[2]

$$r_1 = k_1 x_A, \qquad r_2 = k_2 x_B$$

in which x_j is the *number* of component j molecules in the reactor volume. Note x_j is an integer, unlike c_j of the deterministic model, which is a real number. The reaction probabilities play the role of the rate expressions in the deterministic models. Given the stoichiometry

[2]For the nth-order reaction $n\mathrm{A} \longrightarrow \mathrm{B}$, the number of combinations is

$$\binom{x_A}{n} = \frac{x_A(x_A-1)\cdots(x_A-(n-1))}{n!}$$

and the reaction probabilities, we would like to simulate the expected behavior of the reaction network. One way to accomplish this task is the Gillespie algorithm, which we describe next. The basic idea of the Gillespie algorithm is to: (i) choose randomly the time at which the next reaction occurs, and (ii) choose randomly which reaction occurs at that time. Of course we do not choose arbitrarily. If the total reaction probabilities are large, it is intuitively clear that the time interval until the next reaction should be small, and, if reaction probability r_1 is much larger than r_2, the first reaction is more likely to occur at the next reaction time. The beauty of the Gillespie algorithm is the simple and statistically correct manner in which these two random choices are made.

In a series of papers, Gillespie makes an elegant argument for the use of stochastic simulation in chemical kinetic modeling [7, 9] and provides the following simulation algorithm [7, p.2345].

1. Initialize. Set integer counter n to zero. Set the initial species numbers, $x_j(0), j = 1, \ldots n_s$. Determine stoichiometric matrix $\boldsymbol{\nu}$ and reaction probability laws (rate expressions)

$$r_i = k_i h(x_j)$$

   for all reactions.

2. Compute reaction probabilities, $r_i = k_i h(x_j)$. Compute total reaction probability $r_{tot} = \sum_i r_i$.

3. Select two random numbers, p_1, p_2, from a uniform distribution on the interval $(0, 1)$. Let the time interval until the next reaction be

$$\tilde{t} = -\ln(p_1)/r_{tot} \tag{4.85}$$

   Determine reaction m to take place at this time. The idea here is to partition the interval (0,1) by the relative sizes of each reaction probability and then "throw a dart" at the interval to pick the reaction that occurs. In this manner, all reactions are possible, but the reaction is selected in accord with its probability. See Figure 4.28.

4. Update the simulation time $t(n+1) = t(n) + \tilde{t}$. Update the species numbers for the single occurrence of the mth reaction via

$$x_j(n+1) = x_j(n) + \nu_{mj}, \qquad j = 1, \ldots n_s$$

   Let $n = n + 1$. Return to Step 2.

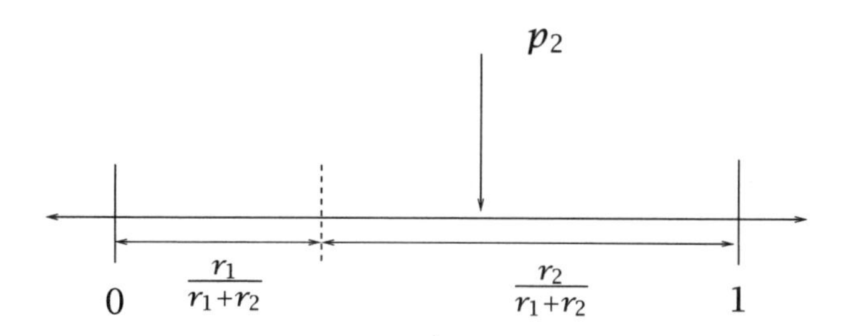

Figure 4.28: Randomly choosing a reaction with appropriate probability; the interval is partitioned according to the relative sizes of the reaction rates; a random number p_2 between zero and one is generated to determine the reaction; in this case, $m = 2$ and the second reaction is selected.

If r_{tot} is the total reaction rate, $r_{\text{tot}}e^{-r_{\text{tot}}\tilde{t}}$ is the probability that a reaction has not occurred during time interval $\tilde{t}$, which leads directly to Equation 4.85 for choosing the time of the next reaction. We will derive this fact in Chapter 8 when we develop the residence-time distribution for a CSTR. Shah, Ramkrishna and Borwanker call this time the "interval of quiescence," and use it to develop a stochastic simulation algorithm for particulate system dynamics rather than chemical kinetics [15].

Figure 4.29 shows the results of this algorithm when starting with $x_A = 100$ molecules. Notice the random aspect of the simulation gives a rough appearance to the number of molecules versus time, which is quite unlike any of the deterministic simulations presented in Section 4.2. In fact, because the *number* of molecules is an integer, the simulation is actually discontinuous with jumps at the reaction times. But in spite of the roughness, we already can make out the classic behavior of the series reaction: loss of starting material A, appearance and then disappearance of the intermediate species B, and slow increase in final product C. Note also that Figure 4.29 is only *one* simulation of the stochastic model. Unlike the deterministic models, if we repeat this simulation, we obtain a different sequence of random numbers and a different simulation. To talk about expected or average behavior of the system, we must perform many of these random simulations and then compute the averages of quantities we wish to report.

Next we explore the effect of increasing the initial number of A molecules on a single simulation [8, p.371]. The results for 1000 and

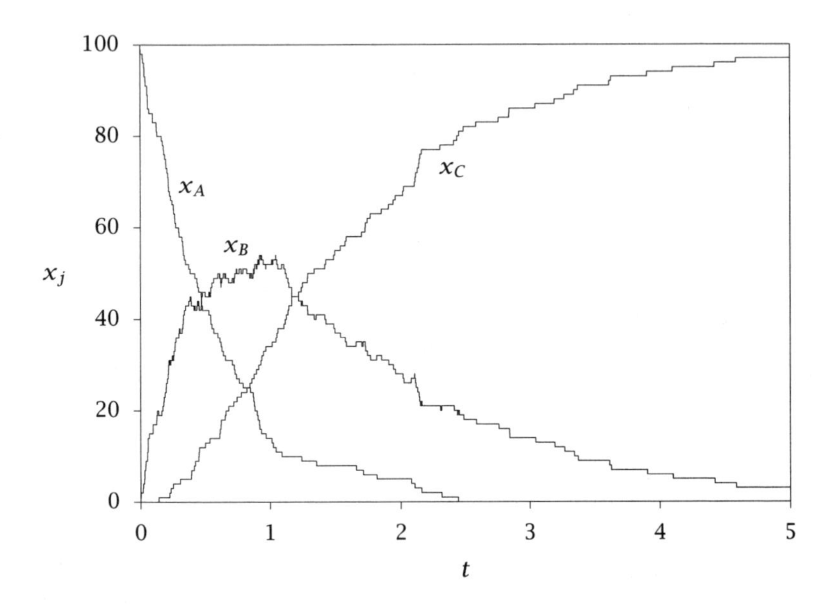

Figure 4.29: Stochastic simulation of the first-order reactions A⟶B⟶C starting with 100 A molecules.

4000 initial A molecules are shown in Figures 4.30 and 4.31, respectively. We see the random fluctuations become less pronounced. Notice that even with only 4000 starting molecules, Figure 4.31 compares very favorably with the deterministic simulation shown in Figure 4.11 of Section 4.2.

Another striking feature of the stochastic approach is the trivial level of programming effort required to make the simulations. In fact, the biggest numerical challenge is producing the random numbers,[3] and many well-developed algorithms are available for that task. The computational time required for performing the stochastic simulation may, however, be large. The solution time depends on the number of simulation steps, and also on whether or not we must repeat the simulations to calculate averages. Usually large numbers of simulation steps are chosen when one has large numbers of initial molecules. If reliable deterministic rate laws are available, at some point it becomes more efficient to use the deterministic models as the number of molecules

[3]It is more accurate to use the term pseudorandom number here to distinguish something we compute from a truly random number.

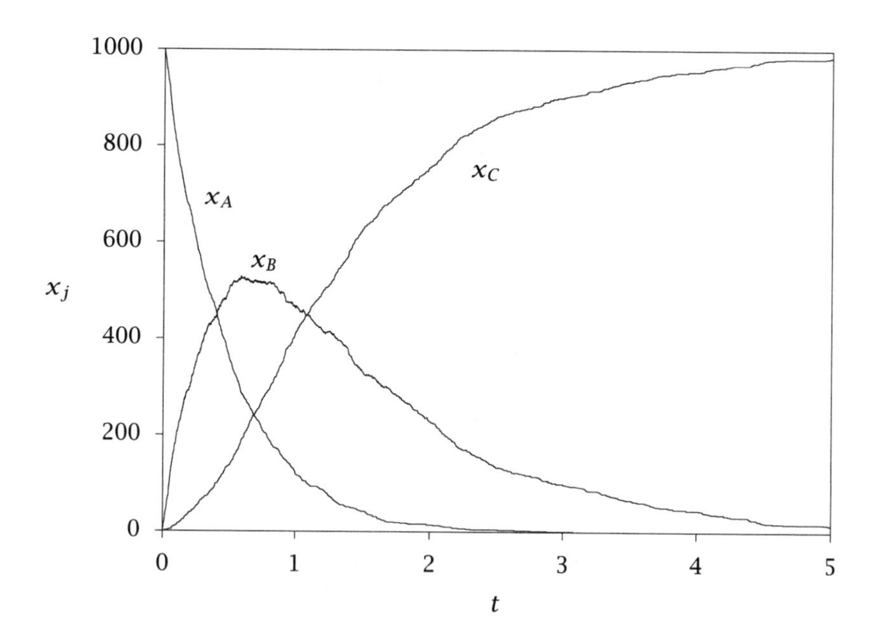

Figure 4.30: Stochastic simulation of the first-order reactions A⟶B⟶C starting with 1000 A molecules.

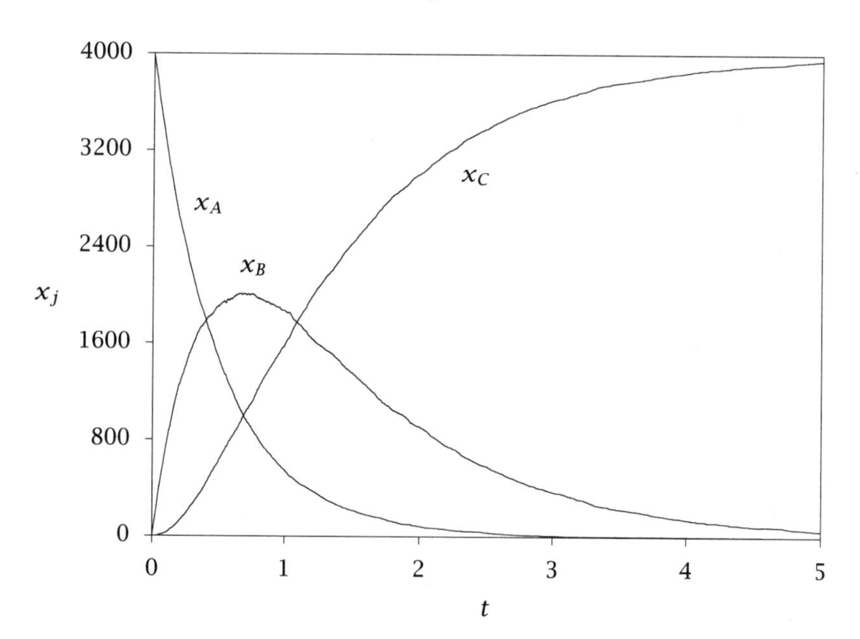

Figure 4.31: Stochastic simulation of the first-order reactions A⟶B⟶C starting with 4000 A molecules.

increases.

But the stochastic approach is invaluable in several ways. It builds a clear intuitive connection between the microscopic and the macroscopic. The microscopic level is characterized by discontinuous, random molecular motion and the probability of collision as the basis for chemical reaction rate. The macroscopic level is characterized by smoothly varying concentrations, and deterministic rate laws and material balances. Watching the transition in Figures 4.29–4.31 and then finally to the deterministic Figure 4.11 is a nice illustration of this connection and provides logical support for the construction of the deterministic rate laws. It is possible to prove that the average of stochastic simulations converges to the deterministic simulation as the number of molecules becomes large, which is known as the thermodynamic limit.

As stressed earlier, the random fluctuations may be an important physical behavior to include in the model. In this situation, the stochastic approach is essential and a deterministic approach cannot be substituted. We illustrate with the hepatitis B virus model introduced in Chapter 1.

Example 4.9: Stochastic versus deterministic simulation of a virus model

Consider the hepatitis B virus model given in Chapter 1.

$$\text{nucleotides} \xrightarrow{\text{cccDNA}} \text{rcDNA} \tag{4.86}$$

$$\text{nucleotides} + \text{rcDNA} \longrightarrow \text{cccDNA} \tag{4.87}$$

$$\text{amino acids} \xrightarrow{\text{cccDNA}} \text{envelope} \tag{4.88}$$

$$\text{cccDNA} \longrightarrow \text{degraded} \tag{4.89}$$

$$\text{envelope} \longrightarrow \text{secreted or degraded} \tag{4.90}$$

$$\text{rcDNA} + \text{envelope} \longrightarrow \text{secreted virus} \tag{4.91}$$

Assume the system starts with a single cccDNA molecule, and no rcDNA and no envelope protein, and use the following rate constants

$$\begin{bmatrix} x_A & x_B & x_C \end{bmatrix}^T = \begin{bmatrix} 1 & 0 & 0 \end{bmatrix}^T \tag{4.92}$$

$$\boldsymbol{k}^T = \begin{bmatrix} 1 & 0.025 & 1000 & 0.25 & 2 & 7.5 \times 10^{-6} \end{bmatrix} (\text{day}^{-1}) \tag{4.93}$$

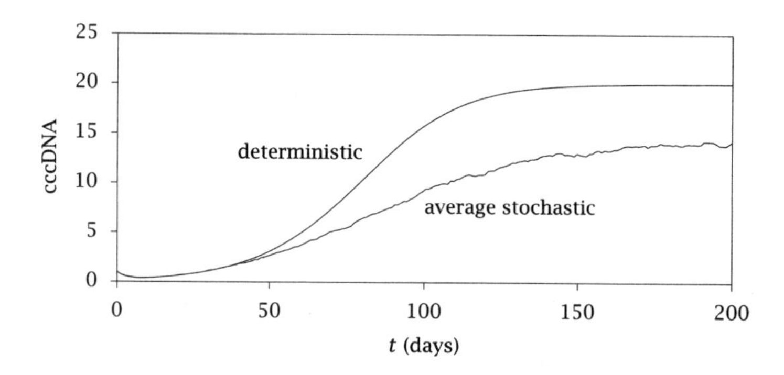

Figure 4.32: Species cccDNA versus time for hepatitis B virus model; deterministic and average stochastic models.

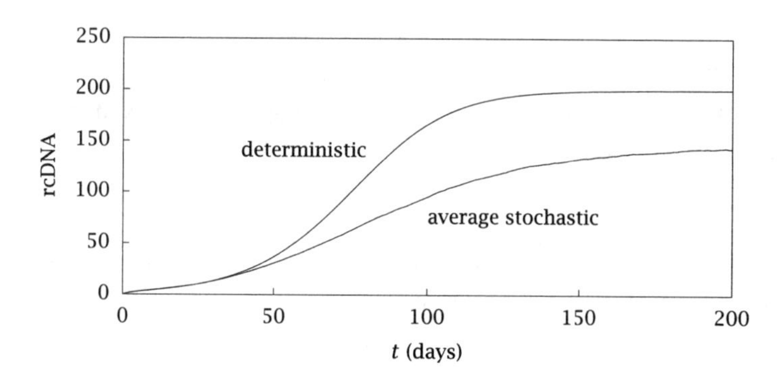

Figure 4.33: Species rcDNA versus time for hepatitis B virus model; deterministic and average stochastic models.

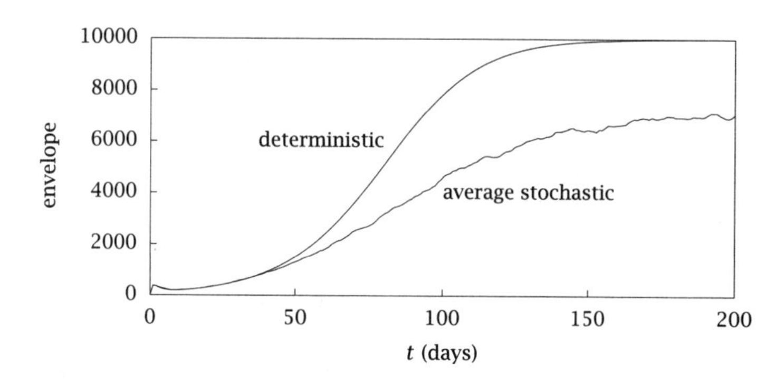

Figure 4.34: Envelope versus time for hepatitis B virus model; deterministic and average stochastic models.

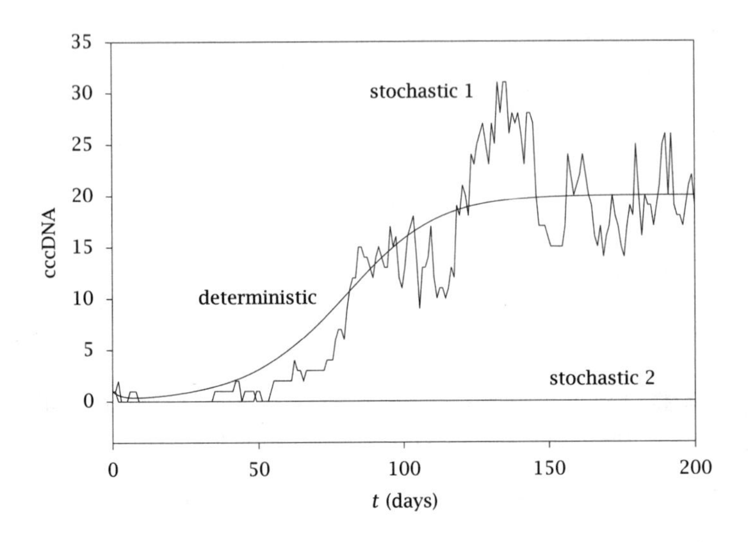

Figure 4.35: Species cccDNA versus time for hepatitis B virus model; two representative stochastic trajectories.

Compare the results of a deterministic simulation to the average of 500 stochastic simulations. If these results are not the same, explain why not.

Solution

The reaction rates and production rates for Reactions 4.86–4.91 are given by

$$\begin{bmatrix} r_1 \\ r_2 \\ r_3 \\ r_4 \\ r_5 \\ r_6 \end{bmatrix} = \begin{bmatrix} k_1 x_A \\ k_2 x_B \\ k_3 x_A \\ k_4 x_A \\ k_5 x_C \\ k_6 x_B x_C \end{bmatrix} \qquad \begin{bmatrix} R_A \\ R_B \\ R_C \end{bmatrix} = \begin{bmatrix} r_2 - r_4 \\ r_1 - r_2 - r_6 \\ r_3 - r_5 - r_6 \end{bmatrix} \tag{4.94}$$

in which A is cccDNA, B is rcDNA, and C is envelope.

Figures 4.32–4.34 show the deterministic model simulation and an average of 500 stochastic simulations. Notice these results are *not* the same, and we should investigate why not. Figure 4.35 shows *two* representative stochastic simulations for only the cccDNA species. Notice the first stochastic simulation does fluctuate around the deterministic simulation as expected. The second stochastic simulation, however, shows complete extinction of the virus. That is another possible steady state for the stochastic model. In fact, it occurs for 125 of the 500 simulations. So the *average* stochastic simulation in Figures 4.32–4.34 consist of 75% trajectories that fluctuate about the deterministic trajectory and 25% trajectories that go to zero. The two types of stochastic trajectories therefore explain why the average stochastic model is not equal to the deterministic model. We should bear this feature in mind when using deterministic models with small numbers of molecules. The interested reader may wish to consult [16] for further discussion of this virus model.

□

4.9 Summary

We have introduced four main reactor types in this chapter: the batch reactor, the continuous-stirred-tank reactor (CSTR), the semi-batch reactor, and the plug-flow reactor (PFR). Table 4.3 summarizes the mole balances for these four reactors. We also have introduced some of the basic reaction-rate expressions:

- first order, irreversible
- first order, reversible
- second order, irreversible
- nth order, irreversible
- two first-order reactions in series
- two first-order reactions in parallel
- two second-order, reversible reactions

We developed the equations required to compute the volume of the reactor if there is a significant volume change upon reaction. We require an equation of state for this purpose. Tables 4.1 and 4.2 describe the appropriate balances for a constant-density mixture, an ideal mixture, and a mixture with a general equation of state.

Several of these simple mass balances with basic rate expressions were solved analytically. In the case of multiple reactions with nonlinear rate expressions (i.e., not first-order reaction rates), the balances must be solved numerically. A high-quality ordinary differential equation (ODE) solver is indispensable for solving these problems. For a

BATCH	$\dfrac{d(c_jV_R)}{dt} = R_jV_R$	(4.95)
constant volume	$\dfrac{dc_j}{dt} = R_j$	(4.96)
CSTR	$\dfrac{d(c_jV_R)}{dt} = Q_fc_{jf} - Qc_j + R_jV_R$	(4.97)
constant density	$\dfrac{dc_j}{dt} = \dfrac{1}{\tau}(c_{jf} - c_j) + R_j$	(4.98)
steady state	$c_j = c_{jf} + R_j\tau$	(4.99)
SEMI-BATCH	$\dfrac{d(c_jV_R)}{dt} = Q_fc_{jf} + R_jV_R$	(4.100)
PFR	$\dfrac{\partial c_j}{\partial t} = -\dfrac{\partial(c_jQ)}{\partial V} + R_j$	(4.101)
steady state	$\dfrac{d(c_jQ)}{dV} = R_j$	(4.102)
constant flowrate	$\dfrac{dc_j}{d\tau} = R_j, \qquad \tau = V/Q_f$	(4.103)

Table 4.3: Summary of mole balances for several ideal reactors.

complex equation of state and nonconstant-volume case, a differential-algebraic equation (DAE) solver may be convenient.

We showed that the PFR achieves higher conversion than the CSTR of the same volume if the reaction rate is an increasing function of a component composition ($n > 0$ for an nth-order rate expression). Conversely, the CSTR achieves higher conversion than the same-volume PFR if the rate is a decreasing function of a component composition ($n < 0$).

Finally, we introduced stochastic simulation to model chemical reactions occurring with *small* numbers of molecules. Each of these random simulation trajectories has a rough appearance and the average of many of these simulations is required to show the expected system behavior. The stochastic model uses basic probability to compute reaction rate. The probability of occurrence of a given reaction is assumed proportional to the number of possible combinations of reactants for the given stoichiometry. Two pseudorandom numbers are chosen to determine: (i) the time of the next reaction and (ii) the reaction that occurs at that time. The smooth behavior of the macroscopic ODE models is recovered by the random simulations in the limit of large numbers of reacting molecules. With small numbers of molecules, however, the average of the stochastic simulation does not have to be equal to the deterministic simulation. We demonstrated this fact with the simple, nonlinear hepatitis B virus model.

Notation

A_c	reactor tube cross-sectional area
A_i	preexponential factor for rate constant i
c_j	concentration of species j
c_{jf}	feed concentration of species j
c_{js}	steady-state concentration of species j
c_{j0}	intial concentration of species j
E_i	activation energy for rate constant i
k_i	reaction rate constant for reaction i
K_i	equilibrium constant for reaction i
M	total mass flow, $N_j M_j$
M	total monomer mass
M_f	total feed mass flow, $N_j M_j$
M_j	molecular weight of species j
n	reaction order
n_j	moles of species j, $V_R c_j$
n_r	number of reactions in the reaction network
n_s	number of species in the reaction network
N_j	molar flow of species j, Qc_j
N_{jf}	feed molar flow of species j, Qc_j
p	uniformly distributed random number on $(0, 1)$
P	pressure
$\widetilde{P}$	total polymer mass
Q	volumetric flowrate
Q_f	feed volumetric flowrate
r_i	reaction rate for ith reaction
r_{tot}	total reaction rate, $\sum_i r_i$
R	gas constant
R_j	production rate for jth species
$\widetilde{R}_P$	mass production rate of polymer
t	time

$\tilde{t}$	time interval in a stochastic simulation
T	temperature
V	reactor volume variable
$\overline{V}_j$	partial molar volume of species j
V_j°	specific molar volume of species j
V_R	reactor volume
ΔV_i	change in volume upon reaction i, $\sum_j \nu_{ij}\overline{V}_j$
x_j	number of molecules of species j in a stochastic simulation
x_j	molar conversion of species j
y_j	mole fraction of gas-phase species j
z	reactor length variable
ε_i	extent of reaction i
ν_{ij}	stoichiometric coefficient for species j in reaction i
ρ	mass density
τ	reactor residence time, V_R/Q_f

4.10 Exercises

Exercise 4.1: Order versus rate of reaction

Liquid A decomposes with nth-order kinetics in a batch reactor

$$\mathrm{A} \longrightarrow \mathrm{B}, \qquad r = kc_A^n$$

The conversion of A reaches 50% in a five-minute run.

What is the order of the reaction if it takes 10 minutes to reach 75% conversion? What is the order of the reaction if it takes 20 minutes to reach 75% conversion? What is the order of the reaction if it takes 30 minutes to reach 75% conversion?

Exercise 4.2: Constant volume versus constant pressure batch reactor

Consider the following two well-mixed, isothermal gas-phase batch reactors for the elementary and irreversible decomposition of A to B,

$$\mathrm{A} \xrightarrow{k} 2\mathrm{B}$$

reactor 1: The reactor volume is held constant (reactor pressure therefore changes).

reactor 2: The reactor pressure is held constant (reactor volume therefore changes).

Both reactors are charged with pure A at 1.0 atm and $k = 0.35\ \mathrm{min}^{-1}$.

(a) What is the fractional decrease in the *concentration* of A in reactors 1 and 2 after five minutes?

(b) What is the *total molar conversion* of A in reactors 1 and 2 after five minutes?

Exercise 4.3: CSTR performance

A liquid-phase reaction

$$A + B \xrightarrow{k} C, \qquad r = kc_B c_A^2$$

takes place in a CSTR of volume V_R in the presence of a large excess of reactant B. Assume the reactor achieves 50% conversion of A at steady state.

(a) What is the steady-state conversion if the original reactor is replaced by two CSTRs of volume $V_R/2$ in series?

(b) What is the conversion if the original reactor is replaced by three CSTRs of volume $V_R/3$ in series?

(c) What is the conversion if the original reactor is replaced by a plug-flow reactor of volume V_R?

Exercise 4.4: Catalyst deactivation in a batch reactor

Consider the irreversible, liquid-phase isomerization reaction carried out in a solvent containing dissolved catalyst at 25°C in a batch reactor

$$A \xrightarrow{k_a} B$$

The *apparent* first-order reaction-rate constant, k_a, decreases with time because of catalyst deterioration. A chemist colleague of yours has studied the catalyst deactivation process and has proposed that it can be modeled by

$$k_a = \frac{k}{1 + k_d t}$$

in which k is the fresh catalyst rate constant and k_d is the deactivation-rate constant.

(a) Write down the mole balance for this reactor.

(b) Solve the mole balance for $c_A(t)$. Sketch your solution.

(c) If it takes two hours to reach 50% conversion and the fresh catalyst has a rate constant of 0.6 hr^{-1}, what is k_d?

(d) How long does it take to reach 75% conversion?

Exercise 4.5: Dynamic CSTR

A CSTR is used to convert A to products B and C via the following liquid-phase reaction

$$A \xrightarrow{k} B + C$$

The reaction is first order in A and irreversible. The tank initially is charged with species A at concentration c_{A0}. At time zero, the feed pump is turned on and delivers constant flowrate, Q_f. The feed concentration of A is c_{Af}, which is also constant. The tank volume is V_R. Liquid density change due to reaction may be neglected.

(a) Write down and solve the dynamic material balance for component A.

(b) Sketch the solution, $c_A(t)$ versus t, for the tank initially filled with solvent, $c_{A0} = 0$. On the same plot, sketch the solution for the tank initially filled with feed, $c_{A0} = c_{Af}$. Clearly label on your plot the initial and steady-state concentrations for both curves.

(c) For a 50 m^3 tank with flowrate of 7 L/s and rate constant $k = 0.02\ \text{min}^{-1}$, what is the steady-state conversion of A?

Exercise 4.6: Linear differential equations and Laplace transforms

Consider the linear, first-order, constant-coefficient differential equation

$$\frac{dx}{dt} = Ax + b$$
$$x(0) = x_0 \qquad (4.104)$$

The Laplace transform is a handy method for solving linear differential equations.

(a) Take the Laplace transform of Equation 4.104 and show

$$\overline{x}(s) = \frac{x_0}{s - A} + \frac{b}{s(s - A)}$$

(b) Invert the transform and show

$$x(t) = x_0 e^{At} - \frac{b}{A}\left[1 - e^{At}\right] \qquad (4.105)$$

You may want to use the following partial fraction formula

$$\frac{1}{s(s - A)} = -\frac{1}{A}\left[\frac{1}{s} - \frac{1}{s - A}\right]$$

(c) Substitute Equation 4.105 into Equation 4.104 and show that it satisfies the differential equation and initial condition.

(d) Identify A, b, so that Equation 4.104 applies to the reversible, first-order reaction case given in Equation 4.13. Substitute these values into Equation 4.105 and verify Equation 4.15.

(e) Identify A, b, so that Equation 4.104 applies to the reversible, first-order reaction case given in Equation 4.41. Substitute these values into Equation 4.105 and verify Equation 4.42.

Exercise 4.7: Multiple reactions in CSTRs in series

The following liquid-phase reactions take place in two identical CSTRs in series,

$$\text{A} \longrightarrow \text{B} + \text{C} \qquad r_1 = k_1 c_A^2$$
$$\text{B} \longrightarrow \text{D} \qquad r_2 = k_2 c_B$$

in which $k_1 = 0.05$ L/mol·min and $k_2 = 0.022\ \text{min}^{-1}$. The feed to the first reactor is pure A, $c_{Af} = 1.5$ mol/L, and the residence time in each reactor is 15 min. Determine the overall conversion of A and the yield of B for the series of reactors. Recall yield is the fraction of reactant converted into a specified product.

Exercise 4.8: Nonconstant density with a liquid-phase reaction

Propylene glycol is produced by the hydrolysis of propylene oxide according to the following reaction

$$\text{propylene oxide} + H_2O \longrightarrow \text{propylene glycol}$$

In the presence of excess water, the reaction has been found to be first-order in propylene oxide

$$r = kc_{PO}$$

and the rate constant is [6]

$$k = k_0 e^{-E_a/RT} \qquad k_0 = 4.71 \times 10^9 \ \text{s}^{-1} \qquad E_a = 18.0 \ \text{kcal/mol}$$

Methanol is added as a solvent, and the reaction is performed in a 1000 L CSTR operating at 60°C. The feed conditions and physical properties are as follows [17]:

Component	Density (g/cm^3)	Mol. wt. (g/mol)	Inlet feedrate (L/hr)
propylene oxide	0.859	58.08	1300
water	1.000	18.02	6600
propylene glycol	1.0361	76.11	0
methanol	0.7914	32.04	1300

Assume the mixture is ideal so that

$$1 = \sum_j c_j V_j^\circ$$

in which $V_j^\circ = M_j/\rho_j^\circ$ are the pure component specific molar volumes. Neglect any change in the pure component densities with temperature in the temperature range 25–60°C.

(a) Compute the steady-state concentrations of all components, Q, and V_R for the following two situations.

1. A float in the top of the tank is used to adjust Q to maintain reactor volume constant at 1000 L.
2. The reactor is initially charged with pure solvent, and a differential pressure measurement is used to adjust Q to maintain constant reactor mass.

Which operation do you recommend, constant volume or constant mass? Look at the conversion of propylene oxide and the total reactor production rate of propylene glycol for the two cases. What are you wasting in constant mass operation?

(b) Resolve the constant reactor volume operation under the assumption that all component densities are equal to the density of water. How much error in the conversion and production rate do you commit under this assumption?

Exercise 4.9: Linear density versus concentration assumption

Consider the important case of liquid-phase reaction in an excess of solvent. Let component $j = s$ be the solvent and consider the solution density at some reference concentrations, $c_j = c_{j0}$,

$$\rho_0 = c_{s0} M_s + \sum_{j \neq s} c_{j0} M_j$$

In this situation, a useful approximation for the density-concentration relationship is the linear form [14]

$$\rho = \rho_0 + \sum_{j \neq s} \left(c_j - c_{j0}\right) M_j$$

or, equivalently, upon substitution of the previous equation

$$\rho = M_s c_{s0} + \sum_{j \neq s} c_j M_j \tag{4.106}$$

(a) Show that even though the density is not assumed constant in Equation 4.106, one still obtains for the reactor volume in a CSTR

$$\frac{dV_R}{dt} = Q_f - Q$$

(b) Show that the approximation error in the density in Equation 4.106 is

$$(c_s - c_{s0})\, M_s$$

in which c_s is the actual solvent concentration in the mixture, which also changes, in general, upon reaction even though the solvent does not take part in any reactions.

Exercise 4.10: Maximizing an intermediate

We considered the following two first-order, irreversible reactions in Reaction 4.28

$$\mathrm{A} \xrightarrow{k_1} \mathrm{B}$$

$$\mathrm{B} \xrightarrow{k_2} \mathrm{C}$$

and derived Equation 4.30 for the intermediate B concentration

$$c_B = c_{B0} e^{-k_2 t} + c_{A0} \frac{k_1}{k_2 - k_1} \left[e^{-k_1 t} - e^{-k_2 t}\right], \qquad k_1 \neq k_2$$

(a) What is the optimal stopping time in order to maximize the concentration of intermediate B starting with pure A?

(b) Check your result against Figure 4.11.

(c) How does this result simplify for the case $k_1 = k_2$?

Exercise 4.11: Some limits

(a) For the second-order, irreversible reaction

$$\mathrm{A} + \mathrm{B} \xrightarrow{k} \mathrm{C} \qquad r = k c_A c_B$$

in the batch reactor, we showed solving the material balance produces Equation 4.24

$$c_A = (c_{A0} - c_{B0})\left[1 - \frac{c_{B0}}{c_{A0}} e^{(c_{B0}-c_{A0})kt}\right]^{-1} \qquad c_{A0} \neq c_{B0}$$

Take the limit $c_{A0} - c_{B0} \to 0$ in this equation and show it produces the other second-order reaction result given in Equation 4.19

$$c_A = \left(\frac{1}{c_{A0}} + kt\right)^{-1}$$

(b) For the n-order, irreversible chemical reaction

$$\text{A} \longrightarrow \text{B} \qquad r = kc_A^n$$

in the batch reactor, we showed solving the material balance produces Equation 4.25

$$\frac{c_A}{c_{A0}} = [1 + (n-1)k_0 t]^{\frac{1}{-n+1}}, \qquad n \neq 1$$

in which

$$k_0 = kc_{A0}^{n-1}$$

Take the limit $n \to 1$ in this equation and show that it produces the first-order reaction result

$$\frac{c_A}{c_{A0}} = e^{-kt}, \qquad n = 1$$

Exercise 4.12: Constant gas flowrate; large or small error?

One of the interesting features of (ideal) gas molecules is that they demand a certain amount of "territory." It doesn't matter if they are heavy or light, big or small — all gas molecules occupy the same amount of space at a given temperature and pressure; hence the ideal gas law.

Consider the following hypothetical reaction taking place in the gas phase in an isothermal, constant pressure PFR with a pure A feed stream

$$\text{A} \longrightarrow b\,\text{B} \tag{4.107}$$

Derive an expression for the tube volume required to achieve molar conversion of A, x_A, for the following two cases.

(a) Ideal gas equation of state, which we call the correct answer, V_{Rc}.

(b) Constant volumetric flowrate, which we call the dubious assumption, V_{Rd}.

(c) Compute the fractional error committed when using the dubious assumption

$$e = \frac{V_{Rc} - V_{Rd}}{V_{Rc}}$$

Your result should be a function of only x_A and b.

(d) Plot e versus x_A for $b = 1, 2, 5, 100$. Note for $b = 100$, Reaction 4.107 is very hypothetical.

(e) If you are unwilling to commit errors larger than 10%, for what b and x_A values must you object to the dubious assumption?

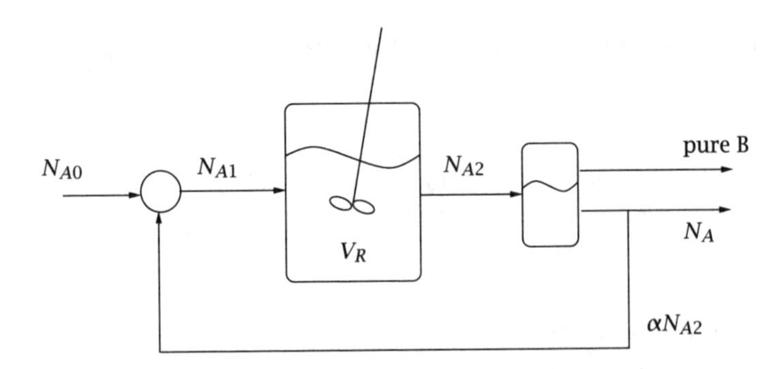

Figure 4.36: An alternative CSTR-separator design.

Bottom line: are you generally making large or small errors when you neglect the change in the number of moles when modeling a gas-phase reaction in a PFR?

Exercise 4.13: Ethane pyrolysis, revisited

Using the reaction chemistry and kinetics presented in Example 4.7, change the feed concentration of NO from 5% to 5 ppm, and examine the effect on the concentration profile for ethane in the PFR at 1050 K. Assume the feed consists only of ethane and NO, and that the pressure and volumetric flowrate are the same as in Example 4.7. Explain the effect of the feed NO concentration on the ethane conversion.

Exercise 4.14: Recycle effects

Consider again Example 4.8, and compute the limiting conversion for a CSTR with recycle for $k\tau = 0.5$. What is the corresponding limiting fractional recycle, α? What happens if one attempts to achieve a higher conversion than this limiting value by recycling more material?

Exercise 4.15: An alternative reactor-separator design

Feinberg and Ellison considered an alternative design for Example 4.8 shown in Figure 4.36. Notice the separator is placed ahead of the recycle split in this design, and after the recycle split in Figure 4.26.

(a) Calculate the flowrate of the recycle stream that achieves the PFR conversion given in Example 4.8.

(b) Compare the two reactor-separator designs. What are the advantages and disadvantages of each design?

Exercise 4.16: Stochastic simulation of a virus model

Perform several random simulations of the virus model given in Equations 4.86–4.93. Perform enough simulations so that some of your trajectories lead to extinction of all three species: cccDNA, rcDNA and envelope protein.

(a) Compare and contrast a simulation that leads to a steady-state infection, and a simulation that leads to extinction.

(b) During the first 25–50 days, is it possible to tell by looking at the trajectory whether it will lead to infection or extinction? About how many days are required before you have a reliable idea of where a trajectory will finish?

(c) Of the trajectories that lead to extinction, what is the longest time required to reach extinction? What fraction of your trajectories lead to extinction?

(d) Which parameters in the model have the largest affect on the behavior?

Exercise 4.17: Stochastic simulation for nonlinear kinetics

Consider the reversible, second-order reaction

$$\mathrm{A} + \mathrm{B} \underset{k_{-1}}{\overset{k_1}{\rightleftharpoons}} \mathrm{C} \qquad r = k_1 c_A c_B - k_{-1} c_C$$

(a) Solve the deterministic material balance for a constant-volume batch reactor with

$$k_1 = 1 \text{ L/mol}\cdot\text{min} \qquad k_{-1} = 1 \text{ min}^{-1}$$
$$c_A(0) = 1 \text{ mol/L} \qquad c_B(0) = 0.9 \text{ mol/L} \qquad c_C(0) = 0 \text{ mol/L}$$

Plot the A, B and C concentrations out to $t = 5$ min.

(b) Compare the result to a stochastic simulation using an initial condition of 400 A, 360 B and zero C molecules. Notice from the units of the rate constants that k_1 should be divided by 400 to compare simulations. Figure 4.37 is a representative comparison for one sequence of pseudorandom numbers.

(c) Repeat the stochastic simulation for an initial condition of 4000 A, 3600 B, zero C molecules. Remember to scale k_1 appropriately. Are the fluctuations noticeable with this many starting molecules?

Exercise 4.18: From microscopic to macroscopic

It is often easier to provide a correct explanation than it is to find the error in someone else's incorrect explanation. Consider the following situation from an instructor's point of view. A student in your class asks you to explain the following difficulty he is having. "I start with the continuity equation for species j,"

$$\frac{\partial c_j}{\partial t} + \nabla \cdot (c_j \boldsymbol{v}_j) - R_j = 0$$

"Since this equation is true at all points in the volume element, I integrate it over the CSTR reactor volume element (as depicted in Figure 4.13),"

$$\int_V \left[\frac{\partial c_j}{\partial t} + \nabla \cdot (c_j \boldsymbol{v}_j) - R_j \right] d\Omega = 0$$

"I then use the divergence theorem that we learned in the Transport Phenomena course to convert the divergence term to a surface integral and obtain,"

$$\int_V \left[\frac{\partial c_j}{\partial t} - R_j \right] d\Omega + \int_S c_j (\boldsymbol{v}_j \cdot \boldsymbol{n}) d\sigma = 0 \qquad (4.108)$$

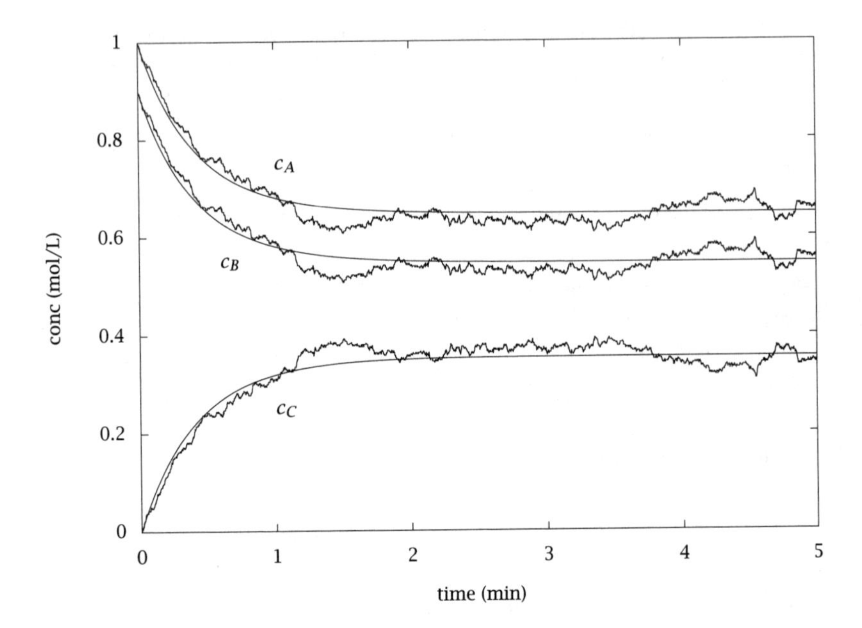

Figure 4.37: Deterministic simulation of reaction A + B $\rightleftharpoons$ C compared to stochastic simulation starting with 400 A molecules.

"It is clear to me that the surface integral picks up the flux of species j into and out of the reactor with the flow streams,"

$$\int_S c_j(\boldsymbol{v}_j \cdot \boldsymbol{n})d\sigma = -Q_f c_{jf} + Qc_j \tag{4.109}$$

"Substituting this into Equation 4.108 gives"

$$\int_V \left[\frac{\partial c_j}{\partial t} - R_j\right] d\Omega - Q_f c_{jf} + Qc_j = 0$$

"Finally since c_j and R_j don't vary over the well-mixed reactor, I simplify to"

$$V_R \frac{dc_j}{dt} = Q_f c_{jf} - Qc_j + R_j V_R$$

"But this doesn't agree with Equation 4.97!" How do you respond to this student? Remember, he does not really want to hear a restatement of why Equation 4.97 is correct; he really wants to know what is wrong with his derivation.

Exercise 4.19: Index and DAEs

We saw in Section 4.5 that it was convenient to specify reactor models for nonconstant-density cases by using DAEs instead of ODEs. Compare, for example, the complexity of the ODE and DAE models appearing in Table 4.2. Some care should be exercised, however, to avoid creating a DAE that is difficult to solve. The DAEs in Table 4.2 are of the following form in which the time derivatives of only the $\boldsymbol{x}$ variables appear explicitly in the model,

$$\frac{d\boldsymbol{x}}{dt} = \boldsymbol{f}(\boldsymbol{x}, \boldsymbol{y}) \tag{4.110}$$

$$\boldsymbol{0} = \boldsymbol{g}(\boldsymbol{x}, \boldsymbol{y}) \tag{4.111}$$

The total moles of the species, n_j, are the $\boldsymbol{x}$ variables and V_R and Q are the $\boldsymbol{y}$ variables.

$$\boldsymbol{x}^T = [n_1 \; n_2 \; \cdots n_{n_s}] \qquad \boldsymbol{y}^T = [V_R \; Q]$$

The index of a DAE is the number of times that the algebraic equations must be differentiated with respect to t to determine $d\boldsymbol{y}/dt$ [3]. An ODE, with no algebraic relations, has index zero. Differentiating Equation 4.111 with respect to time produces

$$\boldsymbol{0} = \frac{\partial \boldsymbol{g}(\boldsymbol{x}, \boldsymbol{y})}{\partial \boldsymbol{x}^T}\frac{d\boldsymbol{x}}{dt} + \frac{\partial \boldsymbol{g}(\boldsymbol{x}, \boldsymbol{y})}{\partial \boldsymbol{y}^T}\frac{d\boldsymbol{y}}{dt}$$

in which $\partial \boldsymbol{g}/\partial \boldsymbol{y}^T$ is a Jacobian matrix with elements

$$\left(\frac{\partial \boldsymbol{g}}{\partial \boldsymbol{y}^T}\right)_{ij} = \frac{\partial g_i}{\partial y_j}$$

If $\partial \boldsymbol{g}/\partial \boldsymbol{y}^T$ is a nonsingular matrix, then we can solve this equation for $d\boldsymbol{y}/dt$ and the DAE has index one. DAEs with index zero and one are generally much easier to solve than those with index two or higher [3]. Notice if the $\boldsymbol{y}$ variables do not appear explicitly in the algebraic equations, for example, then $\partial \boldsymbol{g}/\partial \boldsymbol{y}^T = \boldsymbol{0}$ and the DAE has index two or higher.

Consider the algebraic equations appearing in Table 4.2 for the nonconstant-density cases: 1: reactor volume constant, 2: reactor mass constant, and 3: Q specified. Show that, as expressed in Table 4.2, cases 1 and 2 are high-index DAEs and case 3 is index one. Without differentiating the equation of state, how can case 2 be modified to be index one? Can you find a simple way to modify case 1 to be index one?

Exercise 4.20: Matrix exponential

Consider the general linear kinetic expression written in vector-matrix form

$$\boldsymbol{R}(\boldsymbol{c}) = \boldsymbol{K}\boldsymbol{c}$$

in which $\boldsymbol{R}$ is an n_s-vector of production rates, $\boldsymbol{c}$ is an n_s-vector of concentrations, and $\boldsymbol{K}$ is an $n_s \times n_s$ matrix of rate constants. The batch reactor material balance can be written as

$$\frac{d\boldsymbol{c}}{dt} = \boldsymbol{K}\boldsymbol{c}$$

$$\boldsymbol{c}(0) = \boldsymbol{c}_0 \tag{4.112}$$

The solution to Equations 4.112 can be written as $\boldsymbol{c} = e^{\boldsymbol{K}t}\boldsymbol{c}_0$, in which $e^{\boldsymbol{K}t}$ is the matrix exponential. The matrix exponential can be defined using the power series

$$e^{\boldsymbol{A}} = \boldsymbol{I} + \boldsymbol{A} + \frac{1}{2!}\boldsymbol{A}^2 + \frac{1}{3!}\boldsymbol{A}^3 + \cdots$$

This series converges for all $\boldsymbol{A}$.

(a) Write $\boldsymbol{K}$ for two first-order series reactions in a batch reactor

$$\mathrm{A} \xrightarrow{k_1} \mathrm{B} \qquad \mathrm{B} \xrightarrow{k_2} \mathrm{C}$$

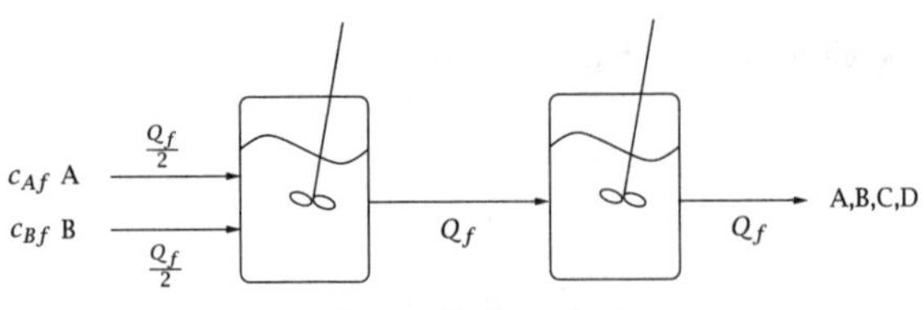

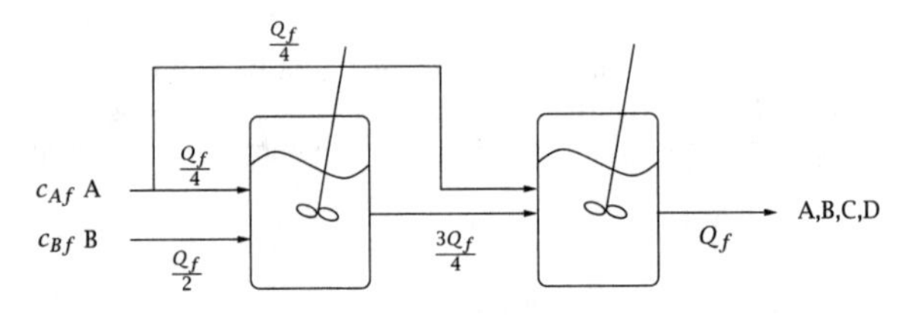

Figure 4.38: Two 2-CSTR reactor configurations with bypass of reactant A in the second configuration.

(b) Write out the series expansion for e^{Kt} and verify the solution derived in Section 4.2.1.

(c) Repeat for two first-order parallel reactions in a batch reactor

$$\mathrm{A} \xrightarrow{k_1} \mathrm{B} \qquad \mathrm{A} \xrightarrow{k_2} \mathrm{C}$$

Exercise 4.21: Reaction probabilities in stochastic kinetics

Consider a stochastic simulation of the following reaction

$$a\,\mathrm{A} + b\,\mathrm{B} \underset{k_{-1}}{\overset{k_1}{\rightleftharpoons}} c\,\mathrm{C} + d\,\mathrm{D}$$

(a) Write out the two reaction probabilities $h_i(x_j), i = 1, -1$ considering the forward and reverse reactions as separate events.

(b) Compare these to the deterministic rate expressions $r_i(c_j), i = 1, -1$ for the forward and reverse reactions considered as elementary reactions. Why are the h_i and r_i expressions different? When do they become close to being the same?

Exercise 4.22: Yield, conversion and reactor configuration

Consider the two liquid-phase, irreversible reactions

$$\mathrm{A} + \mathrm{B} \xrightarrow{k_1} \mathrm{C} \qquad r_1 = k_1 c_A c_B$$

$$2\mathrm{A} \xrightarrow{k_2} \mathrm{D} \qquad r_2 = k_2 c_A^2$$

We wish to achieve a high yield of product C. Consider the two 2-CSTR reactor configurations depicted in Figure 4.38.

(a) Without doing any calculations, which of these configurations do you think gives the higher yield of C? Which gives the higher conversion of A? Explain your reasoning.

$$\text{yield of C} = \frac{\text{moles of C produced}}{\text{moles of A consumed}}$$

$$\text{conversion of A} = \frac{\text{moles of A consumed}}{\text{moles of A fed}}$$

(b) Given the following parameters, list the equations that you will solve to calculate the yield of C and conversion of A at the exit of the two reactor configurations. The only symbols appearing in these equations should be numbers and the unknowns for which you are solving. You should have as many equations as you have unknowns. Clearly indicate which set of equations will be solved for which reactor configuration.

Parameter	Value	Units
V_{R1}	1000	L
V_{R2}	1000	L
Q_f	1000	L/hr
k_1	1	L/mol hr
k_2	1	L/mol hr
c_{Af}	2	mol/L
c_{Bf}	2	mol/L

(c) Solve the equations of Part 4.22b numerically, and compare your numerical results to your answers in Part 4.22a. Explain any differences.

Bibliography

[1] R. B. Bird, W. E. Stewart, and E. N. Lightfoot. *Transport Phenomena.* John Wiley & Sons, New York, second edition, 2002.

[2] W. E. Boyce and R. C. DiPrima. *Elementary Differential Equations and Boundary Value Problems.* John Wiley & Sons, New York, sixth edition, 1997.

[3] K. E. Brenan, S. L. Campbell, and L. R. Petzold. *Numerical Solution of Initial-Value Problems in Differential-Algebraic Equations.* Elsevier Science Publishers, New York, 1989.

[4] M. Feinberg and P. Ellison. General kinetic bounds on productivity and selectivity in reactor-separator systems of arbitrary design: I. Principles. *Ind. Eng. Chem. Res.*, 40(14):3181–3194, 2001.

[5] J. Frank K. Seubold and W. E. Vaughan. Acid-catalyzed decomposition of cumene hydroperoxide. *J. Am. Chem. Soc.*, 75:3790–3792, 1953.

[6] T. Furusawa, H. Nishimura, and T. Miyauchi. Experimental study of a bistable continuous reactor. *J. Chem. Eng. Japan*, 2(1):95–100, 1969.

[7] D. T. Gillespie. Exact stochastic simulation of coupled chemical reactions. *J. Phys. Chem.*, 81:2340–2361, 1977.

[8] D. T. Gillespie. *Markov Processes: An Introduction for Physical Scientists.* Academic Press, New York, 1992.

[9] D. T. Gillespie. A rigorous derivation of the chemical master equation. *Physica A*, 188:404–425, 1992.

[10] O. A. Hougen and K. M. Watson. *Chemical Process Principles. Part Three: Kinetics and Catalysis.* John Wiley & Sons, New York, 1947.

[11] K. J. Laidler and B. W. Wojciechowski. Kinetics and mechanisms of the thermal decomposition of ethane II. The reaction inhibited by nitric oxide. *Proceedings of the Royal Society of London, Math and Physics*, 260A(1300): 103-113, 1961.

[12] O. Levenspiel. *Chemical Reaction Engineering.* John Wiley & Sons, New York, second edition, 1972.

[13] P. R. Pujado and S. Sifniades. Phenol. In J. J. McKetta, editor, *Encyclopedia of Chemical Processing and Design*, volume 35, pages 372–391. Marcel Dekker, Inc., New York, 1990.

[14] T. W. F. Russell and M. M. Denn. *Introduction to Chemical Engineering Analysis.* John Wiley & Sons, New York, 1972.

[15] B. H. Shah, D. Ramkrishna, and J. D. Borwanker. Simulation of particulate systems using the concept of the interval of quiescence. *AIChE J.*, 23(6): 897–904, 1977.

[16] R. Srivastava, L. You, J. Summers, and J. Yin. Stochastic vs. deterministic modeling of intracellular viral kinetics. *J. Theor. Biol.*, 218:309–321, 2002.

[17] R. C. Weast, editor. *CRC Handbook of Chemistry and Physics.* CRC Press, Boca Raton, Florida, 60th edition, 1979.

5
Chemical Kinetics

5.1 Introduction

The purpose of this chapter is to provide a framework for determining the reaction rate given a detailed statement of the reaction chemistry. We use several concepts from the subject of chemical kinetics to illustrate two key points:

1. The stoichiometry of an elementary reaction defines the concentration dependence of the rate expression.

2. The quasi-steady-state assumption (QSSA) and the reaction equilibrium assumption allow us to generate reaction-rate expressions that capture the details of the reaction chemistry with a minimum number of rate constants.

The concepts include:

- the elementary reaction
- Tolman's principle of microscopic reversibility
- elementary reaction kinetics
- the quasi-steady-state assumption
- the reaction equilibrium assumption

This chapter complements Chapter 9 on data analysis and empirical reaction-rate expressions in which the goal is to find a mathematical relation that describes the reaction rate over the range of experimental conditions. The goals here are to develop a chemical kinetics basis for the empirical expression, and to show that kinetic analysis can be used to take mechanistic insight and describe reaction rates from first principles.

In this chapter we also discuss heterogeneous catalytic adsorption and reaction kinetics. Catalysis has a significant impact on the world economy[1] and many important reactions employ catalysts.[2] We describe the kinetic principles that are needed for rate studies and demonstrate how the concepts for homogeneous reactions apply to heterogeneously catalyzed reactions with the added constraint of surface-site conservation. The physical characteristics of catalysts are discussed in Chapter 7.

5.2 Elementary Reactions and Microscopic Reversibility

Stoichiometric statements such as

$$A + B \rightleftharpoons C$$

are used to represent the changes that occur during a chemical reaction. These statements can be interpreted in *two* ways. The reaction statement may represent the change in the relative amounts of species that is observed when the reaction proceeds. Or the reaction statement may represent the actual molecular events that are presumed to occur as the reaction proceeds. The former is referred to as an overall stoichiometry or an overall reaction statement. The latter is referred to as an elementary reaction. The elementary reaction is characterized by a change from reactants to products that proceeds without identifiable intermediate species forming. We show subsequently that for an elementary reaction, the reaction rates for the forward and reverse paths are proportional to the concentration of species taking part in the reaction raised to the absolute value of their stoichiometric coefficients. The reaction order in all species is determined directly from the stoichiometry. Elementary reactions are usually unimolecular or bimolecular because the probability of collision between several species is low and is not observed at appreciable rates. For an overall stoichiometry, on the other hand, any correspondence between the stoichiometric coefficients and the reaction order is purely coincidental.

We use three examples to illustrate overall reaction stoichiometry and elementary reactions. The first involves the mechanism proposed

[1] One-third of material gross national product in the United States involves a catalytic process somewhere in the production chain [4]. The market value of products generated through catalysis has reached about $900 billion per annum in 2003 [1].

[2] Catalysts are used in 90% of the world's chemical processes to produce 60% of its chemical products [2].

for the thermal decomposition of acetone at 900 K to ketene and methyl-ethyl ketone (2-butanone) [35]. The overall reaction can be represented by

$$3CH_3COCH_3 \longrightarrow CO + 2CH_4 + CH_2CO + CH_3COC_2H_5 \qquad (5.1)$$

and is proposed to proceed by the following elementary reactions

$$\begin{aligned}
CH_3COCH_3 &\longrightarrow CH_3 + CH_3CO && (5.2)\\
CH_3CO &\longrightarrow CH_3 + CO && (5.3)\\
CH_3 + CH_3COCH_3 &\longrightarrow CH_4 + CH_3COCH_2 && (5.4)\\
CH_3COCH_2 &\longrightarrow CH_2CO + CH_3 && (5.5)\\
CH_3 + CH_3COCH_2 &\longrightarrow CH_3COC_2H_5 && (5.6)
\end{aligned}$$

In the first elementary reaction, acetone undergoes unimolecular decomposition to a methyl radical and to an acetyl radical. These radicals continue to react: the methyl radical reacts to methane and a third radical (CH_3COCH_2) in elementary Reaction 5.4, and the methyl radical combines with the third radical to produce methyl ethyl ketone in elementary Reaction 5.6. The acetyl radical undergoes unimolecular decomposition to a methyl radical and carbon monoxide in the second elementary reaction. The third radical also undergoes unimolecular decomposition in elementary Reaction 5.5 to ketene and a methyl radical. The thermal decomposition of acetone generates four stable molecules that can be removed from the reaction vessel: CO, CH_4, CH_2CO and $CH_3COC_2H_5$. Three radicals are formed and consumed during the thermal decomposition of acetone: CH_3, CH_3CO and CH_3COCH_2. These three radicals are reaction intermediates and cannot be isolated outside of the reaction vessel.

The overall reaction stoichiometry, Reaction 5.1, is explained as a linear combination of the five reactions in the proposed mechanism. By balancing the species in the overall stoichiometry, one can determine that the overall stoichiometry is produced by adding twice the third reaction to the remaining reactions. If the elementary reactions cannot be combined to form the overall stoichiometry, then the mechanism is not a valid description of the observed stoichiometry.

If we assign the species to the $\boldsymbol{A}$ vector as follows

Species	Formula	Name
A_1	CH_3COCH_3	acetone
A_2	CH_3	methyl radical
A_3	CH_3CO	acetyl radical
A_4	CO	carbon monoxide
A_5	CH_3COCH_2	acetone radical
A_6	CH_2CO	ketene
A_7	CH_4	methane
A_8	$CH_3COC_2H_5$	methyl ethyl ketone

then the stoichiometric matrix is

$$\boldsymbol{\nu} = \begin{bmatrix} -1 & 1 & 1 & 0 & 0 & 0 & 0 & 0 \\ 0 & 1 & -1 & 1 & 0 & 0 & 0 & 0 \\ -1 & -1 & 0 & 0 & 1 & 0 & 1 & 0 \\ 0 & 1 & 0 & 0 & -1 & 1 & 0 & 0 \\ 0 & -1 & 0 & 0 & -1 & 0 & 0 & 1 \end{bmatrix}$$

If we multiply $\boldsymbol{\nu}$ by $\begin{bmatrix} 1 & 1 & 2 & 1 & 1 \end{bmatrix}$ we obtain

$$\begin{bmatrix} -3 & 0 & 0 & 1 & 0 & 1 & 2 & 1 \end{bmatrix}$$

which is the overall stoichiometry given in Reaction 5.1.

The second example involves one of the major reactions responsible for the production of photochemical smog. The overall reaction is

$$2NO_2 + h\nu \longrightarrow 2NO + O_2 \tag{5.7}$$

and one possible mechanism is [15, 13]

$$NO_2 + h\nu \longrightarrow NO + O \tag{5.8}$$
$$O + NO_2 \rightleftharpoons NO_3 \tag{5.9}$$
$$NO_3 + NO_2 \longrightarrow NO + O_2 + NO_2 \tag{5.10}$$
$$NO_3 + NO \longrightarrow 2NO_2 \tag{5.11}$$
$$O + NO_2 \longrightarrow NO + O_2 \tag{5.12}$$

In this reaction mechanism, nitrogen dioxide is activated by absorbing photons and decomposes to nitric oxide and oxygen radicals (elementary Reaction 5.8). As in the previous example, stable molecules are formed, in this case NO and O_2, and radicals, O and NO_3, are generated and consumed during the photochemical reaction. The student should work through the steps to determine the linear combination of the mechanistic steps that produces the overall reaction.

The third example involves the synthesis of methane from synthesis gas, CO and H_2, over a ruthenium catalyst [11]. The overall reaction is

$$3H_2(g) + CO(g) \longrightarrow CH_4(g) + H_2O(g) \tag{5.13}$$

and one possible mechanism is

$$CO(g) + S \rightleftharpoons CO_s \tag{5.14}$$
$$CO_s + S \rightleftharpoons C_s + O_s \tag{5.15}$$
$$O_s + H_2(g) \longrightarrow H_2O(g) + S \tag{5.16}$$
$$H_2(g) + 2S \rightleftharpoons 2H_s \tag{5.17}$$
$$C_s + H_s \rightleftharpoons CH_s + S \tag{5.18}$$
$$CH_s + H_s \rightleftharpoons CH_{2s} + S \tag{5.19}$$
$$CH_{2s} + H_s \rightleftharpoons CH_{3s} + S \tag{5.20}$$
$$CH_{3s} + H_s \longrightarrow CH_4(g) + 2S \tag{5.21}$$

in which subscripts g and s refer to gas phase and adsorbed species, respectively, and S refers to a vacant ruthenium surface site. During the overall reaction, the reagents adsorb (elementary Reactions 5.14 and 5.17), and the products form at the surface and desorb (elementary Reactions 5.16 and 5.21). Adsorbed CO (CO_s) either occupies surface sites or dissociates to adsorbed carbon and oxygen in elementary Reaction 5.15. The adsorbed carbon undergoes a sequential hydrogenation to methyne, methylene, and methyl, all of which are adsorbed on the surface. Hydrogenation of methyl leads to methane in elementary Reaction 5.21. In this example, the overall reaction is twice the fourth reaction added to the remaining reactions.

In each of these examples, the elementary reactions describe the detailed pathway between reactants and products. Numerous factors must be considered in proposing, developing and verifying a mechanism and a complete discussion of these factors is beyond the scope of this text. Keep in mind that the elementary reactions must be possible chemically, that is to say the reactants could form the products (and the products could form the reactants), and the elementary reactions should be kinetically significant, that is to say the reaction contributes to the appearance of intermediates or products. For example, during acetone pyrolysis several radicals are produced and these could recombine in a variety of processes that are chemically possible such as

$$
\begin{aligned}
CH_3 + CH_3COCH_2 &\longrightarrow CH_3COC_2H_5 \\
CH_3 + CH_3 &\longrightarrow C_2H_6 \\
2CH_3COCH_2 &\longrightarrow (CH_3COCH_2)_2
\end{aligned}
$$

However, only Reaction 5.6 is listed in the mechanism presented above because the other two radical recombination reactions occur at rates too small to be significant. Methyl recombination was ruled out because the methyl concentration is anticipated to be much lower than the concentration of CH_3COCH_2; therefore the probability of methyl recombining is much lower than the methyl being consumed in the other steps. Furthermore, the recombination of CH_3COCH_2 can be discounted because the chemical instability of the bond formed in the recombination product $(CH_3COCH_2)_2$ would cause this product to decompose rapidly back into the radical CH_3COCH_2. Laidler [24] and Hill [17] provide a good discussion of some of these kinetic arguments and present additional examples of complex reactions.

One criterion for a reaction to be elementary is that as the reactants transform into the products they do so without forming intermediate species that are chemically identifiable. A second aspect of an elementary reaction is that the reverse reaction also must be possible on energy, symmetry and steric bases, using only the products of the elementary reaction. This reversible nature of elementary reactions is the essence of Tolman's principle of microscopic reversibility [39, p. 699].

> This assumption (at equilibrium the number of molecules going in unit time from state 1 to state 2 is equal to the number going in the reverse direction) should be recognized as a distinct postulate and might be called the principle of microscopic reversibility. In the case of a system in thermodynamic equilibrium, the principle requires not only that the total number of molecules leaving a given quantum state in unit time is equal to the number arriving in that state in unit time, but also that the number leaving by any one particular path, is equal to the number arriving by the reverse of that particular path.

Various models or theories have been postulated to describe the rate of an elementary reaction. Transition-state theory (TST) is reviewed briefly in the next section to describe the flow of systems (reactants $\longrightarrow$ products) over potential-energy surfaces. Using Tolman's principle, the most probable path in one direction must be the most probable path in the opposite direction. Furthermore, the Gibbs energy difference in the two directions must be equal to the overall Gibbs energy difference — the ratio of rate constants for an elementary reaction must be equal to the equilibrium constant for the elementary reaction.

5.3 Elementary Reaction Kinetics

In this section we outline the transition-state theory (TST), which can be used to predict the rate of an elementary reaction. Our purpose is to show how the rate of an elementary reaction is related to the concentration of reactants. The result is

$$r_i = k_i \prod_{j \in \mathcal{R}_i} c_j^{-\nu_{ij}} - k_{-i} \prod_{j \in \mathcal{P}_i} c_j^{\nu_{ij}} \tag{5.22}$$

in which $j \in \mathcal{R}_i$ and $j \in \mathcal{P}_i$ represent the reactant species and the product species in the ith reaction, respectively.

Equation 5.22 forms the basis for predicting reaction rates and is applied to homogeneous and heterogeneous systems. Because of its wide use, the remainder of this section describes the concepts and assumptions that underlie Equation 5.22. Transition-state theory is based on the principles of statistical mechanics and, for the purposes here, you need only an understanding of molecular partition functions at the level presented in undergraduate physical chemistry texts.

Before describing TST it is necessary to develop the concept of the reaction coordinate. We use a two-body reaction example to calculate and illustrate a potential energy surface from which a reaction coordinate diagram is constructed. Elementary reaction kinetics is a rich field [29, 24, 42] and the material outlined here is far from comprehensive — it is intended to provide a framework for developing a physical picture and mathematical model of a chemical reaction.

Molecules containing N atoms require $3N$ Cartesian coordinates to specify the locations of all the nuclei. For convenience, we may use three coordinates to locate the center of mass of the molecule, and three more coordinates to orient a chosen axis passing through the molecule. Only two are needed to describe the orientation if the molecule is diatomic. The remaining $3N - 6$ coordinates are used to specify the relative positions of the atoms. The potential energy $V(r)$

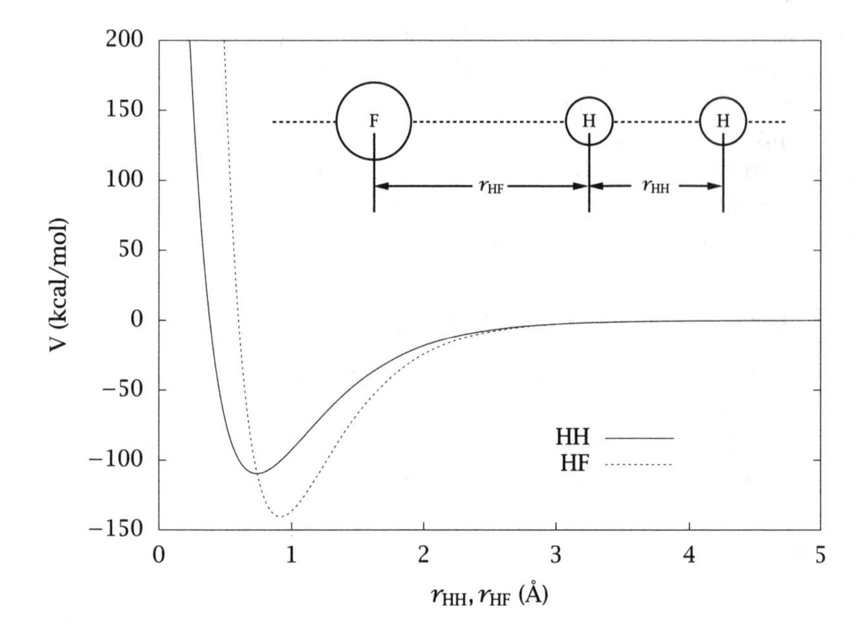

Figure 5.1: Morse potential for H_2 and HF.

of the molecule is related to the relative position of the atoms, and for a diatomic molecule the potential energy can be represented with a Morse function

$$V(r) = D\left[e^{-2\beta r} - 2e^{-\beta r}\right]$$

in which D is the dissociation energy, r is the displacement from the equilibrium bond length, and β is related to the vibrational frequency of the molecule. Figure 5.1 presents the Morse functions for HF and H_2 molecules. The depth of the well is governed by the magnitude of D and the curvature at the bottom is governed by the vibrational frequency. The potential energy is a strong function of the distance of separation between the atoms.

If we increase the number of atoms in the molecule to three, we require three coordinates to describe the internuclear distances. We might choose the two distances from the center atom and the bond angle as the three coordinates. The potential energy is a function of all three coordinate values. A plot of the energy would be a three-dimensional surface, requiring four dimensions to plot the surface. If we fix the bond angle by choosing a collinear molecule, say, only the two distances relative to a central atom are required to describe

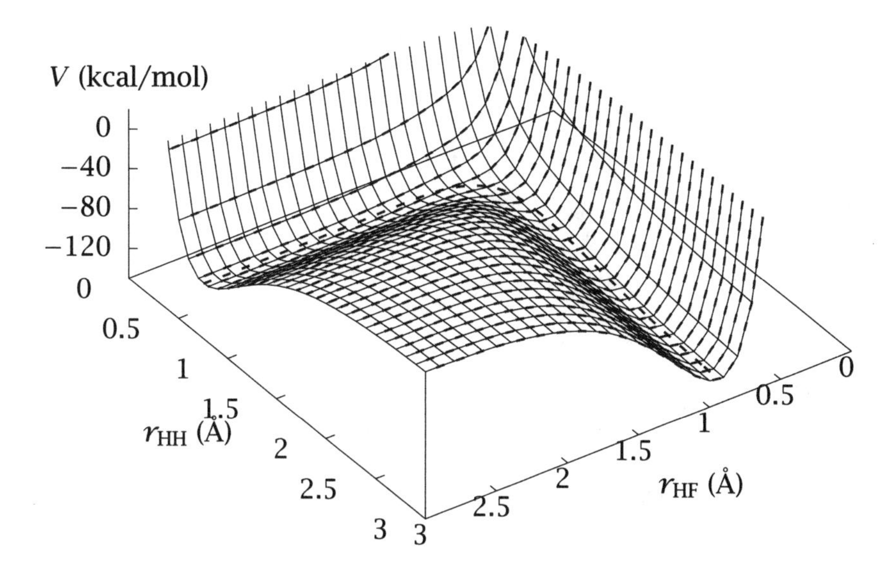

Figure 5.2: Potential-energy surface for the F, H_2 reaction.

the molecule. The potential energy can be expressed as a function of these two distances, and we can view this case as a three-dimensional plot. Figure 5.2 shows a representative view of such a surface for the collinear collision between F and H_2.

$$F + H{-}H \rightleftharpoons \{F{-}H{-}H\} \rightleftharpoons F{-}H + H$$

Several sources provided the data used to calculate this potential-energy surface [32, 34, 33]. Moore and Pearson provide a more detailed discussion of computing potential-energy surfaces [29].

Figure 5.3 gives a contour representation of Figure 5.2. The contour lines are isoenergetic. Figures 5.2 and 5.3 illustrate, for a collinear collision between a diatomic molecule and an atom, the potential-energy surface consists of two valleys connected by a pass. A slice through the energy surface of Figure 5.2 at large values of r_{HF} — in which F atoms exist along with H_2 molecules — reproduces the curve for H_2 in Figure 5.1. A slice at large r_{HH} — in which H atoms exist along with HF molecules — reproduces the curve for HF in Figure 5.1. As the F atom is brought into contact with H_2 in a collinear fashion, a surface of energies

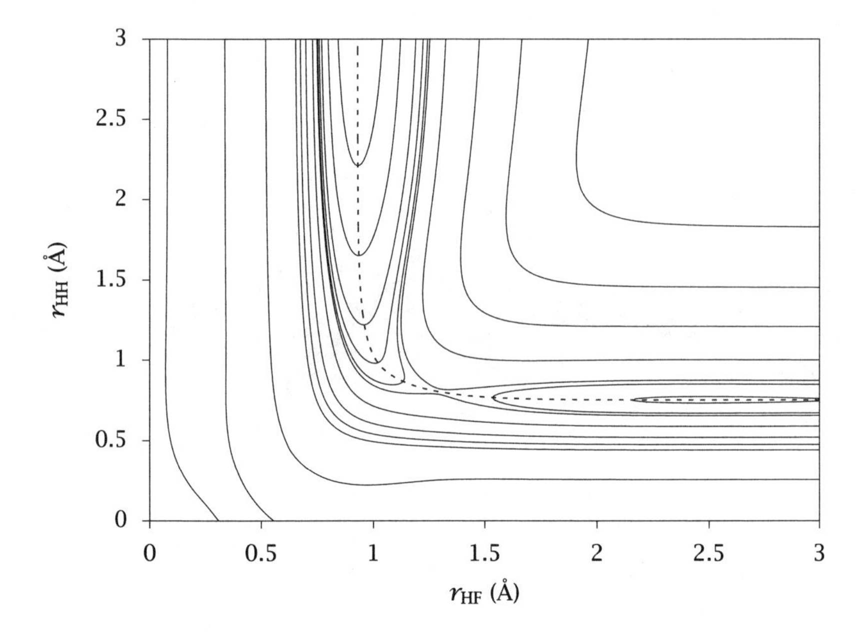

Figure 5.3: Contour representation of the potential-energy surface.

is possible that depends on the values of r_{HF} and r_{HH}. There is a minimum energy path along the valley of decreasing r_{HF} and constant r_{HH}. This path is shown by the dashed line in Figures 5.2 and 5.3. The reaction starts in the H_2 valley at large r_{HF}, proceeds along the minimum energy path and ends in the HF valley at large r_{HH}. Because of the repulsive and attractive forces present during the collision, the maximum value of the energy along the minimum energy path is a saddle point. The dashed line is referred to as the reaction-coordinate pathway, and movement along this pathway is the reaction coordinate, ξ.

Figure 5.4 is a reaction-coordinate diagram, which displays the energy change during this reaction. The reaction coordinate represents travel along the minimum energy path from reactants to products. For this example, there is a one-to-one correspondence between the energy associated with a r_{HF}-r_{HH} coordinate pair in Figure 5.3 and the energy presented in Figure 5.4. The difference in energies between reactants and products is the heat of reaction; in this case it is exothermic by 34 kcal/mol. The barrier height of 4 kcal/mol between reactants and products is used to calculate the activation energy for the reaction.

Potential-energy surfaces such as Figures 5.2 and 5.3 cannot be de-

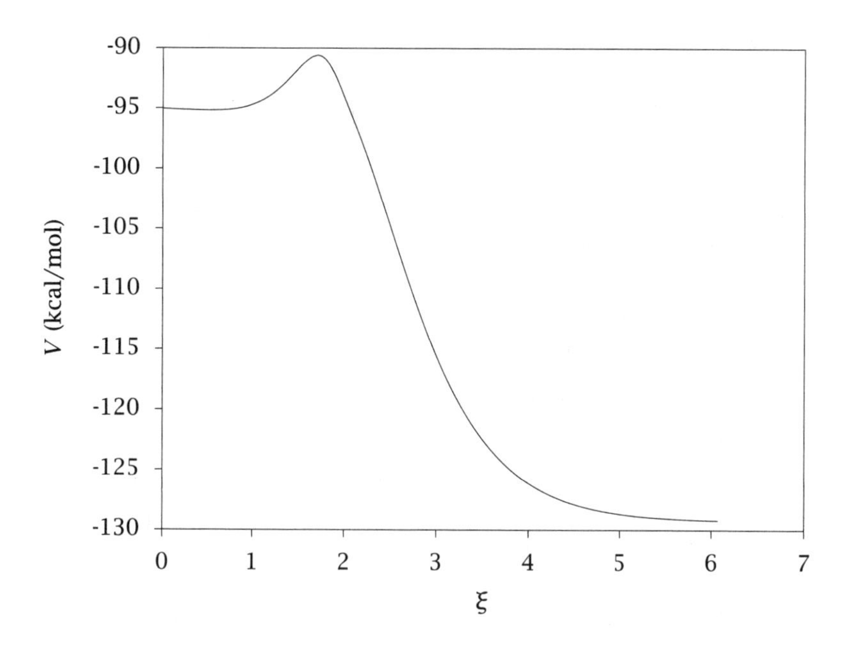

Figure 5.4: Reaction-coordinate diagram.

picted for polyatomic systems with $N > 3$ because more variables would be needed to describe all the relative positions of the nuclei. However, a reaction-coordinate diagram can always be constructed from a minimum energy path along a potential-energy surface. The saddle-point location defines the relative positions of all nuclei in the system, just as in Figure 5.3.

The principle of microscopic reversibility for this elementary reaction implies that the same structure at the saddle point must be realized if the reaction started at HF + H and went in reverse. The molecular structure (the relative positions of F, and two H's) at the saddle point is called the transition-state complex. This transition-state complex is not a chemically identifiable reaction intermediate; it is the arrangement of atoms at the point in energy space between reactants and products. Any change of relative positions of F and the two H's for the transition-state complex (as long as the motions are collinear) results in the complex reverting to reactants or proceeding to products.

Statistical mechanics uses the properties of individual molecules to describe the thermodynamic properties of the bulk system such as the energy, which is the average energy of all molecules in the system.

This energy is found by summing over the product of all possible energies and the probability of finding each energy in the system. The individual probabilities are normalized with a quantity known as the partition function, Q. Therefore, the partition function is a normalization function and we use a microscopic description of the molecules in the system to calculate the partition function. All thermodynamic properties can be expressed in terms of the partition function, including the internal energy, entropy, Helmholtz energy, pressure, enthalpy and Gibbs energy. It is also possible to compute the chemical potential of a component in a mixture from the partition function. This ability to take a microscopic description of the molecules and determine macroscopic properties is exploited to calculate equilibrium constants for a reaction, and rate constants using TST.

The partition function of the system Q is related to the molecular partition function of the individual molecules in the system. In our development of rate constants we make use of the molecular partition functions. The molecular partition function per unit volume for an ideal gas is the product of the translational, rotational, vibrational and electronic energy states in the molecule

$$\left(\frac{q}{V}\right)_j = \left(\frac{q}{V}\right)_{\text{tran}} q_{\text{rot}} q_{\text{vib}} q_{\text{elec}}$$

Table 5.1 lists relationships needed to calculate the various terms of the molecular partition function. Further, the activity for any species j is

$$a_j = \left(\frac{q}{V}\right)_j \frac{k_B T}{f_j^\circ}$$

The molecular partition functions can be used to calculate the equilibrium constant for the reaction between F and H_2. For this gas-phase reaction

$$K = \frac{a_{\text{HF}} a_{\text{H}}}{a_{\text{F}} a_{\text{H}_2}} = \frac{\left(\frac{q}{V}\right)_{\text{HF}} \left(\frac{q}{V}\right)_{\text{H}}}{\left(\frac{q}{V}\right)_{\text{F}} \left(\frac{q}{V}\right)_{\text{H}_2}} \frac{f_{\text{F}}^\circ f_{\text{H}_2}^\circ}{f_{\text{HF}}^\circ f_{\text{H}}^\circ} \tag{5.23}$$

This example serves to illustrate how a microscopic description — the bond distances in HF and H_2, the vibrational frequencies of HF and H_2, the degeneracies of the electronic states of HF, H_2, F and H, and the electronic energy levels of HF, H_2, F and H — can be used to determine a thermodynamic property. This same microscopic description can equally well be used to calculate the enthalpy or any other thermodynamic property of the system.

Component	Units	Expression
translation	length^{-3}	$\left(\frac{q}{V}\right)_{\text{tran}} = \frac{(2\pi m k_B T)^{3/2}}{h^3}$
rotation (linear)	—	$q_{\text{rot}} = \frac{8\pi^2 I k_B T}{\sigma h^2}$
rotation (nonlinear)	—	$q_{\text{rot}} = \frac{8\pi^2 8\pi^3 (I_A I_B I_C)^{1/2} (k_B T)^{3/2}}{\sigma h^3}$
vibration	—	$q_{\text{vib}} = \prod_i^{3N-6} \frac{\exp(-h\nu_i/2k_B T)}{1-\exp(-h\nu_i/k_B T)}$
electronic	—	$q_{\text{elec}} = \sum_i^{\text{states}} g_{ei} \exp(-\epsilon_i/k_B T)$

Table 5.1: Molecular partition function terms.

The reaction-coordinate diagram illustrates the idea that, at some particular orientation of atoms, the reacting molecules (atoms) are in a configuration that either goes forward to products or reverts back to reactants. TST is based on the assumption that the transition-state complex is in equilibrium with the reactants forming it. We use X to designate the transition-state complex, A to denote one reactant, and B to denote the other reactant

$$\text{A} + \text{B} \longrightarrow \text{X}$$

Using the principles of statistical mechanics we can write for this gas-phase reaction

$$K = \frac{a_X}{a_A a_B} = \frac{\left(\frac{q}{V}\right)_X}{\left(\frac{q}{V}\right)_A \left(\frac{q}{V}\right)_B} \frac{f_A^\circ f_B^\circ}{f_X^\circ} \frac{1}{k_B T} \tag{5.24}$$

The concentration of the transition-state complex is required to express the rate as a function of concentration. If activity is defined as

$$a_j = \frac{f_j}{f_j^\circ} = \frac{\phi_j P_j}{f_j^\circ} \tag{5.25}$$

and the equation of state is,

$$P_j V = z k_B T n_j \tag{5.26}$$

then

$$\frac{a_X}{a_A a_B} = \frac{\phi_X}{\phi_A \phi_B} \frac{f_A^\circ f_B^\circ}{f_X^\circ} \frac{1}{z k_B T} \frac{c_X}{c_A c_B} \tag{5.27}$$

Combining Equations 5.24 and 5.27 and solving for c_X gives

$$c_X = \frac{\left(\frac{q}{V}\right)_X}{\left(\frac{q}{V}\right)_A \left(\frac{q}{V}\right)_B} \frac{\phi_A \phi_B z}{\phi_X} c_A c_B \tag{5.28}$$

TST invokes the idea that the rate in the forward direction (H_2 + F $\longrightarrow$ HF + H) is equal to the number of transition-state complexes moving to the right of on the reaction coordinate diagram. For the example worked here, the rate has units of molecules/time·volume. Since movement along the reaction coordinate describes molecules (or atoms) coming together and bonds rearranging, TST assumes that the complex either moves to the right or to the left within the frequency of the single vibration (molecular motion) that best represents the bond to be made (or broken). We refer to this special frequency as ν^*, and we can write

$$r = \nu^* c_X = \nu^* \frac{\left(\frac{q}{V}\right)_X}{\left(\frac{q}{V}\right)_A \left(\frac{q}{V}\right)_B} \frac{\phi_A \phi_B z}{\phi_X} c_A c_B \tag{5.29}$$

The vibrational partition function is

$$q_{\text{vib}} = \prod_{i}^{3N-6} \frac{\exp\left(-h\nu_i / 2k_B T\right)}{1 - \exp\left(-h\nu_i / k_B T\right)}$$

in which $3N-6$ vibrational modes are needed to describe a polyatomic, nonlinear molecule comprised of N atoms ($3N - 5$ modes describe a linear molecule). In the limit of a very weak vibration (small ν_i), the quantity

$$\frac{\exp\left(-h\nu_i / 2k_B T\right)}{1 - \exp\left(-h\nu_i / k_B T\right)} \longrightarrow \frac{k_B T}{h\nu_i} \tag{5.30}$$

This approximation is valid for $h\nu_i / k_B T \leq 0.1$. This limit is sometimes referred to as the classical high-temperature limit. The argument $h\nu / k_B T$ equals 0.1 at a temperature of 28,800 K for a stretching frequency of 2000 cm^{-1}, and it equals 0.1 at a stretching frequency of 20.8 cm^{-1} for a temperature of 300 K. A frequency of 20.8 cm^{-1} is reasonable for a rocking motion or a hindered rotation about a bond.

At this point, we assume a weak bond is made (or broken) as the transition-state complex transforms to products. We then rewrite the molecular partition function for the transition-state complex as

$$\left(\frac{q}{V}\right)_X = \left(\frac{q}{V}\right)_{\text{tran}} q_{\text{rot}} q_{\text{elec}} q^*_{\text{vib}} \frac{\exp\left(-h\nu^*/2k_BT\right)}{1-\exp\left(-h\nu^*/k_BT\right)} \tag{5.31}$$

in which q^*_{vib} now has $3N-7$ (or $3N-6$ linear) vibrational modes remaining because one was used in the argument containing ν^*. At the classical high-temperature limit (shown in Equation 5.30) Equation 5.31 becomes

$$\left(\frac{q}{V}\right)^*_X = \left(\frac{q}{V}\right)_{\text{tran}} q_{\text{rot}} q_{\text{elec}} q^*_{\text{vib}} \frac{k_BT}{h\nu^*}$$

Combining this result with Equation 5.29 gives

$$r = \frac{k_BT}{h} \frac{\left(\frac{q}{V}\right)^*_X}{\left(\frac{q}{V}\right)_A \left(\frac{q}{V}\right)_B} \frac{\phi_A\phi_B z}{\phi_X} c_A c_B \tag{5.32}$$

The rate constant contains all the terms on the right-hand side of Equation 5.32 except for the concentration of reactants,

$$k = \frac{k_BT}{h} \frac{\left(\frac{q}{V}\right)^*_X}{\left(\frac{q}{V}\right)_A \left(\frac{q}{V}\right)_B} \frac{\phi_A\phi_B z}{\phi_X} \tag{5.33}$$

which leads to

$$r = k c_A c_B \tag{5.34}$$

Equation 5.33 can be rewritten as

$$k = k^\circ f(c_j) \tag{5.35}$$

in which

$$k^\circ = \frac{k_BT}{h} \frac{\left(\frac{q}{V}\right)^*_X}{\left(\frac{q}{V}\right)_A \left(\frac{q}{V}\right)_B} \tag{5.36}$$

represents all the terms that are composition independent. For a reaction involving ideal gases, $f(c_j) = 1$.

Most rate expressions use the form of Equation 5.34. The variable k is the rate constant and many texts consider this to be a composition-independent term. The rate "constant" is not strictly constant. It changes with temperature and even composition. The composition dependence is absent for ideal gases and it is reasonable to neglect compositional dependence of k in many situations. The temperature dependence is rigorously found by accounting for the effect of temperature on each term in Equation 5.33. The Arrhenius rate expression is one common way of representing the temperature dependence.

$$k = k_0 \exp(-E_a/k_B T) \tag{5.37}$$

in which k_0 is temperature independent, and E_a is known as the activation energy. Often k_0 varies weakly with temperature

$$k_0 = AT^B$$

with $-2 \leq B \leq 1$. The Arrhenius expression is accurate provided $E_a \gg Bk_B T$.

With Equation 5.32 and a potential-energy surface, it is possible, in principle, to calculate the rate of any elementary reaction. This is a useful result because, as computational chemistry develops better empirical methods for determining the potential-energy surface, it becomes possible to predict reaction rates from first principles and compare them against experimental information or possibly avoid experimental determination of the reaction rate.

TST teaches us that elementary reactions depend on the amount (concentration) of each reactant raised to the absolute value of its stoichiometric coefficient because of the equilibrium assumed between the transition-state complex and the reactants. This allows us to generalize the results illustrated above to all elementary reactions[3] and adopt the convention

$$r_i = k_i \prod_{j \in \mathcal{R}_i} c_j^{-\nu_{ij}} - k_{-i} \prod_{j \in \mathcal{P}_i} c_j^{\nu_{ij}} \tag{5.38}$$

Equation 5.38 cannot be set to zero to define the equilibrium concentration, nor is the ratio of the forward and reverse rate constants equal to the equilibrium constant for the elementary reaction. The equilibrium constant K is given by

$$K = \frac{k_i^\circ}{k_{-i}^\circ} \prod_j^{n_s} \left(\frac{k_B T}{f_j^\circ} \right)^{\nu_{ij}}$$

The equilibrium constant is equal to a ratio of activities. The composition at equilibrium is found by expressing the activities in terms of concentrations.

[3]There is an exception for unimolecular reactions in the low-pressure regime.

Note that k_{-i} may be small compared to k_i, in which case the reverse reaction is insignificant. The reversibility condition, although true in general, may not be important in a particular reaction. Recall also that the production rate of any species j is

$$R_j = \sum_{i=1}^{n_r} \nu_{ij} r_i \tag{5.39}$$

Applying Equations 5.38 and 5.39 to NO_3 in the photochemical smog example gives

$$R_{NO_3} = k_2 c_O c_{NO_2} - k_{-2} c_{NO_3} - k_3 c_{NO_3} c_{NO_2} - k_4 c_{NO_3} c_{NO}$$

Example 5.1: Computing a rate constant for trioxane decomposition

Using TST, predict the value of the rate constant for the unimolecular decomposition of 1,3,5-trioxane at 750 K. The decomposition proceeds by the concerted rupture of three C—O bonds in the ring to form CH_2O.

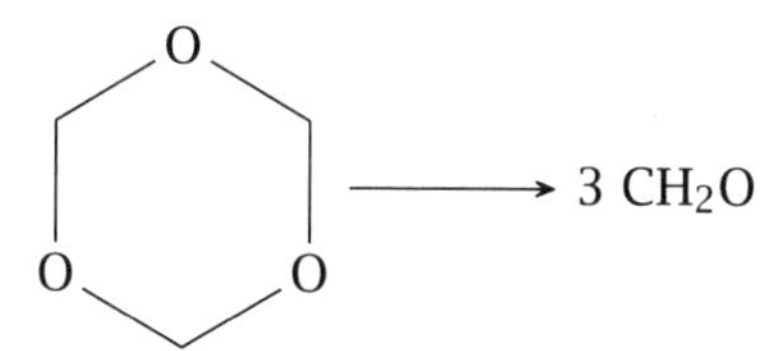

The ring expands during the reaction. The vibrational frequencies and moments of inertia of the molecule and the transition complex are listed in Table 5.2, and the experimentally observed rate constant is [19].

$$k = k_0 e^{-E_a/RT} \qquad k_0 = 10^{15.28 \pm 0.06} \ \text{s}^{-1} \qquad E_a = 47.5 \pm 2.4 \ \text{kcal/mol}$$

Solution

It is reasonable to assume the reacting mixture is ideal and k is given by Equation 5.36.

$$k = \frac{k_B T}{h} \frac{\left(\frac{q}{V}\right)^*_{\text{complex}}}{\left(\frac{q}{V}\right)_{\text{trioxane}}} \tag{5.40}$$

Because the masses of the transition-state complex and the trioxane are equal, $\left(\frac{q}{V}\right)_{\text{tran}}$ is the same for the complex and trioxane molecule. A ratio of the q_{rot} reduces to

$$\frac{(q_{\text{rot}})_{\text{complex}}}{(q_{\text{rot}})_{\text{trioxane}}} = \frac{(\sqrt{I_A I_B I_C})_{\text{complex}}}{(\sqrt{I_A I_B I_C})_{\text{trioxane}}}$$

Parameter	Value
frequencies, cm^{-1}	
trioxane	296, 296, 524, 524, 945, 945, 1070, 1070, 1178, 1178, 1305, 1305, 1410, 1410, 1481, 1481, 2850, 2850, 3025, 3025, 1122, 1242, 1383, 466, 752, 978, 1235, 1495, 2850, 3025
transition complex	100, 100, 200, 200, 945, 945, 1400, 1400, 1178, 1178, 1200, 1200, 1200, 1200, 1481, 1481, 2850, 2850, 3025, 3025, 1000, 1242, 1100, 200, 700, 1200, 1495, 2850, 3025
moments of inertia, amu-Å^2	
trioxane	96.4, 96.4, 173.0
transition complex	125.3, 120.5, 249.2
potential-energy barrier, kcal/mol	51.4

Table 5.2: Parameters for the trioxane reaction.

$$\frac{(q_{\mathrm{rot}})_{\mathrm{complex}}}{(q_{\mathrm{rot}})_{\mathrm{trioxane}}} = \frac{\sqrt{(125.3)(120.5)(249.2)}}{\sqrt{(96.4)(96.4)(96.4)}} = 1.53$$

The vibrational terms use the frequency ν in reciprocal time and we are given the frequency in wave numbers (cm^{-1}). To use wave number units, the following is used in place of $h/k_B T$

$$\frac{hc}{k_B T} = \frac{(6.626 \times 10^{-34}\mathrm{J\ s})(2.998 \times 10^{10}\mathrm{cm/s})}{(1.381 \times 10^{-23}\mathrm{J/K})(750\ \mathrm{K})} = 1.92 \times 10^{-3}\ \mathrm{cm}$$

Using the vibrational frequencies listed in Table 5.2, the vibrational partition functions are

$$q_{\text{vib, complex}} = \prod_{i=1}^{29} \frac{\exp(-1/2(1.92 \times 10^{-3}\text{cm})(\nu_i \text{cm}^{-1}))}{1 - \exp(-(1.92 \times 10^{-3}\text{cm})(\nu_i \text{cm}^{-1}))} = 1.94 \times 10^{-13}$$

$$q_{\text{vib, trioxane}} = \prod_{i=1}^{30} \frac{\exp(-1/2(1.92 \times 10^{-3}\text{cm})(\nu_i \text{cm}^{-1}))}{1 - \exp(-(1.92 \times 10^{-3}\text{cm})(\nu_i \text{cm}^{-1}))} = 3.25 \times 10^{-16}$$

$$\frac{q_{\text{vib, complex}}}{q_{\text{vib, trioxane}}} = 596$$

The reaction frequency is

$$\frac{k_B T}{h} = \frac{(1.381 \times 10^{-23} \text{ J/K})(750 \text{ K})}{6.626 \times 10^{-34} \text{J s}} = 1.56 \times 10^{13} \text{s}^{-1}$$

Assuming degeneracies of unity the electronic terms are

$$q_{\text{elec, complex}} = (1) \exp\left(\frac{-51400 \text{ cal/mol}}{(1.987 \text{ cal/mol K})(750 \text{ K})}\right) = 1.05 \times 10^{-15}$$

$$q_{\text{elec, trioxane}} = 1$$

and the ratio of electronic terms is 1.05×10^{-15}.

Substituting all the individual terms into Equation 5.40 yields

$$k = (1.56 \times 10^{13} \text{ s}^{-1})(1.53)(596)(1.05 \times 10^{-15}) = 14.9 \text{ s}^{-1}$$

This value compares reasonably well to the experimentally determined value of 27.4 s^{-1} at this temperature. A factor of two is not a large disagreement if you consider the temperature dependence of the rate constant. Figure 5.5 compares the experimental and calculated values of the rate constant over the temperature range 700–800 K. □

5.4 Fast and Slow Time Scales

One of the characteristic features of many chemically reacting systems is the widely disparate time scales at which reactions occur. It is not unusual for complex reaction mechanisms to contain rate constants that differ from each other by several orders of magnitude. Moreover, the

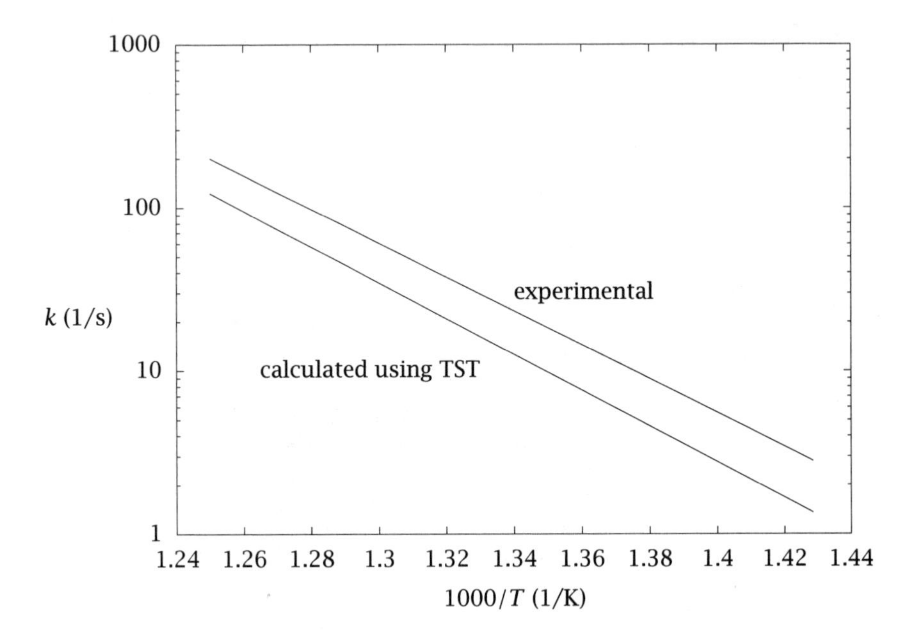

Figure 5.5: Comparison of measured and calculated rate constant versus temperature for trioxane decomposition.

concentrations of highly reactive intermediates may differ by orders of magnitude from the concentrations of relatively stable reactants and products. These widely different time and concentration scales present challenges for accurate estimation of rate constants, measurement of low-concentration species, and even numerical solution of complex models.

On the other hand, these disparate scales often allow us to approximate the complete mechanistic description with simpler rate expressions that retain the essential features of the full problem on the time scale or in the concentration range of interest. Although these approximations were often used in earlier days to allow easier model solution, that is not their primary purpose today. Most models, even stiff differential equation models with fairly disparate time scales, can be solved efficiently with modern ODE solvers. On the other hand, the physical insight provided by these approximations remains valuable. Moreover the reduction of complex mechanisms removes from consideration many parameters that would be difficult to estimate from available data. The next two sections describe two of the most widely used methods of model simplification: the equilibrium assumption and the quasi-steady-state assumption.

5.4.1 The Reaction Equilibrium Assumption

In the reaction equilibrium assumption, we reduce the full mechanism on the basis of fast and slow *reactions.* In a given mechanism consisting of multiple reactions, some reactions may be so much faster than others, that they equilibrate after any displacement from their equilibrium condition. The remaining, slower reactions then govern the rate at which the amounts of reactants and products change. If we take the extreme case in which all reactions except one are assumed at equilibrium, this remaining slow reaction is called the **rate-limiting step**. But the equilibrium approximation is more general and flexible than this one case. We may decompose a full set of reactions into two sets consisting of any number of slow and fast reactions, and make the equilibrium assumption on the set of fast reactions.

We illustrate the main features with the simple series reaction

$$\mathrm{A} \underset{k_{-1}}{\overset{k_1}{\rightleftharpoons}} \mathrm{B}, \qquad \mathrm{B} \underset{k_{-2}}{\overset{k_2}{\rightleftharpoons}} \mathrm{C} \tag{5.41}$$

Assume that the rate constants k_2, k_{-2} are much larger than the rate constants k_1, k_{-1}, so the second reaction equilibrates quickly. By contrast, the first reaction is slow and can remain far from equilibrium for a significant period of time. Because the only other reaction in the network is at equilibrium, the slow, first reaction is called the rate-limiting step. We show that the rate of this slow reaction does indeed determine or limit the rate at which the concentrations change.

It is perhaps clearest to start with the full model and demonstrate what happens if $k_2, k_{-2} \gg k_1, k_{-1}$. Consider the solution to the full model with rate constants $k_1 = 1, k_{-1} = 0.5, k_2 = k_{-2} = 1$ and initial conditions $c_A(0) = 1.0, c_B(0) = 0.4, c_C(0) = 0$ shown in Figure 5.6. Now consider what happens as k_2 and k_{-2} become larger. Figure 5.7 shows the solution for $k_2 = k_{-2} = 10$. Notice the characteristic feature is that equilibrium is now quickly established between species B and C. The time scale for this equilibration is about 0.1 min. The remaining slow part of the system takes about 4 min to reach equilibrium. Figure 5.8 shows the rapid equilibration of c_B and c_C at small times for increasing k_2 holding $K_2 = k_2/k_{-2} = 1$.

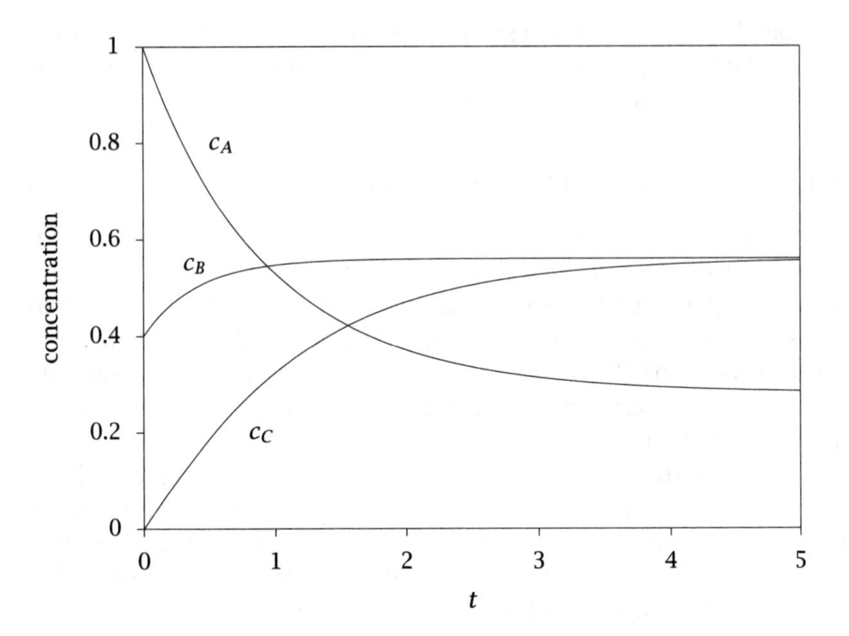

Figure 5.6: Full model solution for $k_1 = 1, k_{-1} = 0.5, k_2 = k_{-2} = 1$.

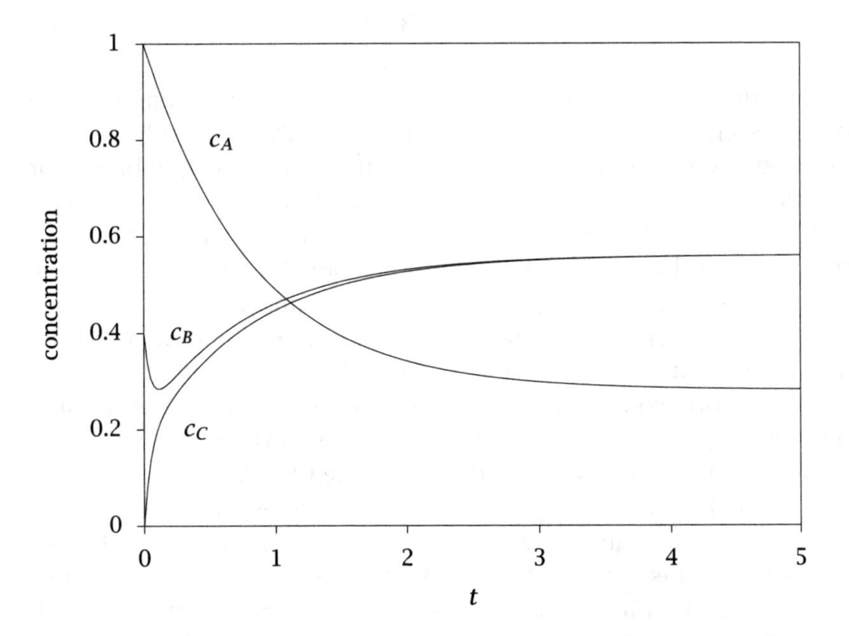

Figure 5.7: Full model solution for $k_1 = 1, k_{-1} = 0.5, k_2 = k_{-2} = 10$.

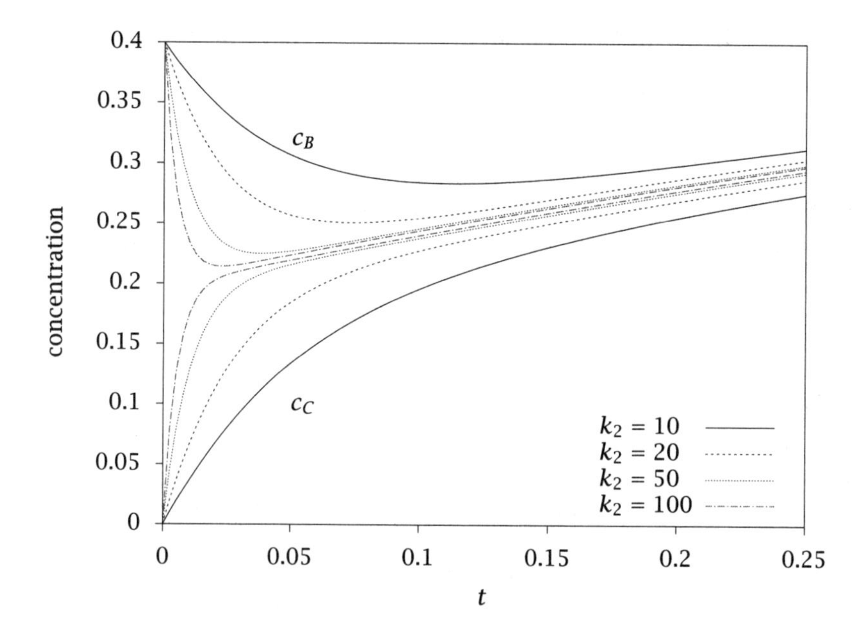

Figure 5.8: Concentrations of B and C versus time for full model with increasing k_2 with $K_2 = k_2/k_{-2} = 1$.

We now analyze the kinetic model. We can express the mole balances in terms of either the extents of the two reactions or the concentrations of the three species. Both approaches are instructive so we examine both, but we start with the two reaction extents because there are only two independent reactions. The full model can be expressed as

$$\frac{d\varepsilon_1}{dt} = r_1 = k_1 c_A - k_{-1} c_B, \qquad \varepsilon_1(0) = 0$$

$$\frac{d\varepsilon_2}{dt} = r_2 = k_2 c_B - k_{-2} c_C, \qquad \varepsilon_2(0) = 0$$

in which, for the batch reactor,

$$c_A = c_{A0} - \varepsilon_1, \qquad c_B = c_{B0} + \varepsilon_1 - \varepsilon_2, \qquad c_C = c_{C0} + \varepsilon_2 \tag{5.42}$$

To describe the fast time-scale behavior displayed in Figure 5.8, let τ denote a fast time variable via

$$\tau = k_{-2} t$$

so a small change in t is equivalent to a large change in τ. If we divide both full model equations by k_{-2} we produce

$$\frac{d\varepsilon_1}{d\tau} = \frac{1}{k_{-2}}\left(k_1 c_A - k_{-1} c_B\right), \qquad \frac{d\varepsilon_2}{d\tau} = K_2 c_B - c_C$$

in which $K_2 = k_2/k_{-2}$ is the equilibrium constant of the fast reaction. Now if we take the limit as $k_{-2} \to \infty$, these equations reduce to

$$\frac{d\varepsilon_1}{d\tau} = 0, \qquad \frac{d\varepsilon_2}{d\tau} = K_2 c_B - c_C$$

which shows the extent of the slow reaction does not change, $\varepsilon_1(\tau) = 0$, over the small time during which the fast reaction equilibrates. We compute the steady state achieved by ε_2 on this fast time scale by substituting Equations 5.42 into the right-hand side of the differential equation for ε_2, and setting the equation to zero, which yields

$$\varepsilon_{1s} = 0, \qquad \varepsilon_{2s} = \frac{K_2}{1+K_2} c_{B0} - \frac{1}{1+K_2} c_{C0}$$

Given this steady state achieved on the fast time scale, we can compute the relaxation on the slow time scale. On the slow time scale, we wish to impose the equilibrium condition $K_2 c_B - c_C = 0$. In terms of extents, this equation is equivalent to

$$\varepsilon_2 = \frac{K_2}{1+K_2}\varepsilon_1 + \frac{K_2}{1+K_2} c_{B0} - \frac{1}{1+K_2} c_{C0} \tag{5.43}$$

This equilibrium relationship takes the place of the differential equation for ε_2 on the slow time scale. Notice the time derivative of ε_2 should not be set to zero on the long time scale because

$$\frac{d\varepsilon_2}{dt} = k_{-2}\left(K_2 c_B - c_C\right)$$

Although the term in parentheses approaches zero for a fast second reaction, k_{-2} approaches infinity, and the product does not go to zero. To determine a differential equation for ε_2, if one is desired, we instead differentiate the equilibrium condition, Equation 5.43, to obtain[4]

$$\frac{d\varepsilon_1}{dt} = r_1 \qquad \frac{d\varepsilon_2}{dt} = \frac{K_2}{1+K_2} r_1$$

[4]Notice we again are choosing between a DAE and an ODE description of the model as in Section 4.5 and Exercise 4.19.

Equations	Fast Time Scale ($\tau = k_{-2}t$)	Slow Time Scale (t)
ODEs	$\frac{d\varepsilon_1}{d\tau} = 0$ $\frac{d\varepsilon_2}{d\tau} = K_2 c_B - c_C$ $\varepsilon_1(0) = 0$ $\varepsilon_2(0) = 0$	$\frac{d\varepsilon_1}{dt} = r_1$ $\frac{d\varepsilon_2}{dt} = \frac{K_2}{1+K_2} r_1$ $\varepsilon_1(0) = 0$ $\varepsilon_2(0) = \varepsilon_{2s} = \frac{K_2 c_{B0} - c_{C0}}{1+K_2}$
DAEs	$\varepsilon_1 = 0$ $\frac{d\varepsilon_2}{d\tau} = K_2 c_B - c_C$ $\varepsilon_2(0) = 0$	$\frac{d\varepsilon_1}{dt} = r_1$ $\varepsilon_2 = \frac{K_2}{1+K_2}\varepsilon_1 + \frac{K_2 c_{B0} - c_{C0}}{1+K_2}$ $\varepsilon_1(0) = 0$
	$c_A = c_{A0} - \varepsilon_1$, $c_B = c_{B0} + \varepsilon_1 - \varepsilon_2$, $c_C = c_{C0} + \varepsilon_2$	

Table 5.3: Fast and slow time-scale models in extents of reactions.

Table 5.3 summarizes the results so far. A network with fast and slow reactions leads to fast and slow time-scale reduced models, shown in the left- and right-hand columns of Table 5.3, respectively. We can express the model either as a set of two ODEs or a DAE consisting of one ODE and one algebraic equation, shown in the top and bottom portions of the table. In either case, the dynamics of the slow time scale are determined only by r_1, verifying that the slow first reaction is the rate-limiting step. The steady state of the fast model becomes the initial condition for the slow model. If we solve the slow time-scale model, we produce the results in Figure 5.9. Notice the concentrations of all components are in excellent agreement after a short time. The big advantage of the reduced model is that values of k_2 and k_{-2}, which are large and hard to estimate from data, are not required. In the reduced model, only their ratio, K_2, appears.

Extents of reaction are convenient variables for a batch reactor, but are inadequate to describe systems with flow terms such as the CSTR. We require species balances for reactors with flow terms, so we reexamine the equilibrium assumption in terms of the concentration variables.

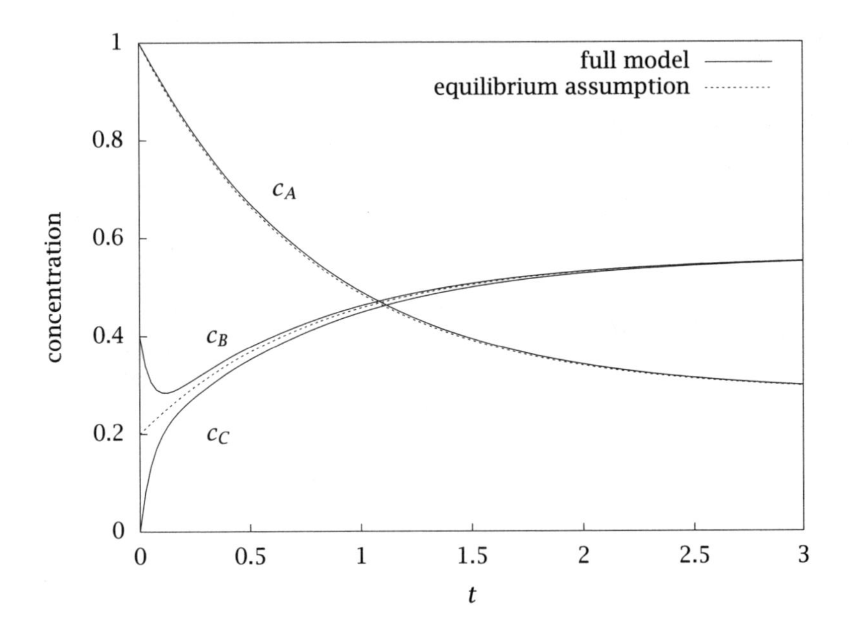

Figure 5.9: Comparison of equilibrium assumption to full model for $k_1 = 1, k_{-1} = 0.5, k_2 = k_{-2} = 10$.

The mole balances for the full model are

$$\frac{dc_A}{dt} = -r_1, \qquad \frac{dc_B}{dt} = r_1 - r_2, \qquad \frac{dc_C}{dt} = r_2$$

We know that only two of these differential equations are independent; we could add the three expressions to deduce the implied algebraic constraint, which could serve as a replacement for any one of the ODEs above

$$\frac{d(c_A + c_B + c_C)}{dt} = 0, \qquad c_A + c_B + c_C = c_{A0} + c_{B0} + c_{C0} \tag{5.44}$$

The short time-scale model is similar to the case with reaction extents. Table 5.4 shows the short time-scale result in the left hand column. Deriving the correct slow time-scale model is less obvious and sometimes leads to confusion. The equilibrium assumption is made by adding the algebraic constraint that $r_2/k_{-2} = 0$ or $K_2 c_B - c_C = 0$. This algebraic equation coupled with Equation 5.44 and the mole balance for component A, in which the second reaction does not appear, then constitute a

Equations	Fast Time Scale ($\tau = k_{-2}t$)	Slow Time Scale (t)
ODEs	$\dfrac{dc_A}{d\tau} = 0, \quad c_A(0) = c_{A0}$ $\dfrac{dc_B}{d\tau} = -(K_2c_B - c_C), \quad c_B(0) = c_{B0}$ $\dfrac{dc_C}{d\tau} = K_2c_B - c_C, \quad c_C(0) = c_{C0}$	$\dfrac{dc_A}{dt} = -r_1, \quad c_A(0) = c_{A0}$ $\dfrac{dc_B}{dt} = \dfrac{1}{1+K_2}r_1, \quad c_B(0) = c_{Bs}$ $\dfrac{dc_C}{dt} = \dfrac{K_2}{1+K_2}r_1, \quad c_C(0) = c_{Cs}$
DAEs	$c_A = c_{A0}$ $\dfrac{dc_B}{d\tau} = -(K_2c_B - c_C), \quad c_B(0) = c_{B0}$ $0 = c_A - c_{A0} + c_B - c_{B0} + c_C - c_{C0}$	$\dfrac{dc_A}{dt} = -r_1, \quad c_A(0) = c_{A0}$ $0 = K_2c_B - c_C$ $0 = c_A - c_{A0} + c_B - c_{B0} + c_C - c_{C0}$
	$c_{Bs} = \dfrac{1}{1+K_2}(c_{B0} + c_{C0}), \quad c_{Cs} = \dfrac{K_2}{1+K_2}(c_{B0} + c_{C0})$	

Table 5.4: Fast and slow time-scale models in species concentrations.

complete set of equations for the slow time scale, shown in the lower, right portion of Table 5.4. If we prefer differential equations, we can add the mole balances for components B and C to *eliminate* r_2

$$\frac{d(c_B + c_C)}{dt} = r_1 \tag{5.45}$$

Notice we have *not* set $r_2 = 0$ to obtain this result. Equation 5.45 involves no approximation. If we differentiate the equilibrium condition, we obtain a second relation, $K_2 dc_B/dt - dc_C/dt = 0$, which allows us to solve for both dc_B/dt and dc_C/dt,

$$\frac{dc_B}{dt} = \frac{1}{1+K_2} r_1 \qquad \frac{dc_C}{dt} = \frac{K_2}{1+K_2} r_1$$

This result also is listed in Table 5.4. Notice that this result is not derivable by setting $r_2 = 0$ in the full model. We see immediately from the slow time-scale differential equations that all concentrations are driven by r_1, again showing the first reaction is the rate-limiting step.

Normally one is interested in the solution to only the slow time-scale model and ignores the small errors at early times. One can also knit the slow and fast time-scale models together, however, and obtain a more accurate reduced model valid at all times. This knitting procedure is known as matching the inner (fast) and outer (slow) solutions in the mathematical theory of singular perturbations [31, 26, 37]. The inverse of the large rate constant plays the role of the perturbation parameter in this theory. The inner (fast) solution is valid in a boundary layer (small time) that decreases in size as the perturbation parameter goes to zero (k_{-2} goes to infinity) as shown in Figure 5.8. The slow and fast time-scale models presented in Tables 5.3 and 5.4 are zero-order terms in a full perturbation series solution. Higher-order corrections also can be computed as shown by O'Malley [31], for example. Although this series solution may be of interest to gain analytical insight into nonlinear kinetic models, the higher-order terms depend on the values of k_2 and k_{-2}. In other words, the higher-order corrections require as much information as the solution of the full model. It is usually simpler to compute numerically the solution to the full model if these rate constants are known. Kumar, Christofides and Daoutidis provide further discussion of the application of the singular perturbation method, and also discuss the resulting control problems arising with chemical reactors having both fast and slow time-scale kinetics [23].

Another interesting extension is the case of larger sets of reactions. The equilibrium assumption for larger sets of reactions is elegantly handled by finding the null space of the stoichiometric matrix of the fast reactions. Ramkrishna and coworkers [14] provide a more complete discussion of this approach.

Finally, if the fast reactions are irreversible, further simplification is possible. In the simple series reactions, let the second reaction be fast and irreversible in Equation 5.41. We start with the slow time-scale model given in Table 5.4 and take the limit as $K_2 \longrightarrow \infty$ giving

$$\begin{aligned} \frac{dc_A}{dt} &= -r_1, & c_A(0) &= c_{A0} \\ \frac{dc_B}{dt} &= 0, & c_B(0) &= 0 \\ \frac{dc_C}{dt} &= r_1, & c_C(0) &= c_{B0} + c_{C0} \end{aligned}$$

In this limit the B disappears, $c_B(t) = 0$, and $r_1 = k_1 c_A$, which describes the irreversible reaction of A directly to C with rate constant k_1

$$\mathrm{A} \xrightarrow{k_1} \mathrm{C} \tag{5.46}$$

We see in the next section that this case is well described also by making the quasi-steady-state assumption on species B.

If we take the second reaction as irreversible in the backward direction, $K_2 = 0$, we obtain from Table 5.4

$$\begin{aligned} \frac{dc_A}{dt} &= -r_1, & c_A(0) &= c_{A0} \\ \frac{dc_B}{dt} &= r_1, & c_B(0) &= c_{B0} + c_{C0} \\ \frac{dc_C}{dt} &= 0, & c_C(0) &= 0 \end{aligned}$$

which describes the reversible reaction between A and B with no second reaction, $c_C(t) = 0$,

$$\mathrm{A} \underset{k_{-1}}{\overset{k_1}{\rightleftharpoons}} \mathrm{B} \tag{5.47}$$

Under the equilibrium assumption, we see that the two series reactions in Equation 5.41 may reduce to A going directly and irreversibly to C, Equation 5.46, or A going reversibly to B, Equation 5.47, depending on the magnitude of the equilibrium constant of the second, fast reaction.

5.4.2 The Quasi-Steady-State Assumption

In the quasi-steady-state assumption, we reduce the full mechanism on the basis of rapidly equilibrating *species* rather than reactions as in the reaction equilibrium assumption. We have seen in Section 5.2 that reaction networks can involve the formation and consumption of intermediate species. In some cases the intermediates are transitory, highly reactive species that are chemically identifiable but unlikely to exist outside the reaction mixture. Examples of these transitory species include atoms, radicals, ions and molecules in excited states. The examples in Section 5.2 contained radicals (CH_3, CH_3CO, CH_3COCH_2 and NO_3) and an atom (O). When the transitory species (reaction intermediates) have certain kinetic properties that we discuss next, their net rate of formation can be set equal to zero, which enables their concentration to be determined in terms of reactants and products using algebraic equations. After solving for these small concentration species, one is then able to construct reaction-rate expressions for the stable reactants and stable products in terms of reactant and product concentrations only.

The idea to set the production rate of a reaction intermediate equal to zero has its origins in the early 1900s [5, 7] and has been termed the Bodenstein-steady-state, pseudo-steady-state, or quasi-steady-state assumption. We use the term quasi-steady-state assumption (QSSA). The intermediate species are referred to as QSSA species [40]. The conditions are: the error in the concentration of the important species (reactants and products) calculated with the QSSA and without the QSSA (the exact solution) be small, and the time domain over which the QSSA is applied be selected to ensure the calculated error is minimized. A number of other qualifying conditions have been applied, which may be valid in certain circumstances, but as pointed out by Turànyi et al. [40, p.172] are not universally applicable.

> There have been several empirical observations or conclusions based on the investigation of small model reaction systems that showed that the rates of consuming reactions of QSSA species are unusually high, that the concentrations, and the net rates of reaction of QSSA species are unusually low, that the induction period is usually short, and that most QSSA species are radicals. These observations are simple consequences of the physical pictures presented above and the error formulas derived from them.

Having stated that certain reactive characteristics (high rates of consumption, short lifetimes, short induction times, low concentrations, radicals, etc.) do not necessarily make a good QSSA species, we still use these characteristics to identify prospective QSSA species, but must then test to determine if these reaction intermediates are QSSA species [40, 29]. The key issue is how well the approximate solution that invokes the QSSA describes the exact solution. A detailed discussion of the error analysis is given by Turànyi et al. [40]. Here we illustrate the features of the QSSA and its application to a QSSA species.

For a spatially homogeneous reaction system, such as a constant-volume batch reactor, we write the following differential equation for each component

$$\frac{dc_j}{dt} = R_j = \sum_{i=1}^{n_r} \nu_{ij} r_i = f(T, c_j) \tag{5.48}$$

A system composed of n_s species leads to n_s differential equations, and the exact solution is found by solving Equations 5.48. The QSSA is applied to k QSSA species, where $k < n_s$. This leads to the following set of algebraic equations for the QSSA species

$$\frac{dc_{js}}{dt} = 0 = \sum_{i=1}^{n_r} \nu_{ij} r_i = g(T, c_s) \tag{5.49}$$

The system of differential-algebraic equations is solved subject to the same initial conditions. We show in the next section how the solution of Equations 5.49 leads to reaction-rate expressions in terms of only the non-QSSA species concentrations, which is the goal. Given today's computing capability, not much has been gained by reducing the set of ODEs required for the model solution. If one knows all the k_i values, it is just as easy to simulate the entire set of ODEs as it is to solve the reduced set of ODEs. If experimental data are being fit to a kinetic model, however, it is advantageous to reduce the number of estimated kinetic parameters. A valuable feature of the QSSA is that it eliminates hard-to-estimate rate constants from the model as demonstrated in Example 5.5.

To illustrate the QSSA we consider two simple elementary reaction schemes. Scheme I is given by

$$\mathrm{A} \xrightarrow{k_1} \mathrm{B} \qquad \mathrm{B} \xrightarrow{k_2} \mathrm{C}$$

and Scheme II is given by

$$\mathrm{A} \underset{k_{-1}}{\overset{k_1}{\rightleftharpoons}} \mathrm{B} \qquad \mathrm{B} \xrightarrow{k_2} \mathrm{C}$$

Let the initial concentration in a constant-volume, isothermal, batch reactor be $c_A = c_{A0}, c_B = c_C = 0$. The exact solution to the set of Equations 5.48 for Scheme I is

$$c_A = c_{A0} e^{-k_1 t} \tag{5.50}$$

$$c_B = c_{A0} \frac{k_1}{k_2 - k_1} \left(e^{-k_1 t} - e^{-k_2 t} \right) \tag{5.51}$$

$$c_C = c_{A0} \frac{1}{k_2 - k_1} \left(k_2 (1 - e^{-k_1 t}) - k_1 (1 - e^{-k_2 t}) \right) \tag{5.52}$$

The exact solution to the set of Equations 5.48 for Scheme II is[5]

$$c_A = c_{A0} \left(\frac{k_1(\alpha - k_2)}{\alpha(\alpha - \beta)} e^{-\alpha t} + \frac{k_1(k_2 - \beta)}{\beta(\alpha - \beta)} e^{-\beta t} \right)$$

$$c_B = c_{A0} \left(\frac{-k_1}{\alpha - \beta} e^{-\alpha t} + \frac{k_1}{\alpha - \beta} e^{-\beta t} \right)$$

$$c_C = c_{A0} \left(\frac{k_1 k_2}{\alpha \beta} + \frac{k_1 k_2}{\alpha(\alpha - \beta)} e^{-\alpha t} - \frac{k_1 k_2}{\beta(\alpha - \beta)} e^{-\beta t} \right)$$

$$\alpha = \frac{1}{2} \left(k_1 + k_{-1} + k_2 + \sqrt{(k_1 + k_{-1} + k_2)^2 - 4 k_1 k_2} \right)$$

$$\beta = \frac{1}{2} \left(k_1 + k_{-1} + k_2 - \sqrt{(k_1 + k_{-1} + k_2)^2 - 4 k_1 k_2} \right)$$

We next examine the effect of increasing k_2 on the concentration of C. Figure 5.10 shows the normalized concentration of C given in Equation 5.52 versus time for different values of k_2/k_1. The curve $c_C/c_{A0} = \left(1 - e^{-k_1 t}\right)$ is shown also (the condition where $k_2/k_1 = \infty$). Figure 5.10 illustrates that as $k_2 \gg k_1$, the exact solution is equivalent to

$$c_C = c_{A0} \left(1 - e^{-k_1 t} \right) \tag{5.53}$$

For $k_2 \gg k_1$ the rate of decomposition of B is much greater than its rate of formation, which is a necessary condition for B to be a QSSA species.

If component B is defined to be a QSSA species, we write

$$\frac{dc_{Bs}}{dt} = 0 = k_1 c_{As} - k_2 c_{Bs}$$

which leads to

$$c_{Bs} = \frac{k_1}{k_2} c_{As} \tag{5.54}$$

[5]The Laplace transform is a good approach for solving these differential equations. Also see Exercise 4.6.

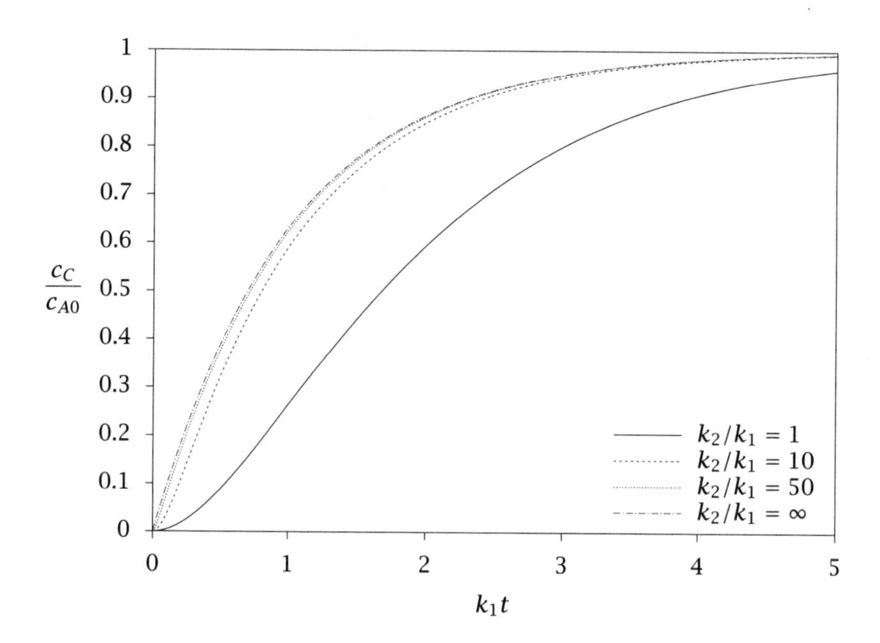

Figure 5.10: Normalized concentration of C versus dimensionless time for the series reaction A → B → C for different values of k_2/k_1.

for Scheme I. Note c_{Bs} is not constant; it changes linearly with the concentration of A. The issue is not if c_B is a constant, but whether or not the B equilibrates quickly to its quasi-steady value.

Substitution of Equation 5.54 into Equation 5.48 for components A and C results in

$$\frac{dc_{As}}{dt} = -k_1 c_{As} \tag{5.55}$$

$$\frac{dc_{Cs}}{dt} = k_1 c_{As} \tag{5.56}$$

for Scheme I. The solutions to Equations 5.55 and 5.56 are presented in Table 5.5 for the initial condition $c_{As} = c_{A0}, c_{Bs} = c_{Cs} = 0$.

Similarly, for Scheme II, when component B is defined to be a QSSA species,

$$c_{Bs} = \frac{k_1}{k_{-1} + k_2} c_{As} \tag{5.73}$$

Substitution of Equation 5.73 into Equation 5.48 for components A and

Scheme I:

$$A \xrightarrow{k_1} B \xrightarrow{k_2} C \tag{5.57}$$

Exact Solution		QSSA Solution	
$c_A = c_{A0} e^{-k_1 t}$	(5.58)	$c_{As} = c_{A0} e^{-k_1 t}$	(5.61)
$c_B = c_{A0} \frac{k_1}{k_2 - k_1} \left(e^{-k_1 t} - e^{-k_2 t}\right)$	(5.59)	$c_{Bs} = c_{A0} \frac{k_1}{k_2} e^{-k_1 t}$	(5.62)
$c_C = c_{A0} \frac{1}{k_2 - k_1} \left(k_2(1 - e^{-k_1 t}) - k_1(1 - e^{-k_2 t})\right)$	(5.60)	$c_{Cs} = c_{A0} \left(1 - e^{-k_1 t}\right)$	(5.63)

Scheme II:

$$A \underset{k_{-1}}{\overset{k_1}{\rightleftharpoons}} B \xrightarrow{k_2} C \tag{5.64}$$

Exact Solution		QSSA Solution	
$c_A = c_{A0} \left(\frac{k_1(\alpha - k_2)}{\alpha(\alpha - \beta)} e^{-\alpha t} + \frac{k_1(k_2 - \beta)}{\beta(\alpha - \beta)} e^{-\beta t} \right)$	(5.65)	$c_{As} = c_{A0} e^{-\frac{k_1 k_2}{k_{-1} + k_2} t}$	(5.70)
$c_B = c_{A0} \left(\frac{-k_1}{\alpha - \beta} e^{-\alpha t} + \frac{k_1}{\alpha - \beta} e^{-\beta t} \right)$	(5.66)	$c_{Bs} = c_{A0} \frac{k_1}{k_{-1} + k_2} e^{-\frac{k_1 k_2}{k_{-1} + k_2} t}$	(5.71)
$c_C = c_{A0} \left(\frac{k_1 k_2}{\alpha\beta} + \frac{k_1 k_2}{\alpha(\alpha - \beta)} e^{-\alpha t} - \frac{k_1 k_2}{\beta(\alpha - \beta)} e^{-\beta t} \right)$	(5.67)	$c_{Cs} = c_{A0} \left(1 - e^{-\frac{k_1 k_2}{k_{-1} + k_2} t} \right)$	(5.72)
$\alpha = \frac{1}{2} \left(k_1 + k_{-1} + k_2 + \sqrt{(k_1 + k_{-1} + k_2)^2 - 4 k_1 k_2} \right)$	(5.68)		
$\beta = \frac{1}{2} \left(k_1 + k_{-1} + k_2 - \sqrt{(k_1 + k_{-1} + k_2)^2 - 4 k_1 k_2} \right)$	(5.69)		

Table 5.5: Exact and QSSA solutions for kinetic Schemes I and II.

C results in

$$\frac{dc_{As}}{dt} = -\frac{k_1 k_2}{k_{-1} + k_2} c_{As} \tag{5.74}$$

$$\frac{dc_{Cs}}{dt} = \frac{k_1 k_2}{k_{-1} + k_2} c_{As} \tag{5.75}$$

The solutions to Equations 5.74 and 5.75 are presented in Table 5.5 for the initial condition $c_{As} = c_{A0}, c_{Bs} = c_{Cs} = 0$.

For the QSSA to be accurate, the error in the predicted concentrations of the reactants and products must be acceptable and the induction time should be small. The results in Table 5.5 permit us to examine the error in the predicted concentration of an important species, the reaction product C, and the induction time to reach the quasi-steady state. Equation 5.60 is used in Figure 5.10 to show c_C/c_{A0} versus dimensionless time ($k_1 t$) for different values of k_2/k_1. The dotted curve shows c_{Cs}/c_{A0} (Equation 5.63). The fractional error between the QSSA and exact values for c_C is shown in Figure 5.11 for much larger ratios of k_2/k_1 and for shorter dimensionless times than are presented in Figure 5.10. Figure 5.11 illustrates several points. As the lifetime of the intermediate product decreases (the reciprocal of k_2), the error decreases and the induction time (dimensionless because we plotted $k_1 t$) to reach an acceptable error, such as 10^{-2}, decreases. Turànyi et al. [40] have shown that the induction period to reach the QSSA is likely to be several times the lifetime of the longest-lived QSSA species. Figure 5.11 illustrates this change in induction time with decreasing lifetime. The induction time should be such that the quasi-steady state is established rapidly (i.e., before a significant amount of reactant A has disappeared). An error of 1% is reached at $k_1 t = 0.1$ for $k_2/k_1 = 10^4$, which corresponds to a conversion of about 10% for A. As k_2/k_1 increases, the amount of A that reacts before an error of 1% is reached decreases as indicated in Figure 5.11.

If component B were a radical or atom, it could be treated as a QSSA species provided k_2/k_1 were large enough because large k_2/k_1 leads to small errors in the predicted concentration of product. In this simple example the net rate of production of the intermediate never reached zero over the values of $k_1 t$ examined, demonstrating it is not necessary for the net rate of production of QSSA species to be zero. At sufficiently large values of k_2/k_1 the intermediate B can be considered a QSSA species because the error in the actual and approximate concentration of the QSSA reaches an acceptable level.

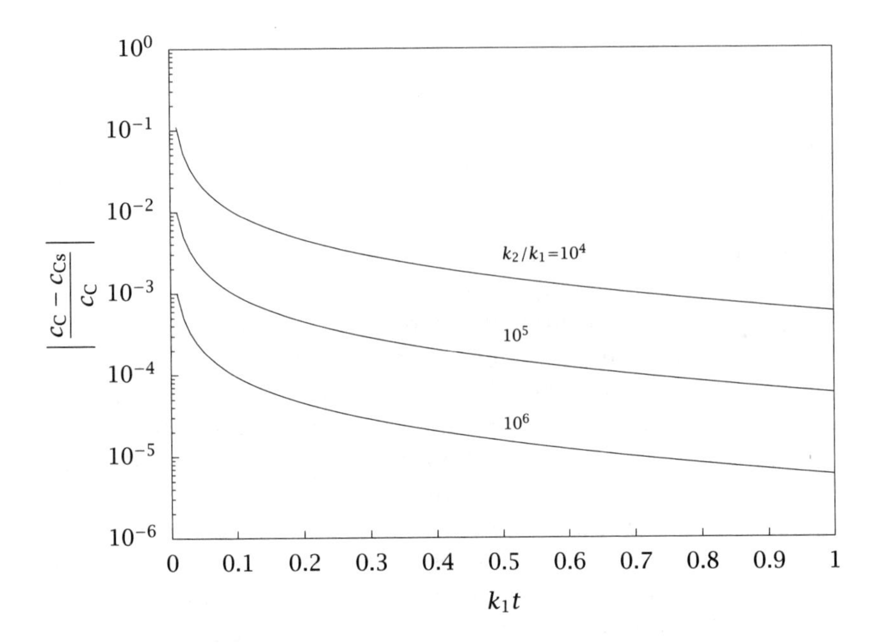

Figure 5.11: Fractional error in the QSSA concentration of C versus dimensionless time for the series reaction A → B → C for different values of k_2/k_1.

Similar concentration plots can be developed for the components in Scheme II to examine the effect of increasing the rate of decomposition of intermediate B relative to its rate of formation. Now both k_{-1} and k_2 need to varied. Figure 5.12 was generated using Equations 5.67 and 5.72 to illustrate the effect of decreasing the lifetime τ of B, here $\tau = (1/k_{-1} + 1/k_2)$. As the lifetime decreases, the error in c_C reaches an acceptable level such as 10^{-2} more rapidly. This effect can be seen by comparing the two curves for $k_{-1} = k_2 = 100k_1$ and $k_{-1} = k_2 = 1000k_1$. Note also that the error at long times does depend on the relative values of k_{-1} and k_2, and there are subtle differences in the induction time for comparable τ. Then provided τ is sufficiently small and k_{-1} and/or k_2 are sufficiently large, we would be justified in letting B be a QSSA species.

In summary, we write for a QSSA species

$$R_j = 0, \qquad j = \text{QSSA species} \tag{5.76}$$

The QSSA was developed for a constant-volume batch reactor, but can

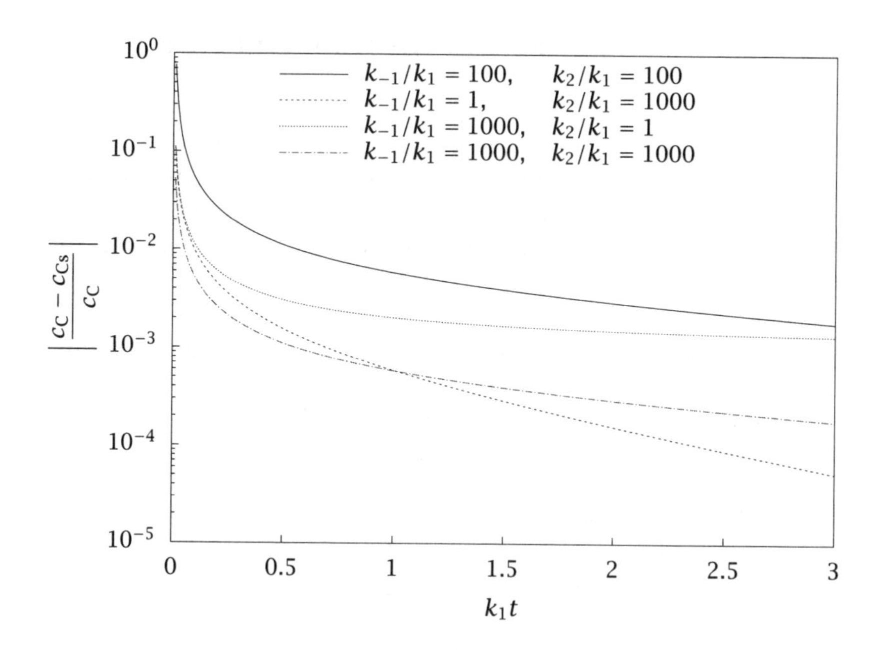

Figure 5.12: Fractional error in the QSSA concentration of C versus dimensionless time for the series-parallel reaction, $A \rightleftharpoons B \rightarrow C$.

be applied equally well to a PFR. For this reason, it is best to define the QSSA based on the production rate of species j, rather than the time derivative of the concentration of species j.

The QSSA is a useful tool in reaction analysis. Material balances for batch and plug-flow reactors are ordinary differential equations. By applying Equation 5.76 to the components that are QSSA species, their material balances become algebraic equations. These algebraic equations can be used to simplify the reaction expressions and reduce the number of equations that must be solved simultaneously. In addition, appropriate use of the QSSA can eliminate the need to know several difficult-to-measure rate constants. The required information can be reduced to ratios of certain rate constants, which can be more easily estimated from data. In the next section we show how the QSSA is used to develop a rate expression for the production of a component from a statement of the elementary reactions, and illustrate the kinetic model simplification that results from the QSSA model reduction.

5.5 Developing Rate Expressions from Complex Mechanisms

In this section we develop rate expressions for the mechanistic schemes presented in Section 5.1. The goal is to determine the rate in terms of measurable quantities, such as the concentrations of the reactants and products, and the temperature. Heterogeneous problems are considered in Section 5.6. The approach taken for homogeneous and heterogeneous reactions is similar, but additional constraints are placed on heterogeneous problems so the synthesis reaction example is presented after discussing these constraints.

The following procedure enables one to develop a reaction-rate expression from a set of elementary reactions.

1. Identify the species that do not appear in equilibrium reactions and write a statement for the rate of production of these species using Equations 5.38 and 5.39.

2. If the rate of production statement developed in step 1 contains the concentration of reaction intermediates that are QSSA species, their concentrations can be found in terms of reactant and product concentrations by writing a necessary number of algebraic statements. These algebraic statements come from applying the QSSA to reaction intermediates. In some cases, either by assumption or because of kinetic insight, an elementary reaction is treated as being at equilibrium.

3. Perform the necessary algebraic manipulations and substitute the resulting intermediate concentrations into the rate of production statement from step 1.

Example 5.2: Production rate of acetone

The thermal decomposition of acetone is represented by the following stoichiometry

$$3CH_3COCH_3 \longrightarrow CO + 2CH_4 + CH_2CO + CH_3COC_2H_5$$

Use the following mechanism to determine the production rate of acetone [35]. You may assume the methyl, acetyl and acetone radicals are

QSSA species.

$$\begin{aligned}
CH_3COCH_3 &\xrightarrow{k_1} CH_3 + CH_3CO && (5.77)\\
CH_3CO &\xrightarrow{k_2} CH_3 + CO && (5.78)\\
CH_3 + CH_3COCH_3 &\xrightarrow{k_3} CH_4 + CH_3COCH_2 && (5.79)\\
CH_3COCH_2 &\xrightarrow{k_4} CH_2CO + CH_3 && (5.80)\\
CH_3 + CH_3COCH_2 &\xrightarrow{k_5} CH_3COC_2H_5 && (5.81)
\end{aligned}$$

Solution

Let the species be designated as:

Species	Formula	Conc.	Name
A_1	CH_3COCH_3	c_{an}	acetone
A_2	CH_3	c_{mr}	methyl radical
A_3	CH_3CO	c_{alr}	acetyl radical
A_4	CO	c_{co}	carbon monoxide
A_5	CH_3COCH_2	c_{anr}	acetone radical
A_6	CH_2CO	c_{ke}	ketene
A_7	CH_4	c_{me}	methane
A_8	$CH_3COC_2H_5$	c_{mek}	methyl ethyl ketone

Write the production rate of acetone (an) using Reactions 5.77 and 5.79

$$R_{an} = -k_1 c_{an} - k_3 c_{an} c_{mr} \tag{5.82}$$

The methyl radical (mr), acetyl radical (alr) and acetone radical (anr) are QSSA species. Apply the QSSA to each of these species

$$R_{mr} = 0 = k_1 c_{an} + k_2 c_{alr} - k_3 c_{an} c_{mr} + k_4 c_{anr} - k_5 c_{mr} c_{anr} \tag{5.83}$$

$$R_{alr} = 0 = k_1 c_{an} - k_2 c_{alr} \tag{5.84}$$

$$R_{anr} = 0 = k_3 c_{an} c_{mr} - k_4 c_{anr} - k_5 c_{mr} c_{anr} \tag{5.85}$$

From Equation 5.84

$$c_{alr} = \frac{k_1}{k_2} c_{an} \tag{5.86}$$

Adding Equations 5.83, 5.84 and 5.85 gives

$$c_{anr} = \frac{k_1}{k_5} \frac{c_{an}}{c_{mr}} \tag{5.87}$$

Inserting the concentrations shown in Equations 5.86 and 5.87 into Equation 5.83 gives

$$k_3 c_{\text{mr}}^2 - k_1 c_{\text{mr}} - \frac{k_1 k_4}{k_5} = 0$$

$$c_{\text{mr}} = \frac{k_1}{2k_3} + \sqrt{\frac{k_1^2}{4k_3^2} + \frac{k_1 k_4}{k_3 k_5}} \tag{5.88}$$

in which the positive sign is chosen to obtain a positive concentration. This result can be substituted into Equation 5.82 to give the rate in terms of measurable species.

$$R_{\text{an}} = -\left(\frac{3}{2}k_1 + \sqrt{\frac{k_1^2}{4} + \frac{k_1 k_3 k_4}{k_5}}\right) c_{\text{an}} \tag{5.89}$$

Equation 5.89 can be simplified to

$$R_{\text{an}} = -k_{\text{eff}} c_{\text{an}} \tag{5.90}$$

by defining an effective rate constant

$$k_{\text{eff}} = \frac{3}{2}k_1 + \sqrt{\frac{k_1^2}{4} + \frac{k_1 k_3 k_4}{k_5}}$$

This simplified form illustrates the rate is first order in the acetone concentration. If, based on kinetic theories, we knew all the individual rate constants, we could calculate the rate of acetone production. Alternately, we can use Equation 5.90 as the basis for designing experiments to determine if the rate of production is first order in acetone and to determine the magnitude of the first-order rate constant, k_{eff}.

□

Example 5.3: Production rate of oxygen

Given the following mechanism for the overall stoichiometry

$$2\text{NO}_2 + h\nu \longrightarrow 2\text{NO} + \text{O}_2 \tag{5.91}$$

derive an expression for the production rate of O_2. Assume atomic O and NO_3 radicals are QSSA species. The production rate should be in terms of the reactant, NO_2, and products, NO and O_2.

$$NO_2 + h\nu \xrightarrow{k_1} NO + O \tag{5.92}$$

$$O + NO_2 \underset{k_{-2}}{\overset{k_2}{\rightleftharpoons}} NO_3 \tag{5.93}$$

$$NO_3 + NO_2 \xrightarrow{k_3} NO + O_2 + NO_2 \tag{5.94}$$

$$NO_3 + NO \xrightarrow{k_4} 2NO_2 \tag{5.95}$$

$$O + NO_2 \xrightarrow{k_5} NO + O_2 \tag{5.96}$$

Solution

From the reaction stoichiometry, the production rate of molecular oxygen is

$$R_{O_2} = k_3 c_{NO_2} c_{NO_3} + k_5 c_{NO_2} c_O \tag{5.97}$$

Applying the QSSA to the reaction intermediates, O and NO_3, gives

$$R_O = 0 = k_1 c_{NO_2} - k_2 c_{NO_2} c_O + k_{-2} c_{NO_3} - k_5 c_{NO_2} c_O \tag{5.98}$$

$$R_{NO_3} = 0 = k_2 c_{NO_2} c_O - k_{-2} c_{NO_3} - k_3 c_{NO_2} c_{NO_3} - k_4 c_{NO} c_{NO_3} \tag{5.99}$$

Adding Equations 5.98 and 5.99

$$0 = k_1 c_{NO_2} - k_3 c_{NO_2} c_{NO_3} - k_4 c_{NO} c_{NO_3} - k_5 c_{NO_2} c_O \tag{5.100}$$

This result can be used to simplify the rate expression. Adding Equations 5.97 and 5.100

$$R_{O_2} = k_1 c_{NO_2} - k_4 c_{NO} c_{NO_3} \tag{5.101}$$

The intermediate NO_3 concentration is found by solving Equation 5.98 for c_O.

$$c_O = \frac{k_1}{k_2 + k_5} + \frac{k_{-2}}{k_2 + k_5} \frac{c_{NO_3}}{c_{NO_2}} \tag{5.102}$$

Substituting Equation 5.102 into Equation 5.99 and rearranging gives

$$c_{NO_3} = \frac{k_1 k_2 c_{NO_2}}{(k_2 k_3 + k_3 k_5) c_{NO_2} + (k_2 k_4 + k_4 k_5) c_{NO} + k_{-2} k_5} \tag{5.103}$$

The rate expression now can be found using Equations 5.101 and 5.103

$$R_{O_2} = k_1 c_{NO_2} - \frac{k_1 k_2 k_4 c_{NO_2} c_{NO}}{(k_2 k_3 + k_3 k_5) c_{NO_2} + (k_2 k_4 + k_4 k_5) c_{NO} + k_{-2} k_5} \tag{5.104}$$

Equation 5.104 is rather complex; under reaction conditions several terms in the denominator are expected to be kinetically unimportant leading to a simple power-law reaction-rate expression. □

Example 5.4: Free-radical polymerization kinetics

Polymers are economically important and many chemical engineers are involved with some aspect of polymer manufacturing during their careers. Polymerization reactions raise interesting kinetic issues because of the long chains that are produced. Consider free-radical polymerization reaction kinetics as an illustrative example. A simple polymerization mechanism is represented by the following set of elementary reactions.

Initiation:

$$\mathrm{I} \xrightarrow{k_1} \mathrm{R_1}$$

Propagation:

$$\begin{aligned}
\mathrm{R_1} + \mathrm{M} &\xrightarrow{k_{p1}} \mathrm{R_2} \\
\mathrm{R_2} + \mathrm{M} &\xrightarrow{k_{p2}} \mathrm{R_3} \\
\mathrm{R_3} + \mathrm{M} &\xrightarrow{k_{p3}} \mathrm{R_4} \\
&\vdots \\
\mathrm{R_j} + \mathrm{M} &\xrightarrow{k_{pj}} \mathrm{R_{j+1}} \\
&\vdots
\end{aligned}$$

Termination:

$$\mathrm{R_m} + \mathrm{R_n} \xrightarrow{k_{tmn}} \mathrm{M_{m+n}}$$

in which M is monomer, M_j is a dead polymer chain of length j, and R_j is a growing polymer chain of length j. In free-radical polymerizations, the initiation reaction generates the free radicals, which initiate the polymerization reactions. An example is the thermal dissociation of benzoyl peroxide as initiator to form a benzyl radical that subsequently reacts with styrene monomer to form the first polymer chain. The termination reaction presented here is a termination of two growing polymer chains by a combination reaction.

Develop an expression for the rate of monomer consumption using the assumptions that $k_{pj} = k_p$ for all j, and $k_{tmn} = k_t$ is independent of the chain lengths of the two terminating chains. You may make the QSSA for all growing polymer radicals.

Solution

From the propagation reactions we can see the rate of monomer consumption is given by

$$R_M = -\sum_{j=1}^{\infty} r_{pj} = -k_p c_M \sum_{j=1}^{\infty} c_{Rj} \tag{5.105}$$

in which the rate of the jth propagation reaction is given by

$$r_{pj} = k_p c_M c_{R_j}$$

Making the QSSA for all polymer radicals, we set their production rates to zero

$$\begin{array}{lcccccl}
 & r_I & - & \boxed{k_p c_{R_1} c_M} & - & k_t c_{R_1} \sum_{j=1}^{\infty} c_{R_j} & = 0 \\
 & \boxed{k_p c_{R_1} c_M} & - & \boxed{k_p c_{R_2} c_M} & - & k_t c_{R_2} \sum_{j=1}^{\infty} c_{R_j} & = 0 \\
 & \boxed{k_p c_{R_2} c_M} & - & \boxed{k_p c_{R_3} c_M} & - & k_t c_{R_3} \sum_{j=1}^{\infty} c_{R_j} & = 0 \\
+ & \vdots & & \vdots & & & \vdots \\
\hline
 & r_I & & & - & k_t \sum_{i=1}^{\infty} c_{R_i} \sum_{j=1}^{\infty} c_{R_j} & = 0
\end{array} \tag{5.106}$$

The first term is the formation of growing chain i by propagation (initiation for $i = 1$), the second term is loss of chain i by propagation with monomer, and the third term is loss of chain i by termination reaction with chains of all other lengths. Notice if we add these equations, propagation terms on each line cancel corresponding entries on the lines directly above and below. It is valid to perform the infinite sum in Equation 5.106 because the polymer concentration for long chains goes to zero, $\lim_{j \to \infty} c_{P_j} = 0$. We solve Equation 5.106 for the total growing polymer concentration and obtain

$$\sum_{j=1}^{\infty} c_{Rj} = \sqrt{r_I / k_t}$$

Substituting this result into Equation 5.105 yields the monomer consumption rate

$$R_M = -k_p c_M \sqrt{r_I / k_t}$$

Reaction	A_0	E(kJ/mol)
1	1.0×10^{17}	356
2	2.0×10^{11}	44
3	3.0×10^{14}	165
4	3.4×10^{12}	28
5	1.6×10^{13}	0

Table 5.6: Ethane pyrolysis kinetics.

From the initiation reaction, the initiation rate is given by $r_I = k_I c_I$, which upon substitution gives the final monomer consumption rate

$$\boxed{R_M = -k_p \sqrt{k_I/k_t} \sqrt{c_I} c_M} \tag{5.107}$$

Notice that this result also provides a mechanistic justification for the production rate used in Example 4.3 in which monomer consumption rate was assumed linear in monomer concentration. □

Example 5.5: Ethane pyrolysis

This example illustrates how to apply the QSSA to a flow reactor. We are interested in determining the effluent concentration from the reactor and in demonstrating the use of the QSSA to simplify the design calculations. Ethane pyrolysis to produce ethylene and hydrogen also generates methane as an unwanted reaction product. The overall stoichiometry for the process is not a simple balance of ethane and the products. The following mechanism and their kinetics have been proposed for ethane pyrolysis [25]

$$\begin{aligned}
C_2H_6 &\xrightarrow{k_1} 2CH_3 \\
CH_3 + C_2H_6 &\xrightarrow{k_2} CH_4 + C_2H_5 \\
C_2H_5 &\xrightarrow{k_3} C_2H_4 + H \\
H + C_2H_6 &\xrightarrow{k_4} H_2 + C_2H_5 \\
H + C_2H_5 &\xrightarrow{k_5} C_2H_6
\end{aligned}$$

The rate constants are listed in Table 5.6 for the elementary reactions, in which $k = A_0 \exp(-E/RT)$. The preexponential factor A_0 has units of s^{-1} or $cm^3/mol\ s$ for first- and second-order reactions, respectively.

The pyrolysis is performed in a 100 cm^3 isothermal PFR, operating at a constant pressure of 1.0 atm and at 925 K. The feed consists of ethane in a steam diluent. The inlet partial pressure of ethane is 50 Torr and the partial pressure of steam is 710 Torr. The feed enters at a flowrate of 35 cm^3/s. The exact solution of this problem uses the methods developed in Chapter 4 and involves solving a set of eight initial-value ODEs of the form

$$\frac{dN_j}{dV} = R_j \tag{5.108}$$

subject to the initial conditions. In this numerical solution, each of the molar flows are computed as the volume increases from 0 to 100, with the concentrations of each component given by

$$c_j = \frac{N_j}{\sum_j N_j} \frac{P}{RT}$$

The results for C_2H_6, C_2H_4 and CH_4 are plotted in Figure 5.13. The H_2 concentration is not shown because it is almost equal to the C_2H_4 concentration. Note the molar flowrate of CH_4 is only 0.2% of the molar flowrate of the other products, C_2H_4 and H_2. The concentrations of the radicals, CH_3, C_2H_5 and H, are on the order of 10^{-6} times the ethylene concentration.

Assuming the radicals CH_3, C_2H_5 and H are QSSA species, develop an expression for the rate of ethylene formation. Verify that this approximation is valid.

Solution

The rate of ethylene formation is

$$R_{C_2H_4} = k_3 c_{C_2H_5} \tag{5.109}$$

Next use the QSSA to relate C_2H_5 to stable reactants and products

$$R_{CH_3} = 0 = 2k_1 c_{C_2H_6} - k_2 c_{CH_3} c_{C_2H_6} \tag{5.110}$$

$$R_H = 0 = k_3 c_{C_2H_5} - k_4 c_H c_{C_2H_6} - k_5 c_H c_{C_2H_5} \tag{5.111}$$

$$R_{C_2H_5} = 0 = k_2 c_{CH_3} c_{C_2H_6} - k_3 c_{C_2H_5} + k_4 c_{C_2H_6} c_H - k_5 c_H c_{C_2H_5} \tag{5.112}$$

Adding Equations 5.110, 5.111 and 5.112 gives

$$c_H = \frac{k_1}{k_5} \frac{c_{C_2H_6}}{c_{C_2H_5}} \tag{5.113}$$

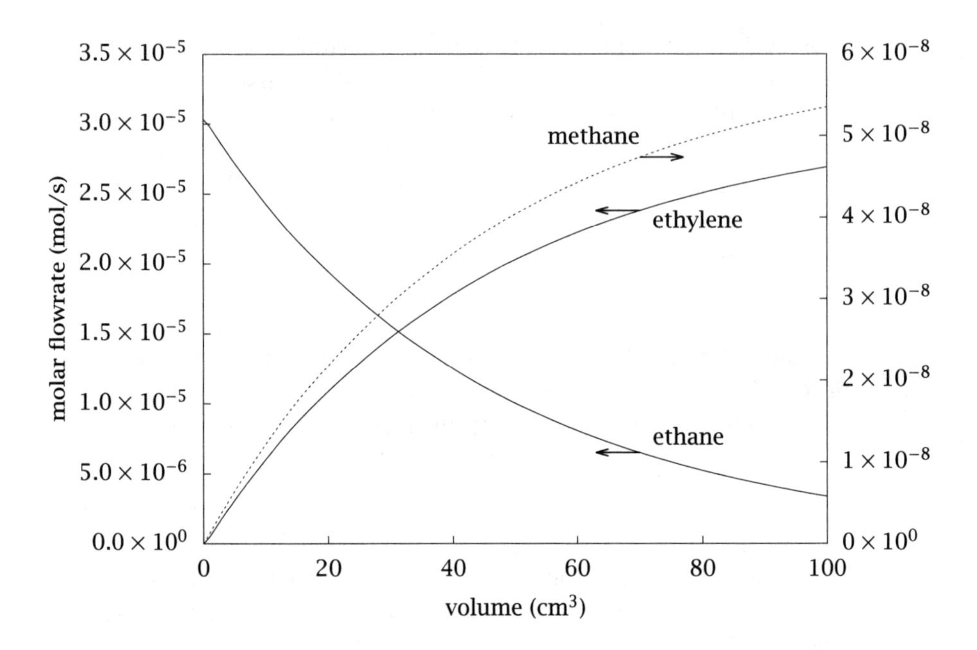

Figure 5.13: Molar flowrates of C_2H_6, C_2H_4 and CH_4 corresponding to the exact solution.

Inserting Equation 5.113 into Equation 5.111 yields

$$0 = k_3 k_5 c_{C_2H_5}^2 - k_4 k_1 c_{C_2H_6}^2 - k_1 k_5 c_{C_2H_6} c_{C_2H_5}$$

$$c_{C_2H_5} = \left(\frac{k_1}{2k_3} + \sqrt{\left(\frac{k_1}{2k_3}\right)^2 + \frac{k_1 k_4}{k_3 k_5}} \right) c_{C_2H_6} \tag{5.114}$$

Finally

$$R_{C_2H_4} = k_3 \left(\frac{k_1}{2k_3} + \sqrt{\left(\frac{k_1}{2k_3}\right)^2 + \frac{k_1 k_4}{k_3 k_5}} \right) c_{C_2H_6} \tag{5.115}$$

which can be rewritten as

$$R_{C_2H_4} = k c_{C_2H_6} \tag{5.116}$$

in which

$$k = k_3 \left(\frac{k_1}{2k_3} + \sqrt{\left(\frac{k_1}{2k_3}\right)^2 + \frac{k_1 k_4}{k_3 k_5}} \right)$$

At 925 K, $k = 0.797\ \mathrm{s}^{-1}$. The student should prove that

$$c_{\mathrm{H}} = \frac{k_1/k_5}{k_1/2k_3 + \sqrt{(k_1/2k_3)^2 + k_1k_4/k_3k_5}} \tag{5.117}$$

$$c_{\mathrm{CH_3}} = \frac{2k_1}{k_2} \tag{5.118}$$

The validity of the QSSA is established by solving the set of ODEs

$$\frac{dN_{\mathrm{C_2H_6}}}{dV} = -r_1 - r_2 - r_4 + r_5 \qquad \frac{dN_{\mathrm{H_2}}}{dV} = r_4$$

$$\frac{dN_{\mathrm{CH_4}}}{dV} = r_2 \qquad \frac{dN_{\mathrm{H_2O}}}{dV} = 0$$

$$\frac{dN_{\mathrm{C_2H_4}}}{dV} = r_3$$

subject to the initial molar flowrates. During the numerical solution the concentrations needed in the rate equations are computed using

$$c_{\mathrm{C_2H_6}} = \left(\frac{N_{\mathrm{C_2H_6}}}{N_{\mathrm{C_2H_6}} + N_{\mathrm{CH_4}} + N_{\mathrm{C_2H_4}} + N_{\mathrm{H_2}} + N_{\mathrm{H_2O}}}\right)\frac{P}{RT} \tag{5.119}$$

and Equations 5.114, 5.117 and 5.118. Note we can neglect H, CH_3 and C_2H_5 in the molar flowrate balance (Equation 5.119) because these components are present at levels on the order of 10^{-6} less than C_2H_6.

Figure 5.14 shows the error in the molar flowrate of ethylene that results from the QSSA. The error is less than 10% after 3.1 cm^3 of reactor volume; i.e., it takes about 3% of the volume for the error to be less than 10%. Similarly, the error is less than 5% after 5.9 cm^3 and less than 1% after 18.6 cm^3. Figure 5.14 provides evidence that the QSSA is valid for sufficiently large reactors, such as 100 cm^3. If the reactor volume were very small (a very short residence time), such as 20 cm^3, the QSSA would not be appropriate.

The reader might question what has been gained in this QSSA example. After all, the full model solution requires the simultaneous solution of eight ODEs and the reduced model requires solving five ODEs. If one had experimental data, such as effluent ethane, ethylene, hydrogen and methane concentrations for different residence times (V_R/Q_f), one would need to fit these data to find the five rate constants for either the full or the reduced model. The results in Figure 5.13, however, demonstrate that CH_4 is a minor product at this temperature and this also would be found in the experimental data. This suggests we can neglect

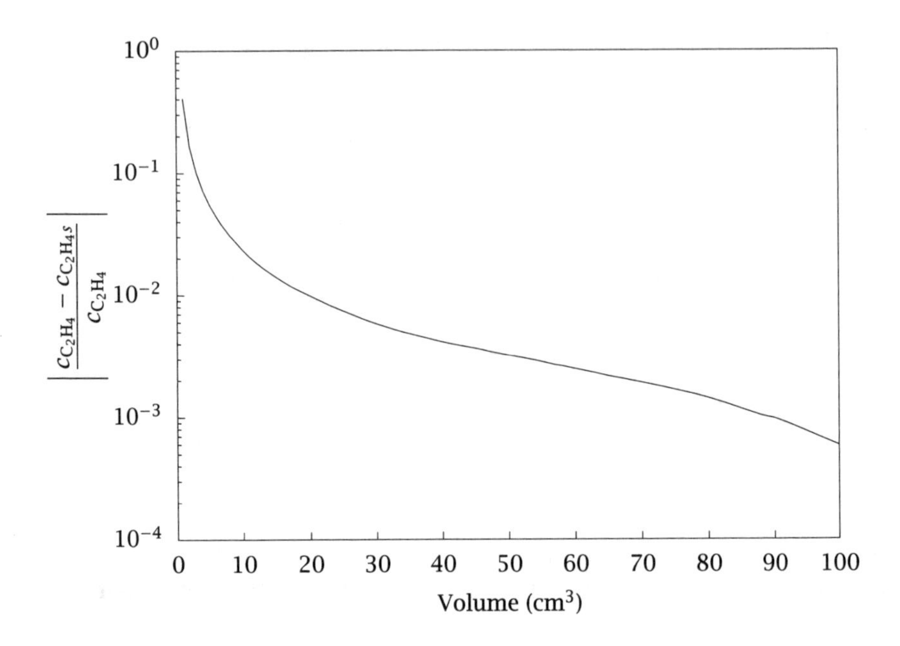

Figure 5.14: Fractional error in the QSSA molar flowrate of C_2H_4 versus reactor volume.

CH_4 in a species balance when computing mole fractions. Therefore the mass action statement

$$C_2H_6 \xrightarrow{k} C_2H_4 + H_2 \tag{5.120}$$

does a reasonable job of accounting for the changes that occur in the PFR during ethane pyrolysis. QSSA analysis predicts the ethane pyrolysis rate should be first order in ethane concentration and the rate constant is $k = 0.797\ s^{-1}$. Using the mass action statement to describe the reaction stoichiometry and the definition of the rate of production of a component in a single reaction

$$R_j = \nu_j r$$

to relate $R_{C_2H_6}$ and R_{H_2} to the result for C_2H_4 found in Equation 5.116, it is possible to solve the problem analytically. Figure 5.15 shows that nearly equivalent predictions are made for the simplified scheme based on the mass action statement, Reaction 5.120 using Equation 5.116 for the rate expression.

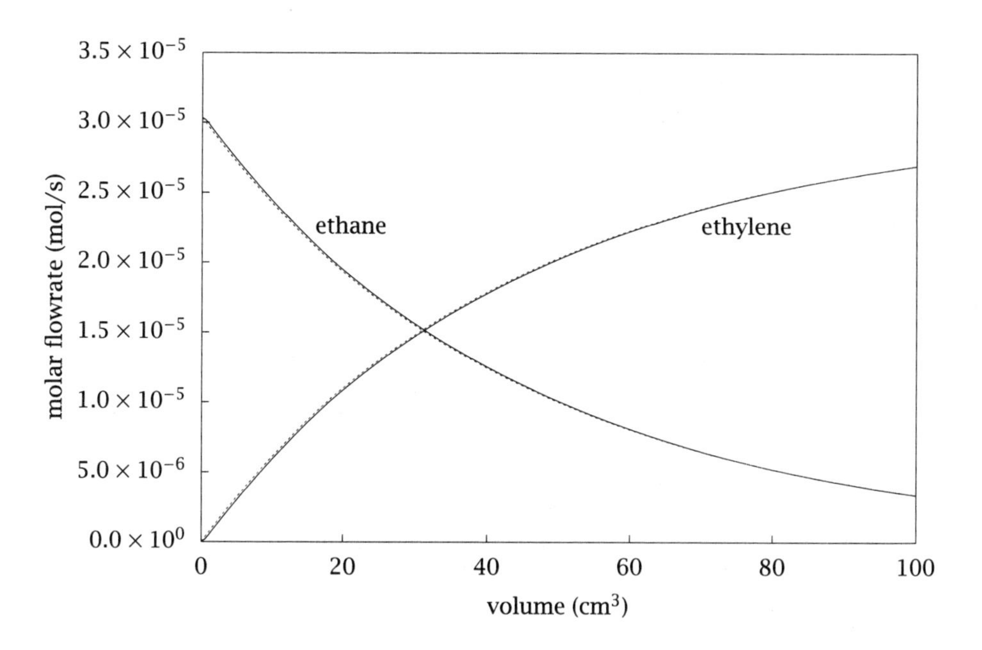

Figure 5.15: Comparison of the molar flowrates of C_2H_6 and C_2H_4 for the exact solution (solid lines) and the simplified kinetic scheme (dashed lines).

This simple example demonstrates that when one has all the kinetic parameters, they should be used because the QSSA buys very little in this case. The reverse situation of knowing a mechanism but not the rate constants could pose a difficult optimization problem when fitting all the rate constants, and the QSSA analysis can provide a framework for simplifying the kinetic expressions against which the data are tested. Here one would expect a first-order expression to describe adequately the appearance of ethylene. □

5.6 Reactions at Surfaces

Many books have been written on this topic and it continues to be an area of intense research. Our intention in this chapter is to provide the student with a simple picture of surfaces and to provide a basis for evaluating heterogeneously catalyzed reaction kinetics and reaction-rate expressions. Keep in mind the goal is to develop a rate expression

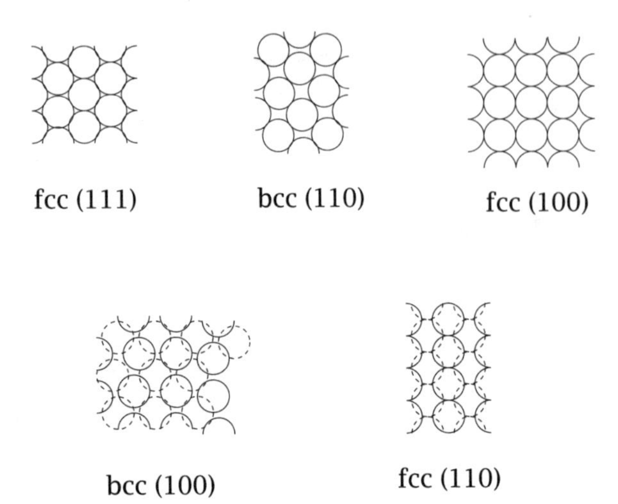

Figure 5.16: Surface and second-layer (dashed) atom arrangements for several low-index surfaces.

from a reaction mechanism. This rate expression can be used to direct the regression of rate data when determining the parameters for a heterogeneously catalyzed reaction. To understand how we construct surface reactions and develop rate expressions, it is necessary to discuss the elementary steps that comprise a surface reaction — adsorption, surface reaction and desorption. We also need to have a model of a surface at which the elementary steps occur. To streamline the presentation, the discussion is limited to gas-solid systems and a single chemical isotherm, the Langmuir isotherm. The interested student can find a more complete and thorough discussion of adsorption, surface reaction and desorption in numerous textbooks [8, 38, 3, 16, 27, 6].

A heterogeneously catalyzed reaction takes place at the surface of a catalyst. Catalysts, their properties, and the nature of catalytic surfaces are discussed in Chapter 7. For this discussion, we approximate the surface as a single crystal with a known surface order. The density of atoms at the low-index planes of transition metals is on the order of 10^{15} cm^{-2}. Figure 5.16 presents the atomic arrangement of low-index surfaces for various metals. This figure illustrates the packing arrangement and different combinations of nearest neighbors that can exist at a surface. For example, an fcc(111) surface atom has six nearest-surface neighbors, an fcc(100) surface atom has four nearest-surface neighbors,

Property	Chemisorption	Physisorption
amount	limited to a monolayer	multilayer possible
specificity	high	none
heat of adsorption	typically > 10 kcal/mol	low (2–5 kcal/mol)
activated	possibly	generally not

Table 5.7: Chemisorption and physisorption properties.

and an fcc(110) surface atom has two nearest-surface neighbors. We find that the interaction between an adsorbing molecule and the surface of a single crystal depends on the surface structure.

To facilitate our discussion of heterogeneous reaction kinetics let us consider the oxidation of CO on Pd, an fcc metal. The following mechanism has been proposed for the oxidation reaction over Pd(111) [12]

$$O_2 + 2S \rightleftharpoons 2O_{ads} \tag{5.121}$$

$$CO + S \rightleftharpoons CO_{ads} \tag{5.122}$$

$$CO_{ads} + O_{ads} \longrightarrow CO_2 + 2S \tag{5.123}$$

where the subscript 'ads' refers to adsorbed species and S refers to vacant surface sites. This simple example illustrates the steps in a catalytic reaction — adsorption of reactants in Reactions 5.121 and 5.122, reaction of adsorbed components in Reaction 5.123, and desorption of products also shown as part of Reaction 5.123.

Adsorption occurs when an incident atom or molecule sticks to the surface. The adsorbing species can be bound weakly to the surface or it can be held tightly to the surface. The manner by which the adsorbed species is held and the properties it exhibits once adsorbed determine the type of adsorption — physical or chemical. The dynamics of the process by which the incident adsorbate finds the adsorption site is used to construct a rate expression and rate constant for the adsorption step from first principles.

A number of criteria have been applied to distinguish between physical and chemical adsorption. In some cases the distinction is not clear. Table 5.7 lists several properties that can be used to distinguish between physisorption and chemisorption. The most distinguishing characteristics are the degree of coverage and the specificity.

Chemisorption: Chemical adsorption occurs when a chemical bond or a partial chemical bond is formed between the surface (adsorbent) and the adsorbate, leading to the specificity of the process. In general, but not always, the formation of a bond limits the coverage to at most one chemisorbed adsorbate for every surface atom. This limit of a single layer or monolayer is exploited later to derive a statement of conservation of sites. The upper limit need not have a one-to-one adsorbate to surface atom stoichiometry; for example, saturation can occur after one-third of all sites are occupied.

Physisorption: Physical adsorption occurs once the partial pressure of the adsorbate is above its saturation vapor pressure. Physisorption is similar to a condensation process and has practically no dependence on the solid surface and the interaction between the adsorbate and adsorbent. Just as water vapor condenses on any cold surface placed in ambient, humid air, a gas such as N_2 condenses on solids maintained at 77 K.

Adsorption is an exothermic process and the magnitude of the heat of adsorption is used to distinguish chemisorption from physisorption. Heats of adsorption greater than 10 kcal/mol are definitely associated with chemisorbed species. Small heats of adsorption (2–5 kcal/mol) do not always indicate physisorption, however. Therefore, it is best to look at more than one property when trying to distinguish between chemisorption and physisorption. Physical adsorption is used to measure the area of high-area oxide catalysts and oxide-supported metal catalysts. Physisorption isotherms and their use are discussed in Chapter 7. The discussion that follows treats only chemisorption.

Figure 5.17 depicts a schematic of a simple adsorption process. Adsorption occurs when an incident CO molecule finds a vacant site. The dynamics of the collision process, accommodation with the surface, and binding to the surface are complex, and the overall rate of this process can change for the same adsorbate on different metals or even different low-index surfaces of the same metal. The flux of CO molecules to the surface (molecules/area·time) is proportional to the number density of incident molecules and the mean kinetic velocity, and is given by

$$F_{CO} = \frac{P_{CO}}{\sqrt{2\pi M_{CO} k_B T}}$$

which, for pressure in torr, equals $3.83 \times 10^{20} P_{CO}$ molecule/s·cm^2 at 300 K. Since most surfaces contain approximately 10^{15} atoms/cm^2, at 2.6×10^{-6} Torr, the number of atoms in a monolayer strikes the surface

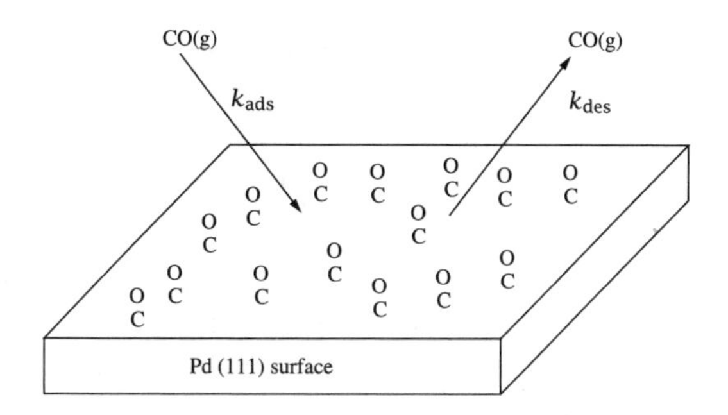

Figure 5.17: Schematic representation of the adsorption/desorption process.

per second. At one atmosphere, 290 million times the number of atoms in a monolayer strike the surface per second.

Not every collision leads to adsorption. The rate of adsorption can be written as the product of the flux times the probability of sticking

$$r_{\text{ads}} = FS(\theta, T)$$

S is known as the sticking coefficient, and is a function of the surface coverage and temperature; the exact functional dependence can be determined once the dynamics of the adsorption process are known [27]. We use the Langmuir adsorption to model the adsorption process.

The Langmuir adsorption model assumes that all surface sites are the same. As long as the incident molecules possess the necessary energy, and entropy considerations are satisfied, adsorption proceeds if a vacant site is available. The Langmuir process also assumes adsorption is a completely random surface process with no adsorbate-adsorbate interactions. Since all sites are equivalent, and adsorbate-adsorbate interactions are neglected, the surface sites can be divided into two groups, vacant and occupied. The surface-site balance is therefore

$$\left\{\begin{array}{c}\text{Total number of}\\ \text{surface sites}\\ \text{per unit area}\end{array}\right\} = \left\{\begin{array}{c}\text{Number of}\\ \text{vacant sites}\\ \text{per unit area}\end{array}\right\} + \left\{\begin{array}{c}\text{Number of}\\ \text{occupied sites}\\ \text{per unit area}\end{array}\right\}$$

or

$$\overline{c}_m = \overline{c}_v + \sum_{j=1}^{n_s} \overline{c}_j \tag{5.124}$$

Dividing by the total number of surface sites leads to the fractional form of the site balance.

$$1 = \frac{\overline{c}_v}{\overline{c}_m} + \sum_{j=1}^{n_s} \frac{\overline{c}_j}{\overline{c}_m} = \theta_v + \sum_{j=1}^{n_s} \theta_j \tag{5.125}$$

Equations 5.124 and 5.125 represent the conservation of sites for non-interacting adsorbates on uniform surfaces. Since active catalysts generally are dispersed on high-surface-area inert carriers, the units for $\overline{c}_m$ often are number of sites per gram of catalyst.

Figure 5.17 shows the model of a surface consisting of vacant sites and sites covered by chemisorbed CO. The CO in the gas phase is adsorbing on the surface and desorbing from the surface. The surface reaction, and the forward rate of adsorption and the reverse rate of desorption, are given by[6]

$$\text{CO} + \text{S} \underset{k_{-1}}{\overset{k_1}{\rightleftharpoons}} \text{CO}_{\text{ads}} \tag{5.126}$$

$$r_{\text{ads}} = k_1 c_{\text{CO}} \overline{c}_v, \qquad r_{\text{des}} = k_{-1} \overline{c}_{\text{CO}}$$

The units of the rate are mol/time·area. The rate expressions follow directly from the reaction mechanism because the reactions are treated as elementary steps. When the adsorption-desorption reactions are in equilibrium, the amount of adsorbed CO is related to the gas-phase partial pressure of CO, an equilibrium constant, and the total number of surface sites. Equating the adsorption and desorption rates at equilibrium gives

$$k_1 c_{\text{CO}} \overline{c}_v = k_{-1} \overline{c}_{\text{CO}}$$

Solving for the surface concentration gives

$$\overline{c}_{\text{CO}} = \frac{k_1}{k_{-1}} c_{\text{CO}} \overline{c}_v = K_1 c_{\text{CO}} \overline{c}_v \tag{5.127}$$

The conservation of sites gives

$$\overline{c}_v = \overline{c}_m - \overline{c}_{\text{CO}} \tag{5.128}$$

[6]This is an oversimplification of the adsorption process. A more acceptable view is to have CO adsorb into a precursor state and move around on the surface until it finds a suitable adsorption site, S. As long as the precursor state is in equilibrium with the gas phase, the Langmuir description of the adsorption process represented by Reaction 5.126 gives the same result as is obtained by a more accurate description of the adsorption and desorption dynamics. Since experimental results are often consistent with the precursor state in equilibrium with the gas phase, we present the Langmuir model.

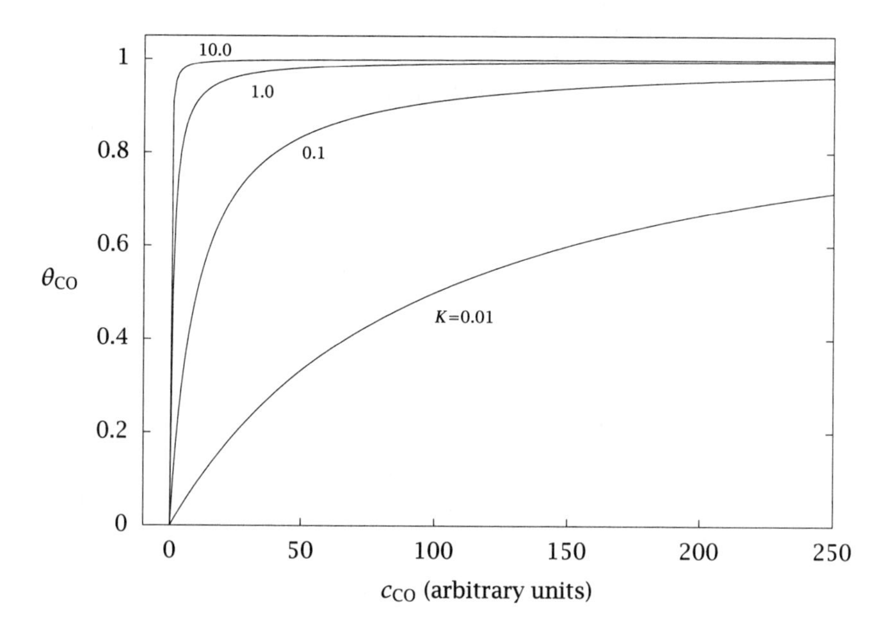

Figure 5.18: Fractional coverage versus adsorbate concentration for different values of the adsorption constant, K.

Substituting Equation 5.128 into Equation 5.127 and rearranging leads to the Langmuir isotherm for single-component, associative adsorption

$$\overline{c}_{CO} = \frac{\overline{c}_m K_1 c_{CO}}{1 + K_1 c_{CO}} \tag{5.129}$$

Dividing both sides of Equation 5.129 by the total concentration of surface sites $\overline{c}_m$ leads to the fractional form of the single component, associative Langmuir adsorption isotherm.

$$\theta_{CO} = \frac{K_1 c_{CO}}{1 + K_1 c_{CO}} \tag{5.130}$$

Figure 5.18 presents the general shape of the fractional coverage as a function of pressure for different values of K. In the limit $K_1 c_{CO} \gg 1$, Equations 5.129 and 5.130 asymptotically approach $\overline{c}_m$ and 1.0, respectively. Since adsorption is exothermic, K_1 decreases with increasing temperature and higher CO partial pressures are required to reach saturation. Figure 5.18 illustrates the effect of K on the coverage. In all cases the surface saturates with the adsorbate; however, the concentration at which this saturation occurs is a strong function of K. Single component adsorption is used to determine the number of active sites, $\overline{c}_m$. The adsorption data can be fit to either the Langmuir isotherm (Equation 5.129) or to the "linear" form of the isotherm. The linear form is obtained by taking the inverse of Equation 5.129 and multiplying by c_{CO}

$$\frac{c_{CO}}{\overline{c}_{CO}} = \frac{1}{K_1 \overline{c}_m} + \frac{c_{CO}}{\overline{c}_m} \tag{5.131}$$

A plot of the linear form provides $1/(K_1 \overline{c}_m)$ for the intercept and $1/\overline{c}_m$ for the slope.

Example 5.6: Fitting Langmuir adsorption constants to CO data

The active area of supported transition metals can be determined by adsorbing carbon monoxide. Carbon monoxide is known to adsorb associatively on ruthenium at 100°C. Use the following uptake data for the adsorption of CO on 10 wt % Ru supported on Al_2O_3 at 100°C to determine the equilibrium adsorption constant and the number of adsorption sites.

P_{CO} (Torr)	100	150	200	250	300	400
CO adsorbed (μmol/g cat)	1.28	1.63	1.77	1.94	2.06	2.21

Solution

Figures 5.19 and 5.20 plot the data for Equations 5.129 and 5.131, respectively. From Figure 5.20 we can estimate using least squares the slope, 0.35 g/μmol, and the zero concentration intercept, 1.822 cm^3/g, and then calculate

$$K_1 = 0.190 \text{ cm}^3/\mu\text{mol}$$
$$\overline{c}_m = 2.89 \ \mu\text{mol/g}$$

□

We now consider multiple-component adsorption. If other components, such as B and D, are assumed to be in adsorption-desorption equilibrium along with CO, we add the following two reactions

$$\text{B} + \text{S} \underset{k_{-2}}{\overset{k_2}{\rightleftharpoons}} \text{B}_{ads}$$
$$\text{D} + \text{S} \underset{k_{-3}}{\overset{k_3}{\rightleftharpoons}} \text{D}_{ads}$$

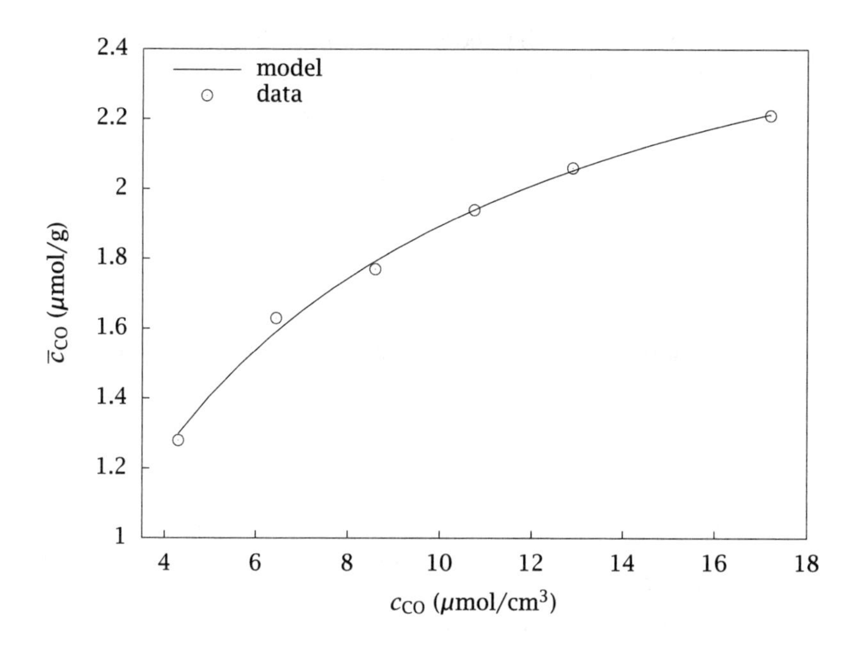

Figure 5.19: Langmuir isotherm for CO uptake on Ru.

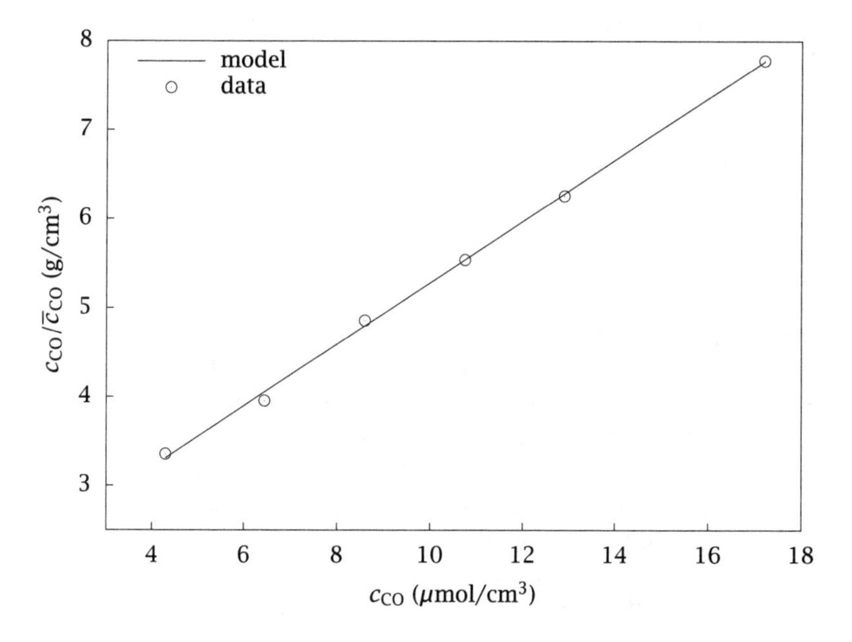

Figure 5.20: Linear form of Langmuir isotherm for CO uptake on Ru.

Following the development that led to Equation 5.127, we can show the equilibrium surface concentrations for B and D are given by

$$\overline{c}_B = K_2 c_B \overline{c}_v \tag{5.132}$$

$$\overline{c}_D = K_3 c_D \overline{c}_v \tag{5.133}$$

The site balance becomes

$$\overline{c}_v = \overline{c}_m - \overline{c}_{\mathrm{CO}} - \overline{c}_B - \overline{c}_D \tag{5.134}$$

Substituting Equation 5.134 into Equation 5.127 and rearranging leads to a different isotherm expression for CO.

$$\overline{c}_{\mathrm{CO}} = \frac{\overline{c}_m K_1 c_{\mathrm{CO}}}{1 + K_1 c_{\mathrm{CO}} + K_2 c_B + K_3 c_D} \tag{5.135}$$

The terms in the denominator of Equation 5.135 have special significance and the magnitude of the equilibrium adsorption constant, when multiplied by the respective concentration, determines the component or components that occupy most of the sites. Each of the terms $K_i c_j$ accounts for sites occupied by species j. From Chapter 3 we know the equilibrium constant is related to the exponential of the heat of adsorption, which is exothermic. Therefore, the more strongly a component is chemisorbed to the surface, the larger the equilibrium constant. Also, if components have very different heats of adsorption, the denominator may be well approximated using only one $K_i c_j$ term.

The CO oxidation example also contains a dissociative adsorption step

$$\mathrm{O}_2 + 2\mathrm{S} \underset{k_{-1}}{\overset{k_1}{\rightleftharpoons}} 2\mathrm{O}_{\mathrm{ads}} \tag{5.136}$$

Adsorption requires the O_2 molecule to find a pair of adsorption sites and desorption requires two adsorbed O atoms to be adjacent. Hill [18] and Kisliuk [21, 22] discuss lattice statistics and the probability of finding pairs of sites in two-dimensional arrays presented by the regular arrangement of surface atoms illustrated in Figure 5.16. Boudart and Djega-Mariadassou [6],and Hayward and Trapnell [16], describe how the probability of finding pairs of sites is used to develop rate expressions on surfaces. When a bimolecular surface reaction occurs, such as dissociative adsorption, associative desorption, or a bimolecular surface reaction, the rate in the forward direction depends on the probability of finding pairs of reaction centers. This probability, in turn, depends

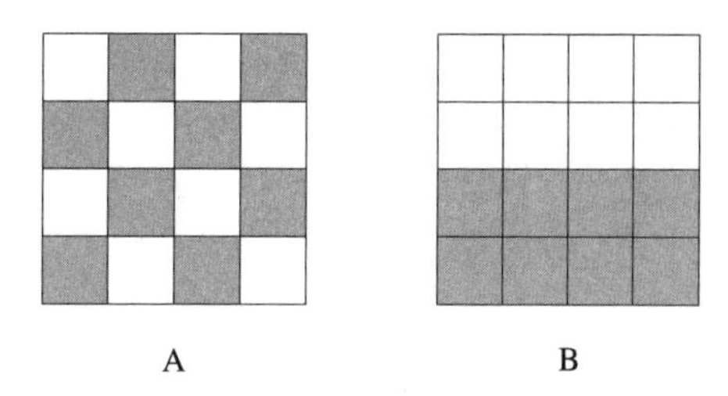

Figure 5.21: Two packing arrangements for a coverage of $\theta = 1/2$.

on whether or not the sites are randomly located, and whether or not the adsorbates are mobile on the surface.

The probability of finding pairs of vacant sites can be seen by considering adsorption onto a checkerboard such as Figure 5.21. Let two sites be adjacent if they share a common line segment (i.e., do not count sharing a vertex as being adjacent) and let θ equal the fraction of surface covered. The checkerboard has a total 24 adjacent site pairs, which can be found by counting the line segments on adjoining squares. Figure 5.21 presents two possibilities where half the sites (the shaded squares) are covered. For Figure 5.21A the probability an incident gas atom striking this surface in a random location hits a vacant site is $1-\theta$ and the probability a gas-phase molecule can dissociate on two adjacent vacant sites is $p = 0$. Figure 5.21B has 10 *vacant* adjacent site pairs. Therefore, the probability of dissociative adsorption is $p = 10/24$. The probability of finding vacant adsorption sites is $0 \leq p \leq 1/2$ for these two examples because the probability depends on the packing arrangement.

For *random* or independent packing of the surface the probability for finding pairs of vacant adjacent sites is $p = (1-\theta)^2$. The probability for random packing can be developed by considering five different arrangements for a site and its four nearest neighbors, with the center site always vacant, as shown in Figure 5.22. The probability of any configuration is $(1-\theta)^n\theta^m$, in which n is the number of vacant sites and m is the number of occupied sites. Arrangement 1 can happen only one way. However, arrangement 2 can happen four ways because the covered site can be in four different locations. Similarly, arrangement 3 can happen six different ways, 4 can happen four ways, and 5 can happen only one way. Therefore, the probability of two adjacent sites being vacant is

$$p_1 + 4p_2 + 6p_3 + 4p_4 + 0p_5 \tag{5.137}$$

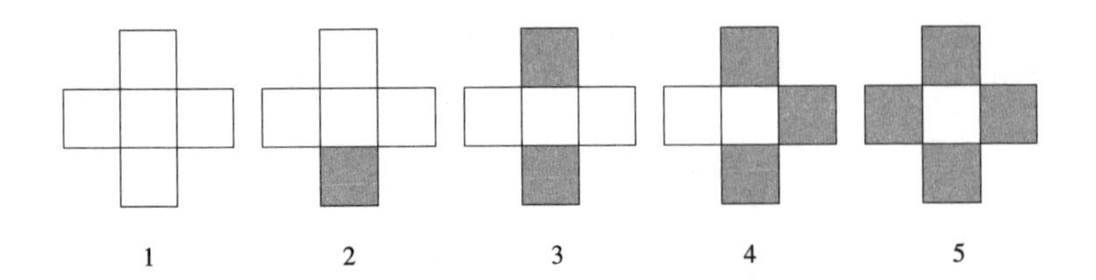

Figure 5.22: Five possible arrangements around a single site.

An incoming gas molecule can hit the surface in one of four positions, with one atom in the center and the other pointing either north, south, east or west. For random molecule orientations, the probabilities that a molecule hits the arrangement with a correct orientation are: p_1, $(3/4)4p_2$, $(1/2)6p_3$ and $(1/4)4p_4$. Therefore, the probability of a successful collision is

$$p = p_1 + 3p_2 + 3p_3 + p_4$$

$$p = (1-\theta)^5 + 3(1-\theta)^4\theta + 3(1-\theta)^3\theta^2 + (1-\theta)^2\theta^3$$

$$p = (1-\theta)^2[(1-\theta)^3 + 3(1-\theta)^2\theta + 3(1-\theta)\theta^2 + \theta^3]$$

Using a binomial expansion, $(x+y)^3 = x^3 + 3x^2y + 3xy^2 + y^3$, the term in brackets can be written as

$$p = (1-\theta)^2[(1-\theta) + \theta]^3$$

or

$$p = (1-\theta)^2$$

Therefore, we use $\theta_v \cdot \theta_v$ to represent the probability of finding pairs of vacant sites for dissociative adsorption. The probability of finding other pair combinations is the product of the fractional coverage of the two types of sites.

Returning to Reaction 5.136

$$O_2 + 2S \underset{k_{-1}}{\overset{k_1}{\rightleftharpoons}} 2O_{ads}$$

and using concentrations of vacant sites instead of fractional coverage, we write the rates of dissociative adsorption and associative desorption of oxygen as

$$r_{ads} = k_1 \bar{c}_v^2 c_{O_2} \tag{5.138}$$

$$r_{des} = k_{-1} \bar{c}_O^2 \tag{5.139}$$

The units on k_1 are such that r_{ads} has the units of mol/time·area. At equilibrium, the rate of adsorption equals the rate of desorption leading to

$$\overline{c}_{\text{O}} = \sqrt{K_1 c_{\text{O}_2}}\, \overline{c}_v \tag{5.140}$$

Combining Equation 5.140 with the site balance, Equation 5.124, and rearranging leads to the Langmuir form for single-component dissociative adsorption

$$\overline{c}_{\text{O}} = \frac{\overline{c}_m \sqrt{K_1 c_{\text{O}_2}}}{1 + \sqrt{K_1 c_{\text{O}_2}}} \tag{5.141}$$

The Langmuir isotherms represented in Equations 5.129 and 5.141 are different because each represents a different gas-surface reaction process. Both share the asymptotic approach to saturation, however, at sufficiently large gas concentration.

Example 5.7: Equilibrium CO and O surface concentrations

Determine the equilibrium CO and O surface concentrations when O_2 and CO adsorb according to

$$\text{CO} + \text{S} \underset{k_{-1}}{\overset{k_1}{\rightleftharpoons}} \text{CO}_{\text{ads}}$$
$$\text{O}_2 + 2\text{S} \underset{k_{-2}}{\overset{k_2}{\rightleftharpoons}} 2\text{O}_{\text{ads}}$$

Solution

At equilibrium the net rates of the two reactions are zero (adsorption rate equals desorption rate)

$$r_1 = 0 = k_1 c_{\text{CO}} \overline{c}_v - k_{-1} \overline{c}_{\text{CO}}$$
$$r_2 = 0 = k_2 c_{\text{O}_2} \overline{c}_v^2 - k_{-2} \overline{c}_{\text{O}}^2$$

Solving for the surface coverages in terms of the concentration of vacant sites gives

$$\overline{c}_{\text{CO}} = K_1 c_{\text{CO}} \overline{c}_v \tag{5.142}$$

$$\overline{c}_{\text{O}} = \sqrt{K_2 c_{\text{O}_2}} \overline{c}_v \tag{5.143}$$

The remaining unknown is $\overline{c}_v$, which can be found using the site balance

$$\overline{c}_m = \overline{c}_v + \overline{c}_{\text{CO}} + \overline{c}_{\text{O}} \tag{5.144}$$

Combining Equations 5.142–5.144 yields

$$\overline{c}_v = \frac{\overline{c}_m}{1 + K_1 c_{CO} + \sqrt{K_2 c_{O_2}}} \tag{5.145}$$

We next substitute the vacant site concentration into Equations 5.142 and 5.143 to give the surface concentrations in terms of gas-phase concentrations and physical constants

$$\overline{c}_{CO} = \frac{\overline{c}_m K_1 c_{CO}}{1 + K_1 c_{CO} + \sqrt{K_2 c_{O_2}}}$$

$$\overline{c}_{O} = \frac{\overline{c}_m \sqrt{K_2 c_{O_2}}}{1 + K_1 c_{CO} + \sqrt{K_2 c_{O_2}}}$$

□

The CO oxidation reaction involves associative adsorption of CO, dissociative adsorption of O_2, and the bimolecular surface reaction

$$CO + S \underset{k_{-1}}{\overset{k_1}{\rightleftharpoons}} CO_{ads}$$

$$O_2 + 2S \underset{k_{-2}}{\overset{k_2}{\rightleftharpoons}} 2O_{ads}$$

$$CO_{ads} + O_{ads} \xrightarrow{k_3} CO_2 + 2S$$

The reaction requires CO_s and O_s to occupy adjacent sites. If this reaction step is essentially irreversible, the rate expression is

$$r_3 = k_3 \overline{c}_{CO} \overline{c}_{O} \tag{5.146}$$

Note this rate expression is based on the probability of finding two dissimilar paired adsorbates.

Example 5.8: Production rate of CO_2

Find the rate of CO_2 production using the reaction mechanism listed above for CO oxidation. Assume the O_2 and CO adsorption steps are at equilibrium. Make a log-log plot of the production rate of CO_2 versus gas-phase CO concentration at constant gas-phase O_2 concentration. What are the slopes of the production rate at high and low CO concentrations?

Solution

The rate of CO_2 production is given by

$$R_{CO_2} = k_3 \overline{c}_{CO} \overline{c}_O \tag{5.147}$$

The surface concentrations for CO and O were determined in the preceding example,

$$\overline{c}_{CO} = \frac{\overline{c}_m K_1 c_{CO}}{1 + K_1 c_{CO} + \sqrt{K_2 c_{O_2}}}$$

$$\overline{c}_O = \frac{\overline{c}_m \sqrt{K_2 c_{O_2}}}{1 + K_1 c_{CO} + \sqrt{K_2 c_{O_2}}}$$

Substituting these concentrations onto Equation 5.147 gives the rate of CO_2 production

$$\boxed{R_{CO_2} = \frac{k_3 \overline{c}_m^2 K_1 c_{CO} \sqrt{K_2 c_{O_2}}}{\left(1 + K_1 c_{CO} + \sqrt{K_2 c_{O_2}}\right)^2}} \tag{5.148}$$

For ease of visualization, we define a dimensionless production rate

$$\tilde{R} = R_{CO_2} / (k_3 \overline{c}_m^2)$$

and plot this production rate versus dimensionless concentrations of CO and O_2 in Figure 5.23. Notice that the production rate goes to zero if either CO or O_2 gas-phase concentration becomes large compared to the other. This feature is typical of competitive adsorption processes and second-order reactions. A large reaction rate in Equation 5.147 requires large surface CO and surface O concentrations. But if one gas-phase concentration is large relative to the other, then that species saturates the surface and the other species surface concentration is small, which causes a small reaction rate. If we hold one of the gas-phase concentrations constant and vary the other, we are taking a slice through the surface in Figure 5.23. Figure 5.24 shows the result when each of the dimensionless gas-phase concentrations is held fixed at 1.0. Notice again that increasing the gas-phase concentration of either reactant increases the rate until a maximum is achieved, and then decreases the rate upon further increase in concentration. High gas-phase concentration of either reactant inhibits the reaction by crowding the other species off the surface. Notice from the plot shown in Figure 5.24 that $\log \tilde{R}$ versus $\log(K_1 c_{CO})$ changes in slope from 1.0 to -1.0 as CO concentration increases. As the concentration of O_2 increases, the slope

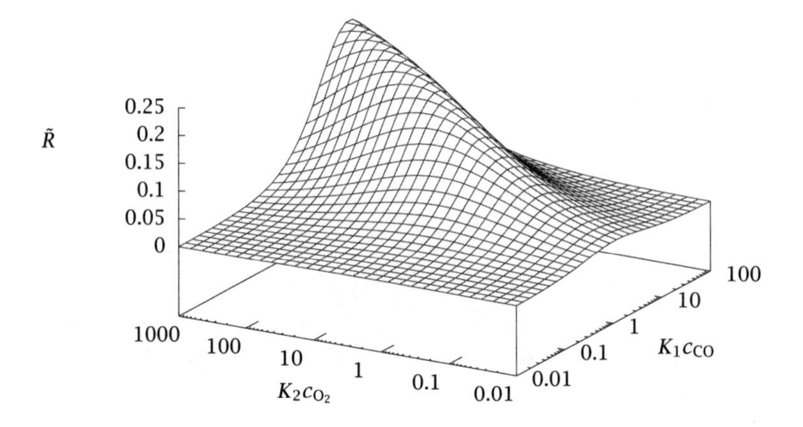

Figure 5.23: Dimensionless CO_2 production rate versus dimensionless gas-phase CO and O_2 concentrations.

changes from 0.5 to −0.5. These values also can be deduced by taking the logarithm of Equation 5.148. □

Rate expressions of the form of Equation 5.148 are known as Hougen-Watson or Langmuir-Hinshelwood kinetics [20]. This form of kinetic expression is often used to describe the species production rates for heterogeneously catalyzed reactions. We complete the section on the kinetics of elementary surface reactions by returning to the methane synthesis reaction listed in Section 5.2. The development proceeds exactly as outlined in Section 5.2. But now it is necessary to add a site-balance expression (Equation 5.124) in Step 3.

Example 5.9: Production rate of methane

Develop a rate expression for the synthesis of methane. The reaction is proposed to proceed as follows over a ruthenium catalyst [11]. The overall reaction is

$$3H_2(g) + CO(g) \longrightarrow CH_4(g) + H_2O(g) \tag{5.149}$$

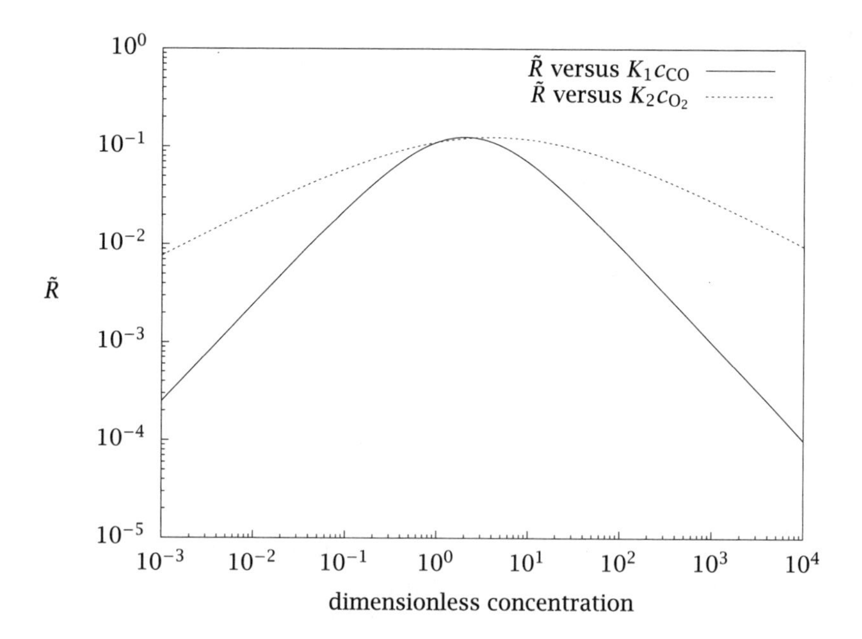

Figure 5.24: Dimensionless CO_2 production rate versus a single dimensionless gas-phase concentration while holding the other fixed at 1.0; slices through Figure 5.23.

and one possible mechanism is

$$\mathrm{CO(g)} + \mathrm{S} \underset{k_{-1}}{\overset{k_1}{\rightleftharpoons}} \mathrm{CO_{ads}} \tag{5.150}$$

$$\mathrm{CO_{ads}} + \mathrm{S} \underset{k_{-2}}{\overset{k_2}{\rightleftharpoons}} \mathrm{C_{ads}} + \mathrm{O_{ads}} \tag{5.151}$$

$$\mathrm{O_{ads}} + \mathrm{H_2(g)} \xrightarrow{k_3} \mathrm{H_2O(g)} + \mathrm{S} \tag{5.152}$$

$$\mathrm{H_2(g)} + 2\mathrm{S} \underset{k_{-4}}{\overset{k_4}{\rightleftharpoons}} 2\mathrm{H_{ads}} \tag{5.153}$$

$$\mathrm{C_{ads}} + \mathrm{H_{ads}} \underset{k_{-5}}{\overset{k_5}{\rightleftharpoons}} \mathrm{CH_{ads}} + \mathrm{S} \tag{5.154}$$

$$\mathrm{CH_{ads}} + \mathrm{H_{ads}} \underset{k_{-6}}{\overset{k_6}{\rightleftharpoons}} \mathrm{CH_{2ads}} + \mathrm{S} \tag{5.155}$$

$$\mathrm{CH_{2ads}} + \mathrm{H_{ads}} \underset{k_{-7}}{\overset{k_7}{\rightleftharpoons}} \mathrm{CH_{3ads}} + \mathrm{S} \tag{5.156}$$

$$\mathrm{CH_{3ads}} + \mathrm{H_{ads}} \xrightarrow{k_8} \mathrm{CH_4(g)} + 2\mathrm{S} \tag{5.157}$$

You may assume the reactions shown to be reversible are at equilibrium and that the surface is dominated by adsorbed CO.

Solution

Let the species be designated as:

Species	Formula	Conc.	Name
A_1	CO(g)	c_{CO}	carbon monoxide
A_2	S	$\overline{c}_v$	vacant site
A_3	CO_{ads}	$\overline{c}_{CO}$	adsorbed carbon monoxide
A_4	C_{ads}	$\overline{c}_C$	adsorbed carbon
A_5	O_{ads}	$\overline{c}_O$	adsorbed atomic oxygen
A_6	H_2(g)	c_{H_2}	hydrogen
A_7	H_2O(g)	c_{H_2O}	water
A_8	H_{ads}	$\overline{c}_H$	adsorbed atomic hydrogen
A_9	CH_{ads}	$\overline{c}_{CH}$	adsorbed methyne
A_{10}	CH_{2ads}	$\overline{c}_{CH_2}$	adsorbed methylene
A_{11}	CH_{3ads}	$\overline{c}_{CH_3}$	adsorbed methyl
A_{12}	CH_4(g)	c_{CH_4}	methane

The rate of methane formation is given by Reaction 5.157.

$$R_{CH_4} = k_8 \overline{c}_H \overline{c}_{CH_3} \tag{5.158}$$

From statements of equilibrium for the reversible reactions

$$\overline{c}_{CO} = K_1 c_{CO} \overline{c}_v \qquad \overline{c}_C = K_2 \frac{\overline{c}_{CO} \overline{c}_v}{\overline{c}_O} \qquad \overline{c}_H = \sqrt{K_4 c_{H_2}} \overline{c}_v$$

$$\overline{c}_{CH} = K_5 \frac{\overline{c}_C \overline{c}_H}{\overline{c}_v} \qquad \overline{c}_{CH_2} = K_6 \frac{\overline{c}_{CH} \overline{c}_H}{\overline{c}_v} \qquad \overline{c}_{CH_3} = K_7 \frac{\overline{c}_{CH_2} \overline{c}_H}{\overline{c}_v}$$

Substituting the expressions for $\overline{c}_{CH_2}$, $\overline{c}_{CH}$, and $\overline{c}_C$ into the expression for $\overline{c}_{CH_3}$ gives

$$\overline{c}_{CH_3} = K_2 K_5 K_6 K_7 \frac{\overline{c}_H^3 \overline{c}_{CO}}{\overline{c}_v^2 \overline{c}_O} \tag{5.159}$$

From the reaction stoichiometry, $R_{CH_4} = R_{H_2O}$ and[7]

$$\overline{c}_O = \frac{k_8}{k_3} \frac{\overline{c}_H \overline{c}_{CH_3}}{c_{H_2}} \tag{5.160}$$

[7]See Exercises 5.9 and 5.10 for a more complete description of when you can legitimately make this assumption.

Substituting Equation 5.160 into Equation 5.159 and rearranging gives

$$\overline{c}_{\text{CH}_3}^2 = \frac{k_3}{k_8} K_2 K_5 K_6 K_7 c_{\text{H}_2} \frac{\overline{c}_{\text{H}}^2 \overline{c}_{\text{CO}}}{\overline{c}_v^2}$$

Substitution of this expression into Equation 5.158 gives

$$R_{\text{CH}_4} = \sqrt{k_3 k_8} \sqrt{K_2 K_5 K_6 K_7} \sqrt{c_{\text{H}_2}} \frac{\overline{c}_{\text{H}}^2 \sqrt{\overline{c}_{\text{CO}}}}{\overline{c}_v} \tag{5.161}$$

We are told the surface is saturated with CO, i.e., $\theta_{\text{CO}} \cong 1$. If only CO adsorbs

$$\theta_3 = \frac{\overline{c}_{\text{CO}}}{\overline{c}_m} = \frac{K_1 c_{\text{CO}}}{1 + K_1 c_{\text{CO}}} \cong 1$$

Therefore $K_1 c_{\text{CO}} \gg 1$ and

$$\overline{c}_v = \frac{\overline{c}_m}{1 + K_1 c_{\text{CO}}} \cong \frac{\overline{c}_m}{K_1 c_{\text{CO}}} \tag{5.162}$$

Substituting Equation 5.162 into the expressions for the hydrogen and carbon monoxide surface concentrations, and combining the resulting expressions with Equation 5.161 leads to

$$R_{\text{CH}_4} = \sqrt{k_3 k_8 K_2 K_5 K_6 K_7} \overline{c}_m{}^{3/2} \frac{K_4}{K_1} \frac{(c_{\text{H}_2})^{3/2}}{c_{\text{CO}}}$$

which simplifies to

$$\boxed{R_{\text{CH}_4} = k_{\text{eff}} \frac{(c_{\text{H}_2})^{3/2}}{c_{\text{CO}}}}$$

□

5.7 Summary

We introduced several concepts in this chapter that are the building blocks for reaction kinetics.

- Most reaction processes consist of more than one elementary reaction. The reaction process can be represented by a mass action statement, such as $C_2H_6 \longrightarrow C_2H_4 + H_2$ in Example 5.5, but this statement does not describe how the reactants convert into the products. The atomistic description of chemical events is found in the elementary reactions. First principles calculations can be used to predict the order of these elementary reactions and the values of their rate constants.

- The reaction order for elementary reactions is determined by the stoichiometry of the reaction

$$r_i = k_i \prod_{j \in \mathcal{R}_i} c_j^{-\nu_{ij}} - k_{-i} \prod_{j \in \mathcal{P}_i} c_j^{\nu_{ij}}$$

- Two assumptions are generally invoked when developing simplified kinetic rate expressions: (i) some of the elementary reactions are slow relative to the others; the fast reactions can be assumed to be at equilibrium. If only one step is slow, this single reaction determines all production rates, and is called the rate-limiting step; and (ii) the rate of formation of highly reactive intermediate species can be set to zero; the concentration of these intermediates is found from the resulting algebraic relations rather than differential equations.

- The chemical steps involved in heterogeneous reactions (adsorption, desorption, surface reaction) are generally treated as elementary reactions. In many cases, one reaction is the slow step and the remaining steps are at equilibrium. For heterogeneous reactions, we add a site balance to account for the vacant and occupied sites that take part in the reaction steps.

Notation

a_j	activity of species j
c_j	concentration of species j
c_{js}	steady-state concentration of j
$\overline{c}_j$	concentration of species j on the catalyst surface
$\overline{c}_m$	total active surface concentration (monolayer coverage concentration)
$\overline{c}_v$	concentration of vacant surface sites
f_j	fugacity of species j
f_j°	standard-state fugacity of pure species j
F_j	flux of species j from the gas phase to the catalyst surface
g_{ei}	degeneracy of the ith electronic energy level
h	Planck's constant
I	moment of inertia for a rigid rotor
I_i	moment of inertia about an axis
k_B	Boltzmann constant
K	reaction equilibrium constant

m	mass of a molecule (atom)
M_j	molecular weight of species j
n_j	moles of species j
N_j	molar flowrate of species j
p	probability of finding adjacent pairs of sites
P	total pressure
P_j	partial pressure of species j
$\mathcal{P}_i$	product species in reaction i
$\left(\frac{q}{V}\right)_j$	molecular partition function of species j
q_{elec}	electronic partition function
q_{rot}	rotational partition function
q_{vib}	vibrational partition function
Q	partition function
Q_f	volumetric flowrate at the reactor inlet
r_i	reaction rate for ith reaction
R	gas constant
R_j	production rate for jth species
$\mathcal{R}_i$	reactant species in reaction i
t	time
T	absolute temperature
$V(r)$	intermolecular potential energy
V_R	reactor volume
z	compressibility factor
ϵ_i	energy of the ith electronic level
ε_i	extent of the ith reaction
θ_j	fractional surface coverage of species j
θ_v	fractional surface coverage of vacant sites
ν_{ij}	stoichiometric number for the jth species in the ith reaction
σ	symmetry number, 1 for a heteronuclear molecule, 2 for homonuclear molecule
τ	lifetime of a component
ϕ_j	fugacity coefficient for species j

5.8 Exercises

Exercise 5.1: Equilibrium assumption and nonlinear kinetics

It is common to add groups to difunctional molecules in two consecutive steps:

$$A + B \underset{k_{-1}}{\overset{k_1}{\rightleftharpoons}} C, \qquad C + B \underset{k_{-2}}{\overset{k_2}{\rightleftharpoons}} D$$

The difunctional starting material (A) combines with the primary reactant (B) to form the monofunctional intermediate (C), which subsequently combines with B to produce the desired final product (D).

(a) Write down the material balances for components A, B, C and D if this reaction takes place in a batch reactor.

(b) Let's assume the first step is very slow compared to the second. Solve for the equilibrium concentration of component D by setting the rate of the second reaction to zero (i.e., the second reaction is at equilibrium compared to the first). Then we can eliminate the ODE for component D in the simplified model and replace it with this algebraic equation.

(c) What are the differential equations for concentrations of components A, B and C in the simplified model? Hint: don't just set $r_2 = 0$ in the full model but eliminate r_2 instead.

(d) A numerical ODE solver like Octave can solve the full set of four ODEs in part 5.1a as easily as the set of three ODEs in part 5.1c.

Why might you prefer the simplified model anyway? Be specific.

Exercise 5.2: Nitrogen dioxide reaction mechanism

The following overall stoichiometry has been observed for the decomposition of gaseous dinitrogen pentoxide to gaseous nitrogen dioxide and oxygen [30].

$$2N_2O_5 \rightleftharpoons 4NO_2 + O_2$$

Consider the following proposed mechanism to explain this stoichiometry.

$$\begin{aligned} N_2O_5 &\rightleftharpoons NO_2 + NO_3 \\ NO_2 + NO_3 &\longrightarrow NO + O_2 + NO_2 \\ NO + NO_3 &\longrightarrow 2NO_2 \end{aligned}$$

(a) Write down the production rate of each species in the mechanism.

(b) After making the quasi-steady-state assumption (QSSA) for the intermediates, write down the production rate of the reactants and products.

(c) How many rate constants would you need to determine experimentally to use the full model to predict the concentrations of N_2O_5, NO_2 and O_2 in the reactor? How many rate constants would you need to determine experimentally to use the QSSA model to predict these same concentrations?

Exercise 5.3: Using the QSSA

Consider the simple series reaction mechanism taking place in a constant-volume batch reactor

$$A \underset{k_{-1}}{\overset{k_1}{\rightleftharpoons}} B, \qquad B \underset{k_{-2}}{\overset{k_2}{\rightleftharpoons}} C$$

(a) Write down the mole balance differential equations for species A, B and C. We call this set of differential equations the full model.

(b) For what values of rate constants would you expect the QSSA to be valid for species B?

(c) Make the QSSA on species B and write down the resulting differential equations for species A and C.

(d) Plot c_A, c_B and c_C versus time for the solution to the full model for initial conditions $c_A(0) = 1$, $c_B(0) = c_C(0) = 0$, and rate constants $k_1 = k_{-2} = 1$, $k_2 = k_{-1} = 20$. You can solve this model analytically or numerically.

(e) Now solve the simplified model, either analytically or numerically. Make three plots, one for each species, showing $c_j(t)$ for the full model and the simplified model. Are you satisfied using the QSSA for this situation? What advantage does the simplified model provide?

Exercise 5.4: Applying the QSSA to develop a rate expression

The thermal decomposition of bis-pentafluorosulfurtrioxide ($SF_5O_3SF_5$)

$$SF_5O_3SF_5 \longrightarrow SF_5O_2SF_5 + \frac{1}{2}O_2$$

is postulated to proceed by the following mechanism for oxygen partial pressures greater than 100 Torr [10].

$$SF_5OOOSF_5 \underset{k_{-1}}{\overset{k_1}{\rightleftharpoons}} SF_5O + SF_5O_2$$

$$SF_5O_2 + SF_5O_2 \overset{k_2}{\longrightarrow} 2SF_5O + O_2$$

$$2SF_5O \overset{k_3}{\longrightarrow} SF_5OOSF_5$$

The first reaction is not at equilibrium and SF_5O and SF_5O_2 are radicals. Develop an expression for the rate of bis-pentafluorosulfurtrioxide decomposition in terms of stable molecules.

Exercise 5.5: Using QSSA to develop a simplified rate expression

Example 4.7 presented elementary reactions for ethane pyrolysis in the presence of NO. Use the data provided in Example 4.7 to answer the following questions.

(a) Apply the quasi-steady-state analysis to these reactions, where HNO, H, and C_2H_5 are reaction intermediates to find the rate of ethylene production in terms of stable molecules.

(b) Detailed analysis of the products in a PFR reveals that the reactions can be represented with the following mass action statement

$$C_2H_6 + NO \xrightarrow{k_{eff}} C_2H_4 + H_2 + NO$$

where

$$r = k_{eff} c_{C_2H_6}$$

$$k_{eff} = k_2 \sqrt{\frac{k_1 k_3 k_{-4}}{k_{-1} k_2 k_4}}$$

and it is necessary to follow only the concentrations of C_2H_6, C_2H_4, H_2 and NO.

Assume the reaction takes place in an isothermal PFR operating at constant pressure (1.0 atm) and constant temperature of 1100 K. The feed to the reactor consists of a mixture of ethane and NO with a molar ratio of 95% ethane and 5% NO. The inlet volumetric flowrate is 600 cm^3/s. Predict the reactor volume required for 98% of the ethane to react, and determine the activation energy for k_{eff}.

(c) Compare the answer for this simplified model that uses k_{eff} with the full solution to this problem in which all components are followed as illustrated in Example 4.7.

Exercise 5.6: Acetaldehyde decomposition mechanism and kinetics

The pyrolysis of acetaldehyde, CH_3CHO, has been proposed to involve the following reactions [35].

$$CH_3CHO \xrightarrow{k_1} CH_3 + CHO$$

$$CH_3 + CH_3CHO \xrightarrow{k_2} CH_4 + CH_3CO$$

$$CH_3CO \xrightarrow{k_3} CH_3 + CO$$

$$2CH_3 \xrightarrow{k_4} C_2H_6$$

In this mechanism CH_3, CH_3CO and CHO are reaction intermediates. The radical CHO undergoes further reactions than are shown, but for simplicity they are ignored here. Assuming the reaction intermediates (CH_3 and CH_3CO) are quasi-steady-state intermediates, determine the rate of methane formation in terms of only stable molecules.

Exercise 5.7: Two candidate reduced models

Consider the simple series reaction mechanism taking place in a constant-volume batch reactor

$$A \underset{k_{-1}}{\overset{k_1}{\rightleftharpoons}} B, \qquad B \underset{k_{-2}}{\overset{k_2}{\rightleftharpoons}} C$$

(a) Write down the mole balance differential equations for species A, B and C. We call this set of differential equations the full model.

(b) For what range of rate constants would you think it is reasonable to assume the second reaction is at equilibrium compared to the first reaction?

(c) We write down two candidate simplified models to try to describe this situation. In the first one, set $r_2 = 0$ and solve for c_C. Then set $r_2 = 0$ in the remaining ODEs for c_A and c_B.

(d) In the second one, solve for c_C as above, but now eliminate r_2 from the differential equations, rather than setting it to zero, and write down differential equations for c_A and c_B.

(e) Solve the three models for the following values (choose any time and concentration units you like). Feel free to solve it analytically or numerically. Plot the concentrations versus time for each model.

$$k_1 = 1 \qquad k_{-1} = 0 \qquad k_2 = 100 \qquad k_{-2} = 200$$
$$c_{A0} = 0.5 \qquad c_{B0} = 0.5 \qquad c_{C0} = 0$$

Which simplified model correctly describes the full model in the limit of a fast second reaction?

What went wrong in the approach that didn't work?

Exercise 5.8: Mechanisms and elementary reactions

(a) What is the important difference between a reaction with an observed stoichiometry and an elementary reaction?

(b) List two simple tests a mechanism must pass to be considered a valid description of an overall stoichiometry.

(c) Name and describe the two major assumptions that are made to simplify the rate expressions for complex reaction networks.

(d) Describe the transition-state concept and give a chemical example.

Exercise 5.9: Disproportionation of cumene

The disproportionation of cumene ($C_6H_5CH(CH_3)_2$) to benzene and propylene

$$\mathrm{C} \longrightarrow \mathrm{B} + \mathrm{P}$$

proceeds by the following catalytic mechanism [9]:

$$\mathrm{C(g)} + \mathrm{X} \underset{k_{-1}}{\overset{k_1}{\rightleftharpoons}} \mathrm{C \cdot X}$$
$$\mathrm{C \cdot X} \underset{k_{-2}}{\overset{k_2}{\rightleftharpoons}} \mathrm{B \cdot X} + \mathrm{P(g)}$$
$$\mathrm{B(g)} + \mathrm{X} \underset{k_{-3}}{\overset{k_3}{\rightleftharpoons}} \mathrm{B \cdot X}$$

in which C, B, P and X represent cumene, benzene, propylene and vacant sites, respectively. The A·X symbol denotes an A molecule adsorbed on the catalyst surface. Propylene does not adsorb on the surface. Develop production-rate expressions for the following two situations.

(a) The surface reaction is rate limiting and irreversible, and the adsorption and desorption of benzene and cumene are at equilibrium. Develop the production-rate expression for propylene.

(b) The adsorption of cumene is rate limiting and irreversible, and the adsorption and desorption of benzene and the surface reaction are at equilibrium. Develop the production-rate expression for cumene.

(c) After glancing at the overall stoichiometry,

$$C \longrightarrow B + P$$

a colleague claims that the production rate of cumene must also be the negative of the production rates of propylene and benzene. Do you agree? Why or why not?

Exercise 5.10: Overall stoichiometry and production rates

Consider again the disproportionation of cumene described in Exercise 5.9.

(a) Assume you are running this reaction in a well-mixed gas-phase CSTR containing a small, solid-catalyst bed. Gas-phase cumene is fed to the reactor, and the effluent gas containing propylene, benzene, and any unreacted cumene is withdrawn. If this reactor achieves steady state, from the CSTR material balances what can you say about the relationships between the steady-state production rates of gas-phase cumene, benzene and propylene?

(b) Assume that you run the experiment while maintaining low concentrations of adsorbed cumene and benzene compared to the gas-phase concentrations. What can you conclude about the production rates of gas-phase cumene, benzene and propylene under these conditions?

(c) Corrigan et al. [9] used initial rates to support the mechanism given in Exercise 5.9. What simple experiments can you do to determine which of the two sets of mechanistic assumptions in Exercise 5.9 better describes the chemistry of cumene disproportionation? Explain how you would interpret the results of your proposed experiments to make this determination.

Exercise 5.11: Fractional coverage for multicomponent adsorption

The conversion of carbon monoxide and hydrogen into synthetic fuels occurs over metals. Carbon monoxide and hydrogen compete for surface sites in the adsorption phase of the synthesis process.

Determine the fraction of surface sites that are covered by CO and atomic hydrogen in the absence of any subsequent reactions. The adsorption reactions are:

$$CO(g) + X \rightleftharpoons CO \cdot X \qquad K_{CO} = 3.8 \times 10^4 \text{ cm}^3/\text{mol}$$

$$H_2(g) + 2X \rightleftharpoons 2H \cdot X \qquad K_{H_2} = 4.9 \times 10^3 \text{ cm}^3/\text{mol}$$

The gas-phase conditions are $P_{CO} = 3$ atm, $P_{H_2} = 6$ atm and $T = 398$ K.

Exercise 5.12: Oxygen adsorption

The data in Table 5.8, which are plotted in Figure 5.25, were collected for adsorption of oxygen on a Pd catalyst. The adsorption is dissociative

$$O_2 + 2X \underset{k_{-1}}{\overset{k_1}{\rightleftharpoons}} 2O \cdot X$$

$c_{O_2} \times 10^3$ (mol/L)	3.05	6.10	9.15	12.2	15.2	61.0	76.2	91.5	107	122	137	152
$\overline{c}_O \times 10^6$ (mol/g·cat)	1.00	1.62	2.30	2.88	2.88	4.51	4.59	4.95	5.21	5.35	5.30	5.59

Table 5.8: Gas-phase oxygen concentration and adsorbed oxygen concentration.

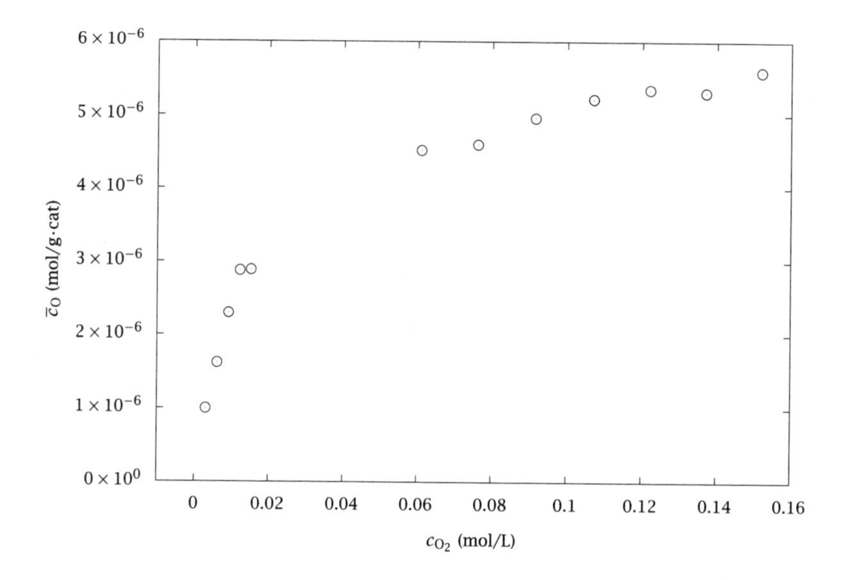

Figure 5.25: Adsorbed oxygen concentration versus gas-phase oxygen concentration.

(a) Derive an expression for the Langmuir isotherm to model the concentration of adsorbed oxygen in terms of the concentration of gas-phase oxygen. How many unknown parameters does your model contain?

(b) Use the experimental data to obtain preliminary estimates of these parameters. You may wish to replot the data as $1/\overline{c}_O$ versus $1/\sqrt{c_{O_2}}$. Don't just plug these numbers into a least-squares calculation without looking at the replotted data first. Table 5.8 provides the numerical values of the data plotted in Figure 5.25.

Does the Langmuir model look reasonable for these data? Explain why or why not.

We revisit these data in Exercise 9.9 after we have developed more general parameter estimation methods in Chapter 9.

Exercise 5.13: Adsorption of propane [8]

Assume we are going to place a large adsorption bed underneath our automobile to store adsorbed propane, which we desorb while driving, to use as a fuel replacement for liquid gasoline. Let's assume that propane adsorbs on the adsorbent material as a Langmuir isotherm,

$$\overline{c} = \frac{\overline{c}_m KP}{1 + KP}$$

in which $\overline{c}$ is the molar concentration of adsorbed propane, $\overline{c}_m$ is the monolayer coverage, K is the adsorption equilibrium constant and P is the gas-phase propane pressure.

Assume the propane "filling station" can charge the bed with a propane supply pressure of $P_2 = 10$ atm and the bed is exhausted at a propane partial pressure of $P_1 = 0.1$ atm. We are interested in selecting an adsorbent material that maximizes the amount of propane that desorbs while changing from P_2 to P_1.

(a) Write down an expression for $\Delta\overline{c} = \overline{c}(P_2) - \overline{c}(P_1)$. What is $\Delta\overline{c}$ for $K = 0$, for $K \to \infty$? Sketch a plot of $\Delta\overline{c}$ versus K.

(b) What is the optimal adsorbent material (K value) and what fraction of the adsorbed propane can be taken off as fuel when using this optimal adsorbent material?

Exercise 5.14: Competitive adsorption and reaction rate

Consider the following heterogeneous reaction in which A and B associatively adsorb on a catalyst surface and undergo reaction to product P, which subsequently desorbs.

$$\mathrm{A} + \mathrm{X} \underset{k_{-1}}{\overset{k_1}{\rightleftharpoons}} \mathrm{A-X}$$

$$\mathrm{B} + \mathrm{X} \underset{k_{-2}}{\overset{k_2}{\rightleftharpoons}} \mathrm{B-X}$$

$$\mathrm{A-X} + \mathrm{B-X} \overset{k_3}{\longrightarrow} \mathrm{P-X} + \mathrm{X}$$

$$\mathrm{P} + \mathrm{X} \underset{k_{-4}}{\overset{k_4}{\rightleftharpoons}} \mathrm{P-X}$$

At 373 K the equilibrium constants for adsorption are

$$K_1 = 190{,}000 \qquad K_2 = 580{,}000 \qquad K_4 = 75{,}000 \qquad \mathrm{cm^3/mol}$$

The heats of adsorption are

$$\Delta H_1 = -20{,}000 \qquad \Delta H_2 = -30{,}000 \quad \Delta H_4 = -12{,}000 \qquad \mathrm{cal/mol}$$

The rate constant for r_3 is

$$k_3 = 7.6 \times 10^{28} \exp(-15000/T) \qquad \frac{(\mathrm{gcat}^2)}{(\mathrm{s})(\mathrm{cm}^3)(\mathrm{mol})}$$

where T is in Kelvin.

(a) Develop a Langmuir-Hinshelwood rate expression (i.e., the rate is expressed in terms of gas-phase components) for the surface reaction when you assume the surface reaction is the slow step and the adsorption and desorption processes are at equilibrium.

---

[8] This exercise came from a seminar given at UT-Austin by Professor Eduardo Glandt on April 18, 1995 [28].

(b) Determine the magnitude of rate of the surface reaction (r_3) at 373 K if the gas contains a 50:50 mixture of A and B at a pressure of 1.0 atm. What is the value of the rate if a 50:10:40 mixture of A:B:P is present instead? $c_m = 0.0008$ mol/g cat.

(c) Examine the effect of changing the composition of the gas in contact with the catalyst on the rate. Plot r_3 for the case where the temperature is 400 K, the total pressure is 1.0 atm, the mole fraction of P is zero, the mole fraction of A is 0.1 and the mole fraction of B is varied between 0 and 0.9. (An inert component that does not adsorb makes up the balance of the gas phase.) Discuss the shape of the curve.

(d) Now repeat part (c) for the 5 and 10 atm total pressure. Why is there a maximum in the rate versus composition curve at higher pressures?

Exercise 5.15: Competitive adsorption

Consider two gas-phase components that adsorb associatively on a heterogeneous catalyst

$$\mathrm{A} + \mathrm{X} \underset{k_{-1}}{\overset{k_1}{\rightleftharpoons}} \mathrm{A-X}, \qquad \mathrm{B} + \mathrm{X} \underset{k_{-2}}{\overset{k_2}{\rightleftharpoons}} \mathrm{B-X}$$

At 373 K the equilibrium constants for adsorption are

$$K_1 = 190{,}000 \qquad K_2 = 580{,}000 \qquad \mathrm{cm^3/mol}$$

The heats of adsorption are

$$\Delta H_1 = -30{,}000 \qquad \Delta H_2 = -20{,}000 \qquad \mathrm{cal/mol}$$

(a) Examine the effect of changing the temperature from 300 to 500 K on the fractional coverage of A and B. Plot θ_A and θ_B versus T when the pressure is 1.0 atm and the gas contains an equimolar mixture of A and B. Comment on the shape of the curves. Why do the curves for the two components have the shapes they do?

(b) Examine the effect of pressure on the fractional coverage of A and B. Plot θ_A and θ_B versus pressure between 0 and 3 atm when the temperature is 373 K and the gas contains an equimolar mixture of A and B. Comment on the shape of the curves. Why does one of the components appear to reach an asymptotic saturation level at a lower pressure than the other one?

(c) Examine the effect of the gas-phase composition on the coverage of A and B. Plot θ_A and θ_B versus the mole fraction of A as it is varied from 0 to 1 when the temperature is 373 K and the pressure is 1.0 atm. Why are the y-axis intercepts different for $y_A = 0$ and $y_A = 1$?

Exercise 5.16: Associative versus dissociative adsorption

The active area of supported group-VIII metals can be determined by adsorbing carbon monoxide. Problems arise with the use of CO because it can adsorb associatively or dissociatively. The type of adsorption is a function of the metal type and the adsorption temperature. The following data describe the adsorption of CO on 10% Ru on Al_2O_3 at 100°C.

P_{CO} (Torr)	CO ads (μmol/g cat)
100	1.28
150	1.63
200	1.77
250	1.94
300	2.06
400	2.21

(a) Test the data to determine if the adsorption is associative or dissociative.

$$CO_g + Ru \rightleftharpoons CO - Ru$$

$$CO_g + 2Ru \rightleftharpoons C - Ru + O - Ru$$

(b) What is the concentration of total surface ruthenium atoms in terms of mol of sites/g of catalyst?

Exercise 5.17: Hougen-Watson kinetics

The irreversible heterogeneous catalytic reaction

$$A + B \longrightarrow C + D$$

was studied over a wide range of partial pressures of reactants A and B, and products C and D. In the experimental observations reported below the partial pressure of one component was varied while holding the partial pressures of the other three components fixed. Use the observations to establish the reaction mechanism. Specifically, which candidate reaction mechanism is consistent with all of the data. Explain your reasoning and justify your answer.

Observations:

Partial pressure varied	Partial pressures fixed	Characteristics of the plot of log rate versus log P varied
A	B, C, D	1. At low P_A the curve has a slope of 1. 2. At intermediate P_A the curve has a positive slope less than 1.
B	A, C, D	1. At low P_B the curve has a positive slope slightly less than 1. 2. At intermediate P_B the curve has a maximum. 3. At high P_B the curve has a slope of −1.
C	A, B, D	1. At low P_C the curve has zero slope. 2. At high P_C the curve has slope −2.

Mechanism 1:

$$A(g) + X \rightleftharpoons A \cdot X$$

$$B(g) + X \rightleftharpoons B \cdot X$$

$$C(g) + X \rightleftharpoons C \cdot X$$

$$D(g) + X \rightleftharpoons D \cdot X$$

$$A \cdot X + B \cdot X \longrightarrow C \cdot X + D \cdot X \qquad \text{rate-limiting step}$$

Mechanism 2:

$$\begin{aligned}
A(g) + X &\rightleftharpoons A \cdot X \\
B(g) + X &\rightleftharpoons B \cdot X \\
C(g) + X &\rightleftharpoons C \cdot X \\
D(g) + X &\rightleftharpoons D \cdot X \\
A(g) + B \cdot X &\longrightarrow C \cdot X + D(g) \qquad \text{rate-limiting step}
\end{aligned}$$

Exercise 5.18: Power-law approximation of Hougen-Watson kinetics

The following catalytic reaction is conducted in a 0.25 cm radius spherical pellet.

$$A + B_2 \longrightarrow C$$

The mechanism for this reaction is shown below. A, B_2 and C are in adsorption-desorption equilibrium with the surface. The bimolecular surface reaction is irreversible and rate limiting.

$$\begin{aligned}
A + \quad S &\underset{k_{-1}}{\overset{k_1}{\rightleftharpoons}} A_{ads} \\
B_2 + \quad 2S &\underset{k_{-2}}{\overset{k_2}{\rightleftharpoons}} 2B_{ads} \\
A_{ads} + B_{ads} &\overset{k_3}{\longrightarrow} C_{ads} + S \\
C + \quad S &\underset{k_{-4}}{\overset{k_4}{\rightleftharpoons}} C_{ads}
\end{aligned}$$

Additional data for this problem follow.

Item	Value	Units
K_1	130,100	cm^3/mol
K_2	6,500	cm^3/mol
K_4	6,440	cm^3/mol
c_A	5.83×10^{-5}	mol/cm^3
c_{B_2}	1.40×10^{-4}	mol/cm^3
c_C	1.17×10^{-5}	mol/cm^3
k_3	7.41×10^{8}	g $cat^2/(mol\ cm^3\ s)$
$\bar{c}_m$	1.8×10^{-5}	mol/g cat

(a) Determine the fractional surface coverages of A, B and C (i.e., θ_A, θ_B and θ_C) at the tabulated conditions.

(b) Assume you can perform experiments that would enable you to determine the rate of the surface reaction as the concentrations of A, B and C are varied plus and minus 50% from the values in the table. You then fit the rate data to the following power-law rate expression

$$r = k c_A^{\alpha} c_{B_2}^{\beta} c_C^{\gamma}$$

What do you expect the values of α, β and γ to be and why?

Exercise 5.19: Ethylene hydrogenation

The hydrogenation of ethylene to ethane is found experimentally to follow the reaction orders given in the following table [36].

Catalyst	Order in H_2	Order in C_2H_4
Rh	0.85	−0.74
Ru	0.95	−0.59
Co	0.55	−0.19

The following steps have been suggested for the catalytic reaction mechanism [41, p.53]. In this mechanism ethylene adsorbs associatively and H_2 adsorbs dissociatively. There is a stepwise addition of adsorbed hydrogen to form an adsorbed ethyl group, $C_2H_{5,ads}$, and then adsorbed ethane, $C_2H_{6,ads}$.

$$\begin{aligned}
C_2H_4 + \quad S \quad &\underset{k_{-1}}{\overset{k_1}{\rightleftharpoons}} C_2H_{4,ads} \\
H_2 + \quad 2S \quad &\underset{k_{-2}}{\overset{k_2}{\rightleftharpoons}} 2H_{ads} \\
C_2H_{4,ads} + \quad H_{ads} \quad &\underset{k_{-3}}{\overset{k_3}{\rightleftharpoons}} C_2H_{5,ads} + S \\
C_2H_{5,ads} + \quad H_{ads} \quad &\underset{k_{-4}}{\overset{k_4}{\rightleftharpoons}} C_2H_{6,ads} + S \\
C_2H_{6,ads} \quad &\underset{k_{-5}}{\overset{k_5}{\rightleftharpoons}} C_2H_6 \quad + S
\end{aligned}$$

where S is a vacant surface site. The rate of the formation of ethane is that of the rate-limiting surface hydrogenation step. There are two choices. For Scheme I, the rate-limiting step is Reaction 3. For Scheme II, the rate-limiting step is Reaction 4.

(a) Develop a rate expression for the rate of ethane formation for Scheme I. Let Reaction 3 be the rate-limiting step and assume all other reactions are at equilibrium. Assume that adsorption from the gas phase follows a Langmuir adsorption isotherm. Neglect the reverse of Reaction 3. When performing a site balance, assume the surface is either vacant or covered with adsorbed ethylene, i.e., the coverage of ethyl, atomic hydrogen and adsorbed ethane are negligible.

(b) Develop a rate expression for the rate of ethane formation for Scheme II. Let Reaction 4 be the rate-limiting step and assume all other reactions are at equilibrium. Assume that adsorption from the gas phase follows a Langmuir adsorption isotherm. Neglect the reverse of Reaction 4. When performing a site balance, assume the surface is either vacant or covered with adsorbed ethylene and ethyl, i.e., the coverage of atomic hydrogen and adsorbed ethane are negligible.

(c) Based on the experimental data listed in the table, is Scheme I or II the more probable mechanism and why?

Bibliography

[1] U.S. climate change technology program: Technology options for the near and long term. A Compendium of Technology Profiles and Ongoing Research and Development at Participating Federal Agencies, page 56, Google Books, 2005.

[2] Recognizing the best in innovation: Breakthrough catalysts. R&D Magazine, page 20, September 2005.

[3] A. W. Adamson. *Physical Chemistry of Surfaces.* John Wiley & Sons, New York, fifth edition, 1990.

[4] A. T. Bell. The impact of nanoscience on heterogeneous catalysis. *Science*, 299:1688–1691, 2003.

[5] M. Bodenstein. Eine Theorie der photochemischen Reaktionsgeschwindigkeiten. *Zeit. physik. Chemie*, 85:329–397, 1913.

[6] M. Boudart and G. Djega-Mariadassou. *Kinetics of Heterogenous Catalytic Reactions.* Princeton University Press, Princeton, NJ, 1984.

[7] D. L. Chapman and L. K. Underhill. The interaction of chlorine and hydrogen. The influence of mass. *J. Chem. Soc. Trans.*, 103:496–508, 1913.

[8] A. Clark. *Theory of Adsorption and Catalysis.* Academic Press, New York, 1970.

[9] T. E. Corrigan, J. C. Garver, H. F. Rase, and R. S. Kirk. Kinetics of catalytic cracking of cumene. *Chem. Eng. Prog.*, 49(11):603–610, 1953.

[10] J. Czarnowski and H. J. Schumacher. The kinetics and the mechanism of the thermal decomposition of bis-pentafluorosulfurtrioxide (SF_5OOOSF_5). *Int. J. Chem. Kinet.*, 11:613–619, 1979.

[11] J. G. Ekerdt and A. T. Bell. Synthesis of hydrocarbons from CO and H_2 over silica-supported Ru: Reaction rate measurements and infrared spectra of adsorbed species. *J. Catal.*, 58:170, 1979.

[12] T. Engel and G. Ertl. In D. P. Woodruff, editor, *The Physics of Solid Surfaces and Heterogeneous Catalysis*, volume four, page 73. Elsevier, Amsterdam, 1982.

[13] H. Gaedtke and J. Troe. Primary processes in the photolysis of NO_2. *Berichte der Bunsen-Gesellschaft für Physikalische Chemie*, 79(2):184–191, 1975.

[14] K. S. Gandhi, R. Kumar, and D. Ramkrishna. Some basic aspects of reaction engineering of precipitation processes. *Ind. Eng. Chem. Res.*, 34(10):3223–3230, 1995.

[15] A. B. Harker and H. S. Johnston. Photolysis of nitrogen dioxide to produce transient O, NO_3 and N_2O_5. *J. Chem. Phys.*, 77(9):1153–1156, 1973.

[16] D. O. Hayward and B. M. W. Trapnell. *Chemisorption.* Butterworths, Washington, D.C., second edition, 1964.

[17] C. G. Hill. *An Introduction to Chemical Engineering Kinetics and Reactor Design.* John Wiley & Sons, New York, 1977.

[18] T. L. Hill. *Introduction to Statistical Mechanics.* Addison-Wesley Publishing Company, Reading, MA, 1960.

[19] S. Hochgreb and F. L. Dryer. Decomposition of 1,3,5-trioxane at 700–800 K. *J. Phys. Chem.*, 96:295–297, 1992.

[20] O. A. Hougen and K. M. Watson. Solid catalysts and reaction rates. *Ind. Eng. Chem.*, 35(5):529–541, 1943.

[21] P. Kisliuk. The sticking probabilities of gases chemisorbed on the surfaces of solids. *J. Phys. Chem. Solids*, 3:95, 1957.

[22] P. Kisliuk. The sticking probabilities of gases chemisorbed on the surfaces of solids - II. *J. Phys. Chem. Solids*, 5:78, 1958.

[23] A. Kumar, P. D. Christofides, and P. Daoutidis. Singular perturbation modeling of nonlinear processes with nonexplicit time-scale multiplicity. *Chem. Eng. Sci.*, 53(8):1491–1504, 1998.

[24] K. J. Laidler. *Chemical Kinetics.* Harper Row, New York, third edition, 1987.

[25] K. J. Laidler and B. W. Wojciechowski. Kinetics and mechanisms of the thermal decomposition of ethane. I. The uninhibited reaction. *Proceedings of the Royal Society of London, Math and Physics*, 260A(1300):91–102, 1961.

[26] C. C. Lin and L. A. Segel. *Mathematics Applied to Deterministic Problems in the Natural Sciences.* Macmillan, New York, 1974.

[27] S. J. Lombardo and A. T. Bell. A review of theoretical models of adsorption, diffusion, desorption, and reaction of gases on metal surfaces. *Surf. Sci. Rep.*, 13:1, 1991.

[28] K. R. Matranga, A. L. Myers, and E. D. Glandt. Storage of natural gas by adsorption on activated carbon. *Chem. Eng. Sci.*, 47(7):1569–1579, 1992.

[29] J. W. Moore and R. G. Pearson. *Kinetics and Mechanism.* John Wiley & Sons, New York, third edition, 1981.

[30] R. G. Ogg Jr. Quasi-unimolecular and quasi-bimolecular steps in complex reactions. *J. Chem. Phys.*, 18(4):572–573, 1950.

[31] R. E. O'Malley Jr. *Singular Perturbation Methods for Ordinary Differential Equations.* Springer Verlag, New York, 1991.

[32] L. Pedersen and R. N. Porter. Modified semiempirical approach to the H_3 potential-energy surface. *J. Chem. Phys.*, 47(11):4751–4757, 1967.

[33] J. C. Polanyi and J. L. Schreiber. Distribution of reaction products (theory). Investigation of an ab initio energy-surface for $F + H_2 \longrightarrow HF + H$. *Chem. Phys. Lett.*, 29(3):319–322, 1974.

[34] L. M. Raff, L. Stivers, R. N. Porter, D. L. Thompson, and L. B. Sims. Semiempirical VB calculation of the (H_2I_2) interaction potential. *J. Chem. Phys.*, 52(7):3449–3457, 1970.

[35] F. O. Rice and K. F. Herzfeld. The thermal decomposition of organic compounds from the standpoint of free radicals. VI. The mechanism of some chain reactions. *J. Am. Chem. Soc.*, 56:284–289, 1934.

[36] G. C. A. Schuit and L. L. van Reijen. The structure and activity of metal-on-silica catalysts. In D. D. Eley, W. G. Frankenberg, and V. I. Komarewsky, editors, *Advances in Catalysis and Related Subjects*, pages 243–317. Academic Press, Inc., New York, tenth edition, 1958.

[37] L. A. Segel and M. Slemrod. The quasi-steady-state assumption: A case study in perturbation. *SIAM Rev.*, 31(3):446–476, 1989.

[38] G. A. Somorjai. *Chemistry in Two Dimensions, Surfaces.* Cornell University Press, Ithaca, New York, 1981.

[39] R. C. Tolman. Duration of molecules in upper quantum states. *J. Phys. Rev.*, 23:693, 1924.

[40] T. Turànyi, A. S. Tomlin, and M. J. Pilling. On the error of the quasi-steady-state approximation. *J. Phys. Chem.*, 97:163, 1993.

[41] R. A. van Santen and J. W. Niemantsverdriet. *Chemical Kinetics and Catalysis.* Plenum Press, New York, 1995.

[42] R. E. Weston and H. A. Schwarz. *Chemical Kinetics.* Prentice Hall, Englewood Cliffs, New Jersey, 1972.

6

The Energy Balance for Chemical Reactors

6.1 General Energy Balance

To specify the rates of reactions in a nonisothermal reactor, we require a model to determine the temperature of the reactor. The temperature is determined by the energy balance for the reactor. We derive the energy balance by considering an arbitrary reactor volume element, shown in Figure 6.1, as we did in deriving the material balance in Chapter 4. The volume element has inlet and outlet streams with mass flowrates m_0 and m_1, respectively. In this chapter, we are again neglecting flux of mass through the surface of the volume element except at stream locations 0 and 1. The molar concentrations of component j in the two streams are given by c_{j0} and c_{j1}, and the total energy mass densities of the streams are denoted by $\hat{E}_0$ and $\hat{E}_1$. The rate of heat

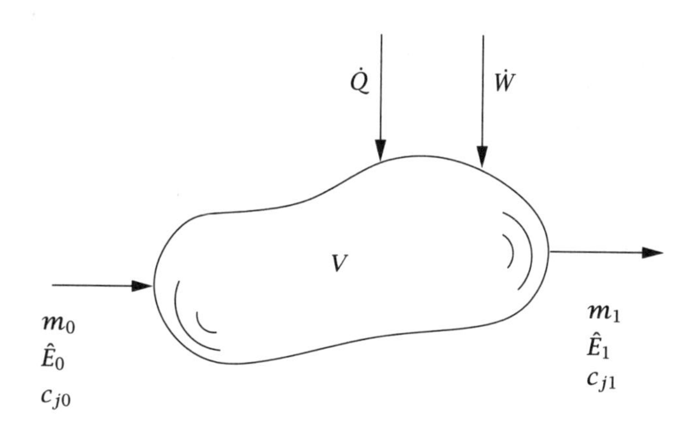

Figure 6.1: Reactor volume element.

added to the system is denoted by $\dot{Q}$, and $\dot{W}$ is the rate of work being done on the system. Be aware that the opposite convention in which $\dot{W}$ points out of the system in Figure 6.1 is also in use. The statement of conservation of energy for this system takes the form,

$$\left\{\begin{array}{c}\text{rate of energy}\\ \text{accumulation}\end{array}\right\} = \left\{\begin{array}{c}\text{rate of energy}\\ \text{entering system}\\ \text{by inflow}\end{array}\right\} - \left\{\begin{array}{c}\text{rate of energy}\\ \text{leaving system}\\ \text{by outflow}\end{array}\right\}$$
$$+ \left\{\begin{array}{c}\text{rate of heat}\\ \text{added to system}\end{array}\right\} + \left\{\begin{array}{c}\text{rate of work}\\ \text{done on system}\end{array}\right\} \tag{6.1}$$

In terms of the defined variables, we can write Equation 6.1 as,

$$\frac{dE}{dt} = m_0\hat{E}_0 - m_1\hat{E}_1 + \dot{Q} + \dot{W} \tag{6.2}$$

in which the hat indicates an energy per unit mass.

6.1.1 Work Term

It is convenient to split the work term into three parts: $\dot{W}_f$, the work done by the flow streams while moving material into and out of the reactor, $\dot{W}_s$, the shaft work being done by stirrers, compressors, etc., and $\dot{W}_b$, the work done when moving the system boundary

$$\underbrace{\dot{W}}_{\text{total work}} = \underbrace{\dot{W}_f}_{\text{flow streams}} + \underbrace{\dot{W}_s}_{\text{shaft work}} + \underbrace{\dot{W}_b}_{\text{boundary work}} \tag{6.3}$$

To calculate the work done by the flow streams, we assume the entering fluid has pressure P_0 and a uniform velocity v_0 normal to the bounding surface, and the exiting stream has pressure P_1 and velocity v_1, and let A_0 and A_1 be the areas on the bounding surface where the streams intersect the boundary as shown in Figure 6.2. In this case $\dot{W}_f$ is [1]

$$\dot{W}_f = v_0 A_0 P_0 - v_1 A_1 P_1 = Q_0 P_0 - Q_1 P_1$$

We also can express the volumetric flowrate as a mass flowrate divided by the density, $Q = m/\rho$

$$\dot{W}_f = m_0 \frac{P_0}{\rho_0} - m_1 \frac{P_1}{\rho_1}$$

[1] We are neglecting any normal viscous forces, which are usually much smaller than the pressure force. Consult Bird for a more detailed development from microscopic considerations [5, 6].

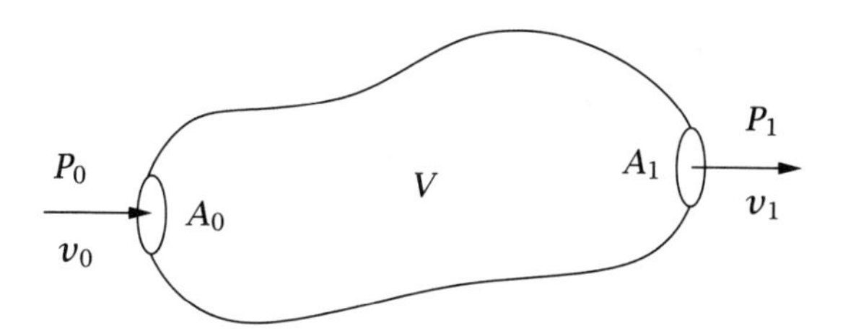

Figure 6.2: Flow streams entering and leaving the volume element.

Note that the units in the product mP/ρ are $(m/t)(m/lt^2)/(m/l^3)$, which gives $(ml^2/t^2)/t$, a unit of energy per time as we expect. The overall rate of work can then be expressed as

$$\dot{W} = \dot{W}_f + \dot{W}_s + \dot{W}_b = m_0 \frac{P_0}{\rho_0} - m_1 \frac{P_1}{\rho_1} + \dot{W}_s + \dot{W}_b \tag{6.4}$$

6.1.2 Energy Terms

The total energy may be regarded as composed of many forms. Obvious contributions to the total energy arise from the internal, kinetic and potential energies.[2]

$$\hat{E} = \hat{U} + \hat{K} + \hat{\Phi} + \cdots$$

For our purposes in this chapter, we consider only these forms of energy. Recalling the definition of enthalpy, $H = U + PV$, or expressed on a per-unit mass basis, $\hat{H} = \hat{U} + P/\rho$, allows us to rewrite Equation 6.2 as

$$\frac{d}{dt}(U + K + \Phi) = m_0\left(\hat{H} + \hat{K} + \hat{\Phi}\right)_0 - m_1\left(\hat{H} + \hat{K} + \hat{\Phi}\right)_1 + \dot{Q} + \dot{W}_s + \dot{W}_b \tag{6.5}$$

6.2 The Batch Reactor

Since the batch reactor has no flow streams Equation 6.5 reduces to

$$\frac{d}{dt}(U + K + \Phi) = \dot{Q} + \dot{W}_s + \dot{W}_b \tag{6.6}$$

[2] In some cases one might need to consider also electrical and magnetic energies. For example, we might consider the motion of charged ionic species between the plates in a battery cell.

In chemical reactors, we normally assume the internal energy is the dominant contribution and neglect the kinetic and potential energies. Normally we neglect the work done by the stirrer, unless the mixture is highly viscous and the stirring operation draws significant power [14]. Neglecting kinetic and potential energies and shaft work yields

$$\frac{dU}{dt} + P\frac{dV_R}{dt} = \dot{Q} \tag{6.7}$$

in which $\dot{W}_b = -PdV_R/dt$. It is convenient to use enthalpy rather than internal energy in the subsequent development. Taking the differential of the definition of enthalpy gives for $V = V_R$

$$dH = dU + PdV_R + V_R dP$$

Forming the time derivatives and substitution into Equation 6.7 gives

$$\frac{dH}{dt} - V_R\frac{dP}{dt} = \dot{Q} \tag{6.8}$$

For *single-phase systems*, we consider the enthalpy as a function of temperature, pressure and number of moles, and express its differential as

$$dH = \left(\frac{\partial H}{\partial T}\right)_{P,n_j} dT + \left(\frac{\partial H}{\partial P}\right)_{T,n_j} dP + \sum_j \left(\frac{\partial H}{\partial n_j}\right)_{T,P,n_k} dn_j \tag{6.9}$$

The first partial derivative is the definition of the heat capacity, C_P. Notice this heat capacity is the extensive heat capacity of the reactor contents. Normally we express this quantity as an intensive heat capacity times the amount of material in the reactor. We can express the intensive heat capacity on either a molar or mass basis. We choose to use the heat capacity on a mass basis, so the total heat capacity can be expressed as

$$C_P = V_R \rho \hat{C}_P$$

The second partial derivative can be expressed as (see Exercise 6.23)

$$\left(\frac{\partial H}{\partial P}\right)_{T,n_j} = V - T\left(\frac{\partial V}{\partial T}\right)_{P,n_j} = V(1 - \alpha T)$$

in which $\alpha = (1/V)(\partial V/\partial T)_{P,n_j}$ is the coefficient of expansion of the mixture. The partial derivatives appearing in the sum in Equation 6.9 are the partial molar enthalpies, $\overline{H}_j$

$$\left(\frac{\partial H}{\partial n_j}\right)_{T,P,n_k} = \overline{H}_j$$

so Equation 6.9 can be written compactly as

$$dH = V_R\rho\hat{C}_P dT + (1-\alpha T)V_R dP + \sum_j \overline{H}_j dn_j \tag{6.10}$$

Forming the time derivatives from this expression and substituting into Equation 6.8 gives

$$V_R\rho\hat{C}_P\frac{dT}{dt} - \alpha T V_R\frac{dP}{dt} + \sum_j \overline{H}_j\frac{dn_j}{dt} = \dot{Q} \tag{6.11}$$

We note that the material balance for the batch reactor is

$$\frac{dn_j}{dt} = R_j V_R = \sum_{i=1}^{n_r} \nu_{ij} r_i V_R, \quad j = 1, \dots, n_s \tag{6.12}$$

which upon substitution into Equation 6.11 yields

$$V_R\rho\hat{C}_P\frac{dT}{dt} - \alpha T V_R\frac{dP}{dt} = -\sum_i \Delta H_{Ri} r_i V_R + \dot{Q} \tag{6.13}$$

in which ΔH_{Ri} is the heat of reaction

$$\Delta H_{Ri} = \sum_j \nu_{ij}\overline{H}_j \tag{6.14}$$

We now consider several special cases. If the reactor operates at constant pressure ($dP/dt = 0$) *or* the fluid is incompressible[3] ($\alpha = 0$), then Equation 6.13 reduces to

Incompressible-fluid or constant-pressure reactor.

$$\boxed{V_R\rho\hat{C}_P\frac{dT}{dt} = -\sum_i \Delta H_{Ri} r_i V_R + \dot{Q}} \tag{6.15}$$

To derive the constant-volume case, we consider the pressure as a function of $T, V (V = V_R), n_j$, and express the pressure differential as

$$dP = \left(\frac{\partial P}{\partial T}\right)_{V,n_j} dT + \left(\frac{\partial P}{\partial V}\right)_{T,n_j} dV + \sum_j \left(\frac{\partial P}{\partial n_j}\right)_{T,V,n_k} dn_j$$

[3] We take incompressible to mean constant density, which is a common practice in fluid mechanics [7, p. 338].

For reactor operation at constant volume, $dV = 0$, and forming time derivatives and substituting into Equation 6.11 gives

$$\left[V_R\rho\hat{C}_P - \alpha TV_R\left(\frac{\partial P}{\partial T}\right)_{V,n_j}\right]\frac{dT}{dt} + \sum_j \left[\overline{H}_j - \alpha TV_R\left(\frac{\partial P}{\partial n_j}\right)_{T,V,n_k}\right]\frac{dn_j}{dt} = \dot{Q}$$

We note that the first term in brackets is $C_V = V_R\rho\hat{C}_V$ (see Exercise 6.23)

$$V_R\rho\hat{C}_V = V_R\rho\hat{C}_P - \alpha TV_R\left(\frac{\partial P}{\partial T}\right)_{V,n_j}$$

The pressure derivative with respect to the moles can be shown to be (see Exercise 6.23)

$$\left(\frac{\partial P}{\partial n_j}\right)_{T,V,n_{k\neq j}} = \frac{\overline{V}_j}{V\kappa_T}$$

in which $\kappa_T = -(1/V)(\partial V/\partial P)_{T,n_j}$ is the isothermal compressibility of the mixture, and $\overline{V}_j$ is the partial molar volume. Substitution of these two thermodynamic relations and the material balance yields the energy balance for the constant-volume batch reactor

Constant-volume reactor.

$$\boxed{V_R\rho\hat{C}_V\frac{dT}{dt} = -\sum_i\left(\Delta H_{Ri} - \frac{\alpha}{\kappa_T}T\Delta V_{Ri}\right)r_iV_R + \dot{Q}} \tag{6.16}$$

in which ΔV_{Ri} is the volume change of reaction

$$\Delta V_{Ri} = \sum_j \nu_{ij}\overline{V}_j$$

If we consider an ideal gas, it is straightforward to calculate $\alpha T = 1$, $\kappa_T P = 1$, and $\Delta V_{Ri} = \bar{\nu}_i(RT/P)$, where $\bar{\nu}_i = \sum_j \nu_{ij}$. Substitution into the constant-volume energy balance gives

Constant-volume reactor, ideal gas.

$$\boxed{V_R\rho\hat{C}_V\frac{dT}{dt} = -\sum_i\left(\Delta H_{Ri} - RT\bar{\nu}_i\right)r_iV_R + \dot{Q}} \tag{6.17}$$

If we consider the constant-volume reactor with incompressible fluid ($\alpha = 0, \hat{C}_V = \hat{C}_P$), Equation 6.16 reduces to Equation 6.15 as it should because Equation 6.15 is valid for any reactor operation with an incompressible fluid. We also notice that, in the constant-pressure case, the same energy balance applies for any fluid mixture (ideal gas, incompressible fluid, etc.), and that this balance is the same as the balance for an incompressible fluid in a constant-volume reactor. Although the same final balances are obtained for these two cases, the physical situations they describe are completely different.

Liquid-phase reactions often are carried out in batch reactors. Equation 6.15 is therefore a common batch-reactor energy balance. Two cases where Equation 6.15 can be simplified include the case of adiabatic operation ($\dot{Q} = 0$) and the case of heat transfer through a jacket or cooling coil

$$\dot{Q} = U^{o}A(T_a - T) \tag{6.18}$$

in which U^{o} is the overall heat-transfer coefficient. Note that Equation 6.15 must be solved simultaneously, and usually numerically, with the batch-reactor material balance, Equation 6.12.

Example 6.1: Constant-pressure versus constant-volume reactors

Consider the following two well-mixed, adiabatic, *gas-phase* batch reactors for the exothermic, elementary decomposition of A to B,

$$\mathrm{A} \xrightarrow{k} 2\mathrm{B} \tag{6.19}$$

Reactor 1: The reactor volume is held constant (reactor pressure therefore changes).

Reactor 2: The reactor pressure is held constant (reactor volume therefore changes).

Both reactors are charged with pure A at 1.0 atm and k has the usual Arrhenius activation energy dependence on temperature,

$$k(T) = k_0 \exp(-E/T)$$

The heat of reaction, ΔH_R, and heat capacity of the mixture, $\hat{C}_P$, may be assumed constant over the composition and temperature range expected.

Write the material and energy balances for these two reactors. Which reactor converts the reactant more quickly?

Solution

Material balance. The material balance for component A in the batch reactor is

$$\frac{d(c_A V_R)}{dt} = R_A V_R$$

Substituting in the reaction-rate expression, $r = k(T)c_A$, and using the number of moles of A, $n_A = c_A V_R$ yields

$$\frac{dn_A}{dt} = -k(T)n_A \tag{6.20}$$

Notice the temperature dependence of $k(T)$ prevents us from solving this differential equation immediately like we did for the isothermal reactor. We must solve it simultaneously with the energy balance, which provides the information for how the temperature changes. The material balance, Equation 6.20, is the same for *both* reactors.

Energy balance. The energy balances for the two reactors are *not* the same. We consider first the constant-volume reactor. For the A → 2B stoichiometry, we substitute the rate expression and $\bar{\nu} = 1$ into Equation 6.17 to obtain

$$C_V \frac{dT}{dt} = -\left(\Delta H_R - RT\right) k n_A$$

in which $C_V = V_R \rho \hat{C}_V$ is the total constant-volume heat capacity.

The energy balance for the constant-pressure case follows from Equation 6.15

$$C_P \frac{dT}{dt} = -\Delta H_R k n_A$$

in which $C_P = V_R \rho \hat{C}_P$ is the total constant-pressure heat capacity. For an ideal gas, we know from thermodynamics that the two total heat capacities are simply related,

$$C_V = C_P - nR \tag{6.21}$$

Comparing the production rates of A and B produces

$$2n_A + n_B = 2n_{A0} + n_{B0}$$

Because there is no B in the reactor initially, subtracting n_A from both sides yields for the total number of moles

$$n = n_A + n_B = 2n_{A0} - n_A$$

Substitution of the above and Equation 6.21 into the constant-volume case yields

$$\frac{dT}{dt} = -\frac{\left(\Delta H_R - RT\right) k n_A}{C_P - (2n_{A0} - n_A)R} \qquad \text{constant volume} \tag{6.22}$$

and the temperature differential equation for the constant-pressure case is

$$\frac{dT}{dt} = -\frac{\Delta H_R k n_A}{C_P} \qquad \text{constant pressure} \tag{6.23}$$

We see by comparing Equations 6.22 and 6.23 that the numerator in the constant-volume case is larger because ΔH_R is negative and the positive RT is subtracted. We also see the denominator is smaller because C_P is positive and the positive nR is subtracted. Therefore the time derivative of the temperature is larger for the constant-volume case. The reaction proceeds more quickly in the constant-volume case. The constant-pressure reactor is expending work to increase the reactor size, and this work results in a lower temperature and slower reaction rate compared to the constant-volume case. □

Example 6.2: Liquid-phase batch reactor

The exothermic elementary *liquid-phase* reaction

$$A + B \xrightarrow{k} C, \qquad r = kc_A c_B$$

is carried out in a batch reactor with a cooling coil to keep the reactor isothermal at 27°C. The reactor is initially charged with equal concentrations of A and B and no C, $c_{A0} = c_{B0} = 2.0$ mol/L, $c_{C0} = 0$.

1. How long does it take to reach 95% conversion?

2. What is the total amount of heat (kcal) that must be removed by the cooling coil when this conversion is reached?

3. What is the maximum *rate* at which heat must be removed by the cooling coil (kcal/min) and at what time does this maximum occur?

4. What is the adiabatic temperature rise for this reactor and what is its significance?

Additional data:

Rate constant, $k = 0.01725$ L/mol·min, at 27°C
Heat of reaction, $\Delta H_R = -10$ kcal/mol A, at 27°C
Partial molar heat capacities, $\overline{C}_{PA} = \overline{C}_{PB} = 20$ cal/(mol K),
$\overline{C}_{PC} = 40$ cal/(mol K)
Reactor volume, $V_R = 1200$ L

Solution

1. Assuming constant density, the material balance for component A is

$$\frac{dc_A}{dt} = -kc_Ac_B$$

The stoichiometry of the reaction, and the material balance for B gives

$$c_A - c_B = c_{A0} - c_{B0} = 0$$

or $c_A = c_B$. Substitution into the material balance for species A gives

$$\frac{dc_A}{dt} = -kc_A^2$$

Separation of variables and integration gives

$$t = \frac{1}{k}\left[\frac{1}{c_A} - \frac{1}{c_{A0}}\right]$$

Substituting $c_A = 0.05c_{A0}$ and the values for k and c_{A0} gives

$$t = 551 \text{ min}$$

2. We assume the incompressible-fluid energy balance is accurate for this liquid-phase reactor. If the heat removal is manipulated to maintain constant reactor temperature, the time derivative in Equation 6.15 vanishes leaving

$$\dot{Q} = \Delta H_R r V_R \tag{6.24}$$

Substituting $dc_A/dt = -r$ and multiplying through by dt gives

$$dQ = -\Delta H_R V_R dc_A$$

Integrating both sides gives

$$Q = -\Delta H_R V_R(c_A - c_{A0}) = -2.3 \times 10^4 \text{ kcal}$$

3. Substituting $r = kc_A^2$ into Equation 6.24 yields

$$\dot{Q} = \Delta H_R k c_A^2 V_R$$

The right-hand side is a maximum in absolute value (note it is a negative quantity) when c_A is a maximum, which occurs for $c_A = c_{A0}$, giving

$$\dot{Q}_{\max} = \Delta H_R k c_{A0}^2 V_R = -828 \text{ kcal/min}$$

4. The adiabatic temperature rise is calculated from the energy balance without the heat-transfer term

$$V_R \rho \hat{C}_P \frac{dT}{dt} = -\Delta H_R r V_R$$

Substituting the material balance $dn_A/dt = -rV_R$ gives

$$V_R \rho \hat{C}_P dT = \Delta H_R dn_A \qquad (6.25)$$

Because we are given the partial molar heat capacities (see Equation 3.48 for the definition of partial molar heat capacity), it is convenient to evaluate the total heat capacity as

$$V_R \rho \hat{C}_P = \sum_{j=1}^{n_s} \overline{C}_{Pj} n_j$$

For a batch reactor, the number of moles can be related to the reaction extent by $n_j = n_{j0} + \nu_j \varepsilon$, so we can express the right-hand side of the previous equation as

$$\sum_{j=1}^{n_s} \overline{C}_{Pj} n_j = \sum_j \overline{C}_{Pj} n_{j0} + \varepsilon \Delta C_P$$

in which $\Delta C_P = \sum_j \nu_j \overline{C}_{Pj}$. If we assume the partial molar heat capacities are independent of temperature and composition we have $\Delta C_P = 0$ and

$$V_R \rho \hat{C}_P = \sum_{j=1}^{n_s} \overline{C}_{Pj} n_{j0}$$

Integrating Equation 6.25 with constant heat capacity gives

$$\Delta T = \frac{\Delta H_R}{\sum_j \overline{C}_{Pj} n_{j0}} \Delta n_A$$

The maximum temperature rise corresponds to complete conversion of the reactants and can be computed from the given data

$$\Delta T_{\max} = \frac{-10 \times 10^3 \text{ cal/mol}}{2(2 \text{ mol/L})(20 \text{ cal/(mol K)})} (0 - 2 \text{ mol/L})$$

$$\Delta T_{\max} = 250 \text{ K}$$

The adiabatic temperature rise indicates the potential danger of a coolant system failure. In this case the reactants contain enough internal energy to raise the reactor temperature by 250 K.

□

6.3 The CSTR

Dynamic Operation

For the continuous-stirred-tank reactor (CSTR), we again assume that the internal energy is the dominant contribution to the total energy and take the entire reactor contents as the volume element. Again we denote the feed stream with flowrate Q_f, density ρ_f, enthalpy $\hat{H}_f$, and component j concentration c_{jf}. The outflow stream is flowing out of a well-mixed reactor and its intensive properties are therefore assumed the same as the reactor contents. Its flowrate is denoted Q. Writing Equation 6.5 for this reactor gives,

$$\frac{dU}{dt} = Q_f\rho_f\hat{H}_f - Q\rho\hat{H} + \dot{Q} + \dot{W}_s + \dot{W}_b \tag{6.26}$$

And if we neglect the shaft work

$$\frac{dU}{dt} + P\frac{dV_R}{dt} = Q_f\rho_f\hat{H}_f - Q\rho\hat{H} + \dot{Q} \tag{6.27}$$

or if we use the enthalpy rather than internal energy

$$\frac{dH}{dt} - V_R\frac{dP}{dt} = Q_f\rho_f\hat{H}_f - Q\rho\hat{H} + \dot{Q} \tag{6.28}$$

As in the batch reactor, for *single-phase systems* we consider the change in enthalpy due to changes in temperature, pressure and the number of moles of component j,

$$dH = V_R\rho\hat{C}_P dT + (1 - \alpha T)V_R dP + \sum_j \overline{H}_j dn_j$$

Substitution into Equation 6.28 gives

$$V_R\rho\hat{C}_P\frac{dT}{dt} - \alpha T V_R\frac{dP}{dt} + \sum_j \overline{H}_j\frac{dn_j}{dt} = Q_f\rho_f\hat{H}_f - Q\rho\hat{H} + \dot{Q} \tag{6.29}$$

The material balance for the CSTR is

$$\frac{dn_j}{dt} = Q_f c_{jf} - Qc_j + \sum_i \nu_{ij} r_i V_R \tag{6.30}$$

Substitution into Equation 6.29 and rearrangement yields

$$V_R\rho\hat{C}_P\frac{dT}{dt} - \alpha T V_R\frac{dP}{dt} = -\sum_i \Delta H_{Ri} r_i V_R + \sum_j c_{jf}Q_f(\overline{H}_{jf} - \overline{H}_j) + \dot{Q} \tag{6.31}$$

Again, a variety of important special cases may be considered. These are listed in Table 6.8 in the chapter summary. A common case is the liquid-phase reaction, which usually is well approximated by the incompressible-fluid equation,

$$V_R \rho \hat{C}_P \frac{dT}{dt} = -\sum_i \Delta H_{Ri} r_i V_R + \sum_j c_{jf} Q_f (\overline{H}_{jf} - \overline{H}_j) + \dot{Q} \qquad (6.32)$$

In the next section we consider further simplifying assumptions that require less thermodynamic data and yield useful approximations.

Steady-State Operation

If the CSTR is at steady state, the time derivatives in Equations 6.30 and 6.31 can be set to zero yielding,

$$Q_f c_{jf} - Q c_j + \sum_i \nu_{ij} r_i V_R = 0 \qquad (6.33)$$

$$-\sum_i \Delta H_{Ri} r_i V_R + \sum_j c_{jf} Q_f (\overline{H}_{jf} - \overline{H}_j) + \dot{Q} = 0 \qquad (6.34)$$

Equations 6.33 and 6.34 provide $n_s + 1$ algebraic equations that can be solved simultaneously to obtain the steady-state concentrations and temperature in the CSTR. Note that the heats of reaction ΔH_{Ri} are evaluated at the reactor temperature and composition.

For a liquid-phase reaction, there is a final approximation that is often useful. If the heat capacity of the liquid phase does not change significantly with composition or temperature, possibly because of the presence of a large excess of a nonreacting solvent, and we neglect the pressure effect on enthalpy, which is normally small for a liquid, we obtain

$$\overline{H}_{jf} - \overline{H}_j = \overline{C}_{Pj}(T_f - T)$$

Substitution into Equation 6.34 gives

$$-\sum_i r_i \Delta H_{Ri} V_R + Q_f \rho_f \hat{C}_P (T_f - T) + \dot{Q} = 0 \qquad (6.35)$$

Example 6.3: Temperature control in a CSTR

An aqueous solution of species A undergoes the following elementary reaction in a 2000 L CSTR as depicted in Figure 6.3.

$$\mathrm{A} \underset{k_{-1}}{\overset{k_1}{\rightleftharpoons}} \mathrm{R} \qquad \Delta \mathrm{H_R} = -18 \text{ kcal/mol}$$

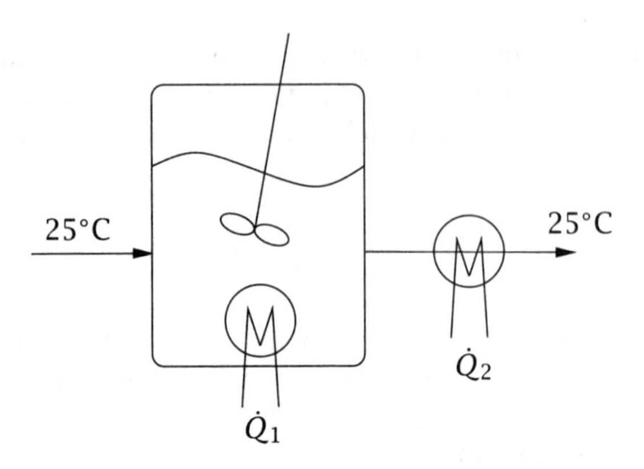

Figure 6.3: CSTR and heat exchangers.

The feed concentration, C_{Af}, is 4 mol/L and feed flowrate, Q_f, is 250 L/min. The reaction-rate constants have been determined experimentally

$$k_1 = 3 \times 10^7 e^{-5838/T} \text{ min}^{-1}$$
$$K_1 = 1.9 \times 10^{-11} e^{9059/T}$$

1. At what temperature must the reactor be operated to achieve 80% conversion?

2. What are the heat duties of the two heat exchangers if the feed enters at 25°C and the product is to be withdrawn at this temperature? The heat capacity of feed and product streams can be approximated by the heat capacity of water, $\hat{C}_P = 1$ cal/(g K).

Solution

1. The steady-state material balances for components A and R in a constant-density CSTR are

$$Q(c_{Af} - c_A) - rV_R = 0$$
$$Q(c_{Rf} - c_R) + rV_R = 0$$

Adding these equations and noting $c_{Rf} = 0$ gives

$$c_R = c_{Af} - c_A$$

Substituting this result into the rate expression gives

$$r = k_1(c_A - \frac{1}{K_1}(c_{Af} - c_A))$$

Substitution into the material balance for A gives

$$Q(c_{Af} - c_A) - k_1\left(c_A - \frac{1}{K_1}(c_{Af} - c_A)\right)V_R = 0 \tag{6.36}$$

If we set $c_A = 0.2c_{Af}$ to achieve 80% conversion, we have one equation and one unknown, T, because k_1 and K_1 are given functions of temperature. Solving this equation numerically gives

$$T = 334 \text{ K}$$

A word of caution is in order here. Because the reaction is reversible, we do not know if 80% conversion is achievable for *any* temperature when we attempt to solve Equation 6.36. It may be valuable to first make a plot of the conversion as a function of reactor temperature. If we solve Equation 6.36 for c_A, we have

$$c_A = \frac{Q/V_R + k_1/K_1}{Q/V_R + k_1(1 + 1/K_1)} c_{Af}$$

or for $x_A = 1 - c_A/c_{Af}$

$$x_A = \frac{k_1}{Q/V_R + k_1(1 + 1/K_1)} = \frac{k_1\tau}{1 + k_1\tau(1 + 1/K_1)}$$

Figure 6.4 displays x_A versus T and we see that the conversion 80% is just reachable at 334 K, and that for any conversion lower than this value, there are two solutions.

2. A simple calculation for the heat-removal rate required to bring the reactor outflow stream from 334 K to 298 K gives

$$\begin{aligned}\dot{Q}_2 &= Q_f \rho \hat{C}_P \Delta T \\ &= (250 \text{ L/min})(1000 \text{ g/L})(1 \text{ cal/(g K)})(298 - 334 \text{ K}) \\ &= -9 \times 10^3 \text{ kcal/min}\end{aligned}$$

Applying Equation 6.35 to this reactor gives

$$\begin{aligned}\dot{Q}_1 &= k_1\left(c_A - \frac{1}{K_1}(c_{Af} - c_A)\right)\Delta H_R V_R - Q_f \rho \hat{C}_P (T_f - T) \\ &= -5.33 \times 10^3 \text{ kcal/min}\end{aligned}$$

□

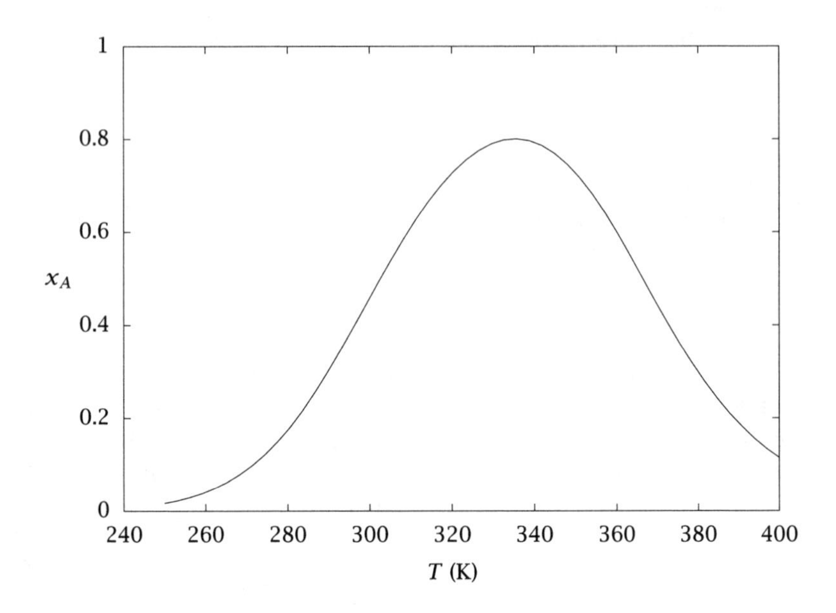

Figure 6.4: Conversion of A versus reactor temperature.

6.3.1 Steady-State Multiplicity

The coupling of the material and energy balances for the CSTR can give rise to some surprisingly complex and interesting behavior. Even the steady-state solution of the material and energy balances holds some surprises. In this section we explore the fact that the steady state of the CSTR is not necessarily unique. As many as three steady-state solutions to the material and energy balances may exist for even the simplest kinetic mechanisms. This phenomenon is known as steady-state multiplicity.

We introduce this topic with a simple example [24]. Consider an adiabatic, constant-volume CSTR with the following elementary reaction taking place in the liquid phase

$$\mathrm{A} \xrightarrow{k} \mathrm{B}$$

We wish to compute the steady-state reactor conversion and temperature. The data and parameters are listed in Table 6.1.

The material balance for component A is

$$\frac{d(c_A V_R)}{dt} = Q_f c_{Af} - Q c_A + R_A V_R$$

Parameter	Value	Units
T_f	298	K
T_m	298	K
$\hat{C}_P$	4.0	kJ/kg K
c_{Af}	2.0	kmol/m^3
k_m	0.001	min^{-1}
E	8.0×10^3	K
ρ_f	10^3	kg/m^3
ΔH_R	-3.0×10^5	kJ/kmol
U^o	0	

Table 6.1: Parameter values for multiple steady states.

The production rate is given by

$$R_A = -k(T)c_A$$

For the steady-state reactor with constant-density, liquid-phase streams, the material balance simplifies to

$$\boxed{0 = c_{Af} - (1 + k\tau)c_A} \tag{6.37}$$

Equation 6.37 is one nonlinear algebraic equation in two unknowns: c_A and T. The temperature appears in the rate-constant function,

$$k(T) = k_m e^{-E(1/T - 1/T_m)}$$

A second equation is provided by the energy balance, and, in this example, we assume the heat capacity of the mixture is constant and independent of composition and temperature. With these assumptions, the steady-state energy balance reduces to

$$0 = -kc_A \Delta H_R V_R + Q_f \rho_f \hat{C}_P (T_f - T) + U^o A(T_a - T)$$

Dividing through by V_R and noting $U^o = 0$ for the adiabatic reactor gives

$$\boxed{0 = -kc_A \Delta H_R + \frac{C_{Ps}}{\tau}(T_f - T)} \tag{6.38}$$

in which $C_{Ps} = \rho_f \hat{C}_P$, a heat capacity per volume. The solution of Equations 6.37 and 6.38 for c_A and T provide the steady-state CSTR solution. The parameters appearing in the problem are: c_{Af}, T_f, τ, C_{Ps}, k_m, T_m, E, ΔH_R. We wish to study this solution as a function of one of these parameters, τ, the reactor residence time. Consider the heat of reaction taking values in one of three cases.

Thermoneutral reaction. Assume the heat of reaction is zero, $\Delta H_R = 0$. Equation 6.38 then implies $T = T_f$ for all residence times. In this case, we already have shown that

$$c_A = \frac{c_{Af}}{1 + k\tau} \qquad x = \frac{k\tau}{1 + k\tau}$$

The x and T values for this case are plotted versus residence time as the $\Delta H_R = 0$ lines in Figures 6.5 and 6.6. This case is much as we expect, the conversion of A increases monotonically with residence time because the reactant molecules have more time to react at longer residence times.

Endothermic reaction. Let $\Delta H_R = +5 \times 10^4$ kJ/kmol. Because the heat of reaction is positive, the steady-state temperature decreases with increasing conversion and residence time. The lower temperature decreases the rate constant and one has to operate the reactor at significantly longer residence times to achieve the same conversion as the isothermal case as shown in Figure 6.5.

Exothermic reaction. The solution for the exothermic reaction is plotted in Figures 6.5 and 6.6, for $\Delta H_R = -5, -10, -20, -30 \times 10^4$ kJ/kmol. In these cases, temperature increases with conversion and hence residence time. The rate constant increases significantly with temperature, and higher conversions are achieved at smaller residence times as the exothermicity increases. Note that if the heat of reaction is more exothermic than -10 kJ/kmol, there is a range of residence times in which there is not one but several steady-state solutions, three solutions in this case. The reactor is said to exhibit **steady-state multiplicity** for these values of residence time. For intermediate values of the heat of reaction, one can see the formation of the s-shaped steady-state temperature curves. We also should note that we have assumed the reaction mixture remains in the liquid phase at these temperatures and the reactor pressure.

The points at which the steady-state curves turn are known as **ignition** and **extinction** points. Figures 6.7 and 6.8 show the $\Delta H_R = -3 \times 10^5$ kJ/kmol case again, without a log scale on the residence-time axis. Consider a reactor at steady state for a small value of residence time, 10 min, at low conversion of A and low temperature. If the feed flowrate were decreased slightly (τ increased), there would be a small upset and the reactor would increase in conversion and temperature as it approached the new steady state at the new residence time. Consider the situation at the ignition point, however. Let the reactor be at steady

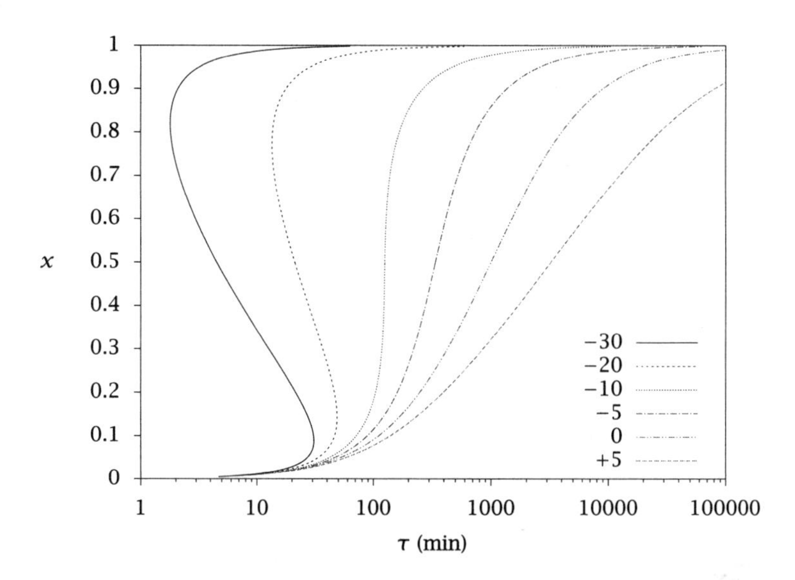

Figure 6.5: Steady-state conversion versus residence time for different values of the heat of reaction ($\Delta H_R \times 10^{-4}$ kJ/kmol).

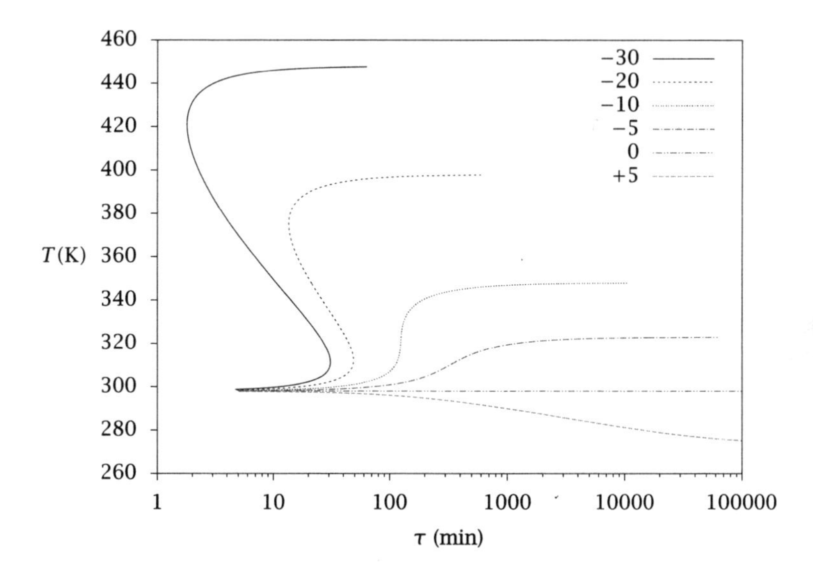

Figure 6.6: Steady-state temperature versus residence time for different values of the heat of reaction ($\Delta H_R \times 10^{-4}$ kJ/kmol).

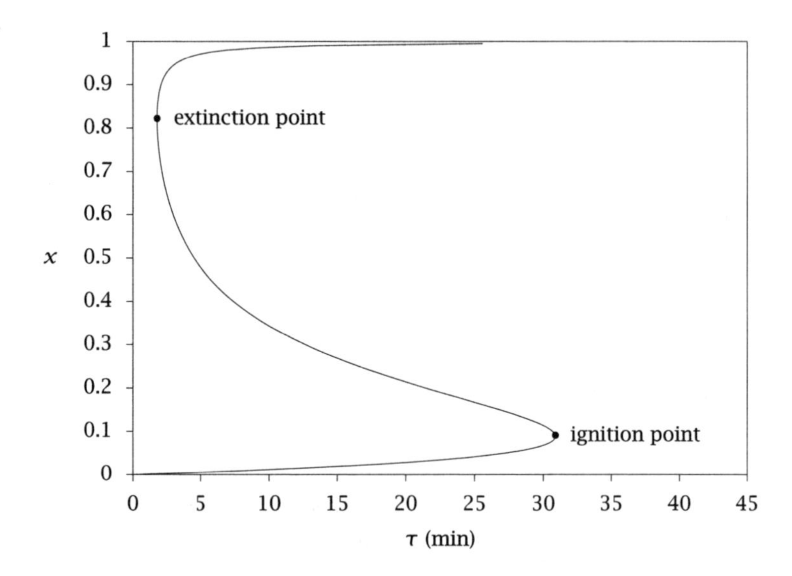

Figure 6.7: Steady-state conversion versus residence time for $\Delta H_R = -3 \times 10^5$ kJ/kmol; ignition and extinction points.

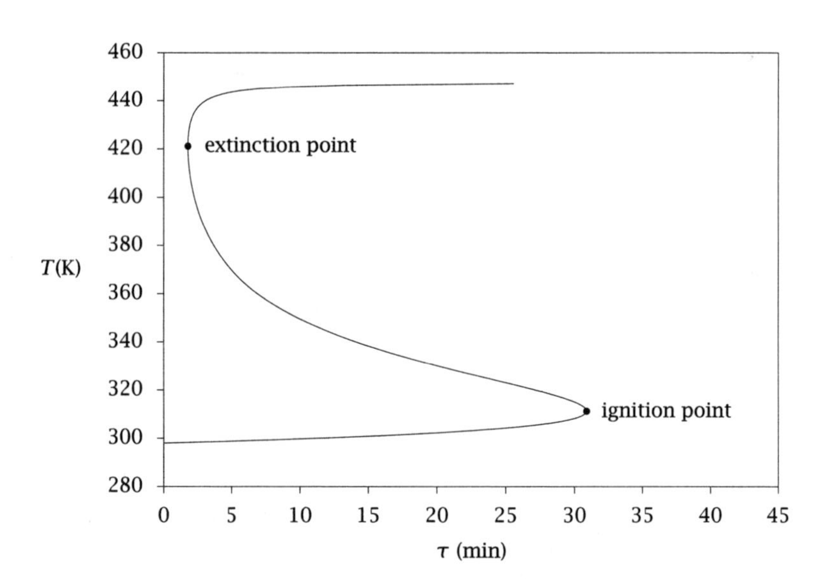

Figure 6.8: Steady-state temperature versus residence time for $\Delta H_R = -3 \times 10^5$ kJ/kmol; ignition and extinction points.

state for $\tau = 30.9$ min at $x = 0.09$ and $T = 311$ K. If there is a small decrease in feed flowrate so the residence time is 32 min, there is no steady state near the temperature and concentration of the reactor. A large release of heat occurs and the reactor ignites and moves to the steady state near $x = 1$ and $T = 448$ K. We can compute this maximal temperature corresponding to complete conversion as follows.

$$T_{\max} - T_f = -\frac{c_{A0}\Delta H_R}{C_{Ps}} = 150 \text{ K}$$

A reactor operating near the extinction point can exhibit the opposite phenomenon. A small increase in feed flowrate causes the residence time to decrease enough so that no steady-state solution exists near the current temperature and concentration. A rapid drop in temperature and increase in concentration of A occurs as the reactor approaches the new steady state. The curves of low temperature and low conversion connecting to the ignition point are known as the **lower branch** of solutions. The curves connecting the extinction point to the high temperature and conversion at large residence time are known as the **upper branch**. The curves connecting the ignition and extinction points together are called the **middle branch**. The points on the middle branch, while satisfying the steady-state equations, are **unstable** solutions. An unstable steady-state solution has the property that small perturbations to a reactor operating at this steady state causes the reactor to leave this solution and approach either the steady state on the upper or lower branch. The choice of which branch the reactor approaches depends on the exact nature of the perturbation the reactor experiences. This situation is much the same as a pencil initially balanced on end on a desk. Which direction the pencil falls is not easily predicted and depends in a sensitive way on the actual perturbation the pencil experiences. The upper and lower branches, on the other hand, are **stable** steady states. The reactor returns to these steady states after small perturbations.

6.3.2 Stability of the Steady State

We next discuss why some steady states are stable and others are unstable. This discussion comes in two parts. First we present a plausibility argument and develop some physical intuition by constructing and examining van Heerden diagrams [26]. In the next section we present a rigorous mathematical argument, which has wide applicability in analyzing the stability of any system described by differential equations.

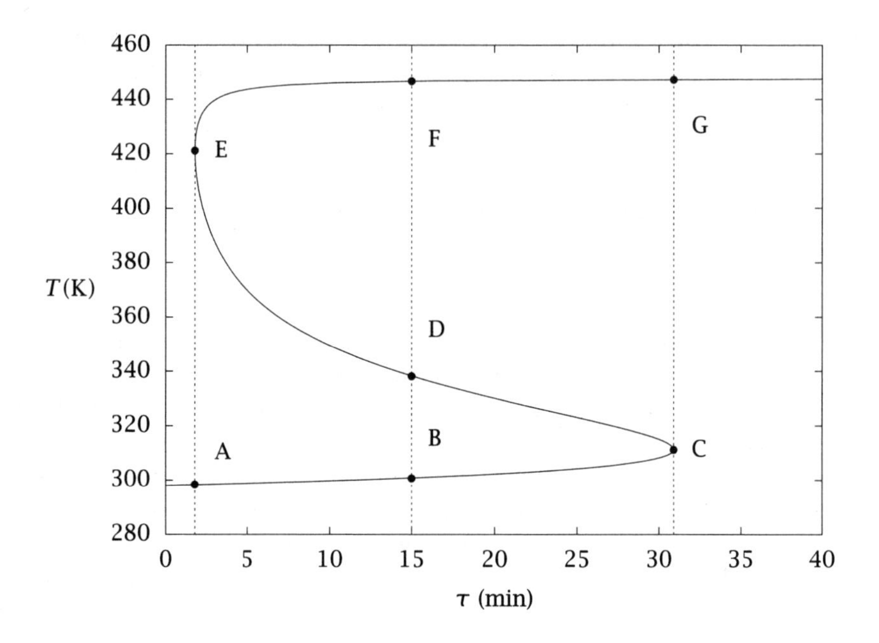

Figure 6.9: Steady-state temperature versus residence time for $\Delta H_R = -3 \times 10^5$ kJ/kmol.

Point	A	B	C	D	E	F	G
τ(min)	1.79	15	30.9	15	1.79	15	30.9
T(K)	298.2	300.5	311	338	421	446.5	447

Table 6.2: Selected values of steady-state temperatures and residence times.

Reactor Stability — Plausibility Argument

Consider again the steady-state temperature versus the residence-time curve shown in Figure 6.9. We have labeled seven points on this curve: A, B and C on the lower branch; D on the middle branch; and E, F and G on the upper branch. The numerical values at these points are listed in Table 6.2. If we first substitute the solution of the mass balance, Equation 6.37, for c_A into the energy balance, Equation 6.38, we obtain a single equation for the single unknown T

$$0 = \underbrace{-\frac{k}{1+k\tau} c_{Af} \Delta H_R}_{\dot{Q}_g} + \underbrace{\frac{C_{Ps}}{\tau}(T_f - T)}_{-\dot{Q}_r} \tag{6.39}$$

The first term, the reaction rate times the negative of the heat of reaction, we call the heat-generation rate, $\dot{Q}_g$. The second term is the enthalpy difference between the inflow and outflow streams, which we call the heat-removal rate, $\dot{Q}_r$,

$$\dot{Q}_g = -\frac{k(T)}{1+k(T)\tau} c_{Af} \Delta H_R, \qquad \dot{Q}_r = \frac{C_{Ps}}{\tau}(T - T_f)$$

in which we emphasize the temperature dependence of the rate constant,

$$k(T) = k_m e^{-E(1/T - 1/T_m)}$$

Obviously we have a steady-state solution when these two quantities are equal. Consider plotting these two functions as T varies. The heat-removal rate is simply a straight line with slope C_{Ps}/τ. The heat-generation rate is a nonlinear function that is asymptotically constant at low temperatures ($k(T)$ much less than one) and high temperatures ($k(T)$ much greater than one). These two functions are plotted for $\tau = 1.79$ min in Figure 6.10. Notice the two intersections of the heat-generation and heat-removal functions corresponding to steady states A and E. If we decrease the residence time slightly, the slope of the heat-removal line increases and the intersection corresponding to point A shifts slightly. Because the two curves are just tangent at point E, however, the solution at point E disappears, another indicator that point E is an extinction point.

We also can make a plausibility argument on the stability of steady-state A. If we were to increase the reactor temperature slightly, we would be to the right of point A in Figure 6.10. To the right of A we notice that the heat-removal rate is larger than the heat-generation rate. That causes the reactor to cool, which moves the temperature back to the left. In other words, the system responds by resisting our applied perturbation. Similarly, consider a decrease to the reactor temperature. To the left of point A, the heat-generation rate is larger than the heat-removal rate causing the reactor to heat up and move back to the right. Point A is a stable steady state because small perturbations are rejected by the system. Using this argument we see the situation at

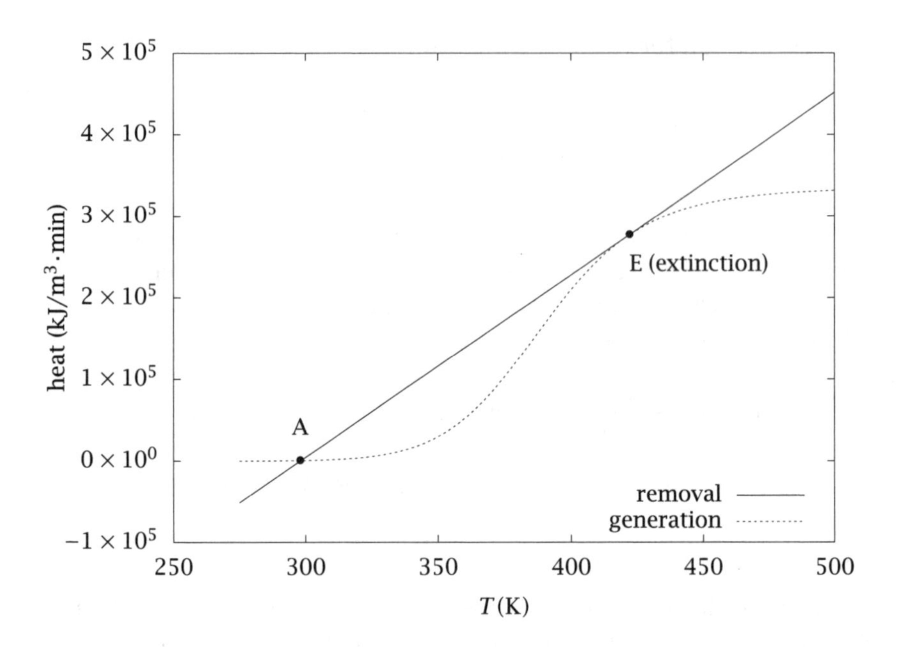

Figure 6.10: Rates of heat generation and removal for $\tau = 1.79$ min.

point E is quite different from the stability perspective. Increasing the temperature causes the system to cool, but decreasing the temperature also causes the system to cool. The system tends to increase the size of a cooling perturbation and does not return to the steady state. That is characteristic of an unstable solution.

Consider next the points on the middle branch. Figure 6.11 displays the heat-generation and heat-removal rates for points B, D and F, $\tau = 15$ min. Point B on the lower branch is stable as was point A. Point F on the upper branch also is stable because the slope of the heat-generation rate is smaller than the heat-removal rate at point F, just as at point A. At point D, however, the slope of the heat-generation rate is larger than the heat-removal rate. For point D, increasing temperature causes heat generation to be larger than heat removal, and decreasing temperature causes heat generation to be smaller than heat removal. Both of these perturbations are amplified by the system at point D, and this solution is unstable. All points on the middle branch are similar to point D.

Figure 6.12 displays the heat-generation and heat-removal rates for $\tau = 30.9$ min. Notice that point G on the upper branch is stable and

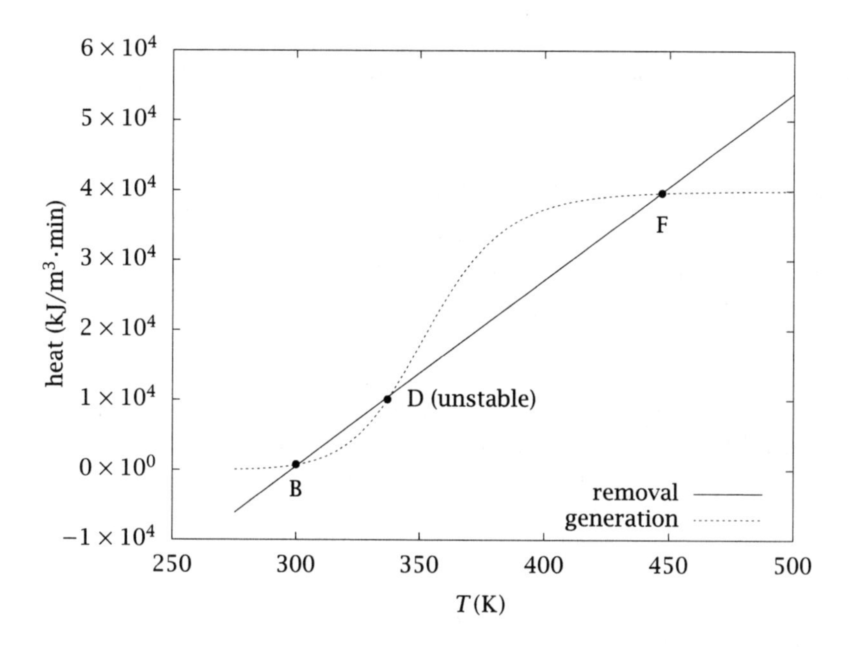

Figure 6.11: Rates of heat generation and removal for $\tau = 15$ min.

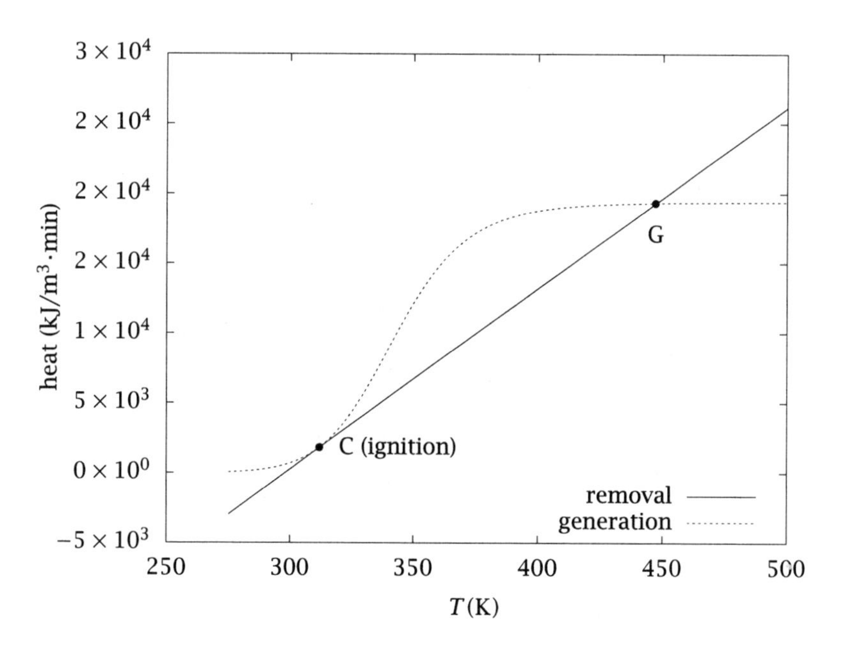

Figure 6.12: Rates of heat generation and removal for $\tau = 30.9$ min.

point C, the ignition point, is similar to extinction point E, perturbations in one direction are rejected, but in the other direction they are amplified.

Reactor Stability — Rigorous Argument

The stability discussion of the previous section accompanying the van Heerden diagrams is useful and provides insight, particularly in simple situations. In this section we develop proper arguments that determine stability rigorously in complex situations where physical intuition may not be as readily forthcoming. We also develop analysis tools that tell us more than just whether a steady state is stable or unstable; we also gain insight on the dynamic response of the system to perturbations.

Consider again the dynamic model of the reactor

$$\frac{dc_A}{dt} = \frac{c_{Af} - c_A}{\tau} - kc_A \tag{6.40}$$

$$\frac{dT}{dt} = \frac{U^oA}{V_R C_{Ps}}(T_a - T) + \frac{T_f - T}{\tau} - \frac{\Delta H_R}{C_{Ps}} kc_A \tag{6.41}$$

For convenience we summarize these two equations as the following nonlinear differential equations

$$\begin{aligned} \frac{dc_A}{dt} &= f_1(c_A, T) \\ \frac{dT}{dt} &= f_2(c_A, T) \end{aligned} \tag{6.42}$$

Given some values of parameters, we define steady-state solutions to these equations, c_{As}, T_s, as satisfying

$$\begin{aligned} 0 &= f_1(c_{As}, T_s) \\ 0 &= f_2(c_{As}, T_s) \end{aligned}$$

Imagine the reactor is initially at this steady state and at $t = 0$ we perturb the temperature and concentration by small amounts. We would like to know whether or not the system returns to the steady state after this initial condition perturbation. If so, we call the steady-state solution (asymptotically) stable. If not, we call the steady state unstable. Obviously we can solve numerically the nonlinear differential equations to answer this question, but then we answer the question on a case-by-case basis. By linearizing the nonlinear differential equations, we can gain further insight without resorting to full numerical solution.

Consider the Taylor series expansion of the nonlinear functions f_1, f_2 around the steady state

$$f_1(c_A, T) = f_1(c_{As}, T_s) + \frac{\partial f_1}{\partial c_A}(c_A - c_{As}) + \frac{\partial f_1}{\partial T}(T - Ts)$$
$$+ \text{ higher-order terms}$$
$$f_2(c_A, T) = f_2(c_{As}, T_s) + \frac{\partial f_2}{\partial c_A}(c_A - c_{As}) + \frac{\partial f_2}{\partial T}(T - Ts)$$
$$+ \text{ higher-order terms}$$

Define deviation variables, ϵ_1, ϵ_2, that measure how far the system is from the steady state

$$\epsilon_1 = c_A - c_{As}$$
$$\epsilon_2 = T - T_s$$

We can differentiate these deviation variables to see that they also satisfy Equation 6.42. Substituting the Taylor series expansion for f_1, f_2 and neglecting the higher-order terms, which are valid when the perturbations are small, produce the following approximate linear equations for the deviation variables

$$\frac{d\epsilon_1}{dt} = \frac{\partial f_1}{\partial c_A}\epsilon_1 + \frac{\partial f_1}{\partial T}\epsilon_2$$
$$\frac{d\epsilon_2}{dt} = \frac{\partial f_2}{\partial c_A}\epsilon_1 + \frac{\partial f_2}{\partial T}\epsilon_2$$

If we define the Jacobian matrix of f_1, f_2 with respect to c_A, T

$$\boldsymbol{J} = \begin{bmatrix} \frac{\partial f_1}{\partial c_A} & \frac{\partial f_1}{\partial T} \\ \frac{\partial f_2}{\partial c_A} & \frac{\partial f_2}{\partial T} \end{bmatrix} \tag{6.43}$$

we can compactly summarize the linear deviation variable equations in the convenient notation

$$\frac{d\boldsymbol{\epsilon}}{dt} = \boldsymbol{J\epsilon} \tag{6.44}$$

For our case, taking partial derivatives shows the Jacobian is

$$\boldsymbol{J} = \begin{bmatrix} -\frac{1}{\tau} - k & -kc_{As}\frac{E}{T_s^2} \\ -\frac{\Delta H_R}{C_{Ps}}k & -\frac{U^oA}{V_R C_{Ps}} - \frac{1}{\tau} - \frac{\Delta H_R}{C_{Ps}}kc_{As}\frac{E}{T_s^2} \end{bmatrix} \tag{6.45}$$

If Equation 6.44 governed exactly, rather than approximately, the evolution of the perturbations, we would be essentially finished. Linear differential equations can be solved analytically. The important feature for us to note is that the solution contains linear combinations of the terms $e^{\lambda t}$ in which λ are the eigenvalues of $\boldsymbol{J}$. In our case the 2×2 $\boldsymbol{J}$ matrix has two eigenvalues, which may be real or complex. If complex, they come as complex conjugate pairs. Stability of the linear equations is determined by the real parts of the eigenvalues. In fact,

> *The solution to the linear model, Equation 6.44, is asymptotically stable if and only if the eigenvalues of $\boldsymbol{J}$ have strictly negative real parts.*

We also know that if the eigenvalues are complex, the solution to Equation 6.44 contains oscillatory terms due to the relation

$$e^{(a+bi)t} = e^{at} \left(\cos(bt) + i \sin(bt)\right)$$

and the period of these oscillatory terms is

$$T = 2\pi / b \tag{6.46}$$

Finally, this linear problem is important to us only because it can be connected to the solution of the full nonlinear problem. Although precise statements of this connection require more mathematics than we can assume, the general result may be summed up for "nice" systems by saying [12]

> *For small initial perturbations from the steady state, the local behavior of the nonlinear model, Equation 6.42, is well described by the corresponding behavior of the linear model, Equation 6.44.*

The interested reader should see the first chapter in Guckenheimer and Holmes [12] for an introduction into the mathematical details and important restrictions on this general statement. For the models analyzed in this text, this general statement is valid. This connection forms the basis for all linear stability analysis.

We next use this result to analyze the multiple steady states of the previous section. In Figure 6.13 we plot the eigenvalues of $\boldsymbol{J}$ versus the steady-state reactor temperature and label the points A–G as indicated in Table 6.2 and shown in Figure 6.9. Both eigenvalues are real for all values of residence time. Notice that the steady state loses stability at point C, the ignition point, as one of the eigenvalues changes sign from negative to positive. Stability is regained on the upper branch at

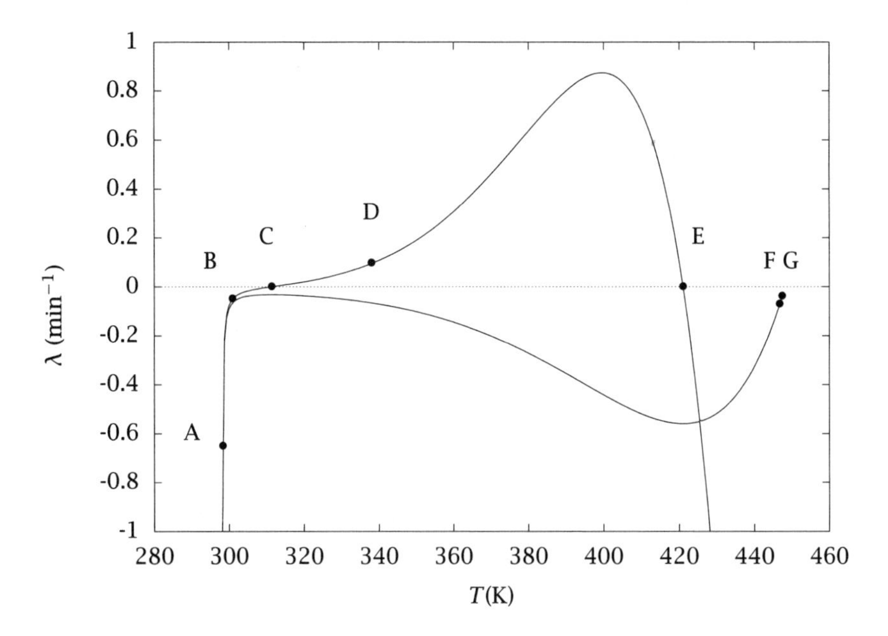

Figure 6.13: Eigenvalues of the Jacobian matrix versus reactor temperature.

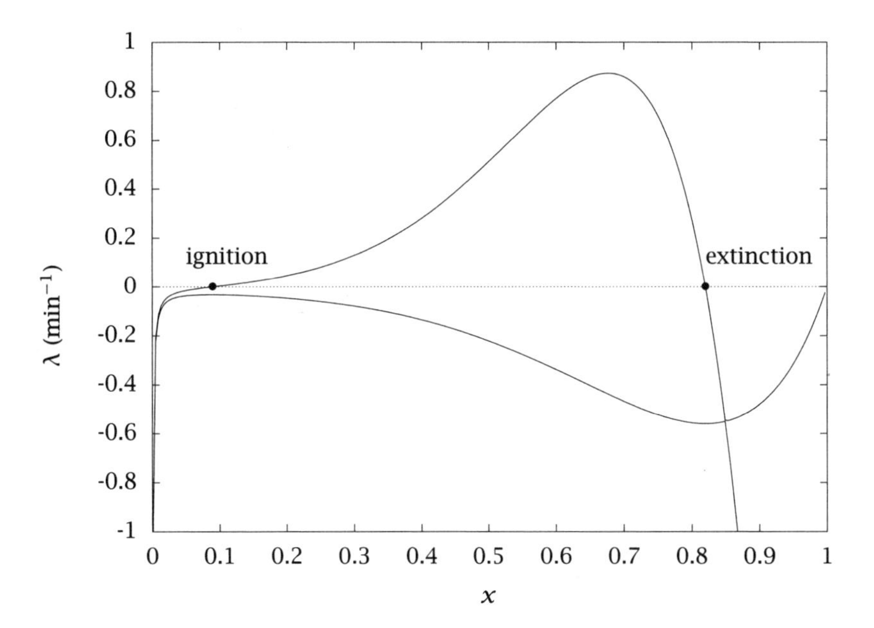

Figure 6.14: Eigenvalues of the Jacobian matrix versus reactor conversion.

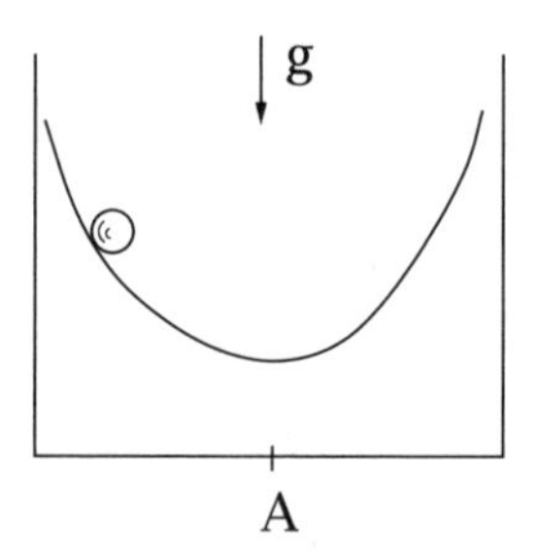

Figure 6.15: Marble on a track in a gravitational field; point A is the unique, stable steady state.

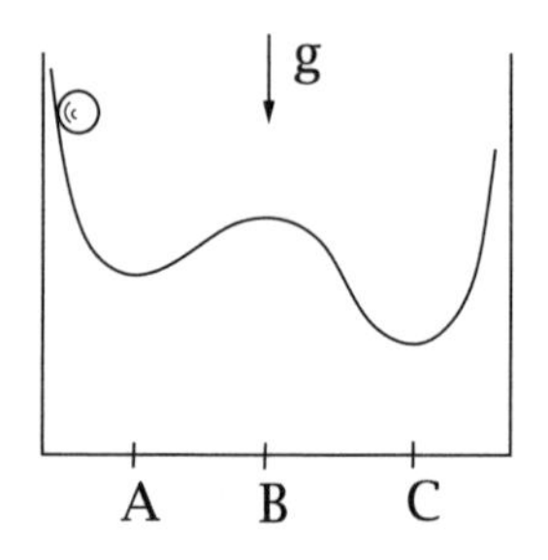

Figure 6.16: Marble on a track with three steady states; points A and C are stable, and point B is unstable.

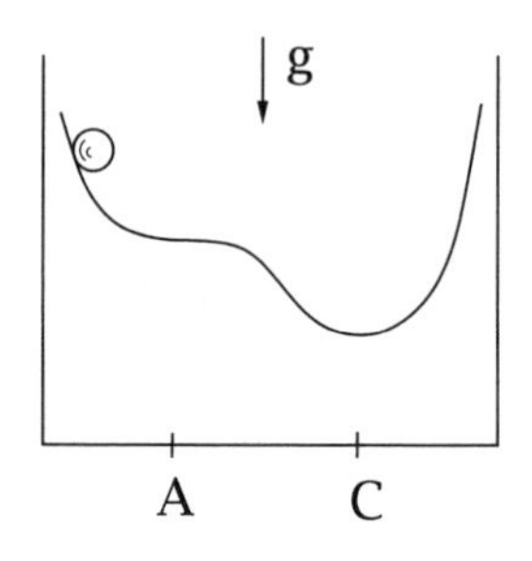

Figure 6.17: Marble on a track with an ignition point (A) and a stable steady state (C).

point E, the extinction point, as the eigenvalue changes sign back from positive to negative. This analysis validates the plausibility argument and shows the entire middle branch is unstable. Figure 6.14 provides another view of the same phenomenon, in which the eigenvalues are plotted versus the conversion instead of temperature; the ignition and extinction points are labeled.

Students may find it helpful to draw an analogy between the chemical reactor with multiple steady states and simple mechanical systems that exhibit the same behavior. Consider a marble on a track in a gravitational field as depicted in Figure 6.15. Based on our physical experience with such systems we conclude immediately that the system has a single steady state, position A, which is asymptotically stable. If we expressed Newton's laws of motion for this system, and linearized the model at point A, we would expect to see eigenvalues with negative real part and nonzero imaginary part because the system exhibits a decaying oscillation back to the steady-state position after a perturbation. The oscillation decays because of the friction between the marble and the track.

Now consider the track depicted in Figure 6.16. Here we have three steady states, the three positions where the tangent curve to the track has zero slope. This situation is analogous to the chemical reactor with multiple steady states. The steady states A and C are obviously stable and B is unstable. Perturbations from point B to the right are attracted to steady-state C and perturbations to the left are attracted to steady-state A. The significant difference between the reactor and marble systems is that the marble decays to steady state in an oscillatory fashion, and the reactor, with its zero imaginary eigenvalues, returns to the steady state without overshoot or oscillation. Now consider the track depicted in Figure 6.17. We have flattened the track between points A and B in Figure 6.16 so there is just a single point of zero slope, now marked point A. Point A now corresponds to a reactor ignition point as shown in Figures 6.7 and 6.8. Small perturbations push the marble over to the only remaining steady state, point C, which remains stable.

6.3.3 Sustained Oscillations, Limit Cycles

We next explore the fact that the dynamic behavior of the CSTR can be more complicated than multiple steady states with ignition, extinction and hysteresis. In fact, at a given operating condition, all steady states may be unstable and the reactor may exhibit sustained oscillations or limit cycles. Consider the same simple kinetic scheme as in the previous section,

$$A \xrightarrow{k} B$$

Param.	Value	Units
T_f	298	K
T_a	298	K
T_m	298	K
$\hat{C}_P$	4.0	kJ/(kg K)
c_{Af}	2.0	kmol/m^3
$k_m(T_m)$	0.004	min^{-1}
E	1.5×10^4	K
ρ	10^3	kg/m^3
ΔH_R	-2.2×10^5	kJ/kmol
$U^o A/V_R$	340	kJ/(m^3 min K)

Table 6.3: Parameter values for limit cycles.

but with the following parameter values. Notice that the activation energy in Table 6.3 is significantly larger than in Table 6.1. If we compute the solutions to the steady-state mass and energy balances, Equations 6.37 and 6.38, with these new values of parameters, we obtain the results displayed in Figures 6.18 and 6.19. If we replot these results using a log scaling to stretch out the x axis, we obtain the results in Figures 6.20 and 6.21. Notice the steady-state solution curve has become

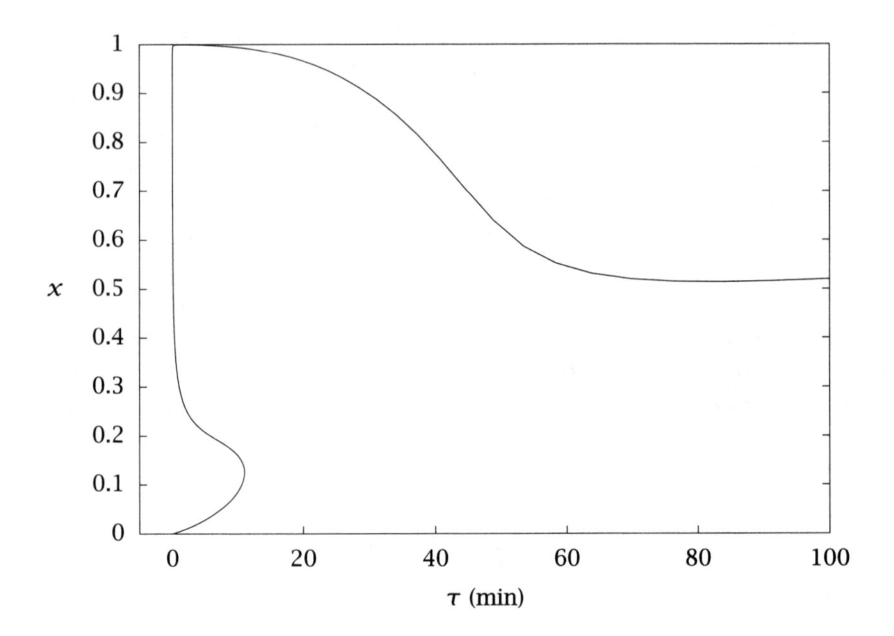

Figure 6.18: Steady-state conversion versus residence time.

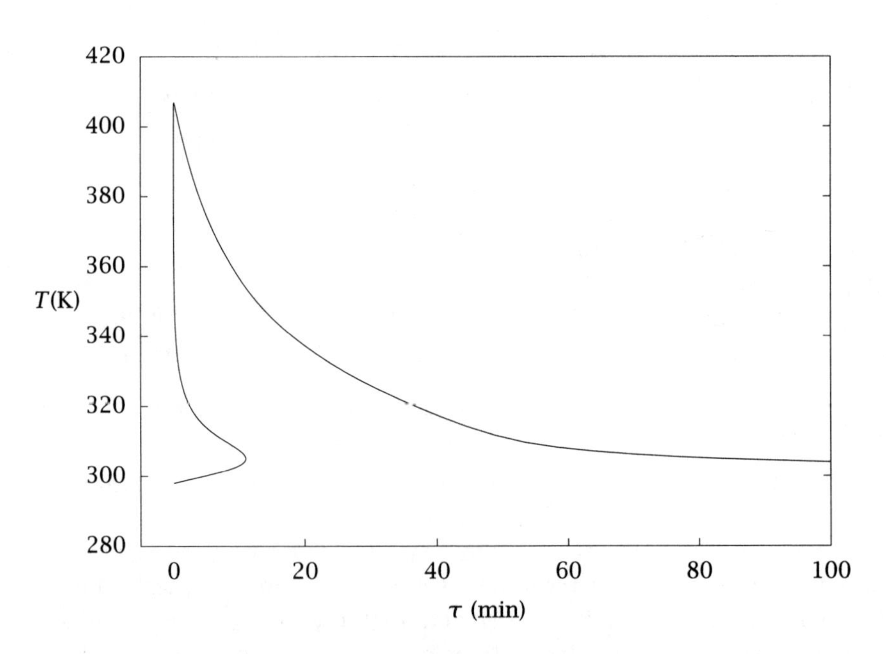

Figure 6.19: Steady-state temperature versus residence time.

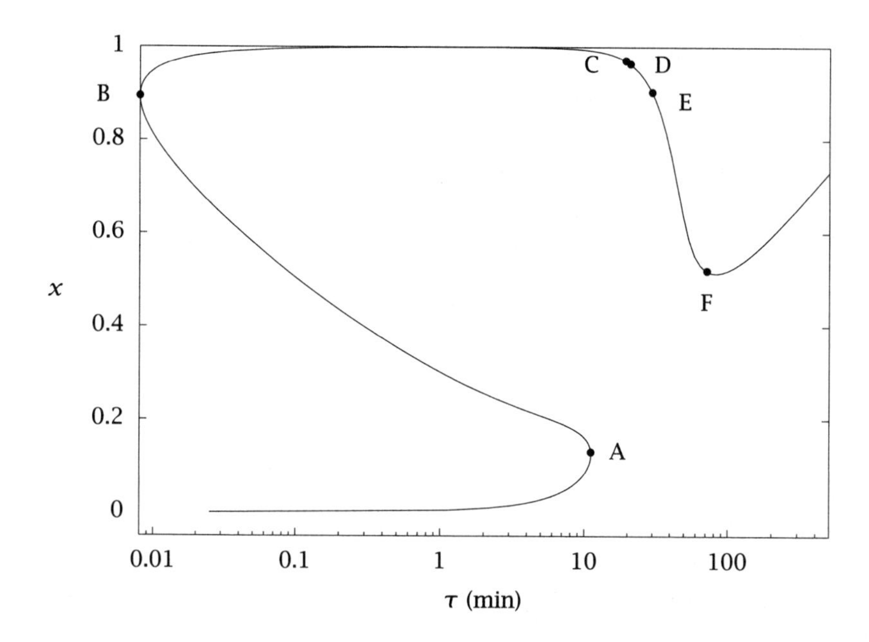

Figure 6.20: Steady-state conversion versus residence time — log scale.

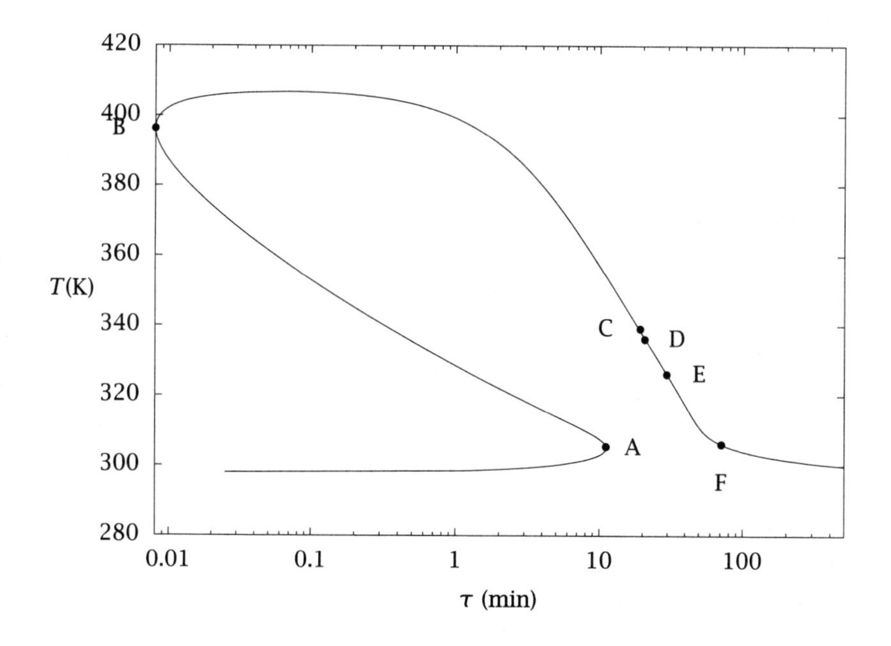

Figure 6.21: Steady-state temperature versus residence time — log scale.

Point	τ(min)	x	T(K)	Re(λ)(min^{-1})	Im(λ)(min^{-1})
A	11.1	0.125	305	0	0
B	0.008	0.893	396	0	0
C	19.2	0.970	339	−0.218	0
D	20.7	0.962	336	−0.373	0
E	29.3	0.905	327	0	0.159
F	71.2	0.519	306	0	0.0330

Table 6.4: Steady state and eigenvalue with largest real part at selected points in Figures 6.20 and 6.21.

a bit deformed and the simple s-shaped multiplicities in Figures 6.7 and 6.8 have taken on a mushroom shape with the new parameters. We have labeled points A–F on the steady-state curves in Figures 6.20 and 6.21. Table 6.4 summarizes the locations of these points in terms of the residence times, the steady-state conversions and temperatures, and the eigenvalue of the Jacobian with largest real part. Now we take a walk along the steady-state solution curve and examine the dynamic behavior of the system.

From the beginning of the curve through points A and B, the two eigenvalues are real. Their values are plotted in Figure 6.22 versus the conversion. As in the s-shaped multiplicity of the previous section, we see that an eigenvalue becomes positive at point A, which is again an ignition point. Figure 6.23 shows a magnified view. The middle branch between A and B is unstable and the upper branch after point B is stable again as the positive eigenvalue passes through zero and becomes negative.

As we continue along the upper branch, some surprising things happen. The two eigenvalues remain real and negative as we pass through point C and then merge into the same value at point D. Figure 6.24 shows the real and imaginary parts of the eigenvalues from points C–F. Notice the two eigenvalues approach each other along the real axis until they become identical at point D. After point D the imaginary parts of the eigenvalues are no longer zero, and we have a complex pair of eigenvalues. The two eigenvalues are complex conjugates; they have the same real part and opposite imaginary parts so the eigenvalue curves from D to F are mirror images. But the real parts of the eigenvalues remain negative from point D to E and the steady states along this portion of the upper branch are stable. At point E, however, the pair

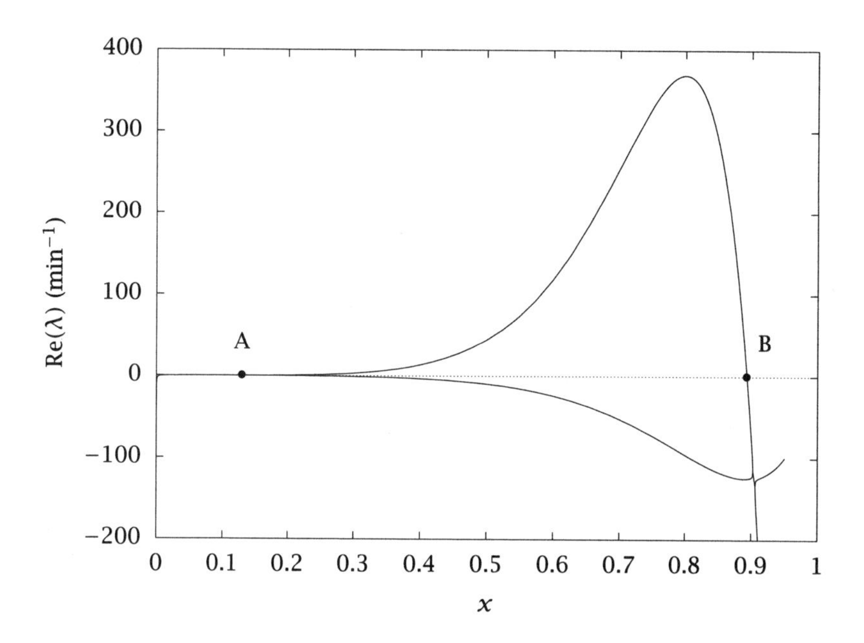

Figure 6.22: Eigenvalues of the Jacobian matrix versus reactor conversion in the region of steady-state multiplicity.

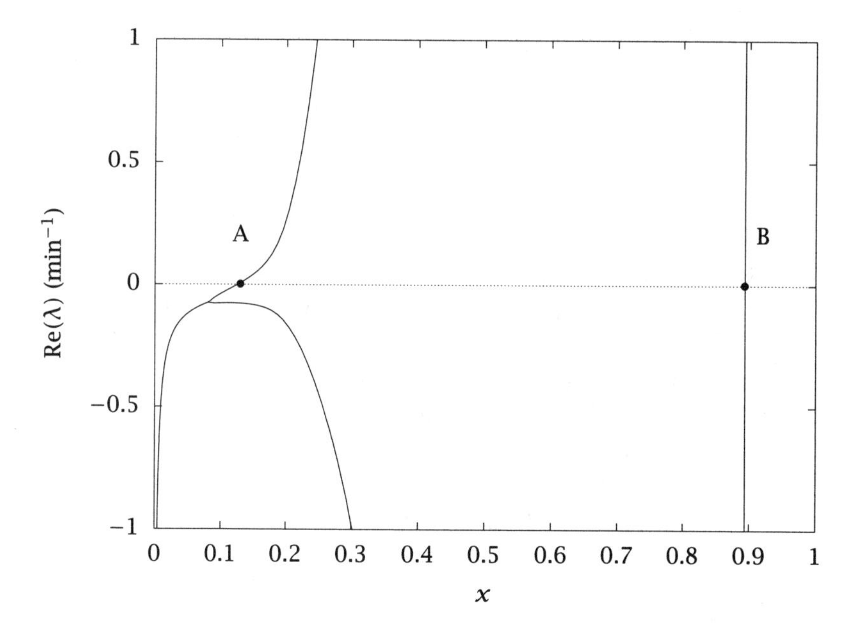

Figure 6.23: Eigenvalues of the Jacobian matrix versus reactor conversion in the region of steady-state multiplicity—magnified y axis.

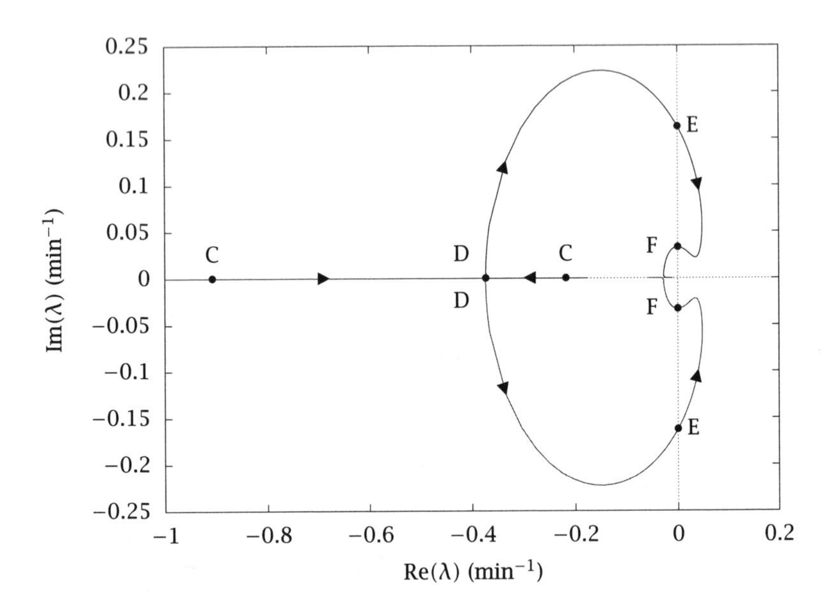

Figure 6.24: Real and imaginary parts of the eigenvalues of the Jacobian matrix near points C–F.

of eigenvalues cross back into the right half of the complex plane and have positive real parts. From point E to F, the eigenvalue analysis indicates the upper branch is unstable. Notice from Figures 6.18 and 6.19, however, that there is no lower-branch steady state for these values of residence time. It appears we have unique steady states and they are unstable. So a pressing issue now is what the reactor does after it is perturbed from these unique and unstable steady states.

Let us see what simulations can show us. A residence time of $\tau =$ 35 min is between points E and F as shown in Table 6.4. Solving the dynamic mass and energy balances with this value of residence time produces Figure 6.25. We see that the solution does not approach the steady state but oscillates continuously. These oscillations are sustained; they do not damp out at large times. Notice also that the amplitude of the oscillation is large, more than 80 K in temperature and 50% in conversion.

We can obtain another nice view of this result if we plot the conversion versus the temperature rather than both of them versus time. This kind of plot is known as a phase plot or phase portrait and is shown in

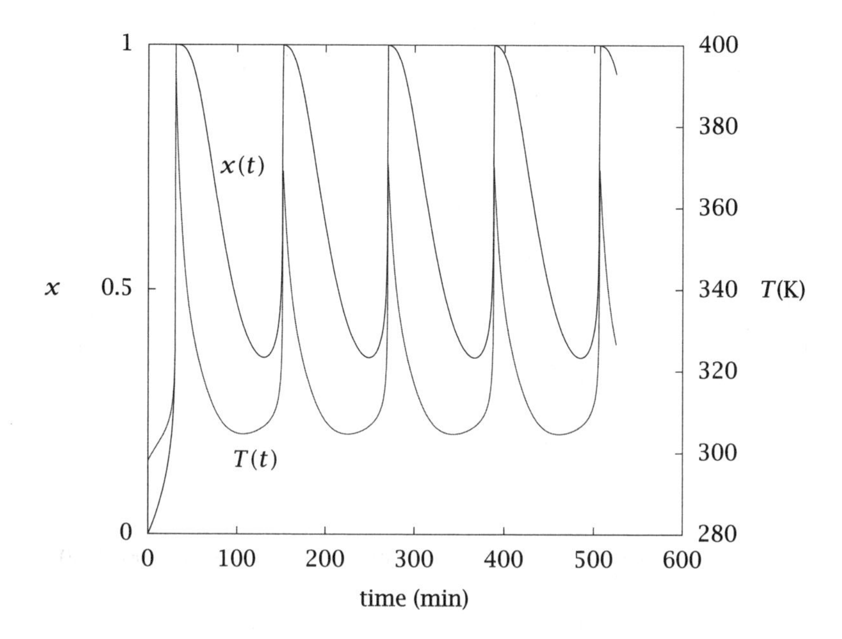

Figure 6.25: Conversion and temperature vs. time for $\tau = 35$ min.

Figure 6.26. Time increases as we walk along the phase plot; the reactor ignites, then slowly decays, ignites again, and eventually winds onto the steady limit cycle shown in Figure 6.26. Figure 6.27 explores the effect of initial conditions. The trajectory starting with the feed temperature and concentration as in Figure 6.26 is shown again. The trajectory starting in the upper left of the figure has the feed temperature and zero A concentration as its initial condition. Several other initial conditions inside the limit cycle are shown also, including starting the reactor at the unstable steady state. All of these initial conditions wind onto the same final limit cycle. We say that the limit cycle is a global *attractor* because all initial conditions wind onto this same solution.

If we decrease the residence time to $\tau = 30$ min, we are close to point E, where the stability of the upper steady state changes. A simulation at this residence time is shown in Figure 6.28. Notice the amplitude of these oscillations is much smaller, and the shape is more like a pure sine wave. Near point E in Figure 6.24, where the transition to instability occurs, the amplitude of the emerging oscillation is small and the period of the oscillation is well predicted by Equation 6.46 ($T = 2\pi/b$). Table 6.4 shows the imaginary part of the eigenvalue at point E is 0.159

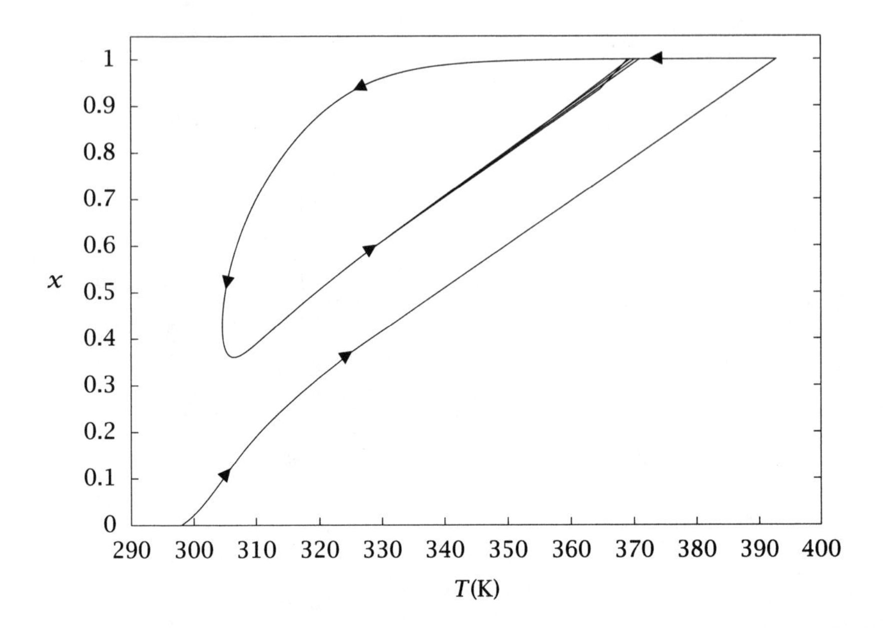

Figure 6.26: Phase portrait of conversion versus temperature for feed initial condition; $\tau = 35$ min.

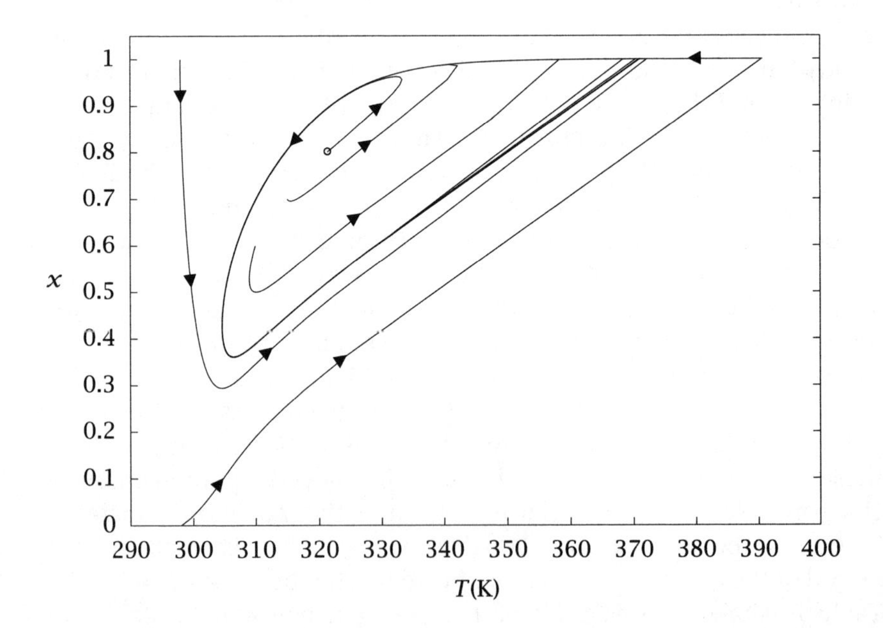

Figure 6.27: Phase portrait of conversion versus temperature for several initial conditions; $\tau = 35$ min.

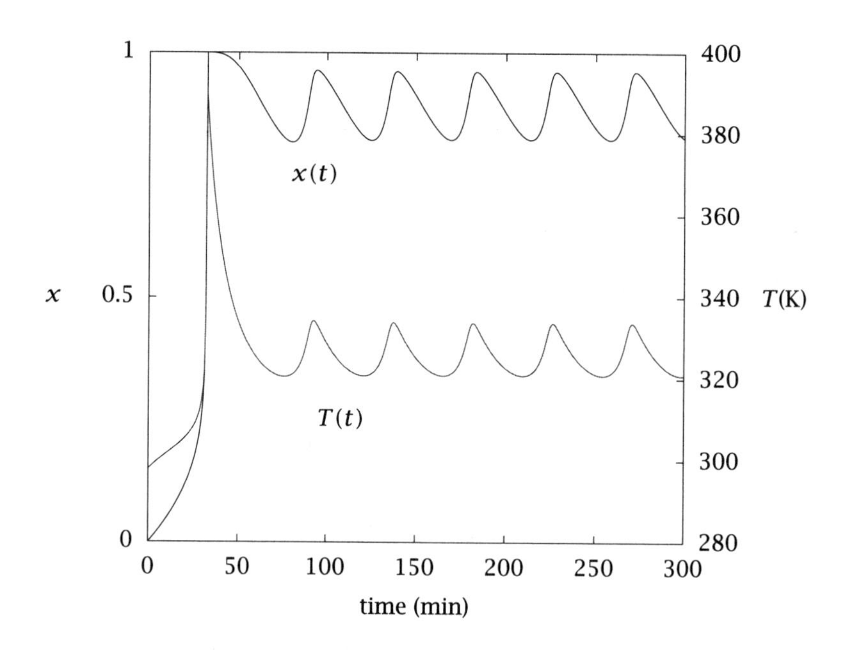

Figure 6.28: Conversion and temperature vs. time for $\tau = 30$ min.

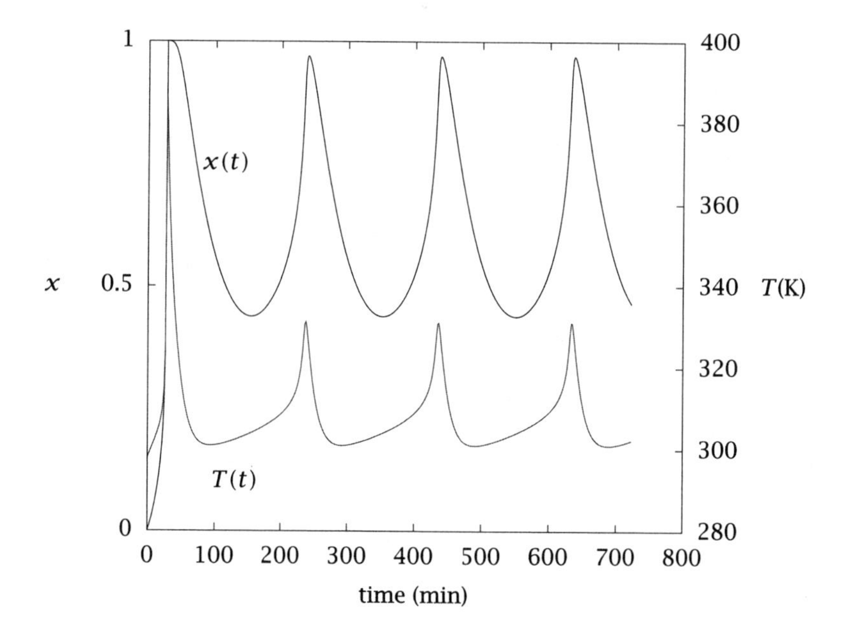

Figure 6.29: Conversion and temperature vs. time for $\tau = 72.3$ min.

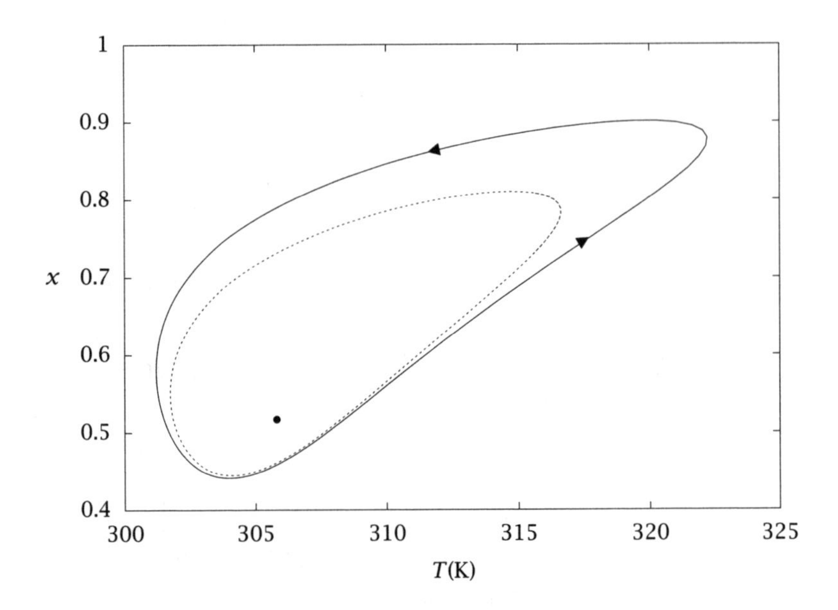

Figure 6.30: Phase portrait of conversion versus temperature at τ = 73.1 min showing stable and unstable limit cycles, and a stable steady state.

so the predicted period is

$$T = 2\pi/0.159 \text{ min} = 40 \text{ min}$$

which is in good agreement with the simulation shown in Figure 6.28.

As we pass point F, the eigenvalues in Figure 6.24 cross back into the left-half plane and the steady state is again stable. A simulation near point F is shown in Figure 6.29. Notice, in contrast to point E, the amplitude of the oscillations is not small near point F. Indeed, even though Figure 6.24 indicates the steady state is stable for $\tau \geq 72.3$ min, that does not mean that limit-cycle solutions are not possible in this part of the parameter space. To see how limit cycles can remain after the steady state regains its stability, consider Figure 6.30, constructed for $\tau = 73.1$ min. The figure depicts the stable steady state, indicated by a solid circle, surrounded by an unstable limit cycle, indicated by the dashed line.[4] The unstable limit cycle is in turn surrounded by a

[4]This curve was constructed by reversing time in the integration of the mass and energy balances, which stabilizes the unstable limit cycle.

stable limit cycle. Note that all initial conditions outside of the stable limit cycle would converge to the stable limit cycle from the outside. All initial conditions in the region between the unstable and stable limit cycles would converge to the stable limit cycle from the inside. Finally, all initial conditions inside the unstable limit cycle are attracted to the stable steady state. We see that the unstable limit cycle is the boundary between the regions of attraction of the stable steady state and stable limit cycle. We also have a quantitative measure of a perturbation capable of knocking the system from the steady state onto a periodic solution. Exercise 6.12 explores the regions of attraction of the steady state and the stable limit cycle in further detail.

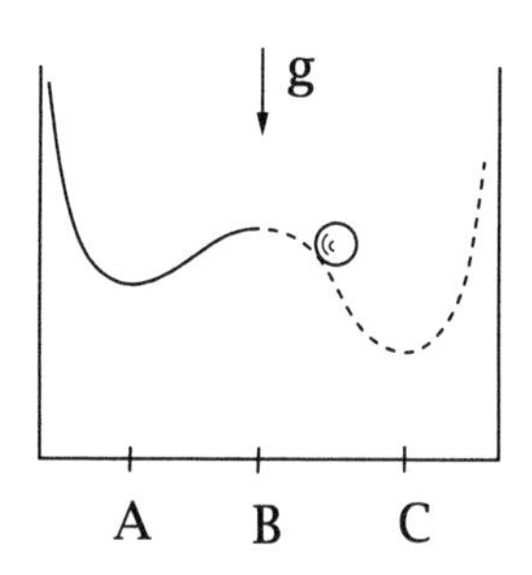

Figure 6.31: Marble on a track with three steady states; dashed line indicates frictionless track; point C is not asymptotically stable and is surrounded by limit cycles.

We may modify our simple mechanical system to illustrate somewhat analogous limit-cycle behavior. Consider the marble and track system depicted in Figure 6.31. We have three steady states; steady-state A is again stable and steady-state B is unstable as in Figure 6.16. At this point we cheat thermodynamics a bit to achieve the desired behavior. Imagine the track consists of a frictionless material to the right of point B. Although in violation of the second law, we can either use our imagination or consider extremely low friction material. Without friction in the vicinity of point C, the steady state is not asymptotically stable. Perturbations from point C do not return to the steady state but continually oscillate. The analogy is not perfect because a single limit cycle does not surround the unstable point C as in the chemical reactor shown in Figures 6.27 and 6.30. The amplitude of the oscillation in the marble system depends on the initial perturbation and a continuous family of limit cycles surrounds the unstable steady-state C. But the analogy may prove helpful in demystifying these kinds of reactor behaviors. The reader may also want to consider why we were compelled to violate the second law to achieve sustained oscillations in the simple mechanical system but the reactor can continually oscillate without such violation. Consider what is the essential difference between these thermodynamic systems.

Further Reading on CSTR Dynamics and Stability

Consideration of the coupled mass and energy balances for the CSTR have led to possible behaviors that may seem surprisingly complex for even the simplest kinetic mechanism, an irreversible first-order reaction. Just because these behaviors are possible does not mean that they are normally observed in reactor operation for something as simple as A goes to B.

Complex kinetics, such as polymerization kinetics [19, 22], lead to complex types of behavior, even under isothermal operation, i.e., *without* the nonlinearity introduced by the rate constant's exponential dependence on temperature. Bailey [4] summarizes some cases of experimentally observed oscillations in CSTRs including chlorination of methyl chloride [8], hydrolysis of acetyl chloride [3], the reaction between methanol and hydrogen chloride [13], and the reaction between sodium thiosulfate and hydrogen peroxide [9].

In addition to the multiple steady states and sustained oscillations, chemical reactors are a rich source of other types of behavior including deterministic chaos [18, 17, 23, 15]. The chemical engineer should be aware of the complex behavior that is possible with simple nonlinear CSTR models, especially if confronted with apparently complex operating data.

6.4 The Semi-Batch Reactor

The development of the semi-batch reactor energy balance follows directly from the CSTR energy balance derivation by setting $Q = 0$. The main results are summarized in Table 6.9 at the end of this chapter. Note in particular that Equations 6.81–6.82 in the semi-batch reactor Table 6.9 are identical to the corresponding Equations 6.72–6.73 in the CSTR Table 6.8.

It is good practice to enumerate the unknowns and applicable equations when setting up a model for the nonisothermal semi-batch reactor. As discussed in Chapter 4, for a single-phase system, the molar concentrations of the components, and the temperature and the pressure specify all intensive variables of the reactor. If we use the reactor volume as the single extensive variable, then we have $n_s + 3$ unknowns. As in Chapter 4, we have n_s equations from the material balances and one equation of state. The energy balance of this chapter provides an additional equation. Finally, we must specify the reactor pressure or some other system constraint that determines the reactor pressure.

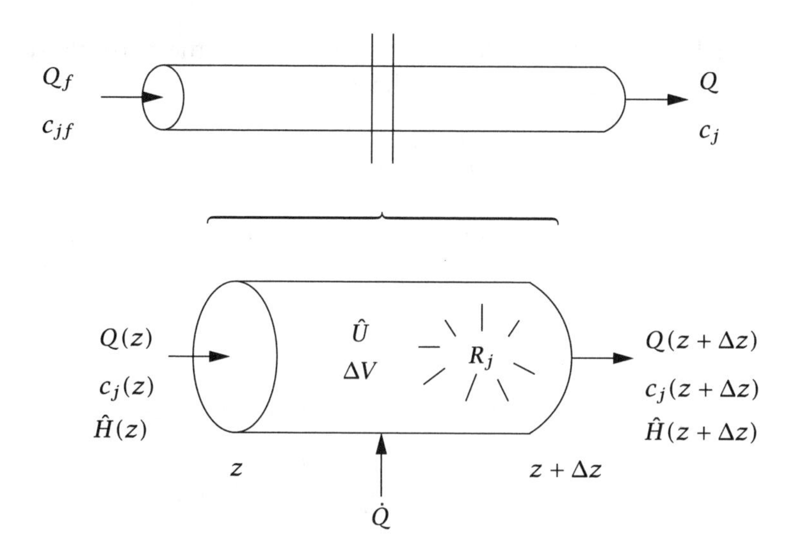

Figure 6.32: Plug-flow reactor volume element.

As in Section 4.5 of Chapter 4, if the reactor outflow term is nonzero as well, then an additional statement of reactor operation, such as constant reactor volume or mass, is required to determine Q, which then appears in the material and energy balances.

6.5 The Plug-Flow Reactor

To derive an energy balance for the plug-flow reactor (PFR), consider the volume element in Figure 6.32. If we write Equation 6.5 for this element and neglect kinetic and potential energies and shaft work, we obtain

$$\frac{\partial}{\partial t}(\rho \hat{U} A_c \Delta z) = m\hat{H}|_z - m\hat{H}|_{z+\Delta z} + \dot{Q}$$

in which A_c is the cross-sectional area of the tube, R is the tube outer radius, and $\dot{Q}$ is the heat transferred through the wall, normally expressed using an overall heat-transfer coefficient

$$\dot{Q} = U^o 2\pi R \Delta z (T_a - T)$$

Dividing by $A_c \Delta z$ and taking the limit $\Delta z \to 0$, gives

$$\frac{\partial}{\partial t}(\rho \hat{U}) = -\frac{1}{A_c}\frac{\partial}{\partial z}(Q\rho\hat{H}) + \dot{q}$$

in which $\dot{q} = (2/R)U^o(T_a - T)$ and we express the mass flowrate as $m = Q\rho$. In the steady state, we have

$$\frac{d}{dV}(Q\sum_j c_j\overline{H}_j) = \dot{q}$$

in which we use the volume instead of length of the reactor, $V = A_c z$, and partial molar enthalpy to express the enthalpy per volume,

$$\rho\hat{H} = \sum_j c_j\overline{H}_j$$

The enthalpy term can be broken apart into two terms

$$\frac{d}{dV}(Q\sum_j c_j\overline{H}_j) = \sum_j \left[Qc_j\frac{d\overline{H}_j}{dV} + \overline{H}_j\frac{d}{dV}(Qc_j)\right] \tag{6.47}$$

We treat the $\sum_j c_j d\overline{H}_j$ term by deriving the Gibbs-Duhem equation for enthalpy as follows. We have $H = \sum_j n_j\overline{H}_j$ because $\overline{H}_j$ is a partial molar enthalpy (see also Exercise 3.6 in Chapter 3). Its differential is therefore

$$dH = \sum_j \overline{H}_j dn_j + n_j d\overline{H}_j$$

Subtracting this expression from Equation 6.10, which is valid for single-phase systems,

$$dH = V\rho\hat{C}_P dT + (1 - \alpha T)VdP + \sum_j \overline{H}_j dn_j$$

yields a Gibbs-Duhem equation for enthalpy

$$0 = V\rho\hat{C}_P dT + (1 - \alpha T)VdP - \sum_j n_j d\overline{H}_j$$

We express an intensive version of this result by dividing by V and rearranging to obtain

$$\boxed{\sum_j c_j d\overline{H}_j = \rho\hat{C}_P dT + (1 - \alpha T)dP} \qquad \text{Gibbs-Duhem equation} \tag{6.48}$$

We can form the derivative of Equation 6.48 with reactor volume to obtain

$$\sum_j c_j\frac{d\overline{H}_j}{dV} = \rho\hat{C}_P\frac{dT}{dV} + (1 - \alpha T)\frac{dP}{dV} \tag{6.49}$$

The second term in Equation 6.47 is expressed using the material balance for the PFR

$$\frac{d}{dV}(Qc_j) = R_j = \sum_i \nu_{ij} r_i$$

Substitution of this result and Equation 6.49 into Equation 6.47 then gives

$$\boxed{Q\rho\hat{C}_P \frac{dT}{dV} + Q(1 - \alpha T)\frac{dP}{dV} = -\sum_i \Delta H_{Ri} r_i + \dot{q}} \qquad (6.50)$$

In unpacked tubes, the pressure drop is usually negligible, and for an ideal gas, $\alpha T = 1$. For both of these cases, we have

Ideal gas, or neglect pressure drop.

$$\boxed{Q\rho\hat{C}_P \frac{dT}{dV} = -\sum_i \Delta H_{Ri} r_i + \dot{q}} \qquad (6.51)$$

If the fluid is incompressible, $\alpha = 0$, and Equation 6.50 reduces to

Incompressible fluid.

$$\boxed{Q\rho\hat{C}_P \frac{dT}{dV} + Q\frac{dP}{dV} = -\sum_i \Delta H_{Ri} r_i + \dot{q}} \qquad (6.52)$$

Equation 6.51 is the usual energy balance for PFRs in this chapter. The next chapter considers packed-bed reactors in which the pressure drop may be significant. Equation 6.50 is more appropriate in that situation unless one can safely make the ideal-gas assumption.

Example 6.4: PFR and interstage cooling

Consider the reversible, gas-phase reaction

$$\mathrm{A} \underset{k_{-1}}{\overset{k_1}{\rightleftharpoons}} \mathrm{B}$$

The reaction is carried out in two long, adiabatic, plug-flow reactors with an interstage cooler between them as shown in Figure 6.33. (Refer to the discussion in Section 1.4.2 for more on interstage cooling.) The feed consists of component A diluted in an inert N_2 stream, $N_{Af}/N_{If} = 0.1, N_{Bf} = 0$, and $Q_f = 10{,}000$ ft^3/hr at $P_f = 180$ psia and $T_f = 830$°R. Since the inert stream is present in such excess, we assume that the heat

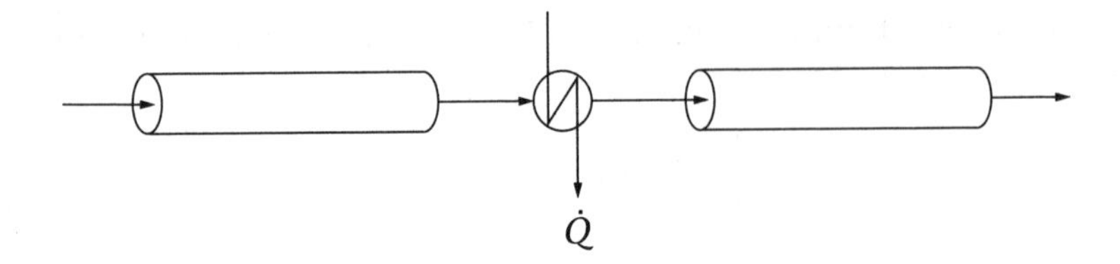

Figure 6.33: Tubular reactors with interstage cooling.

capacity of the mixture is equal to the heat capacity of nitrogen and is independent of temperature for the temperature range we expect. The heat of reaction is $\Delta H_R = -5850$ BTU/lbmol and can be assumed constant. The value of the equilibrium constant is $K = k_1/k_{-1} = 1.5$ at the feed temperature.

1. Write down the mole and energy balances that would apply in the reactors. Make sure all variables in your equations are expressed in terms of T and N_A. What other assumptions did you make?

2. If the reactors are long, we may assume that the mixture is close to equilibrium at the exit. Using the mole balance, express N_A at the exit of the first reactor in terms of the feed conditions and the equilibrium constant, K.

3. Using the energy balance, express T at the exit of the first reactor in terms of the feed conditions and N_A.

4. Notice we have two equations and two unknowns because K is a strong function of T. Solve these two equations numerically and determine the temperature and conversion at the exit of the first reactor. Alternatively, you can substitute the material balance into the energy balance to obtain one equation for T. Solve this equation to determine the temperature at the exit of the first reactor. What is the conversion at the exit of the first reactor?

5. Assume that economics dictate that we must run this reaction to 70% conversion to make a profit. How much heat must be removed in the interstage cooler to be able to achieve this conversion at the exit of the second reactor? What are the temperatures at the inlet and outlet of the second reactor?

6. How would you calculate the *actual* conversion achieved for two PFRs of *specified* sizes (rather than "long" ones) with this value of $\dot{Q}$?

Solution

1. The steady-state molar flow of A is given by the PFR material balance

$$\frac{dN_A}{dV} = R_A = -r \tag{6.53}$$

and the rate expression for the reversible reaction is given by

$$r = k_1 c_A - k_{-1} c_B = (k_1 N_A - k_{-1} N_B)/Q$$

The molar flow of B is given by $dN_B/dV = r$, so we conclude

$$N_B = N_{A_f} + N_{Bf} - N_A = N_{Af} - N_A$$

If we assume the mixture behaves as an ideal gas at these conditions, $c = P/RT$ or

$$Q = \frac{RT}{P} \sum_{j=1}^{n_S} N_j$$

The material balance for inert gives $dN_I/dV = 0$, so we have the total molar flow is $\sum_{j=1}^{n_s} N_j = N_{Af} + N_{If}$ and the volumetric flowrate is

$$Q = \frac{RT}{P}(N_{Af} + N_{If})$$

and the reaction rate is

$$r = \frac{P}{RT}\left(\frac{k_1 N_A - k_{-1}(N_{Af} - N_A)}{N_{Af} + N_{If}}\right)$$

which is in terms of T and N_A. The adiabatic PFR energy balance for an ideal gas is given by

$$\frac{dT}{dV} = -\frac{\Delta H_R}{Q\rho \hat{C}_P} r \tag{6.54}$$

2. For long reactors, $r = 0$ or

$$k_1 N_A - k_{-1}(N_{Af} - N_A) = 0$$

Dividing by k_{-1} and solving for N_A gives

$$N_A = \frac{N_{Af}}{1 + K_1}$$

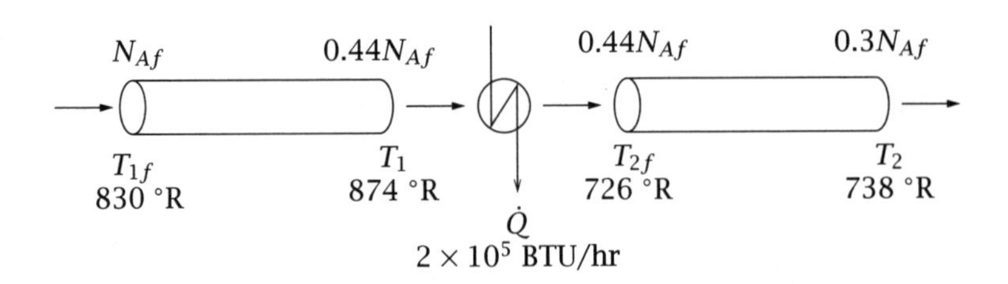

Figure 6.34: Temperatures and molar flows for tubular reactors with interstage cooling.

3. Substituting $r = -dN_A/dV$ into the energy balance and multiplying through by dV gives

$$dT = \frac{\Delta H_R}{Q\rho \hat{C}_P} dN_A$$

The term $Q\rho = m$ in the denominator is the mass flowrate, which is constant and equal to the feed mass flowrate. If we assume the heat of reaction and the heat capacity are weak functions of temperature and composition, we can perform the integral yielding

$$T_1 - T_{1f} = \frac{\Delta H_R}{m\hat{C}_P}(N_A - N_{Af})$$

4. $T_1 - 830 + 80.1 \left(\frac{1}{1 + 0.0432e^{2944/T_1}} - 1 \right) = 0,$

   $T_1 = 874°\text{R}, \qquad x = 0.56$

5. We compute the equilibrium T_2 to achieve 70% conversion, T_{2f} from T_2, and the interstage cooler duty from T_1 and T_{2f}

$$T_2 = 738°\text{R} \qquad T_{2f} = 726°\text{R} \qquad \dot{Q} = 200,000 \text{ BTU/hr}$$

6. Integrate Equations 6.53 and 6.54.

The results are summarized in Figure 6.34. □

6.5.1 Plug-Flow Reactor Hot Spot and Runaway

For exothermic, gas-phase reactions in a PFR, the heat release generally leads to the formation of a reactor hot spot, a point along the reactor length at which the temperature profile achieves a maximum. If the reaction is highly exothermic, the temperature profile can be very sensitive to parameters, and a small increase in the inlet temperature or reactant feed concentration, for example, can lead to large changes in the temperature profile. A sudden, large increase in the reactor temperature due to a small change in feed conditions is known as reactor runaway. Reactor runaway is highly dangerous, and operating conditions are normally chosen to keep reactors far from the runaway condition. The following example, oxidation of o-xylene to phthalic anhydride, illustrates the PFR hotspot and reactor runaway.

Example 6.5: Oxidation of o-xylene to phthalic anhydride

The gas-phase oxidation of o-xylene to phthalic anhydride

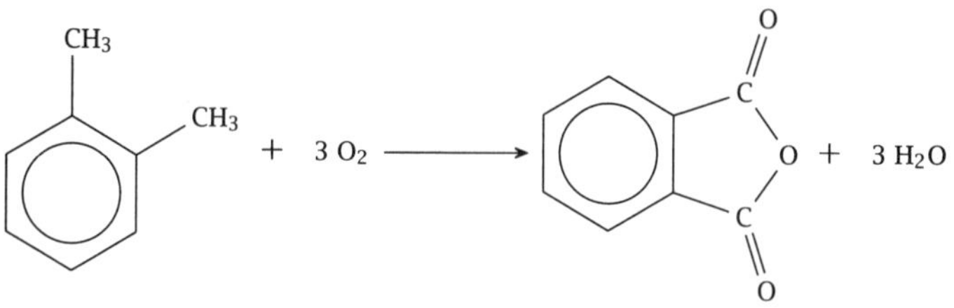

CH_3 CH_3 + 3 O_2 ⟶ (C=O, O, C=O) O + 3 H_2O

is highly exothermic. The reaction is carried out in PFR tube bundles with molten salt circulating as the heat transfer fluid [11]. The o-xylene is mixed with air before entering the PFR. The reaction rate is limited by maintaining a low concentration of hydrocarbon in the feed. The mole fraction of o-xylene is less than 2%.

Under these conditions, the large excess of oxygen leads to a pseudo-first-order rate expression

$$r = k_m \exp\left[-E\left(\frac{1}{T} - \frac{1}{T_m}\right)\right] c_x$$

in which c_x is the o-xylene concentration. The operating pressure is atmospheric. Calculate the temperature and o-xylene composition profiles. The kinetic parameters are adapted from Van Welsenaere and Froment and given in Table 6.5 [27].

Solution

If we assume constant thermochemical properties, an ideal gas mixture, and express the mole and energy balances in terms of reactor length,

Parameter	Value	Units
k_m	2.0822	s^{-1}
T_a	625	K
T_m	625	K
P_f	1.0	atm
l	1.5	m
R	0.0125	m
$\hat{C}_P$	0.992	kJ/(kg K)
U^o	0.373	kJ/(m^2 s K)
y_{xf}	0.019	
E	1.3636×10^4	K
ΔH_R	-1.284×10^6	kJ/kmol
$Q\rho$	2.6371×10^{-3}	kg/s

Table 6.5: PFR operating conditions and parameters for o-xylene example.

we obtain

$$\frac{dN_x}{dz} = -A_c r$$
$$\frac{dT}{dz} = -\beta r + \gamma(T_a - T)$$
$$r = k\frac{P}{RT}\frac{N_x}{N}$$

in which

$$\beta = \frac{\Delta H_R A_c}{Q\rho\hat{C}_P}, \qquad \gamma = \frac{2\pi R U^o}{Q\rho\hat{C}_P}$$

and the total molar flow is constant and equal to the feed molar flow because of the stoichiometry. Figure 6.35 shows the molar flow of o-xylene versus reactor length for several values of the feed temperature. The corresponding temperature profile is shown in Figure 6.36. We see a hotspot in the reactor for each feed temperature. Notice the hotspot temperature increases and moves down the tube as we increase the feed temperature. Finally, notice if we increase the feed temperature above about 625 K, the temperature spikes quickly to a large value and all of the o-xylene is converted by $z = 0.6$ m, which is a classic example of reactor runaway. To avoid this reactor runaway, we must maintain the feed temperature below a safe value. This safe value obviously also depends on how well we can control the composition and temperature in the feed stream. Tighter control allows us to operate safely at higher

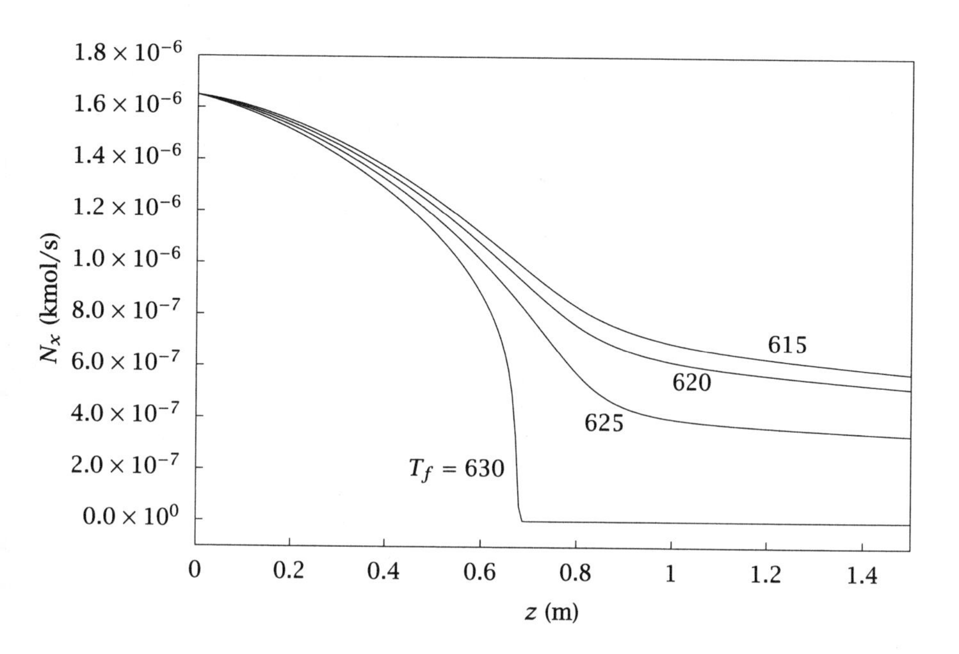

Figure 6.35: Molar flow of o-xylene versus reactor length for different feed temperatures.

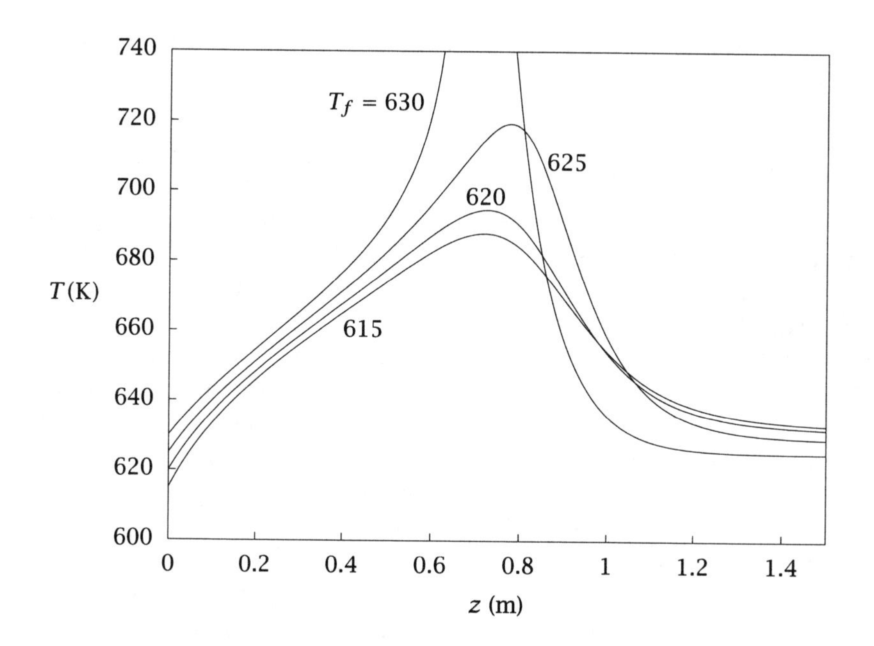

Figure 6.36: Reactor temperature versus length for different feed temperatures.

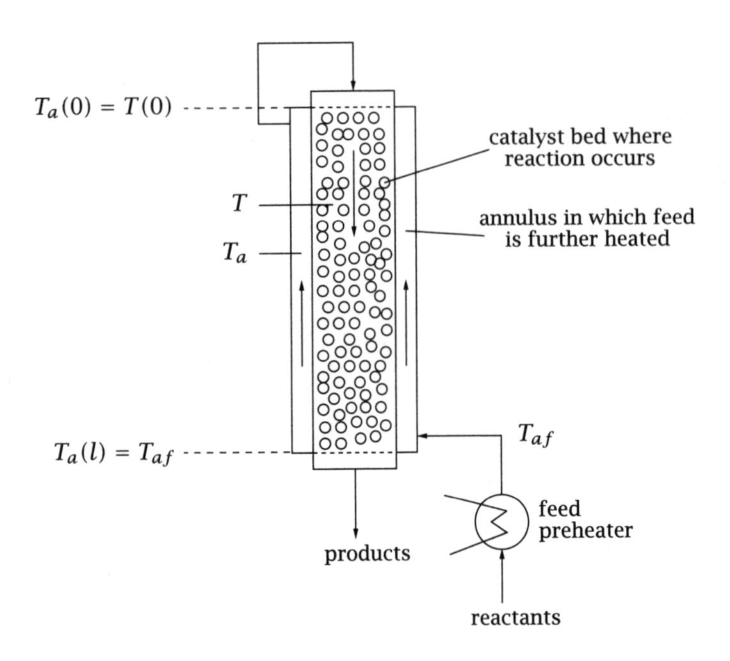

Figure 6.37: Autothermal plug-flow reactor; the heat released by the exothermic reaction is used to preheat the feed.

feed temperatures and feed o-xylene mole fractions, which increases the production rate. □

6.5.2 The Autothermal Plug-Flow Reactor

In many applications, it is necessary to heat a feed stream to achieve a reactor inlet temperature having a high reaction rate. If the reaction also is exothermic, we have the possibility to lower the reactor operating cost by heat integration. The essential idea is to use the heat released by the reaction to heat the feed stream. As a simple example of this concept, consider the heat integration scheme depicted in Figure 6.37 [1]. This reactor configuration is known as an autothermal plug-flow reactor. The reactor system is an annular tube. The feed passes through the outer region and is heated through contact with the hot reactor wall. The feed then enters the inner reaction region, which is filled with the catalyst, and flows countercurrently to the feed stream. The heat released due to reaction in the inner region is used to heat the feed in the outer region. When the reactor is operating at steady state, no external heat is required to preheat the feed. Of course, during the reactor start up, external heat must be supplied to ignite the reactor.

Although recycle of energy can offer greatly lower operating costs, the dynamics and control of these reactors may be complex. We next examine an ammonia synthesis example to show that multiple steady states are possible. Ammonia synthesis is also interesting because of its large impact on the early development of the chemical engineering discipline. Quoting Aftalion [2, p. 101]

> While physicists and chemists were linking up to understand the structure of matter and giving birth to *physical chemistry*, another discipline was emerging, particularly in the United States, at the beginning of the twentieth century, that of *chemical engineering* ... it was undoubtedly the synthesis of ammonia by BASF, successfully achieved in 1913 in Oppau, which forged the linking of chemistry with physics and engineering as it required knowledge in areas of analysis, equilibrium reactions, high pressures, catalysis, resistance of materials, and design of large-scale apparatus.

Example 6.6: Ammonia synthesis

Calculate the steady-state conversion for the synthesis of ammonia using the autothermal process shown in Figure 6.37 [26]. A rate expression for the reaction

$$\mathrm{N_2 + 3H_2 \underset{k_{-1}}{\overset{k_1}{\rightleftharpoons}} 2NH_3}$$

over an iron catalyst at 300 atm pressure is suggested by Temkin [21]

$$r = k_{-1}/RT\left[K^2(T)\frac{P_N P_H^{3/2}}{P_A} - \frac{P_A}{P_H^{3/2}}\right]$$

in which P_N, P_H, P_A are the partial pressures (divided by 1.0 atm) of nitrogen, hydrogen and ammonia, respectively, and K is the equilibrium constant for the reaction forming one mole of ammonia. For illustration, we assume the thermochemical properties are constant and the gases form an ideal-gas mixture. More accurate thermochemical properties and a more accurate equation of state do not affect the fundamental behavior predicted by the reactor model.

The steady-state material balance for the ammonia is

$$\frac{dN_A}{dV} = R_A = 2r$$
$$N_A(0) = N_{Af}$$

and the other molar flows are calculated from

$$N_N = N_{Nf} - 1/2(N_A - N_{Af})$$
$$N_H = N_{Hf} - 3/2(N_A - N_{Af})$$

If we assume an ideal gas in this temperature and pressure range, the volumetric flowrate is given by

$$Q = \frac{RT}{P}(N_A + N_N + N_H)$$

The energy balance for the reactor is the usual

$$Q\rho\hat{C}_P \frac{dT}{dV} = -\Delta H_R r + \dot{q} \tag{6.55}$$

in which $\dot{q}$ is the heat transfer taking place between the reacting fluid and the cold feed

$$\dot{q} = \frac{2}{R} U^o (T_a - T)$$

The material balances for the feed-heating section are simple because reaction does not take place without the catalyst. Without reaction, the molar flow of all species are constant and equal to their feed values and the energy balance for the feed-heating section is

$$Q_a \rho_a \hat{C}_{Pa} \frac{dT_a}{dV_a} = -\dot{q} \tag{6.56}$$
$$T_a(0) = T_{af}$$

in which the subscript a represents the fluid in the feed-heating section. Notice the heat terms are of opposite signs in Equations 6.56 and 6.55. If we assume the fluid properties do not change significantly over the temperature range of interest, and switch the direction of integration in Equation 6.56 using $dV_a = -dV$, we obtain

$$Q\rho\hat{C}_P \frac{dT_a}{dV} = \dot{q} \tag{6.57}$$
$$T_a(V_R) = T_{af} \tag{6.58}$$

Parameter	Value	Units
P	300	atm
Q_f	0.05713	m^3/s
x_{Af}	0.015	
x_{Nf}	0.985(1/4)	
x_{Hf}	0.985(3/4)	
A_c	1	m^2
l	12	m
T_{af}	323	K
$\gamma = \dfrac{2\pi RU^o}{Q\rho\hat{C}_P}$	0.5	1/m
$\beta = \dfrac{\Delta H_R A_c}{Q\rho\hat{C}_P}$	-2.342	m^2 s K/mol
ΔG°	-4.25×10^3	cal/mol
ΔH°	-1.2×10^4	cal/mol
k_{-10}	7.794×10^{11}	atm/s
E_{-1}/R	2×10^4	K

Table 6.6: Parameter values for Example 6.6; heat of reaction and mixture heat capacity assumed constant.

Finally we require a boundary condition for the reactor energy balance, which we have from the fact that the heating fluid enters the reactor at $z = 0, T(0) = T_a(0)$. Combining these balances and boundary conditions and converting to reactor length in place of volume gives the model

$$\boxed{\begin{aligned} \frac{dN_A}{dz} &= 2A_c r & N_A(0) &= N_{Af} \\ \frac{dT}{dz} &= -\beta r + \gamma(T_a - T) & T(0) &= T_a(0) \\ \frac{dT_a}{dz} &= \gamma(T_a - T) & T_a(l) &= T_{af} \end{aligned}} \tag{6.59}$$

in which

$$\beta = \frac{\Delta H_R A_c}{Q\rho\hat{C}_P} \qquad \gamma = \frac{2\pi RU^o}{Q\rho\hat{C}_P}$$

Equation 6.59 is a boundary-value problem, rather than an initial-value problem, because T_a is specified at the exit of the reactor. A simple solution strategy is to guess the reactor inlet temperature, solve the model to the exit of the reactor, and then compare the computed feed preheat temperature to the specified value, T_{af}. This strategy is known as a shooting method. We guess the missing values required to produce an initial-value problem. We solve the initial-value problem, and

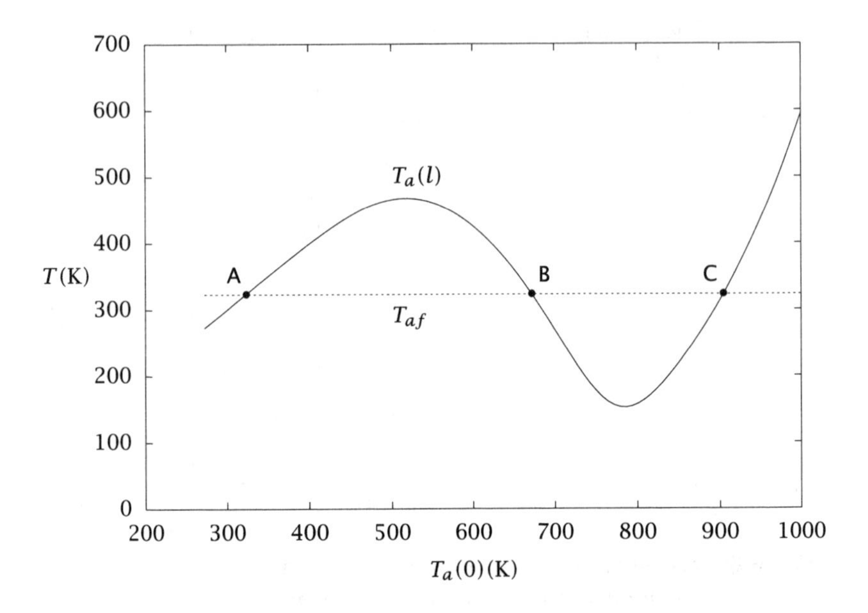

Figure 6.38: Coolant temperature at reactor outlet versus temperature at reactor inlet, $T_a(l)$ versus $T_a(0)$; intersection with coolant feed temperature T_{af} indicates three steady-state solutions (A,B,C).

then iterate on the guessed values until we match the specified boundary conditions. We will see more about boundary-value problems and shooting methods when we treat diffusion in Chapter 7.

Solution

Figure 6.38 shows the results for the parameter values listed in Table 6.6, which are based on those used by van Heerden [26]. The feed consists of 1.5% ammonia and 98.5% stoichiometric mixture of nitrogen and hydrogen. For given values of $T_a(0)$, we solve the initial-value problem, Equation 6.59, and plot the resulting $T_a(V_R)$ as the solid line in Figure 6.38. The intersection of that line with the feed temperature $T_{af} = 323$ K indicates a steady-state solution. Notice three steady-state solutions are indicated in Figure 6.38 for these values of parameters. The profiles in the reactor for these three steady states are shown in Figures 6.39 and 6.40. It is important to operate at the upper steady state so that a reasonably large production of ammonia is achieved. □

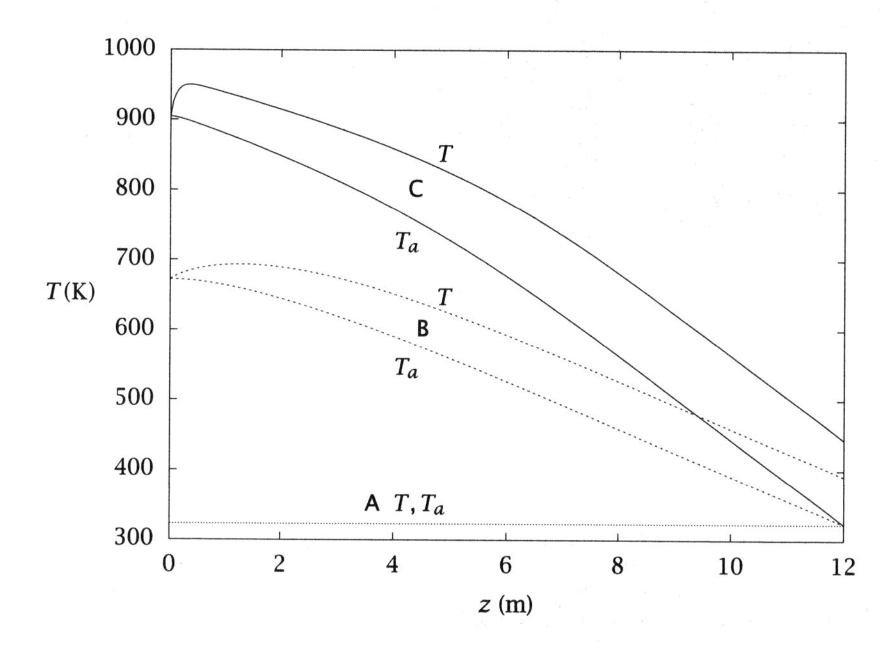

Figure 6.39: Reactor and coolant temperature profiles versus reactor length; lower (A), unstable middle (B), and upper (C) steady states.

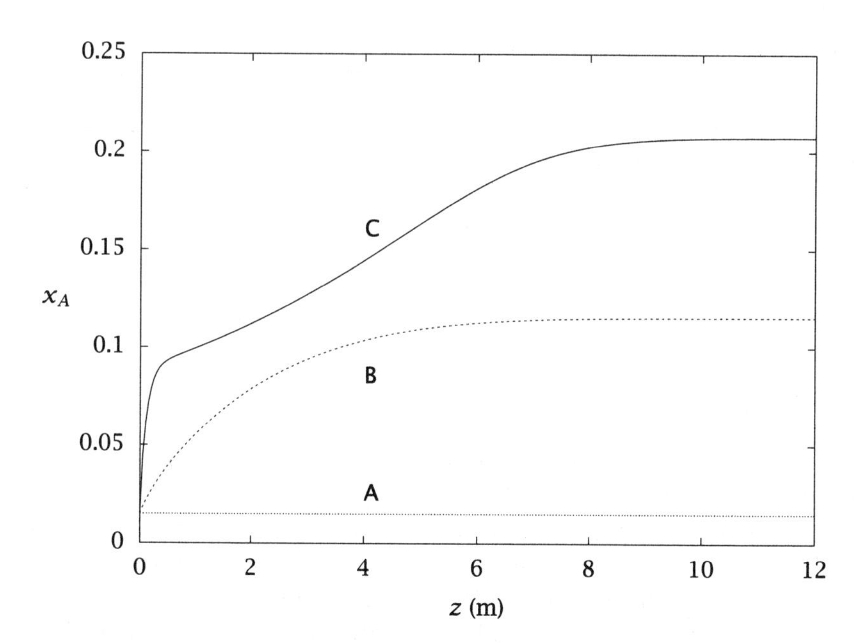

Figure 6.40: Ammonia mole fraction versus reactor length; lower (A), unstable middle (B), and upper (C) steady states.

Neglect kinetic and potential energies

$$\frac{dU}{dt} = \dot{Q} + \dot{W}_s + \dot{W}_b \tag{6.60}$$

Neglect shaft work

$$\frac{dU}{dt} + P\frac{dV_R}{dt} = \dot{Q} \tag{6.61}$$

$$\frac{dH}{dt} - V_R\frac{dP}{dt} = \dot{Q} \tag{6.62}$$

Single phase

$$V_R\rho\hat{C}_P\frac{dT}{dt} - \alpha T V_R\frac{dP}{dt} + \sum_j \overline{H}_j\frac{dn_j}{dt} = \dot{Q} \tag{6.63}$$

$$V_R\rho\hat{C}_P\frac{dT}{dt} - \alpha T V_R\frac{dP}{dt} = -\sum_i \Delta H_{Ri} r_i V_R + \dot{Q} \tag{6.64}$$

---

a. Incompressible-fluid or constant-pressure reactor

$$V_R\rho\hat{C}_P\frac{dT}{dt} = -\sum_i \Delta H_{Ri} r_i V_R + \dot{Q} \tag{6.65}$$

---

b. Constant-volume reactor

$$V_R\rho\hat{C}_V\frac{dT}{dt} = -\sum_i \left(\Delta H_{Ri} - \frac{\alpha}{\kappa_T} T\Delta V_{Ri}\right) r_i V_R + \dot{Q} \tag{6.66}$$

---

b.1 Constant-volume reactor, ideal gas

$$V_R\rho\hat{C}_V\frac{dT}{dt} = -\sum_i \left(\Delta H_{Ri} - RT\bar{\nu}_i\right) r_i V_R + \dot{Q} \tag{6.67}$$

Table 6.7: Energy balances for the batch reactor.

6.6 Summary

Tables 6.7–6.10 summarize the important energy balances for the batch, continuous-stirred-tank, semi-batch, and plug-flow reactors. In contrast to the material balance, which is reasonably straightforward, choosing the proper energy balance requires some care. It is unwise to select an energy balance from a book without carefully considering the assumptions that have been made in the derivation of that particular

Neglect kinetic and potential energies

$$\frac{dU}{dt} = Q_f \rho_f \hat{H}_f - Q\rho\hat{H} + \dot{Q} + \dot{W}_s + \dot{W}_b \tag{6.68}$$

Neglect shaft work

$$\frac{dU}{dt} + P\frac{dV_R}{dt} = Q_f \rho_f \hat{H}_f - Q\rho\hat{H} + \dot{Q} \tag{6.69}$$

$$\frac{dH}{dt} - V_R\frac{dP}{dt} = Q_f \rho_f \hat{H}_f - Q\rho\hat{H} + \dot{Q} \tag{6.70}$$

Single phase

$$V_R\rho\hat{C}_P\frac{dT}{dt} - \alpha T V_R\frac{dP}{dt} + \sum_j \overline{H}_j\frac{dn_j}{dt} = Q_f \rho_f \hat{H}_f - Q\rho\hat{H} + \dot{Q} \tag{6.71}$$

$$V_R\rho\hat{C}_P\frac{dT}{dt} - \alpha T V_R\frac{dP}{dt} = -\sum_i \Delta H_{Ri} r_i V_R + \sum_j c_{jf} Q_f(\overline{H}_{jf} - \overline{H}_j) + \dot{Q} \tag{6.72}$$

a. Incompressible-fluid or constant-pressure reactor

$$V_R\rho\hat{C}_P\frac{dT}{dt} = -\sum_i \Delta H_{Ri} r_i V_R + \sum_j c_{jf} Q_f(\overline{H}_{jf} - \overline{H}_j) + \dot{Q} \tag{6.73}$$

b. Constant-volume reactor

$$V_R\rho\hat{C}_V\frac{dT}{dt} = -\sum_i \left(\Delta H_{Ri} - \frac{\alpha}{\kappa_T}T\Delta V_{Ri}\right) r_i V_R + \sum_j c_{jf} Q_f(\overline{H}_{jf} - \overline{H}_j) + \frac{\alpha}{\kappa_T}T\sum_j \overline{V}_j(c_{jf}Q_f - c_jQ) + \dot{Q} \tag{6.74}$$

b.1 Constant-volume reactor, ideal gas

$$V_R\rho\hat{C}_V\frac{dT}{dt} = -\sum_i \left(\Delta H_{Ri} - RT\bar{\nu}_i\right) r_i V_R + \sum_j c_{jf} Q_f(\overline{H}_{jf} - \overline{H}_j) + RT\sum_j (c_{jf}Q_f - c_jQ) + \dot{Q} \tag{6.75}$$

c. Steady state, constant $\hat{C}_P$, $P = P_f$

$$-\sum_i \Delta H_{Ri} r_i V_R + Q_f\rho_f\hat{C}_P(T_f - T) + \dot{Q} = 0 \tag{6.76}$$

Table 6.8: Energy balances for the CSTR.

Neglect kinetic and potential energies

$$\frac{dU}{dt} = Q_f \rho_f \hat{H}_f + \dot{Q} + \dot{W}_s + \dot{W}_b \tag{6.77}$$

Neglect shaft work

$$\frac{dU}{dt} + P\frac{dV_R}{dt} = Q_f \rho_f \hat{H}_f + \dot{Q} \tag{6.78}$$

$$\frac{dH}{dt} - V_R\frac{dP}{dt} = Q_f \rho_f \hat{H}_f + \dot{Q} \tag{6.79}$$

Single phase

$$V_R \rho \hat{C}_P \frac{dT}{dt} - \alpha T V_R \frac{dP}{dt} + \sum_j \overline{H}_j \frac{dn_j}{dt} = Q_f \rho_f \hat{H}_f + \dot{Q} \tag{6.80}$$

$$V_R \rho \hat{C}_P \frac{dT}{dt} - \alpha T V_R \frac{dP}{dt} = -\sum_i \Delta H_{Ri} r_i V_R + \sum_j c_{jf} Q_f (\overline{H}_{jf} - \overline{H}_j) + \dot{Q} \tag{6.81}$$

a. Incompressible-fluid or constant-pressure reactor

$$V_R \rho \hat{C}_P \frac{dT}{dt} = -\sum_i \Delta H_{Ri} r_i V_R + \sum_j c_{jf} Q_f (\overline{H}_{jf} - \overline{H}_j) + \dot{Q} \tag{6.82}$$

a.1 Constant $\hat{C}_P$

$$V_R \rho \hat{C}_P \frac{dT}{dt} = -\sum_i \Delta H_{Ri} r_i V_R + Q_f \rho_f \hat{C}_P (T_f - T) + \dot{Q} \tag{6.83}$$

Table 6.9: Energy balances for the semi-batch reactor.

energy balance. Adding to the confusion, many books do not state clearly what assumptions have been made in deriving various energy balances, and some books list energy balances that are simply incorrect. See Denn [10] for a thorough discussion of common errors in energy balances and a list of books containing incorrect energy balances.

Nonisothermal reactor design requires the simultaneous solution of the appropriate energy balance and the species material balances. For the batch, semi-batch, and steady-state plug-flow reactors, these balances are sets of initial-value ODEs that must be solved numerically. In very limited situations (constant thermodynamic properties, single

Neglect kinetic and potential energies and shaft work
$$\frac{\partial}{\partial t}(\rho \hat{U}) = -\frac{1}{A_c}\frac{\partial}{\partial z}(Q\rho \hat{H}) + \dot{q} \qquad (6.84)$$
Heat transfer with an overall heat-transfer coefficient
$$\dot{q} = \frac{2}{R}U^o(T_a - T) \qquad (6.85)$$
Steady state
$$\frac{d}{dV}(Q\rho \hat{H}) = \dot{q} \qquad (6.86)$$
Single phase
$$Q\rho \hat{C}_P \frac{dT}{dV} + Q(1-\alpha T)\frac{dP}{dV} = -\sum_i \Delta H_{Ri} r_i + \dot{q} \qquad (6.87)$$
a. Neglect pressure drop, or ideal gas
$$Q\rho \hat{C}_P \frac{dT}{dV} = -\sum_i \Delta H_{Ri} r_i + \dot{q} \qquad (6.88)$$
b. Incompressible fluid
$$Q\rho \hat{C}_P \frac{dT}{dV} + Q\frac{dP}{dV} = -\sum_i \Delta H_{Ri} r_i + \dot{q} \qquad (6.89)$$

Table 6.10: Energy balances for the plug-flow reactor.

reaction, adiabatic), one can solve the energy balance to get an algebraic relation between temperature and concentration or molar flowrate.

The nonlinear nature of the energy and material balances can lead to multiple steady-state solutions. Steady-state solutions may be unstable, and the reactor can exhibit sustained oscillations. These reactor behaviors were illustrated with exothermic CSTRs and autothermal tubular reactors.

Notation

A	heat transfer area
A_c	reactor tube cross-sectional area
A_i	preexponential factor for rate constant i
c_j	concentration of species j
c_{jf}	feed concentration of species j
c_{js}	steady-state concentration of species j
c_{j0}	initial concentration of species j
C_P	constant-pressure heat capacity
$\overline{C}_{Pj}$	partial molar heat capacity of species j
C°_{Pj}	specific heat capacity of pure species j
C_{Ps}	heat capacity per volume
$\hat{C}_P$	constant-pressure heat capacity per mass
$\hat{C}_V$	constant-volume heat capacity per mass
ΔC_P	heat capacity change on reaction, $\Delta C_P = \sum_j \boldsymbol{\nu}_j \overline{C}_{Pj}$
E	activation energy (divided by the gas constant)
E_i	activation energy for rate constant i
E_k	total energy of stream k
$\hat{E}_k$	total energy per mass of stream k
ΔG°	Gibbs energy change on reaction at standard conditions
$\overline{H}_j$	partial molar enthalpy
$\hat{H}$	enthalpy per unit mass
ΔH_{Ri}	enthalpy change on reaction, $\Delta H_{Ri} = \sum_j \nu_{ij} \overline{H}_j$
ΔH°	enthalpy change on reaction at standard conditions
k_i	reaction rate constant for reaction i
k_m	reaction rate constant evaluated at mean temperature T_m
K_i	equilibrium constant for reaction i
$\hat{K}$	kinetic energy per unit mass
l	tubular reactor length
m_k	total mass flow of stream k
n	reaction order
n_j	moles of species j, $V_R c_j$
n_r	number of reactions in the reaction network
n_s	number of species in the reaction network
N_j	molar flow of species j, Qc_j
N_{jf}	feed molar flow of species j, Qc_j
P	pressure
P_j	partial pressure of component j
$\dot{q}$	heat transfer rate per volume for tubular reactor, $\dot{q} = \frac{2}{R} U^{o} (T_a - T)$

Q	volumetric flowrate
Q_f	feed volumetric flowrate
$\dot{Q}$	heat transfer rate to reactor, usually modeled as $\dot{Q} = U^o A(T_a - T)$
r_i	reaction rate for ith reaction
r_{tot}	total reaction rate, $\sum_i r_i$
R	gas constant
R_j	production rate for jth species
t	time
T	temperature
T_a	temperature of heat transfer medium
T_m	mean temperature at which k is evaluated
U^o	overall heat transfer coefficient
$\hat{U}$	internal energy per mass
v_k	velocity of stream k
V	reactor volume variable
$\overline{V}_j$	partial molar volume of species j
V_j°	specific molar volume of species j
V_R	reactor volume
ΔV_i	change in volume upon reaction i, $\sum_j \nu_{ij}\overline{V}_j$
$\dot{W}$	rate work is done on the system
x_j	number of molecules of species j in a stochastic simulation
x_j	molar conversion of species j
y_j	mole fraction of gas-phase species j
z	reactor length variable
α	coefficient of expansion of the mixture, $\alpha = (1/V)(\partial V/\partial T)_{P,n_j}$
ε_i	extent of reaction i
κ_T	isothermal compressibility of the mixture, $\kappa_T = -(1/V)(\partial V/\partial P)_{T,n_j}$
ν_{ij}	stoichiometric coefficient for species j in reaction i
$\bar{\nu}_i$	$\sum_j \nu_{ij}$
ρ	mass density
ρ_k	mass density of stream k
τ	reactor residence time, $\tau = V_R/Q_f$
$\hat{\Phi}$	potential energy per mass

6.7 Exercises

Exercise 6.1: Batch energy balance with heat transfer

Consider Example 6.2 with the following additional data:

$$k_m = 0.01725 \text{ L/mol·min}$$
$$T_m = 300 \text{ K}$$
$$E_a/R = 2660 \text{ K}$$

so the temperature dependence of the rate constant is given by

$$k = k_m \exp\left(-\frac{E_a}{R}\left(\frac{1}{T} - \frac{1}{T_m}\right)\right)$$

(a) What is the adiabatic temperature rise for the reactor?

(b) How long does it take to reach 95% conversion if the reactor operates isothermally at 27°C?

(c) How long does it take to reach 95% conversion if the reactor operates adiabatically? Plot c_A and T versus time for this case. Put in enough points so we can see a smooth curve.

(d) Plot c_A and T versus time for the nonadiabatic case with heat exchange:

$$U^o A/V_R = 0.01 \text{ kcal/(min L K)}$$

and the temperature of the heat transfer fluid is $T_a = 27°\text{C}$.

(e) Assume the batch is ruined if the temperature exceeds 350 K during the run. What value of heat-transfer coefficient ($U^o A/V_R$) should your design achieve so that this temperature is not exceeded. How long does it take to reach 95% conversion with your design? How should you operate the reactor if you want to speed things up but cannot violate the 350 K limit?

Exercise 6.2: Batch energy balance with constrained heat transfer

Consider Example 6.2 with the following additional data:

$$k_m = 0.01725 \text{ L/mol·min}$$
$$T_m = 300 \text{ K}$$
$$E_a/R = 2660 \text{ K}$$

so the temperature dependence of the rate constant is given by

$$k = k_m \exp\left(-\frac{E_a}{R}\left(\frac{1}{T} - \frac{1}{T_m}\right)\right)$$

The heat-transfer equipment accidentally has been underdesigned by using the average rather than the maximum required rate of heat transfer

$$\dot{Q} = \frac{-2.3 \times 10^4}{551} = -41.7 \text{ kcal/min} \tag{6.90}$$

which is much less in absolute value than the required -828 kcal/min calculated in Example 6.2. Assume the control system can only remove heat at this constant rate.

Component	$H_f^{298\text{ K}}$ kcal/mol	A	B ($\times 10^2$)	C ($\times 10^5$)	D ($\times 10^9$)
Cl_2	0.00	6.432	0.8082	−0.9241	3.695
C_3H_6	4.88	0.866	5.602	−2.771	5.266
C_3H_5Cl	−0.15	0.604	7.277	−5.442	17.42
HCl	−22.06	7.235	−0.172	0.2976	−0.931
$1,2{-}C_3H_6Cl_2$	−39.60	2.496	8.729	−6.219	18.49

Table 6.11: Thermodynamic data for allyl chloride example.

(a) What is the maximum temperature achieved in this reactor and when does this maximum occur?

(b) How long does it take for the control system to return the temperature to 27°C?

(c) When does the reactor achieve 95% conversion?

Exercise 6.3: CSTR energy balance with multiple reactions

Allyl chloride is to be produced in a 0.83 ft^3 CSTR [20]

$$Cl_2 + C_3H_6 \xrightarrow{k_1} C_3H_5Cl + HCl$$
$$Cl_2 + C_3H_6 \xrightarrow{k_2} C_3H_6Cl_2$$

The feed is a 4:1 molar ratio of propylene to chlorine and it enters at a feed rate of 0.85 lbmol/hr, 2.0 atm of pressure, and 392°F. The reactor pressure may be assumed constant.

The rate constants have units of lbmol/(hr ft^3 atm^2) and are

$$k_1 = 2.06 \times 10^5 \ \exp \frac{-27200}{RT}$$
$$k_2 = 11.7 \ \exp \frac{-6860}{RT}$$

where T is in degrees Rankine and R is in Btu/(lbmol°R). The rate expressions are

$$r_1 = k_1 P_{C_3H_6} P_{Cl_2}$$
$$r_2 = k_2 P_{C_3H_6} P_{Cl_2}$$

The thermodynamic data for this reaction are listed in Table 6.11 [16]. The partial molar heat capacities can be calculated using

$$\overline{C}_{Pj} = A_j + B_j T + C_j T^2 + D_j T^3 \qquad \text{cal/mol K}$$

(a) Compute the molar flowrates of Cl_2, C_3H_5Cl and $C_3H_6Cl_2$, and the reactor temperature for adiabatic operation. You need to use an algebraic equation solver for this problem, a reasonable first guess on the temperature is 1200°R.

(b) Now consider an approximate solution along the lines presented by Smith [20] in which the heat of reaction is computed at the inlet temperature and assumed constant. Compare the heat of reaction in Part 6.3a with the value at the inlet temperature, and comment on the differences in the molar flowrates and the temperatures in Parts 6.3a and 6.3b.

Exercise 6.4: CSTR stability

An adiabatic CSTR with a first-order, liquid-phase reaction

$$A \longrightarrow B, \qquad r = kc_A$$

operates at the conditions shown below

Parameter	Value	Units
T_f	298	K
c_{Af}	3	kmol/m^3
Q_f	60×10^{-6}	m^3/s
ΔH_R	-2.09×10^8	J/kmol
$\hat{C}_P$	4.19×10^3	J/(kg K)
ρ	10^3	kg/m^3
V_R	18×10^{-3}	m^3
k	$4.48 \times 10^6 \exp(-7550/T)$	s^{-1}; T in K

Find the steady-state temperatures and conversions. How many steady states can you find? Which steady states are stable and which are unstable?

Exercise 6.5: CSTR energy balance

The liquid-phase reactions

$$A \longrightarrow B$$
$$A \longrightarrow C$$

are each first order in the concentration of A. The feed to a nonisothermal CSTR contains pure A at 45 °C and 5 mol/L. Additional information for the reactions and the reactor are provided below. We want to operate this reactor such that the selectivity to B is greater than the selectivity to C, i.e., the rate at which B forms is greater than the rate at which C forms.

Parameter	Value	Units
ΔH_{r1}	$-12{,}000$	cal/mol
ΔH_{r2}	$-15{,}000$	cal/mol
V	1000	L
Q_f	100	L/min
ρ_f	0.932	g/cm^3
$\hat{C}_{pf}$	0.22	cal/(g K)
k_1	$3.16 \times 10^{14} \exp(-12{,}500/T)$	min^{-1}; T in K
k_2	$2.52 \times 10^9 \exp(-8{,}500/T)$	min^{-1}; T in K

(a) Defining the selectivity as $S = R_B/R_C$, determine the rate at which heat is transferred to the surroundings to achieve a selectivity of five.

(b) Determine the rate at which heat is transferred to the surroundings to achieve a selectivity of four.

(c) Explain why the answer to Part A or Part B is greater.

Exercise 6.6: Finding the CSTR steady state

The first-order reaction

$$A \xrightarrow{k} B \qquad r = k_0 e^{-E/T} c_A$$

takes place in a CSTR. Determine the effluent temperature(s) and concentration(s) of A for a reactor operating at the conditions listed in the following table.

Parameter	Value	Units
E	7550	K
A	3600	cm^2
T_a	312	K
T_f	298	K
c_{Af}	3	$kmol/m^3$
U^o	0.225	$J/(cm^2\ s\ K)$
ΔH_R	-8.09×10^8	J/kmol
k_0	4.48×10^6	1/s
ρ	10^3	kg/m^3
$\hat{C}_P$	4.19×10^3	J/(kg K)
V_R	18×10^{-3}	m^3
Q_f	60×10^{-6}	m^3/s

Exercise 6.7: Testing a CSTR operating condition for stability

The reaction

$$A \xrightarrow{k} B \qquad r = k_0 e^{-E/T} c_A$$

is carried out in a CSTR. Find the three steady states corresponding to the conditions in the following table. Determine whether each of these three steady states is stable or unstable.

Parameter	Value	Units
E	7550	K
T_f	298	K
c_{Af}	3	$kmol/m^3$
U^o	0	
ΔH_R	-2.09×10^8	J/kmol
k_0	4.48×10^6	1/s
$\hat{C}_P$	4.19×10^3	J/(kg K)
ρ	10^3	kg/m^3
V_R	18×10^{-3}	m^3
Q_f	60×10^{-6}	m^3/s

Exercise 6.8: CSTR ignition

Consider the CSTR parameter values listed in Table 6.1 and the steady-state x, T versus residence time plots in Figures 6.7 and 6.8.

(a) Write down the dynamic CSTR equations for the adiabatic reactor. How many parameters are there and what are their values? Choose consistent units.

(b) Choose a residence time of $\tau = 10$ min. According to Figures 6.7 and 6.8, how many steady-state solutions are there at this residence time? What are the stable (c_A, T) steady-state values and what are the unstable ones? You can just read these off the plots.

Next simulate the dynamic reactor. Choose as initial conditions for c_A and T, values near the lower-branch and upper-branch steady states. Simulate the reactor for a time long enough for it to reach steady state. Check that the dynamic simulation converges to a steady state consistent with Figures 6.7 and 6.8 (this is your program debugging check). Once you are sure the program is working correctly simulate the reactor starting from $c_A = c_{Af}, T = T_f$. Increase only the initial temperature and simulate again. How warm can the reactor be started before it ignites and reaches the upper branch steady state instead of the lower branch. Hand in plots of $c_A(t)$ and $T(t)$ starting one degree below and above this critical temperature.

(c) If it were advantageous to operate at an unstable steady state, how might you modify the reactor design to stabilize the steady state? Sketch your control scheme. Be sure to show and label clearly your sensor and actuator (if you are interested you can simulate the controlled reactor and see what controller gain is required to achieve stability).

(d) Change the residence time to $\tau = 32$ min. How many steady-state solutions are there for this residence time. Start the reactor from the feed conditions and simulate for 1000 min. Does anything unusual happen? Describe what the reactor is doing. Does your simulation point out any safety issues? Can you think of a less violent method to start up the reactor? What values of (c_A, T) are you trying to avoid during start up?

Exercise 6.9: Removing heat in a CSTR

The liquid-phase reaction

$$A \longrightarrow B$$

$$r = kc_A$$

is to be carried out in a nonisothermal, nonadiabatic CSTR operating at 95 °C. The feed contains pure A at 45 °C and 5 mol/L. Using the additional information provided below determine the temperature of the cooling fluid medium.

Parameter	Value	Units
E	12,500	K
ΔH_R	-3700	cal/mol
V_R	1000	L
ρ_f	932	g/L
Q_f	100	L/min
U^o	0.54	cal/(min cm^2 K)
$\hat{C}_P$	0.22	cal/g K
k_0	3.16×10^{14}	min^{-1}
A	1.8×10^4	cm^2

Exercise 6.10: CSTR and energy balance

Consider the reaction

$$A \xrightarrow{k} B$$

in a liquid-phase, constant-volume CSTR with heating/cooling system. The heat capacity and density of the mixture can be assumed independent of temperature and composition. The following data and parameters are known.

Parameter	Value	Units
E	7554	K
T_f	298	K
$\hat{C}_{Ps}$	4.19	kJ/(kg K)
c_{Af}	3.0	kmol/m^3
V_R	0.018	m^3
k_0	4.48×10^6	s^{-1}
ΔH_R	-2.09×10^5	kJ/kmol
ρ	10^3	kg/m^3
Q_f	6.0×10^{-5}	m^3/s

(a) At what temperature must the reactor be operated to achieve 80% conversion?

(b) What is the heat duty on the heating/cooling system when operating at this steady state and are you heating or cooling the reactor, i.e., what is the sign of the heat duty?

Exercise 6.11: CSTR stability

(a) For the reaction of A to products, sketch the conversion of A versus mean reactor residence time for an isothermal CSTR.

(b) Draw a separate sketch of conversion of A and reactor temperature versus residence time for an exothermic reaction in an adiabatic CSTR that exhibits steady-state multiplicity.

(c) Describe an ignition and extinction point and locate them on the sketch of the CSTR with steady-state multiplicity. What is special about these points? What does hysteresis mean? What are the practical operational problems associated with trying to operate a CSTR near one of these points. Is it possible to operate near one of these points? If not, why not? If so, how do you overcome the operational problems described above?

Exercise 6.12: Attractors in the CSTR

Show which initial conditions are attracted to which solutions in Figure 6.30.

Exercise 6.13: Semi-batch reactor balance

Derive the CSTR energy balance given by Equation 6.72 in Table 6.8 by making the assumptions listed in the table. Now derive the semi-batch reactor Equation 6.81 in Table 6.9. Why are these two energy balances identical even though they apply to different reactor types?

Exercise 6.14: Adiabatic plug-flow reactor

Consider the elementary, gas-phase isomerization reaction shown below:

$$A(g) \underset{k_{-1}}{\overset{k_1}{\rightleftharpoons}} B(g)$$

This isomerization is to be carried out in an *adiabatic* PFR. Only A and B are present in the reactor feed stream. The feed rate, Q_f, concentrations, c_{Af} and c_{Bf}, temperature, T_f, and pressure, P_f, are known at the inlet to the PFR. Pressure drop in the reactor can be neglected.

(a) What does the stoichiometry tell you about the relationship between the molar flux of B, N_B, and the molar flux of A, N_A, at any point in the reactor?

(b) Write down the mole balance for component A. Remove any reference to c_B in this equation with the result from 6.14a. Also write down an expression to evaluate the volumetric flowrate, Q, because it is not constant.

(c) Starting from the general energy balance equation for a PFR given below, determine the appropriate energy balance equation for the PFR in this problem. Be sure that all variables in both the mole and energy balances can be evaluated in terms of your dependent variables, T and N_A

$$Q\rho\hat{C}_P \frac{dT}{dV} = -\Delta H_R r + \frac{2}{R} U^o (T_a - T)$$

(d) Compute the equilibrium concentration and conversion of component A that could be achieved if the reactor operated *isothermally* at the inlet temperature given below. The molar ratio of A to B in the feed is 4:1. Assume ideal gas behavior.

$T_f = 830°\text{R}$; $P_f = 180$ psia ; $K = k_1/k_{-1} = 1.5$; $R = 10.731$ psi·ft^3/lbmol°R

(e) Calculate the minimal inlet temperature of an *adiabatic* reactor required to achieve the equilibrium conversion of component A determined in 6.14d. Assume the heat capacities of A and B are the same and the heat capacity and heat of reaction are constant at the values below.

$\Delta H_R = 5850$ BTU/lbmol ; $C_P = 20.0$ BTU/lbmol°R; $Q_f = 10,000$ ft^3/hr

Exercise 6.15: PFR energy balance with a single reaction

The gas-phase reaction

$$A + B \longrightarrow D + E$$

is carried out in a PFR. The reaction-rate expression is

$$r = kP_A P_B$$

in which

$$k = 2.06 \times 10^5 \exp(-13,700/T)$$

The units for r are lbmol/hr ft^3, P is in atm, and T is in °R. The feed is a 4:1 molar ratio of A and B and enters at 500°F, at a feed rate of 0.85 lbmol/hr and 2 atm pressure.

Assuming constant thermodynamic properties, determine the length of a 2-in inner-diameter (ID) tubular reactor that operates adiabatically and provides 93% conversion of B. The heat of reaction is $\Delta H_r = -48,000$ Btu/lbmol and the mixture heat capacity is $C_P = 110$ Btu/lbmol°R.

Exercise 6.16: Temperature and the equilibrium limit

The following exothermic, reversible, gas-phase reaction is to be conducted in an adiabatic PFR

$$\mathrm{A} \rightleftharpoons \mathrm{B} \qquad r = k\left(c_A - \frac{c_B}{K}\right)$$

You may assume the heat of reaction ($\Delta H_r = -20,000$ cal/mol) and the partial molar heat capacities ($\overline{C}_{PA} = \overline{C}_{PB} = 50$ cal/(mol K)) are constant. The equilibrium constant at 298 K is $K = 10^5$.

Determine the maximum possible conversion of A if the feed contains pure A at a temperature of 300 K.

Exercise 6.17: PFR energy balance with multiple reactions

Allyl chloride is to be produced in a 25-ft long 2-in ID tube operating as a PFR [20]. The feed is a 4:1 molar ratio of propylene to chlorine and it enters at a feed rate of 0.85 lbmol/hr, 2 atm of pressure, and 392°F. The reactor pressure may be assumed constant.

$$\mathrm{Cl_2 + C_3H_6 \xrightarrow{k_1} C_3H_5Cl + HCl}$$

$$\mathrm{Cl_2 + C_3H_6 \xrightarrow{k_2} C_3H_6Cl_2}$$

The rate constants have units of lbmol/(hr ft^3 atm^2) and are

$$k_1 = 2.06 \times 10^5 \exp\left(-27200/(RT)\right)$$

$$k_2 = 11.7 \times \exp\frac{-6860}{RT}$$

in which T is in °R and R is in Btu/(lbmol°R). The rate expressions are

$$r_1 = k_1 P_{\mathrm{C_3H_6}} P_{\mathrm{Cl_2}}$$

$$r_2 = k_2 P_{\mathrm{C_3H_6}} P_{\mathrm{Cl_2}}$$

The thermodynamic data for this reaction are listed in Table 6.11[16]. The partial molar heat capacities are expressed in units of cal/(mol K) and can be calculated using

$$\overline{C}_{Pj} = A_j + B_j T + C_j T^2 + D_j T^3$$

Two modes of operation have been considered for this PFR, adiabatic and constant wall temperature. A constant wall temperature of 392°F can be realized by boiling ethylene glycol on the outer surface of the reactor wall. The inside heat-transfer coefficient is 5 Btu/hr ft^2°F for a feed rate of 0.85 lbmol/hr and 2 atm total pressure.

(a) Compute the molar flowrates of Cl_2, C_3H_5Cl and $C_3H_6Cl_2$, and the reactor temperature as a function of reactor length for adiabatic operation. Plot these dependent variables versus reactor length.

(b) Compute the molar flowrates of Cl_2, C_3H_5Cl and $C_3H_6Cl_2$, and the reactor temperature as a function of reactor length for constant wall temperature operation. Plot these dependent variables versus reactor length. What do you notice when you compare the extent of product formation and reactor temperature for both cases? Why do you think this happens?

(c) Now compare the exact solutions developed above to the approximate solution presented in Smith [20] where the component heat capacities and the heat of reaction are evaluated at the inlet temperature and assumed constant. Plot the molar flowrates of C_3H_5Cl and $C_3H_6Cl_2$ for the exact and and approximate solutions on different graphs, one for adiabatic and one for nonadiabatic operation. Prepare similar plots for the temperature. Comment on the errors this approximation introduces. Is it sensible that the error is greater for adiabatic operation? Why?

Exercise 6.18: The adiabatic limit in a PFR

We wish to convert the CO in a waste stream to CO_2 by reaction in an adiabatic bundle of empty tubes.

$$CO + \frac{1}{2}O_2 \rightleftharpoons CO_2$$

The waste stream is a mixture of CO and air, and is mixed with a pure air stream to achieve 100% excess air to CO. This stream enters the reactor tube bundle at 100 °C. You want to calculate the maximum temperature that might be reached in the tube to select appropriate material of construction for the tube bundle.

(a) What is the maximum outlet temperature that could be achieved? What assumptions do you make to compute this temperature?

(b) If this temperature is too high for the materials on hand, what simple process design changes can lower the outlet temperature?

(c) What would you do next if you wanted a more accurate estimate of the temperature of the effluent from the reactor?

Additional information.

The mean pure component heat capacities and heats of formation at 25 °C are as follows.

Component	C_{Pj}° cal/mol °C	$-\Delta H_{jf}^\circ$ kcal/mol
CO	6.98	26.42
CO_2	12.9	94.05
O_2	8.35	0
N_2	7.92	0

Exercise 6.19: Using a simplified PFR energy balance

Consider the following reaction in an adiabatic, constant-pressure PFR

$$Cl_2 + C_3H_6 \xrightarrow{k_1} C_3H_5Cl + HCl$$

The rate constant has units of lbmol/(hr ft^3 atm^2) and is given by

$$k_1 = 2.06 \times 10^5 \exp(-27200/RT)$$

where T is in °R and R is in Btu/(lbmol°R). The rate expression is

$$r_1 = k_1 P_{C_3H_6} P_{Cl_2}$$

The feed is a 4:1 molar ratio of propylene to chlorine and it enters at a feed rate of 0.85 lbmol/hr, 2.0 atm pressure, and 392°F (852°R). The reactor temperature must be maintained below 880°R. The heat of reaction is

$$\Delta H_R = -48{,}000 \qquad \text{Btu/lbmol}$$

and the heat capacity of the reactor feed mixture is

$$C_P = 110 \qquad \text{Btu/lbmol°R}$$

If you assume ΔH_R and C_P are constant and do not change as an inert is added to the feed, determine the minimum inert flowrate (in lbmol/hr) that must be added to the feed to keep the reactor temperature below 880°R when the conversion of chlorine is 95%.

Exercise 6.20: The effect of heat transfer in an autothermal reactor

Consider what happens in the ammonia synthesis system described in Example 6.6 if we decrease the heat-transfer parameter, γ. We wish to find the value of γ such that the middle and upper steady states coalesce.

(a) For a range of $T_a(0)$ values, solve the initial-value problem, Equation 6.59, and plot the resulting $T_a(l)$ as in Figure 6.38. Use the parameter values given in Table 6.6.

(b) Decrease the heat-transfer rate by 20% and repeat this calculation. Continue to decrease γ until you produce Figure 6.41. What is the critical value of γ that corresponds to Figure 6.41?

(c) What happens if you operate the autothermal reactor with a value of γ less than the critical value found in part 6.20b?

Exercise 6.21: Moving volume element in PFR

In Section 6.5 we derived the energy balance for the PFR by considering a fixed volume element. Consider an adiabatic PFR $\dot{Q} = 0$, for which we showed

$$\frac{\partial}{\partial t}(\rho \hat{U} A_c \Delta z) = m\hat{H}|_z - m\hat{H}|_{z+\Delta z} + \underbrace{\dot{W}_b}_{\text{zero}} \tag{6.91}$$

For the fixed volume element we have convection terms because material streams through the boundary, but no work term for moving the boundary because the element is fixed in space. Dividing by Δz and taking the limit leads to the energy balance

$$\frac{\partial(\rho \hat{U})}{\partial t} + \frac{\partial(v \rho \hat{H})}{\partial z} = 0 \tag{6.92}$$

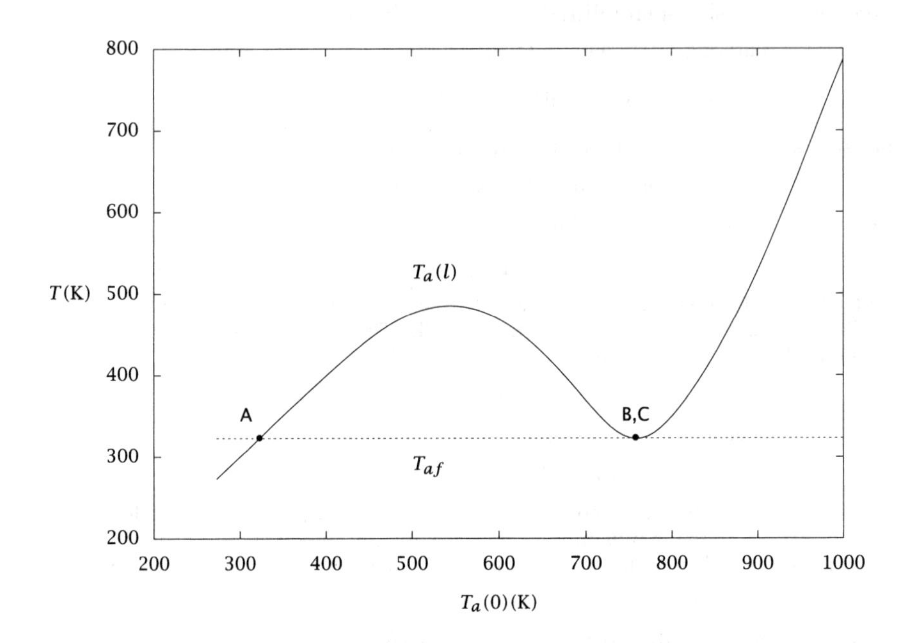

Figure 6.41: Coolant temperature at reactor outlet versus temperature at reactor inlet, $T_a(l)$ versus $T_a(0)$, at the critical value of heat-transfer coefficient; upper and middle steady-state solutions have coalesced.

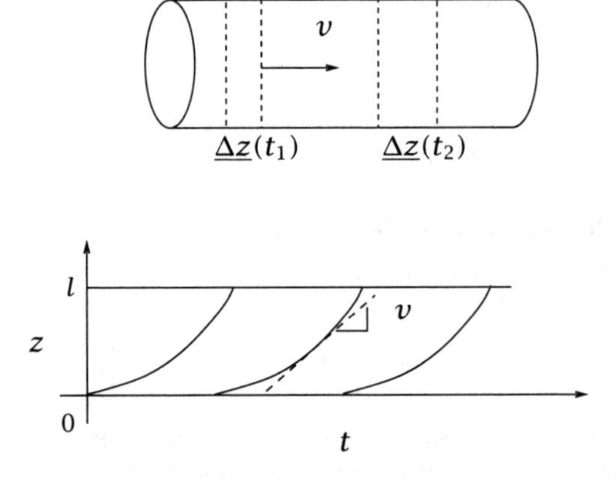

Figure 6.42: Volume element moving with the PFR fluid velocity.

in which v is the PFR fluid velocity.

Consider a volume element that moves with the fluid velocity as depicted in Figure 6.42. The location of the moving element traces out one of the lines shown in the (z, t) plane. These lines are known as the characteristic lines for Equation 6.92. Notice also the two circular ends of the element move with different velocities if the fluid density is not constant, so this volume element may expand or contract as it moves down the reactor. Applying Equation 6.91 to this moving element produces

$$\frac{\partial}{\partial t}(\underline{\rho \hat{U}} A_c \underline{\Delta z}) = \underbrace{\underline{m\hat{H}}|_z}_{\text{zero}} - \underbrace{\underline{m\hat{H}}|_{z+\Delta z}}_{\text{zero}} + \underline{\dot{W}_b} \tag{6.93}$$

in which the underline represents the quantities in the moving element. Notice for this choice of volume element the convection terms are zero, but the work term for moving the boundary is nonzero. Show that we still arrive at the same energy balance, Equation 6.92, starting with Equation 6.93 for the moving element. It is not difficult to add the heat term to this development as well. The moral of the story is that the choice of volume element is based on convenience; that choice does not affect the final energy balance.

Hints:

1. First show
$$\underline{\dot{W}_b} = -A_c \frac{\partial (Pv)}{\partial z} \underline{\Delta z}$$
by considering the work done while moving the two ends of the element.

2. Differentiate the left-hand side of Equation 6.93 to obtain
$$\frac{\partial}{\partial t}(\underline{\rho \hat{U}} A_c \underline{\Delta z}) = A_c \left[\frac{\partial (\underline{\underline{\rho \hat{U}}})}{\partial t} \underline{\Delta z} + \underline{\rho \hat{U}} \frac{\partial \underline{\underline{\Delta z}}}{\partial t} \right]$$

3. Compute the change in the volume element size due to the change in the fluid velocity with position, and show
$$\frac{\partial \underline{\underline{\Delta z}}}{\partial t} = \frac{\partial v}{\partial z} \underline{\Delta z}$$
Use the chain rule to show for $\underline{\rho \hat{U}} = \rho \hat{U}(z, t)$
$$\frac{\partial (\underline{\underline{\rho \hat{U}}})}{\partial t} = \frac{\partial (\rho \hat{U})}{\partial t} + \frac{\partial (\rho \hat{U})}{\partial z} \frac{\partial z}{\partial t} \tag{6.94}$$
in which $\partial z / \partial t = v$ is the fluid velocity because we consider the moving element. The left-hand side of Equation 6.94 is the rate of change of internal energy as we move with the fluid along a characteristic line in Figure 6.42. In the transport literature this derivative also is called the material or substantial derivative. The right-hand side contains the usual partial derivatives in which we hold either z or t constant.

4. Combine these results to obtain Equation 6.92.

Exercise 6.22: CSTR steady-state isola

The first-order reaction

$$\text{A} \longrightarrow \text{B}$$

has reaction-rate expression

$$r = k_0 e^{-E/T} c_A$$

in which the units for the rate are kmol/m^3 s, c_A is in kmol/m^3, and T is in K. Consider the reactor operating conditions listed in the table in Exercise 6.6.

Calculate the steady-state conversion and temperature as a function of reactor residence time over the range $0 < \tau < 800$ s. Also use your steady-state plot to check your answer to Exercise 6.6.

Hint: this reactor has multiple steady states. Be aware that steady-state multiplicity may be more complex than the simple s-shaped curves shown in Figures 6.5 and 6.6 of Section 6.3.1 and Figures 6.18–6.21 of Section 6.3.3. Read the discussion of isola multiplicity provided by Uppal, Ray and Poore [25] before deciding on a computational approach for this problem.

Exercise 6.23: Useful thermodynamic identities

Establish the following thermodynamic results, which are useful in the derivations of energy balances in the chapter.

(a)

$$\left(\frac{\partial S}{\partial P}\right)_{T,n_j} = -\alpha V$$

Hint: consider the roles of entropy and volume in the Gibbs energy.

(b)

$$\left(\frac{\partial P}{\partial n_j}\right)_{T,V,n_{k\neq j}} = \frac{\overline{V}_j}{V\kappa_T}$$

Hint: use Euler's cyclic chain rule.

(c)

$$\left(\frac{\partial H}{\partial P}\right)_{T,n_j} = V(1 - \alpha T)$$

(d)

$$C_P = C_V + \alpha T V \left(\frac{\partial P}{\partial T}\right)_{V,n_j}$$

(e)

$$\left(\frac{\partial P}{\partial T}\right)_{V,n_j} = \frac{\alpha}{\kappa_T}$$

Bibliography

[1] R. A. Adomaitis and A. Cinar. The bifurcation behavior of an autothermal packed bed reactor. *Chem. Eng. Sci.*, 43(4):887–898, 1988.

[2] F. Aftalion. *A History of the International Chemical Industry.* Chemical Heritage Press, Philadelphia, second edition, 2001. Translated by Otto Theodor Benfey.

[3] G. P. Baccaro, N. Y. Gaitonde, and J. M. Douglas. An experimental study of oscillating reactors. *AIChE J.*, 16(2):249–254, March 1970.

[4] J. E. Bailey. Periodic phenomena. In L. Lapidus and N. R. Amundson, editors, *Chemical Reactor Theory. A Review,* pages 758–813. Prentice-Hall, Inc., Englewood Cliffs, New Jersey, 1977.

[5] R. B. Bird. The equations of change and the macroscopic mass, momentum, and energy balances. *Chem. Eng. Sci.*, 6:123–131, 1957.

[6] R. B. Bird. Viewpoints on transport phenomena. *Korean J. Chem. Eng.*, 15 (2):105–123, 1998.

[7] R. B. Bird, W. E. Stewart, and E. N. Lightfoot. *Transport Phenomena.* John Wiley & Sons, New York, second edition, 2002.

[8] S. F. Bush. The measurement and prediction of sustained temperature oscillations in a chemical reactor. *Proc. Roy. Soc. A.*, 309:1–26, 1969.

[9] M. Chang and R. A. Schmitz. An experimental study of oscillatory states in a stirred reactor. *Chem. Eng. Sci.*, 30:21–34, 1975.

[10] M. M. Denn. *Process Modeling.* Pitman Publishing, 1986.

[11] G. F. Froment and K. B. Bischoff. *Chemical Reactor Analysis and Design.* John Wiley & Sons, New York, second edition, 1990.

[12] J. Guckenheimer and P. Holmes. *Nonlinear Oscillations, Dynamical Systems, and Bifurcations of Vector Fields.* Springer Verlag, New York, 1983.

[13] M. D. Hancock and C. N. Kenney. Instabilities in two-phase reactors. In *Proceedings of the fifth European Symposium on Chemical Reaction Engineering, Amsterdam,* pages B3–47–B3–58. Elsevier Publishing Company, 1972.

[14] L. S. Henderson. Stability analysis of polymerization in continuous stirred-tank reactors. *Chem. Eng. Prog.*, pages 42–50, March 1987.

[15] M. Marek and I. Schreiber. *Chaotic Behavior of Deterministic Dissipative Systems.* Cambridge University Press, New York, 1991.

[16] B. E. Poling, J. M. Prausnitz, and J. P. O'Connell. *Properties of Gases and Liquids.* McGraw-Hill, New York, 2001.

[17] L. F. Razon, S.-M. Chang, and R. A. Schmitz. Chaos during the oxidation of carbon monoxide over platinum — experiments and analysis. *Chem. Eng. Sci.*, 41:1561–1576, 1986.

[18] R. A. Schmitz, K. R. Graziani, and J. L. Hudson. Experimental evidence of chaotic states in the Belousov–Zhabotinskii reaction. *J. Chem. Phys.*, 67: 3040–3044, 1977.

[19] F. J. Schork and W. H. Ray. The dynamics of the continuous emulsion polymerization of methylmethacrylate. *J. Appl. Polym. Sci.*, 34:1259–1276, 1987.

[20] J. M. Smith. *Chemical Engineering Kinetics.* McGraw-Hill, New York, third edition, 1981.

[21] M. Temkin and V. Pyzhev. Kinetics of ammonia synthesis on promoted iron catalysts. *Acta Physicochimica U.R.S.S.*, 12(3):327–356, 1940.

[22] F. Teymour and W. H. Ray. The dynamic behavior of continuous polymerization reactors — V. Experimental investigation of limit-cycle behavior for vinyl acetate polymerization. *Chem. Eng. Sci.*, 47(15/16):4121–4132, 1992.

[23] F. Teymour and W. H. Ray. The dynamic behavior of continuous polymerization reactors — VI. Complex dynamics in full-scale reactors. *Chem. Eng. Sci.*, 47(15/16):4133–4140, 1992.

[24] A. Uppal, W. H. Ray, and A. B. Poore. On the dynamic behavior of continuous stirred tank reactors. *Chem. Eng. Sci.*, 29:967–985, 1974.

[25] A. Uppal, W. H. Ray, and A. B. Poore. The classification of the dynamic behavior of continuous stirred tank reactors — influence of reactor residence time. *Chem. Eng. Sci.*, 31:205–214, 1976.

[26] C. van Heerden. Autothermic processes. Properties and reactor design. *Ind. Eng. Chem.*, 45(6):1242–1247, 1953.

[27] R. J. Van Welsenaere and G. F. Froment. Parametric sensitivity and runaway in fixed bed catalytic reactors. *Chem. Eng. Sci.*, 25:1503–1516, 1970.

7

Fixed-Bed Catalytic Reactors

7.1 Introduction

The analysis and rational design of reactors requires the simultaneous solution of material and energy balances. For a steady-state, tubular, plug-flow reactor, we showed in Chapter 6 that the molar flows and temperature are governed by the following differential equations

$$\frac{d(c_jQ)}{dV} = R_j = \sum_{i=1}^{n_r} \nu_{ij} r_i, \qquad j = 1, 2, \ldots, n_s \tag{7.1}$$

$$Q\rho\hat{C}_P \frac{dT}{dV} = -\sum_{i=1}^{n_r} \Delta H_{Ri} r_i + \frac{2}{R} U^o (T_a - T) \tag{7.2}$$

A rate expression, r_i, as a function of concentration and temperature, for each of the n_r reactions is required to complete the problem description. Equations 7.1 and 7.2 can be used for any reaction within the reactor as long as the plug-flow conditions are satisfied. These balances are applicable to reactions that are homogeneous or heterogeneous; the heterogeneous case is referred to as a fixed-bed reactor.

In a fixed-bed reactor the catalyst pellets are held in place and do not move with respect to a fixed reference frame. Material and energy balances are required for both the fluid, which occupies the interstitial region between catalyst particles, and the catalyst particles, in which the reactions occur. For heterogeneously catalyzed reactions, the effects of intraparticle transport on the rate of reaction must be considered. Catalytic systems operate somewhere between two extremes: kinetic control, in which mass and energy transfer are very rapid; and intraparticle transport control, in which the reaction is very rapid. Separate material and energy balances are needed to describe the concentration and temperature profile inside the catalyst pellet. The concentrations of each component and the temperature in the pellet are related to the concentrations and temperature in the bulk fluid through the boundary conditions that complete the problem description. Numerical methods are used to solve the resulting coupled differential equations, except in some special cases.

This chapter's goal is to describe the reaction at any location within the reactor. In certain cases, analytical solutions to the transport of mass and energy within the catalyst pellet can be used to develop algebraic relationships that couple the rate of reaction within the pellet to the temperature and concentration that exist in the bulk fluid. Even though the applicability of these relationships is limited, deriving analytical solutions develops intuition and clearly illustrates the coupling between the pellet and the fluid.

Figure 7.1 presents several views of the fixed-bed reactor. Equation 7.1 applies to the bulk fluid in the interstitial volume. The species production rates in the bulk fluid are *essentially zero.* That is the reason we are using a catalyst. Essentially all reaction occurs within the catalyst particles. The fluid in contact with the external surface of the catalyst pellet is denoted with subscript s. When we need to discuss both fluid and pellet concentrations and temperatures (see Section 7.7.1), we use a tilde on the variables within the catalyst pellet.

During any catalytic reaction the following steps occur:

1. transport of reactants and energy from the bulk fluid up to the catalyst pellet exterior surface,

2. transport of reactants and energy from the external surface into the porous pellet,

3. adsorption, chemical reaction, and desorption of products at the catalytic sites,

4. transport of products from the catalyst interior to the external surface of the pellet, and

5. transport of products into the bulk fluid.

The coupling of transport processes with chemical reaction can lead to concentration and temperature gradients within the pellet, between the surface and the bulk, or both. For the remainder of this section, the pellet, surface and bulk temperatures are assumed to be equal.

The five steps just listed describe the simultaneous processes that occur in the fixed-bed reactor. Usually one or at most two of these

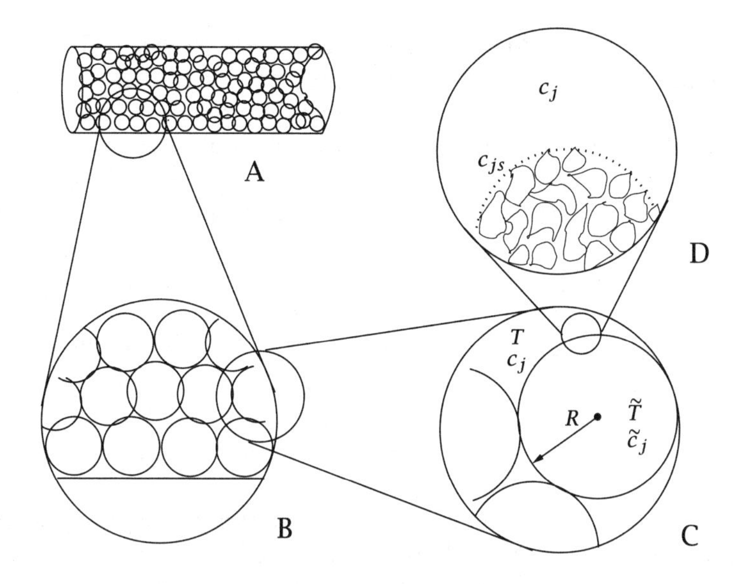

Figure 7.1: Expanded views of a fixed-bed reactor.

steps are rate limiting and act to influence the overall rate of reaction in the pellet. The other steps are inherently faster than the slow step(s) and can accommodate any change in the rate of the slow step. The system is *intraparticle transport controlled* if step 2 is the slow process (sometimes referred to as diffusion limited). For *kinetic or reaction control,* step 3 is the slowest process. Finally, if step 1 is the slowest process, the reaction is said to be *externally transport controlled.*

In this chapter, we model the system on the scale of Figure 7.1C. The problem is solved for one pellet by averaging the microscopic processes that occur on the scale of Figure 7.1D over the volume of the pellet or over a solid surface volume element. This procedure requires an effective diffusion coefficient, D_j, to be identified that contains information about the physical diffusion process and pore structure.

7.2 Catalyst Properties

The chemical steps for catalytic reactions: adsorption, desorption and surface reaction, are covered in Chapter 5. These processes occur at a fluid-solid interface and the rates scale directly with the total number of surface sites, $\overline{c}_m$. Table 7.1 lists some of the important commercial catalysts and their uses [12]. References [12, 13, 14, 15] provide an excellent introduction to catalysis and catalytic processes. To make a catalytic process commercially viable, the number of sites per unit reactor volume should be such that the rate of product formation is on the order of 1 mol/L·hour [22]. In the case of metal catalysts, the metal is generally dispersed onto a high-area oxide such as alumina. Metal oxides also can be dispersed on a second carrier oxide such as vanadia supported on titania, or it can be made into a high-area oxide. These carrier oxides can have surface areas ranging from 0.05 m^2/g to greater than 100 m^2/g. The carrier oxides generally are pressed into shapes or extruded into pellets. The following shapes are frequently used in applications: 20–100 μm diameter spheres for fluidized-bed reactors, 0.3–0.7 cm diameter spheres for fixed-bed reactors, 0.3–1.3 cm diameter cylinders with a length-to-diameter ratio of 3–4, and up to 2.5 cm diameter hollow cylinders or rings.

Figure 7.1D shows a schematic representation of the cross section of a single pellet. The solid density is denoted ρ_s. The pellet volume consists of both void and solid. The pellet void fraction (or porosity) is denoted by ϵ and

$$\epsilon = \rho_p V_g$$

in which ρ_p is the effective particle or pellet density and V_g is the pore volume. The pore structure is a strong function of the preparation method, and catalysts can have pore volumes (V_g) ranging from 0.1–1 cm^3/g pellet. The pores can be the same size or there can be a bimodal distribution with pores of two different sizes, a large size to facilitate transport and a small size to contain the active catalyst sites. Pore sizes can be as small as molecular dimensions (several Ångströms) or as large as several millimeters.

Total catalyst area is generally determined using a physically adsorbed species, such as N_2. The procedure was developed in the 1930s by Brunauer, Emmett and and Teller [7], and the isotherm they developed is referred to as the BET isotherm. In the physisorption process multiple layers are allowed to form. The BET isotherm model treats the first layer differently from all subsequent layers and leads to the following expression

$$v = \frac{v_m c P}{(P_0 - P)\,(1 + (c-1)P/P_0)} \tag{7.3}$$

in which c is a parameter related to the heats of adsorption in the first layer and subsequent layers, P_0 is the normal saturation vapor pressure of the adsorbate at the adsorption temperature, and v is the

Catalyst	Reaction
Metals (e.g., Ni, Pd, Pt, as powders or on supports) or metal oxides (e.g., Cr_2O_3)	C=C bond hydrogenation, e.g., olefin + $H_2 \longrightarrow$ paraffin
Metals (e.g., Cu, Ni, Pt)	C=O bond hydrogenation, e.g., acetone + $H_2 \longrightarrow$ isopropanol
Metal (e.g., Pd, Pt)	Complete oxidation of hydrocarbons, oxidation of CO
Fe (supported and promoted with alkali metals)	$3H_2 + N_2 \longrightarrow 2NH_3$
Ni	$CO + 3H_2 \longrightarrow CH_4 + H_2O$ (methanation)
Fe or Co (supported and promoted with alkali metals)	$CO + H_2 \longrightarrow$ paraffins + olefins + H_2O + CO_2 (+ other oxygen-containing organic compounds) (Fischer-Tropsch reaction)
Cu (supported on ZnO, with other components, e.g., Al_2O_3)	$CO + 2H_2 \longrightarrow CH_3OH$
Re + Pt (supported on η-Al_2O_3 or γ-Al_2O_3 promoted with chloride)	Paraffin dehydrogenation, isomerization and dehydrocyclization
Solid acids (e.g., SiO_2-Al_2O_3, zeolites)	Paraffin cracking and isomerization
γ-Al_2O_3	Alcohol $\longrightarrow$ olefin + H_2O
Pd supported on acidic zeolite	Paraffin hydrocracking
Metal-oxide-supported complexes of Cr, Ti or Zr	Olefin polymerization, e.g., ethylene $\longrightarrow$ polyethylene
Metal-oxide-supported oxides of W or Re	Olefin metathesis, e.g., 2 propylene $\rightarrow$ ethylene + butene
Ag(on inert support, promoted by alkali metals)	Ethylene + 1/2 $O_2 \rightarrow$ ethylene oxide (with $CO_2 + H_2O$)
V_2O_5 or Pt	$2\ SO_2 + O_2 \rightarrow 2\ SO_3$
V_2O_5 (on metal oxide support)	Naphthalene + $9/2O_2 \rightarrow$ phthalic anhydride + $2CO_2$ +$2H_2O$
Bismuth molybdate	Propylene + $1/2O_2 \rightarrow$ acrolein
Mixed oxides of Fe and Mo	$CH_3OH + O_2 \rightarrow$ formaldehyde (with $CO_2 + H_2O$)
Fe_3O_4 or metal sulfides	$H_2O + CO \rightarrow H_2 + CO_2$

Table 7.1: Industrial reactions over heterogeneous catalysts. This material is used by permission of John Wiley & Sons, Inc., Copyright ©1992 [12].

volume of gas adsorbed at a particular pressure P. Uptake data are fit to Equation 7.3 to determine c and v_m, in which v_m is the volume of gas that corresponds to a monolayer of the entire solid. One then assumes each adsorbed molecule in the volume v_m covers a particular area, i.e., the footprint of the molecule at the adsorption temperature. For N_2 at 77 K the area of each molecule is 0.162 nm^2. The total area per gram of catalyst is denoted by S_g.

7.2.1 Effective Diffusivity

We define a diffusion coefficient that takes into account the random distribution of solid and void as one moves from the exterior to the interior of the pellet, so that the balance can be written over the entire pellet. This idea leads to the concept of an effective diffusivity. Starting at any pore opening in Figure 7.1D and moving inward along a straight line to the center, one quickly encounters solid. Inside, any void space transport can be described by the diffusion of a substance under a concentration gradient. The diffusion rate is controlled by the microscopic details of the transport in the voids.

Section 7.2 discussed pore sizes and pore size distributions. Assume here that all pores are straight cylinders and that the pellet has a single pore size distribution. The average radius, r_a, of the pores is then

$$r_a = \frac{2V_g}{S_g} \tag{7.4}$$

The type of diffusion in the pore, *bulk* or Knudsen, depends on whether the diffusing species collide more often with each other or with the pore wall surface. Liquid-phase diffusion is described by liquid-liquid intermolecular collisions. For a gas-phase molecule the mean free path, λ, is

$$\lambda = \frac{k_B T}{\sqrt{2}\pi\sigma^2 P}$$

and it has a value of approximately 10^{-5} cm at 1.0 atm and room temperature for simple gases. When $\lambda < r_a$ by an order of magnitude, bulk flow dominates and diffusion is described by collisions in the gas phase. When $\lambda > r_a$ by an order of magnitude, Knudsen flow dominates and diffusion is described by collisions with the pore walls. Finally, when r_a is approximately equal to the dimensions of the diffusing reactants, transport is described by surface or configurational diffusion. We only consider the first two in this text.

A number of correlations have been developed for bulk flow that describe binary interactions. One such correlation is the Chapman-Enskog relation,

$$D_{AB} = 0.0018583 \frac{\sqrt{T^3 \left(\frac{1}{M_A} + \frac{1}{M_B}\right)}}{P\sigma_{AB}^2 \Omega_{D,AB}}$$

which predicts the diffusivity in cm^2/s for T in K, P in atm and σ_{AB} in Ångström units. In multicomponent systems (n_s species), all the binary interactions must be considered, as well as the contributions relative molar fluxes make to the transport of each component. The correct and complete treatment requires the use of Stefan-Maxwell relations. For example, when only one reaction is present for a multicomponent system, the diffusivity of species j in the mixture, D_{jm}, can be found using

$$\frac{1}{D_{jm}} = \left[\sum_{k \neq j}^{n_s} \frac{1}{D_{jk}} \left(y_k + y_j \frac{\nu_k}{\left|\nu_j\right|}\right)\right] \frac{1}{1 + \delta_j y_j}$$

in which

$$\delta_j = \frac{\sum_1^{n_s} \nu_k}{\left|\nu_j\right|}$$

in which ν_k and y_k represent the stoichiometric coefficient and mole fraction of component k, respectively. This one limiting case illustrates that the diffusion coefficient can change with temperature, pressure and composition.

For Knudsen flow the diffusivity of species j is

$$D_{jK} = \frac{8r_a}{3} \sqrt{\frac{RT}{2\pi M_j}} \tag{7.5}$$

Substitution of the gas constant gives

$$D_{jK} = 9.7 \times 10^3 r_a \sqrt{\frac{T}{M_j}}$$

which predicts the diffusivity in cm^2/s for T in K and M_j in g/mol.

Bulk or Knudsen flow dominate when r_a and λ vary by at least an order of magnitude. A combined diffusivity is needed when r_a and λ are of the same order of magnitude. Again for the single reaction

$$\frac{1}{D_{jc}} = \sum_{k=1}^{n_s} \frac{1}{D_{jk}} \left(y_k - y_j \frac{N_k}{N_j}\right) + \frac{1}{D_{jK}} \tag{7.6}$$

in which N_k is the molar flux of component k in the pore. For binary systems (components A and B) with equimolar counter diffusion, Equation 7.6 reduces to

$$\frac{1}{D_{Ac}} = \frac{1}{D_{AB}} + \frac{1}{D_{AK}} \tag{7.7}$$

Equations 7.6 and 7.7 can be used to describe diffusion down a straight cylindrical pore. A porous solid does not consist of straight cylindrical pores, each having the same length and radius. Models for pore structure have been proposed that describe the pore size distribution and orientation as a function of location within the pellet [2]. These microscopic descriptions can be used to predict the porosity, pore size distribution, pore volume and pore area, all of which can be measured experimentally.

We also can use these microscopic models to describe diffusion radially inward, accounting for the changing pore shape and the presence of void or solid as a function of any position (r, θ, ϕ) within the pellet. The texts by Aris [2] and Petersen [18] illustrate the approach. In short, the rate of mass transfer is solved rigorously in a pellet at the level depicted in Figure 7.1D to give an equation of the form

$$I_j = -D_j 4\pi r^2 \frac{dc_j}{dr} \xi \tag{7.8}$$

in which D_j refers to either D_{jm} or D_{jK} and ξ is an integral over the bounding surface of a dimensionless concentration gradient for j that depends on the microscopic description of the pore surfaces. As the microscopic description changes, ξ changes; however, this same ξ would apply to other diffusing species as well. Now if one considers the solid to be an equivalent homogeneous medium, which is the level depicted in Figure 7.1C, then the same rate can described in terms of an effective diffusion coefficient, D_e.

$$I_j = -D_{ej} 4\pi r^2 \frac{dc_j}{dr} \tag{7.9}$$

Comparison of Equations 7.8 and 7.9 leads to

$$D_{ej} = D_j \xi$$

The integral ξ is generally represented as

$$\xi = \frac{\epsilon}{\tau}$$

Catalyst	ϵ	τ
100–110μm powder packed into a tube	0.416	1.56
pelletized Cr_2O_3 supported on Al_2O_3	0.22	2.5
pelletized boehmite alumina	0.34	2.7
Girdler G-58 Pd on alumina	0.39	2.8
Haldor-Topsøe MeOH synthesis catalyst	0.43	3.3
0.5% Pd on alumina	0.59	3.9
1.0% Pd on alumina	0.5	7.5
pelletized Ag/8.5% Ca alloy	0.3	6.0
pelletized Ag	0.3	10.0

Table 7.2: Porosity and tortuosity factors for diffusion in catalysts.

in which τ is the tortuosity factor. The use of the ratio ϵ/τ has no real physical significance. It is done to remind us that as a diffusing molecule travels radially inward it encounters solid and void but it diffuses mainly in the void regions. Similarly, the molecule cannot travel in a straight line from the exterior to the center point, rather it must follow a tortuous path that effectively adds to the distance of travel.

Rigorous models such as the parallel, cross-linked, pore model can be used to predict the value of τ. If the pores are modeled as straight cylindrical tubes of various lengths, one finds $\tau = 3$. Table 7.2 lists experimental values for τ [8]. The values of τ generally range from about 2 to 7. As more anisotropy in the pore distribution is introduced by the manufacturing process, τ increases. Because of this variability in τ, it should be determined experimentally for the catalyst of interest. Using effective transport properties to model transport in pores and other microstructures has been effective in many contexts. For example, Saltzman and Langer [20] find effective diffusion coefficients to model protein transport in aqueous, constricted macropores.

7.3 The General Balances in the Catalyst Particle

In this section we consider the mass and energy balances that arise with diffusion in the solid catalyst particle when considered at the scale of Figure 7.1C. Consider the volume element depicted in Figure 7.2. Assume a fixed laboratory coordinate system in which the velocities are defined and let $\boldsymbol{v}_j$ be the velocity of species j giving rise to molar flux $\boldsymbol{N}_j$

$$\boldsymbol{N}_j = c_j \boldsymbol{v}_j, \qquad j = 1, 2, \ldots, n_s$$

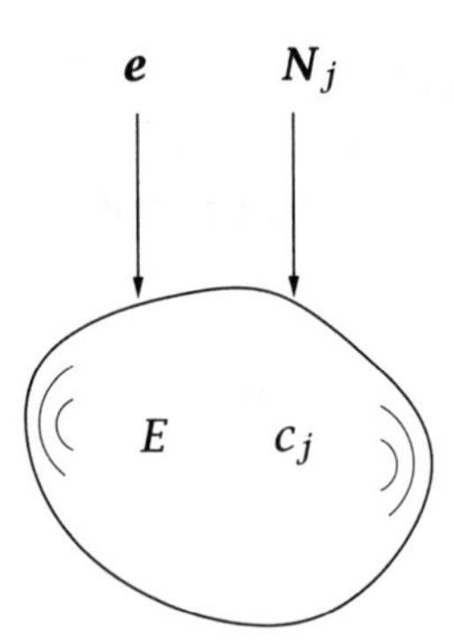

Figure 7.2: Volume element in a stationary solid containing energy density E and molar concentrations c_j, with energy flux $\boldsymbol{e}$ and mass fluxes $\boldsymbol{N}_j$.

Let E be the total energy within the volume element and $\boldsymbol{e}$ be the flux of total energy through the bounding surface due to all mechanisms of transport. The conservation of mass and energy for the volume element implies

$$\frac{\partial c_j}{\partial t} = -\nabla \cdot \boldsymbol{N}_j + R_j, \qquad j = 1, 2, \ldots, n_s \tag{7.10}$$

$$\frac{\partial E}{\partial t} = -\nabla \cdot \boldsymbol{e} \tag{7.11}$$

in which R_j accounts for the production of species j due to chemical reaction. As in Chapter 6, we consider the internal energy to be the dominant contribution to total energy, and neglect kinetic, potential and other forms of energy, so $E = U$.

Next we consider the fluxes. Since we are considering the diffusion of mass in a stationary, solid particle, we assume the mass flux is well approximated by

$$\boldsymbol{N}_j = -D_j \nabla c_j, \qquad j = 1, 2, \ldots, n_s$$

in which D_j is an effective diffusivity for species j. Notice we are neglecting a variety of other possible mechanisms of mass transport, including the flux of species j due to gradients in the other species concentrations, and the flux of species j due to temperature gradients (thermal diffusion), pressure gradients, and gravitational or other external fields (forced diffusion). Bird, Stewart and Lightfoot [5, pp. 767–768] provide further discussion of these contributions to the mass flux.

We approximate the total energy flux by

$$e = -\hat{k}\nabla T + \sum_j N_j \overline{H}_j$$

This expression accounts for the transfer of heat by conduction, in which $\hat{k}$ is the effective thermal conductivity of the solid, and transport of energy due to the mass diffusion. We have neglected any viscous heating effects, and the flux of energy due to concentration gradients (Dufour energy) and radiation. So the energy balance becomes

$$\frac{\partial \left(\rho \hat{U}\right)}{\partial t} = \nabla \cdot \hat{k}\nabla T - \sum_j \left[(\nabla \cdot N_j)\overline{H}_j + N_j \cdot \nabla \overline{H}_j\right] \tag{7.12}$$

In this chapter, we are concerned mostly with the steady state. Setting the time derivative to zero in Equation 7.10 produces

$$0 = -\nabla \cdot N_j + R_j \tag{7.13}$$

Substituting Equation 7.13 into Equation 7.12, using the definition of the heat of reaction, and setting the time derivative to zero, yields

$$0 = \nabla \cdot \hat{k}\nabla T - \sum_i \Delta H_{Ri} r_i - \sum_j N_j \cdot \nabla \overline{H}_j$$

Finally, we often assume that the diffusivity, thermal conductivity and partial molar enthalpies are independent of temperature and composition to produce the following coupled mass and energy balances for the steady-state problem

$$0 = D_j \nabla^2 c_j + R_j, \qquad j = 1, 2, \ldots, n_s \tag{7.14}$$

$$0 = \hat{k}\nabla^2 T - \sum_i \Delta H_{Ri} r_i \tag{7.15}$$

In multiple-reaction, nonisothermal problems, we must solve these equations numerically, so the assumption of constant transport and thermodynamic properties is driven by the lack of data, and not analytical convenience.

7.4 Single Reaction in an Isothermal Particle

We start with the simplest cases and steadily remove restrictions and increase the generality. In this section, therefore, we consider a single reaction taking place in an isothermal particle. We start with the spherical particle, first-order reaction, and neglect the external mass-transfer resistance. Next we consider other catalyst shapes, then other reaction orders, and then other kinetic expressions such as the Hougen-Watson kinetics of Chapter 5. We end this section by considering the effects of finite external mass transfer.

7.4.1 First-Order Reaction in a Spherical Particle

In this section we consider a single, irreversible, first-order reaction occurring in an isothermal spherical pellet

$$\mathrm{A} \xrightarrow{k} \mathrm{B}, \qquad r = kc_A$$

Substituting the production rate into Equation 7.14, expressing the equation in spherical coordinates, and assuming pellet symmetry in θ and ϕ coordinates gives

$$D_A \frac{1}{r^2} \frac{d}{dr}\left(r^2 \frac{dc_A}{dr}\right) - kc_A = 0 \qquad (7.16)$$

in which D_A is the effective diffusivity in the pellet for species A. As written here, the first-order rate constant k has units of inverse time. Be aware that the units for a heterogeneous reaction rate constant are sometimes expressed per mass or per area of catalyst. In these cases, the reaction rate expression includes the conversion factors, catalyst density or catalyst area, as illustrated in Example 7.1. We require two boundary conditions for Equation 7.16. In this section we assume the concentration at the outer boundary of the pellet, c_{As}, is known, and the symmetry of the spherical pellet implies the vanishing of the derivative at the center of the pellet.[1] Therefore the two boundary conditions for Equation 7.16 are

$$c_A = c_{As}, \qquad r = R$$

$$\frac{dc_A}{dr} = 0 \qquad r = 0$$

At this point we can obtain better insight by converting the problem into dimensionless form. Equation 7.16 has two dimensional quantities, length and concentration. We might naturally choose the sphere

[1] Some may prefer to use other boundary conditions at $r = 0$, such as boundedness of c_A or dc_A/dr, or a zero source/sink condition. These boundary conditions are equivalent for the problems considered here.

radius R as the length scale, but we will find that a better choice is to use the pellet's volume-to-surface ratio. For the sphere, this characteristic length is

$$a = \frac{V_p}{S_p} = \frac{\frac{4}{3}\pi R^3}{4\pi R^2} = \frac{R}{3}$$

The only concentration appearing in the problem is the surface concentration in the boundary condition, so we use that quantity to nondimensionalize the concentration

$$\overline{r} = \frac{r}{a}, \qquad \overline{c} = \frac{c_A}{c_{As}}$$

Dividing through by the various dimensional quantities produces

$$\frac{1}{\overline{r}^2}\frac{d}{d\overline{r}}\left(\overline{r}^2\frac{d\overline{c}}{d\overline{r}}\right) - \Phi^2\overline{c} = 0 \tag{7.17}$$

$$\overline{c} = 1 \qquad \overline{r} = 3$$

$$\frac{d\overline{c}}{d\overline{r}} = 0 \qquad \overline{r} = 0$$

in which Φ is given by

$$\boxed{\Phi = \sqrt{\frac{ka^2}{D_A}} \qquad \frac{\text{reaction rate}}{\text{diffusion rate}} \qquad \text{Thiele modulus}} \tag{7.18}$$

The single dimensionless group appearing in the model is referred to as the Thiele number or Thiele modulus in recognition of Thiele's pioneering contribution in this area [21].[2] The Thiele modulus quantifies the ratio of the reaction rate to the diffusion rate in the pellet.

Equation 7.18 can be considered as a special case of a more general definition

$$\Phi_j = \sqrt{\frac{\left|R_{js}\right| a^2}{c_{js}D_j}} = \sqrt{\frac{kc_{As}a^2}{c_{As}D_A}} = \sqrt{\frac{ka^2}{D_A}} \tag{7.19}$$

We now wish to solve Equation 7.17 with the given boundary conditions. Because the reaction is first order, the model is linear and we

[2] In his original paper, Thiele used the term *modulus* to emphasize that this then unnamed dimensionless group was positive. Later when Thiele's name was assigned to this dimensionless group, the term modulus was retained. Thiele number would seem a better choice, but the term Thiele modulus has become entrenched.

can derive an analytical solution. It is often convenient in spherical coordinates to consider the variable transformation

$$\overline{c}(\overline{r}) = \frac{u(\overline{r})}{\overline{r}} \tag{7.20}$$

Substituting this relation into Equation 7.17 provides a simpler differential equation for $u(\overline{r})$,

$$\frac{d^2u}{d\overline{r}^2} - \Phi^2 u = 0 \tag{7.21}$$

with the transformed boundary conditions

$$u = 3 \qquad \overline{r} = 3$$
$$u = 0 \qquad \overline{r} = 0$$

The boundary condition $u = 0$ at $\overline{r} = 0$ ensures that $\overline{c}$ is finite at the center of the pellet. The student may recall from the differential equations course that the solution to Equation 7.21 is

$$u(\overline{r}) = c_1 \cosh \Phi\overline{r} + c_2 \sinh \Phi\overline{r} \tag{7.22}$$

This solution is analogous to the sine and cosine solutions if one replaces the negative sign with a positive sign in Equation 7.21. These functions are shown in Figure 7.3. Some of the properties of the hyperbolic functions are

$$\cosh r = \frac{e^r + e^{-r}}{2} \qquad \frac{d\cosh r}{dr} = \sinh r$$
$$\sinh r = \frac{e^r - e^{-r}}{2} \qquad \frac{d\sinh r}{dr} = \cosh r$$
$$\tanh r = \frac{\sinh r}{\cosh r}$$

The constants c_1 and c_2 are determined by the boundary conditions. Substituting Equation 7.22 into the boundary condition at $\overline{r} = 0$ gives $c_1 = 0$, and applying the boundary condition at $\overline{r} = 3$ gives $c_2 = 3/\sinh 3\Phi$. Substituting these results into Equations 7.22 and 7.20 gives the solution to the model

$$\overline{c}(\overline{r}) = \frac{3}{\overline{r}}\frac{\sinh \Phi\overline{r}}{\sinh 3\Phi} \tag{7.23}$$

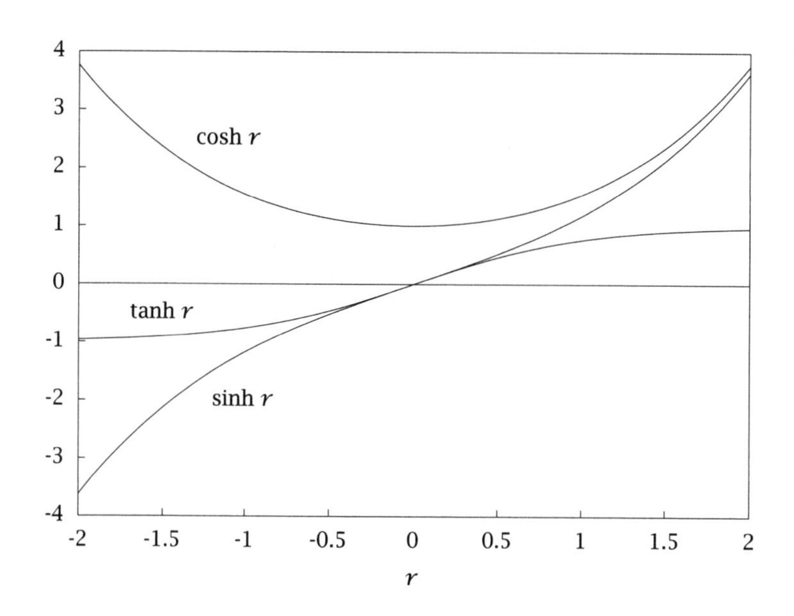

Figure 7.3: Hyperbolic trigonometric functions sinh, cosh and tanh.

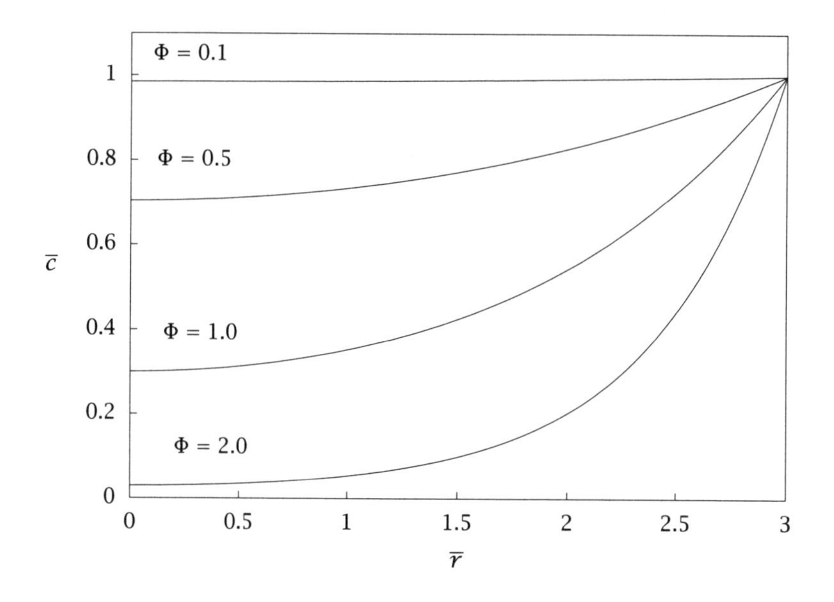

Figure 7.4: Dimensionless concentration versus dimensionless radial position for different values of the Thiele modulus.

Figure 7.4 displays this solution for various values of the Thiele modulus. Note for small values of Thiele modulus, the reaction rate is small compared to the diffusion rate, and the pellet concentration becomes nearly uniform. For large values of Thiele modulus, the reaction rate is large compared to the diffusion rate, and the reactant is converted to product before it can penetrate very far into the pellet.

We now calculate the pellet's overall production rate given this concentration profile. We can perform this calculation in two ways. The first and more direct method is to integrate the local production rate over the pellet volume. The second method is to use the fact that, at steady state, the rate of consumption of reactant within the pellet is equal to the rate at which material fluxes through the pellet's exterior surface. The two expressions are

$$R_{Ap} = \frac{1}{V_p} \int_0^R R_A(r) 4\pi r^2 dr \qquad \text{volume integral} \tag{7.24}$$

$$R_{Ap} = -\frac{S_p}{V_p} D_A \left. \frac{dc_A}{dr} \right|_{r=R} \qquad \text{surface flux (assumes steady state)} \tag{7.25}$$

in which the local production rate is given by $R_A(r) = -kc_A(r)$. We use the direct method here and leave the other method as an exercise. Substituting the local production rate into Equation 7.24 and converting the integral to dimensionless radius gives

$$R_{Ap} = -\frac{kc_{As}}{9} \int_0^3 \overline{c}(\overline{r}) \overline{r}^2 d\overline{r}$$

Substituting the concentration profile, Equation 7.23, and changing the variable of integration to $x = \Phi \overline{r}$ gives

$$R_{Ap} = -\frac{kc_{As}}{3\Phi^2 \sinh 3\Phi} \int_0^{3\Phi} x \sinh x dx$$

The integral can be found in a table or derived by integration by parts to yield finally

$$R_{Ap} = -kc_{As} \frac{1}{\Phi} \left[\frac{1}{\tanh 3\Phi} - \frac{1}{3\Phi} \right] \tag{7.26}$$

It is instructive to compare this actual pellet production rate to the rate in the absence of diffusional resistance. If the diffusion were arbitrarily fast, the concentration everywhere in the pellet would be equal to the surface concentration, corresponding to the limit $\Phi = 0$. The pellet rate for this limiting case is simply

$$R_{As} = -kc_{As} \tag{7.27}$$

We define the effectiveness factor, η, to be the ratio of these two rates

$$\eta \equiv \frac{R_{Ap}}{R_{As}}, \qquad \text{effectiveness factor} \tag{7.28}$$

The effectiveness factor is a dimensionless pellet production rate that measures how effectively the catalyst is being used. For η near unity, the entire volume of the pellet is reacting at the same high rate because the reactant is able to diffuse quickly through the pellet. For η near zero, the pellet reacts at a low rate. The reactant is unable to penetrate significantly into the interior of the pellet and the reaction rate is small in a large portion of the pellet volume. The pellet's diffusional resistance is large and this resistance lowers the overall reaction rate. We can substitute Equations 7.26 and 7.27 into the definition of effectiveness factor to obtain for the first-order reaction in the spherical pellet

$$\eta = \frac{1}{\Phi}\left[\frac{1}{\tanh 3\Phi} - \frac{1}{3\Phi}\right] \tag{7.29}$$

Figures 7.5 and 7.6 display the effectiveness factor versus Thiele modulus relationship given in Equation 7.29. The log-log scale in Figure 7.6 is particularly useful, and we see the two asymptotic limits of Equation 7.29. At small Φ, $\eta \approx 1$, and at large Φ, $\eta \approx 1/\Phi$. Figure 7.6 shows that the asymptote $\eta = 1/\Phi$ is an excellent approximation for the spherical pellet for $\Phi \geq 10$. For large values of the Thiele modulus, the rate of reaction is much greater than the rate of diffusion, the effectiveness factor is much less than unity, and we say the pellet is *diffusion limited.* Conversely, when the diffusion rate is much larger than the reaction rate, the effectiveness factor is near unity, and we say the pellet is *reaction limited.*

Example 7.1: Using the Thiele modulus and effectiveness factor

The first-order, irreversible reaction ($A \longrightarrow B$) takes place in a 0.3 cm radius spherical catalyst pellet at $T = 450\,\text{K}$. At 0.7 atm partial pressure of A, the pellet's production rate is -2.5×10^{-5} mol/(g s). Determine the production rate at the same temperature in a 0.15 cm radius spherical pellet. The pellet density is $\rho_p = 0.85\ \text{g/cm}^3$. The effective diffusivity of A in the pellet is $D_A = 0.007\ \text{cm}^2/\text{s}$.

Solution

We can use the production rate and pellet parameters for the 0.3 cm pellet to find the value for the rate constant k, and then compute the Thiele

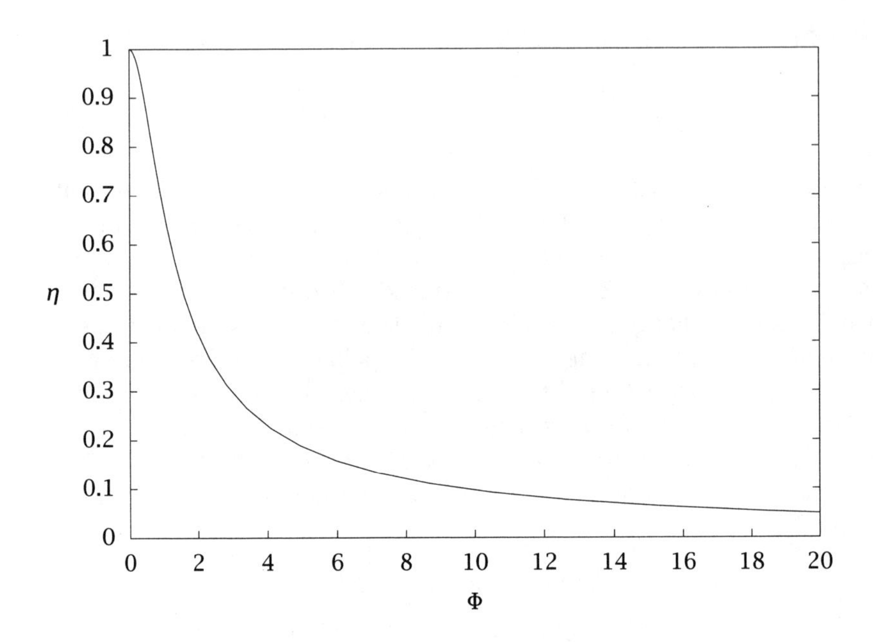

Figure 7.5: Effectiveness factor versus Thiele modulus for a first-order reaction in a sphere.

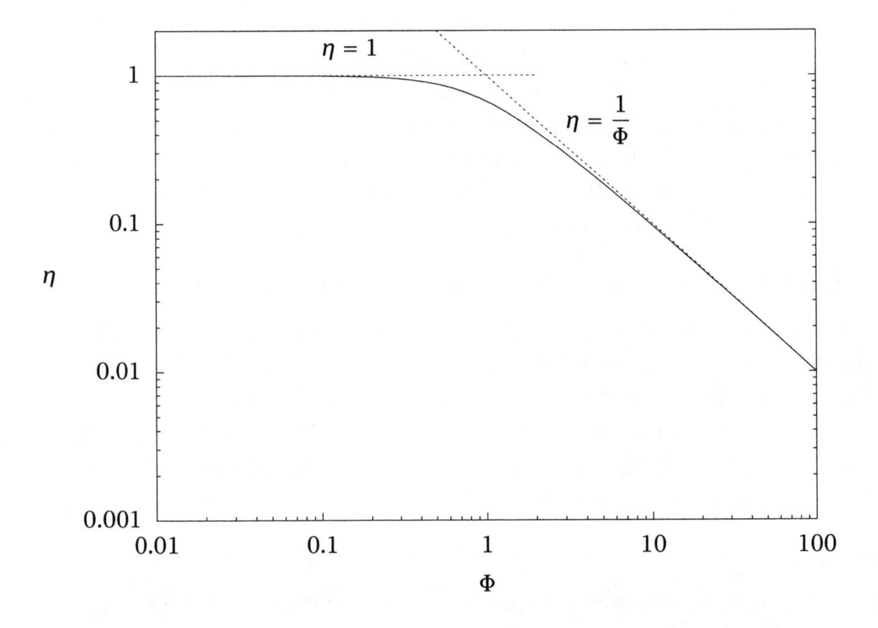

Figure 7.6: Effectiveness factor versus Thiele modulus for a first-order reaction in a sphere (log-log scale).

modulus, effectiveness factor and production rate for the smaller pellet.

We have three unknowns, k, Φ, η, and the following three equations

$$R_{Ap} = -\eta k c_{As} \tag{7.30}$$

$$\Phi = \sqrt{\frac{ka^2}{D_A}} \tag{7.31}$$

$$\eta = \frac{1}{\Phi}\left[\frac{1}{\tanh 3\Phi} - \frac{1}{3\Phi}\right] \tag{7.32}$$

The production rate is given in the problem statement. Solving Equation 7.31 for k, and substituting that result and Equation 7.32 into 7.30, give one equation in the unknown Φ

$$\Phi\left[\frac{1}{\tanh 3\Phi} - \frac{1}{3\Phi}\right] = -\frac{R_{Ap}a^2}{D_A c_{As}} \tag{7.33}$$

The surface concentration and pellet production rates are given by

$$c_{As} = \frac{0.7 \text{ atm}}{\left(82.06\frac{\text{cm}^3 \text{ atm}}{\text{mol K}}\right)(450 \text{ K})} = 1.90 \times 10^{-5} \text{mol/cm}^3$$

$$R_{Ap} = \left(-2.5 \times 10^{-5}\frac{\text{mol}}{\text{g s}}\right)\left(0.85\frac{\text{g}}{\text{cm}^3}\right) = -2.125 \times 10^{-5}\frac{\text{mol}}{\text{cm}^3 \text{ s}}$$

Substituting these values into Equation 7.33 gives

$$\Phi\left[\frac{1}{\tanh 3\Phi} - \frac{1}{3\Phi}\right] = 1.60$$

This equation can be solved numerically yielding the Thiele modulus

$$\Phi = 1.93$$

Using this result, Equation 7.31 gives the rate constant

$$k = 2.61 \text{ s}^{-1}$$

The smaller pellet is half the radius of the larger pellet, so the Thiele modulus is half as large or $\Phi = 0.964$, which gives $\eta = 0.685$. The production rate is therefore

$$R_{Ap} = -0.685\left(2.6\text{s}^{-1}\right)\left(1.90 \times 10^{-5}\text{mol/cm}^3\right) = -3.38 \times 10^{-5}\frac{\text{mol}}{\text{cm}^3 \text{ s}}$$

We see that decreasing the pellet size increases the production rate by almost 60%. Notice that this type of increase is possible only when the pellet is in the diffusion-limited regime. □

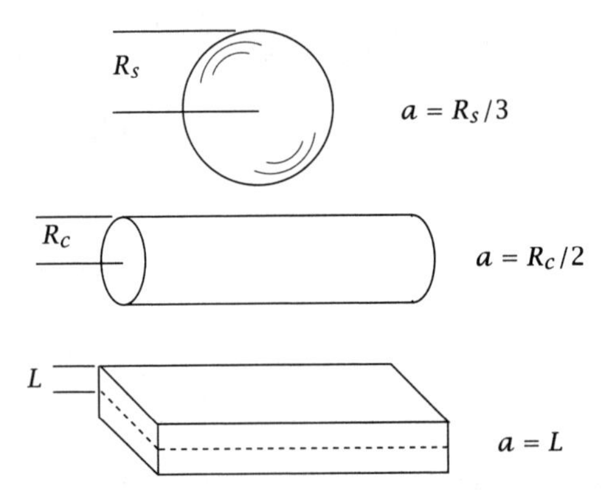

Figure 7.7: Characteristic length a for sphere, semi-infinite cylinder and semi-infinite slab.

7.4.2 Other Catalyst Shapes: Cylinders and Slabs

Here we consider the cylinder and slab geometries in addition to the sphere covered in the previous section. To have a simple analytical solution, we must neglect the end effects; we therefore consider in addition to the sphere of radius R_s, the semi-infinite cylinder of radius R_c, and the semi-infinite slab of thickness $2L$, depicted in Figure 7.7. We can summarize the reaction-diffusion mass balance for these three geometries by

$$D_A \frac{1}{r^q} \frac{d}{dr}\left(r^q \frac{dc_A}{dr}\right) - kc_A = 0 \tag{7.34}$$

in which

$$\begin{array}{ll} q = 2 & \text{sphere} \\ q = 1 & \text{cylinder} \\ q = 0 & \text{slab} \end{array}$$

The associated boundary conditions are

$$c_A = c_{As} \qquad \begin{cases} r = R_s & \text{sphere} \\ r = R_c & \text{cylinder} \\ r = L & \text{slab} \end{cases}$$

$$\frac{dc_A}{dr} = 0 \qquad r = 0 \quad \text{all geometries}$$

The characteristic length a is again best defined as the volume-to-surface ratio, which gives for these geometries

$$a = \frac{R_s}{3} \qquad \text{sphere}$$
$$a = \frac{R_c}{2} \qquad \text{cylinder}$$
$$a = L \qquad \text{slab}$$

The dimensionless form of Equation 7.34 is

$$\frac{1}{\overline{r}^q}\frac{d}{d\overline{r}}\left(\overline{r}^q\frac{d\overline{c}}{d\overline{r}}\right) - \Phi^2\overline{c} = 0 \tag{7.35}$$

$$\overline{c} = 1 \qquad \overline{r} = q+1$$
$$\frac{d\overline{c}}{d\overline{r}} = 0 \qquad \overline{r} = 0$$

in which the boundary conditions for all three geometries can be compactly expressed in terms of q.

The effectiveness factor for the different geometries can be evaluated using the integral and flux approaches, Equations 7.24–7.25, which lead to the two expressions

$$\eta = \frac{1}{(q+1)^q}\int_0^{q+1}\overline{c}\overline{r}^q d\overline{r} \tag{7.36}$$
$$\eta = \frac{1}{\Phi^2}\left.\frac{d\overline{c}}{d\overline{r}}\right|_{\overline{r}=q+1} \tag{7.37}$$

We have already solved Equations 7.35 and 7.36 (or 7.37) for the sphere, $q = 2$. Analytical solutions for the slab and cylinder geometries also can be derived. See Exercise 7.1 for the slab geometry. The results are summarized in Table 7.3. Note that I_0 and I_1 are modified Bessel functions of the first kind of orders zero and one, respectively.

The effectiveness factors versus Thiele modulus for the three geometries are plotted in Figure 7.8. Although the functional forms listed in Table 7.3 appear quite different, we see in Figure 7.8 that these solutions are quite similar. The effectiveness factor for the slab is largest, the cylinder is intermediate, and the sphere is the smallest at all values of Thiele modulus. The three curves have identical small Φ and large Φ asymptotes. The maximum difference between the effectiveness factors of the sphere and the slab η is about 16%, and occurs at $\Phi = 1.6$. For $\Phi < 0.5$ and $\Phi > 7$, the difference between all three effectiveness factors is less than 5%.

Sphere	$\eta = \dfrac{1}{\Phi}\left[\dfrac{1}{\tanh 3\Phi} - \dfrac{1}{3\Phi}\right]$	(7.38)
Cylinder	$\eta = \dfrac{1}{\Phi}\dfrac{I_1(2\Phi)}{I_0(2\Phi)}$	(7.39)
Slab	$\eta = \dfrac{\tanh\Phi}{\Phi}$	(7.40)

Table 7.3: Closed-form solution for the effectiveness factor versus Thiele modulus for the sphere, semi-infinite cylinder, and semi-infinite slab.

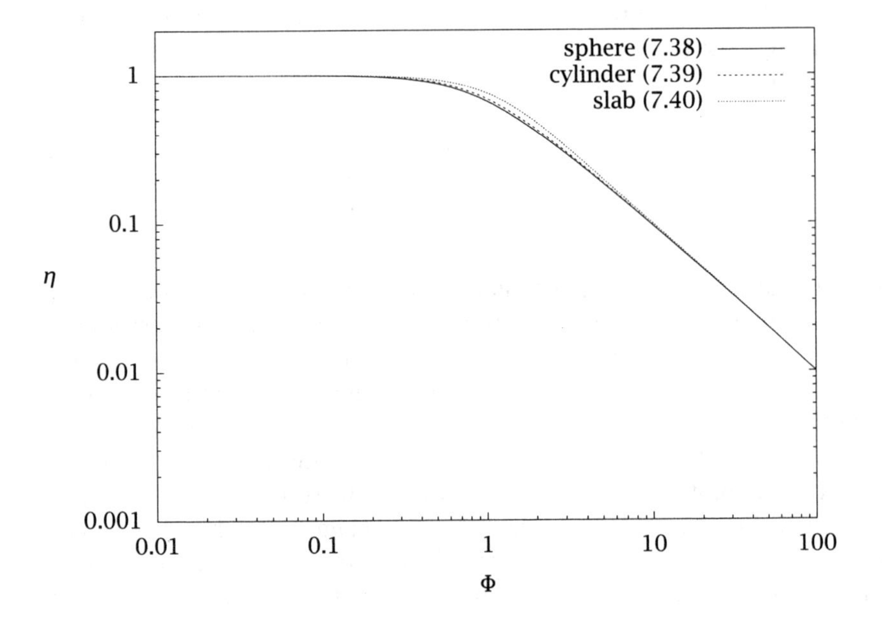

Figure 7.8: Effectiveness factor versus Thiele modulus for the sphere, cylinder and slab.

7.4.3 Other Reaction Orders

For reactions other than first order, the reaction-diffusion equation is nonlinear and numerical solution is required. We will see, however, that many of the conclusions from the analysis of the first-order reaction case still apply for other reaction orders. We consider nth-order, irreversible reaction kinetics

$$\mathrm{A} \xrightarrow{k} \mathrm{B}, \qquad r = kc_A^n$$

The reaction-diffusion equation and boundary conditions for this case are

$$D_A \frac{1}{r^q} \frac{d}{dr}\left(r^q \frac{dc_A}{dr}\right) - kc_A^n = 0 \tag{7.41}$$

We might naturally define the Thiele modulus as in Equation 7.19, but the results for various reaction orders have a common asymptote if we instead define

$$\Phi = \sqrt{\frac{n+1}{2} \frac{kc_{As}^{n-1} a^2}{D_A}} \qquad \text{Thiele modulus } n\text{th-order reaction} \tag{7.42}$$

$$\frac{1}{\overline{r}^q} \frac{d}{d\overline{r}}\left(\overline{r}^q \frac{d\overline{c}}{d\overline{r}}\right) - \frac{2}{n+1}\Phi^2 \overline{c}^n = 0$$

$$\overline{c} = 1 \qquad \overline{r} = q+1$$

$$\frac{d\overline{c}}{d\overline{r}} = 0 \qquad \overline{r} = 0$$

$$\eta = \frac{1}{(q+1)^q} \int_0^{q+1} \overline{c}^n \overline{r}^q d\overline{r}$$

$$\eta = \frac{n+1}{2} \frac{1}{\Phi^2} \left.\frac{d\overline{c}}{d\overline{r}}\right|_{\overline{r}=q+1}$$

Figure 7.9 shows the effect of reaction order for $n \geq 1$ in a spherical pellet. As the reaction order increases, the effectiveness factor decreases. Notice that the definition of Thiele modulus in Equation 7.42 has achieved the desired goal of giving all reaction orders a common asymptote at high values of Φ. Figure 7.10 shows the effectiveness factor versus Thiele modulus for reaction orders less than unity. Notice the discontinuity in slope of the effectiveness factor versus Thiele modulus that occurs when the order is less than unity. Recall from

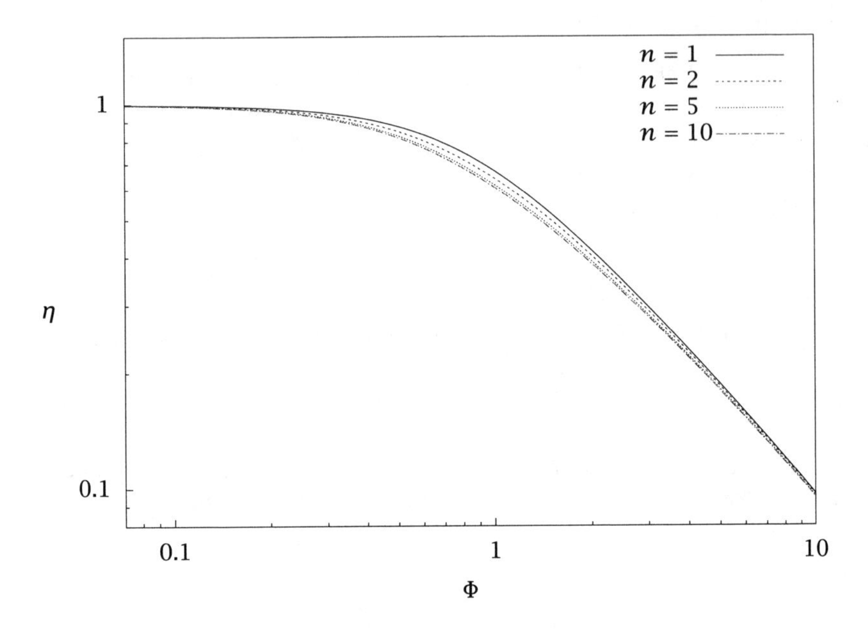

Figure 7.9: Effectiveness factor versus Thiele modulus in a spherical pellet; reaction orders greater than unity.

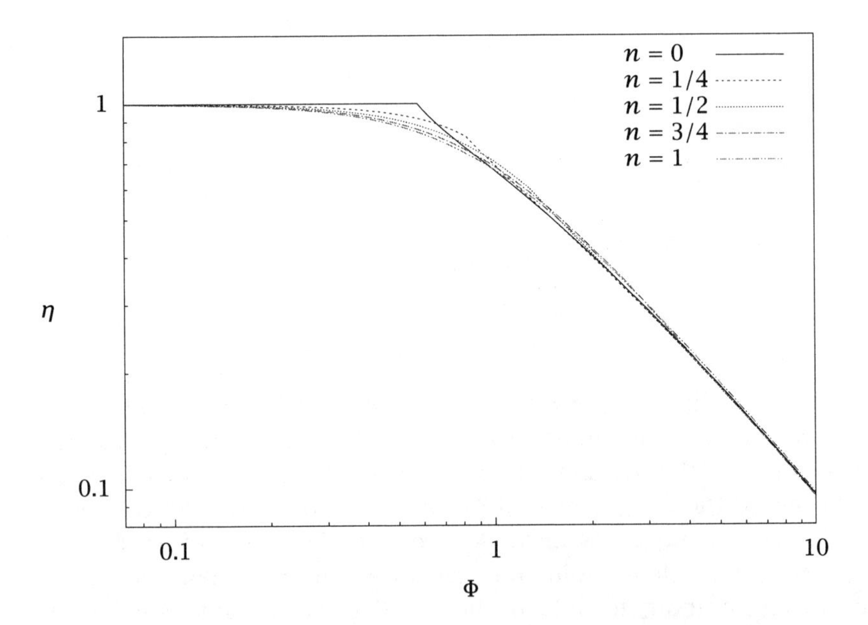

Figure 7.10: Effectiveness factor versus Thiele modulus in a spherical pellet; reaction orders less than unity.

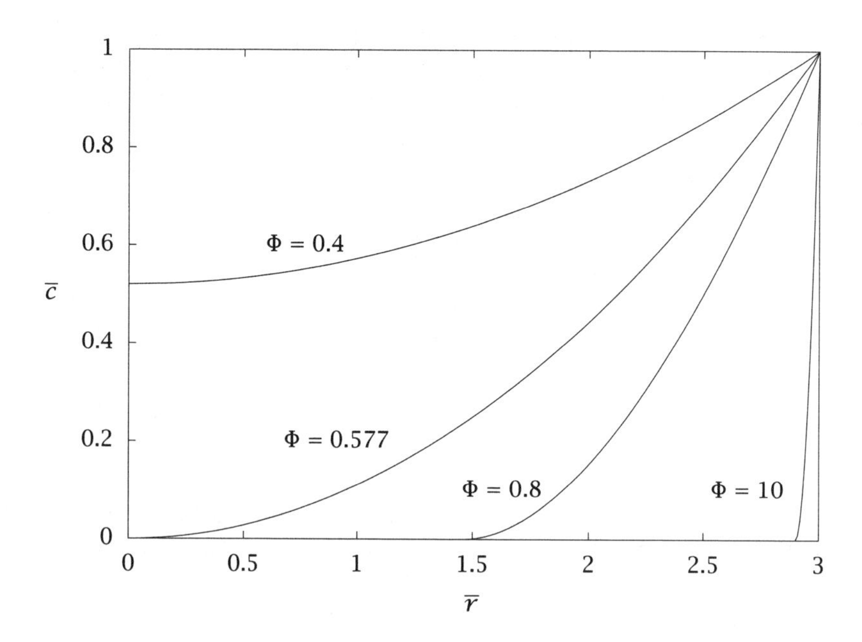

Figure 7.11: Dimensionless concentration versus radius for zero-order reaction ($n = 0$) in a spherical pellet ($q = 2$); for large Φ the inner region of the pellet has zero A concentration.

the discussion in Chapter 4 that if the reaction order is less than unity in a batch reactor, the concentration of A reaches zero in finite time. Equation 4.27 shows the discontinuity that is introduced into the mass balance when the concentration reaches zero, and Figure 4.8 displays the discontinuity in the slope of concentration versus time. In the reaction-diffusion problem in the pellet, the same kinetic effect causes the discontinuity in η versus Φ. For large values of Thiele modulus, the diffusion is slow compared to reaction, and the A concentration reaches zero at some nonzero radius inside the pellet. For orders less than unity, an inner region of the pellet has identically zero A concentration. Figure 7.11 shows the reactant concentration versus radius for the zero-order reaction case in a sphere at various values of Thiele modulus. For $\Phi = 0.577$, the A concentration reaches zero just at the center of the pellet. Notice that the discontinuity in slope of η versus Φ in Figure 7.10 occurs at this same Φ value. For larger values of Φ, the A concentration reaches zero at finite radius. This radius increases with increase in Thiele modulus so that for $\Phi = 10$, only a small outer shell of the pellet has nonzero A concentration. The same qualitative picture holds for all values of reaction order less than one.

7.4.4 Hougen-Watson Kinetics

Given the discussion in Section 5.6 of adsorption and reactions on catalyst surfaces, it is reasonable to expect our best catalyst rate expressions may be of the Hougen-Watson form. In this section we study an example reaction mechanism of this form. Consider the following reaction and rate expression

$$\text{A} \longrightarrow \text{products} \qquad r = k\overline{c}_m \frac{K_A c_A}{1 + K_A c_A}$$

This expression arises when gas-phase A adsorbs onto the catalyst surface and the reaction is first order in the adsorbed A concentration. If we consider the slab catalyst geometry, the mass balance is

$$D_A \frac{d^2 c_A}{dr^2} - k\overline{c}_m \frac{K_A c_A}{1 + K_A c_A} = 0$$

and the boundary conditions are

$$c_A = c_{As} \qquad r = L$$

$$\frac{dc_A}{dr} = 0 \qquad r = 0$$

We would like to study the effectiveness factor for these kinetics. First we define dimensionless concentration and length as before to arrive at the dimensionless reaction-diffusion model

$$\frac{d^2\overline{c}}{d\overline{r}^2} - \tilde{\Phi}^2 \frac{\overline{c}}{1 + \phi\overline{c}} = 0 \tag{7.43}$$

$$\overline{c} = 1 \qquad \overline{r} = 1$$

$$\frac{d\overline{c}}{d\overline{r}} = 0 \qquad \overline{r} = 0 \tag{7.44}$$

in which we now have *two* dimensionless groups

$$\tilde{\Phi} = \sqrt{\frac{k\overline{c}_m K_A a^2}{D_A}}, \qquad \phi = K_A c_{As} \tag{7.45}$$

We use the tilde to indicate $\tilde{\Phi}$ is a good first guess for a Thiele modulus for this problem, but we will find a better candidate subsequently. The new dimensionless group ϕ represents a dimensionless adsorption constant. The effectiveness factor is calculated from

$$\eta = \frac{R_{Ap}}{R_{As}} = \frac{-(S_p/V_p)D_A \, dc_A/dr|_{r=a}}{-k\overline{c}_m K_A c_{As}/(1 + K_A c_{As})}$$

which becomes upon definition of the dimensionless quantities

$$\eta = \frac{1+\phi}{\tilde{\Phi}^2} \left. \frac{d\overline{c}}{d\overline{r}} \right|_{\overline{r}=1} \tag{7.46}$$

Now we wish to define a Thiele modulus so that η has a common asymptote at large Φ for all values of ϕ. This goal was accomplished for the nth-order reaction as shown in Figures 7.9 and 7.10 by including the factor $(n+1)/2$ in the definition of Φ given in Equation 7.42. We now turn our attention to finding the right Thiele modulus for the Hougen-Watson kinetics problem. First we calculate η for large $\tilde{\Phi}$.

Asymptotic behavior for large $\tilde{\Phi}$. As a first step we change the independent variable from $\overline{r}$ to

$$z = \tilde{\Phi}(1 - \overline{r}) \tag{7.47}$$

in order to remove $\tilde{\Phi}$ from the differential equation [18]. Then we take the limit $\tilde{\Phi} \to \infty$ to obtain the model describing the asymptotic case. Changing variables in the reaction-diffusion model, Equations 7.43–7.44, and taking the limit produces

$$\frac{d^2\overline{c}}{dz^2} - \frac{\overline{c}}{1+\phi\overline{c}} = 0 \tag{7.48}$$

$$\overline{c} = 1 \qquad z = 0$$

$$\frac{d\overline{c}}{dz} = 0 \qquad z = \infty$$

The effectiveness factor is obtained by changing variables in Equation 7.46, producing

$$\eta = -\frac{1+\phi}{\tilde{\Phi}} \left. \frac{d\overline{c}}{dz} \right|_{z=0} \tag{7.49}$$

We need to perform one integral of Equation 7.48, without resorting to numerical integration, to evaluate the effectiveness factor.

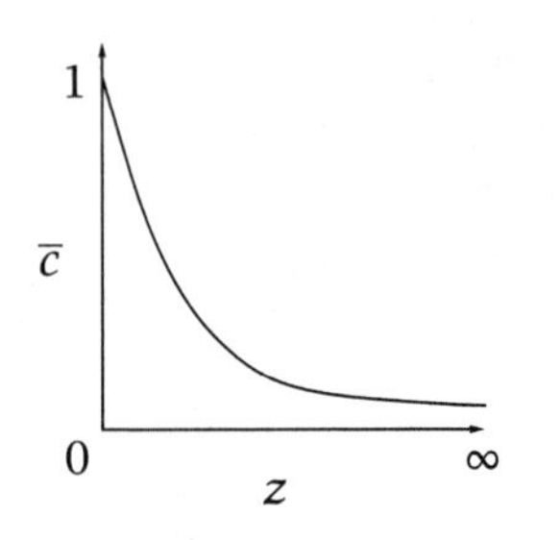

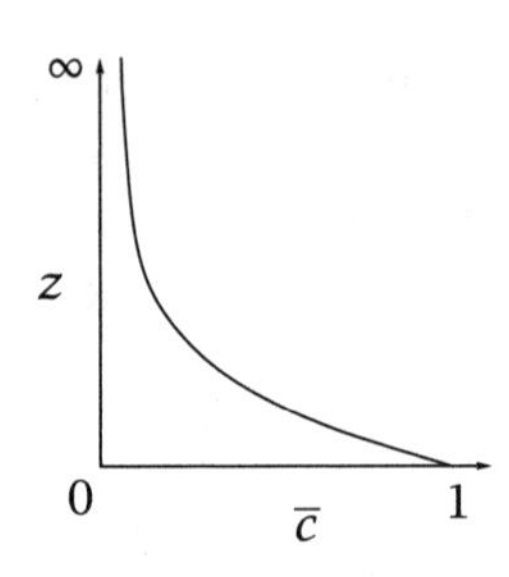

Figure 7.12: Dimensionless concentration $\overline{c}$ versus z for large $\tilde{\Phi}$; consider the function $z(\overline{c})$ in place of $\overline{c}(z)$.

Define $p(z) = d\overline{c}/dz$ to be the desired first derivative. The only nonobvious step in this section is the next one. We make the following variable transformation; instead of considering p as a function of z, define a transformation that makes it a function of $\overline{c}$. Figure 7.12 illustrates what we have in mind. The variable transformation is

$$p(z) = p(z(\overline{c})) = \tilde{p}(\overline{c})$$

We now use the familiar chain rule of differentiation

$$\frac{dp}{dz} = \frac{d\tilde{p}}{d\overline{c}}\frac{d\overline{c}}{dz} = \frac{d\tilde{p}}{d\overline{c}}p(z) = \frac{d\tilde{p}}{d\overline{c}}\tilde{p}$$

and substitute the mass balance, Equation 7.48, to obtain

$$\frac{d\tilde{p}}{d\overline{c}}\tilde{p} = \frac{\overline{c}}{1+\phi\overline{c}}$$

This equation can be separated to yield

$$\int_0^{\tilde{p}} \tilde{p}\,d\tilde{p} = \int_0^{\overline{c}} \frac{\overline{c}}{1+\phi\overline{c}}d\overline{c}$$

Performing the integral on the left-hand side and solving for $\tilde{p}$ gives

$$\tilde{p} = -\left[2\int_0^{\overline{c}} \frac{\overline{c}}{1+\phi\overline{c}}d\overline{c}\right]^{1/2}$$

in which we choose the correct sign to correspond to Figure 7.12. Performing the integral on the right-hand side yields

$$\tilde{p} = -\left[\frac{2}{\phi}\left(\overline{c} - \frac{1}{\phi}\ln(1+\phi\overline{c})\right)\right]^{1/2} \tag{7.50}$$

Evaluating Equation 7.50 at $\overline{c} = 1$, which corresponds to $z = 0$ (see Figure 7.12), and substituting the result into Equation 7.49 produces

$$\eta = \frac{1}{\tilde{\Phi}} \left(\frac{1+\phi}{\phi} \right) \sqrt{2\left(\phi - \ln(1+\phi)\right)} \tag{7.51}$$

Equation 7.51 tells us what we need to know. We wish to redefine the Thiele modulus so that this equation reads simply

$$\eta = \frac{1}{\Phi}$$

We can accomplish this with a simple rescaling of the Thiele modulus via

$$\Phi = \left(\frac{\phi}{1+\phi} \right) \frac{1}{\sqrt{2\left(\phi - \ln(1+\phi)\right)}} \, \tilde{\Phi}$$

So, in summary, we have the following two dimensionless groups for this problem

$$\Phi = \left(\frac{\phi}{1+\phi} \right) \sqrt{\frac{k\overline{c}_m K_A a^2}{2D_A\left(\phi - \ln(1+\phi)\right)}}, \qquad \phi = K_A c_{As} \tag{7.52}$$

Obviously it does not pay to try to guess an appropriate Thiele modulus for these more complex rate expressions. An asymptotic analysis as presented here is required to find the appropriate scaling. This idea appears to have been discovered independently by *three* chemical engineers in 1965. To quote from Aris [2, p. 113]

> This is the essential idea in three papers published independently in March, May and June of 1965; see Bischoff [6], Aris [1] and Petersen [19]. A more limited form was given as early as 1958 by Stewart in Bird, Stewart and Lightfoot [4, p. 338].

The payoff for this analysis is shown in Figures 7.13 and 7.14. If we use our first guess for the Thiele modulus, Equation 7.45, we obtain Figure 7.13 in which the various values of ϕ have different asymptotes. Using the Thiele modulus defined in Equation 7.52, we obtain the results in Figure 7.14. Figure 7.14 displays things more clearly. First, notice from the mass balance, Equation 7.48, that the dimensionless reaction rate is

$$\frac{\overline{c}}{1+\phi\overline{c}} \qquad \begin{array}{l}\text{dimensionless reaction rate} \\ \text{Hougen-Watson kinetics}\end{array}$$

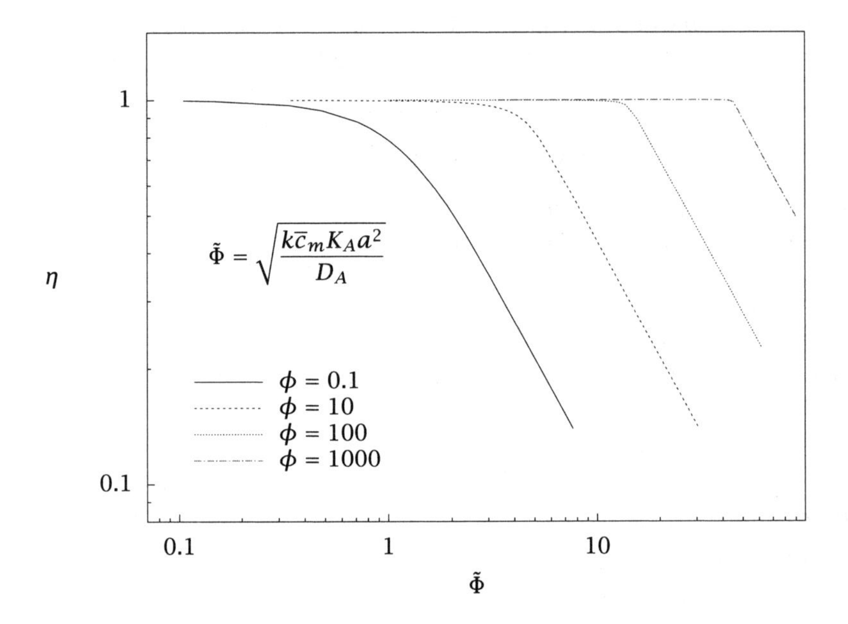

Figure 7.13: Effectiveness factor versus an inappropriate Thiele modulus in a slab; Hougen-Watson kinetics.

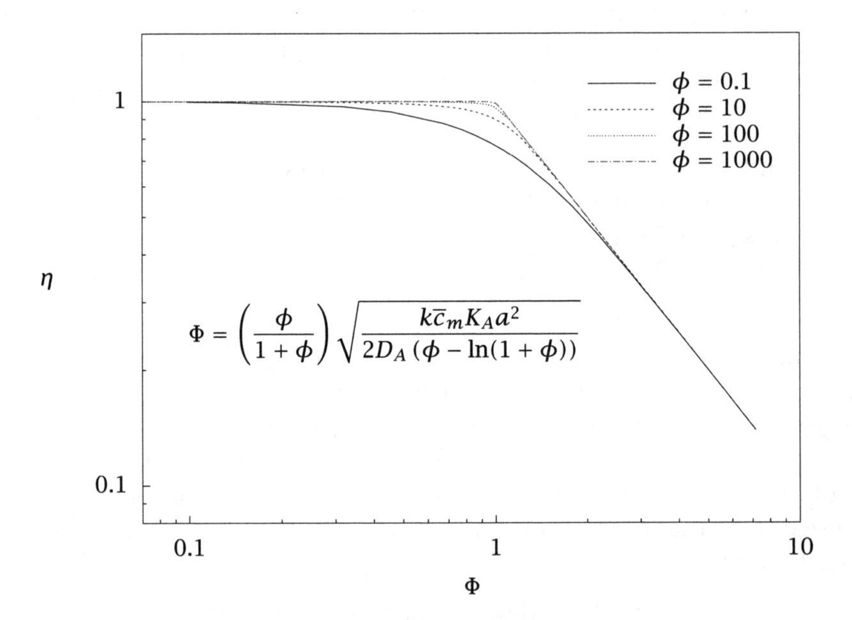

Figure 7.14: Effectiveness factor versus appropriate Thiele modulus in a slab; Hougen-Watson kinetics.

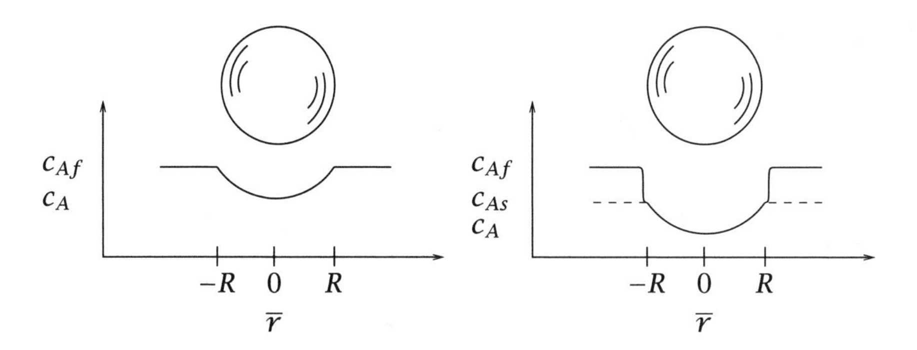

Figure 7.15: Effect of external mass transfer on pellet surface concentration.

and, as we have seen in Chapter 5, the apparent order of the reaction changes depending on the size of the dimensionless adsorption constant ϕ. For small ϕ, we expect first-order reaction behavior, and for large ϕ, we expect zero-order reaction behavior. Comparing Figure 7.14 to Figure 7.10 we see this trend. Recall that Figure 7.10 corresponds to spherical geometry, and we are using the slab geometry in this section, which causes the small differences in the effectiveness factor plots. Notice that we approximate the zero-order reaction discontinuity in the slope of η versus Φ for large ϕ. Again we see that as long as we choose an appropriate Thiele modulus, we can approximate the effectiveness factor for all values of ϕ with the first-order reaction. The largest approximation error occurs near $\Phi = 1$, and if $\Phi > 2$ or $\Phi < 0.2$, the approximation error is negligible.

7.4.5 External Mass Transfer

If the mass-transfer rate from the bulk fluid to the exterior of the pellet is not high, then the boundary condition

$$c_A(r = R) = c_{Af}$$

is not satisfied. Figure 7.15 provides a depiction of the concentration profile under low and high external mass-transfer rates. If the external mass-transfer rate is low, the concentrations in the bulk fluid and external catalyst surface are significantly different. To obtain a simple model of the external mass transfer, we replace the boundary condition above with a flux boundary condition

$$D_A \frac{dc_A}{dr} = k_m \left(c_{Af} - c_A \right), \qquad r = R \tag{7.53}$$

in which k_m is the external mass-transfer coefficient. If we multiply Equation 7.53 by $a/c_{Af}D_A$, we obtain the dimensionless boundary condition

$$\frac{d\overline{c}}{d\overline{r}} = B\left(1 - \overline{c}\right), \qquad \overline{r} = 3 \tag{7.54}$$

in which

$$B = \frac{k_m a}{D_A} \tag{7.55}$$

is the Biot number or dimensionless mass-transfer coefficient.

Summarizing, for finite external mass transfer, the dimensionless model and boundary conditions are

$$\frac{1}{\overline{r}^2}\frac{d}{d\overline{r}}\left(\overline{r}^2 \frac{d\overline{c}}{d\overline{r}}\right) - \Phi^2 \overline{c} = 0 \tag{7.56}$$

$$\frac{d\overline{c}}{d\overline{r}} = B\left(1 - \overline{c}\right) \qquad \overline{r} = 3$$

$$\frac{d\overline{c}}{d\overline{r}} = 0 \qquad \overline{r} = 0$$

The solution to the differential equation satisfying the center boundary condition can be derived as in Section 7.4 to produce

$$\overline{c}(\overline{r}) = \frac{c_2}{\overline{r}} \sinh \Phi \overline{r}$$

in which c_2 is the remaining unknown constant. Evaluating this constant using the external boundary condition gives

$$\overline{c}(\overline{r}) = \frac{3}{\overline{r}} \frac{\sinh \Phi \overline{r}}{\sinh 3\Phi + \left(\Phi \cosh 3\Phi - (\sinh 3\Phi)/3\right)/B} \tag{7.57}$$

The effectiveness factor can again be derived by integrating the local reaction rate or computing the surface flux, and the result is

$$\eta = \frac{1}{\Phi}\left[\frac{1/\tanh 3\Phi - 1/(3\Phi)}{1 + \Phi\left(1/\tanh 3\Phi - 1/(3\Phi)\right)/B}\right] \tag{7.58}$$

in which

$$\eta = \frac{R_{Ap}}{R_{Ab}}$$

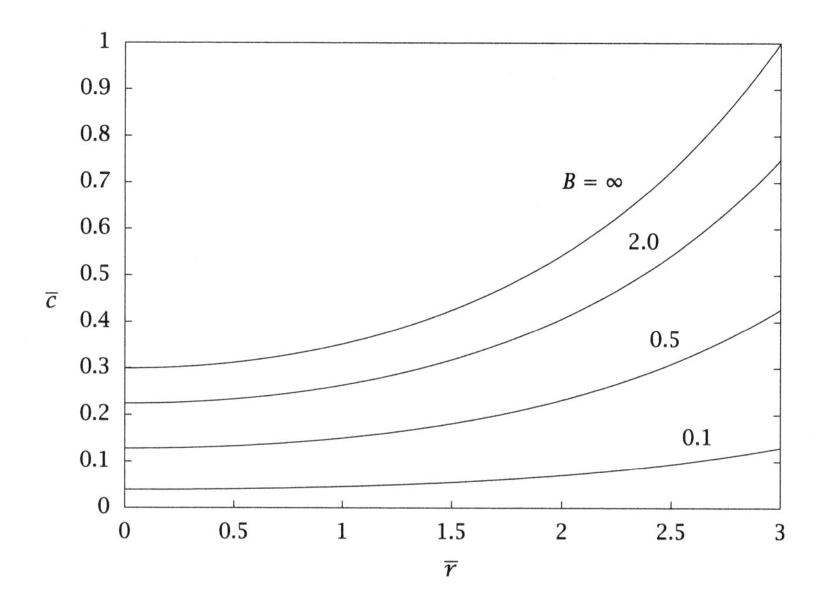

Figure 7.16: Dimensionless concentration versus radius for different values of the Biot number; first-order reaction in a spherical pellet with $\Phi = 1$.

Notice we are comparing the pellet's reaction rate to the rate that would be achieved if the pellet reacted at the *bulk fluid concentration* rather than the pellet exterior concentration as before.

Figure 7.16 plots Equation 7.57 for different values of B. When $B = \infty$, Equation 7.57 reduces to Equation 7.23, and the curve for $\Phi = 1.0$ in Figure 7.4 is the same as the curve for $B = \infty$ in Figure 7.16. With decreasing B, corresponding to slower external mass transfer, the concentration profile in the pellet becomes more uniform *and* the dimensionless surface concentration decreases. The lower concentration leads to lower reaction rates. Therefore one normally designs the reactor and chooses operating conditions, such as large gas velocities, to enhance external mass transfer and make the Biot number large.

Figure 7.17 shows the effect of the Biot number on the effectiveness factor or total pellet reaction rate. Notice that the slope of the log-log plot of η versus Φ has a slope of negative *two* rather than negative one as in the case without external mass-transfer limitations ($B = \infty$). Figure 7.18 shows this effect in more detail. If B is small, the log-log

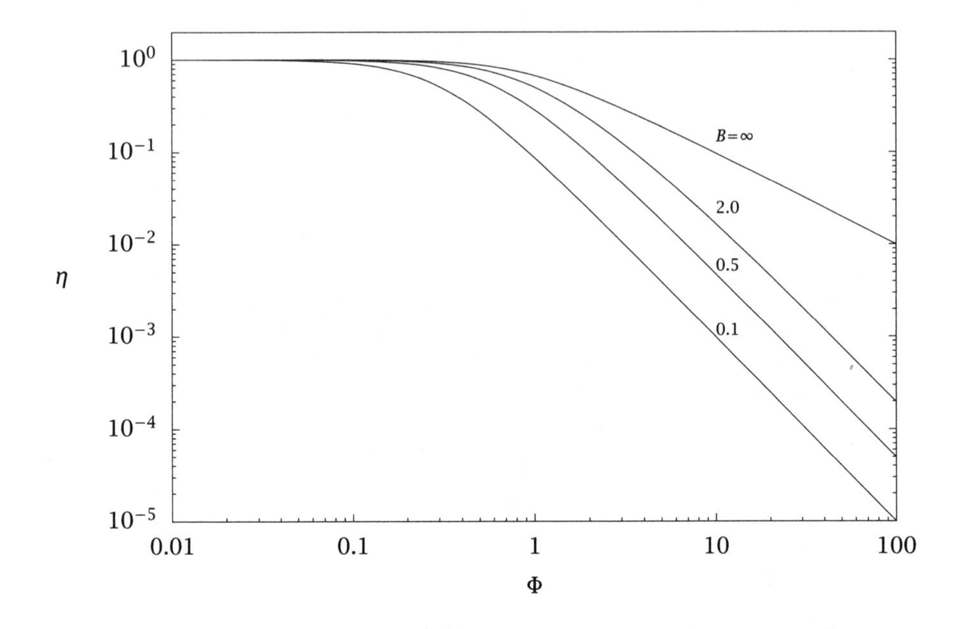

Figure 7.17: Effectiveness factor versus Thiele modulus for different values of the Biot number; first-order reaction in a spherical pellet.

plot corners with a slope of negative two at $\Phi = \sqrt{B}$. If B is large, the log-log plot first corners with a slope of negative one at $\Phi = 1$, then it corners again and decreases the slope to negative two at $\Phi = \sqrt{B}$. Both mechanisms of diffusional resistance, the diffusion within the pellet and the mass transfer from the fluid to the pellet, show their effect on pellet reaction rate by changing the slope of the effectiveness factor by negative one.[3] Given the value of the Biot number, one can easily sketch the straight line asymptotes shown in Figure 7.18. Then, given the value of the Thiele modulus, one can determine the approximate concentration profile, and whether internal diffusion or external mass transfer or both limit the pellet reaction rate. The possible cases are summarized in Table 7.4.

[3]Students who are taking or already have taken the process control course might want to think about what happens to the amplitude of a signal as it passes through a series of first-order processes.

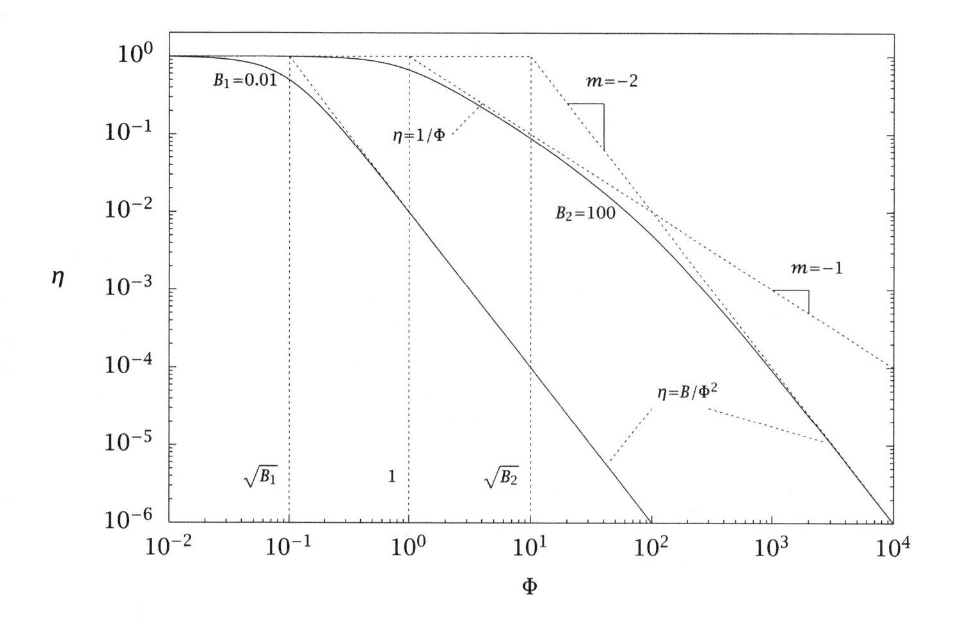

Figure 7.18: Asymptotic behavior of the effectiveness factor versus Thiele modulus; first-order reaction in a spherical pellet.

Biot number	Thiele modulus	Mechanism controlling pellet reaction rate
$B < 1$	$\Phi < \sqrt{B}$ $\sqrt{B} < \Phi < 1$ $1 < \Phi$	reaction external mass transfer both external mass transfer and internal diffusion
$1 < B$	$\Phi < 1$ $1 < \Phi < B$ $B < \Phi$	reaction internal diffusion both internal diffusion and external mass transfer

Table 7.4: The controlling mechanisms for pellet reaction rate given finite rates of internal diffusion and external mass transfer.

7.4.6 Observed versus Intrinsic Kinetic Parameters

We often need to determine a reaction order and rate constant for some catalytic reaction of interest. Assume the following nth-order reaction takes place in a catalyst particle

$$\mathrm{A} \longrightarrow \mathrm{B}, \qquad r_1 = kc_A^n$$

We call the values of k and n the intrinsic rate constant and reaction order to distinguish them from what we may estimate from data. The typical experiment is to change the value of c_A in the bulk fluid, measure the rate r_1 as a function of c_A, and then find the values of the parameters k and n that best fit the measurements. We explain this procedure in much more detail in Chapter 9. Here we show only that one should exercise caution with this estimation if we are measuring the rates with a solid catalyst. The effects of reaction, diffusion and external mass transfer may all manifest themselves in the measured rate. We express the reaction rate as

$$r_1 = \eta k c_{Ab}^n \tag{7.59}$$

We also know that at steady state, the rate is equal to the flux of A into the catalyst particle

$$r_1 = k_{mA}(c_{Ab} - c_{As}) = D_A \left.\frac{dc_A}{dr}\right|_{r=R} \tag{7.60}$$

We now study what happens to our experiment under different rate-limiting steps.

Reaction limited. First assume that both the external mass transfer and internal pellet diffusion are fast compared to the reaction. Then $\eta = 1$, and we would estimate the intrinsic parameters correctly in Equation 7.59

$$k_{\mathrm{ob}} = k$$
$$n_{\mathrm{ob}} = n$$

Everything goes according to plan when we are reaction limited.

Diffusion limited. Next assume that the external mass transfer and reaction are fast, but the internal diffusion is slow. In this case we have

$\eta = 1/\Phi$, and using the definition of Thiele modulus and Equation 7.59

$$r_1 = k_{ob} c_{As}^{(n+1)/2} \tag{7.61}$$

$$k_{ob} = \frac{1}{a}\sqrt{\frac{2}{n+1}D_A} \quad \sqrt{k} \tag{7.62}$$

$$n_{ob} = (n+1)/2 \tag{7.63}$$

So we see two problems. The rate constant we estimate, k_{ob}, varies as the square root of the intrinsic rate constant, k. The diffusion has affected the measured rate of the reaction and disguised the rate constant. We even obtain an incorrect reaction order: the first-order reaction is correct, but a second-order reaction appears 3/2 order, and so on.

Also consider what happens if we vary the temperature and try to determine the reaction's activation energy. Let the temperature dependence of the diffusivity, D_A, be represented also in Arrhenius form, with E_{diff} the activation energy of the diffusion coefficient. Let E_{rxn} be the intrinsic activation energy of the reaction. The observed activation energy from Equation 7.62 is

$$E_{ob} = \frac{E_{diff} + E_{rxn}}{2}$$

so both activation energies show up in our estimated activation energy. Normally the temperature dependence of the diffusivity is much smaller than the temperature dependence of the reaction, $E_{diff} \ll E_{rxn}$, so we would estimate an activation energy that is one-half the intrinsic value.

Mass transfer limited. Finally, assume the reaction and diffusion are fast compared to the external mass transfer. Then we have $c_{Ab} \gg c_{As}$ and Equation 7.60 gives

$$r_1 = k_{mA} c_{Ab}$$

If we vary c_{Ab} and measure r_1, we would find the mass-transfer coefficient instead of the rate constant, and a first-order reaction instead of the true reaction order

$$k_{ob} = k_{mA}$$

$$n_{ob} = 1$$

Normally, mass-transfer coefficients also have fairly small temperature dependence compared to reaction rates, so the observed activation energy would be almost zero, independent of the true reaction's activation energy.

The moral to this story is that mass transfer and diffusion resistances disguise the reaction kinetics. We can solve this problem in two ways. First, we can arrange the experiment so that mass transfer and diffusion are fast and do not affect the estimates of the kinetic parameters. We can accomplish this by making the catalyst particles small (Φ small) and using a flow reactor with high fluid flowrates to ensure a high mass-transfer coefficient. If this experimental design is impractical or too expensive, we can alternatively model the effects of the mass transfer and diffusion, and estimate the parameters D_A and k_{mA} simultaneously with k and n. We develop techniques in Chapter 9 to handle this more complex estimation problem.

7.5 Nonisothermal Particle Considerations

We now consider situations in which the catalyst particle is not isothermal. Given an exothermic reaction, for example, if the particle's thermal conductivity is not large compared to the rate of heat release due to chemical reaction, the temperature rises inside the particle. We wish to explore the effects of this temperature rise on the catalyst performance. We have already written the general mass and energy balances for the catalyst particle in Section 7.3. Consider the single-reaction case, in which we have $R_A = -r$ and Equations 7.14 and 7.15 reduce to

$$D_A \nabla^2 c_A = r$$

$$\hat{k} \nabla^2 T = \Delta H_R r$$

We can eliminate the reaction term between the mass and energy balances to produce

$$\nabla^2 T = \frac{\Delta H_R D_A}{\hat{k}} \nabla^2 c_A$$

which relates the conversion of the reactant to the rise (or fall) in temperature. Because we have assumed constant properties, we can integrate this equation twice to give the relationship between temperature and A concentration

$$T - T_s = \frac{-\Delta H_R D_A}{\hat{k}} (c_{As} - c_A) \tag{7.64}$$

We now consider a first-order reaction and assume the rate constant has an Arrhenius form,

$$k(T) = k_s \exp\left[-E\left(\frac{1}{T} - \frac{1}{T_s}\right)\right]$$

in which T_s is the pellet exterior temperature, and we assume fast external mass transfer. Substituting Equation 7.64 into the rate constant expression gives

$$k(T) = k_s \exp\left[\frac{E}{T_s}\left(1 - \frac{T_s}{T_s + \Delta H_R D_A (c_A - c_{As})/\hat{k}}\right)\right]$$

It now simplifies matters to define dimensionless concentration and temperature, and three dimensionless parameters

$$\overline{c} = \frac{c_A}{c_{As}} \qquad \overline{T} = \frac{T - T_s}{T_s}$$

$$\gamma = \frac{E}{T_s} \qquad \beta = \frac{-\Delta H_R D_A c_{As}}{\hat{k} T_s} \qquad \tilde{\Phi}^2 = \frac{k(T_s)}{D_A} a^2$$

in which γ is a dimensionless activation energy, β is a dimensionless heat of reaction, and $\tilde{\Phi}$ is the usual Thiele modulus. Again we use the tilde to indicate we will find a better Thiele modulus subsequently. With these variables, we can express the rate constant as

$$k(T) = k_s \exp\left[\frac{\gamma\beta(1-\overline{c})}{1+\beta(1-\overline{c})}\right]$$

We then substitute the rate constant into the mass balance, and assume a spherical particle to obtain the final dimensionless model

$$\frac{1}{\overline{r}^2}\frac{d}{d\overline{r}}\left(\overline{r}^2 \frac{d\overline{c}}{d\overline{r}}\right) = \tilde{\Phi}^2 \overline{c} \exp\left(\frac{\gamma\beta(1-\overline{c})}{1+\beta(1-\overline{c})}\right)$$

$$\overline{c} = 1 \qquad \overline{r} = 3$$

$$\frac{d\overline{c}}{d\overline{r}} = 0 \qquad \overline{r} = 0 \qquad (7.65)$$

Equation 7.65 is sometimes called the Weisz-Hicks problem in honor of Weisz and Hicks's outstanding paper in which they computed accurate numerical solutions to this problem [23]. Given the solution to Equation 7.65, we can compute the effectiveness factor for the nonisothermal pellet using the usual relationship

$$\eta = \frac{1}{\tilde{\Phi}^2} \left.\frac{d\overline{c}}{d\overline{r}}\right|_{\overline{r}=3}$$

If we perform the same asymptotic analysis of Section 7.4.4 on the Weisz-Hicks problem, we find, however, that the appropriate Thiele

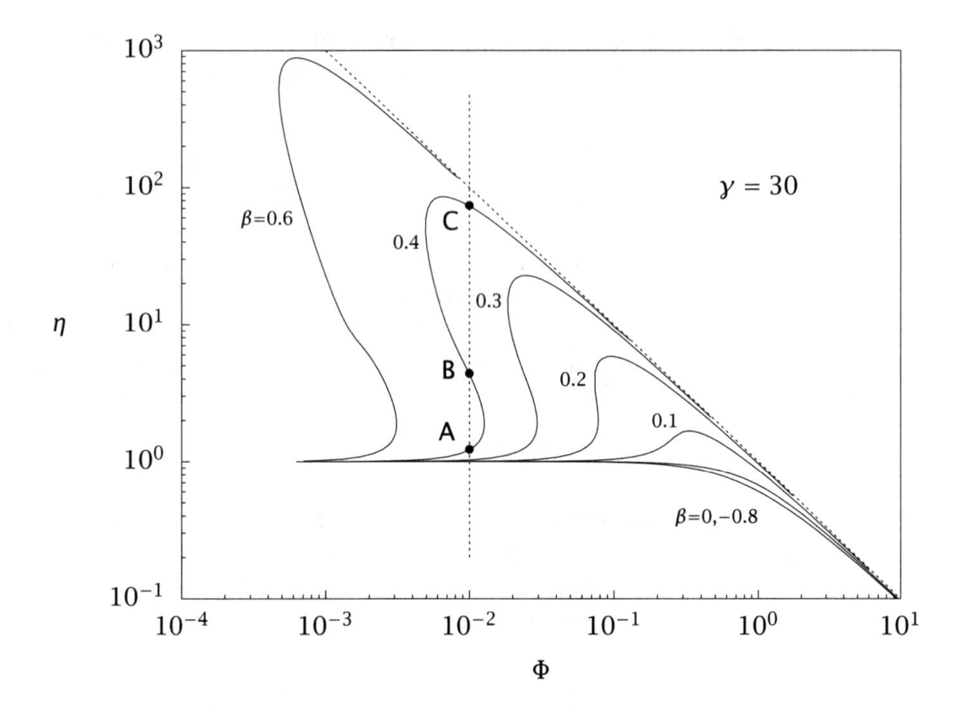

Figure 7.19: Effectiveness factor versus normalized Thiele modulus for a first-order reaction in a nonisothermal spherical pellet.

modulus for this problem is

$$\Phi = \tilde{\Phi}/I(\gamma, \beta), \qquad I(\gamma, \beta) = \left[2 \int_0^1 c \exp\left(\frac{\gamma\beta(1-c)}{1+\beta(1-c)} \right) dc \right]^{1/2} \tag{7.66}$$

The normalizing integral $I(\gamma, \beta)$ can be expressed as a sum of exponential integrals [2] or evaluated by quadrature.

Figure 7.19 shows the effectiveness factor versus Thiele modulus for activation energy $\gamma = 30$ and a variety of heats of reaction, β. Note that Φ is well chosen in Equation 7.66 because the large Φ asymptotes are the same for all values of γ and β. The first interesting feature of Figure 7.19 is that the effectiveness factor is *greater than unity* for some values of the parameters. Notice that feature is more pronounced as we increase the exothermic heat of reaction. If we consider what is happening within the pellet, this effect may not be too surprising. For the highly exothermic case, the pellet's interior temperature is significantly *higher* than the exterior temperature T_s. The rate constant inside the pellet is therefore much larger than the value at the exterior, k_s. Even though the concentration is lower inside the pellet because A is consumed, the increase in the rate constant is more significant and the product $r = kc_A$ is *larger* inside the pellet. Because the effectiveness factor compares the actual rate in the pellet to the rate at the surface conditions, it is possible for the effectiveness factor to exceed unity in a nonisothermal pellet, which we see in Figure 7.19.

A second striking feature of the nonisothermal pellet is that multiple steady states are possible. Consider the case $\Phi = 0.01$, $\beta = 0.4$ and $\gamma = 30$ shown in Figure 7.19. The effectiveness factor has three possible values for this case. We show in Figures 7.20 and 7.21 the solution to Equation 7.65 for this case. The three temperature and concentration profiles correspond to an ignited steady state (C), an extinguished steady state (A), and an unstable intermediate steady state (B). As we showed in Chapter 6, whether we achieve the ignited or extinguished steady state in the pellet depends on how the reactor is started. Aris provides further discussion of these cases and shows that many steady-state solutions are possible in some cases [3, p. 51]. For realistic values of the catalyst thermal conductivity, however, the pellet can often be considered isothermal and the energy balance can be neglected [17]. Multiple steady-state solutions in the particle may still occur in practice, however, if there is a large external heat transfer resistance.

7.6 Multiple Reactions

As the next step up in complexity, we consider the case of multiple reactions. Some analytical solutions are available for simple cases with multiple reactions, and Aris provides a comprehensive list [2], but the scope of these is limited. We focus on numerical computation as a general method for these problems. Indeed, we find that even numerical solution of some of these problems is challenging for two reasons. First, steep concentration profiles often occur for realistic parameter values, and we wish to compute these profiles accurately. It is not unusual for species concentrations to change by 10 orders of magnitude within the pellet for realistic reaction and diffusion rates. Second, we are solving boundary-value problems because the boundary conditions are provided at the center and exterior surface of the pellet. Boundary-value problems (BVPs) are generally much more difficult to solve than

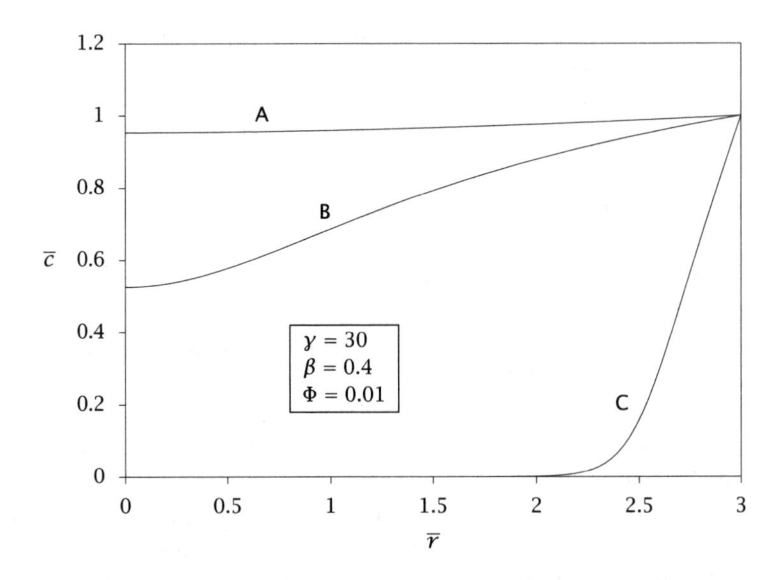

Figure 7.20: Dimensionless concentration versus radius for the non-isothermal spherical pellet: lower (A), unstable middle (B), and upper (C) steady states.

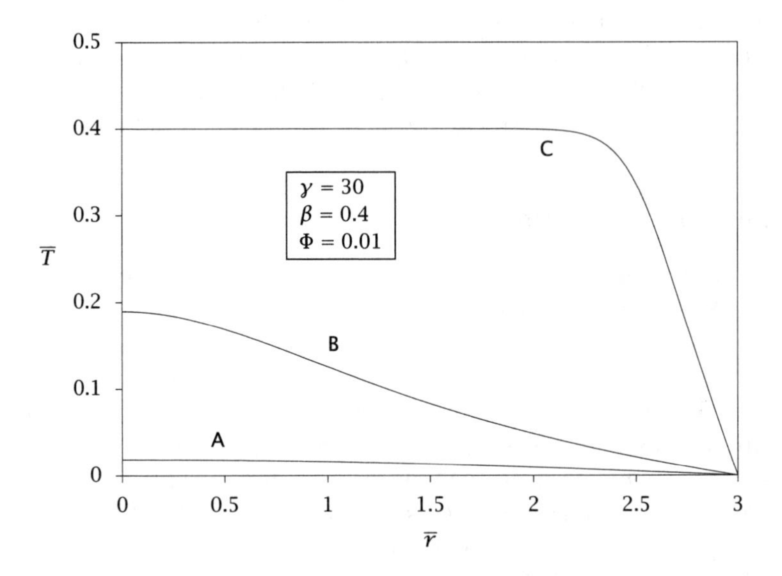

Figure 7.21: Dimensionless temperature versus radius for the non-isothermal spherical pellet: lower (A), unstable middle (B), and upper (C) steady states.

initial-value problems (IVPs).

A detailed description of numerical methods for this problem is out of place here. We use the collocation method, which is described in more detail in Appendix A. The next example involves five species, two reactions with Hougen-Watson kinetics, and both diffusion and external mass-transfer limitations.

Example 7.2: Catalytic converter

Consider the oxidation of CO and a representative volatile organic such as propylene in a automobile catalytic converter containing spherical catalyst pellets with particle radius 0.175 cm. The particle is surrounded by a fluid at 1.0 atm pressure and 550 K containing 2% CO, 3% O_2 and 0.05% (500 ppm) C_3H_6. The reactions of interest are

$$\mathrm{CO} + \frac{1}{2}\mathrm{O_2} \longrightarrow \mathrm{CO_2}$$

$$\mathrm{C_3H_6} + \frac{9}{2}\mathrm{O_2} \longrightarrow 3\mathrm{CO_2} + 3\mathrm{H_2O}$$

with rate expressions given by Oh et al. [16]

$$r_1 = \frac{k_1 c_{\mathrm{CO}} c_{\mathrm{O_2}}}{(1 + K_{\mathrm{CO}} c_{\mathrm{CO}} + K_{\mathrm{C_3H_6}} c_{\mathrm{C_3H_6}})^2}$$

$$r_2 = \frac{k_2 c_{\mathrm{C_3H_6}} c_{\mathrm{O_2}}}{(1 + K_{\mathrm{CO}} c_{\mathrm{CO}} + K_{\mathrm{C_3H_6}} c_{\mathrm{C_3H_6}})^2}$$

The rate constants and the adsorption constants are assumed to have Arrhenius form. The parameter values are given in Table 7.5 [16]. The mass-transfer coefficients are taken from DeAcetis and Thodos [9]. The pellet may be assumed to be isothermal. Calculate the steady-state pellet concentration profiles of all reactants and products.

Solution

We solve the steady-state mass balances for the three reactant species,

$$D_j \frac{1}{r^2} \frac{d}{dr}\left(r^2 \frac{dc_j}{dr}\right) = -R_j$$

with the boundary conditions

$$\frac{dc_j}{dr} = 0 \qquad r = 0$$

$$D_j \frac{dc_j}{dr} = k_{mj}\left(c_{jf} - c_j\right) \qquad r = R$$

Parameter	Value	Units	Parameter	Value	Units
P	1.013×10^5	N/m^2	k_{10}	7.07×10^{19}	$cm^3/mol \cdot s$
T	550	K	k_{20}	1.47×10^{21}	$cm^3/mol \cdot s$
R	0.175	cm	K_{CO0}	8.099×10^6	cm^3/mol
E_1	13,108	K	$K_{C_3H_60}$	2.579×10^8	cm^3/mol
E_2	15,109	K	D_{CO}	0.0487	cm^2/s
E_{CO}	−409	K	D_{O_2}	0.0469	cm^2/s
$E_{C_3H_6}$	191	K	$D_{C_3H_6}$	0.0487	cm^2/s
c_{COf}	2.0 %		k_{mCO}	3.90	cm/s
c_{O_2f}	3.0 %		k_{mO_2}	4.07	cm/s
$c_{C_3H_6f}$	0.05 %		$k_{mC_3H_6}$	3.90	cm/s

Table 7.5: Kinetic and mass-transfer parameters for the catalytic converter example.

$j = \{CO, O_2, C_3H_6\}$. The model is solved using the collocation method. The reactant concentration profiles are shown in Figures 7.22 and 7.23. Notice that O_2 is in excess and both CO and C_3H_6 reach very low values within the pellet. The log scale in Figure 7.23 shows that the concentrations of these reactants change by seven orders of magnitude. Obviously the consumption rate is large compared to the diffusion rate for these species. The external mass-transfer effect is noticeable, but not dramatic.

The product concentrations could simply be calculated by solving their mass balances along with those of the reactants. Because we have only two reactions, however, the concentrations of the products are computable from the stoichiometry and the mass balances. If we take the following mass balances

$$\begin{aligned}
D_{CO}\nabla^2 c_{CO} &= -R_{CO} &&= r_1 \\
D_{C_3H_6}\nabla^2 c_{C_3H_6} &= -R_{C_3H_6} &&= r_2 \\
D_{CO_2}\nabla^2 c_{CO_2} &= -R_{CO_2} &&= -r_1 - 3r_2 \\
D_{H_2O}\nabla^2 c_{H_2O} &= -R_{H_2O} &&= -3r_2
\end{aligned}$$

and form linear combinations to eliminate the reaction-rate terms for the two products, we obtain

$$\begin{aligned}
D_{CO_2}\nabla^2 c_{CO_2} &= -D_{CO}\nabla^2 c_{CO} - 3D_{C_3H_6}\nabla^2 c_{C_3H_6} \\
D_{H_2O}\nabla^2 c_{H_2O} &= -3D_{C_3H_6}\nabla^2 c_{C_3H_6}
\end{aligned}$$

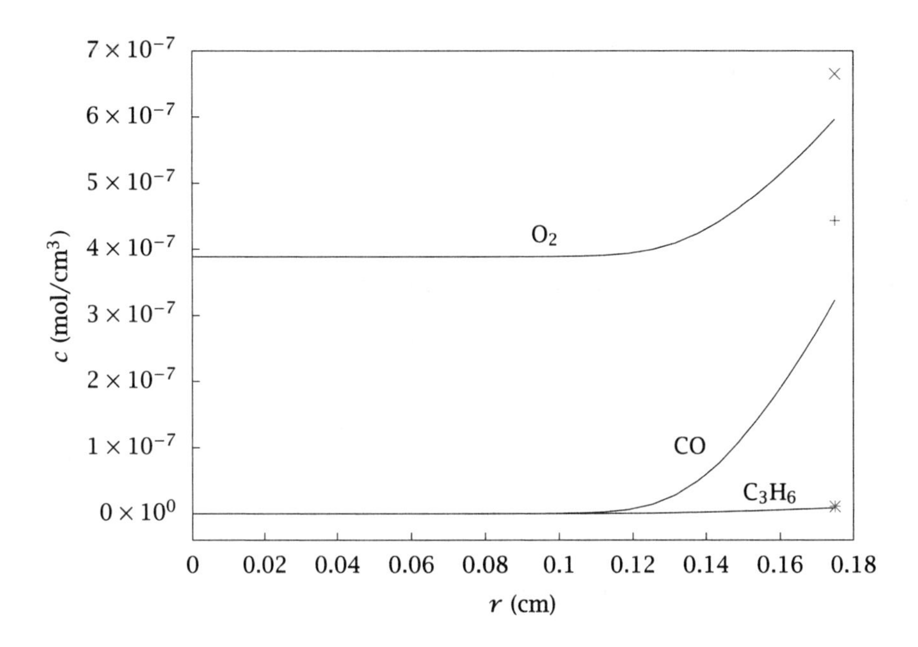

Figure 7.22: Concentration profiles of reactants; fluid concentration of O_2 (×), CO (+), C_3H_6 (∗).

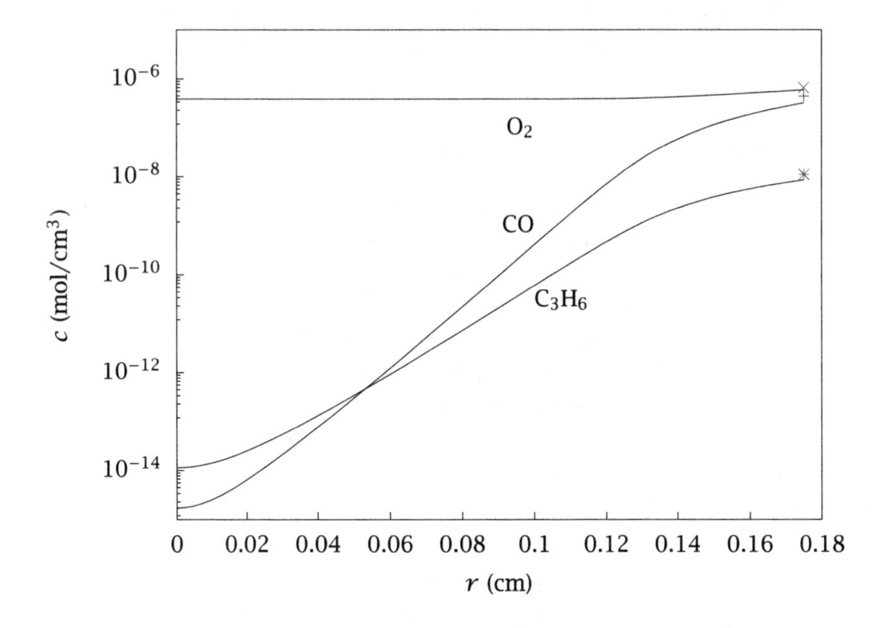

Figure 7.23: Concentration profiles of reactants (log scale); fluid concentration of O_2 (×), CO (+), C_3H_6 (∗).

Because the diffusivities are assumed constant, we can integrate these once on $(0, r)$ to obtain for the products

$$D_{CO_2} \frac{dc_{CO_2}}{dr} = -D_{CO} \frac{dc_{CO}}{dr} - 3D_{C_3H_6} \frac{dc_{C_3H_6}}{dr}$$

$$D_{H_2O} \frac{dc_{H_2O}}{dr} = -3D_{C_3H_6} \frac{dc_{C_3H_6}}{dr}$$

The exterior boundary condition can be rearranged to give

$$c_j - c_{jf} = -\frac{D_j}{k_{mj}} \frac{dc_j}{dr}$$

Substituting in the relationships for the products gives

$$c_{CO_2} = c_{CO_2 f} + \frac{1}{k_{mCO_2}} \left[D_{CO} \frac{dc_{CO}}{dr} + 3D_{C_3H_6} \frac{dc_{C_3H_6}}{dr} \right]$$

$$c_{H_2O} = c_{H_2O f} + \frac{1}{k_{mH_2O}} \left[3D_{C_3H_6} \frac{dc_{C_3H_6}}{dr} \right]$$

The right-hand sides are available from the solution of the material balances of the reactants. Plotting these results for the products gives Figure 7.24. We see that CO_2 is the main product. Note the products flow out of the pellet, unlike the reactants shown in Figures 7.22 and 7.23, which are flowing into the pellet. □

7.7 Fixed-Bed Reactor Design

Given our detailed understanding of the behavior of a single catalyst particle, we now are prepared to pack a tube with a bed of these particles and solve the fixed-bed reactor design problem. In the fixed-bed reactor, we keep track of two phases. The fluid-phase streams through the bed and transports the reactants and products through the reactor. The reaction-diffusion processes take place in the solid-phase catalyst particles. The two phases communicate to each other by exchanging mass and energy at the catalyst particle exterior surfaces. We have constructed a detailed understanding of all these events, and now we assemble them together.

7.7.1 Coupling the Catalyst and Fluid

We make the following assumptions:

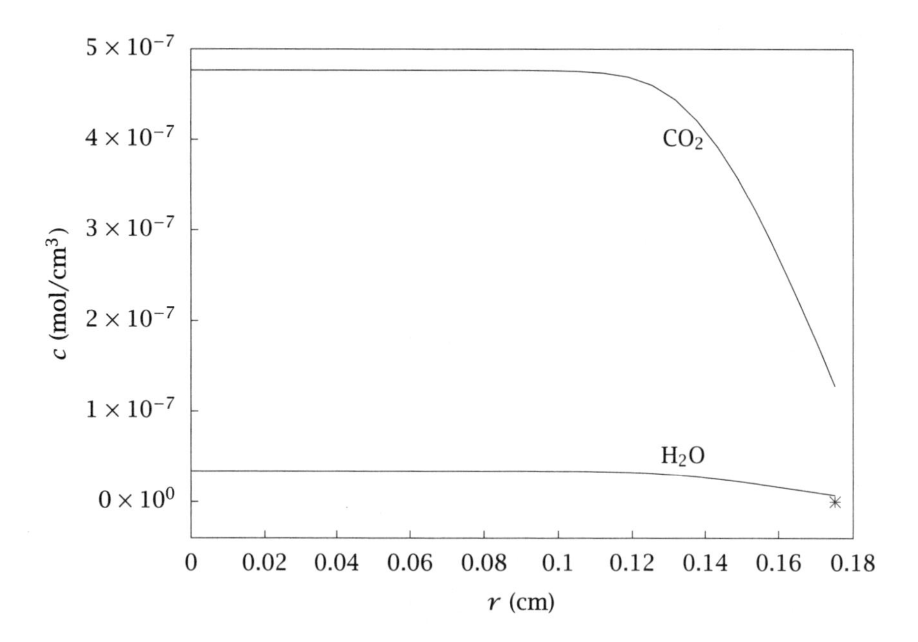

Figure 7.24: Concentration profiles of the products; fluid concentration of CO_2 (×), H_2O (+).

1. Uniform catalyst pellet exterior. Particles are small compared to the length of the reactor.

2. Plug flow in the bed, no radial profiles.

3. Neglect axial diffusion in the bed.

4. Steady state.

Fluid. In the fluid phase, we track the molar flows of all species, the temperature and the pressure. We can no longer neglect the pressure drop in the tube because of the catalyst bed. We use an empirical correlation to describe the pressure drop in a packed tube, the well-known Ergun equation [10]. Therefore, we have the following differential equations for the fluid phase

$$\frac{dN_j}{dV} = R_j \tag{7.67}$$

$$Q\rho\hat{C}_p\frac{dT}{dV} = -\sum_i \Delta H_{Ri} r_i + \frac{2}{R}U^o(T_a - T) \tag{7.68}$$

$$\frac{dP}{dV} = -\frac{(1-\epsilon_B)}{D_p\epsilon_B^3}\frac{Q}{A_c^2}\left[150\frac{(1-\epsilon_B)\mu_f}{D_p} + \frac{7}{4}\frac{\rho Q}{A_c}\right] \tag{7.69}$$

The fluid-phase boundary conditions are provided by the known feed conditions at the tube entrance

$$N_j = N_{jf}, \qquad z = 0$$
$$T = T_f, \qquad z = 0$$
$$P = P_f, \qquad z = 0$$

Catalyst particle. Inside the catalyst particle, we track the concentrations of all species and the temperature. We neglect any pressure effect inside the catalyst particle. We have the following differential equations for the catalyst particle

$$D_j\frac{1}{r^2}\frac{d}{dr}\left(r^2\frac{d\tilde{c}_j}{dr}\right) = -\tilde{R}_j \tag{7.70}$$

$$\hat{k}\frac{1}{r^2}\frac{d}{dr}\left(r^2\frac{d\tilde{T}}{dr}\right) = \sum_i \Delta H_{Ri}\tilde{r}_i \tag{7.71}$$

The boundary conditions are provided by the mass-transfer and heat-transfer rates at the pellet exterior surface, and the zero-slope conditions at the pellet center

$$\frac{d\tilde{c}_j}{dr} = 0 \qquad r = 0 \tag{7.72}$$

$$D_j\frac{d\tilde{c}_j}{dr} = k_{mj}(c_j - \tilde{c}_j) \qquad r = R \tag{7.73}$$

$$\frac{d\tilde{T}}{dr} = 0 \qquad r = 0 \tag{7.74}$$

$$\hat{k}\frac{d\tilde{T}}{dr} = k_T(T - \tilde{T}) \qquad r = R \tag{7.75}$$

Coupling equations. Finally, we equate the production rate R_j experienced by the fluid phase to the production rate inside the particles, which is where the reaction takes place. Analogously, we equate the enthalpy change on reaction experienced by the fluid phase to the enthalpy change on reaction taking place inside the particles. These expressions are given below

$$\underbrace{R_j}_{\text{rate } j \text{ / vol}} = -\underbrace{(1-\epsilon_B)}_{\text{vol cat / vol}} \underbrace{\frac{S_p}{V_p} D_j \left.\frac{d\tilde{c}_j}{dr}\right|_{r=R}}_{\text{rate } j \text{ / vol cat}} \tag{7.76}$$

$$\underbrace{\sum_i \Delta H_{Ri} r_i}_{\text{rate heat / vol}} = \underbrace{(1-\epsilon_B)}_{\text{vol cat / vol}} \underbrace{\frac{S_p}{V_p} \hat{k} \left.\frac{d\tilde{T}}{dr}\right|_{r=R}}_{\text{rate heat / vol cat}} \tag{7.77}$$

Notice we require the bed porosity to convert from the rate per volume of particle to the rate per volume of reactor. The bed porosity or void fraction, ϵ_B, is defined as the volume of voids per volume of reactor. The volume of catalyst per volume of reactor is therefore $1 - \epsilon_B$. This information can be presented in a number of equivalent ways. We can easily measure the density of the pellet, ρ_p, and the density of the bed, ρ_B. From the definition of bed porosity, we have the relation

$$\rho_B = (1 - \epsilon_B)\rho_p$$

or if we solve for the volume fraction of catalyst

$$1 - \epsilon_B = \rho_B/\rho_p$$

Figure 7.25 shows the particles and fluid, and summarizes the coupling relations between the two phases.

Equations 7.67–7.77 provide the full packed-bed reactor model given our assumptions. We next examine several packed-bed reactor problems that can be solved without solving this full set of equations. Finally, we present Example 7.7, which requires numerical solution of the full set of equations.

Example 7.3: First-order, isothermal fixed-bed reactor

Use the rate data presented in Example 7.1 to find the fixed-bed reactor volume and the catalyst mass needed to convert 97% of A. The feed to the reactor is pure A at 1.5 atm at a rate of 12 mol/s. The 0.3 cm pellets are to be used, which leads to a bed density $\rho_B = 0.6$ g/cm^3. Assume the reactor operates isothermally at 450 K and that external mass-transfer limitations are negligible.

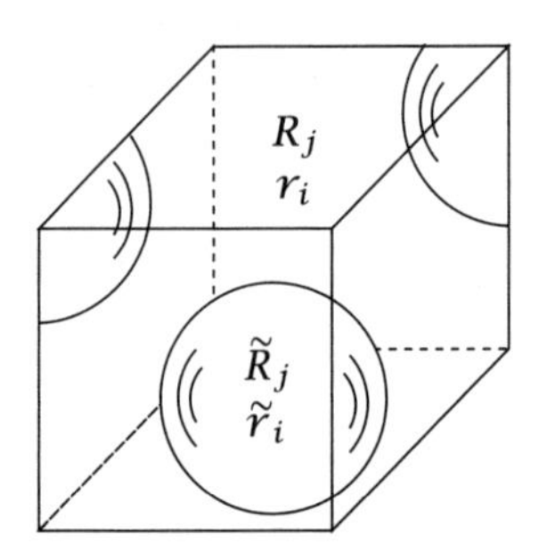

Mass

$$R_j = (1 - \epsilon_B)\widetilde{R}_{jp}$$

$$\widetilde{R}_{jp} = -\frac{S_p}{V_p} D_j \left. \frac{d\widetilde{c}_j}{dr} \right|_{r=R}$$

Energy

$$\sum_i \Delta H_{Ri} r_i = (1 - \epsilon_B) \sum_i \Delta H_{Ri} \widetilde{r}_{ip}$$

$$\sum_i \Delta H_{Ri} \widetilde{r}_{ip} = \frac{S_p}{V_p} \hat{k} \left. \frac{d\widetilde{T}}{dr} \right|_{r=R}$$

Figure 7.25: Fixed-bed reactor volume element containing fluid and catalyst particles; the equations show the coupling between the catalyst particle balances and the overall reactor balances.

Solution

We solve the fixed-bed design equation

$$\frac{dN_A}{dV} = R_A = -(1 - \epsilon_B)\eta k c_A$$

between the limits N_{Af} and $0.03N_{Af}$, in which c_A is the A concentration in the fluid. For the first-order, isothermal reaction, the Thiele modulus is independent of A concentration, and is therefore independent of axial position in the bed

$$\Phi = \frac{R}{3}\sqrt{\frac{k}{D_A}} = \frac{0.3\text{cm}}{3}\sqrt{\frac{2.6\text{s}^{-1}}{0.007\text{cm}^2/\text{s}}} = 1.93$$

The effectiveness factor is also therefore a constant

$$\eta = \frac{1}{\Phi}\left[\frac{1}{\tanh 3\Phi} - \frac{1}{3\Phi}\right] = \frac{1}{1.93}\left[1 - \frac{1}{5.78}\right] = 0.429$$

In Chapter 4, Equation 4.70, we express the concentration of A in terms of molar flows for an ideal-gas mixture

$$c_A = \frac{P}{RT}\left(\frac{N_A}{N_A + N_B}\right)$$

The total molar flow is constant due to the reaction stoichiometry so $N_A + N_B = N_{Af}$ and we have

$$c_A = \frac{P}{RT}\frac{N_A}{N_{Af}}$$

Substituting these values into the material balance, rearranging and integrating over the volume gives

$$V_R = -\left(\frac{1}{1-\epsilon_B}\right)\left(\frac{RTN_{Af}}{\eta kP}\right)\int_{N_{Af}}^{0.03N_{Af}} \frac{dN_A}{N_A}$$

$$V_R = -\left(\frac{0.85}{0.6}\right)\frac{(82.06)(450)(12)}{(0.429)(2.6)(1.5)}\ln(0.03) = 1.32\times 10^6 \text{cm}^3$$

and

$$W_c = \rho_B V_R = \frac{0.6}{1000}\left(1.32\times 10^6\right) = 789 \text{ kg}$$

We see from this example that if the Thiele modulus and effectiveness factors are constant, finding the size of a fixed-bed reactor is no more difficult than finding the size of a plug-flow reactor. □

Example 7.4: Mass-transfer limitations in a fixed-bed reactor

Reconsider Example 7.3 given the following two values of the mass-transfer coefficient

$$k_{m1} = 0.07 \text{ cm/s}$$
$$k_{m2} = 1.4 \text{ cm/s}$$

Solution

First we calculate the Biot numbers from Equation 7.55 and obtain

$$B_1 = \frac{(0.07)(0.1)}{(0.007)} = 1$$
$$B_2 = \frac{(1.4)(0.1)}{(0.007)} = 20$$

Inspection of Figure 7.17 indicates that we expect a significant reduction in the effectiveness factor due to mass-transfer resistance in the first case, and little effect in the second case. Evaluating the effectiveness factors with Equation 7.58 indeed shows

$$\eta_1 = 0.165$$
$$\eta_2 = 0.397$$

which we can compare to $\eta = 0.429$ from the previous example with no mass-transfer resistance. We can then calculate the required catalyst mass from the solution of the previous example without mass-transfer limitations, and the new values of the effectiveness factors

$$W_{c1} = \left(\frac{0.429}{0.165}\right)(789) = 2051 \text{ kg}$$
$$W_{c2} = \left(\frac{0.429}{0.397}\right)(789) = 852 \text{ kg}$$

As we can see, the first mass-transfer coefficient is so small that more than twice as much catalyst is required to achieve the desired conversion compared to the case without mass-transfer limitations. The second mass-transfer coefficient is large enough that only 8% more catalyst is required. □

Example 7.5: Second-order, isothermal fixed-bed reactor

Estimate the mass of catalyst required in an isothermal fixed-bed reactor for the second-order, heterogeneous reaction.

$$\text{A} \xrightarrow{k} \text{B}$$

$$r = kc_A^2 \qquad k = 2.25 \times 10^5 \text{cm}^3/\text{mol s}$$

The gas feed consists of A and an inert, each with molar flowrate of 10 mol/s, the total pressure is 4.0 atm and the temperature is 550 K. The desired conversion of A is 75%. The catalyst is a spherical pellet with a radius of 0.45 cm. The pellet density is $\rho_p = 0.68 \text{ g/cm}^3$ and the bed density is $\rho_B = 0.60 \text{ g/cm}^3$. The effective diffusivity of A is $0.008 \text{ cm}^2/\text{s}$ and may be assumed constant. You may assume the fluid and pellet surface concentrations are equal.

Solution

We solve the fixed-bed design equation

$$\frac{dN_A}{dV} = R_A = -(1 - \epsilon_B)\eta k c_A^2$$
$$N_A(0) = N_{Af} \tag{7.78}$$

between the limits N_{Af} and $0.25N_{Af}$. We again express the concentration of A in terms of the molar flows

$$c_A = \frac{P}{RT}\left(\frac{N_A}{N_A + N_B + N_I}\right)$$

As in the previous example, the total molar flow is constant and we know its value at the entrance to the reactor

$$N_T = N_{Af} + N_{Bf} + N_{If} = 2N_{Af}$$

Therefore,

$$c_A = \frac{P}{RT}\frac{N_A}{2N_{Af}} \tag{7.79}$$

Next we use the definition of Φ for nth-order reactions given in Equation 7.42

$$\Phi = \frac{R}{3}\left[\frac{(n+1)kc_A^{n-1}}{2D_A}\right]^{1/2} = \frac{R}{3}\left[\frac{(n+1)k}{2D_A}\left(\frac{P}{RT}\frac{N_A}{2N_{Af}}\right)^{n-1}\right]^{1/2} \tag{7.80}$$

Substituting in the parameter values gives

$$\Phi = 9.17\left(\frac{N_A}{2N_{Af}}\right)^{1/2} \tag{7.81}$$

For the second-order reaction, Equation 7.81 shows that Φ varies with the molar flow, which means Φ and η vary along the length of the reactor as N_A decreases. We are asked to estimate the catalyst mass needed to achieve a conversion of A equal to 75%. So for this particular example, Φ decreases from 6.49 to 3.24. As shown in Figure 7.9, we can *approximate* the effectiveness factor for the second-order reaction using the analytical result for the first-order reaction, Equation 7.38,

$$\eta = \frac{1}{\Phi}\left[\frac{1}{\tanh 3\Phi} - \frac{1}{3\Phi}\right] \tag{7.82}$$

Summarizing so far, to compute N_A versus V_R, we solve one differential equation, Equation 7.78, in which we use Equation 7.79 for c_A, and Equations 7.81 and 7.82 for Φ and η. We march in V_R until $N_A = 0.25N_{Af}$. The solution to the differential equation is shown in Figure 7.26. The required reactor volume and mass of catalyst are:

$$V_R = 361 \text{ L}, \qquad W_c = \rho_B V_R = 216 \text{ kg}$$

As a final exercise, given that Φ ranges from 6.49 to 3.24, we can make the large Φ approximation

$$\eta = \frac{1}{\Phi} \tag{7.83}$$

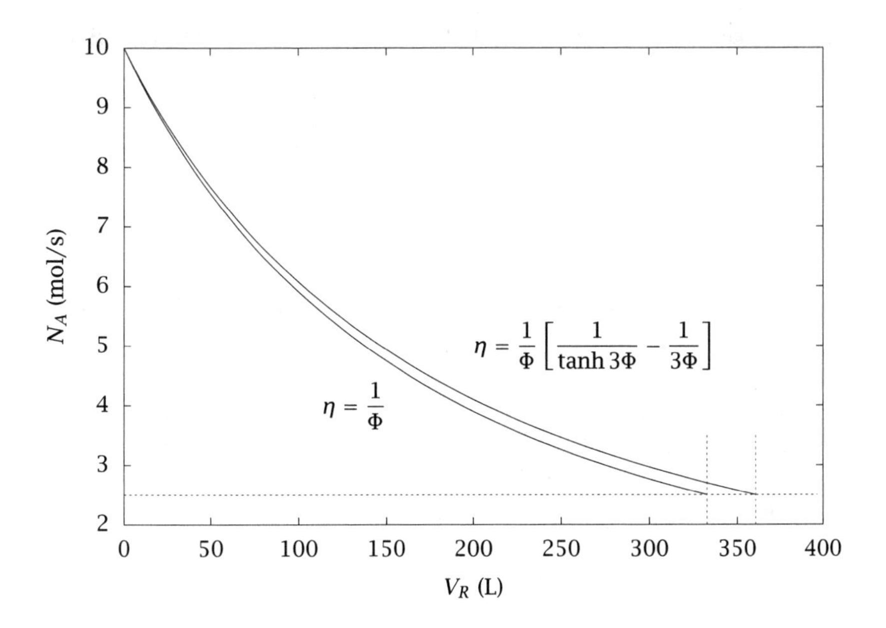

Figure 7.26: Molar flow of A versus reactor volume for second-order, isothermal reaction in a fixed-bed reactor.

to obtain a closed-form solution. If we substitute this approximation for η, and Equation 7.80 into Equation 7.78 and rearrange we obtain

$$\frac{dN_A}{dV} = \frac{-(1-\epsilon_B)\sqrt{2kD_A/3}\,(P/RT)^{3/2}}{(R/3)(2N_{Af})^{3/2}} N_A^{3/2}$$

Separating and integrating this differential equation gives

$$V_R = \frac{4\left[(1-x_A)^{-1/2}-1\right]N_{Af}(R/3)}{(1-\epsilon_B)\sqrt{kD_A/3}\,(P/RT)^{3/2}} \tag{7.84}$$

Large Φ approximation

The results for the large Φ approximation also are shown in Figure 7.26. Notice from Figure 7.9 that we are slightly overestimating the value of η using Equation 7.83, so we underestimate the required reactor volume. The reactor size and the percent change in reactor size are

$$V_R = 333 \text{ L}, \qquad \Delta = -7.7\%$$

Given that we have a result valid for all Φ that requires solving only a single differential equation, one might question the value of this closed-form solution. One advantage is purely practical. We may not have a computer available. Instructors are usually thinking about in-class examination problems at this juncture. The other important advantage is insight. It is not readily apparent from the differential equation what would happen to the reactor size if we double the pellet size, or halve the rate constant, for example. Equation 7.84, on the other hand, provides the solution's dependence on all parameters. As shown in Figure 7.26 the approximation error is small. Remember to check that the Thiele modulus is large for the entire tube length, however, before using Equation 7.84. □

Example 7.6: Hougen-Watson kinetics in a fixed-bed reactor

The following reaction converting CO to CO_2 takes place in a catalytic, fixed-bed reactor operating isothermally at 838 K and 1.0 atm

$$CO + \frac{1}{2}O_2 \longrightarrow CO_2$$

The following rate expression and parameters are adapted from a different model given by Oh et al. [16]. The rate expression is assumed to be of the Hougen-Watson form

$$r = \frac{k c_{CO} c_{O_2}}{1 + K c_{CO}} \qquad \text{mol/s cm}^3 \text{ pellet}$$

The constants are provided below

$$k = 1.3828 \times 10^{19} \exp(-13{,}500/T) \text{ cm}^3/\text{mol s}$$
$$K = 8.099 \times 10^{6} \exp(409/T) \text{ cm}^3/\text{mol}$$
$$D_{CO} = 0.0487 \text{ cm}^2/\text{s}$$

in which T is in Kelvin. The spherical catalyst pellet radius is 0.1 cm, and the densities are $\rho_p = 0.68, \rho_B = 0.60$ g/cm^3. The feed to the reactor consists of 16.7 mol% CO, 83.3 mol% O_2, and zero CO_2, with volumetric flowrate $Q_f = 792$ L/s. Find the reactor volume required to achieve 95% conversion of the CO.

Solution

Given the reaction stoichiometry and the excess of O_2, we can neglect the change in c_{O_2} and approximate the reaction as pseudo-first order in CO

$$r = \frac{k' c_{CO}}{1 + K c_{CO}} \qquad \text{mol/s cm}^3 \text{ pellet}$$
$$k' = k c_{O_2 f}$$

which is of the form analyzed in Section 7.4.4. We can write the mass balance for the molar flow of CO,

$$\frac{dN_{CO}}{dV} = -(1 - \epsilon_B)\eta r(c_{CO})$$

in which c_{CO} is the fluid CO concentration. From the reaction stoichiometry, we can express the remaining molar flows in terms of N_{CO}

$$N_{O_2} = N_{O_2 f} + 1/2(N_{CO} - N_{COf})$$
$$N_{CO_2} = N_{COf} - N_{CO}$$
$$N = N_{O_2 f} + 1/2(N_{CO} + N_{COf})$$

The concentrations follow from the molar flows assuming an ideal-gas mixture

$$c_j = \frac{P}{RT}\frac{N_j}{N}$$

To decide how to approximate the effectiveness factor shown in Figure 7.14, we evaluate $\phi = K_{CO} c_{CO}$, at the entrance and exit of the fixed-bed reactor. With ϕ evaluated, we compute the Thiele modulus given in Equation 7.52 and obtain

$$\phi = 32.0 \qquad \Phi = 79.8, \qquad \text{entrance}$$
$$\phi = 1.74 \qquad \Phi = 326, \qquad \text{exit}$$

It is clear from these values and Figure 7.14 that $\eta = 1/\Phi$ is an excellent approximation for this reactor. Substituting this equation for η into the mass balance and solving the differential equation produces the results shown in Figure 7.27. The concentration of O_2 is nearly constant, which justifies the pseudo-first-order rate expression. Reactor volume

$$V_R = 233 \text{ cm}^3$$

is required to achieve 95% conversion of the CO. Recall that the volumetric flowrate varies in this reactor so conversion is based on molar flow, not molar concentration. Figure 7.28 shows how Φ and ϕ vary with position in the reactor. □

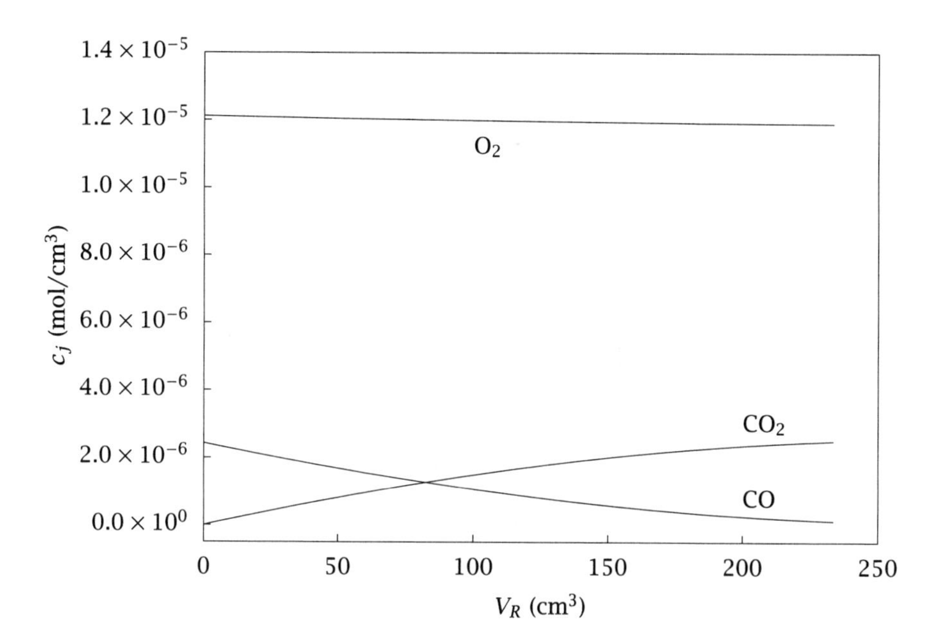

Figure 7.27: Molar concentrations versus reactor volume.

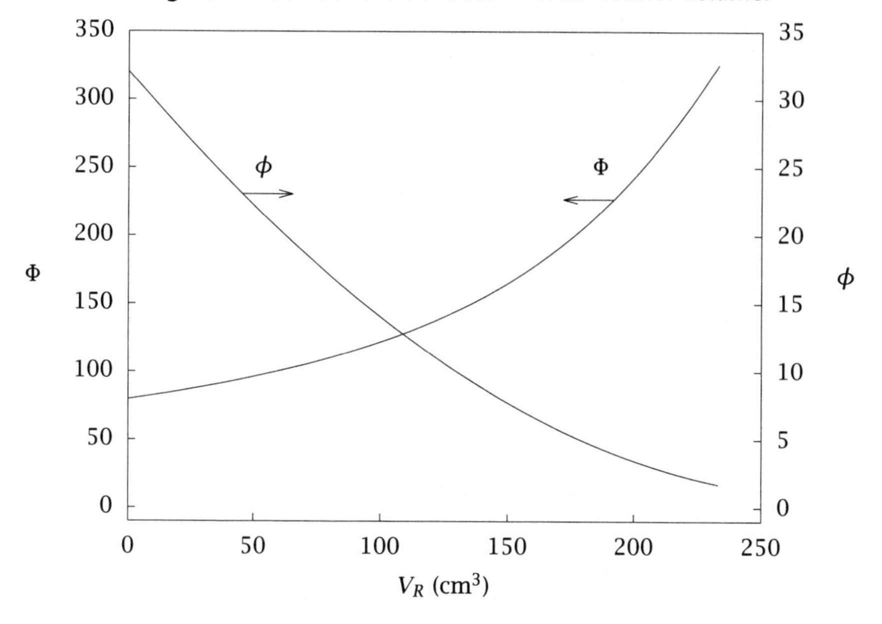

Figure 7.28: Dimensionless equilibrium constant and Thiele modulus versus reactor volume. Values indicate $\eta = 1/\Phi$ is a good approximation for entire reactor.

Parameter	Value	Units
P_f	2.02×10^5	N/m^2
T_f	550	K
R_t	5.0	cm
u_f	75	cm/s
T_a	325	K
U^o	5.5×10^{-3}	cal/(cm^2 Ks)
ΔH_{R1}	-67.63×10^3	cal/(mol CO)
ΔH_{R2}	-460.4×10^3	cal/(mol C_3H_6)
$\hat{C}_p$	0.25	cal/(g K)
μ_f	0.028×10^{-2}	g/(cm s)
ρ_B	0.51	g/cm^3
ρ_p	0.68	g/cm^3

Table 7.6: Feed flowrate and heat-transfer parameters for the fixed-bed catalytic converter.

In the previous examples, we have exploited the idea of an effectiveness factor to reduce fixed-bed reactor models to the same form as plug-flow reactor models. This approach is useful and solves several important cases, but this approach is also limited and can take us only so far. In the general case, we must contend with multiple reactions that are not first order, nonconstant thermochemical properties, and nonisothermal behavior in the pellet and the fluid. For these cases, we have no alternative but to solve numerically for the temperature and species concentrations profiles in both the pellet and the bed. As a final example, we compute the numerical solution to a problem of this type.

We use the collocation method to solve the next example, which involves five species, two reactions with Hougen-Watson kinetics, both diffusion and external mass-transfer limitations, and nonconstant fluid temperature, pressure and volumetric flowrate.

Example 7.7: Multiple-reaction, nonisothermal fixed-bed reactor

Evaluate the performance of the catalytic converter in converting CO and propylene. Determine the amount of catalyst required to convert 99.6% of the CO and propylene. The reaction chemistry and pellet mass-transfer parameters are given in Table 7.5. The feed conditions and heat-transfer parameters are given in Table 7.6.

Solution

The fluid balances govern the change in the fluid concentrations, temperature and pressure. The pellet concentration profiles are solved with the collocation approach. The pellet and fluid concentrations are coupled through the mass-transfer boundary condition. The fluid concentrations are shown in Figure 7.29. A bed volume of 1098 cm^3 is required to convert the CO and C_3H_6. Figure 7.29 also shows that oxygen is in slight excess.

The reactor temperature and pressure are shown in Figure 7.30. The feed enters at 550 K, and the reactor experiences about a 130 K temperature rise while the reaction essentially completes; the heat losses then reduce the temperature to less than 500 K by the exit. The pressure drops from the feed value of 2.0 atm to 1.55 atm at the exit. Notice the catalytic converter exit pressure of 1.55 atm must be large enough to account for the remaining pressure drops in the tail pipe and muffler.

In Figures 7.31 and 7.32, the pellet CO concentration profile at several reactor positions is displayed. The feed profile, marked by ① in Figure 7.32, is similar to the one shown in Figure 7.23 of Example 7.2 (the differences are caused by the different feed pressures). We see that as the reactor heats up, the reaction rates become large and the CO is rapidly converted inside the pellet. By 490 cm^3 in the reactor, the pellet exterior CO concentration has dropped by two orders of magnitude, and the profile inside the pellet has become very steep. As the reactions go to completion and the heat losses cool the reactor, the reaction rates drop. At 890 cm^3, the CO begins to diffuse back into the pellet. Finally, the profiles become much flatter near the exit of the reactor.

It can be numerically challenging to calculate rapid changes and steep profiles inside the pellet. The good news, however, is that accurate pellet profiles are generally *not* required for an accurate calculation of the overall pellet reaction rate. The reason is that when steep profiles are present, essentially all of the reaction occurs in a thin shell near the pellet exterior. We can calculate accurately down to concentrations on the order of 10^{-15} as shown in Figure 7.32, and by that point, essentially zero reaction is occurring, and we can calculate an accurate overall pellet reaction rate. It is always a good idea to vary the numerical approximation in the pellet profile, by changing the number of collocation points, to ensure convergence in the fluid profiles. □

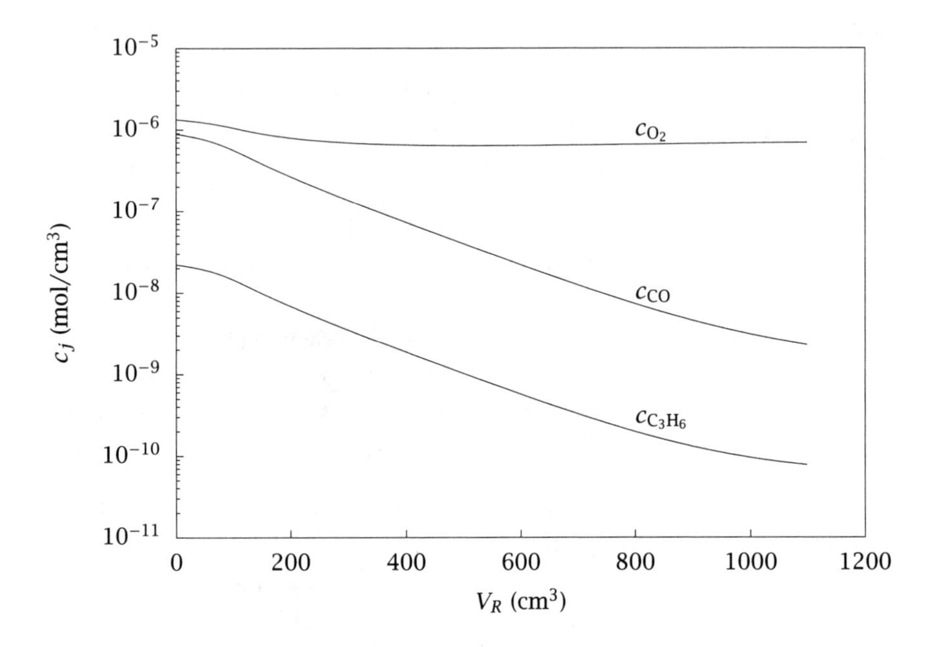

Figure 7.29: Fluid molar concentrations versus reactor volume.

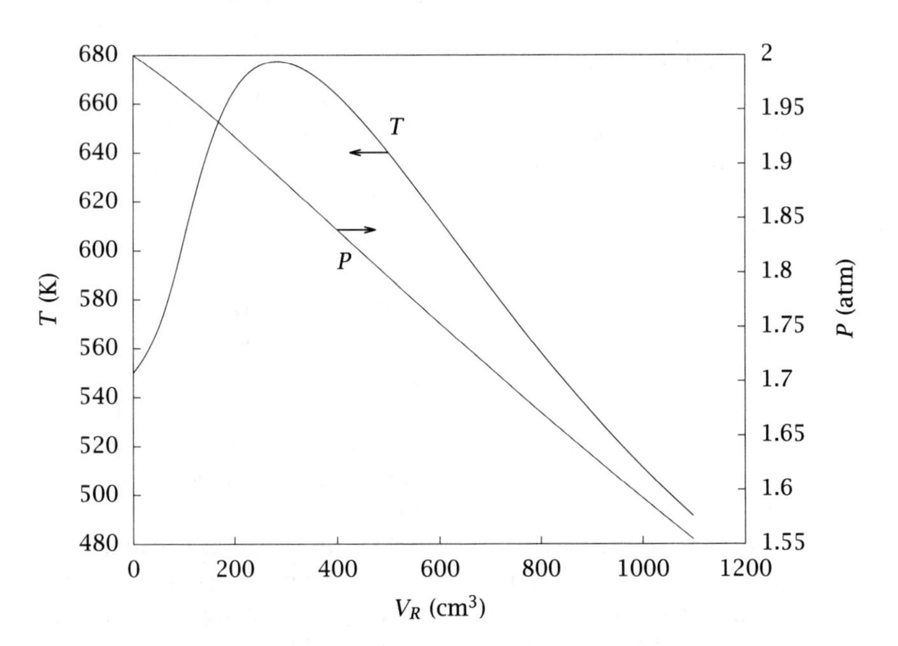

Figure 7.30: Fluid temperature and pressure versus reactor volume.

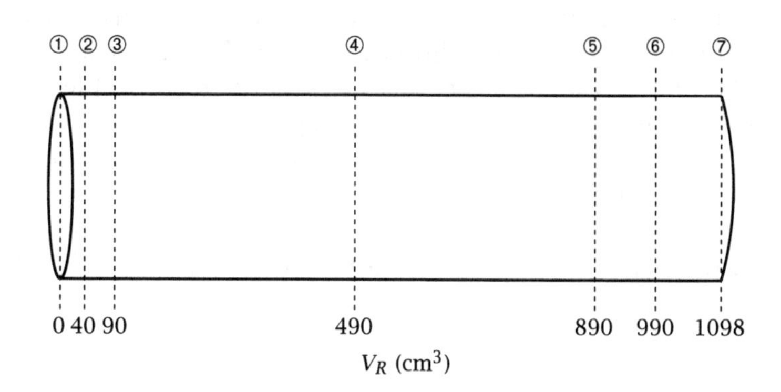

Figure 7.31: Reactor positions for pellet profiles.

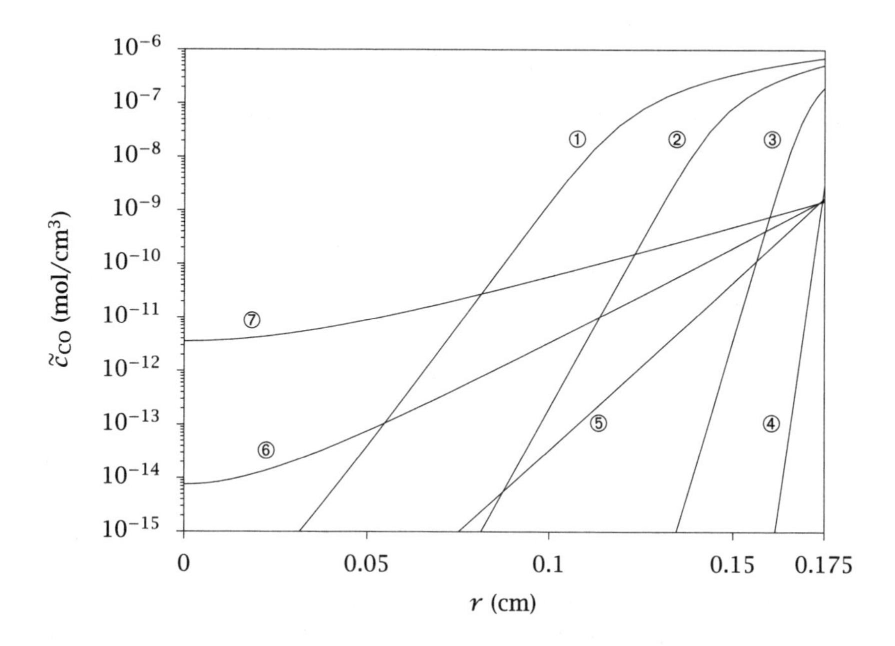

Figure 7.32: Pellet CO profiles at several reactor positions.

7.7.2 Logarithmic Transformation

As seen in Figure 7.32, the concentration profiles inside the catalyst pellet can change by many orders of magnitude during a fixed-bed simulation at realistic reactor conditions. Global polynomials are not well suited for approximating functions with these steep profiles. Since concentration is always non-negative, a logarithmic transformation is a useful way to overcome this difficulty and obtain accurate solutions with modest computation time.

The concentration inside the catalyst pellet is found by solving the material balance, Equation 7.70, and the boundary conditions, Equations 7.72 and 7.73, repeated here

$$D_j \frac{1}{r^2} \frac{d}{dr} \left(r^2 \frac{d\widetilde{c}_j}{dr} \right) = -\widetilde{R}_j \tag{7.85}$$

$$\frac{d\widetilde{c}_j}{dr} = 0 \qquad r = 0 \tag{7.86}$$

$$D_j \frac{d\widetilde{c}_j}{dr} = k_{mj}(c_j - \widetilde{c}_j) \qquad r = R \tag{7.87}$$

Consider the logarithmic transformation of concentration

$$w_j = \ln(\widetilde{c}_j) \qquad \widetilde{c}_j = e^{w_j} \tag{7.88}$$

Differentiating the transformation twice and substituting into Equations 7.85–7.87 allows us to express the model in terms of the transformed variable w_j

$$\frac{d^2 w_j}{dr^2} + \frac{dw_j}{dr} \left[\frac{dw_j}{dr} + \frac{2}{r} \right] = -\frac{1}{D_j} \widetilde{R}_j e^{-w_j} \tag{7.89}$$

$$\frac{dw_j}{dr} = 0 \qquad r = 0 \tag{7.90}$$

$$D_j \frac{dw_j}{dr} = k_{mj}(e^{-w_j} c_j - 1) \qquad r = R \tag{7.91}$$

As shown in the semi-log plot of Figure 7.32, polynomials provide an excellent fit to w_j in contrast to $\widetilde{c}_j$. In preparing Figure 7.32 with the untransformed model, for example, 200 collocation points are required to achieve an accurate solution of the fluid balances, and the pellet profiles are still not accurately computed below about $\widetilde{c}_j = 10^{-15}$. Solving the transformed model, on the other hand, only 40 collocation points are required and the pellet profiles are computed accurately down to

$\tilde{c}_j = 10^{-90}$. In addition, the transformed problem can be solved approximately 20 times faster than the original problem. For better accuracy *and* faster computation, using Equations 7.89–7.91 for the pellet concentration balances is highly recommended in fixed-bed reactor simulations with large reaction rates.

7.8 Summary

This chapter treated the fixed-bed reactor, a tubular reactor packed with catalyst pellets. We started with a general overview of the transport and reaction events that take place in the fixed-bed reactor: transport by convection in the fluid; diffusion inside the catalyst pores; and adsorption, reaction and desorption on the catalyst surface. We summarized the transport properties of the catalyst particles, and described bulk and Knudsen diffusion phenomena.

In order to simplify the model, we assumed an effective diffusivity could be used to describe diffusion in the catalyst particles. We next presented the general mass and energy balances for the catalyst particle. Next we solved a series of reaction-diffusion problems in a single catalyst particle. These included:

- Single reaction in an isothermal pellet. This case was further divided into a number of special cases.
    - First-order, irreversible reaction in a spherical particle.
    - Reaction in a semi-infinite slab and cylindrical particle.
    - nth order, irreversible reaction.
    - Hougen-Watson rate expressions.
    - Particle with significant external mass-transfer resistance.
- Single reaction in a nonisothermal pellet.
- Multiple reactions.

For the single-reaction cases, we performed dimensional analysis and found a dimensionless number, the Thiele modulus, which measures the rate of production divided by the rate of diffusion of some component. A complete analysis of the first-order reaction in a sphere suggested a general approach to calculate the production rate in a pellet in terms of the rate evaluated at the pellet exterior surface conditions. This motivated the definition of the pellet effectiveness factor, which is a function of the Thiele modulus.

For the single-reaction, nonisothermal problem, we solved the so-called Weisz-Hicks problem, and determined the temperature and concentration profiles within the pellet. We showed the effectiveness factor can be greater than unity for this case. Multiple steady-state solutions also are possible for this problem, but for realistic values of the catalyst thermal conductivity, the pellet often can be considered isothermal and the energy balance can be neglected. Multiple steady-state solutions in the particle may occur in practice, however, if there is a large external heat-transfer resistance.

For complex reactions involving many species, we cannot use the simple Thiele modulus and effectiveness factor approach, and must solve numerically the complete reaction-diffusion problem. These problems are challenging because of the steep pellet profiles that are possible.

Finally, we showed several ways to couple the mass and energy balances over the fluid flowing through a fixed-bed reactor to the balances within the pellet. For simple reaction mechanisms, we were still able to use the effectiveness factor approach to solve the fixed-bed reactor problem. For complex mechanisms, we solved numerically the full problem given in Equations 7.67–7.77. We solved the reaction-diffusion problem in the pellet coupled to the mass and energy balances for the fluid, and we used the Ergun equation to calculate the pressure in the fluid.

Notation

a	characteristic pellet length, V_p/S_p
A_c	reactor cross-sectional area
B	Biot number for external mass transfer
c	constant for the BET isotherm
c_j	concentration of species j
c_{js}	concentration of species j at the catalyst surface
$\overline{c}$	dimensionless pellet concentration
$\overline{c}_m$	total number of active surface sites
D_{AB}	binary diffusion coefficient
D_j	effective diffusion coefficient for species j
D_{jK}	Knudsen diffusion coefficient for species j
D_{jm}	diffusion coefficient for species j in the mixture
D_p	pellet diameter
E_{diff}	activation energy for diffusion

E_{obs}	experimental activation energy
E_{rxn}	intrinsic activation energy for the reaction
ΔH_{Ri}	heat of reaction i
I_j	rate of transport of species j into a pellet
I_0	modified Bessel function of the first kind, zero order
I_1	modified Bessel function of the first kind, first order
k_e	effective thermal conductivity of the pellet
k_{mj}	mass-transfer coefficient for species j
k_n	nth-order reaction rate constant
L	pore length
M_j	molecular weight of species j
n_r	number of reactions in the reaction network
N	total molar flow, $\sum_j N_j$
N_j	molar flow of species j
P	pressure
Q	volumetric flowrate
r	radial coordinate in catalyst particle
r_a	average pore radius
r_i	rate of reaction i per unit reactor volume
r_{obs}	observed (or experimental) rate of reaction in the pellet
r_{ip}	total rate of reaction i per unit catalyst volume
$\overline{r}$	dimensionless radial coordinate
R	spherical pellet radius
R	gas constant
R_j	production rate of species j
R_{jf}	production rate of species j at bulk fluid conditions
R_{jp}	total production rate of species j per unit catalyst volume
R_{js}	production rate of species j at the pellet surface conditions
S_g	BET area per gram of catalyst
S_p	external surface area of the catalyst pellet
T	temperature
$\overline{T}$	dimensionless temperature
T_f	bulk fluid temperature
T_s	pellet surface temperature
u_f	feed gas velocity
U^o	overall heat-transfer coefficient
v	volume of gas adsorbed in the BET isotherm
v_m	volume of gas corresponding to an adsorbed monolayer
V	reactor volume coordinate

V_g	pellet void volume per gram of catalyst
V_p	volume of the catalyst pellet
V_R	reactor volume
W_c	total mass of catalyst in the reactor
y_j	mole fraction of species j
z	position coordinate in a slab
ϵ	porosity of the catalyst pellet
ϵ_B	fixed-bed porosity or void fraction
η	effectiveness factor
λ	mean free path
μ_f	bulk fluid viscosity
ν_{ij}	stoichiometric number for the jth species in the ith reaction
ξ	integral of a diffusing species over a bounding surface
ρ	bulk fluid density
ρ_B	reactor bed density
ρ_p	overall catalyst pellet density
ρ_s	catalyst solid-phase density
σ	hard sphere collision radius
τ	tortuosity factor
Φ	Thiele modulus
$\Omega_{D,AB}$	dimensionless function of temperature and the intermolecular potential field for one molecule of A and one molecule of B

7.9 Exercises

Exercise 7.1: Isothermal slab with first-order kinetics

A first-order, irreversible reaction

$$A \xrightarrow{k} 2B$$

takes place in a catalyst pellet with the slab geometry depicted in Figure 7.33. Assume that the pellet height, h, and length, l, are much greater than the width, w. The catalyst exterior is assumed to be at constant species A concentration, c_{As}. Species A has an effective diffusivity, D_A, in the pellet and the catalyst is isothermal.

(a) The steady-state material balance for component A is

$$D_A \nabla^2 c_A + R_A = 0 \tag{7.92}$$

In rectangular coordinates, this becomes

$$D_A \left(\frac{\partial^2 c_A}{\partial x^2} + \frac{\partial^2 c_A}{\partial y^2} + \frac{\partial^2 c_A}{\partial z^2} \right) + R_A = 0$$

What happens to the terms $\partial^2 c_A / \partial y^2$ and $\partial^2 c_A / \partial z^2$ if h and l are much greater than w? Why? Make this simplification and substitute in the production rate for A for the final dimensional model. What are the boundary conditions for the model?

(b) Now consider writing the model in dimensionless form. What is the appropriate characteristic length, a? What is the characteristic concentration? Nondimensionalize the model and boundary conditions, and show the final dimensionless model is

$$\frac{d^2\overline{c}_A}{d\overline{x}^2} - \Phi^2\overline{c}_A = 0$$

and the boundary conditions are

$$\overline{c}_A = 1 \qquad \overline{x} = 1$$
$$\overline{c}_A = 1 \qquad \overline{x} = -1$$

What is the definition of the Thiele modulus for this problem?

(c) What is the solution of this model?

(d) Perform the integration to compute the effectiveness factor and show

$$\eta = \frac{\tanh\Phi}{\Phi}$$

(e) Plot $\overline{c}_A(\overline{x}), 0 \leq \overline{x} \leq 1$, for $\Phi = 1/2, 1, 2, 5, 10$. Plot $\overline{c}_A(\overline{r})$ for the sphere for the same Φ values. Comment on the differences and similarities.

(f) Plot $\eta(\Phi), 0.1 \leq \Phi \leq 10$ on a log-log scale for the sphere and slab in the same figure. Comment on the differences and similarities. What is the largest error that you would commit by using the sphere in place of the slab geometry and for what Φ value does it occur?

Exercise 7.2: Pellet profile of product and reactant

Consider the same reaction and slab geometry of Exercise 7.1,

$$\text{A} \xrightarrow{k} 2\text{B}$$

Now let's compute the product concentration profile in the pellet, c_B. The catalyst exterior is assumed to be at constant species B concentration, c_{Bs}, and species B has an effective diffusivity, D_B, which may be different from D_A.

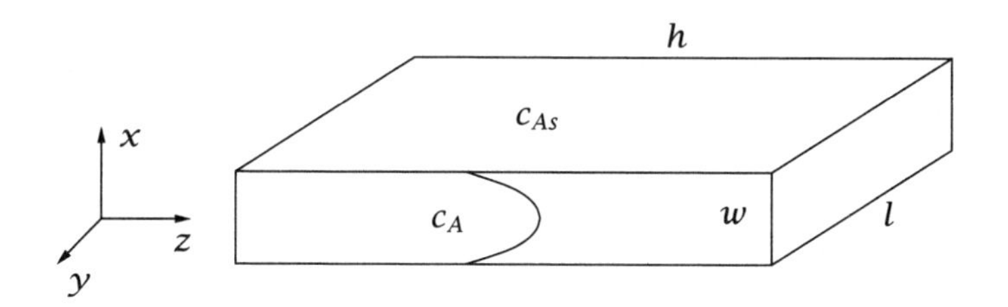

Figure 7.33: Catalyst pellet with slab geometry.

(a) The steady-state material balance for component B is

$$D_B \nabla^2 c_B + R_B = 0$$

For the given stoichiometry, $R_B = -2R_A$. Using Equation 7.92, show that

$$2\left(D_A \nabla^2 c_A\right) + \left(D_B \nabla^2 c_B\right) = 0$$

(b) Solve this equation by performing two integrals and show

$$c_B = c_{Bs} + 2\frac{D_A}{D_B}(c_{As} - c_A)$$

(c) Let's compare the A and B concentrations. Assume the diffusivities are equal and that $c_{As} = 1$ and $c_{Bs} = 0$. Plot $c_A(x)$ and $c_B(x)$ for $\Phi = 10$. Comment on the differences between c_A and c_B.

(d) Now assume $D_B = 10D_A$ and replot c_A and c_B. Explain what happens to the product B concentration with this larger value of D_B.

Exercise 7.3: Slab with finite external mass transfer

Consider the same reaction and slab geometry of Exercise 7.1

$$\mathrm{A} \xrightarrow{k} 2\mathrm{B}$$

but include the mass-transfer resistance.

(a) Write the dimensionless material balance and define the Thiele modulus and Biot number for this situation.

(b) Solve the model and find $\overline{c}_A(\overline{x})$. Plot this result for the same values of Biot number as in Figure 7.16 and compare the results.

(c) Compute the effectiveness factor versus Thiele modulus, plot this result and compare to Figure 7.17. Comment on the effect of pellet geometry on the overall reaction rate.

Exercise 7.4: Thiele modulus and a second-order reaction

The following second-order catalytic reaction is currently carried out in a fixed-bed reactor operating isothermally at 650 K.

$$\mathrm{A} \xrightarrow{k} \mathrm{B} \qquad k = 4.54 \times 10^7 \frac{\mathrm{cm}^3}{\mathrm{mol\ s}}$$

Pure A is fed to the reactor and a conversion of 93% is achieved at a pressure of 2.0 atm. The catalyst vendor has changed fabrication procedures and has verified that the intrinsic rate constant listed above is unchanged. Using the following data, determine if the mass of catalyst needs to be changed for the new catalyst to maintain the same production rate of B. The pressure drop in the tube is negligible for both catalysts.

Property	Old catalyst	New catalyst
effective diffusivity	0.0095 $\mathrm{cm^2/s}$	0.0072 $\mathrm{cm^2/s}$
particle density	1.75 $\mathrm{g/cm^3}$	1.79 $\mathrm{g/cm^3}$
bed density	0.84 $\mathrm{g/cm^3}$	0.96 $\mathrm{g/cm^3}$
shape	sphere (0.635 cm diam)	cylinder (0.48 cm diam, 0.79 cm length)

Exercise 7.5: Thiele modulus and Hougen-Watson kinetics

The following catalytic reaction is conducted in a 0.25-cm radius spherical pellet.

$$\mathrm{A} + \mathrm{B_2} \longrightarrow \mathrm{C}$$

The mechanism for this reaction is shown below. A, B_2 and C are in adsorption-desorption equilibrium with the surface. The bimolecular surface reaction is irreversible and rate limiting.

$$\begin{aligned}
\mathrm{A} + \quad \mathrm{S} &\underset{k_{-1}}{\overset{k_1}{\rightleftharpoons}} \mathrm{A_{ads}} \\
\mathrm{B_2} + \quad 2\mathrm{S} &\underset{k_{-2}}{\overset{k_2}{\rightleftharpoons}} 2\mathrm{B_{ads}} \\
\mathrm{A_{ads}} + \mathrm{B_{ads}} &\overset{k_3}{\longrightarrow} \mathrm{C_{ads}} + \mathrm{S} \\
\mathrm{C} + \quad \mathrm{S} &\underset{k_{-4}}{\overset{k_4}{\rightleftharpoons}} \mathrm{C_{ads}}
\end{aligned}$$

Additional data needed for this problem are provided in the following table.

Parameter	Value	Units
K_1	90,100	$\mathrm{cm^3/mol}$
K_2	6,500	$\mathrm{cm^3/mol}$
K_4	64,400	$\mathrm{cm^3/mol}$
T	523	K
D_{eA}	0.045	$\mathrm{cm^2/s}$
c_A	5.83×10^{-5}	$\mathrm{mol/cm^3}$
c_{B_2}	1.40×10^{-4}	$\mathrm{mol/cm^3}$
c_C	1.17×10^{-5}	$\mathrm{mol/cm^3}$
k_3	7.41×10^{8}	$\mathrm{g\ cat^2/mol{\cdot}cm^3{\cdot}s}$
c_m	1.8×10^{-5}	mol/g cat

(a) Determine if the concentration of A changes appreciably within the pellet from the value it has at the pellet surface.

(b) Estimate the concentration of A at the center of the pellet.

Exercise 7.6: Thiele modulus and first-order kinetics

A catalytic reaction that is first order in the concentration of A

$$\mathrm{A} \longrightarrow \mathrm{B}$$

is carried out in a spherical pellet with a 0.20-cm radius. The effective diffusivity of A in the pellet is 0.015 $\mathrm{cm^2/s}$. The rate of reaction at 398 K is 2.63×10^{-5} $\mathrm{mol/cm^3{\cdot}s}$ when the concentration of A is 3.25×10^{-5} $\mathrm{mol/cm^3}$. The intrinsic activation energy is 20 kcal/mol.

Assuming D_A is independent of temperature, what is the rate at 448 K for the same concentration of A?

Exercise 7.7: Thiele modulus and rate data

The second-order catalytic reaction

$$A \longrightarrow B$$

$$r = kc_A^2 \qquad \text{mol/s·g}$$

is carried out over a spherical catalyst pellet. The reactant and product are gases at the reaction conditions. A series of experiments was performed using 0.70-cm diameter catalyst beads. At 150°C and 1.0–3.0 atm of pure A, a plot of log rate versus log c_A was a straight line of slope 3/2. The following table lists additional information.

T(°C)	R_A (mol/s·g)	P_A (atm)
150	-6.17×10^{-4}	1.0
200	-2.37×10^{-3}	1.0

Estimate the value of W_c/N_{Af} required to achieve 50% conversion of A at 200°C in a fixed-bed reactor using the same catalyst but with a bead diameter of 1.0 cm. Pure A is fed to the reactor at a total pressure of 1.0 atm. Assume the diffusion process is dominated by Knudsen diffusion, and the diffusion coefficient is constant.

Exercise 7.8: First-order reaction in a fixed-bed reactor

The following first-order catalytic reaction is conducted in a fixed-bed reactor. The reactant and product are gases.

$$A \longrightarrow B$$

The intrinsic rate constant is $5.74 \times 10^{13} \exp(-38{,}000/RT)$ s^{-1}, in which the units of R are cal/mol K, and the units of T are K. The feed consists of pure A at 1.0 atm, 630 K, at molar flowrate 0.5 mol/s. The catalyst is a cylindrical-shaped pellet with a radius of 0.35 cm and a length of 0.5 cm. The catalyst pellet density $\rho_p = 0.84$ g/cm^3. The reactor bed density $\rho_B = 0.52$ g/cm^3. At 630 K the effective diffusivity in the pellet is 1.40×10^{-3} cm^2/s. Determine the mass of catalyst required to achieve 90% conversion of A if the reactor operates isothermally.

Exercise 7.9: A hollow-cylinder catalyst pellet in a fixed-bed reactor

Estimate the mass of catalyst required in an isothermal fixed-bed reactor for the second-order reaction.

$$A \xrightarrow{k_1} B$$

$$r = k_1 c_A^2 \qquad k_1 = 1.21 \times 10^5 \frac{\text{cm}^3}{\text{mol} \cdot \text{s}}$$

The feed is pure A at a molar flowrate of 15 mol/s, the total pressure is 3.0 atm and the temperature is 550 K. The reactor achieves 90% conversion of A. The catalyst is a hollow cylinder with an outer diameter of 1.3 cm, an inner diameter of 0.5 cm, and a length of 0.7 cm. The pellet density $\rho_p = 0.73$ g/cm^3 and the bed density $\rho_B = 0.58$ g/cm^3. The effective diffusivity of A is 0.008 cm^2/s and may be assumed constant. You may assume the bulk and pellet surface concentrations are equal.

Exercise 7.10: Catalyst size and production rate

The following second-order catalytic reaction is carried out in a fixed-bed reactor operating isothermally at 650 K.

$$A \xrightarrow{k} B \qquad k = 5.15 \times 10^5 \frac{\text{cm}^6}{\text{s}\cdot\text{g}\cdot\text{mol}}$$

Pure A is fed to the reactor with a nominal inlet pressure of 2.0 atm. The desired production rate per tube is 18.84 kg/hr of B. If you choose to operate above the nominal inlet pressure, an additional compressor has to be installed upstream of the reactor. The existing tubes are 100-cm long and 20-cm^2 cross-sectional area. The remaining A in the product stream must be separated in a second unit for recycle. For the separation unit to run efficiently, the conversion in the reactor must be at least 93% and the outlet pressure must be at least 1.5 atm. If the reactor effluent pressure is below 1.5 atm, an additional compressor must be installed downstream of the reactor.

The current catalyst has reached the end of its useful lifetime and needs to be replaced. Your project is to make a design change with the new catalyst to improve the reactor efficiency. In particular you are considering changing the catalyst size. The catalyst vendor has told you that they easily can make spherical pellets with diameters of 0.075 cm, 0.15 cm, and 0.30 cm. You also have been assured that the effective diffusivity, bed density, and reaction rate constant do not vary among the catalysts in this size range. One of your team members wants to use the smallest diameter catalyst to minimize the total mass of catalyst required. Another team member wants to use the largest diameter catalyst to minimize the pressure drop.

You have been assigned to model the reactor and decide which is the best choice. Some of the design constraints that you must satisfy include the production rate, the minimal final conversion, the maximal tube length (it is fine to use reactor length less than 100 cm, one simply leaves the end of the tube empty of catalyst), the maximal inlet pressure, and the minimal outlet pressure.

Can you meet these design constraints with all three of the catalysts? Once you have met the constraints, you can optimize the reactor operation over the remaining design decision variables. What is your final choice of catalyst size, and at what nominal inlet pressure will you run the reactor? Include a plot of the pressure, conversion, and Thiele modulus versus reactor length for your final design.

Please write a brief report discussing how the design variables (catalyst size, inlet pressure, bed length) affect both the capital and operating costs of this process. What additional information would you require to perform a better design?

Additional data for the catalyst and reaction are listed in the following table.

Parameter	Value	Units
effective diffusivity	0.00375	cm^2/s
particle density	1.75	g/cm^3
bed density	1.12	g/cm^3
molecular weight of A	100	kg/kmol
viscosity of gas	10^{-4}	g/cm·s

Exercise 7.11: Mass transfer with reaction

(a) Draw a sketch of the effectiveness factor versus the Thiele modulus for a spherical pellet. Be sure to label clearly your axes and put as many numbers on the plot as you can.

(b) Choose a value of $\Phi = 1$ and draw a sketch of $c_A(r)$ and $c_B(r)$ versus r for an irreversible, first-order reaction, $A \longrightarrow B$, in an isothermal pellet. Make this sketch as accurate as possible. If you are using dimensionless variables, be sure to indicate what quantities you have used to make the variables dimensionless.

(c) Comment on the differences and similarities between the effectiveness factor versus Thiele modulus plot for slabs and cylinders compared to the spherical case. How do you nondimensionalize your length for these other two geometries.

(d) Provide a concise explanation of the reactor design and operation trade-off involved in choosing between a catalyst pellet that gives a diffusion-controlled reactor versus one that gives a kinetically controlled reactor.

Exercise 7.12: Mass transfer with reaction for reversible kinetics

Consider the *reversible* first-order elementary reaction in an isothermal solid catalyst slab of thickness w surrounded by a fluid

$$A \underset{k_{-1}}{\overset{k_1}{\rightleftharpoons}} B$$

The external mass-transfer rate to the outer surface of the catalyst is very large and the fluid has A and B concentrations c_{Af} and c_{Bf}, respectively.

(a) Write down the steady-state material balance for species A and B considering the reaction and diffusion in the slab. Species A has effective diffusivity D_A and species B has effective diffusivity D_B, which may not be the same. What are the boundary conditions for these two material balances. Write down the reaction rate expression.

(b) Add the two material balances together and find an expression for c_B in terms of c_A, c_{Af}, c_{Bf}, and $D = D_A/D_B$.

(c) At equilibrium, the rate of reaction is zero. Find an expression for the equilibrium A concentration, c_{Ae}, in terms of D, c_{Af}, c_{Bf}, and $K = k_1/k_{-1}$.

(d) Define dimensionless concentration and distance via

$$\overline{c} = \frac{c_A - c_{Ae}}{c_{Af}}, \qquad \overline{x} = \frac{x}{w/2}$$

and show the material balance for A in dimensionless variables is

$$\frac{d^2\overline{c}}{d\overline{x}^2} - \Phi^2\overline{c} = 0$$

What is the Thiele modulus for this problem? How does it differ from the irreversible case? What are the dimensionless boundary conditions?

(e) Solve your model and show

$$\overline{c} = \left(1 - c_{Ae}/c_{Af}\right)\frac{\cosh \Phi\overline{x}}{\cosh \Phi}$$

Exercise 7.13: Reaction-limited and diffusion-limited fixed-bed reactors

A second-order catalytic reaction

$$A \longrightarrow B$$

is carried out in an isothermal fixed-bed reactor containing spherical catalyst pellets. The exit conversion is 85%. The reactant and product are gases. What happens to the required bed length to achieve 85% conversion if the feed pressure is doubled and the feed volumetric flowrate is held constant? You may assume that the flowrate is small enough that the pressure drop across the bed is unimportant. Consider the following two cases.

(a) What is the new required bed length if the overall pellet reaction rate is controlled by the intrinsic reaction rate?

(b) What is the new required bed length if the overall pellet reaction rate is controlled by the pellet diffusion rate?

You may assume the entire bed length is either in the intrinsic reaction rate or pellet diffusion rate-limited regimes and that the external mass-transfer resistance between the fluid and the solid catalyst is negligible.

Exercise 7.14: Second-order reaction in a fixed-bed reactor

The following second-order gas-phase catalytic reaction is conducted in an isothermal fixed-bed reactor:

$$A \longrightarrow B$$

The intrinsic rate constant is 15.0 L/mol·s at 800 K. The feed is 15,000 L/hr of pure A at 10.0 atm and 800 K. Spherical catalyst pellets of radius 1.2 cm are used to pack the reactor. The catalyst has a pellet density of 0.90 g/cm^3, and the reactor bed density is 0.52 g/cm^3. The effective diffusivity of A inside the catalyst pellet is 2.06×10^{-4} cm^2/s.

Calculate the mass of catalyst necessary to achieve 85% conversion of A.

Exercise 7.15: Using experimental data from a fixed-bed reactor

The irreversible catalytic reaction

$$A \longrightarrow B$$

is first order in the concentration of A. Use the data in the following table to determine the mass of catalyst required for 95% conversion of A in a fixed-bed reactor that can be modeled as a PFR. The experimental rate was measured in a 0.25-cm diameter spherical catalyst pellet at 1.0 atm of A and 510°C. The fixed-bed reactor you are asked to size is isothermal (510°C), constant pressure (1.0 atm), and the feed enters at $N_{Af} = 1.5$ mol/s.

Name	Symbol	Value	Units
experimental rate	—	1.04×10^{-4}	mol/(s·cm^3)
catalyst radius	R_p	0.125	cm
effective diffusivity of A	D_A	0.007	cm^2/s
catalyst pellet density	ρ_p	0.85	g/cm^3
catalyst bed density	ρ_B	0.55	g/cm^3

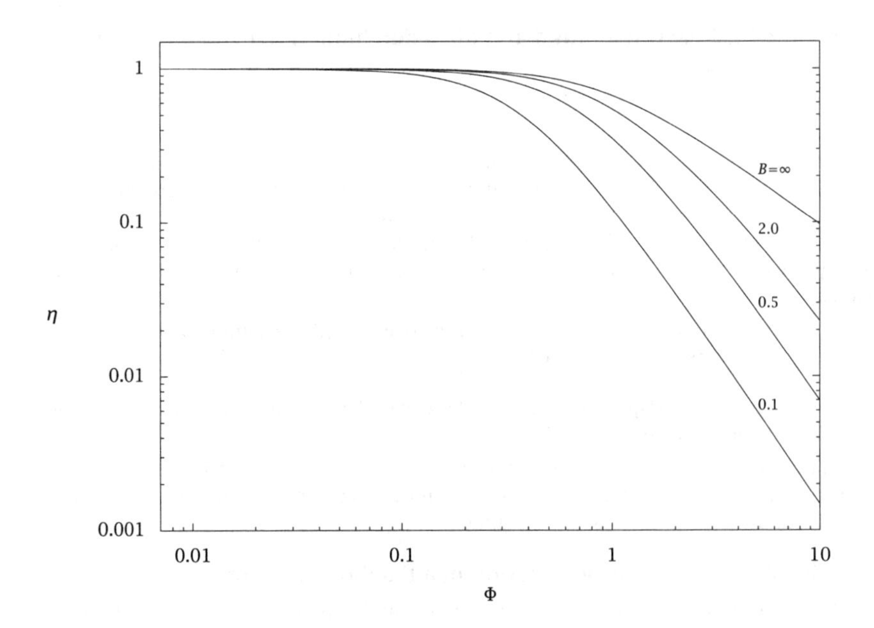

Figure 7.34: Effectiveness factor versus Thiele modulus for different values of the Biot number; second-order reaction in a cylindrical pellet.

Exercise 7.16: External mass-transfer effects on a second-order reaction

Consider the irreversible, second-order reaction

$$A \longrightarrow B, \qquad r = kc_A^2$$

taking place in a semi-infinite cylindrical pellet. Compute the effectiveness factor versus Thiele modulus for the following values of the Biot number, $B = \infty$, 2, 0.5, 0.1. Compare your calculation to Figure 7.34.

Exercise 7.17: Hougen-Watson kinetics

Consider the Hougen-Watson kinetics of Section 7.4.4. Using Equation 7.52, verify the following limits

$$\Phi = \tilde{\Phi}, \qquad \phi \text{ small} \tag{7.93}$$

$$\Phi = \frac{\tilde{\Phi}}{\sqrt{2\phi}}, \qquad \phi \text{ large} \tag{7.94}$$

Verify the result in Equation 7.94 using Figures 7.13 and 7.14, and $\phi = 1000$.

Exercise 7.18: General kinetics

Consider the large Φ reaction-diffusion equation, Equation 7.48, with general kinetic expression

$$\frac{d^2\overline{c}}{dz^2} - f(\overline{c}) = 0 \tag{7.95}$$

$$\overline{c} = 1 \qquad z = 0$$
$$\frac{d\overline{c}}{dz} = 0 \qquad z = \infty$$

in which $f(\overline{c})$ is the dimensionless reaction rate. Notice the Hougen-Watson and nth-order reaction kinetics are special cases of this more general form.

(a) Apply the change of variables of Section 7.4.4 and show the solution is [4, p. 338][18]

$$z(\overline{c}) = \int_{\overline{c}}^{1} \frac{d\overline{c}'}{\sqrt{2\int_0^{\overline{c}'} f(\overline{c}'')d\overline{c}''}} \tag{7.96}$$

(b) Verify Equation 7.95 is satisfied by $z(\overline{c})$ by taking two derivatives with respect to z of Equation 7.96.

Hint: recall the Leibniz rule for differentiating an integral in which the limits of integration are variable.

$$\frac{d}{dt}\int_{a(t)}^{b(t)} f(x,t)dx = -\frac{da}{dt}f(a,t) + \frac{db}{dt}f(b,t) + \int_a^b \frac{\partial f}{\partial t}dx$$

Exercise 7.19: Multiple-reaction, nonisothermal fixed-bed reactor

Consider again the catalytic converter problem of Examples 7.2 and 7.7. Farrauto and Bartholomew provide a table of exhaust gas velocities for various driving conditions [11]. The exhaust pipe gas velocities range from 0.54–25.2 m/s as the engine operation varies from idling to full load. We used 0.75 m/s in Example 7.7, which corresponds to a low engine load.

In order to get reasonable pressure drops at these higher engine loads, let's assume the catalyst bed porosity is $\epsilon_B = 0.4$, and the inlet pressure and temperature are $P_f = 1.5$ atm, $T_f = 570$ K. Assume the cross-section of the catalytic converter bed has an area four times as large as the exhaust pipe's. Therefore, the velocity in the entrance to the converter is 1/4 the velocity in the exhaust pipe.

(a) Calculate the fluid and pellet profiles in the bed for $u_f = 5$ m/s exhaust pipe velocity at the entrance to the catalytic converter. How much catalyst is required to reach 97.0% conversion of CO and C_3H_6 under the new engine operation? What is the pressure drop for this case?

(b) Using the bed size from the previous part, resolve the problem with the idling speed, $u_f = 0.75$ m/s exhaust pipe velocity at the entrance to the catalytic converter. What is the conversion CO and C_3H_6 at the end of the bed? What is the pressure drop across the bed?

Exercise 7.20: Logarithmic transformation

Derive the transformed pellet material balance and boundary conditions, Equations 7.89–7.91, given the original model, Equations 7.85–7.87, and the transformation

$$w_j = \ln(\widetilde{c}_j) \qquad \widetilde{c}_j = e^{w_j} \tag{7.97}$$

Exercise 7.21: Hougen-Watson kinetics with inerts in the feed

Let's see the effect of adding inert N_2 to the feed in Example 7.6. The following reaction converting CO to CO_2 takes place in a catalytic, fixed-bed reactor operating isothermally at 838 K and 1.0 atm

$$CO + \frac{1}{2}O_2 \longrightarrow CO_2$$

The following rate expression and parameters are adapted from a different model given by Oh et al. [16]. The rate expression is assumed to be of the Hougen-Watson form

$$r = \frac{k c_{CO} c_{O_2}}{1 + K c_{CO}} \qquad \text{mol/s cm}^3 \text{ pellet}$$

The constants are provided below

$$k = 1.3828 \times 10^{19} \exp(-13{,}500/T) \text{ cm}^3\text{/mol s}$$
$$K = 8.099 \times 10^{6} \exp(409/T) \text{ cm}^3\text{/mol}$$
$$D_{CO} = 0.0487 \text{ cm}^2\text{/s}$$

in which T is in Kelvin. The catalyst pellet radius is 0.1 cm.

In this case, the feed to the reactor consists of 2 mol% CO, 10 mol% O_2, zero CO_2 and 88 mol% inerts, with total volumetric flowrate $Q_f = 792$ L/s. Find the reactor volume required to achieve 95% conversion of the CO. Do you expect it to be larger or smaller than the value of $V_R = 233$ cm^3 found in Example 7.6? Why?

Bibliography

[1] R. Aris. A normalization for the Thiele modulus. *Ind. Eng. Chem. Fundam.*, 4:227, 1965.

[2] R. Aris. *The Mathematical Theory of Diffusion and Reaction in Permeable Catalysts. Volume I: The Theory of the Steady State.* Clarendon Press, Oxford, 1975.

[3] R. Aris. *The Mathematical Theory of Diffusion and Reaction in Permeable Catalysts. Volume II: Questions of Uniqueness, Stability, and Transient Behavior.* Clarendon Press, Oxford, 1975.

[4] R. B. Bird, W. E. Stewart, and E. N. Lightfoot. *Notes on Transport Phenomena.* John Wiley & Sons, New York, 1958.

[5] R. B. Bird, W. E. Stewart, and E. N. Lightfoot. *Transport Phenomena.* John Wiley & Sons, New York, second edition, 2002.

[6] K. B. Bischoff. Effectiveness factors for general reaction rate forms. *AIChE J.*, 11:351, 1965.

[7] S. Brunauer, P. H. Emmett, and E. Teller. Adsorption of gases in multimolecular layers. *J. Am. Chem. Soc.*, 60:309–319, 1938.

[8] J. B. Butt. *Reaction Kinetics and Reactor Design.* Prentice Hall, Englewood Cliffs, New Jersey, 1980.

[9] J. DeAcetis and G. Thodos. Flow of gases through spherical packings. *Ind. Eng. Chem.*, 52(10):1003–1006, 1960.

[10] S. Ergun. Fluid flow through packed columns. *Chem. Eng. Prog.*, 48(2): 89–94, 1952.

[11] R. J. Farrauto and C. H. Bartholomew. *Fundamentals of Industrial Catalytic Processes.* Blackie Academic and Professional, London, UK, 1997.

[12] B. C. Gates. *Catalytic Chemistry.* John Wiley & Sons, New York, 1992.

[13] B. E. Leach, editor. *Industrial Catalysis*, volume one. Academic Press, New York, 1983.

[14] B. E. Leach, editor. *Industrial Catalysis*, volume two. Academic Press, New York, 1983.

[15] B. E. Leach, editor. *Industrial Catalysis*, volume three. Academic Press, New York, 1984.

[16] S. H. Oh, J. C. Cavendish, and L. L. Hegedus. Mathematical modeling of catalytic converter lightoff: Single-pellet studies. *AIChE J.*, 26(6):935–943, 1980.

[17] C. J. Pereira, J. B. Wang, and A. Varma. A justification of the internal isothermal model for gas-solid catalytic reactions. *AIChE J.*, 25(6):1036–1043, 1979.

[18] E. E. Petersen. *Chemical Reaction Analysis*. Prentice Hall, Englewood Cliffs, New Jersey, 1965.

[19] E. E. Petersen. A general criterion for diffusion influenced chemical reactions in porous solids. *Chem. Eng. Sci.*, 20:587–591, 1965.

[20] W. M. Saltzman and R. Langer. Transport rates of proteins in porous materials with known microgeometry. *Biophys. J.*, 55:163–171, 1989.

[21] E. W. Thiele. Relation between catalytic activity and size of particle. *Ind. Eng. Chem.*, 31(7):916–920, 1939.

[22] P. B. Weisz. Zeolites - new horizons in catalysts. *Chem. Tech.*, 3:498, 1973.

[23] P. B. Weisz and J. S. Hicks. The behaviour of porous catalyst particles in view of internal mass and heat diffusion effects. *Chem. Eng. Sci.*, 17: 265–275, 1962.

8
Mixing in Chemical Reactors

8.1 Introduction

The three main reactor types developed thus far — batch, continuous-stirred-tank, and plug-flow reactors — are useful for modeling many complex chemical reactors, and to this point we have neglected a careful treatment of the fluid flow pattern within the reactor. In this chapter we explore some of the limits of this approach and develop methods to address and overcome some of the more obvious limitations.

Scope of problem. The general topic of mixing, even in the restricted context of chemical reactors, is an impossibly wide one to treat comprehensively. Obviously the phases of matter that are mixed play a large role and a comprehensive treatment would consider: liquid/liquid, liquid/solid, liquid/gas, and gas/solid mixing. Even in the contacting of two liquid phases, the miscibility of the two phases plays a large role in determining whether or not mixing occurs. We could also consider the mixing of different solid particles, in which particle size plays a large role. In this chapter, we will restrict ourselves to fluid-phase systems.

One natural approach to describing mixing is to solve the equations of motion of the fluid. In fluid systems, the type of fluid flow is obviously important, and we should consider both laminar and turbulent flow, and various mechanisms of diffusion (molecular diffusion, eddy diffusion). Using fluid mechanics to describe all cases of interest is a difficult problem, both from the modeling and computational perspectives. Rapid developments in computational fluid dynamics (CFD), however, make this approach increasingly attractive [2].

A second, classical approach to describing mixing is to use simple tests to experimentally probe the system of interest. These empirical testing approaches do not use any of the structure of the equations of motion, but they can provide some rough features of the mixing taking

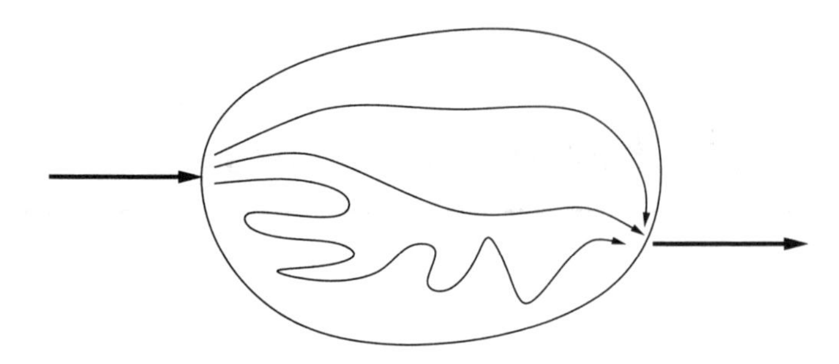

Figure 8.1: Arbitrary reactor with steady flow profile.

place in the system under study. In this chapter we first develop this classical approach, and find out what kinds of understanding it can provide. We also identify some of the limitations of this approach. Nauman and Buffham provide a more in-depth treatment of many of the classical topics covered in this chapter, and provide many further citations to the research literature [24].

Convection and diffusion. We should also mention one point to bear in mind when thinking about how convection and diffusion affect mixing. One might intuitively expect that to enhance mixing and reduce spatial variation in concentration, one should seek conditions that maximize the rate of diffusion. Although this notion is correct for mixing on the finest length scales, it is generally much more important in macroscopic scale processes to decrease variations on the larger length scales. Mixing in this regime is enhanced primarily by improving the convection, and diffusion plays only a small role. In simple terms, one does not expect to appreciably decrease the time required to mix the cream in one's coffee by increasing the coffee temperature (enhanced diffusion); one instead turns a spoon a few times (enhanced convection). On the finest length scales, mixing is accomplished readily for small molecules by the random process of molecular diffusion; in fact, the random molecular motions are the only effective mixing processes taking place on the finest length scales.

8.2 Residence-Time Distribution

8.2.1 Definition

Consider an arbitrary reactor with single feed and effluent streams depicted in Figure 8.1. Without solving for the entire flow field, which might be quite complex, we would like to characterize the flow pattern established in the reactor at steady state. The residence-time distribution (RTD) of the reactor is one such characterization or measure of the flow pattern. Imagine we could slip some inert tracer molecules into the feed stream and could query these molecules on their exit from the reactor as to how much time they had spent in the reactor. We assume that we can add a small enough amount of tracer in the feed so that we do not disturb the established flow pattern. Some of the tracer molecules might happen to move in a very direct path to the exit; some molecules might spend a long time in a poorly mixed zone before finally finding their way to the exit. Due to their random motions as well as convection with the established flow, which itself might be turbulent, we would start recording a distribution of residence times and we would create the residence-time probability density or residence-time distribution. If the reactor is at steady state, and after we had collected sufficient residence-time statistics, we expect the residence-time distribution to also settle down to a steady function. Let $p(\theta)$ represent the probability density or residence-time distribution, and $P(\theta)$ the cumulative residence-time distribution so

$p(\theta)d\theta$,	probability that a feed molecule spends time θ to $\theta + d\theta$ in the reactor
$P(\theta)$,	probability that a feed molecule spends time zero to θ in the reactor

The two versions of the probability function obviously contain the same information and are related by

$$P(\theta) = \int_0^\theta p(\theta')d\theta', \qquad p(\theta) = \frac{dP(\theta)}{d\theta}$$

Some prefer to use exclusively $P(\theta)$ because if $P(\theta)$ is not differentiable, then even defining $p(\theta)$ is somewhat problematic. We sidestep this issue by introducing the impulse or delta function to describe $p(\theta)$ in the case where $P(\theta)$ has jump discontinuities.

8.2.2 Measuring the RTD

As a thought experiment to define the RTD, querying tracer molecules on their exit from the reactor is a fine concept. But we plan to actually measure the RTD, so we require an implementable experiment with actual measurements. We cannot measure the time spent by a particular tracer molecule in the reactor; to us, all tracer molecules are identical. We can measure concentration of tracer molecules in the effluent, however, and that will prove sufficient to measure the RTD. Imagine an experiment in which we measure the concentration of tracer in the feed and effluent streams over some time period, while the reactor maintains a steady flow condition. From the definition of the RTD in the previous section, the effluent tracer concentration at some time t is established by the combined exit of many tracer molecules with many different residence times. The concentration of molecules that enter the reactor at time t' and spend time $t - t'$ in the reactor before exiting is given by $c_f(t')p(t - t')dt'$. These molecules are the ones leaving the reactor at time t that establish effluent concentration $c_e(t)$, so we have

$$c_e(t) = \int_{-\infty}^{t} c_f(t')p(t - t')dt' \tag{8.1}$$

The inlet and outlet concentrations are connected through this convolution integral with the residence-time distribution. If we conduct the experiment so that the feed tracer concentration is zero before an initial time $t = 0$, then the integral reduces to

$$c_e(t) = \int_{0}^{t} c_f(t')p(t - t')dt', \qquad c_f(t) = 0, t \leq 0 \tag{8.2}$$

Notice we can change the variable of integration in Equation 8.2 to establish an equivalent representation

$$c_e(t) = \int_{0}^{t} c_f(t - t')p(t')dt' \tag{8.3}$$

which is sometimes a convenient form. This connection between the inlet and outlet concentrations, and the RTD, allows us to determine the RTD by measuring only tracer concentrations. We next describe some of the convenient experiments to determine the RTD.

Step response. Although we can in principle use any feed concentration time function to determine the RTD, some choices are convenient for ease of data analysis. One of these is the step response. In the step-response experiment, at time zero we abruptly change the feed tracer concentration from steady value c_0 to steady value c_f. For convenience we assume $c_0 = 0$. Exercise 8.4 shows that we can easily remove this assumption using deviation variables. Because the feed concentration is constant at c_f after time zero, we can take it outside the integral in

Equation 8.3 and obtain

$$c_e(t) = c_f \int_0^t p(t')dt' = c_f P(t)$$

So for a step-response experiment, the effluent concentration versus time provides immediately the cumulative form of the residence-time distribution

$$P(\theta) = c_e(\theta)/c_f, \qquad \text{step response} \tag{8.4}$$

Pulse and impulse responses. An impulse response is an idealized experiment, but is a useful concept. As we will see it provides the RTD directly rather than the cumulative RTD. To motivate the impulse-response experiment, imagine we abruptly change the inlet tracer concentration from zero to a large value and return it to zero after a short time as sketched in Figure 8.2. Such a test is called a pulse test. The pulse test is no more difficult to implement than the step test; it is merely two step changes in feed concentration in rapid succession. In some ways it is a superior test to the step response, because by returning the tracer concentration to zero, we use less tracer in the experiment and we cause less disruption of the normal operation of the reactor.

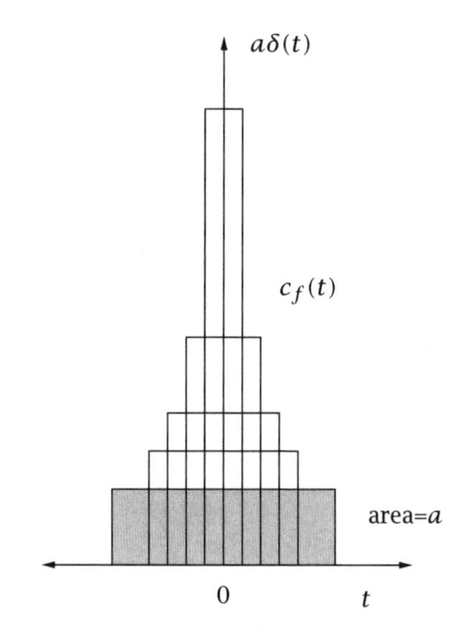

Figure 8.2: Family of pulses and the impulse or delta function.

The impulse response is an idealized limit of the pulse response. Consider a family of pulse tests as sketched in Figure 8.2 of shorter and shorter duration Δt. We maintain constant total tracer addition by spiking the feed with higher and higher concentrations so that the product $c_f \Delta t = a$ is constant. The impulse response is the limit of this experiment as $\Delta t \longrightarrow 0$. We call this limiting feed concentration versus time function the delta function, $a\delta(t)$. It is also called the Dirac delta function or an impulse, hence the name, impulse response. The constant a is the amplitude of the delta function. The mathematical problem with this approach is that the limit is not a well-defined function. The values at all points other than $t = 0$ go to zero as we take narrow enough pulses, and the value at $t = 0$ is unbounded. But mathematicians have taken a bold step and defined this new kind of function by what it *does* rather than its numerical values. The main property of the delta function is that, because it is so narrowly focused, it extracts the value of an integrand at a point in the interval of integration,

$$\int_{-\infty}^{\infty} g(t)\delta(t)dt = g(0), \qquad \text{all } g(t) \tag{8.5}$$

$$\int_{-\infty}^{\infty} \delta(t)dt = 1, \qquad \text{normalized} \tag{8.6}$$

So if we can approximate $c_f(t) = a\delta(t)$, then we have from Equation 8.1

$$c_e(t) = a\int_{-\infty}^{t} \delta(t')p(t-t')dt' = ap(t)$$

So for an experiment approximating an impulse, the effluent concentration versus time provides the residence-time distribution directly

$$p(\theta) = c_e(\theta)/a, \qquad \text{impulse response} \tag{8.7}$$

We do not require any sophisticated mathematics to use these generalized functions such as the delta function. We introduce them here because they make our work simpler, not more complicated. You will also find these functions useful in the process control course when you analyze the dynamics and feedback control of processes. Moreover, empirical tests, such as the step- and impulse-response experiments, provide a general method for modeling the dynamic behavior of many processes. Using these methods to determine a reactor RTD is just one small example of their uses in chemical process modeling.

Arbitrary feed concentration experiment. It may be inconvenient or impossible to perform a step or impulse test on a reactor. In some situations it is possible to measure a low-concentration impurity in the feed and effluent streams, and construct the RTD from those measurements. In these cases the feed concentration $c_f(t)$ is an arbitrary, but measured and known, function of time. One must take care that the noise in the measurements does not obscure the RTD, and replication of experiment and assessment of uncertainty in the RTD is highly recommended. Shinnar [28] provides a detailed discussion for further reading on the science and art of determining and using RTDs to solve industrial reactor problems.

8.2.3 Continuous-Stirred-Tank Reactor (CSTR)

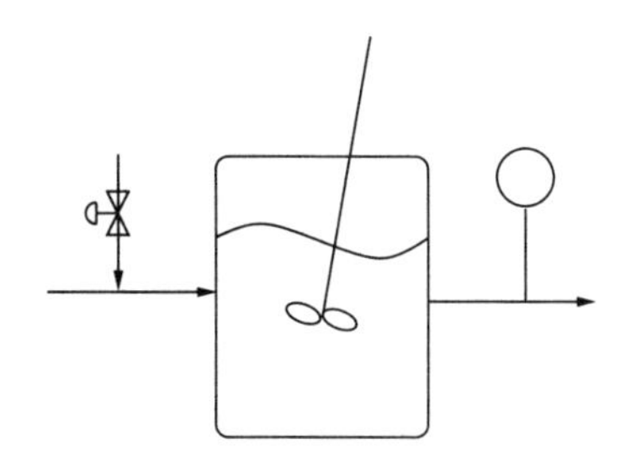

Figure 8.3: CSTR and tracer experiment.

We next examine again the well-stirred reactor introduced in Chapter 4. Consider the following step-response experiment: a clear fluid with flowrate Q_f enters a well-stirred reactor of volume V_R as depicted in Figure 8.3. At time zero we start adding a small flow of a tracer to the feed stream and measure the tracer concentration in the effluent stream. We assume we keep the flow of the tracer small enough that we do not disturb the existing flow pattern in the reactor. We expect to see a continuous change in the concentration of the effluent stream until, after a long time, it matches the concentration of the feed stream. We are by now experts on solving this type of problem, especially in this simple situation without chemical reaction to complicate matters. Assuming constant density, the differential equation governing the concentration of dye in the reactor follows from Equation 4.37

$$\frac{dc}{dt} = \frac{Q_f}{V_R}(c_f - c), \qquad c(0) = 0 \tag{8.8}$$

in which c is the concentration of the dye in the reactor and effluent stream. In Chapter 4, we named the parameter $\tau = V_R/Q_f$ the "mean residence time". We show subsequently that τ is indeed the mean of the RTD for a CSTR. We already derived the solution to Equation 8.8, where

$$c(t) = (1 - e^{-t/\tau})c_f$$

so we have immediately

$$P(\theta) = 1 - e^{-\theta/\tau}$$

which upon differentiation gives

$$p(\theta) = \frac{1}{\tau}e^{-\theta/\tau}, \qquad \text{CSTR residence-time distribution}$$

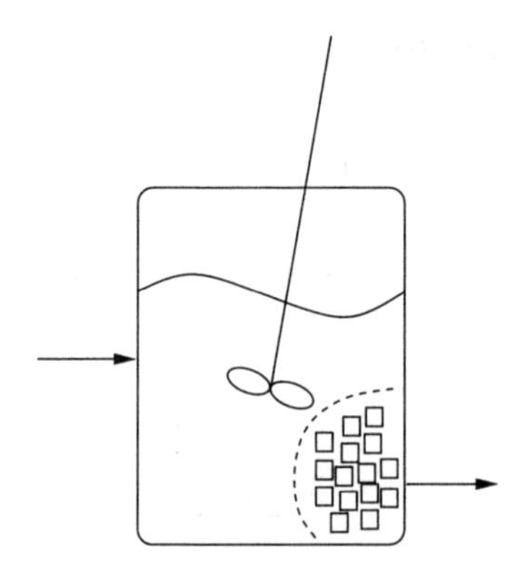

Figure 8.4: CSTR and volume elements in a game of chance.

It is often helpful to have two different views of the same result, so we next consider the well-mixed reactor as a game of chance and rederive the RTD for the CSTR. Imagine we divide the reactor into a number of volume elements, depicted in Figure 8.4. At each time instant, some group of volume elements is withdrawn from the reactor, and new volume elements enter from the feed stream. We assign a probability law to model which volume elements are withdrawn from the reactor. The well-mixed assumption is equivalent to the statement that all volume elements are equally likely to be withdrawn from the reactor. One can view the mixing as quickly (instantaneously in the case of perfect mixing) randomizing the locations of the volume elements so that each one is equally likely to be nearest to the exit when the next element is withdrawn in the effluent stream. We also see why perfect mixing is a physical idealization; it is not possible to instantaneously randomize the locations of the volume elements with finite power to the stirrer.

We have n volume elements, each with volume $V_n = V_R/n$. In time interval Δt, a volume of $Q_f \Delta t$ leaves with the effluent stream. The number of elements leaving in the time interval is therefore $Q_f \Delta t / V_n$ out of n, and the fraction of elements withdrawn is

$$f = \frac{Q_f \Delta t}{V_R} = \frac{\Delta t}{\tau}$$

Now consider the probability that a particular volume element is still in the reactor after i time intervals. The probability of removal is f, so the probability of survival is $1 - f$. After i trials, each of which is independent, the probability of survival is the product $(1 - f)^i$. Therefore the probability that an element has residence (survival) time $\theta = i\Delta t$ is

$$\tilde{p}(\theta) = \left(1 - \frac{\Delta t}{\tau}\right)^{\theta/\Delta t} \tag{8.9}$$

We now take the limit as $\Delta t \to 0$. Recall from calculus

$$\lim_{x \to 0} (1 + ax)^{1/x} = e^a$$

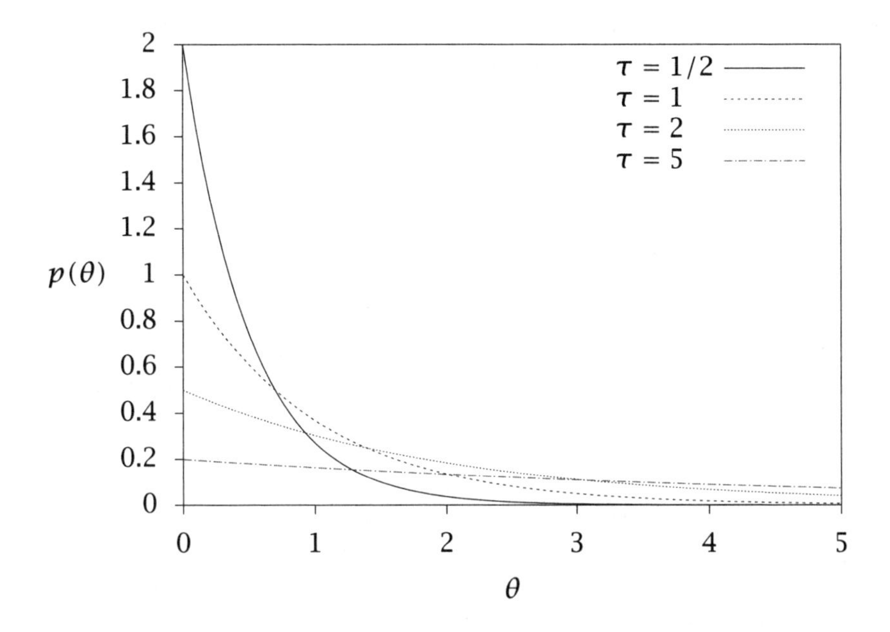

Figure 8.5: CSTR residence-time distribution.

Taking the limit and using this result in Equation 8.9 gives

$$\tilde{p}(\theta) = e^{-\theta/\tau}$$

which, after normalization, is again the residence-time distribution of the CSTR. The normalization constant is computed by integration

$$\int_0^\infty \tilde{p}(\theta)\, d\theta = \int_0^\infty e^{-\theta/\tau}\, d\theta = \tau$$

so the residence-time distribution is

$$p(\theta) = \frac{1}{\tau} e^{-\theta/\tau} \tag{8.10}$$

which is plotted in Figure 8.5 for a variety of mean residence times. We next compute the mean of this distribution. An integration by parts produces

$$\begin{aligned}
\overline{\theta} &= \int_0^\infty \theta\, p(\theta)\, d\theta = \frac{1}{\tau}\int_0^\infty \theta e^{-\theta/\tau}\, d\theta \\
&= \frac{1}{\tau}\left[-\tau\theta\, e^{-\theta/\tau} - (\tau)^2 e^{-\theta/\tau}\right]\Big|_0^\infty \\
&= \tau
\end{aligned}$$

and we have established that the mean of the RTD for the CSTR is indeed $\tau = V_R/Q_f$. We can therefore compactly write Equation 8.10 as

$$p(\theta) = \frac{1}{\overline{\theta}} e^{-\theta/\overline{\theta}} \tag{8.11}$$

Notice the exponential distribution tells us that it is unlikely for a volume element to remain in this reactor for long because at each instant there is a constant probability that the element is withdrawn in the effluent. It is difficult to win for long in this kind of game of chance, which explains the exponential decrease in $p(\theta)$. In fact, zero is the most likely residence time, a classic and slightly counter-intuitive case: the most likely value of θ, the $\max_\theta p(\theta)$ occurs at $\theta = 0$, which is different from the mean of the distribution, $\overline{\theta} = V_R/Q_f$. Again, if the mixing is not perfect, this maximum does not occur at $\theta = 0$ because it requires finite time for a volume element entering with the feed to be convected to the exit. Moreover, our derivation of the RTD of the CSTR shows the following general principle: given an event with constant probability of occurrence, the *time* until the next occurrence of the event is distributed as a decreasing exponential function. This principle is used, for example, to choose the time of the next reaction in the stochastic simulation of chemical kinetics in Chapter 4.

8.2.4 Plug-Flow Reactor (PFR) and Batch Reactor

The simple flow pattern in the PFR produces a simple residence-time distribution. Consider a step test in which the reactor is initially free of tracer and we increase the feed tracer concentration from zero to c_f at time zero. As shown in Figure 8.6, the tracer travels in a front that first reaches the reactor exit at time $t = l/v$, in which v is the velocity of the axial flow and l is the reactor length. From these physical considerations, we can write the reactor tracer concentration immediately

$$c(t,z) = \begin{cases} 0, & z - vt > 0 \\ c_f, & z - vt < 0 \end{cases} \tag{8.12}$$

For $z - vt > 0$, the tracer front has not reached location z at time t so the tracer concentration is zero. For $z - vt < 0$, the front has passed location z at time t and the tracer concentration is equal to the feed value c_f.

It is convenient to introduce the unit step or Heaviside function to

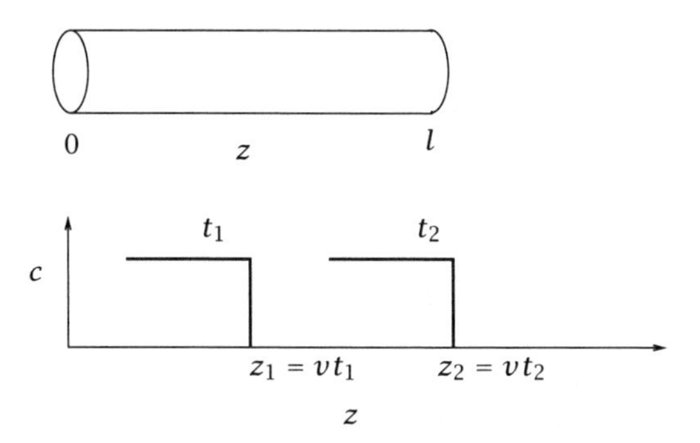

Figure 8.6: Plug-flow reactor with moving front of tracer.

summarize this result. The Heaviside function is defined as follows

$$H(t) = \begin{cases} 0, & t < 0 \\ 1, & t > 0 \end{cases} \tag{8.13}$$

Because we are armed with the delta function, we can even consider differentiating this discontinuous function to obtain the relationship

$$\frac{dH(t)}{dt} = \delta(t) \tag{8.14}$$

Equation 8.12 can then be summarized compactly by

$$c(t,z) = c_f H(t - z/v)$$

so the effluent tracer concentration is given by

$$c_e(t) = c_f H(t - l/v) = c_f H(t - V_R/Q_f)$$

and the integrated from of the residence-time distribution is therefore

$$P(\theta) = H(\theta - V_R/Q_f)$$

We can differentiate this result using Equation 8.14 to obtain

$$p(\theta) = \delta(\theta - V_R/Q_f)$$

In other words, all tracer molecules spend exactly the same time V_R/Q_f in the reactor. The mean of this distribution is then also V_R/Q_f, which is verified by using Equation 8.5

$$\overline{\theta} = \int_0^\infty \theta\delta(\theta - V_R/Q_f)d\theta = V_R/Q_f$$

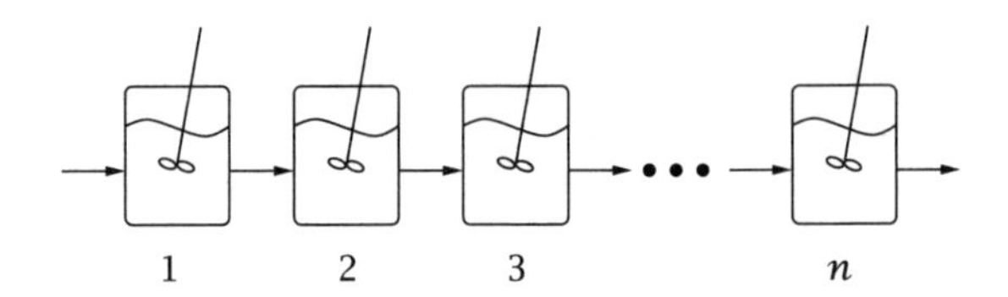

Figure 8.7: Series of n equal-sized CSTRs with residence time τ/n.

which shows that V_R/Q_f is the mean residence time for the PFR as well as the CSTR, even though the residence-time distributions of these two reactors are quite different. We can compactly summarize the RTD for the plug-flow reactor by

$$\boxed{\begin{array}{ll} p(\theta) = \delta(\theta - \overline{\theta}) & \text{PFR and} \\ P(\theta) = H(\theta - \overline{\theta}) & \text{batch reactors} \end{array}} \tag{8.15}$$

Likewise, the RTD for a batch reactor is immediate. All material is charged to the reactor at time $t = 0$ and remains in the reactor until the final batch time, which we may call $\overline{\theta}$. Then Equations 8.15 also apply to the batch reactor.

8.2.5 CSTRs in Series

Consider dividing the volume of a single CSTR into n equal-sized CSTRs in series as shown in Figure 8.7. If the single CSTR has volume V_R and residence time $\tau = V_R/Q_f$, each of the CSTRs in series has volume V_R/n and residence time τ/n. If we solve for the effluent concentration after an impulse at time zero, we can show

$$p(\theta) = \left(\frac{n}{\tau}\right)^n \frac{\theta^{n-1}}{(n-1)!} e^{-n\theta/\tau} \tag{8.16}$$

which is plotted in Figure 8.8 for $\tau = 2$. Notice the residence-time distribution becomes more narrowly focused at τ as n increases. In fact Equation 8.16 for large n is another approximation for $\delta(\theta - \tau)$. If we integrate Equation 8.16 we obtain

$$P(\theta) = \frac{\gamma(n, n\theta/\tau)}{\Gamma(n)} \tag{8.17}$$

in which the gamma function is defined as

$$\Gamma(n) = \int_0^\infty t^{(n-1)} e^{-t} dt \tag{8.18}$$

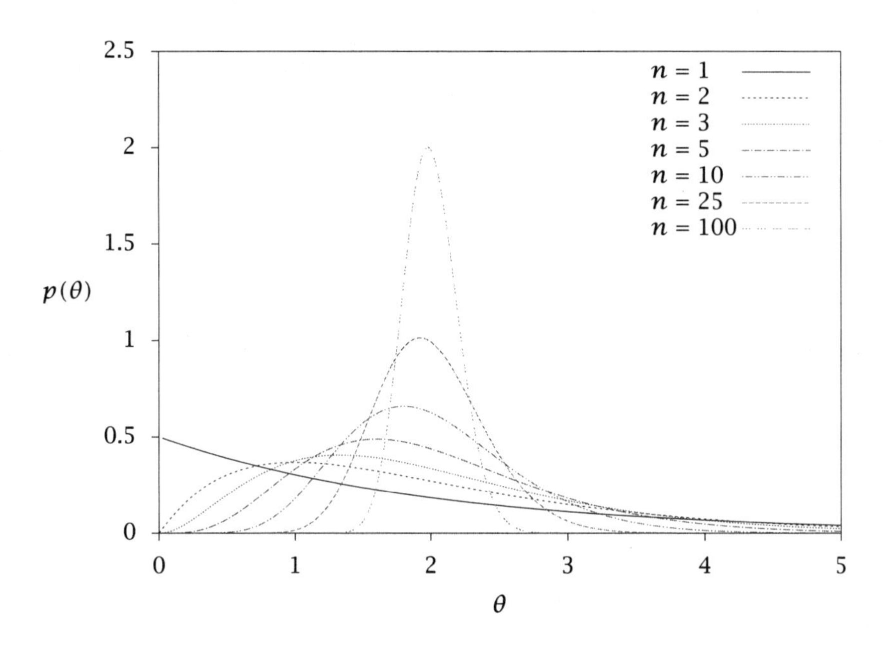

Figure 8.8: Residence-time distribution $p(\theta)$ versus θ for n CSTRs in series, $\tau = 2$.

Notice the gamma function is a generalization of the familiar factorial and for n an integer,

$$\Gamma(n) = (n-1)!$$

The incomplete gamma function is defined by the same integrand on the finite interval

$$\gamma(n, x) = \int_0^x t^{(n-1)} e^{-t} dt \tag{8.19}$$

You can perform integration by parts on Equation 8.16 to obtain a series representation

$$P(\theta) = 1 - \left(1 + \frac{n\theta/\tau}{1!} + \frac{(n\theta/\tau)^2}{2!} + \cdots + \frac{(n\theta/\tau)^{n-1}}{(n-1)!}\right) e^{-n\theta/\tau} \tag{8.20}$$

Evaluating the incomplete gamma function is numerically better than summing this series when n is large. Figure 8.9 shows Equation 8.17 for a range of n values.

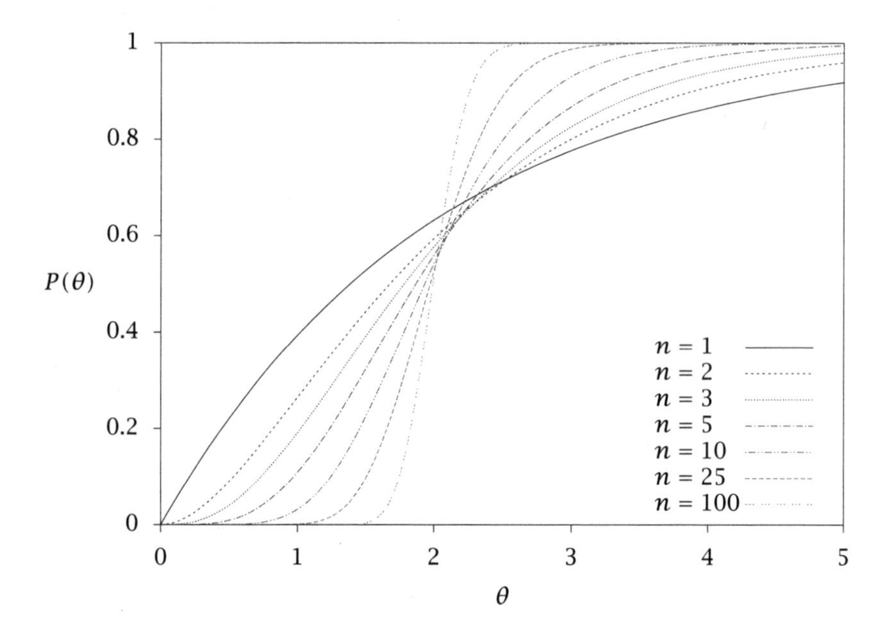

Figure 8.9: $P(\theta)$ versus θ for n CSTRs in series, $\tau = 2$.

8.2.6 Dispersed Plug Flow

Ideal plug flow may not be realized in many packed-bed reactors. We develop next a model that allows for deviations from plug flow. As shown in Figure 8.6, in the PFR a change in the feed concentration travels as a perfect front. But measurements in actual packed-bed reactors may show a fair amount of spreading or dispersion as this front travels through the reactor. The dispersed plug-flow model given in Equation 8.21 allows for this spreading phenomenon. A diffusion-like term has been added to the mass balance. The coefficient, D_l, is called the dispersion coefficient. Note that this term is not meant to represent true molecular diffusion caused by the random motion of the molecules. Molecular diffusion *in the direction of the convected flow* is always negligible in packed bed reactors operating at reasonable flowrates. The spreading of concentration fronts is caused by other phenomena, such as Taylor dispersion [5, p.643].

$$\underbrace{\frac{\partial c}{\partial t}}_{\text{accumulation}} = \underbrace{-v\frac{\partial c}{\partial z}}_{\text{convection}} + \underbrace{D_l\frac{\partial^2 c}{\partial z^2}}_{\text{diffusion}} \tag{8.21}$$

Given diffusion in the tube, the inlet boundary condition is no longer just the feed condition, $c(0) = c_f$, that we used in the PFR. To derive the boundary condition at the inlet, we write a material balance over a small region containing the entry point, and consider diffusion and convection terms. The diffusion term introduces a second-order derivative in Equation 8.21, so we now require two boundary conditions. We specify a zero slope condition at the tube exit.

$$
\begin{aligned}
vc|_{0^-} &= vc|_{0^+} - D_l \left.\frac{\partial c}{\partial z}\right|_{0^+} & z &= 0 \\
\frac{\partial c}{\partial z} &= 0 & z &= l
\end{aligned}
\tag{8.22}
$$

These two boundary conditions have become known as Danckwerts boundary conditions [11], but they were derived at least 45 years prior to Danckwerts in a classic paper by Langmuir [22]. Further discussion of the applicability and some alternatives to the Danckwerts boundary conditions are given by Bischoff [6], Levenspiel [23, p.272], and Parulekar and Ramkrishna [27].

It is now convenient to introduce a coordinate moving with a scaled velocity

$$
x = \frac{z - vt}{\sqrt{4D_l t}} = \frac{\overline{z} - t/\tau}{\sqrt{4Dt/\tau}}
$$

in which

$$
D = \frac{D_l}{vl} = \frac{D_l \tau}{l^2}, \qquad \text{dimensionless dispersion number}
$$

Transforming variables from z and t in Equation 8.21 to x gives

$$
\frac{d^2 c}{dx^2} + 2x \frac{dc}{dx} = 0 \tag{8.23}
$$

Rather than use the Danckwerts boundary conditions, we can approximate the behavior with the following simplified boundary conditions

$$
\begin{aligned}
c &= 1, & x &= -\infty \\
c &= 0, & x &= \infty
\end{aligned}
\tag{8.24}
$$

These boundary conditions correspond to stating that $c = 0$ for z, t values corresponding to arbitrarily long times before the step change arrives, and that $c = 1$, for z, t values corresponding to arbitrarily long times after the step change has passed. See Exercise 8.10 for comparing the dispersed PFR RTD with Danckwerts boundary conditions to the one we calculate here with the simplified boundary conditions.

The solution to Equation 8.23 with boundary conditions listed in Equations 8.24 is

$$c(x) = 1/2\left[1 - \frac{2}{\sqrt{\pi}}\int_0^x e^{-t^2}dt\right]$$

The integral can be expressed in terms of the error function, which is defined as

$$\operatorname{erf}(x) = \frac{2}{\sqrt{\pi}}\int_0^x e^{-t^2}dt \tag{8.25}$$

Substituting in the original variables and setting $\overline{z} = 1$ to obtain the response at the reactor outlet as a function of time gives

$$c(t, \overline{z} = 1) = 1/2\left[1 - \operatorname{erf}\left(\frac{1 - t/\tau}{\sqrt{4Dt/\tau}}\right)\right]$$

and we have calculated the cumulative RTD for the dispersed PFR with simplified boundary conditions

$$P(\theta) = 1/2\left[1 - \operatorname{erf}\left(\frac{1 - \theta/\tau}{\sqrt{4D\theta/\tau}}\right)\right] \tag{8.26}$$

Equation 8.26 is plotted in Figure 8.10 for $\tau = 2$ and various dispersion numbers D. We can differentiate Equation 8.26 to obtain the dispersed plug-flow RTD

$$p(\theta) = \frac{1}{4\tau\sqrt{\pi D}}\frac{1 + \tau/\theta}{\sqrt{\theta/\tau}}\exp\left(-\left(\frac{1 - \theta/\tau}{\sqrt{4D\theta/\tau}}\right)^2\right) \tag{8.27}$$

This RTD is plotted in Figure 8.11.

The dispersion number, D, is related to another dimensionless group, the mass-transfer analog of the inverse of the Péclet number,

$$\text{Pe} = \frac{vl}{D_A}, \qquad \frac{1}{\text{Pe}} = \frac{D_A}{vl}$$

which measures the rate of diffusion compared to the rate of convection. The key difference is the Péclet number contains the molecular diffusivity, D_A, and the dispersion number contains the effective axial dispersion coefficient, D_l. Levenspiel makes a compelling case that these two quantities have different origins and motivations and deserve different names. To further complicate matters, the inverse of

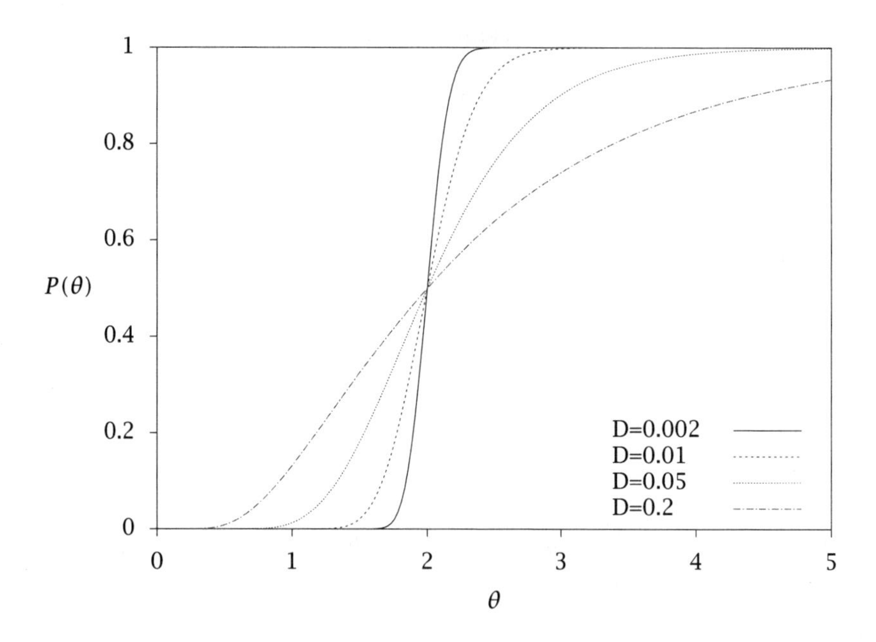

Figure 8.10: $P(\theta)$ versus θ for plug flow with dispersion number D, $\tau = 2$.

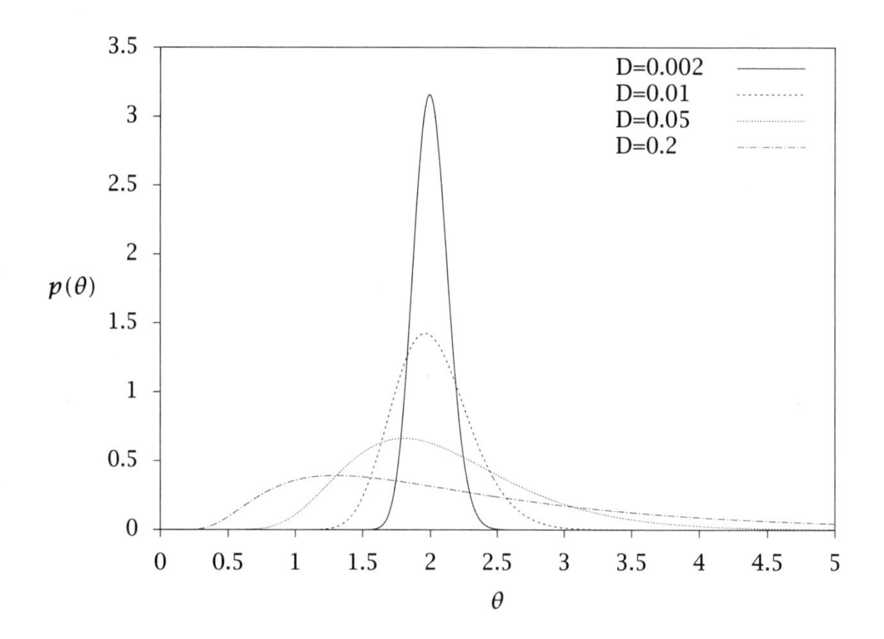

Figure 8.11: Residence-time distribution $p(\theta)$ versus θ for plug flow with dispersion number D, $\tau = 2$.

Parameter	Value	Units
k	0.5	L/mol·min
c_{Af}	1	mol/L
D_{Al}	0.01	m^2/min
v	0.5	m/min
l	1	m

Table 8.1: Mass-transfer and kinetic parameters for Example 8.1.

the Péclet number is often called the Bodenstein number in European literature. Weller [32] provides an interesting discussion of the history of the literature on the Bodenstein number, which does not appear to have been defined or used by Bodenstein, but was defined and used by Langmuir [22].

Dispersed plug flow with reaction. We modify Equation 8.21 for dispersed plug flow to account for chemical reaction,

$$\frac{\partial c_j}{\partial t} = -v\frac{\partial c_j}{\partial z} + D_{jl}\frac{\partial^2 c_j}{\partial z^2} + R_j \tag{8.28}$$

Danckwerts boundary conditions, as given in Equations 8.22, can be applied without change.

Up to this point in the text, we have solved *exclusively steady-state* profiles in tubular reactors. Obviously tubular reactors experience a start-up transient like every other reactor, and this time-dependent behavior is also important and interesting. Calculating the transient tubular-reactor behavior involves solving the partial differential equation (PDE), Equation 8.28, rather than the usual ODE for the steady-state profile. Appendix A describes the method we use for this purpose, which is called orthogonal collocation.

Example 8.1: Transient start-up of a PFR

Compute the transient behavior of the dispersed plug-flow reactor for the isothermal, liquid-phase, second-order reaction

$$2\text{A} \longrightarrow \text{B}, \qquad r = kc_A^2$$

The reactor is initially filled with solvent. The kinetic and reactor parameters are given in Table 8.1.

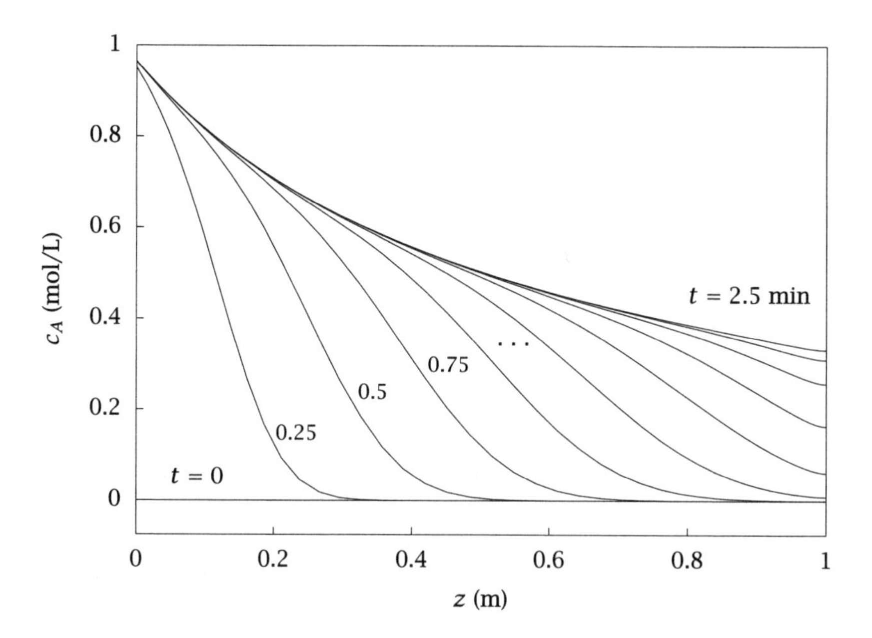

Figure 8.12: Start-up of the tubular reactor; $c_A(t, z)$ versus z for various times, $0 \le t \le 2.5$ min, $\Delta t = 0.25$ min.

Solution

The mass balance for component A is

$$\frac{\partial c_A}{\partial t} = -v \frac{\partial c_A}{\partial z} + D_{Al} \frac{\partial^2 c_A}{\partial z^2} - 2kc_A^2$$

The boundary conditions are

$$v c_{Af} = v c_A|_{0^+} - D_{Al} \left. \frac{\partial c_A}{\partial z} \right|_{0^+}, \qquad z = 0$$

$$\frac{\partial c_A}{\partial z} = 0, \qquad z = l$$

Finally, an initial condition is required

$$c_A(t, z) = 0, \qquad t = 0$$

Figure 8.12 shows the transient profiles. We see the reactor initially has zero A concentration. The feed enters the reactor and the A concentration at the inlet rises rapidly. Component A is transported by

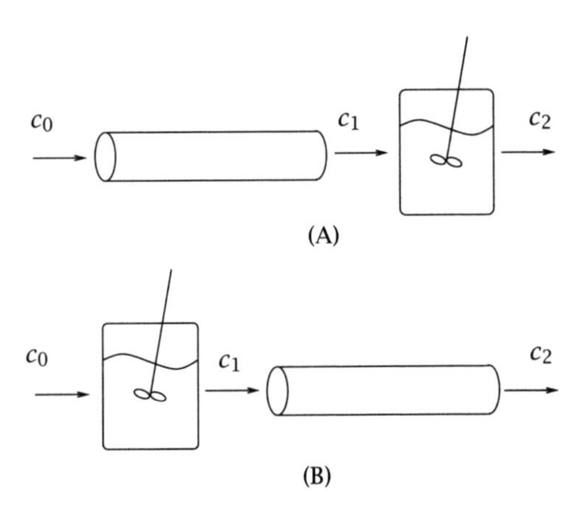

Figure 8.13: PFR followed by CSTR (A), and CSTR followed by PFR (B).

convection and diffusion down the reactor, and the reaction consumes the A as it goes. After about $t = 2.5$ min, the concentration profile has reached its steady value. Given the low value of dispersion in this problem, the steady-state profile is close to the steady-state PFR profile for this problem. □

Example 8.2: Order matters

Consider two arrangements of a PFR and CSTR of equal volume in series as shown in Figure 8.13. What are the residence-time distributions for the two reactor systems? What are the overall conversions for the two systems for the reaction

$$\mathrm{A} \longrightarrow \mathrm{B} \qquad r = kc_A^2$$

with second-order, irreversible kinetics?

Solution

Consider a unit step test for the CSTR–PFR arrangement. The outlet concentration for the CSTR is given by

$$c_1(t) = 1 - \exp(-t/\tau)$$

That feed concentration to the PFR is then simply delayed by τ time units to give for the CSTR–PFR arrangement.

$$P(\theta) = (1 - \exp(-(\theta - \tau)/\tau))\, H(\theta - \tau)$$

Next consider a unit step test into the PFR–CSTR arrangement. For this case the intermediate stream is given by a delayed step input

$$c_1(t) = H(t - \tau)$$

With this feed into the CSTR, the effluent is merely the CSTR response to a unit step change after we shift the starting time of the step forward τ time units,

$$c_2(t) = (1 - \exp(-(t - \tau)/\tau))\, H(t - \tau)$$

so again for this case

$$P(\theta) = (1 - \exp(-(\theta - \tau)/\tau))\, H(\theta - \tau)$$

and the two residence-time distributions are equal.

The steady-state conversions for both arrangements are also simply calculated. For a single CSTR, the steady-state inlet and outlet concentrations are related by

$$c_o/c_i = \frac{-1 + \sqrt{1 + 4k\tau c_i}}{2k\tau c_i} \equiv C(c_i) \tag{8.29}$$

For a single PFR, the inlet and outlet concentrations are related by

$$c_o/c_i = \frac{1}{1 + k\tau c_i} \equiv \mathcal{P}(c_i) \tag{8.30}$$

So we wish to compare $\mathcal{P}(C(c_0))$ for the CSTR–PFR case and $C(\mathcal{P}(c_0))$ for PFR–CSTR case. Because we are not even told $k\tau c_0$, we check over a range of values. Figure 8.14 displays the result. We see that the conversions are *not* the same and that *the PFR–CSTR gives higher conversion* (lower outlet concentration) than the CSTR–PFR for all values of $k\tau c_0$ for a second-order reaction. □

8.3 Limits of Reactor Mixing

We have seen in the previous section that complete knowledge of the reactor residence-time distribution is insufficient to predict the reactor performance. Although we have characterized completely the time tracer molecules spend in the reactor, we have not characterized their

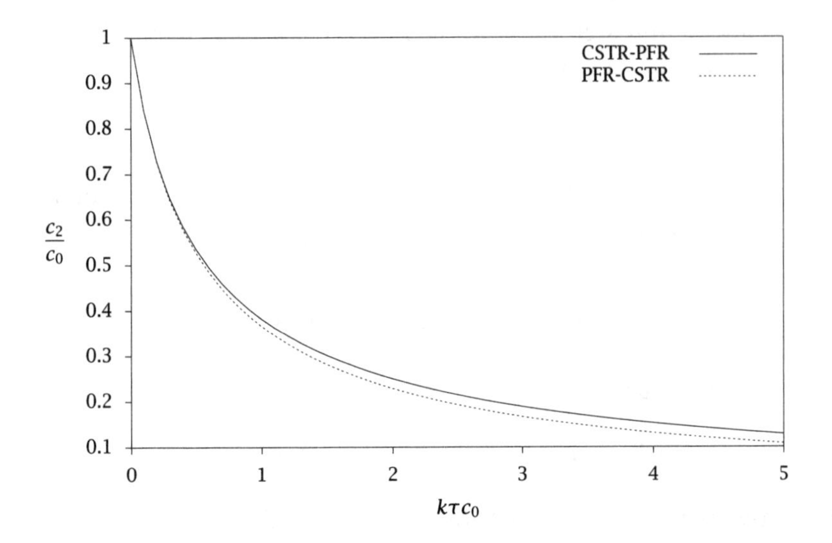

Figure 8.14: Comparison of the effluent concentrations for the two cases shown in Figure 8.13.

surrounding environment in the reactor during this time.[1] In the literature these two effects are sometimes termed **macromixing:** the distribution of residence times of molecules moving with the flow; and **micromixing:** the exchange of material between different volume elements during their residence times. Although we may find it instructive to separate these two phenomena in the simple reactor mixing models under discussion, in actual reactors this separation may be unrealistic. Accurate prediction of reactor performance may require solution or approximate solution of the equations of motion for the fluid, including material transport due to diffusion.

In defense of the simple mixing models, however, they do provide another important insight. We can determine the limits of achievable mixing consistent with a measured reactor residence-time distribution. These mixing limits do provide some insight into the limits of achiev-

[1] If someone were to characterize your learning in this course by measuring your hours spent in the classroom (RTD), they would hopefully obtain a positive correlation between learning and residence time. But we would naturally want to evaluate the environment inside the classroom during these hours if we were going to make more accurate predictions of learning. We would want to know if the instructor was prepared for lecture and saying reasonable things, if the students were attentive or asleep, and so on.

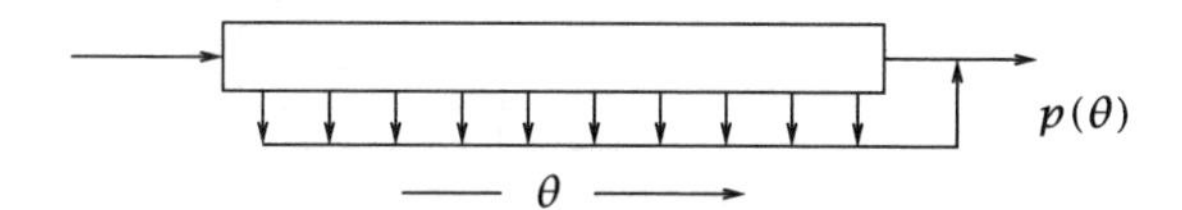

Figure 8.15: Completely segregated flow as a plug-flow reactor with side exits; outlet flows adjusted to achieve given RTD.

able reactor performance, although this connection remains an active area of research as discussed in section 8.4.

8.3.1 Complete Segregation

Imagine we know a reactor RTD, $p(\theta)$, either through direct measurement in a tracer experiment or solution of the equations of motion or some other means. We know from Example 8.2 that this constraint does not tell us the complete state of mixing in the reactor. We define next the two extreme limits of mixing consistent with the given RTD. These ideas were laid out in influential papers by Danckwerts and Zwietering. The first limit is called complete segregation; it is the limit of no mixing between volume elements. We can realize this limit by considering the ideal reactor depicted in Figure 8.15. As we progress down the plug-flow reactor, the residence time θ of the material reaching that location increases. We can imagine withdrawing from the reactor at each location or θ a fraction of the flow corresponding to the required RTD value $p(\theta)$, although this might be difficult to achieve in practice. A PFR with this removal rate then has the specified RTD. No material in two volume elements with different residence times is ever exchanged because the plug flow has zero backmixing. This last point is perhaps more clear if we redraw the reactor configuration as an equivalent bank of PFRs of different lengths without side exits, as in Figure 8.16.A [31]. Each tube has a single θ value according to its length. We feed the fraction $p(\theta)$ of the total flow into each tube of residence time θ so as to achieve the given RTD for the composite reactor system. This reactor system is called completely segregated because there is no exchange of material between the various tubes. Each tube acts as its own private reactor that processes material for a given amount of time and then discharges it to be mixed with the other reactors at the exit.

It is a simple matter to predict the behavior of this completely segregated reactor. We assume a single reaction and constant density throughout the following discussion. Each tube of specified length or

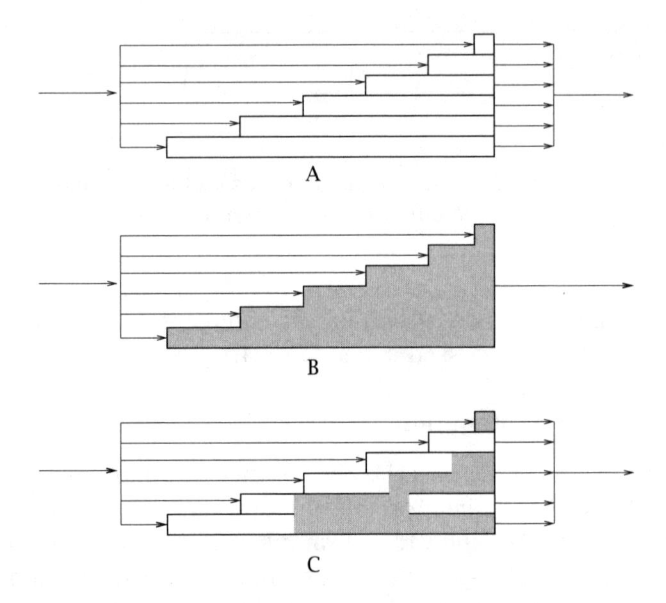

Figure 8.16: Alternate representation of completely segregated flow (A), maximum mixed flow (B), and an intermediate mixing pattern (C).

volume V can be assigned a residence time according to $\theta = V/Q$. Let $c(\theta)$ represent the concentration of a reactant in a volume element that has spent time θ in the reactor. Because the reactor is segregated, each tube satisfies the constant density PFR equation, Equation 4.103,

$$\frac{dc}{d\theta} = R(c), \qquad c(0) = c_f \tag{8.31}$$

The concentration of the effluent stream is then simply computed by multiplying the concentration of each tube by the fraction of the total feed passing through that tube

$$c_s = \int_0^\infty p(\theta)c(\theta)d\theta \tag{8.32}$$

in which $c(\theta)$ comes from the solution of Equations 8.31. It is often convenient to remove the explicit integration required by Equation 8.32. Let $c_s(\theta)$ represent the effect of combining streams with residence times less than or equal to θ, so

$$c_s(\theta) = \int_0^\theta p(\theta)c(\theta)d\theta$$

From this definition it is clear that $c_s(\theta)$ satisfies the following differential equation and initial condition

$$\frac{dc_s}{d\theta} = p(\theta)c(\theta), \qquad c_s(0) = 0$$

and the reactor effluent concentration is the limit of $c_s(\theta)$ as $\theta \longrightarrow \infty$. We can combine the two differential equations for convenient numerical solution of the segregated case

$$\boxed{\begin{aligned} \frac{dc}{d\theta} &= R(c) & c(0) &= c_f \\ \frac{dc_s}{d\theta} &= p(\theta)c(\theta) & c_s(0) &= 0 \end{aligned}} \tag{8.33}$$

Notice that this is an initial-value problem, but, in general, we require the solution at $\theta = \infty$ to determine the effluent concentration of the reactor. Differential equations on semi-infinite domains are termed singular, and require some care in their numerical treatment as we discuss next. On the other hand, if the residence-time distribution is zero beyond some maximum residence time, $\theta_{\max}$, then it is straightforward to integrate the initial-value problem on $0 \leq \theta \leq \theta_{\max}$.

Numerical solution. We can solve Equation 8.33 as an initial-value problem as written with an ODE solver. Because of the semi-infinite domain, we would need to check the solution for a sequence of increasingly large θ values and terminate the ODE solver when the value of $c_s(\theta)$ stops changing. Alternatively, we can map the semi-infinite domain onto a finite domain and let the ODE solver do the work for us. Many transformations are possible, such as $z = \exp(-\theta)$, but experience suggests a strongly decreasing function like the exponential causes the right-hand side to go to infinity at $z = 1$, and we simply exchange one singularity for another. A more gentle transformation and its inverse are

$$z = \frac{\theta}{1+\theta}, \qquad \theta = \frac{z}{1-z}$$

Using this change of variable, we rewrite the derivative as

$$\frac{dc}{d\theta} = \frac{dc}{dz}\frac{dz}{d\theta} = (1-z)^2\frac{dc}{dz}$$

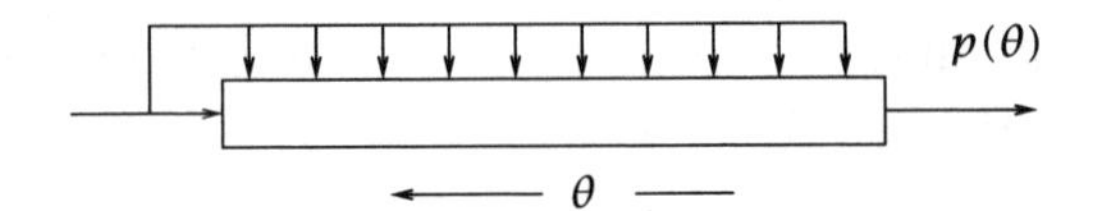

Figure 8.17: Maximum mixed flow as a plug-flow reactor with side entrances; inlet flows adjusted to achieve a given RTD.

Using this result, we transform Equation 8.33 to

$$\boxed{\begin{aligned} \frac{dc}{dz} &= \frac{R(c)}{(1-z)^2} & \qquad c(0) &= c_f \\ \frac{dc_s}{dz} &= \frac{p(z/(1-z))\,c}{(1-z)^2} & \qquad c_s(0) &= 0 \end{aligned}} \tag{8.34}$$

Most modern ODE solvers allow the user to specify critical stopping values. These are values of the variable of integration beyond which the ODE solver will not step. We would specify $z = 1$ as a critical value because the right-hand side is not defined past $z = 1$. At the value $z = 1$, we would specify the right-hand sides are zero because the reaction will have equilibrated at $z = 1, \theta = \infty$ so $R(c) = 0$, and $p(\theta) = 0$ at $\theta = \infty$. Again, some care must be taken because the denominators are also going to zero. If the ODE solver terminates successfully, that usually indicates the transformation was successful. It is useful to plot $c(z)$ to make sure the $z = 1$ end does not exhibit some unusual behavior.

8.3.2 Maximum Mixedness

We realize the opposite mixing limit, maximum mixedness, by reversing the flow in the segregated reactor as shown in Figure 8.17 [34]. The feed stream is distributed along the length of the PFR and injected at the appropriate rate at various side entrances corresponding to different θ "locations" to achieve the required RTD. Notice that because the flow has been reversed compared to the segregated case, the θ locations increase from zero at the exit of the tube to large values at the entrance to the tube. We allow an infinitely long tube if we wish to allow RTDs such as the CSTR defined on a semi-infinite domain. Reactors with these specified sidestream addition policies are conceptually important in understanding recent research on achievable reactor performance as discussed in Section 8.4.

Consider the equivalent representation of maximum mixedness in Figure 8.16.B. The shading means that the material at these locations is

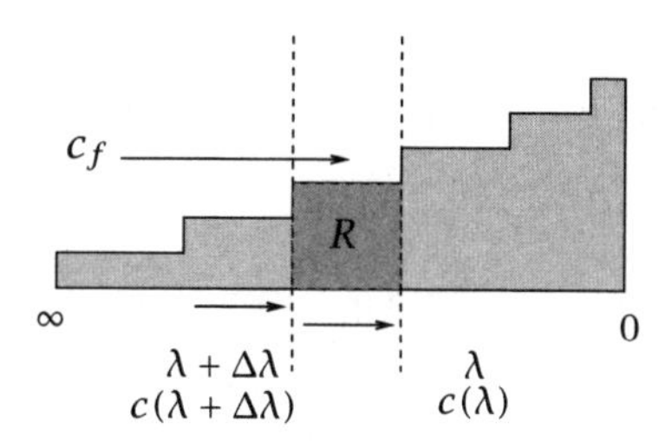

Figure 8.18: Volume element in the state of maximum mixedness.

completely mixed with the material from the other tubes at these same locations [31]. Notice that we have aligned the exits of the tubes in Figure 8.16. Therefore when we mix material between tubes, we are mixing material not with common time spent in the reactor but rather with a common time-to-go λ. Indeed, the mixing process at a given location is mixing material with different times spent in the reactor. It is not possible to mix material with different times-to-go without changing their exit times for the following reason. If we mix two groups of molecules with different times-to-go, λ_1, λ_2, the mixture must later be separated again so that the molecules may exit the reactor at the specified times λ_1 and λ_2. Such a separation is not possible because molecules are not distinguishable based on their times-to-go in the reactor. We face the equivalent problem if we mix two identical, pure-component gases initially on opposite sides of a partition. After mixing, we have no process to separate them again based on their initial locations because the molecules are identical. Such a separation process would violate the second law of thermodynamics.

We next derive the differential equation that governs the maximum mixedness reactor [34]. Consider an expanded view of the reactor in a state of maximum mixedness shown in Figure 8.16.B. As depicted in Figure 8.18, feed with flowrate $Qp(\theta)d\theta$ is added to each tube having residence time θ to achieve the desired RTD. So if we consider a volume element to be a mixed section of the composite reactor between times-to-go λ and $\lambda + \Delta\lambda$, the material balance for this element consists of the following terms:

entering at $\lambda + \Delta\lambda$: $\quad Q\left(\int_{\lambda+\Delta\lambda}^{\infty} p(\lambda')d\lambda'\right) c(\lambda+\Delta\lambda)$

leaving at λ: $\quad Q\left(\int_{\lambda}^{\infty} p(\lambda')d\lambda'\right) c(\lambda)$

feed addition: $\quad Q\,(p(\lambda)\Delta\lambda)\, c_f$

production: $\quad \underbrace{Q\left(\int_{\lambda}^{\infty} p(\lambda')d\lambda'\right)\Delta\lambda}_{\text{volume of reactor element}} \; \underbrace{R(c(\lambda))}_{\text{rate per volume}}$

Considering the reactor is at steady state, we set the accumulation to zero and obtain

$$c(\lambda+\Delta\lambda)\int_{\lambda+\Delta\lambda}^{\infty} p(\lambda')d\lambda' - c(\lambda)\int_{\lambda}^{\infty} p(\lambda')d\lambda' + c_f p(\lambda)\Delta\lambda + R(c)\Delta\lambda \int_{\lambda}^{\infty} p(\lambda')d\lambda' = 0$$

We can combine the first two integral terms and divide by $\Delta\lambda$ to obtain

$$\frac{c(\lambda+\Delta\lambda)-c(\lambda)}{\Delta\lambda}\int_{\lambda+\Delta\lambda}^{\infty} p(\lambda')d\lambda' - c(\lambda)p(\lambda) + c_f p(\lambda) + R(c)\int_{\lambda}^{\infty} p(\lambda')d\lambda' = 0$$

Taking the limit as $\Delta\lambda \to 0$ and rearranging gives

$$\frac{dc}{d\lambda} = \frac{p(\lambda)}{\int_{\lambda}^{\infty} p(\lambda')d\lambda'}\left(c(\lambda) - c_f\right) - R(c)$$

Equivalently we can express the integral in terms of the integrated form of the RTD and write

$$\boxed{\frac{dc}{d\lambda} = \frac{p(\lambda)}{1-P(\lambda)}\left(c(\lambda) - c_f\right) - R(c)} \tag{8.35}$$

We wish to calculate the reactor effluent concentration, which is given by $c(\lambda)$ at $\lambda = 0$. As in the segregated reactor case, this first-order differential equation is singular; we wish to integrate from $\lambda = \infty$, the entrance to the longest tube, to the combined tube exits at $\lambda = 0$. A boundary condition is required at $\lambda = \infty$. For c to remain bounded as $\lambda \longrightarrow \infty$, we stipulate the boundary condition

$$\frac{dc}{d\lambda} = 0, \qquad \lambda = \infty$$

Provided we know the limit $p(\lambda)/(1 - P(\lambda))$ as $\lambda \longrightarrow \infty$, we can solve Equation 8.35 directly for the boundary condition on c at $\lambda = \infty$; we call this value c_∞. Note that $c_\infty \neq c_f$.

Numerical solution. We wish to transform the $\lambda \in (\infty, 0)$ interval into $z \in (0, 1)$. The analogous transformation to the segregated reactor is

$$z = \frac{1}{1+\lambda}, \qquad \lambda = \frac{1-z}{z}$$

The derivative becomes

$$\frac{dc}{d\lambda} = \frac{dc}{dz}\frac{dz}{d\lambda} = -z^2\frac{dc}{dz}$$

in which the minus sign arises because we are changing the direction when integrating in the transformed z variable. Equation 8.35 then becomes

$$\boxed{\frac{dc}{dz} = -\frac{1}{z^2}\left[\frac{p((1-z)/z)}{1 - P((1-z)/z)}\left(c - c_f\right) - R(c)\right] \qquad c(0) = c_\infty} \tag{8.36}$$

and we integrate from $z = 0$ to $z = 1$. Again, a critical stopping value should be set at $z = 1$ to avoid an undefined right-hand side. We set the right-hand side to zero at $z = 0$ because we determined the value of c_∞ such that the bracketed term in Equation 8.36 was zero. Again, care should be exercised at $z = 0$ because the denominator goes to zero at $z = 0 (\lambda = \infty)$. Plotting $c(z)$ and examining the $z = 0$ end for unusual behavior is recommended.

Example 8.3: Two CSTRs in series

We illustrate the results of these sections with an example taken from Zwietering [34].

Given the RTD of two equal-sized CSTRs in series for a single, second-order, irreversible reaction, compute the reactor effluent concentration for the following cases: segregated flow, maximum mixedness and two ideal CSTRs.

Solution

The residence-time distribution for two CSTRs in series is given by Equations 8.16 and 8.20 for $n = 2$,

$$p(\theta) = \frac{4\theta}{\tau^2}e^{-2\theta/\tau}$$

$$1 - P(\theta) = (1 + 2\theta/\tau)e^{-2\theta/\tau}$$

in which $\tau = V_R/Q_f$ and V_R is the total volume of the CSTRs. The balance for the maximum mixedness case becomes

$$\frac{dc}{d\lambda} = \frac{4\lambda}{\tau(\tau+2\lambda)}(c - c_f) + kc^2$$

Defining dimensionless variables, $\overline{c} = c/c_0$ and $\overline{\lambda} = \lambda/\tau$, the equation becomes

$$\frac{d\overline{c}}{d\overline{\lambda}} = \frac{4\overline{\lambda}}{2\overline{\lambda}+1}(\overline{c} - 1) + K\overline{c}^2$$

in which $K = kc_0\tau$. Notice that all the physical constants of the reactor combine into the single dimensionless constant K. If we apply the zero slope condition at $\overline{\lambda} = \infty$, we obtain the quadratic equation

$$2(\overline{c}_\infty - 1) + K\overline{c}_\infty^2 = 0$$

which can be solved for $\overline{c}_\infty$. Again we have an equation on a semi-infinite interval, which we can transform via

$$z = \frac{1}{1+\overline{\lambda}} \qquad \overline{\lambda} = \frac{1-z}{z}$$

in which $\overline{\lambda} \in (\infty, 0)$ is transformed to $z \in (0, 1)$. The transformed derivative satisfies

$$\frac{d\overline{c}}{d\overline{\lambda}} = \frac{d\overline{c}}{dz}\frac{dz}{d\overline{\lambda}} = -z^2\frac{d\overline{c}}{dz}$$

so the final differential equation is

$$\frac{d\overline{c}}{dz} = -\frac{1}{z^2}\left[\frac{4(1-z)/z}{2(1-z)/z+1}(\overline{c} - 1) + K\overline{c}^2\right]$$

$$\overline{c}(0) = \overline{c}_\infty$$

The effluent of the maximum mixed reactor is given by the solution $\overline{c}(z)$ at $z = 1$. Figure 8.19 displays the solution to this differential equation for a range of K values. □

Intermediate conditions of mixing. Weinstein and Adler [31] also proposed an interesting general conceptual mixing model by allowing a general mixing pattern between the various tubes as depicted in Figure 8.16.C.

The segregated reactor depicted in Figure 8.15 and Figure 8.16.A is sometimes referred to as late mixing or mixing as late as possible. The material remains segregated until it reaches the common exit where

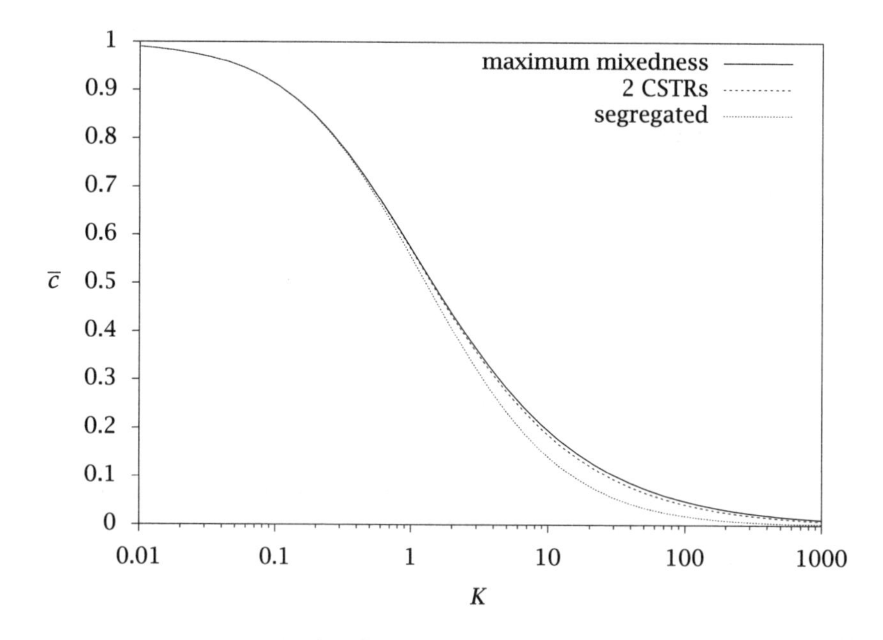

Figure 8.19: Dimensionless effluent concentration $\overline{c} = c/c_0$ versus dimensionless rate constant $K = kc_0\tau$ for second-order reaction; the RTD for all cases is given by 2 CSTRs in series.

the segregated streams are finally mixed in the reactor effluent. The situation depicted by the maximum mixedness reactor of Figure 8.17 and Figure 8.16.B is sometimes called early mixing. The material is mixed at the earliest possible times as it travels down the tubes; no segregated streams remain to be combined at the reactor exit.

Consider again the two reactors in Example 8.2. The conceptual mixing pattern is sketched in Figure 8.20. The reactors have identical RTDs. Comparing these two reactor configurations, the reactor with the CSTR preceding the PFR is in the condition of maximum mixedness because the CSTR is the condition of maximum mixedness and the feed to the PFR is therefore well mixed, so the different portions of the RTD in the PFR section have identical compositions, and could be considered well mixed or segregated. The PFR preceding the CSTR is not in the condition of maximum mixedness, nor is it segregated. As shown in Figure 8.20, it displays an intermediate state of mixing, similar to case C in Figure 8.16. We show in Section 8.4 that because the reac-

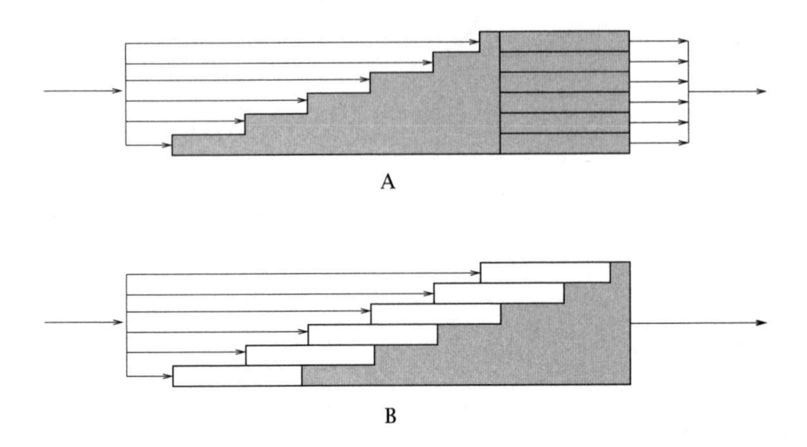

Figure 8.20: CSTR followed by PFR (A) and PFR followed by CSTR (B) as examples of complete and partial mixing; RTDs for the two configurations are equal.

tion rate is second order, complete mixing gives the lowest conversion possible consistent with the given RTD. This conclusion is consistent with the calculation performed in Example 8.2 in which the CSTR–PFR arrangement had lower conversion than the PFR–CSTR.

8.3.3 Mass Transfer and Limits of Reactor Mixing

Consider the following physical picture to help in our understanding of segregated flow and maximum mixedness. Figure 8.21 shows the classic situation in which we mix two liquid-phase feed streams in a stirred tank for the second-order reaction

$$A + B \longrightarrow C$$

We model the action of the stirrer as shearing the fluid A stream into small, uniformly sized "particles" of component A dispersed in the continuous phase containing component B dissolved in a solvent. The size of the A "particles" is one measure of how well the stirrer is working. This physical picture, although idealized, is motivated by several types of real reactors, such as suspension and emulsion polymerization reactors. Ottino provides a well-illustrated discussion of the detailed results of fluid shear [26, pp.1–17]. We assume these "particles" of component A move rapidly about the reactor with the fluid flow. We

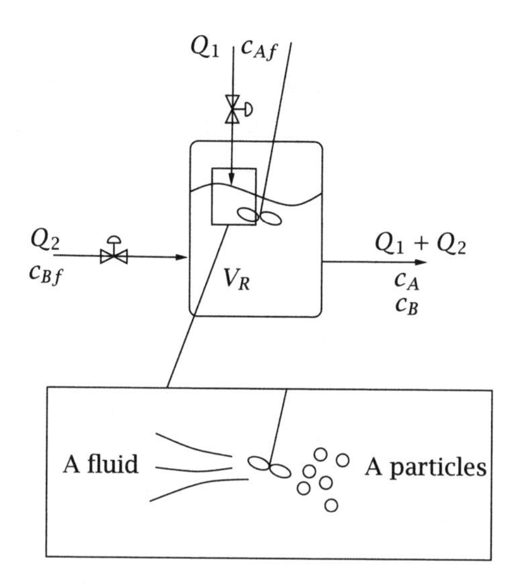

Figure 8.21: Adding two liquid-phase feed streams to a stirred tank; the stirrer is modeled as shearing the liquid A feed stream into small, uniformly sized particles of A and randomly distributing them in the continuous phase containing B.

Parameter	Value	Units
k	1	L/mol·min
k_{mA}	1.67×10^{-4}	cm/min
k_{mB}	1.67×10^{-4}	cm/min
$\alpha = Q_1/Q_2$	1	
c_{Af}	1	mol/L
c_{Bf}	1	mol/L
$\tau = V_R/(Q_1 + Q_2)$	10	min

Table 8.2: Mass-transfer and kinetic parameters for micromixing problem.

therefore have an ideal CSTR residence-time distribution; if we inject tracer with the A stream or the B stream, we would see the classic step response for the CSTR. In other words, the macromixing is excellent, and determining the residence-time distribution would not indicate anything except an ideally mixed reactor.

Now we model the micromixing. Let the mass transfer between the particles and the continuous phase be described by a mass-transfer coefficient, so the mass balance for components A and B inside the particles is given by

$$\begin{aligned} V\frac{dc_A}{d\theta} &= k_{mA}(\overline{c}_A - c_A)S - kc_Ac_BV \qquad c_A(0) = c_{Af} \\ V\frac{dc_B}{d\theta} &= k_{mB}(\overline{c}_B - c_B)S - kc_Ac_BV \qquad c_B(0) = 0 \end{aligned} \tag{8.37}$$

in which θ is the time the particle has been in the reactor, V and S are the particle volume and area, respectively, and k_{mA} and k_{mB} are the A and B mass-transfer coefficients. The variables $\overline{c}_A$ and $\overline{c}_B$ are the continuous-phase concentrations of A and B. The initial conditions follow from the fact that the particles are initially formed from the pure A feed stream. Only as θ increases do they have time to communicate with the continuous phase. To determine the A and B concentrations in the continuous phase, we write the overall, steady-state mass balances for both components

$$\begin{aligned} 0 &= Q_1c_{Af} - Q_1\int_0^\infty c_A(\theta)p(\theta)d\theta - Q_2\overline{c}_A - \frac{V_R}{1+\alpha}\left[\alpha\int_0^\infty kc_Ac_Bp(\theta)d\theta + k\overline{c}_A\overline{c}_B\right] \\ 0 &= Q_2c_{Bf} - Q_1\int_0^\infty c_B(\theta)p(\theta)d\theta - Q_2\overline{c}_B - \frac{V_R}{1+\alpha}\left[\alpha\int_0^\infty kc_Ac_Bp(\theta)d\theta + k\overline{c}_A\overline{c}_B\right] \end{aligned} \tag{8.38}$$

We use orthogonal collocation on $z = \theta/(1+\theta)$ to solve Equations 8.37 simultaneously with Equations 8.38 [30]. Orthogonal collocation is described briefly in Appendix A. The kinetic and mass-transfer parameters are given in Table 8.2 We compute the total A and B concentration in the effluent by summing over both particle and continuous phases

$$\begin{aligned} c_{At} &= \frac{\alpha}{1+\alpha}\int_0^\infty c_A(\theta)p(\theta)d\theta + \frac{1}{1+\alpha}\overline{c}_A \\ c_{Bt} &= \frac{\alpha}{1+\alpha}\int_0^\infty c_B(\theta)p(\theta)d\theta + \frac{1}{1+\alpha}\overline{c}_B \end{aligned}$$

We next study the effect of particle size. Figure 8.22 shows c_{At} for particle sizes ranging from 0.1 μm to 1.0 cm. We see that if the stirrer is able to produce A particles of 1.0 μm or less, then the reactor is essentially in the state of maximum mixedness, or, equivalently, operates as an ideally mixed CSTR. At the other extreme, if the A particles are larger than about 1.0 mm, then the reactor operates essentially as a segregated-flow reactor. Segregated flow essentially reduces the reaction rate to zero because the A and B species cannot come into contact.

Figure 8.23 provides a detailed look inside the particles for $r = 1, 10$ and 100 μm. For $r = 1$ μm, the A and B concentrations in the particles rapidly change from the feed values to the continuous phase values as they spend time in the reactor. This equilibration with the continuous phase is rapid because the particles are small, the total surface area and rate of mass transfer are therefore large. This case is close to maximum mixedness. For $r = 100$ μm, the particles are 100 times larger, and the total surface area and rate of mass transfer are small. Therefore, these particles remain at the inlet feed conditions for a large time. They are washed out of the reactor before they can produce hardly any reaction rate. This case corresponds to essentially complete segregation.

Summarizing, this example is instructive for two reasons. First the residence-time distribution corresponds to a perfect CSTR regardless of particle size. Residence-time distribution measures the reactor macromixing, which is excellent. The particle size governs the micromixing. Small particles have large mass-transfer rates and equilibrate with the continuous phase and the particles in the reactor with different ages leading to the case of maximum mixedness. Large particles have small mass-transfer rates and do not exchange much material with the continuous phase nor therefore with particles of other ages. This case corresponds to segregated flow, which leads to essentially zero rate of reaction. Particles of intermediate size then describe the reactors in intermediate states of mixing.

8.4 Limits of Reactor Performance

8.4.1 A Single Convex (Concave) Reaction Rate

To generalize the results of Examples 8.2 and 8.3, we define convex and concave functions. As presented in the introductory calculus course, the simplest version pertains to functions having at least two derivatives. In that case, a function is convex (concave upward) if its second derivative is everywhere greater than or equal to zero. A function is concave (concave downward) if its second derivative is everywhere less than or equal to zero, as shown in Figure 8.24

$$\frac{d^2 f(x)}{dx} \geq 0, \qquad f \text{ convex}$$

$$\frac{d^2 f(x)}{dx} \leq 0, \qquad f \text{ concave}$$

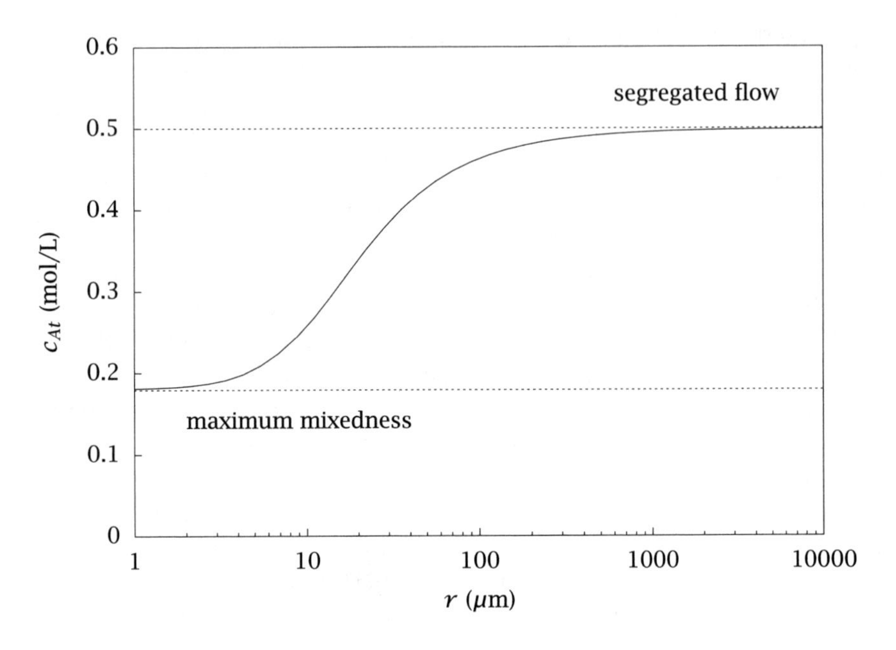

Figure 8.22: Total concentration of A in the reactor effluent versus particle size.

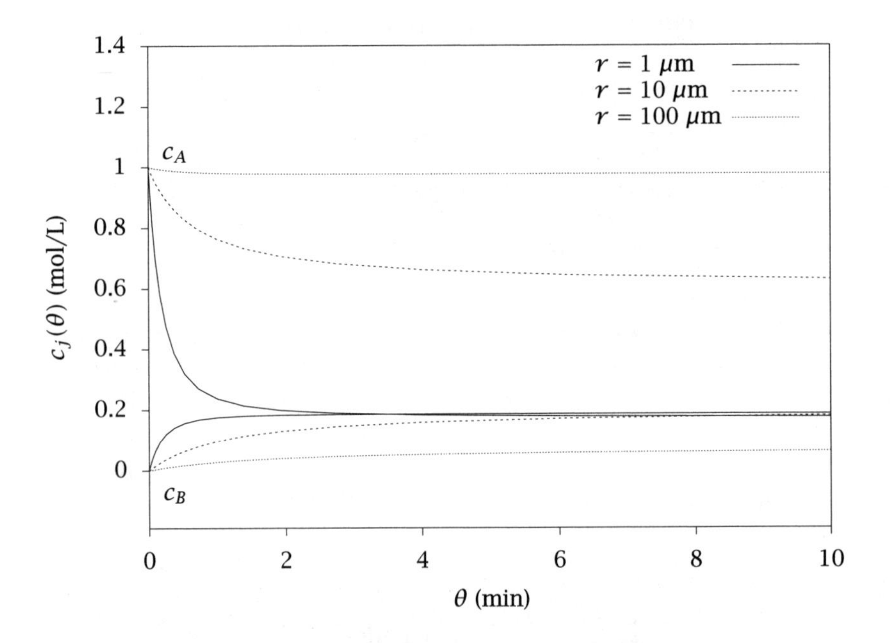

Figure 8.23: Particle concentrations of A and B versus particle age for three different-sized particles.

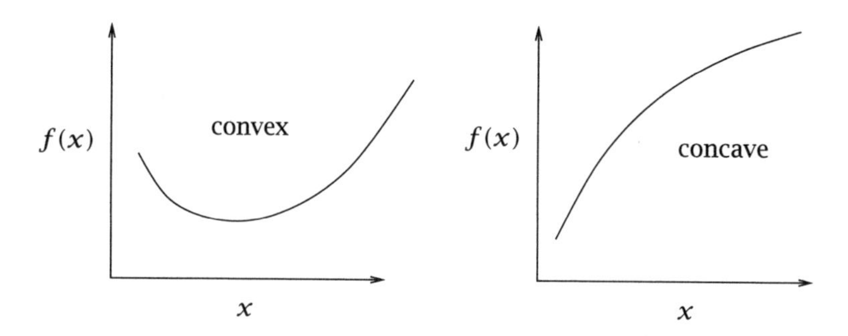

Figure 8.24: Differentiable convex and concave functions.

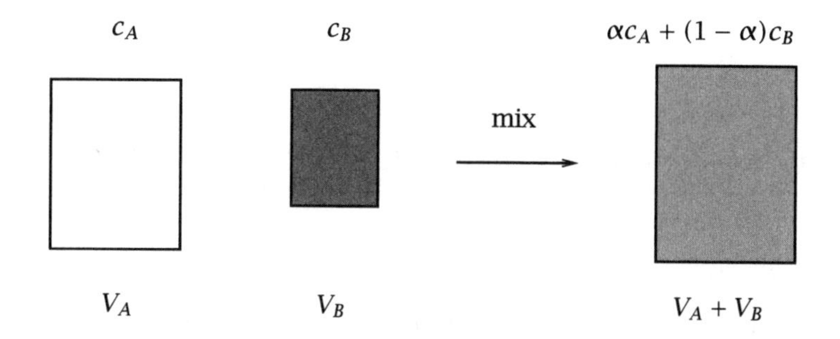

Figure 8.25: Two volume elements before and after mixing.

For example, the nth-order reaction-rate expression $r = c^n$, is *convex if* $n \geq 1$ and *concave if* $n \leq 1$. Note that first-order rate functions are both convex and concave.

The general result for the single reaction is

> Given a single reaction with convex (concave) reaction rate expression, the highest (lowest) conversion for a given RTD is achieved by the segregated reactor and the lowest (highest) conversion is achieved by the maximally mixed reactor.

This nonobvious result is a significant generalization of the numerical Examples 8.2 and 8.3, and Exercise 8.6, and requires justification. The argument presented next first appeared in Chauhan et al. [10]; Nauman and Buffham [24] also provide a detailed discussion.

Step 1. To start, consider the two volume elements shown in Figure 8.25. Note that in this discussion c_A and c_B represent concentration of the *same reactant species* in volume elements or tubes A and B. When

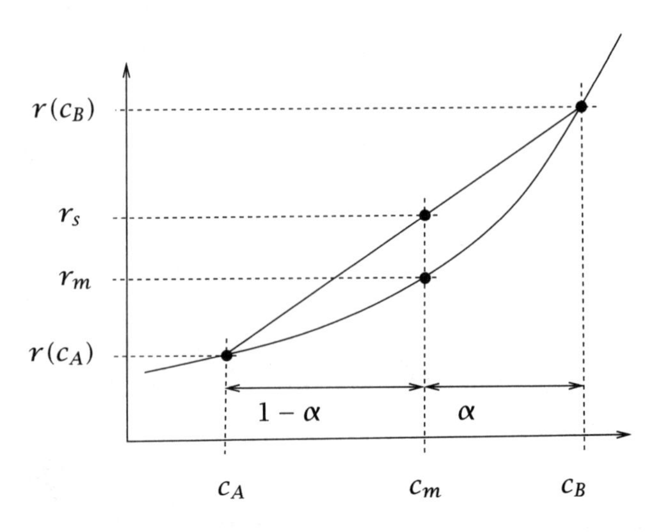

Figure 8.26: Convex rate expression and the effect of mixing; rate of the mean (r_m) is less than the mean of the rate (r_s).

the volume elements are segregated the total reaction rate r_s is simply

$$r_s = r(c_A)V_A + r(c_B)V_B$$

so that the segregated rate per volume is

$$r_s = \alpha r(c_A) + (1-\alpha)r(c_B), \qquad 0 \leq \alpha \leq 1$$

in which α is the volume fraction of element A

$$\alpha = \frac{V_A}{V_A + V_B}$$

On the other hand, if we mix the contents, the concentration is

$$c_m = \frac{c_A V_A + c_B V_B}{V_A + V_B} = \alpha c_A + (1-\alpha)c_B$$

The total reaction rate per volume after mixing is therefore

$$r_m = r(c_m)$$

As shown in Figure 8.26, for all c_A, c_B and α, *if we mix the two volume elements, we lower the overall reaction rate.* The opposite conclusion

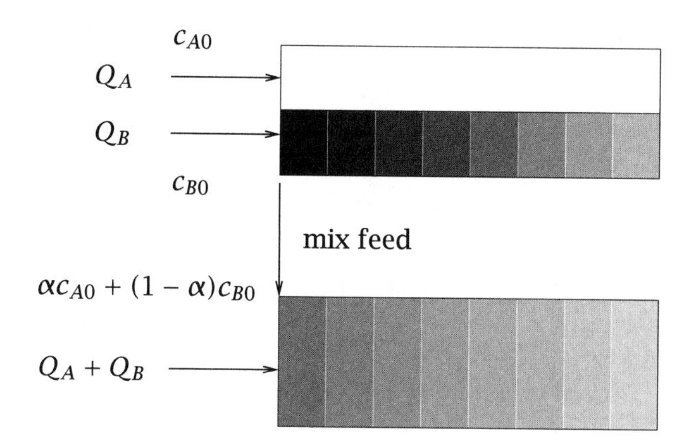

Figure 8.27: Two tubes before and after mixing the entering feed; averaging the two segregated tubes produces c_s; the mixed feed tube produces c_m.

applies if we have a concave rate expression. The rate of the mean r_m is less than the mean of the rate r_s for convex reactions, or

$$r(\alpha c_A + (1-\alpha)c_B) \leq \alpha r(c_A) + (1-\alpha)r(c_B), \qquad \text{all } c_A, c_B, 0 \leq \alpha \leq 1$$

This result is the key to understanding what happens in the general reactor. In fact, this statement can be taken as the definition of convexity (Exercise 8.13).

Step 2. Now consider two tubes as shown in Figure 8.27, which we may choose to mix or maintain segregated as material proceeds down their lengths. Again assume a single reaction takes place and the reaction-rate expression is a convex function of a single reactant species. Without loss of generality assume the stoichiometric coefficient for the limiting species is negative one. For constant density, the material balances for the segregated tubes are

$$\frac{dc_A}{d\theta} = -r(c_A), \qquad c_A(0) = c_{A0}$$
$$\frac{dc_B}{d\theta} = -r(c_B), \qquad c_B(0) = c_{B0}$$

in which $\theta = V/Q_f$. We can track the mean concentration for the segregated case c_s by simply summing the molar flows for tubes A and B divided by the total flow

$$c_s = \alpha c_A + (1-\alpha)c_B \tag{8.39}$$

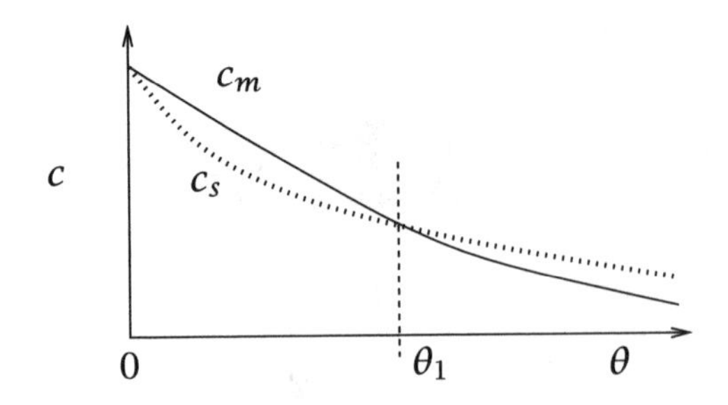

Figure 8.28: Mean segregated and mixed concentrations versus θ; curves crossing at θ_1 is a contradiction.

in which α is now the flowrate fraction in tube A

$$\alpha = \frac{Q_A}{Q_A + Q_B}$$

We also can write a differential equation for c_s by simply differentiating Equation 8.39

$$\frac{dc_s}{d\theta} = -\left[\alpha r(c_A) + (1-\alpha) r(c_B)\right], \qquad c_s(0) = \alpha c_{A0} + (1-\alpha) c_{B0} \tag{8.40}$$

Consider now the mixed case. If the tubes are mixed at some point, which we may call $\theta = 0$, then the material balance for the concentration after that point is

$$\frac{dc_m}{d\theta} = -r(c_m), \qquad c_m(0) = \alpha c_{A0} + (1-\alpha) c_{B0} \tag{8.41}$$

Our goal now is to show $c_m \geq c_s$ for all reactor positions, θ, and all feed concentrations and flowrates, c_{A0}, c_{B0} and α. We know at $\theta = 0$

$$\frac{dc_m}{d\theta} = -\left[r(\alpha c_{A0} + (1-\alpha) c_{B0})\right] \geq -\left[\alpha r(c_A) + (1-\alpha) r(c_B)\right] = \frac{dc_s}{d\theta}$$

If the initial derivatives have this relationship we know, for at least some small distance down the tube, $c_m \geq c_s$ as shown in Figure 8.28. How do we know, however, that the curves do not cross each other at some later time? Assume this crossing can happen as shown in Figure 8.28, and we establish a contradiction. Let θ_1 be the first such crossing time. At θ_1, c_A and c_B have some well-defined values and $c_s = \alpha c_A + (1-\alpha) c_B$. We have assumed that $c_m = c_s$ at θ_1 so the differential equation for c_m, Equation 8.41, gives

$$\frac{dc_m}{d\theta} = -\left[r(\alpha c_A + (1-\alpha)c_B)\right], \qquad \theta = \theta_1$$

The differential equation for c_s still applies and Equation 8.40 gives

$$\frac{dc_s}{d\theta} = -\left[\alpha r(c_A) + (1-\alpha)r(c_B)\right], \qquad \theta = \theta_1$$

Comparing the right-hand sides of these two differential equations and using the convexity of $r(c)$, we conclude

$$\frac{dc_m}{d\theta} \geq \frac{dc_s}{d\theta}, \qquad \theta = \theta_1$$

But this relationship contradicts the assumption that the c_s and c_m curves cross each other. Therefore there can be no time θ_1 at which the curves cross and we conclude

$$c_m(\theta) \geq c_s(\theta), \qquad \text{all } \theta$$

This argument and result apply equally well for all c_{A0}, c_{B0} and α.

Step 3. Finally, consider a segregated reactor with arbitrary residence-time distribution as depicted in Figure 8.16.C. We select any pair of tubes, mix them, make the same argument that we made in Step 2, and replace the segregated tubes with mixed tubes that achieve lower conversion than the original system. We continue in this fashion, and after we pairwise mix all the segregated tubes with mixed tubes, we achieve the reactor of maximum mixedness in Figure 8.16.B. and the lowest possible conversion. Note that this pairing and mixing procedure does not affect the RTD.

8.4.2 The General Case

One might expect that the limits of reactor mixing determine directly the limits of reactor performance for more general kinetic schemes as well as the single convex or concave rate expression of the last section. Unfortunately nature is more subtle. We present next an example that dispels this notion, and then discuss what is known about the limits of reactor performance. This example is based on one presented by Glasser, Hildebrandt and Godorr [16]. Levenspiel [23] shows how to find the optimal reactor configuration for this type of example.

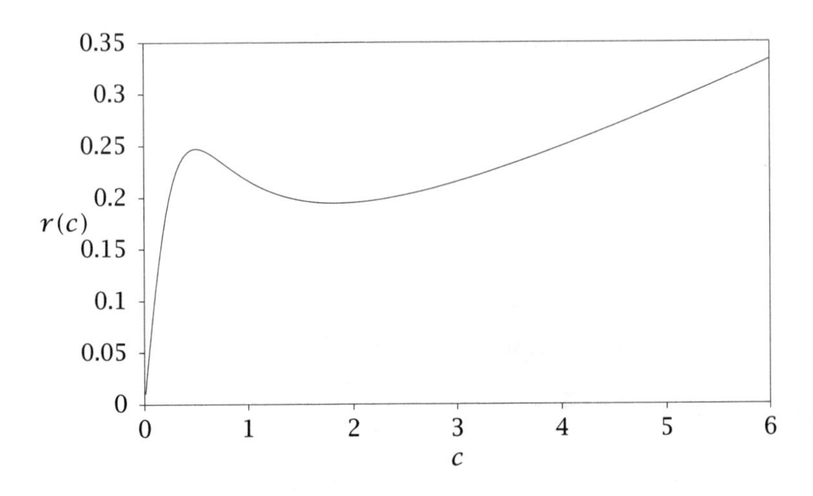

Figure 8.29: Reaction rate versus concentration of limiting reactant; rate expression is neither convex nor concave.

Example 8.4: Optimal is neither segregated nor maximally mixed

Consider the rate expression

$$r(c) = \frac{c}{1 + 5c^2} + 0.05c \tag{8.42}$$

which is plotted in Figure 8.29. For a feed concentration of 5, find the combination of CSTRs and PFRs that achieve 95% conversion with the smallest total reactor volume. Determine the RTD for this reactor configuration. What conversion is achieved in a segregated reactor with this RTD? What conversion is achieved in a maximally mixed reactor with this RTD?

Solution

As we mentioned in Chapter 4, the smallest volume can be achieved with a series combination of CSTRs and PFRs. First we plot the inverse of the rate as shown in Figure 8.30. Then we find any minima in the inverse rate function and construct CSTRs from those values until we intersect the inverse rate curve. In the remaining sections of the curve where the inverse rate is a decreasing function of concentration, we use PFRs. Examining the plotted $1/r$ function in Figure 8.30, we see the optimal configuration is a PFR-CSTR-PFR; this configuration is sketched in Figure 8.30. We can calculate the sizes of the reactors as follows. We

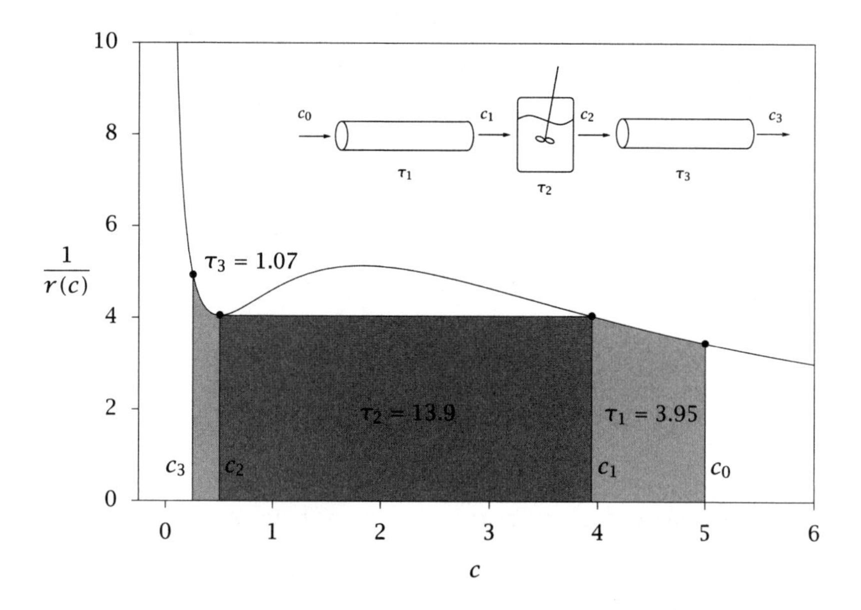

Figure 8.30: Inverse of reaction rate versus concentration; optimal sequence to achieve 95% conversion is PFR-CSTR-PFR.

know from the problem statement that $c_0 = 5, c_3 = 0.25$. We next find the point where $dr(c)/dc = 0$. Notice these are also the places where $d(1/r(c))/dc = 0$. Setting the derivative of Equation 8.42 to zero gives a quadratic equation with two roots: 0.501 and 1.83. We choose the one corresponding to the minimum in $1/r$, which gives

$$c_2 = 0.501, \qquad 1/r(c_2) = 4.045$$

Next we find the concentration c_1 such that $1/r(c_1) = 1/r(c_2)$. This results in a cubic equation, which we solve numerically. Then the residence time is given by $\tau_2 = 1/r(c_2)(c_1 - c_2)$ which gives

$$c_1 = 3.94, \qquad \tau_2 = 13.9$$

To size the PFRs we simply use the PFR design equation and obtain

$$\tau_1 = -\int_{c_0}^{c_1} \frac{1}{r(c)} dc = 3.95, \qquad \tau_3 = -\int_{c_2}^{c_3} \frac{1}{r(c)} dc = 1.07$$

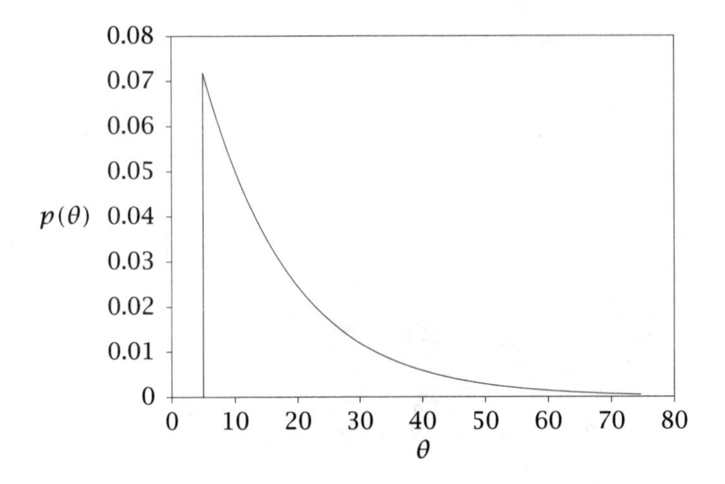

Figure 8.31: Residence-time distribution for the optimal reactor configuration.

These results are displayed in Figure 8.30. Because we have a series of CSTRs and PFRs, we can write the RTD immediately

$$p(\theta) = \frac{1}{\tau_2} \exp\left[-\frac{\theta - (\tau_1 + \tau_3)}{\tau_2}\right] H(\theta - (\tau_1 + \tau_3))$$

which is plotted in Figure 8.31. With the RTD in hand, we can compute both the maximally mixed, Equation 8.36, and segregated, Equation 8.34, reactor cases. The results of those two calculations are summarized in the following table.

Reactor	Conversion
optimal	0.95
segregated	0.68
maximally mixed	0.75

We see that these two mixing limits do *not* bound the performance of the actual reactor sequence with the given RTD. In fact, they are off by more than 20%. Even with a single reaction, if the rate expression is neither convex nor concave, we cannot bound the performance of an actual reactor between the segregated and maximally mixed mixing limits. □

The attainable region. The primary use of the classical mixing models, such as the segregated reactor and the maximally mixed reactor, is to build insight into the effects of mixing on reactor behavior under the constraint of a fixed, and presumably measurable, RTD. As we have seen in Example 8.4, however, if we are mainly interested in determining bounds on achievable reactor states (conversions, yields, etc.), these simple mixing models are insufficient. In this section we would like to provide a brief overview of what is known about finding sharp bounds on reactor performance. The general problem can be defined in this way.

> Given a feed stream of known composition and a set of chemical reactions with known rate expressions, determine the set of all possible steady-state species concentrations that can be achieved by *any* combination of chemical reactors.

This set was proposed by Horn almost 40 years ago and named the **attainable region** [18]. Because the set is defined for *all possible reactor combinations*, it seems conceptually difficult to formulate a procedure by which we can calculate this set. We should also note that by considering all reactor combinations, we are also considering all possible residence-time distributions, which is a considerable generalization from the single RTD that was considered in the mixing discussion in previous sections. In spite of the seeming difficulty in finding the attainable region, excellent, recent research progress has been made. Feinberg provides a nice summary overview of the history and many recent developments [12].

Glasser and Hildebrandt revived recent interest in this problem [17, 16]. Feinberg and Hildebrandt [14] characterized the boundary of the attainable region, which is of importance because it bounds the possible steady-state concentrations. They showed, for example, that the extreme points of the attainable region boundary are made up entirely of plug-flow reactor trajectories. They also showed that combinations of PFRs, CSTRs, and what are called differential side-stream reactors (PFRs with addition of feed along the side of the tube), provide the means to find all of the attainable region extreme points.

In addition to properties and conceptual characterization of the attainable region, researchers have proposed computational procedures to approximate the attainable region and solve reactor synthesis problems. Some of these are based on proposing a superstructure of reactor types and numbers, and optimizing an objective function among the possible reactors [20]. Because the superstructure does not enumerate all possibilities, the solution may not be close to the true attainable region. A person skilled in reactor design may be able to choose reactor numbers and types well and overcome this general difficulty on specific reaction networks of interest.

Some computational methods are based on finding the boundary of the attainable region using the reactor types that characterize the attainable region extreme points. Hybrid methods involving superstructures and geometric considerations have also been proposed [21]. Biegler, Grossmann and Westerberg provide an overview in Chapter 13 of their text [4].

Burri, Wilson and Manousiouthakis recently proposed an infinite dimensional state-space approach (IDEAS) that requires only PFRs, CSTRS and mixing [8]. The advantage of this approach is that one solves only convex, linear optimization problems. The disadvantage is the problems are infinite dimensional and require a finite dimensional approximation for calculation. A full analysis of the convergence properties of the finite dimensional approximation is not yet available, but the approach shows promise on numerical examples. Kauchali et al. have proposed another linear programming approach [19]. Finally, Abraham and Feinberg recently proposed bounding the attainable region from the outside using hyperplanes [1].

If we wish to allow separation as well as chemical reaction, and almost all industrial designs would fall into this category, then the problem switches from a pure reactor synthesis problem to a reactor-separator synthesis problem. The CSTR equivalence principle of Chapter 4 is an example of the strikingly simple and general results that recently have been achieved for the reactor-separator synthesis problem [13].

Forecasting is always risky business, but given the rapid pace of recent progress, it seems likely that new and highly useful results on pure reactor and reactor-separator synthesis problems will be forthcoming. These ideas and results may have immediate industrial impact, and certainly fall within the scope of the course in reactor analysis and design.

8.5 Examples in Which Mixing is Critical

Returning to the topic of mixing, we would like to close the chapter by presenting a few more chemical mechanisms for which reactor mixing

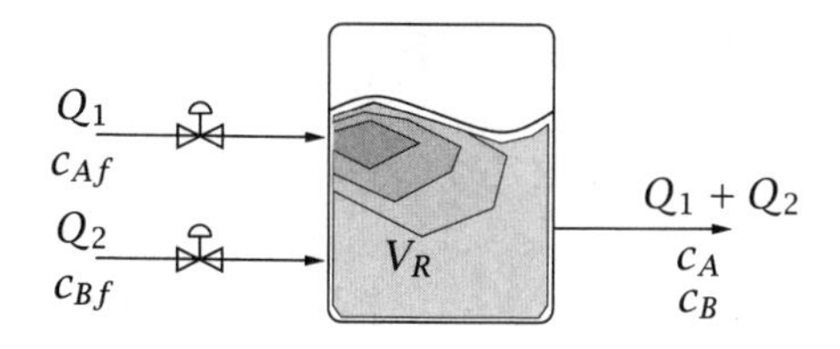

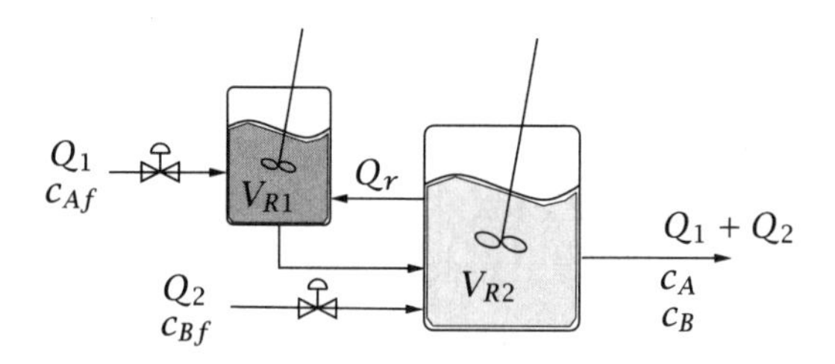

Figure 8.32: Imperfect mixing (top reactor) leads to formation of an A-rich zone, which is modeled as a small CSTR feeding a second CSTR (bottom two reactors).

can play a critical role.

Example 8.5: Mixing two liquid-phase streams in a stirred tank

A classic mixing problem arises when we must bring two liquid-phase feed streams together to perform the second-order reaction

$$\mathrm{A} + \mathrm{B} \xrightarrow{k_1} \mathrm{C} \qquad r_1 = k_1 c_A c_B \tag{8.43}$$

in the presence of the undesirable side reaction

$$\mathrm{A} \xrightarrow{k_2} \mathrm{D} \qquad r_2 = k_2 c_A^n \tag{8.44}$$

If the rate of the second degradation reaction is fast compared to the rate of mixing of the two feed streams, we can anticipate problems. To understand what happens in this situation, consider the mixing model depicted in Figure 8.32. Component A is assumed to be the limiting reactant. It is added at a low flowrate to a CSTR that contains an excess of reactant B. In the top figure we depict the ideal-mixing case in which the rate of mixing is arbitrarily fast compared to the rate of either reaction. But this ideal mixing may be impossible to achieve if the reaction rates are reasonably large. So in the bottom figure, we model

Parameter	Value	Units
c_{Af}	1	mol/L
c_{Bf}	1	mol/L
k_1	1	L/mol·min
k_2	2	L/mol·min
n	2	
$\tau_1 = V_{R1}/Q_2$	1	min
$\tau_2 = V_{R2}/Q_2$	2	min
$\tau = V_R/Q_2$		
$= \tau_1 + \tau_2$	3	min
$\alpha = Q_1/Q_2$	0.1	
$\rho = Q_r/Q_2$	varies	

Table 8.3: Reactor and kinetic parameters for feed-mixing example.

the formation of an A-rich zone near the feed entry point. This small CSTR exchanges mass with a larger reactor that contains the excess of reactant B. We can vary the recycle flowrate between the two CSTRs, Q_r, and the sizes of the two reactors, V_{R1} and V_{R2}, to vary the degree of mixing. For large Q_r, we expect the two-reactor mixing model to approach the single, ideally mixed CSTR.

As discussed in Chapter 4, the conversion and yield are the usual quantities of interest in competing parallel reactions of the type given in Reactions 8.43 and 8.44. We assume the density of this liquid-phase system is constant, and define the overall conversion of reactant A and yield of desired product C as follows:

$$x_A = \frac{Q_1 c_{Af} - (Q_1 + Q_2)c_A}{Q_1 c_{Af}} \qquad y_C = \frac{(Q_1 + Q_2)c_C}{Q_1 c_{Af} - (Q_1 + Q_2)c_A}$$

Given the parameters and rate constants in Table 8.3, calculate x_A and y_C versus Q_r for the two-reactor mixing model shown in Figure 8.32, and compare the result to the single, well-mixed reactor. Then calculate the residence-time distribution $P(\theta)$ for tracer injected with the A feed stream for the two models. Discuss whether or not the residence-time distribution is a reliable indicator for problems with yield in the imperfectly mixed reactor.

Solution

The steady-state mass balance for the single, well-mixed CSTR is

$$0 = Q_1 c_{Af} - (Q_1 + Q_2)c_A - (k_1 c_A c_B + k_2 c_A^n)V_R$$
$$0 = Q_2 c_{Bf} - (Q_1 + Q_2)c_B - k_1 c_A c_B V_R$$

Defining the following parameters

$$\alpha = \frac{Q_1}{Q_2} \qquad \tau = \frac{V_R}{Q_2} \qquad \rho = \frac{Q_r}{Q_2}$$

allows us to write these as

$$0 = \alpha c_{Af} - (1+\alpha)c_A - (k_1 c_A c_B + k_2 c_A^n)\tau$$
$$0 = c_{Bf} - (1+\alpha)c_B - k_1 c_A c_B \tau$$

We can solve numerically the two equations for the two unknowns c_A, c_B. The concentration of C in the outflow is determined from the change in the concentration of B,

$$(Q_1 + Q_2)c_C = Q_2 c_{Bf} - (Q_1 + Q_2)c_B$$

Using this relationship and the defined parameters gives for conversion and yield,

$$x_A = \frac{\alpha c_{Af} - (1+\alpha)c_A}{\alpha c_{Af}} \qquad y_C = \frac{c_{Bf} - (1+\alpha)c_B}{\alpha c_{Af} - (1+\alpha)c_A}$$

For the two-reactor system, we write mass balances for each reactor. Let $c_{A1}, c_{A2}, c_{B1}, c_{B2}$ be the unknown A and B concentrations in the two-reactors, respectively. The mass balances are

Reactor 1:

$$0 = Q_1 c_{Af} - (Q_1 + Q_r)c_{A1} + Q_r c_{A2} - (k_1 c_{A1} c_{B1} + k_2 c_{A1}^2)V_{R1}$$
$$0 = -(Q_1 + Q_r)c_{B1} + Q_r c_{B2} - k_1 c_{A1} c_{B1} V_{R1}$$

Reactor 2:

$$0 = (Q_1 + Q_r)c_{A1} - Q_r c_{A2} - (Q_1 + Q_2)c_{A2} - (k_1 c_{A2} c_{B2} + k_2 c_{A2}^2)V_{R2}$$
$$0 = Q_2 c_{Bf} + (Q_1 + Q_r)c_{B1} - Q_r c_{B2} - (Q_1 + Q_2)c_{B2} - k_1 c_{A2} c_{B2} V_{R2}$$

We can summarize this case using the previously defined variables as four equations in four unknowns

$$0 = \alpha c_{Af} - (\alpha + \rho)c_{A1} + \rho c_{A2} - (k_1 c_{A1} c_{B1} + k_2 c_{A1}^2)\tau_1$$
$$0 = -(\alpha + \rho)c_{B1} + \rho c_{B2} - k_1 c_{A1} c_{B1} \tau_1$$
$$0 = (\alpha + \rho)c_{A1} - \rho c_{A2} - (1 + \alpha)c_{A2} - (k_1 c_{A2} c_{B2} + k_2 c_{A2}^2)\tau_2$$
$$0 = c_{Bf} + (\alpha + \rho)c_{B1} - \rho c_{B2} - (1 + \alpha)c_{B2} - k_1 c_{A2} c_{B2} \tau_2$$

Figures 8.33 and 8.34 show the yield and conversion for the two cases as a function of Q_r. The conversion is not adversely affected by the poor mixing. In fact, the conversion in the two-reactor system is higher than the single, well-mixed reactor. Notice, however, that at low values of Q_r, which corresponds to poor mixing at the feed location, the yield changes from *more than 90% to less than 15%.* Low yield is a qualitatively different problem than low conversion. If the conversion is low, we can design a separation system to remove the unreacted A and recycle it, or use it as feed in a second reactor. With low yield, however, the A has been irreversibly converted to an undesired product D. The raw material is lost and cannot be recovered. It is important to diagnose the low yield as a reactor mixing problem, and fix the problem at the reactor. A yield loss cannot be recovered by downstream processing.

Next we compute the outcome of injecting a unit step change in a tracer in the A feed stream. We solve the transient CSTR balances and calculate the tracer concentration at the outlet. Because the tracer does not take part in any reactions, this can be done analytically or numerically. The result is shown in Figure 8.35. We see the familiar single CSTR step response. For the two-reactor mixing model, when $\rho = 0$, which corresponds to the poorest mixing and lowest yield, the step test does reliably indicate the poor mixing. At the end of this chapter and also in Chapter 9 we show how to use this step response to determine the best value of ρ to model the mixing. When ρ is reasonably large, $Q_r = Q_2$, and the single CSTR and two-reactor cases have similar yields and step responses.

Notice in all three step responses, the tracer concentration reaches only $c_{Is} = 0.091 = \alpha/(1 + \alpha)$ because we inject tracer in only one of the two feed streams. □

This example is one of the classic sets of reactions in which mixing has a significant impact on the reactor performance and the product yield. It deserves careful study because it builds intuition and leads us to ask good questions when confronted with more complex cases.

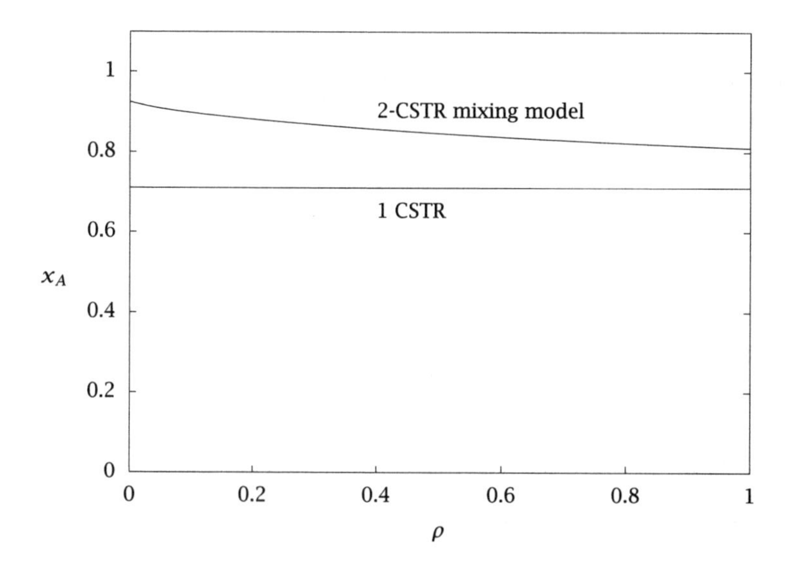

Figure 8.33: Conversion of reactant A for single, ideal CSTR, and as a function of internal flowrate, $\rho = Q_r/Q_2$, in a 2-CSTR mixing model.

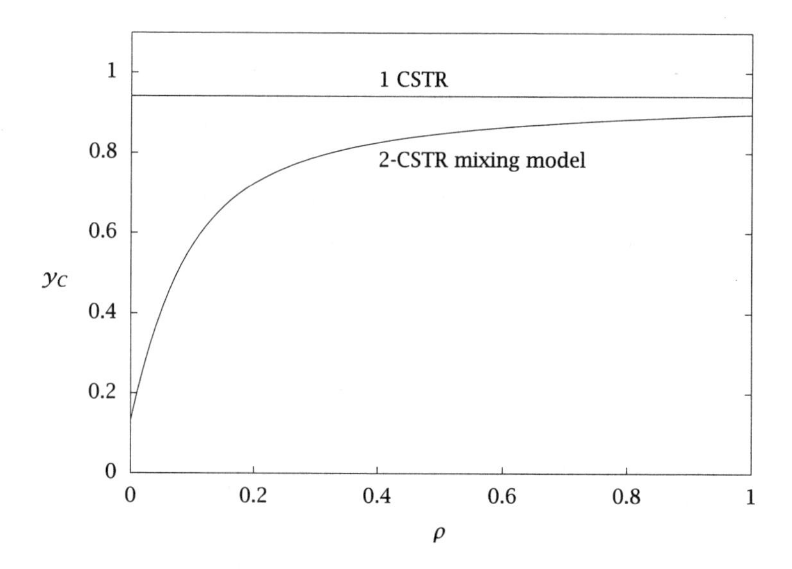

Figure 8.34: Yield of desired product C for single, ideal CSTR, and as a function of internal flowrate, $\rho = Q_r/Q_2$, in a 2-CSTR mixing model.

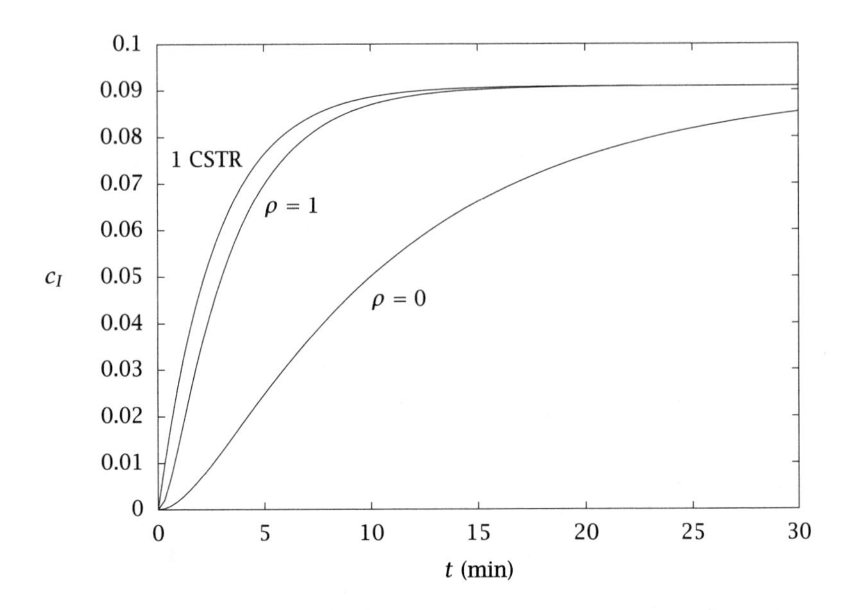

Figure 8.35: Step response for single, ideal CSTR, and 2-CSTR mixing model with $\rho = 0, 1$.

For example, Villa et al. [29] discuss similar issues that arise in more complex polymerization reaction engineering problems.

Example 8.6: Maximizing yield in dispersed plug flow

Consider the following two liquid-phase reactions in which B is the desired product

$$\mathrm{A} \xrightarrow{k_1} \mathrm{B}, \qquad r_1 = k_1 c_A$$

$$2\mathrm{B} \xrightarrow{k_2} \mathrm{C}, \qquad r_2 = k_2 c_B^2$$

The second reaction can represent the first step in a polymerization process of species B, which is undesirable in this case.

Because the second reaction is second order in B, it is desirable to keep the *average* B concentration in the reactor low, to avoid yield losses, but achieve high B concentration near the reactor exit to maximize the production rate. Intuitively the CSTR is a bad choice, because it maintains the same B concentration everywhere in the reactor. A PFR

Parameter	Value	Units
k_1	1	min^{-1}
k_2	1	L/mol·min
c_{Af}	1	mol/L
c_{Bf}	0	
v	1	m/min
l	0.5	m
D_l	varies	m^2/min

Table 8.4: Parameters for the dispersed PFR example.

should offer higher yield. The B concentration is low near the tube entrance, and increases to its maximum value at the tube exit if we choose the right length or residence time. If we make the tube too long, however, the B is largely converted to C and the yield is again low. In this case, yield is adversely affected by mixing.

Calculate the steady-state conversion of A and yield of B versus PFR length for the kinetic and reactor parameters in Table 8.4. What is an appropriate reactor length to maximize yield of B? Study the effect of dispersion. Approximately how large can the dispersion number be before the advantages of the PFR over the CSTR are lost?

Solution

The steady-state mass balances for components A and B are

$$v\frac{dc_A}{dz} - D_l\frac{d^2c_A}{dz^2} = R_A \qquad v\frac{dc_B}{dz} - D_l\frac{d^2c_B}{dz^2} = R_B$$

in which

$$R_A = -k_1c_A \qquad R_B = k_1c_A - 2k_2c_B^2$$

and we have assumed the dispersion numbers of both species are the same, $D_{Al} = D_{Bl} = D_l$. Because the fluid is a liquid, we assume the velocity is constant. We use Danckwerts boundary conditions for both species

$$\begin{aligned} vc_{jf} &= vc_j(0) - D_l\frac{dc_A}{dz}(0), \qquad && z = 0 \\ \frac{dc_j}{dz} &= 0, && z = l \end{aligned}$$

$j = (A, B)$. Given the concentrations, and because the flowrate is constant, the conversion and yield are

$$x_A = \frac{c_{Af} - c_A}{c_{Af}} \qquad y_B = \frac{c_B}{c_{Af} - c_A}$$

Figures 8.36 and 8.37 show the conversion of A and yield of B versus tube length for a tube designed to maximize the yield of B. A tube length of about 0.5 m is appropriate. As the length increases above this value, the conversion of A increases, but the yield of B drops rapidly, defeating the main purpose of using a PFR. For the kinetic parameters chosen, the CSTR yield can be improved by about 8% with a PFR. As shown in Figure 8.36, the high-dispersion PFR is essentially a CSTR, and achieves $y_B = 0.79$. The PFR with $D = 0.001$ achieves $y_B = 0.87$. We see that the dispersion number must be kept less than about 0.1 to maintain this advantage in yield. □

8.6 Summary

In this chapter we generalized the two flow assumptions of the idealized reactor models: the perfect mixing assumption of the batch reactor and CSTR, and the plug-flow assumption of the PFR. We defined the residence-time distribution (RTD) of a reactor, and showed how to measure the RTD with simple tracer experiments such as the step test, pulse test and (idealized) impulse test. The RTD gives a rough measure of the flow pattern in the reactor, but it does not determine completely the reactor performance. Indeed, reactors with different flow patterns, and therefore different performances, may have identical RTDs. We showed the CSTR has an exponential RTD. The derivation of the RTD of the CSTR also illustrated the following general principle: given an event with constant probability of occurrence, the time until the next occurrence of the event is distributed as a decreasing exponential function. This principle was used, for example, to choose the time of the next reaction in the stochastic simulations of Chapter 4.

The residence-time distribution of the PFR was shown to be arbitrarily sharp because all molecules spend identical times in the PFR. We introduced the delta function to describe this arbitrarily narrow RTD. We added a dispersion term to the PFR equations to model the spread of the RTD observed in actual tubular reactors. We computed the full, transient behavior of the dispersed plug-flow model, and displayed the evolution of the concentration profile after a step change in the feed concentration.

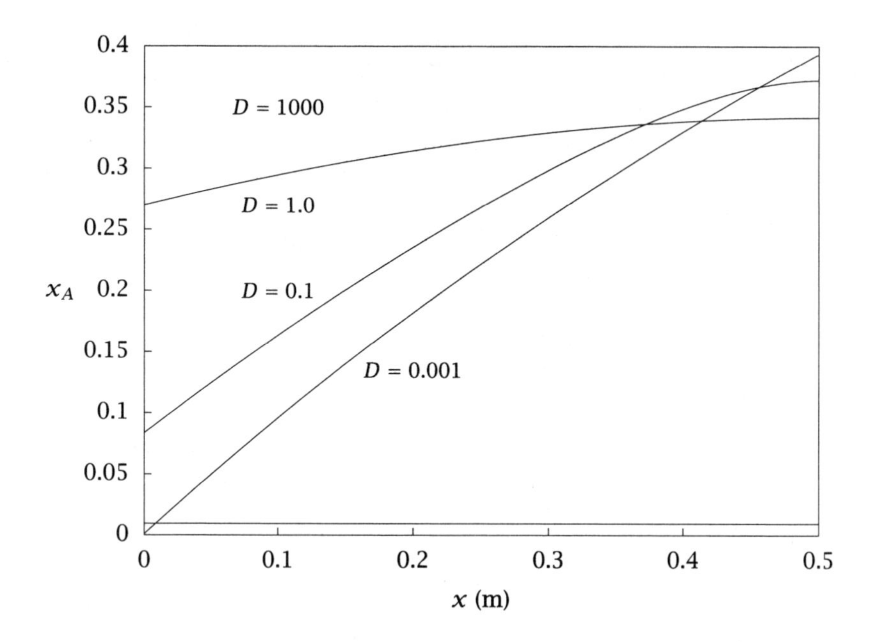

Figure 8.36: Conversion of reactant A versus reactor length for different dispersion numbers.

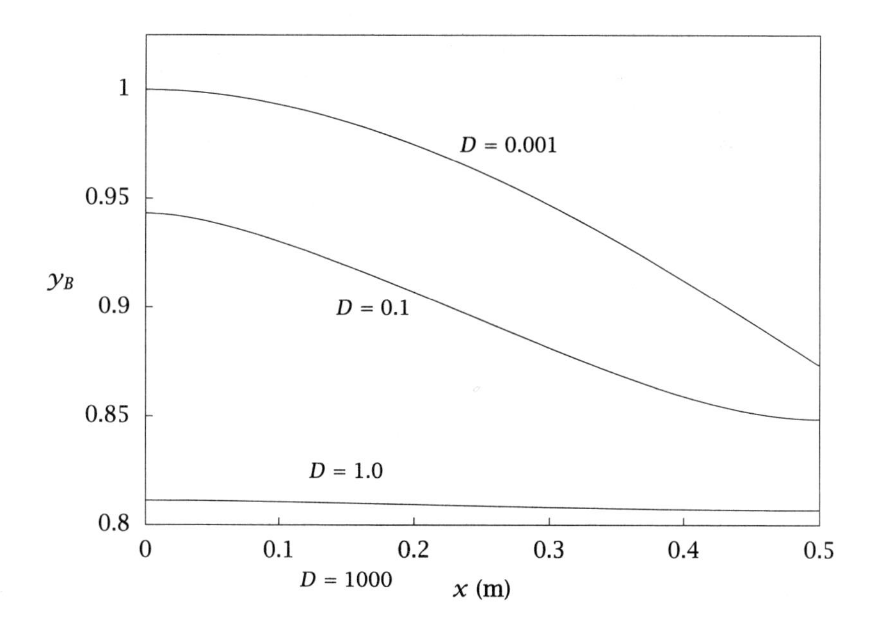

Figure 8.37: Yield of desired product B versus reactor length for different dispersion numbers.

We then examined the limits of reactor mixing consistent with a given RTD. The two limits are segregated flow and maximum mixedness. We showed how a physical process such as mass transfer between a continuous phase and a particle phase can approach segregated flow for large particles (small mass-transfer rates) and can approach maximum mixedness for small particles (high mass-transfer rates).

We also showed that the mixing limits bound the possible reactor behavior for the case of a single, convex reaction-rate expression. For more general reaction networks, however, the mixing limits do not bound the reactor performance. For the general reaction network, recent research on the attainable region has started to shed light on the possible reactor performance. If we consider separation as well as reaction, rather general results such as the CSTR equivalence principle of Chapter 4 have recently been discovered. These topics seem ripe for further research progress.

Next we discussed two contrasting cases in which mixing plays a critical role. In the mixing of two liquid reactants, we showed that formation of a poorly mixed zone can lead to significant yield losses. By contrast, for the kinetics of the second example, good mixing leads to yield losses; in this example the reactor should be designed to approach segregated flow.

Finally, the recent progress in the area of computational fluid dynamics (CFD) gives us reason to believe that direct solution of the equations of motion for the fluid will be a tractable approach for designing reactors and evaluating their performance [2]. It seems reasonable to expect the classical RTD methods and simple flow models to complement the computationally intensive CFD methods. CFD methods may be used to validate simpler mixing models. These validated, simple mixing models may continue to play important roles in reactor analysis, design and optimization. Hybrid models that combine aspects of mixing models with CFD calculations are also finding application, particularly in bioreactor modeling in which fluid shear has important effects on both cell damage and mass transfer rates [3].

Notation

c_e	effluent concentration in RTD measurement
c_f	feed concentration
c_j	concentration of species j
c_m	concentration in a maximally mixed flow model
c_s	concentration in a segregated flow model
c_∞	concentration boundary condition in maximum mixedness model
C	outflow of CSTR, Equation 8.29
D	dimensionless dispersion number, $D = D_l \tau / l^2$
D_A	molecular diffusivity
D_{jl}	dispersion coefficient for species j
D_l	dispersion coefficient
$\text{erf}(x)$	error function, Equation 8.25
$H(x)$	Heaviside or unit step function, Equation 8.13
k_{mj}	mass-transfer coefficient
l	tubular reactor length
n	number of CSTRs in a mixing model
$p(\theta)$	RTD, probability that molecule spends time θ to $\theta + d\theta$ in reactor
$P(\theta)$	cumulative RTD, probability that molecule spends time zero to θ in the reactor
$\mathcal{P}$	outflow of the PFR, Equation 8.30
Pe	Péclet number, $\text{Pe} = vl/D_A$
Q	volumetric flowrate
Q_f	feed volumetric flowrate
r	particle radius in mixing model
r	reaction rate of (single) reaction
R_j	production rate of species j
v	fluid axial velocity
V_R	reactor volume
x_j	molar conversion of component j
y_j	yield of species j
z	reactor axial coordinate
$\gamma(n, x)$	incomplete gamma function of order n, Equation 8.19
$\Gamma(n)$	gamma function of n, Equation 8.18
$\delta(x)$	delta or impulse function, Equations 8.5 and 8.14, and Figure 8.2
θ	residence time of tracer molecule in reactor
$\overline{\theta}$	mean residence time, $\overline{\theta} = \int_0^\theta \theta' p(\theta') d\theta'$
λ	time-to-go before molecule exits reactor
τ	V_R/Q_f

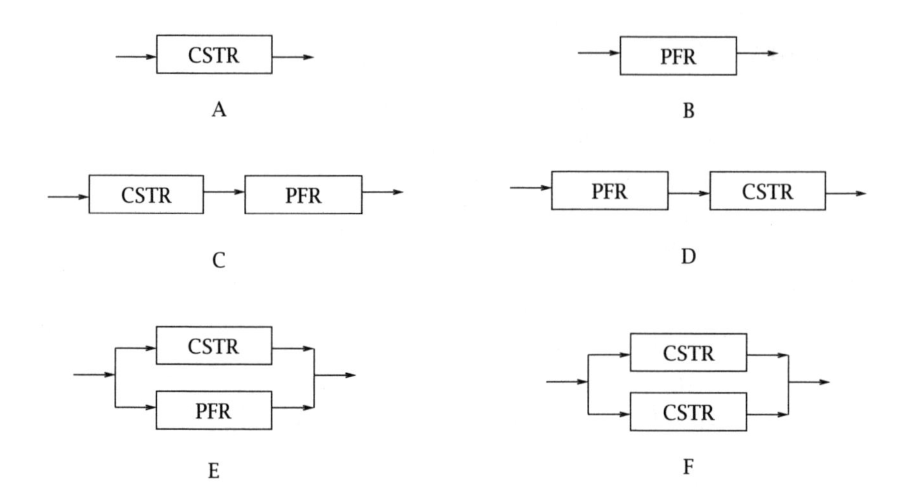

Figure 8.38: Reactor configurations subjected to a step test in tracer concentration.

8.7 Exercises

Exercise 8.1: Configuration to RTD

Make a qualitative sketch of the integrated form of the residence-time distribution $P(\theta)$ by considering a step change in tracer concentration in each of the reactor configurations shown in Figure 8.38. Be sure to mark the mean of the RTD on your sketches. The total flow into each configuration is the same and each individual reactor has the same volume. Is each of the answers in A–F unique?

Exercise 8.2: RTD to configuration

Consider the feed and effluent tracer concentrations from a pulse test shown in Figure 8.39. Draw the simplest reactor configuration consisting of PFRs and CSTRs similar to those shown in Figure 8.38 that you would use to model the reactor based on these test data. Determine $\tau = V_R/Q_f$ for each reactor in your configuration.

Is this configuration unique? Why or why not?

Exercise 8.3: More equivalent RTDs with different conversions

Prepare a plot like that in Figure 8.14 for the following reaction rates

$$r = kc, \qquad r = kc^{1/2}, \qquad r = k\frac{Kc}{1+Kc}$$

Which configuration achieves higher conversion for these cases, CSTR–PFR or PFR–CSTR?

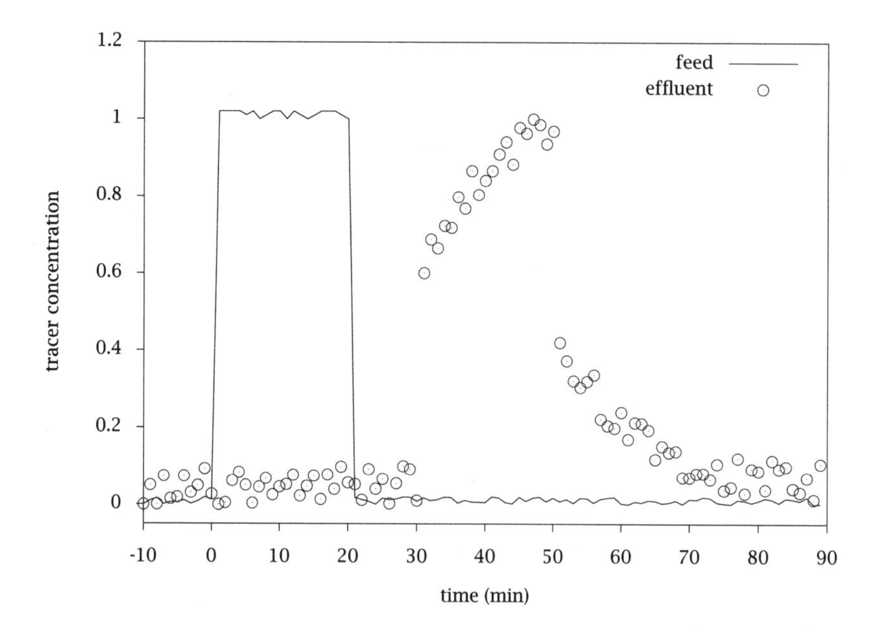

Figure 8.39: Tracer concentrations in the feed and effluent streams versus time.

Exercise 8.4: Deviation variables

Consider again the RTD relationship, Equation 8.1

$$c_e(t) = \int_{-\infty}^{t} c_f(t')p(t-t')dt'$$

Assume the feed tracer concentration is not initially zero but some steady concentration c_0. As in the step-response experiment, at time zero the feed is abruptly changed to a new steady value $c_f \neq c_0$. Consider a dimensionless deviation variable for effluent tracer concentration.

$$\overline{c}_e(t) = \frac{c_e(t) - c_0}{c_f - c_0}$$

(a) Show that the results of Section 8.2.2 still apply in this situation. Namely, the RTD is easily determined from the effluent concentration via

$$P(\theta) = \overline{c}_e(\theta)$$

(b) For a constant-density CSTR, what differential equation and initial condition does the deviation variable $\overline{c}_e(t)$ satisfy?

Exercise 8.5: Predicting performance

The residence-time distribution is tabulated below for a flow reactor that has a feed rate of 16.2 ft^3/min. Under the feed conditions (c_{Af} = 0.02 lbmol/ft^3 and c_{Bf} = 2.34

lbmol/ft^3) the reaction

$$\mathrm{A} + \mathrm{B} \xrightarrow{k_1} \mathrm{C} \qquad r = k_1 c_{\mathrm{A}}$$

is essentially first order in the concentration of A and the rate constant is 0.23 min^{-1} at the temperature of the reactor. Determine the effluent concentration of reactant A when the total feed to the reactor is 16.2 ft^3/min.

θ (min)	$p(\theta)$ (min^{-1})	θ (min)	$p(\theta)$ (min^{-1})	θ (min)	$p(\theta)$ (min^{-1})
0	0	18.5	0.0604	37.0	0.0026
3.7	0.0005	22.2	0.0437	40.7	0.0010
7.4	0.0120	25.9	0.0259	44.4	0.0004
11.1	0.0415	29.6	0.0133	48.1	0.0001
14.8	0.0627	33.3	0.0061	51.8	0.0000

Exercise 8.6: Three CSTRs and maximum mixedness

(a) Given feed flowrate Q and total reactor volume V, let $\tau = V/Q$ and express the residence-time distribution for three equal-sized CSTRs in series.

(b) Consider second-order kinetics, $r = kc^2$, with feed concentration c_0. Define dimensionless variables

$$\overline{c} = c/c_0, \qquad x = \lambda/\tau, \qquad K = kc_0\tau$$

and show the maximum mixedness case for the three-CSTR RTD is described by

$$\frac{d\overline{c}}{dx} = K\overline{c}^2 + \frac{27x^2}{9x^2 + 6x + 2}(\overline{c} - 1), \qquad d\overline{c}/dx = 0 \text{ for } x = \infty \tag{8.45}$$

What "initial" value $\overline{c}(\infty)$ replaces the derivative boundary condition in Equation 8.45? Solve Equation 8.45 for a range of K values and plot $\overline{c}(0)$ versus K.

(c) Compare your results above with Zwietering's results for this problem, Tables 1 and 2 in Zwietering [34]. Why do you suppose Zwietering does not report values for the maximum mixedness case for $K = 3$ and $K = 50$? Did you experience any difficulties computing these cases? What typographical error do you find in the published tables? See also the results in Example 8.3.

Exercise 8.7: RTD for laminar-flow reactor

Calculate the residence-time distribution (RTD) for a tubular reactor undergoing steady, laminar flow (Hagen-Poiseuille flow). The velocity profile for Hagen-Poiseuille flow is [5, p. 51]

$$v(r) = \frac{2Q_f}{\pi R^2}\left[1 - \left(\frac{r}{R}\right)^2\right]$$

in which v is the axial velocity and R is the tube radius. Plot this RTD and compare to the RTD for the plug-flow reactor.

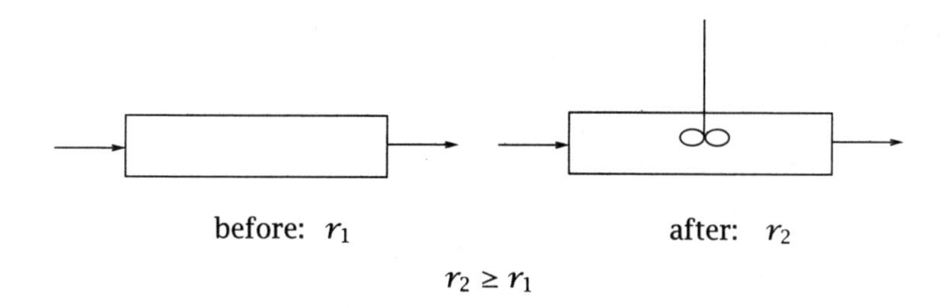

Figure 8.40: Reactor before and after stirrer is started; overall reaction rates before and just after mixing are r_1 and r_2, respectively.

Exercise 8.8: Other forms of the rate expression

(a) Establish that $r_m \leq r_s$ if the rate expression has the form

$$r(c) = \sum_i \alpha_i c^{\gamma_i} \qquad \alpha_i \geq 0 \quad \gamma_i \geq 1$$

(b) Establish that $r_m \geq r_s$ if the rate expression has the form

$$r(c) = \sum_i \beta_i c^{\delta_i} \qquad \beta_i \geq 0 \quad 0 \leq \delta_i \leq 1$$

Exercise 8.9: Mixing it up

Another perplexed student shows up at your office hours and complains, "Look, this mixing stuff doesn't make any sense." You and he discuss the simple case of a single, irreversible reaction with a concave rate expression,

$$\mathrm{A} \longrightarrow \mathrm{B}, \qquad r = kc_A^n$$

with $n \leq 1$. Displaying impeccable logic, he continues, "Assume we have a PFR operating at steady state at some overall conversion. If we turn on a stirrer and suddenly mix the contents of the PFR (shown in Figure 8.40), you claimed in class that the reactor's overall reaction rate will suddenly *increase* as well. But you also told us in Chapter 4 that for these same kinetics, the PFR has a higher steady-state conversion than a CSTR of the same volume. Both of these statements cannot be true!"

(a) Resolve this seeming contradiction by drawing a sketch of the overall reaction rate and the effluent concentration of the reactor versus time from the time just before the stirrer is started until the reactor comes to its new steady state with the stirrer running.

(b) Draw the same sketch if the reaction rate is convex.

(c) Calculate the exact results for the cases $k = 1$; $\tau = 1$; $c_0 = 1$; $n = 1/2, 1, 2$ (in appropriate units); and compare the numerical results to your sketches. Make the sketches before you perform the computation if you want to check your understanding of the issues.

Exercise 8.10: RTD for dispersed PFR with Danckwerts BCs

(a) Calculate the RTD for the dispersed PFR using the Danckwerts boundary conditions by solving numerically Equation 8.21 subject to Equations 8.22. You may wish to try the same approach used in solving Example 8.1.

(b) Compare your calculated values to Figures 8.10 and 8.11 for the same values of the dispersion number and mean residence time. These figures were calculated for the dispersed PFR with the simplified boundary conditions given in Equations 8.24. For what values of the dispersion number are the RTDs similar, and for what values of dispersion number are the RTDs different?

(c) Because the model is linear, the RTD of the dispersed PFR can also be solved analytically. Solve Equation 8.21 subject to Equations 8.22 for a unit step change in the feed using either Laplace transforms or an eigenfunction expansion. Compare your result with that given by Otake and Kunigita [25] (see also the discussion by Westerterp, Van Swaaij and Beenackers [33, p.187], Brenner [7, p. 231] and Carslaw and Jaeger [9, pp.114-119]). Compare your analytical solution to your previous calculation. Which approach do you prefer?

Exercise 8.11: Fitting a reactor model to a step response

Consider the 2-CSTR reactor model of Example 8.5.

(a) Write the mass balances for an inert component that is added to the Q_1 feed stream depicted in the bottom of Figure 8.32 and show

$$\begin{aligned} \frac{dc_1}{dt} &= \frac{1}{\tau_1}\left[\alpha c_{1f} - (\alpha+\rho)c_1 + \rho c_2\right] \\ \frac{dc_2}{dt} &= \frac{1}{\tau_2}\left[(\alpha+\rho)c_1 - (1+\alpha+\rho)c_2\right] \end{aligned} \tag{8.46}$$

in which c_1 and c_2 are the inert component concentrations in the first and second reactors, respectively.

A unit-step change in inert concentration in the feed is made to the first reactor, and the outlet concentration is shown in Figure 8.41.

(b) Adjust ρ by hand, solve Equations 8.46, and compare to the data. The other parameters are given in Table 8.3. Iterate until you have a pretty good fit to the data. What is the best value of ρ?

(c) Predict the conversion of A and yield of C for this reactor.

Exercise 8.12: Ordering PFRs and CSTRs

Consider a series of CSTRs and PFRs with a single nth-order reaction taking place in which

$$r(c) = kc^n$$

We showed in Example 8.2 that for a convex reaction-rate expression, the PFR should precede the CSTR to maximize conversion. The reason is shown in Figure 8.20. The CSTR–PFR configuration is a maximally mixed reactor, which has the lowest conversion for a convex rate expression.

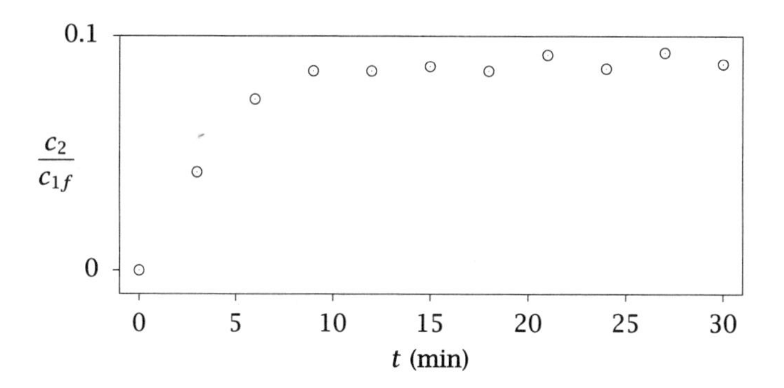

Figure 8.41: Effluent concentration versus time after unit step change in the first reactor.

t (min)	0	3	6	9	12	15	18	21	24	27	30
c_2/c_{1f}	0.00	0.042	0.073	0.085	0.085	0.087	0.085	0.092	0.086	0.093	0.088

(a) Consider now a series of four reactors: two CSTRs and two PFRs, which has six possible configurations: PPCC, PCPC, PCCP, CPPC, CPCP, and CCPP. For $n = 2$, we should be able to safely reason that the lowest conversion is achieved with series: CCPP, because that is also a maximally mixed reactor. Rank order all other configurations from lowest to highest conversion. Be careful with the ordering of PCCP and CPPC. Which of these achieves higher conversion?

(b) Repeat for $n = 1/2$. We know in this case the series CCPP should achieve the highest conversion because the rate expression is concave. Is your order in the two parts completely reversed or not? If not, can you explain why not?

(c) Can you guess the solution for a series of six reactors: three CSTRs and three PFRs? Note there are 20 possible reactor configurations in this case. Can you suggest a general result for ordering PFRs and CSTRs given a single, convex or concave rate expression? The authors are not aware of a result of this type in the research literature.

Exercise 8.13: Convex and concave functions

Convexity is a fundamental notion in the theory of optimization. Convexity can be defined to apply to functions that are not differentiable such as cases B and C in Figure 8.42. We say $f(x)$ is a convex function if

$$f(\alpha x + (1-\alpha)y) \leq \alpha f(x) + (1-\alpha) f(y), \qquad \text{all } x, y, 0 \leq \alpha \leq 1$$

In chemical reactor problems, the results stated for convex rate expressions apply to functions like $r = c^n$ in which $n \geq 1$, but also apply to rate expressions like that shown in Figure 8.42, which may be only piecewise differentiable and not monotone.

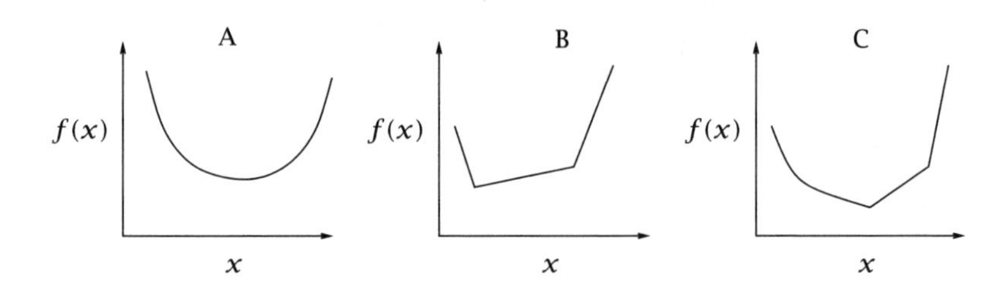

Figure 8.42: Some convex functions; differentiability is not required.

(a) Establish the following fact about convex functions, which will prove useful. Given f is convex and some value of x, then there exists a constant m such that

$$f(y) - f(x) \geq m(y - x), \qquad \text{for every } y \tag{8.47}$$

Hint: without loss of generality assume $x < y$ and consider an arbitrary z such that $x < z \leq y$. Use the definition of convexity of f to show

$$f(y) - f(x) \geq \left(\frac{f(z) - f(x)}{z - x} \right) (y - x)$$

We now choose m to minimize the function $(f(z) - f(x))/(z - x)$ over z to establish the result.

(b) Find m graphically for each of the functions sketched in Figure 8.42. Notice m depends on the value of x.

Exercise 8.14: Jensen's inequality

Jensen's inequality [15] states that given a convex function R, arbitrary function c and probability density $p(\theta)$,

$$\int_0^\infty R\left(c(\theta)\right) p(\theta) d\theta \geq R\left(\int_0^\infty c(\theta) p(\theta) d\theta \right) \tag{8.48}$$

In terms of the chemical reactor, given a convex production rate expression, the mean production rate is greater than or equal to the production rate evaluated at the mean concentration, i.e., mixing the reactor contents lowers or does not change the overall production rate.

(a) Derive Jensen's inequality. Hint: Let

$$\alpha = \int_0^\infty c(\theta) p(\theta) d\theta$$

and use Equation 8.47 to show

$$R(c(\theta)) - R(\alpha) \geq m(c(\theta) - \alpha), \qquad \text{for every } c(\theta)$$

Multiply by $p(\theta)$ and integrate to obtain Jensen's inequality.

(b) What analogous inequality holds when R is concave?

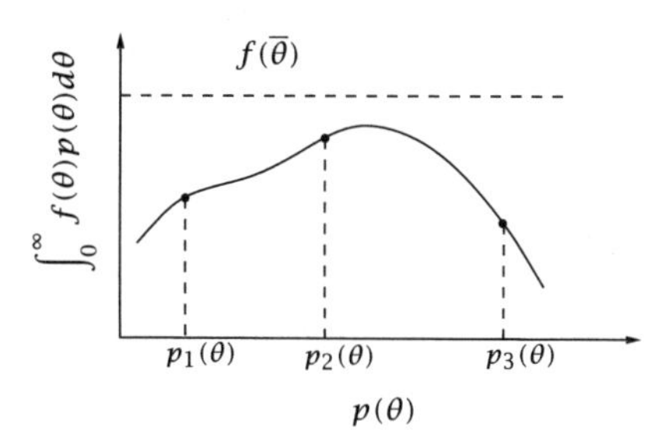

Figure 8.43: Value of $\int_0^\infty f(\theta)p(\theta)d\theta$ as the function $p(\theta)$ varies.

Exercise 8.15: Optimization of a residence-time distribution function

Let f be a concave function and $p(\theta)$ a residence-time distribution with mean

$$\overline{\theta} = \int_0^\infty \theta p(\theta) d\theta \tag{8.49}$$

Consider maximizing f over *all possible RTDs* having the specified mean $\overline{\theta}$

$$\max_{p(\theta)} \int_0^\infty f(\theta)p(\theta)d\theta \tag{8.50}$$

(a) Show the maximum is achieved for

$$p(\theta) = \delta(\theta - \overline{\theta})$$

i.e., the maximum is achieved when the residence-time distribution is arbitrarily narrow at the specified mean value.

Hint: Use Jensen's inequality and choose $c(\theta) = \theta$ to show

$$f(\overline{\theta}) \geq \int_0^\infty f(\theta)p(\theta)d\theta, \qquad \text{for all } p(\theta) \tag{8.51}$$

which provides an upper bound for the maximum that is independent of $p(\theta)$ as sketched in Figure 8.43.[2] By examining Equation 8.51, for what $p(\theta)$ is this upper bound achieved? For what f is this maximum unique? See also Nauman and Buffham [24, p.162].

(b) What analogous optimization problem can you solve when f is a convex function?

[2]Notice Figure 8.43 is merely a representation of the true situation because the abscissa is meant to depict a set of *functions*, not a set of scalar values as in an ordinary graph.

Bibliography

[1] T. K. Abraham and M. Feinberg. Kinetic bounds on attainability in the reactor synthesis problem. *Ind. Eng. Chem. Res.*, 43(2):449–457, 2004.

[2] A. Bakker, A. H. Haidari, and E. M. Marshall. Design reactors via CFD. *Chem. Eng. Prog.*, pages 30–39, December 2001.

[3] F. Bezzo, S. Macchietto, and C. C. Pantelides. General hybrid multi-zonal/CFD approach for bioreactor modeling. *AIChE J.*, 49(8):2133–2148, August 2003.

[4] L. T. Biegler, I. E. Grossmann, and A. W. Westerberg. *Systematic Methods of Chemical Process Design.* Prentice Hall PTR, Upper Saddle River, New Jersey, 1997.

[5] R. B. Bird, W. E. Stewart, and E. N. Lightfoot. *Transport Phenomena.* John Wiley & Sons, New York, second edition, 2002.

[6] K. B. Bischoff. A note on boundary conditions for flow reactors. *Chem. Eng. Sci.*, 16(1-2):131–133, 1961.

[7] H. Brenner. The diffusion model of longitudinal mixing in beds of finite length. Numerical values. *Chem. Eng. Sci.*, 17:229–243, 1962.

[8] J. F. Burri, S. D. Wilson, and V. I. Manousiouthakis. Infinite dimensional state-space approach to reactor network synthesis: Application to attainable region construction. *Comput. Chem. Eng.*, 26(6):849–862, 2002.

[9] H. S. Carslaw and J. C. Jaeger. *Conduction of Heat in Solids.* Oxford University Press, Oxford, second edition, 1959.

[10] S. P. Chauhan, J. P. Bell, and R. J. Adler. On optimum mixing in continuous homogeneous reactors. *Chem. Eng. Sci.*, 27:585–591, 1972.

[11] P. V. Danckwerts. Continuous flow systems: Distribution of residence times. *Chem. Eng. Sci.*, 2:1–13, 1953.

[12] M. Feinberg. Toward a theory of process synthesis. *Ind. Eng. Chem. Res.*, 41(16):3751–3761, 2002.

[13] M. Feinberg and P. Ellison. General kinetic bounds on productivity and selectivity in reactor-separator systems of arbitrary design: I. Principles. *Ind. Eng. Chem. Res.*, 40(14):3181–3194, 2001.

[14] M. Feinberg and D. Hildebrandt. Optimal reactor design from a geometric viewpoint — I. Universal properties of the attainable region. *Chem. Eng. Sci.*, 52(10):1637-1665, 1997.

[15] W. Feller, editor. *An Introduction to Probability Theory and Its Applications: Volume II.* John Wiley & Sons, New York, second edition, 1971.

[16] D. Glasser, D. Hildebrandt, and S. Godorr. The attainable region for segregated, maximum mixed and other reactor models. *Ind. Eng. Chem. Res.*, 33:1136-1144, 1994.

[17] D. Hildebrandt and D. Glasser. The attainable region and optimal reactor structures. *Chem. Eng. Sci.*, 45:2161-2168, 1990.

[18] F. J. M. Horn. Attainable and non-attainable regions in chemical reaction technique. In *Proceedings of the third European Symposium on Chemical Reaction Engineering*, pages 293-302, London, UK, 1964. Pergamon Press.

[19] S. Kauchali, W. C. Rooney, L. T. Biegler, D. Glasser, and D. Hildebrandt. Linear programming formulations for attainable region analysis. *Chem. Eng. Sci.*, 57(11):2015-2028, 2002.

[20] A. C. Kokossis and C. A. Floudas. Synthesis of isothermal reactor-separator-recycle systems. *Chem. Eng. Sci.*, 46(5/6):1361-1383, 1991.

[21] A. Lakshmanan and L. T. Biegler. Synthesis of optimal chemical reactor networks. *Ind. Eng. Chem. Res.*, 35:1344-1353, 1996.

[22] I. Langmuir. The velocity of reactions in gases moving through heated vessels and the effect of convection and diffusion. *J. Am. Chem. Soc.*, 30 (11):1742-1754, 1908.

[23] O. Levenspiel. *Chemical Reaction Engineering.* John Wiley & Sons, New York, third edition, 1999.

[24] E. B. Nauman and B. A. Buffham. *Mixing in Continuous Flow Systems.* John Wiley & Sons, New York, 1983.

[25] T. Otake and E. Kunigita. *Kagaku Kogaku*, 22:144, 1958.

[26] J. M. Ottino. *The kinematics of mixing: stretching, chaos and transport.* Cambridge University Press, Cambridge, 1989.

[27] S. J. Parulekar and D. Ramkrishna. Analysis of axially dispersed systems with general boundary conditons —I. Formulation. *Chem. Eng. Sci.*, 39 (11):1571-1579, 1984.

[28] R. Shinnar. Use of residence- and contact-time distributions. In J. J. Carberry and A. Varma, editors, *Reactor Design in Chemical Reaction and Reactor Engineering*, pages 63–150. Marcel Dekker, Inc., New York, 1986.

[29] C. M. Villa, J. O. Dihora, and W. H. Ray. Effects of imperfect mixing on low-density polyethylene reactor dynamics. *AIChE J.*, 44(7):1646–1656, 1998.

[30] J. V. Villadsen and W. E. Stewart. Solution of boundary-value problems by orthogonal collocation. *Chem. Eng. Sci.*, 22:1483–1501, 1967.

[31] H. Weinstein and R. J. Adler. Micromixing effects in continuous chemical reactors. *Chem. Eng. Sci.*, 22:65–75, 1967.

[32] S. W. Weller. Langmuir as chemical engineer... or, from Danckwerts to Bodenstein and Damköhler. *Chem. Eng. Ed.*, 28:262–264, 1994.

[33] K. R. Westerterp, W. P. M. van Swaaij, and A. A. C. M. Beenackers. *Chemical Reactor Design and Operation.* John Wiley & Sons, New York, 1984.

[34] T. N. Zwietering. The degree of mixing in continuous flow systems. *Chem. Eng. Sci.*, 11:1–15, 1959.

9

Parameter Estimation for Reactor Models

Much of this book is devoted to constructing and explaining the fundamental principles governing chemical reactor behavior. As we have seen, these principles are efficiently expressed as mathematical models. Drawing quantitative conclusions from these principles requires not only the models, however, but the values of the parameters in the models as well. In most practical applications, many of the parameters in these models are not known or available beforehand. Evaluating the model parameters therefore usually requires something outside the scope of the theory: experimental data. In this chapter we consider the issues involved in gathering data and estimating parameters from data with the expressed purpose of identifying a chemical reactor model. In the process, we develop convenient and reliable tools that aid us in extracting as much information as possible from the expensive and time-consuming process of building experimental facilities, designing experiments, and making careful measurements.

9.1 Experimental Methods

9.1.1 Analytical Probes for Concentration

Experimental design includes specifying what variables to measure and how best to measure them. Included in the list of variables are reactor volume, inlet flowrates, temperature, inlet (initial) concentrations of one or more components, and effluent (final) concentrations of one or more components. Concentration or molar flowrate are the dependent composition variables in the design equations, and reaction rates are generally specified in terms of component concentrations. Whether the reaction is homogeneous or heterogeneous, solution of the material balance requires knowledge of the fluid-phase concentrations, so we focus here on the important methods for measuring component concentrations in liquid-phase and gas-phase mixtures. You are likely to encounter these methods in research and analytical services laboratories.

To determine concentration, a unique chemical or physical attribute of the molecule under study is measured. For example, the vibrational frequency can be probed by infrared spectroscopy, the thermal conductivity can be probed by gas chromatography, or the mass/charge ratio for an ionized molecule can be probed by mass spectroscopy. The first example can be applied to the mixture while the second example requires the components to be separated before measuring the thermal conductivity. The final example samples the mixture and performs the mass discrimination as part of the analytical technique.

Analytical methods require calibration and proper selection of conditions to ensure the components of interest can be detected and their amounts quantified. Often only a few components out of an entire mixture can be monitored, and it is necessary to select the ones that provide the most information about the progress of the reactions. In most situations, the composition is measured only at discrete sample times, t_i. The general problem is to collect enough of the proper information so that the experimental data can be tested against appropriate kinetic and process models.

Infrared spectroscopy exploits the absorption of electromagnetic radiation associated with molecular motion of chemical bonds, in particular stretching, bending and rocking motions. The energies of these vibrational motions are in the infrared region of the electromagnetic spectrum (2.5–20 μm; 500–4000 cm^{-1}) [19, 3, 23]. Examples include the C—H asymmetric stretching mode of a methyl group (CH_3) at 2962 cm^{-1}, the C—C stretching mode of ethylene at 1623 cm^{-1}, the O—H stretching mode of methanol at 3682 cm^{-1}, the R branch of the C—O stretching mode of carbon monoxide at 2170 cm^{-1}, the C—O asymmetric stretching mode of carbon dioxide at 2350 cm^{-1}, and the C—O stretching mode of acetone at 1718 cm^{-1}. Infrared spectroscopy can be applied to gas, liquid and solid samples, and it is both a qualitative technique to identify the presence of molecules and a quantitative technique that can establish the concentration of the molecules.

The quantitative aspects are based on Beer's law,

$$A = \epsilon b c$$

in which A is the measured absorbance, ϵ is the molar absorptivity, b

is the path length and c is the concentration of the absorbing species. The liquid-phase or gas-phase sample is placed in a fixed-path-length cell. For gases, this cell may be simply a tube with infrared transparent windows affixed to each end. The intensity of a particular wavelength of light passing through the sample is recorded and compared to the intensity recorded when the cell is empty or filled with an inert substance. The infrared spectrometer reports the intensity difference as an absorbance. With proper calibration using standard mixtures, the absorbance intensity can be converted into a concentration using Beer's law because the path length and molar absorptivity are fixed variables. Infrared spectroscopy can be used as a continuous monitoring technique by passing the gas or liquid through the cell as it exits a flow reactor.

Infrared spectrometers vary by the means used to discriminate wavelengths. Dispersive instruments use a monochrometer and a grating to select the particular wavelength of interest or range of wavelengths that are scanned. Fourier transform spectrometers use an interferometer with a movable mirror to modulate the light source before it is incident on the sample, and all wavelengths are sampled simultaneously. The signal is recorded as a function of the mirror displacement, and after averaging multiple measurements and transforming the interferograms, the spectrometer reports absorbance versus wavelength.

Because infrared spectroscopy is an optical technique, the sample must be transparent in the spectral region of interest. Infrared spectroscopy can be used for heterogeneous samples and adsorbates on catalyst surfaces [11, 17]. If the sample cannot be made thin enough to be transparent, special probing techniques such as attenuated total internal reflectance can be used to monitor the concentration.

Gas chromatography is the workhorse of routine analysis [14, 18, 21, 16] because it can be used for simple gases, such as separating the constituents in air, as well as complex hydrocarbon mixtures, such as the components in gasoline. Gas chromatography can resolve concentrations in the range of parts per billion to tens of percent. Figure 9.1 presents a simplified schematic of a gas chromatograph. It consists of an injector, column and detector; each section has a separate temperature zone. The gas/liquid sample to be analyzed is injected as a pulse input in the injector and it flows along with the carrier gas through the column, where it is separated into components, and then flows to the detector. Most gases, and liquids that vaporize at temperatures less than about 350°C, can be analyzed by gas chromatography. More spe-

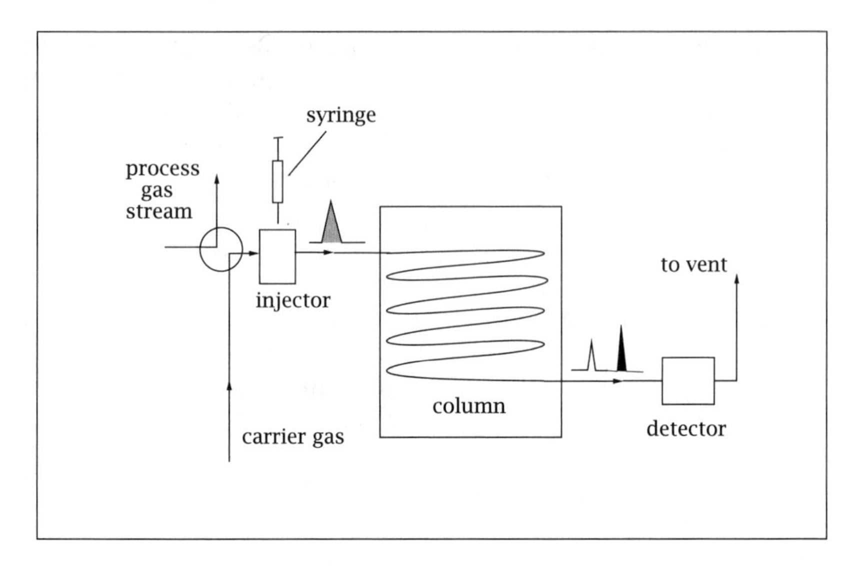

Figure 9.1: Gas chromatograph schematic.

cialized chromatography techniques, such as high-pressure liquid chromatography, extend the basic chromatographic technique to very high molecular weight compounds or to thermally unstable compounds.

The detector is typically a thermal conductivity detector or a flame ionization detector. The thermal conductivity detector is just a heated wire that has a constant current passed through it. The wire cools as a carrier gas flows over it and a steady wire temperature is established as a base-line response. If the gas flowing past the wire changes composition and therefore thermal conductivity, the wire temperature changes. The voltage drop across the wire and the current through the wire are related via Ohm's law, $V = IR$, to the wire resistance, which is temperature dependent. The thermal conductivity detector then operates by converting the voltage drop change to a temperature change as the gas mixture flowing past the wire changes from carrier gas only to carrier gas plus sample. A carrier gas is selected, typically He or N_2, that has a thermal conductivity significantly different from any component in the mixture. The detector identifies the *thermal conductivity* of the gas mixture, not the composition of the gas mixture. The thermal conductivity can be related to the composition by calibrating with carrier gas and samples of known composition.

The flame ionization detector operates by blending the column effluent with H_2 and burning it with air in an ionizing flame. The ions are

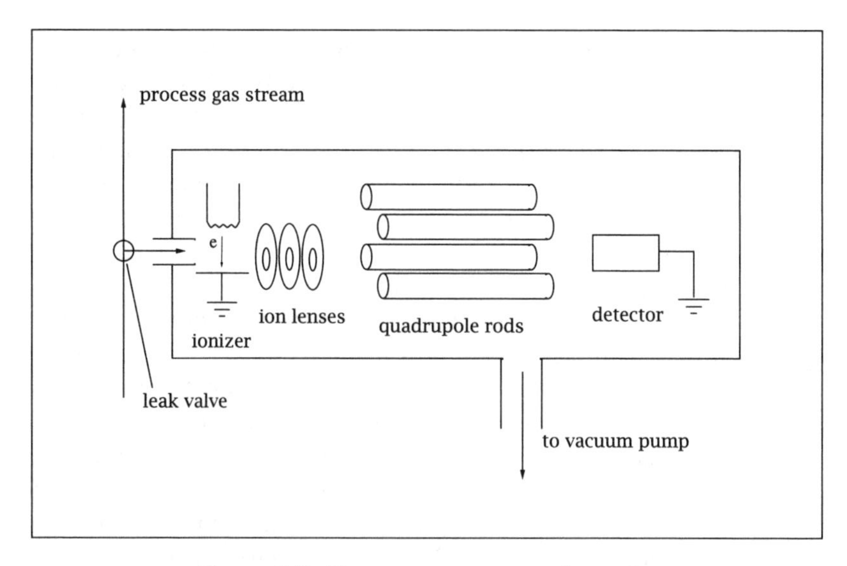

Figure 9.2: Mass spectrometer schematic.

counted with an electrometer. A flame ionization detector works best with hydrocarbons. The electrometer merely records the presence of ions. The greater the concentration of a particular component in the flame, the greater the signal. The flame ionization detector does not identify the composition of the gas in the flame. Separation of the injected sample is accomplished by column selection. The column is a long, narrow tube that is either filled with a packing such as a porous adsorbent, or contains a thin coating of a high-molecular-weight liquid on the inner tube surface that acts as the adsorbent. As the injected pulse travels along the column length, the components interact with the adsorbent to differing degrees and are eventually separated into pulses of almost pure components. The science of gas chromatography lies in designing and selecting a column that separates the constituents, perhaps sorting them by boiling point or polarity. Standard mixtures are used to establish the elution order and time of elution for gases in a particular column.

Mass spectroscopy can be used to determine the partial pressure of gases in a mixture. Figure 9.2 presents a simple schematic of a mass spectrometer for gas monitoring. The mass spectrometer is maintained in a high-vacuum chamber at pressures less than 10^{-6} Torr. The gas to be analyzed flows past a special valve, called a leak valve, that permits a small fraction of the flowing gas stream to enter the vacuum chamber.

A steady state is established in which the amount of gas leaked into the chamber is pumped away. Pumping speeds of different molecules may vary; a predictable relationship between the relative mole fractions in the flowing gas stream and the relative mole fractions in the vacuum chamber can be established [20]. The mass spectrometer measures the partial pressures of the molecules inside the vacuum chamber. As configured in Figure 9.2, the mass spectrometer generates a continuous measure of the gas-phase composition.

The spectrometer has four major components: ionizer, focusing lens system, quadrupole section and detector. The spectrometer operates on the principal that ions can be steered into the detector or diverted away from the detector. The ionizer and lens system produce, collimate and impart a constant kinetic energy to the ions. The quadrupole rods have a varying direct current (dc) potential and oscillating radio frequency (rf) field applied across pairs of rods that causes all ions except those with a particular mass/charge (m/z) ratio to follow an unstable path and collide with the rods. Depending on the instantaneous dc voltage, only ions with a single m/z follow a stable path to the detector, which is at a large negative potential (−2 to −4 kiloelectronvolt (keV)). The ions received at the detector induce a current to flow in the detector.

Various ionization techniques can be used [14, 13, 9]; the most common is electron impact ionization. Electrons (typically 70 eV) emitted from a very hot filament collide with molecules in the ionizer region. The electron impact ionization process ionizes the molecule or atom

$$\mathrm{M} + \mathrm{e}^- \longrightarrow \mathrm{M}^+ + 2\mathrm{e}^-$$

and ionizes and fragments the molecule into a number of ions, M_j^+, of mass less than M.

$$\mathrm{M} + \mathrm{e}^- \longrightarrow \mathrm{M}_j^+ + 2\mathrm{e}^- + \text{neutral fragments}$$

The relative ratios of the M_j^+ fragments can be used to identify the structure of M. The current generated by M^+ and all M_j^+ fragments at the detector is used to establish the partial pressure of each component in the vacuum chamber.

9.1.2 Experimental Reactors for Kinetics Studies

Time and position are the independent variables, and concentration, volumetric flowrate and temperature are the dependent variables. Ultimately, the experimental data are fit against a model for the reactor in which the data were collected. This process becomes more challenging when the reactor is nonisothermal. Therefore, one often tries to design experimental reactors and select reaction conditions that permit isothermal operation. The effect of temperature is most easily explored by operating the isothermal reactor at different temperatures.

The reactor models presented in Chapters 4–7 were based on specific flow assumptions, such as the well-mixed reactor and plug flow reactors. Experimental reactors must retain these flow attributes if we wish to use the simple material balances for estimating model parameters.

Starting and quenching the reaction can present formidable challenges. The experiment should be designed to ensure that the reaction takes place only in the reactor and under the conditions assumed in the reactor model. For example, if a reaction has a large activation energy, it is necessary to operate the reaction at elevated temperatures to achieve a measurable rate. If using a flow reactor, the reactants must be preheated and injected into the reactor without any reaction occurring prior to the reactor inlet. In an isothermal batch reactor, the reactor contents must be heated rapidly to the desired temperature and held at constant temperature. Unless steps are taken, the reactions may continue in the samples withdrawn from the reactor. If the reaction has a low activation energy, lowering the temperature does not quench the reaction. It may prove necessary to dilute the sample or add components to cause side reactions to quench a reaction.

Most experimental reactors are small to ensure large surface-to-volume ratio for good temperature regulation, and to minimize the chemical inventory needed to conduct the experiments. As illustrated in Figures 1.6 and 1.8, flow and batch reactors with volumes of several hundred nanoliters have been fabricated out of silicon wafers and used for organic synthesis screening studies.

Liquid-phase reactions. A batch reactor, CSTR or PFR could be used for liquid-phase studies. Many times, the choice of a batch reactor is made by pumping considerations. Continuous, steady and nonpulsating flow may be difficult to realize for liquids in a laboratory setting. Accurate and large-volume positive displacement pumps are costly, and small centrifugal pumps many not be available to handle the reactants. In the simplest case, a batch reactor may be nothing more than a beaker with a magnet stir bar for agitation. The situation becomes more complex if temperature regulation, efficient mixing, and isolation of the contents from ambient gases are important factors. Samples are typically withdrawn as aliquots at preset time intervals via syringe and saved for subsequent analysis. Figure 1.2 illustrates the batch reactor internals required to control the temperature and withdraw samples.

Gas-phase reactions. The reactor of choice for gas-phase reactions is the plug-flow reactor. The batch reactor and CSTR are rarely used for gas-phase studies because it is difficult to achieve ideal mixing of gases. The isothermal PFR is usually nothing more complicated than an open tube maintained at a constant temperature. Different tube diameters and lengths are used to change the reactor volume as an experimental variable. Gas flows are routinely and reproducibly regulated with mass flow controllers or manually with metering valves and flow meters; this enables a wide range of volumetric flowrates to be examined as an experimental variable. Sampling is best accomplished by directing the gas effluent through a gas sampling valve so that precise samples can be injected automatically into a gas chromatograph.

It is quite common in experimental studies to operate a PFR as a so-called **differential reactor**. A differential reactor is a PFR in which the reactants experience only a small (differential) change in the extents of the reactions. The differential reactor is modeled using a limiting form of the PFR material balance design equation and generates data of the form rate versus concentration. Rate versus concentration data may prove useful in certain kinetic parameter studies. The PFR design equation is

$$\frac{d}{dV}N_j = R_j = \sum_i \nu_{ij} r_i \tag{9.1}$$

If we assume the molar flows change by only a small amount, we can approximate the derivative with a finite difference formula

$$\frac{N_j - N_{jf}}{V_R} \approx R_{jf} = \sum_i \nu_{ij} r_{if} \tag{9.2}$$

$$N_j \approx N_{jf} + R_{jf} V_R \tag{9.3}$$

in which R_{jf} and r_{if} are the production rates and reaction rates evaluated at the feed conditions. Now we account for the change in volumetric flowrate. (Exercise 9.7 explores what happens if we neglect the volumetric flowrate change.) If we assume an ideal gas, we have $Q = NRT/P$, in which N is the total molar flow, and summing Equation 9.3 produces

$$N = N_f + \sum_i r_{if} \bar{\nu}_i V_R$$

$$Q = Q_f + \frac{RT}{P} \sum_i \bar{\nu}_i r_{if} V_R$$

Substituting the Q relation into Equation 9.3 and rearranging produces

$$\sum_i (\nu_{ij} - y_j \bar{\nu}_i) r_{if} = \frac{Q_f}{V_R} (c_j - c_{jf}) \tag{9.4}$$

in which y_j is the mole fraction of component j in the effluent. Experimentally, one measures c_j in the effluent stream, which determines the right-hand side of Equation 9.4. One then solves a least-squares problem to determine all of the r_{if}. By varying c_{jf}, it is possible to measure efficiently the reaction rates over a range of concentrations.

For example, if we have the single reaction A $\longrightarrow$ 2B, and we measure both c_A and c_B, using Equation 9.4, we obtain two estimates of the single reaction rate

$$r = \frac{Q_f}{V_R} \frac{(c_{Af} - c_A)}{-(1 + y_A)} \qquad r = \frac{Q_f}{V_R} \frac{(c_{Bf} - c_B)}{(2 - y_B)} \tag{9.5}$$

and the least-squares solution is equivalent to taking the average.

Heterogeneously catalyzed reactions. Macroscopic fluid models are combined with microscopic transport models in the catalyst particles to describe how concentration changes with time and position in a catalytic reactor. Special considerations must be given to the selection of experimental temperature and catalyst particle size to minimize (and hopefully eliminate) internal transport limitations on the catalytic reaction rate. The next requirement is that the flow pattern in the reactor is accurately represented by the well-mixed or plug-flow assumption. The subsequent discussion applies to gas-phase reactants.

Commercial experimental reactors that achieve ideal mixing are available. These reactors either hold the catalyst in a basket arrangement and circulate the reacting gases through the bed or they spin the bed at a high revolution rate. The design achieves a sufficiently high linear velocity across the catalyst bed and volumetric flowrate in the reactor vessel so that the contents of the fluid phase can be assumed to be well mixed. The high internal circulation rate and linear velocities ensure uniform temperatures in the reactor, and special seals permit operation at pressures exceeding 100 atm. These well-mixed reactors can be treated as batch or CSTR, depending on how they are operated.

A second type of reactor is the plug-flow, fixed-bed reactor. The reacting gas must move through the reactor as a plug, with a flat velocity profile. Previous studies of transport in porous media have led to characteristic groups that gauge how well the fluid flow can be characterized as plug-like [8]. These groups and their values include the particle Reynolds number

$$\mathrm{Re} = \frac{D_p m}{S \mu \epsilon_B} \geq 30$$

the axial aspect ratio

$$\frac{L}{D_p} \geq 30$$

and the radial aspect ratio

$$\frac{D_t}{D_p} \geq 10$$

The Reynolds number ensures turbulent flow and with it effective radial mixing. The axial aspect ratio ensures that axial dispersion is minimal. The radial aspect ratio ensures that channeling does not occur. Channeling refers to the situation in which the fluid close to the reactor walls travels faster than the fluid at the center of the tube. When these three dimensionless conditions are satisfied, one can usually model the reactor as a PFR. The velocity profiles are complex, however, and broad generalizations should be used with caution [28].

9.1.3 Characterizing Catalysts and Surfaces

A wide variety of analytical probes are used to study, characterize and monitor catalysts and catalyst surfaces. Our intent here is to discuss some of the more common and routine techniques. Much more detail and many more techniques can be found in specialized books [33, 32, 27, 12]. A catalyst functions through the highly specific interactions the active sites have with the reactants. The catalyst might be a metal dispersed on an inert carrier, a polycrystalline or amorphous mixture of metal oxides, or a zeolite (a crystalline and highly porous oxide). The experimentalist is typically interested in the catalyst composition, structure of the catalyst, distribution of active sites, presence of poisons/impurities after the catalyst has been used, and number of active sites — parameters that influence the catalytic activity.

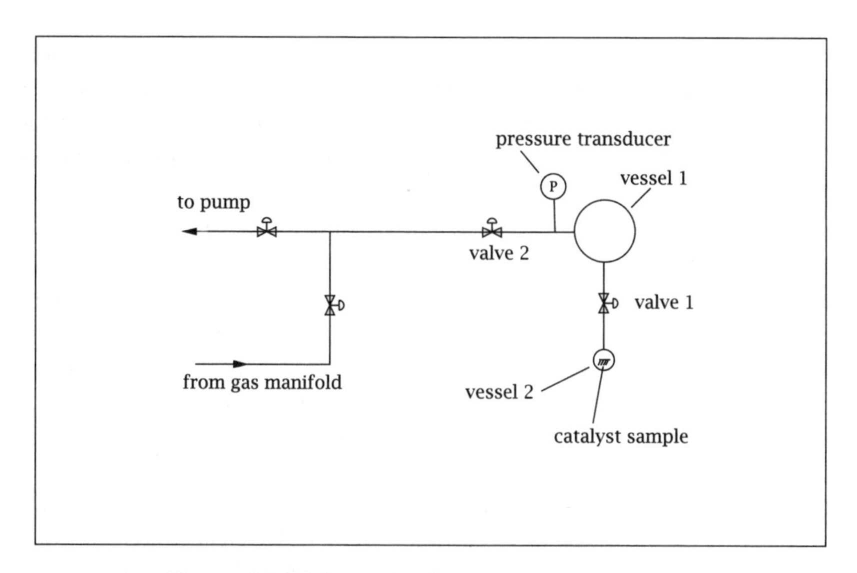

Figure 9.3: Volumetric chemisorption apparatus.

Active site densities can be determined using adsorption isotherms as discussed in Chapter 5. The equipment is straightforward. Figure 9.3 presents a schematic view of the essential components of a chemisorption apparatus. The apparatus includes a pressure transducer for accurately measuring the pressure, a gas-mixing manifold for introducing the adsorbate gas, a vacuum pump for establishing a low base pressure and removing all adsorbates from the catalyst, and two vessels of known volumes (V_1 and V_2) separated by a valve. The catalyst is placed in Vessel 2 and this section of the apparatus is maintained at constant temperature. An experiment proceeds by adding the catalyst and evacuating the entire apparatus until nothing is adsorbed on the catalyst and the pressure is below the detection level of the pressure transducer. Valve 1 is closed, Valve 2 is opened, and gas is admitted to Vessel 1. Valve 2 is closed, and the pressure and temperature of Vessel 1 are recorded. Valve 1 is then opened and gas both expands into Vessel 2 and adsorbs on the catalyst. The pressure is noted after reaching a steady value. Using an equation of state, such as the ideal gas law, the temperatures and volumes of Vessels 1 and 2, and the pressure before and after admitting gas to Vessel 2, one can determine the moles of adsorbed gas. The experiment continues by closing Valve 1, opening Valve 2 and refilling Vessel 1, and then closing Valve 2. The pressure in Vessel 1 is noted before and after opening

Valve 1. This sequence of filling and expanding is repeated until there is no change in the pressure upon expanding the gas into Vessel 2. This procedure generates the number of moles adsorbed as a function of pressure, which can be used to determine the number of active sites and the adsorption equilibrium constant.

Crystallinity and particle size can be determined by X-ray diffraction. Supported metals and metal oxides generally adopt their native packing when they form crystallites sufficiently large to define a structure. A diffraction pattern identifies the atomic packing structure, provided the material is crystalline, and is in sufficiently large particles ($\geq$ 10 nm) and has a diffraction pattern that is not obscured by the support. X-ray diffraction is used to assess the presence of amorphous versus crystalline phases and changes in the relative amounts of different crystalline phases with thermal treatment. A given crystal has a unique diffraction pattern. As particle size decreases, the diffraction pattern becomes diffuse and the peaks broaden. This phenomenon is termed **line broadening** and it can be used to determine particle size [33].

Particle size also can be established using transmission electron microscopy. In this technique, a mono-energetic beam of high-energy electrons is directed through a thin, solid specimen ($\leq$ 200 nm). As the electrons travel through the sample, they interact with the atomic constituents of the sample and scatter. This scattering process leads to contrast images being projected on a detector. Generally, it is easy to discern a heavy element from lighter elements, such as Pt dispersed on Al_2O_3. The heavier element produces a different intensity image. It is routine to resolve less than 0.2 nm and image individual atoms. The particle size is determined by measuring the size of the image and working back from the magnification used in the experiment. A sufficient number of particles must be measured to give a statistically meaningful average particle size. This technique can be applied before and after reaction to determine if the particles have changed size during the reaction.

Composition can be determined by dissolving the catalyst in a suitable liquid, injecting the liquid solution into an inductively coupled plasma, and analyzing the ions generated in the plasma with a mass spectrometer. The inductively coupled plasma source is highly efficient at ionizing the injected sample, and this measurement has very high elemental sensitivity and is quantitative. The digestion technique is destructive.

Composition also can be determined in a nondestructive manner using electrons or soft X-ray sources to excite the core levels of the atoms. Secondary or primary electrons are ejected from the sample at energies that are characteristic of the elements. These ejected electrons have a mean free path that is energy dependent and can only exit the sample if they are formed in the top 0.5–10 nm of the sample surface. The ejected electron techniques identify what is present; it is difficult to make these techniques quantitative.

Auger electron spectroscopy uses a focused electron beam (3–5 keV) to eject an inner atomic level electron and create a singly ionized excited atom. Electrons from other energy states in the atom rapidly fill the vacancy. The energy released in this de-excitation electronic transition can be transferred to a third electron. If the binding energy of this third electron is less than the energy transferred to it, it is ejected as an Auger electron. The energy of this ejected electron is monitored with an Auger electron spectrometer. Elements have discrete electron energy levels with distinguishable binding energies, so the energy of the ejected Auger electrons provide a means of identifying the elements that are present [10].

X-ray photoelectron spectroscopy uses soft X-rays (Mg $K\alpha$ at 1253.6 eV and Al $K\alpha$ at 1486.6 eV) to eject core-level electrons from the unknown sample. The kinetic energy of the ejected electrons is monitored and related back to the binding energy. The binding energy is used to identify the elements present [22]. Since the binding energy of inner-shell electrons typically changes with the valence of the element, X-ray photoelectron spectroscopy can be used to follow oxidation or reduction of the catalyst as well as identify the presence of elements.

9.2 Data Modeling and Analysis

Having provided the context in which we perform experiments and make measurements, we now turn to the issue of how we extract the information contained in the data. The two large questions we must address are: what model *structure* is appropriate to describe the reacting system of interest, and, having selected a structure, what model *parameters* best represent the data we have collected, and how certain are we about these parameter values. The first question has occupied us up to this point; it has been the central focus of Chapters 1–8. Indeed, understanding the fundamental principles and models that explain the many kinds of chemical reaction and reactor behaviors is one of the essential goals of the chemical engineering education.

In this section, we focus finally on the second question, how to estimate model parameters from data. This problem is also important, and has a distinguished place in the history of science and engineering. Accurate prediction of the motions of the planets based on astronomical measurements was one of the early motivating problems of parameter estimation. Solving this problem led Gauss to invent the least-squares method in the late 1700s. Gauss's summary of this effort more than 175 years ago remains valid today:

> One of the most important problems in the application of mathematics to the natural sciences is to choose the best of these many combinations, i.e., the combination that yields values of the unknowns that are least subject to errors.
>
> Theory of the Combination of Observations Least Subject to Errors. C. F. Gauss, 1821 [15, p.31].

9.2.1 Review of the Normal Distribution

Probability and statistics provide one useful set of tools to model the uncertainty in experimental data. It is appropriate to start with a brief review of the normal distribution, which plays a central role in analyzing data. The normal or Gaussian distribution is ubiquitous in applications. It is characterized by its mean, m, and variance, σ^2, and is given by

$$p(x) = \frac{1}{\sqrt{2\pi\sigma^2}} \exp\left(-\frac{1}{2}\frac{(x-m)^2}{\sigma^2}\right) \tag{9.6}$$

Figure 9.4 shows the univariate normal with mean zero and unit variance. We adopt the following notation to write Equation 9.6 more compactly

$$x \sim N(m, \sigma^2)$$

which is read "the random variable x is distributed as a normal with mean m and variance σ^2." Equivalently, the probability density $p(x)$ for random variable x is given by Equation 9.6.

For distributions in more than one variable, we let $\boldsymbol{x}$ be an n_p-vector and the generalization of the normal is

$$p(\boldsymbol{x}) = \frac{1}{(2\pi)^{n_p/2}\,|\boldsymbol{P}|^{1/2}} \exp\left[-\frac{1}{2}(\boldsymbol{x}-\boldsymbol{m})^T\boldsymbol{P}^{-1}(\boldsymbol{x}-\boldsymbol{m})\right]$$

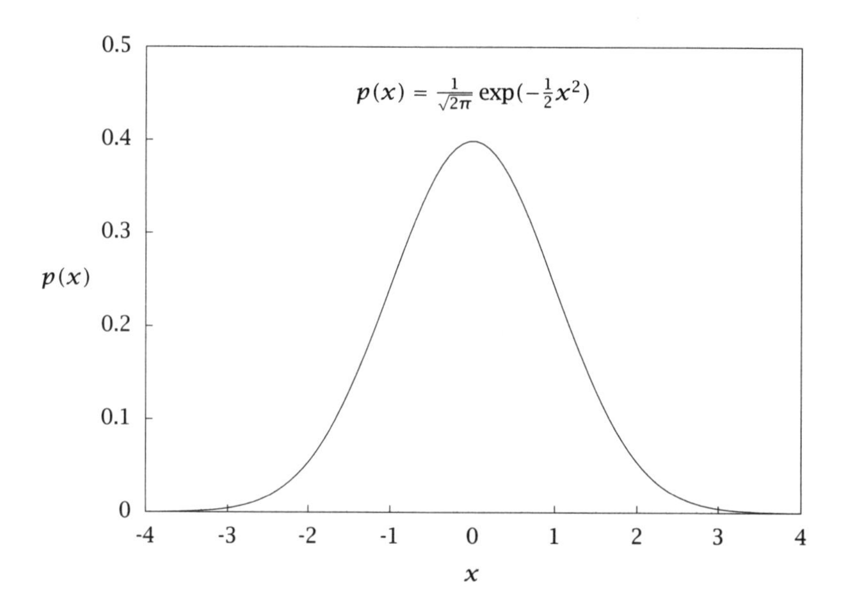

Figure 9.4: Univariate normal with zero mean and unit variance.

in which the n_p-vector $\boldsymbol{m}$ is the mean and the $n_p \times n_p$-matrix $\boldsymbol{P}$ is called the covariance matrix. The notation $|\boldsymbol{P}|$ denotes determinant of $\boldsymbol{P}$. We also can write for the random variable $\boldsymbol{x}$ vector

$$\boldsymbol{x} \sim N(\boldsymbol{m}, \boldsymbol{P})$$

The matrix $\boldsymbol{P}$ is a real, symmetric matrix. Figure 9.5 displays a multivariate normal for

$$\boldsymbol{P}^{-1} = \begin{bmatrix} 3.5 & 2.5 \\ 2.5 & 4.0 \end{bmatrix}$$

As displayed in Figure 9.5, lines of constant probability in the multivariate normal are lines of constant

$$(\boldsymbol{x} - \boldsymbol{m})^T \boldsymbol{P}^{-1} (\boldsymbol{x} - \boldsymbol{m})$$

To understand the geometry of lines of constant probability (ellipses in two dimensions, ellipsoids or hyperellipsoids in three or more dimensions) we examine the eigenvalues and eigenvectors of the $\boldsymbol{P}$ matrix.

9.2.2 Eigenvalues and Eigenvectors

An eigenvector of a matrix $\boldsymbol{A}$ is a nonzero vector $\boldsymbol{v}$ such that when multiplied by $\boldsymbol{A}$, the resulting vector points in the same direction as $\boldsymbol{v}$,

$$p(\boldsymbol{x}) = \exp\left(-1/2\left(3.5x_1^2 + 2(2.5)x_1x_2 + 4.0x_2^2\right)\right)$$

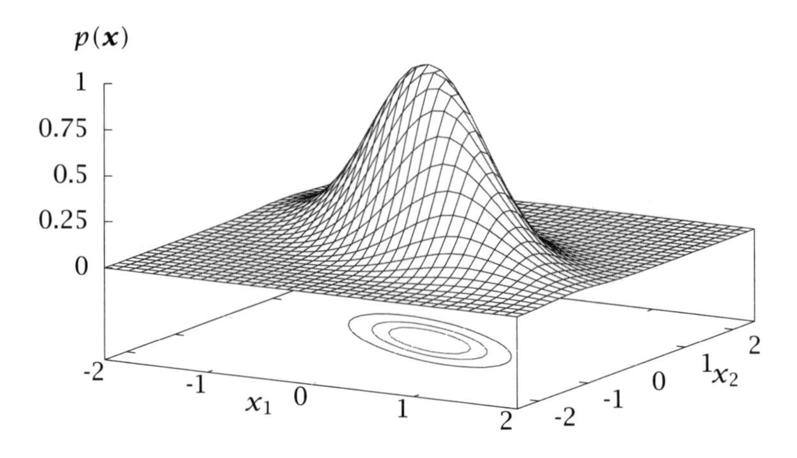

Figure 9.5: Multivariate normal for $n_p = 2$.

and only its magnitude is rescaled. The rescaling factor is known as the corresponding eigenvalue λ of $\boldsymbol{A}$. Therefore the eigenvalues and eigenvectors satisfy the relation

$$\boldsymbol{A}\boldsymbol{v} = \lambda\boldsymbol{v}, \qquad \boldsymbol{v} \neq \boldsymbol{0}$$

We normalize the eigenvectors so

$$\boldsymbol{v}^T\boldsymbol{v} = \sum_i v_i^2 = 1$$

The eigenvectors show us the orientation of the ellipse given by the normal distribution. Consider the ellipse in the two-dimensional $\boldsymbol{x}$ coordinates given by the quadratic

$$\boldsymbol{x}^T\boldsymbol{A}\boldsymbol{x} = b$$

If we march along a vector $\boldsymbol{x}$ pointing in the eigenvector $\boldsymbol{v}$ direction, we calculate how far we can go in this direction until we hit the ellipse $\boldsymbol{x}^T\boldsymbol{A}\boldsymbol{x} = b$. Substituting $\alpha\boldsymbol{v}$ for $\boldsymbol{x}$ in this expression yields

$$(\alpha\boldsymbol{v}^T)\boldsymbol{A}(\alpha\boldsymbol{v}) = b$$

$$\boldsymbol{x}^T \boldsymbol{A}\boldsymbol{x} = b$$
$$\boldsymbol{A}\boldsymbol{v}_i = \lambda_i \boldsymbol{v}_i$$

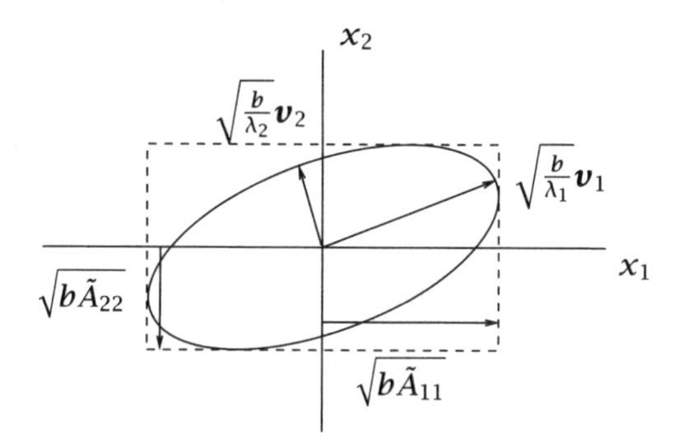

Figure 9.6: The geometry of quadratic form $\boldsymbol{x}^T \boldsymbol{A}\boldsymbol{x} = b$.

Using the fact that $\boldsymbol{A}\boldsymbol{v} = \lambda \boldsymbol{v}$ for the eigenvector gives

$$\alpha^2 \lambda \boldsymbol{v}^T \boldsymbol{v} = b$$

because the eigenvectors are of unit length, we solve for α and obtain

$$\alpha = \sqrt{\frac{b}{\lambda}}$$

which is shown in Figure 9.6. Each eigenvector of $\boldsymbol{A}$ points along one of the axes of the ellipse. The eigenvalues show us how stretched the ellipse is in each eigenvector direction.

If we want to put simple bounds on the ellipse, then we draw a box around it as shown in Figure 9.6. Notice the box contains much more area than the corresponding ellipse and we have lost the correlation between the elements of $\boldsymbol{x}$. This loss of information means we can put different tangent ellipses of quite different shapes inside the same box. The size of the bounding box is given by

$$\text{length of } i\text{th side} = \sqrt{b\tilde{A}_{ii}}$$

in which

$$\tilde{A}_{ii} = (i, i) \text{ element of } \boldsymbol{A}^{-1}$$

Figure 9.6 displays these results: the eigenvectors are aligned with the ellipse axes and the eigenvalues scale the lengths. The lengths of the sides of the box that is tangent to the ellipse are proportional to the square root of the diagonal elements of $\boldsymbol{A}^{-1}$.

9.2.3 Least-Squares Estimation

Consider again the problem of fitting a straight line to data

$$y_i = mx_i + b$$

in which y_i is the measurement at x_i, $i = 1, \dots n_d$ and n_d is the number of data points. Using matrix vector notation, we can write the equation for all the data as

$$\boldsymbol{y} = \boldsymbol{X\theta}$$

in which the parameters to be estimated are placed in the $\boldsymbol{\theta}$ vector

$$\boldsymbol{\theta} = \begin{bmatrix} m \\ b \end{bmatrix}$$

and the $\boldsymbol{y}$ vector and $\boldsymbol{X}$ matrix are given by

$$\boldsymbol{y} = \begin{bmatrix} y_1 \\ y_2 \\ \vdots \\ y_{n_d} \end{bmatrix} \qquad \boldsymbol{X} = \begin{bmatrix} x_1 & 1 \\ x_2 & 1 \\ \vdots & \vdots \\ x_{n_d} & 1 \end{bmatrix}$$

We do not expect the best fit line to pass through all the data points, so we modify the model to account for measurement error[1]

$$\boldsymbol{y} = \boldsymbol{X\theta} + \boldsymbol{e} \tag{9.7}$$

in which $\boldsymbol{e}$ is a random variable. We model the measurement error as a normal distribution with mean $\boldsymbol{0}$ and variance σ^2.

$$\boldsymbol{e} \sim N(\boldsymbol{0}, \sigma^2\boldsymbol{I}) \tag{9.8}$$

The best estimate of $\boldsymbol{\theta}$ in a least-squares sense is given by

$$\hat{\boldsymbol{\theta}} = (\boldsymbol{X}^T\boldsymbol{X})^{-1}\boldsymbol{X}^T\boldsymbol{y} \tag{9.9}$$

a formula that you have used often. However, we also can examine the distribution of parameter estimates given the observed measurements corrupted by the measurement errors. Imagine we create replicate datasets by drawing measurement errors $\boldsymbol{e}$ from the distribution

[1]Notice the model *structure* is usually in error also, e.g., the true relationship between y and x may be nonlinear, variables other than x may be required to predict y, and so on. The procedure outlined here lumps structural error into $\boldsymbol{e}$ as well, but structural error is not accounted for correctly in this way. If the structure is in serious doubt, one may pose instead model discrimination tests to choose between competing models with different structures [30, 31].

given in Equation 9.8. For each dataset we apply Equation 9.9 and produce a parameter estimate. The distribution of measurement errors creates a distribution of parameter estimates. In fact, for models linear in the parameters, we can show the parameter estimates also are normally distributed (see also Exercise 9.14)

$$\hat{\boldsymbol{\theta}} \sim N(\boldsymbol{\theta}, \boldsymbol{P})$$

in which the mean is the true value of the parameters and the covariance is

$$\boldsymbol{P} = \sigma^2 (\boldsymbol{X}^T \boldsymbol{X})^{-1}$$

We also can calculate the parameter "confidence intervals." We merely compute the size of the ellipse containing a given probability of the multivariate normal. That can be shown to be the chi-squared probability function [4, p. 116]. Given the number of estimated parameters, n_p, and the confidence level, α, then

$$\frac{(\boldsymbol{\theta} - \hat{\boldsymbol{\theta}})^T \boldsymbol{X}^T \boldsymbol{X} (\boldsymbol{\theta} - \hat{\boldsymbol{\theta}})}{\sigma^2} \leq \chi^2(n_p, \alpha) \tag{9.10}$$

The χ^2 distribution is tabulated in many statistics handbooks [5] and is available in many computing environments.

To illustrate the ideas we examine a classic problem: how to estimate the preexponential factor and activation energy of a rate constant.

Example 9.1: Estimating the rate constant and activation energy

Assume a reaction rate has been measured at several different temperatures in the range 300 K $\leq T \leq$ 500 K. Estimate the preexponential factor and activation energy of the rate constant, and quantify your uncertainty in the estimated parameters.

Solution

We first model the rate (rate constant) as

$$k = k_0 \exp(-E/T) \tag{9.11}$$

in which k_0 is the preexponential factor and E is the activation energy, scaled by the gas constant.

Figure 9.7 shows a typical experiment. Notice that the measurement of the rate constant is somewhat noisy, a likely outcome if we

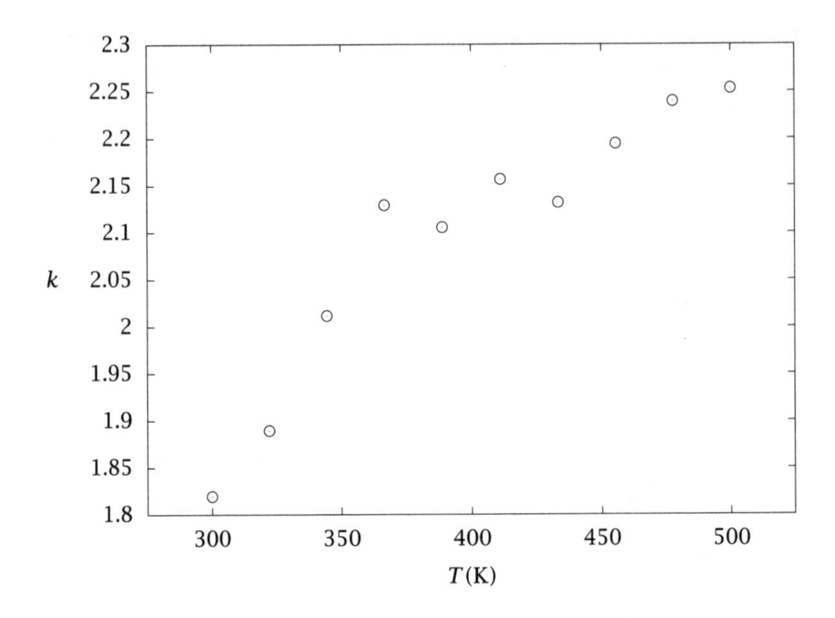

Figure 9.7: Measured rate constant at several temperatures.

differentiate the concentration data to obtain the rate. To make the estimation problem linear, we transform the data by taking the logarithm of Equation 9.11

$$\ln k = \ln k_0 - E/T$$

The transformed data are shown in Figure 9.8. To generate these data we assume the model is correct and suppress the units of the parameter values

$$\ln k_0 = 1, \qquad E = 100$$

The measurements of $\ln k$ are corrupted with normally distributed errors having variance 0.001,

$$e \sim N(0, 0.001)$$

We apply Equation 9.9 to estimate the parameter using the transformed model, so $x_i = 1/T_i$ and $y_i = \ln k_i$. To quantify the uncertainty, imagine we replicate the experiment. Figure 9.9 shows several more experiments. Each experiment that we perform allows us to estimate the slope and intercept. Then we can plot the distribution of parameters. Figure 9.10 shows the parameter estimates for 500 replicated experiments. Notice the points are clustering in an elliptical shape.

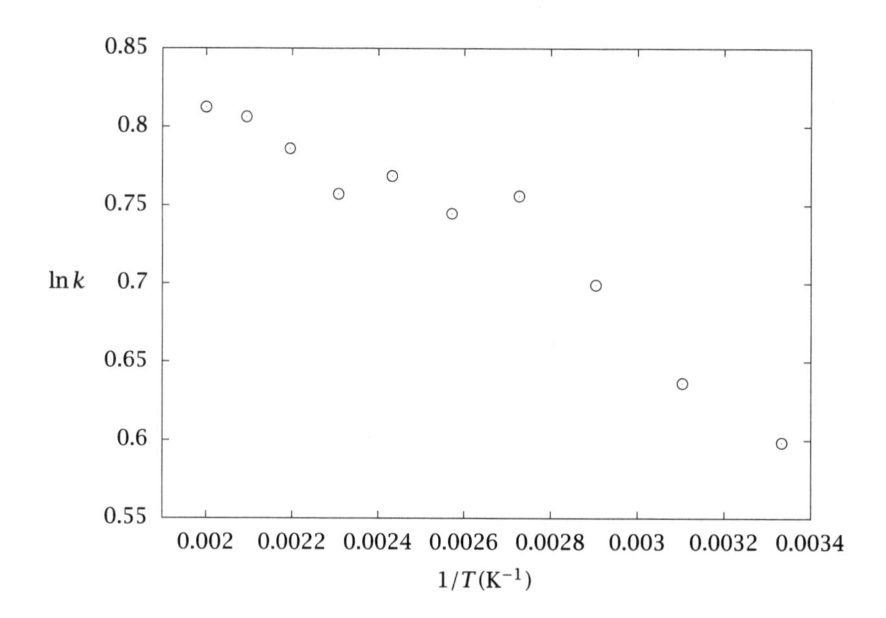

Figure 9.8: Transformed data set, $\ln k$ versus $1/T$.

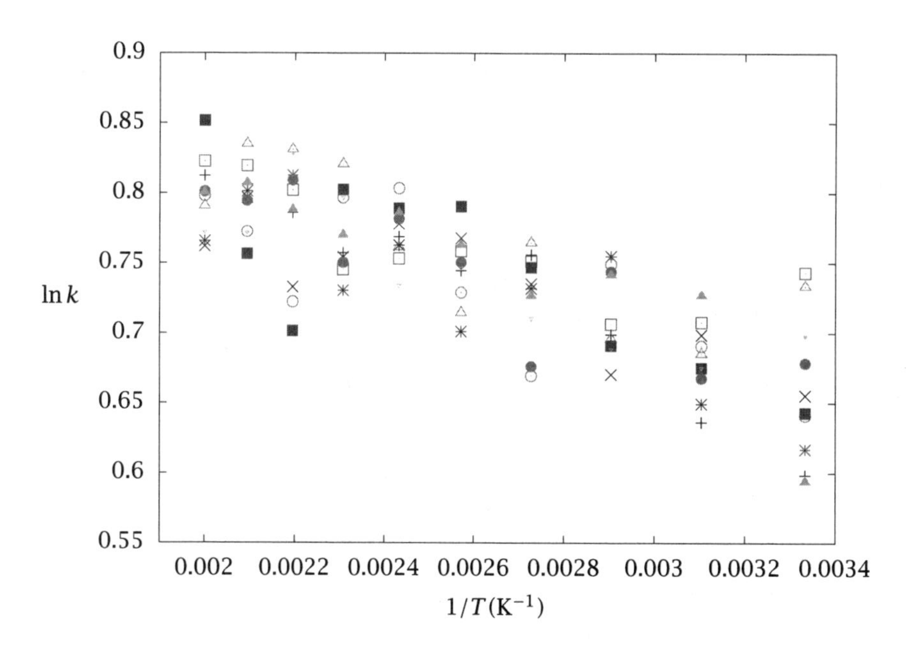

Figure 9.9: Several replicate data sets, $\ln k$ versus $1/T$.

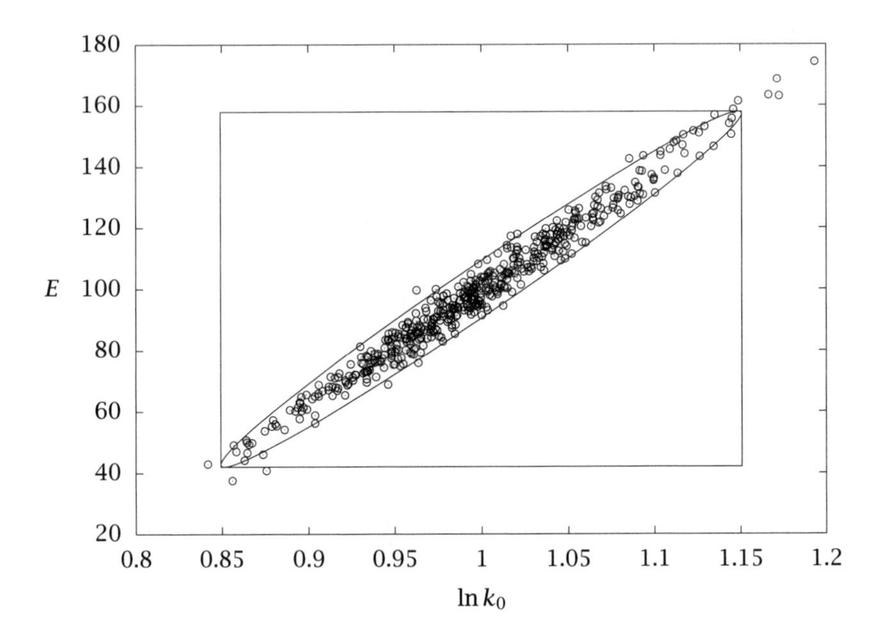

Figure 9.10: Distribution of estimated parameters.

We construct the 95% confidence interval from Equation 9.10. In this problem $n_p = 2$ and $\alpha = 0.95$, and we obtain from a statistics table $\chi^2(2, 0.95) = 5.99$, so we plot

$$\frac{(\boldsymbol{\theta} - \widehat{\boldsymbol{\theta}})^T \boldsymbol{X}^T \boldsymbol{X} (\boldsymbol{\theta} - \widehat{\boldsymbol{\theta}})}{0.001} \leq 5.99 \tag{9.12}$$

This ellipse is also shown in Figure 9.10. In fact, 24 out of the 500 points, or 4.8% of the estimated parameters, lie outside this ellipse, which indicates Equation 9.12 is fairly accurate with this many random experimental trials. Note one often sees parameters reported with plus/minus limits. For this problem, one might report

$$\begin{bmatrix} \ln k_0 \\ E \end{bmatrix} = \begin{bmatrix} 1 \\ 100 \end{bmatrix} \pm \begin{bmatrix} 0.15 \\ 60 \end{bmatrix}$$

But notice these limits are misleading. The rectangle in Figure 9.10 does *not* indicate the strong correlation between the parameters. The ellipse is much more informative in this case.

Next we show how a simple reparameterization of the rate constant can reduce the parameter correlation. Consider the mean of the tem-

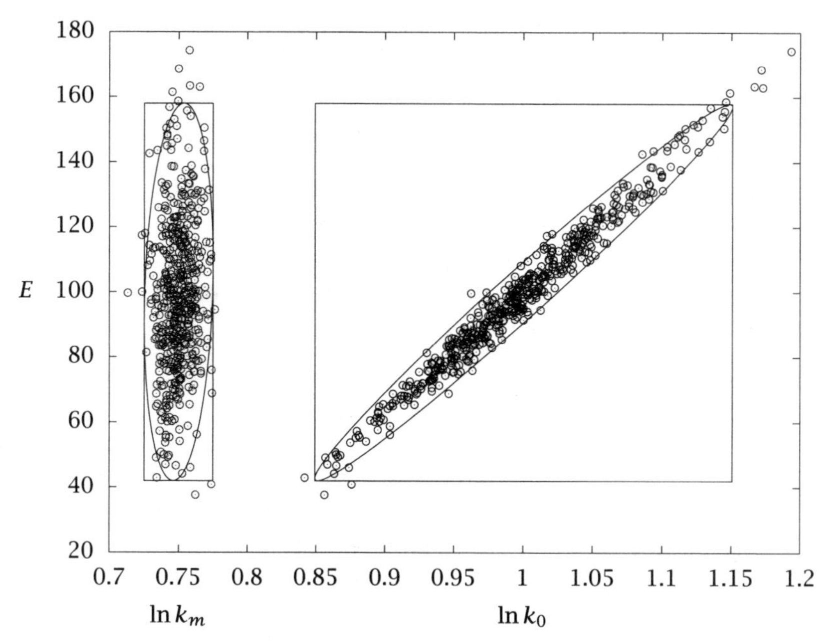

Figure 9.11: Reducing parameter correlation by centering the data.

peratures at which data were collected, and reparameterize the rate constant as in Chapter 6,

$$k = k_m \exp(-E(1/T - 1/T_m)) \tag{9.13}$$

We may wish to consider the mean of temperature or inverse temperature for T_m. Here we let T_m be the mean temperature

$$T_m = \frac{300 + 500}{2} = 400 \text{ K}$$

Both Equations 9.11 and 9.13 are two parameter models and we can convert between them using $k_0 = k_m \exp(E/T_m)$. But in Equation 9.13 we estimate the rate constant at the mean temperature in contrast to infinite temperature ($1/T = 0$) in Equation 9.11.

If we estimate E and k_m in place of k_0 we obtain the results shown in Figure 9.11. Notice that the correlation between k_m and E is much reduced compared to k_0 and E. In fact, reporting confidence limits

$$\begin{bmatrix} \ln k_m \\ E \end{bmatrix} = \begin{bmatrix} 0.75 \\ 100 \end{bmatrix} \pm \begin{bmatrix} 0.025 \\ 60 \end{bmatrix}$$

is an accurate representation of the true 95% confidence interval ellipse.

□

9.2.4 Least Squares with Unknown Variance

In the previous problem, we assumed the variance in the error was known to us. It is often the case that we do not know the measurement error variance, but must estimate it also from the data. In this case, it can be shown that the distribution of parameter estimates is a multivariate t-distribution (instead of a normal distribution) using the same least-squares estimate as before

$$\hat{\boldsymbol{\theta}} = (\boldsymbol{X}^T\boldsymbol{X})^{-1}\boldsymbol{X}^T\boldsymbol{y} \tag{9.14}$$

For the confidence intervals, we compute the size of the ellipse containing a given probability of the multivariate t-distribution. That can be shown to be an F probability function [4, p. 117]. Given the number of estimated parameters, n_p, the confidence level, α, and the number of data points, n_d,

$$\frac{(\boldsymbol{\theta} - \hat{\boldsymbol{\theta}})^T\boldsymbol{X}^T\boldsymbol{X}(\boldsymbol{\theta} - \hat{\boldsymbol{\theta}})}{s^2} \leq n_p F(n_p, n_d - n_p, \alpha) \tag{9.15}$$

defines the confidence interval ellipse. The F distribution is also tabulated in statistics handbooks [5] and available in computing environments. The sample variance

$$s^2 = \frac{1}{n_d - n_p}(\boldsymbol{y} - \boldsymbol{X}\hat{\boldsymbol{\theta}})^T(\boldsymbol{y} - \boldsymbol{X}\hat{\boldsymbol{\theta}}) \tag{9.16}$$

is the estimate of the unknown error variance. Notice the number of data points, n_d, shows up in the confidence interval when the error variance is unknown. In the limit of large n_d, the F distribution converges to the χ^2

$$\lim_{n_d \to \infty} n_p F(n_p, n_d - n_p, \alpha) = \chi^2(n_p, \alpha)$$

and the confidence intervals given in Equations 9.10 and 9.15 are the same. The sample variance also converges to the error variance in the limit of large number of data points.

On the other hand, in engineering problems we often must contend with few data points, because they are difficult or expensive to obtain. Figure 9.12 shows the values of χ^2 and $n_p F$ as a function of the number of data points for 95% confidence limits, and the cases of two and five

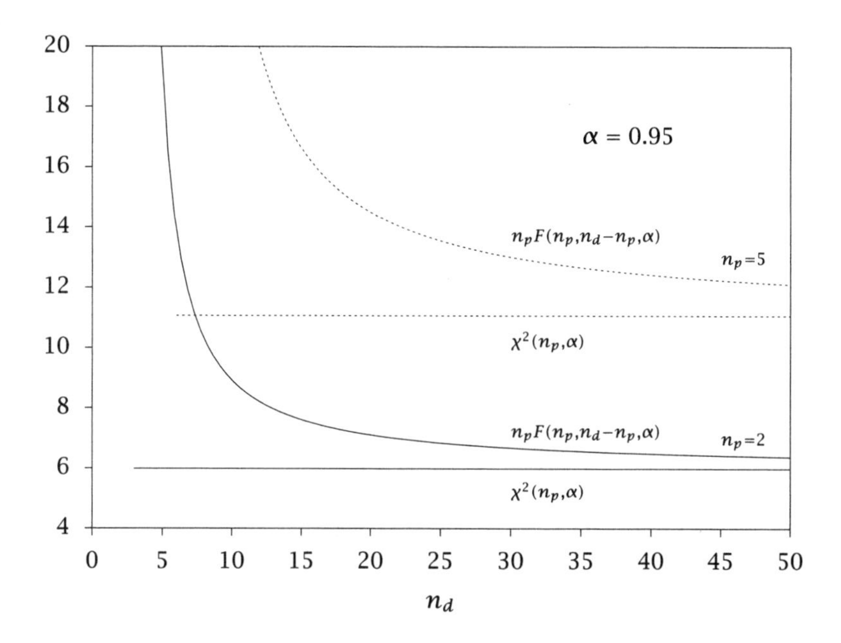

Figure 9.12: Values of χ^2 and F versus the number of data points when estimating 2 and 5 parameters; the larger values of F show information loss (larger parameter uncertainty) when estimating the measurement variance from the data.

estimated parameters. Notice that n_pF is much larger than χ^2 at small n_d. With a small number of data points, estimating the variance as well as the parameters from the data inflates significantly the confidence intervals. Figure 9.12 quantifies this effect; we can see that if we have about 10 times as many data points as parameters, $n_d > 10n_p$, this effect is rather small.

Although we should expect larger confidence intervals when n_d is small, that may not be what happens with a particular dataset as we show with the next example.

Example 9.2: Unknown measurement variance and few data points

Consider the data shown in Figure 9.13 with two unknown parameters and only 10 data points. The measurement errors are drawn from a normal distributed with zero mean and variance $\sigma^2 = 10^{-3}$. Compute the best estimates of activation energy and mean rate constant and the

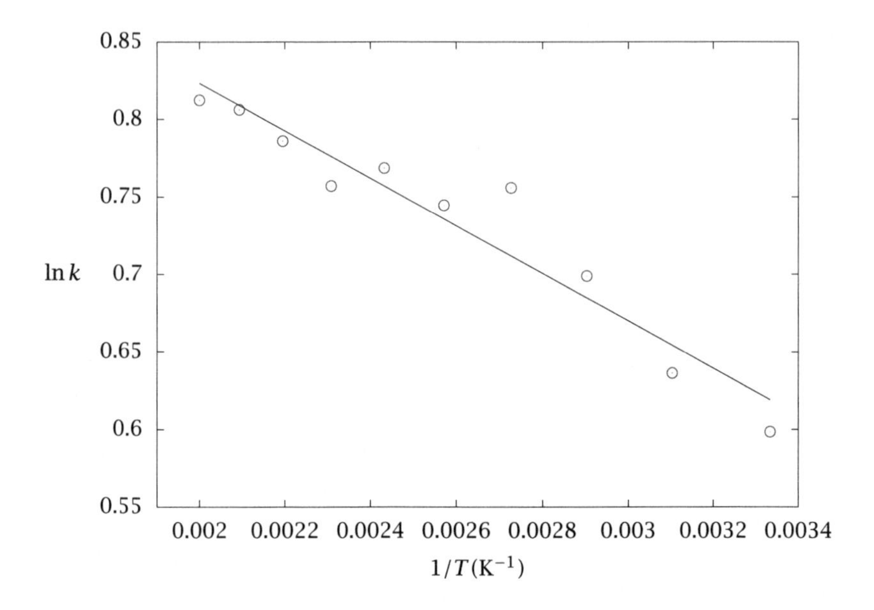

Figure 9.13: Parameter estimates with only 10 data points.

95% confidence intervals for the cases of known and unknown measurement variance.

Solution

We calculate $\hat{\boldsymbol{\theta}}$ from Equation 9.14 and the sample variance from Equation 9.16 and obtain

$$\hat{\boldsymbol{\theta}} = \begin{bmatrix} 0.747 \\ 153 \end{bmatrix}, \qquad s^2 = 0.000454$$

We construct the 95% confidence interval from Equations 9.10 and 9.15. We obtain from a statistics table $F(2, 8, 0.95) = 4.46$ so the two confidence intervals are given by

$$(\boldsymbol{\theta} - \hat{\boldsymbol{\theta}})^T \boldsymbol{X}^T \boldsymbol{X} (\boldsymbol{\theta} - \hat{\boldsymbol{\theta}}) \leq (0.001)(5.99), \qquad \text{known variance}$$

$$(\boldsymbol{\theta} - \hat{\boldsymbol{\theta}})^T \boldsymbol{X}^T \boldsymbol{X} (\boldsymbol{\theta} - \hat{\boldsymbol{\theta}}) \leq (0.000454)(2)(4.46), \qquad \text{unknown variance}$$

The two 95% confidence ellipses are shown in Figure 9.14. Notice that although $n_p F$ is 50% *larger* than χ^2, our 95% confidence interval for the unknown measurement variance case is *smaller* than the known measurement variance case.

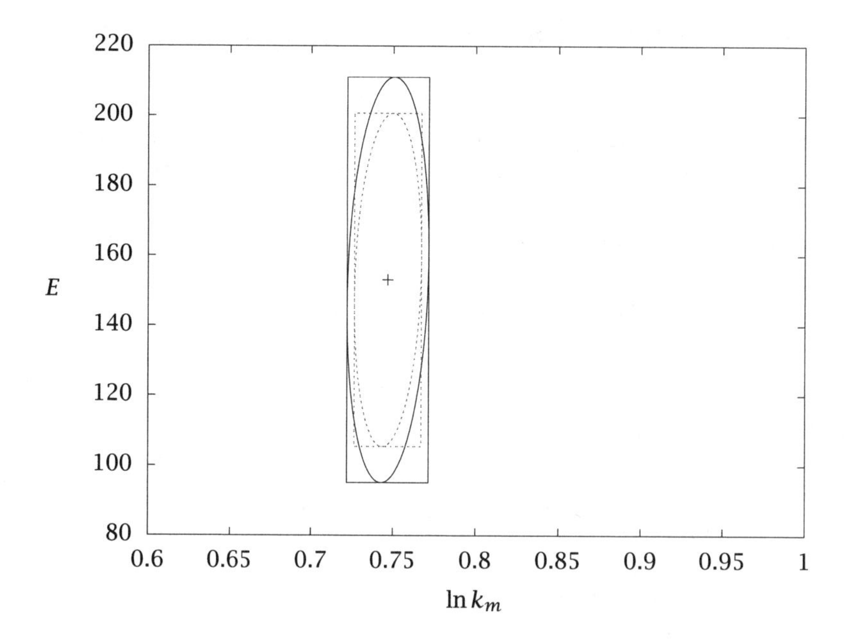

Figure 9.14: Confidence intervals with known (solid line) and unknown (dashed line) error variance.

Can you resolve this apparent contradiction? What experiment can you propose in which we would find the confidence interval with estimated measurement variance to be about 50% larger than known measurement variance in agreement with the statistics in Figure 9.12? □

9.2.5 Nonlinear Least Squares

In the previous two examples we transformed the model by taking logarithms to obtain a linear estimation problem. In many situations we do not want to make such a transformation, or such transformations simply do not exist. For example, the measurement error may be distributed normally in the original variables and the transformation may distort this distribution and subsequent parameter estimates and confidence intervals.

If we decide to treat the estimation problem using the nonlinear model, the problem becomes more challenging. The parameter estimation becomes a nonlinear optimization that must be solved numerically instead of a linear matrix inversion that can be solved analytically as in Equation 9.9. Moreover, the confidence intervals become more difficult to compute, and they lose their strict probabilistic interpretation as α-level confidence regions. As we will see, however, the "approximate" confidence intervals remain very useful in nonlinear problems. The numerical challenges for nonlinear models can be addressed successfully in many reaction engineering problems if we employ high-quality numerical software.

Consider the nonlinear model

$$y_i = h(x_i, \boldsymbol{\theta}), \qquad i = 1, \ldots n_d \tag{9.17}$$

and the least-squares objective

$$\Phi(\boldsymbol{\theta}) = \sum_{i=1}^{n_d} \left(\tilde{y}_i - h(x_i, \boldsymbol{\theta})\right)^2$$

We obtain the parameter estimates by solving the optimization problem

$$\min_{\boldsymbol{\theta}} \Phi(\boldsymbol{\theta})$$

subject to Equation 9.17. Call the solution to this problem $\hat{\boldsymbol{\theta}}$. We now consider the function $\Phi(\boldsymbol{\theta})$ near the optimum. Expanding in a second-order multivariable Taylor series gives

$$\Phi(\boldsymbol{\theta}) \approx \Phi(\hat{\boldsymbol{\theta}}) + (\nabla\Phi)^T \Big|_{\boldsymbol{\theta}=\hat{\boldsymbol{\theta}}} (\boldsymbol{\theta} - \hat{\boldsymbol{\theta}}) + \frac{1}{2}(\boldsymbol{\theta} - \hat{\boldsymbol{\theta}})^T \boldsymbol{H}\Big|_{\boldsymbol{\theta}=\hat{\boldsymbol{\theta}}} (\boldsymbol{\theta} - \hat{\boldsymbol{\theta}}) \tag{9.18}$$

in which the gradient of the objective function is the vector of first derivatives of Φ with respect to the model parameters

$$(\nabla\Phi)_j = \frac{\partial\Phi}{\partial\theta_j}$$

and the Hessian of the objective function is the matrix of second derivatives

$$H_{kj} = \frac{\partial^2\Phi}{\partial\theta_k \partial\theta_j}$$

The gradient is zero at the optimum (see also Figure 3.5 of Chapter 3), which allows us to rearrange Equation 9.18 to give

$$\Phi(\boldsymbol{\theta}) - \Phi(\hat{\boldsymbol{\theta}}) \approx \frac{1}{2}(\boldsymbol{\theta} - \hat{\boldsymbol{\theta}})^T \boldsymbol{H}\Big|_{\boldsymbol{\theta}=\hat{\boldsymbol{\theta}}} (\boldsymbol{\theta} - \hat{\boldsymbol{\theta}})$$

Therefore, lines of constant objective function are approximately quadratic functions as shown in Figure 9.6 and we use $\boldsymbol{H}$ as the $\boldsymbol{A}$ matrix. This quadratic approximation using the Hessian matrix evaluated at the optimum is accurate if we are in the neighborhood of the optimal parameter values. We also can obtain order-of-magnitude confidence intervals using the relation

$$(\boldsymbol{\theta} - \hat{\boldsymbol{\theta}})^T \boldsymbol{H}\Big|_{\boldsymbol{\theta}=\hat{\boldsymbol{\theta}}} (\boldsymbol{\theta} - \hat{\boldsymbol{\theta}}) \leq 2s^2 n_p F(n_p, n_d - n_p, \alpha) \tag{9.19}$$

in which s^2 is again the sample variance

$$s^2 = \frac{1}{n_d - n_p} \sum_{i=1}^{n_d} \left(\tilde{y}_i - h(x_i, \hat{\boldsymbol{\theta}})\right)^2 = \frac{\Phi(\hat{\boldsymbol{\theta}})}{n_d - n_p}$$

These confidence intervals are exact only if the model is linear, in which case $\boldsymbol{H} = 2\boldsymbol{X}^T\boldsymbol{X}$. The intervals should be checked occasionally with Monte Carlo simulations when the model is nonlinear. We illustrate this check in Example 9.4.

Example 9.3: Fitting single and multiple adsorption experiments

To illustrate the use of nonlinear models, we study an adsorption experiment. The system studied was the adsorption of H_2 on a Pd catalyst with SiO_2 support [25, 24]. The adsorption is assumed dissociative. Both the catalyst and the support adsorb H_2 so the adsorption isotherm model is

$$\mathrm{H_2} + 2\mathrm{X_P} \rightleftharpoons 2\mathrm{H}\cdot\mathrm{X_P}$$
$$\mathrm{H_2} + 2\mathrm{X_S} \rightleftharpoons 2\mathrm{H}\cdot\mathrm{X_S}$$

in which X_P represents a Pd vacant site and X_S represents a SiO_2 vacant site. We can apply the methods of Section 5.6 to derive the surface coverage of H atoms

$$\overline{c}_H = \frac{\overline{c}_{mP}\sqrt{K_P}\sqrt{c_{\mathrm{H_2}}}}{1 + \sqrt{K_P}\sqrt{c_{\mathrm{H_2}}}} + \frac{\overline{c}_{mS}\sqrt{K_S}\sqrt{c_{\mathrm{H_2}}}}{1 + \sqrt{K_S}\sqrt{c_{\mathrm{H_2}}}}$$

It is known that the adsorption constant on the Pd is large, so we may reduce the model to

$$\overline{c}_H = \overline{c}_{mP} + \frac{\overline{c}_{mS}\sqrt{K_S}\sqrt{c_{\mathrm{H_2}}}}{1 + \sqrt{K_S}\sqrt{c_{\mathrm{H_2}}}} \tag{9.20}$$

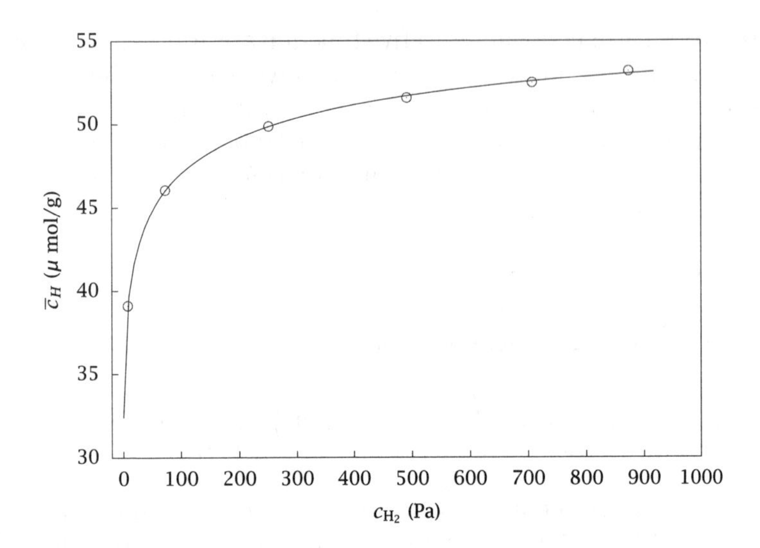

Figure 9.15: Model fit to a single adsorption experiment.

and we have three parameters to estimate from data,

$$\boldsymbol{\theta}^T = \begin{bmatrix} \sqrt{K_S} & \overline{c}_{mS} & \overline{c}_{mP} \end{bmatrix}$$

As the measure of fit to the data we choose a least-squares objective,

$$\Phi = \sum_i \left(\tilde{\overline{c}}_{Hi} - \overline{c}_{Hi} \right)^2$$

and propose the optimization problem

$$\min_{\boldsymbol{\theta}} \Phi(\boldsymbol{\theta})$$

subject to Equation 9.20 to determine the parameter estimates.

Solution

Figure 9.15 shows the best fit to a single adsorption experiment. The parameters and the 95% confidence intervals are given by

$$\hat{\boldsymbol{\theta}} = \begin{bmatrix} \sqrt{K_S} \\ \overline{c}_{mS} \\ \overline{c}_{mP} \end{bmatrix} = \begin{bmatrix} 0.127 \\ 26.1 \\ 32.4 \end{bmatrix} \pm \begin{bmatrix} 0.048 \\ 1.5 \\ 2.5 \end{bmatrix}$$

We can see that the model fit to the data is excellent, and the parameters are determined with fairly tight confidence intervals.

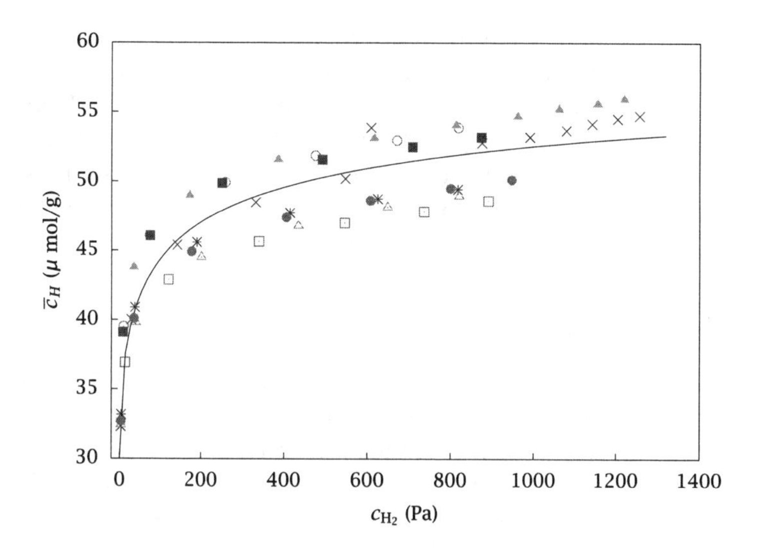

Figure 9.16: Model fit to all adsorption experiments.

For this catalyst sample, additional adsorption experiments are available. After the first adsorption experiment, reaction studies were performed with the catalyst sample. Then the catalyst was regenerated with heat treatment, and a second adsorption experiment was performed, followed by an additional reaction study, and so on. The additional adsorption data are shown in Figure 9.16. Notice these are not replicate experiments, because the adsorption experiments are performed after different reaction studies have been performed, and the catalyst regeneration step does not necessarily return the catalyst to exactly the same condition. Estimating the parameters by fitting all of these data simultaneously produces the solid line in Figure 9.16 and the following parameter values and confidence intervals

$$\hat{\boldsymbol{\theta}} = \begin{bmatrix} \sqrt{K_S} \\ \overline{c}_{mS} \\ \overline{c}_{mP} \end{bmatrix} = \begin{bmatrix} 0.0895 \\ 30.4 \\ 30.1 \end{bmatrix} \pm \begin{bmatrix} 0.096 \\ 4.9 \\ 6.2 \end{bmatrix}$$

The total coverage estimated here, 60.5 μmol/g, compares favorably to the apparent saturation limit or total uptake value of 55 μmol/g reported by Natal-Santiago et al. [24, p.157].

Notice because the reaction studies have introduced significant variability in the data, the confidence intervals for the parameter values are significantly larger than those for the single adsorption experiment.

Both sets of estimated parameters are valid, but they have different meanings. If the model is intended to represent the catalyst in its native state, one might use the estimated values from a single adsorption experiment on a freshly prepared sample. If the model is intended to represent the "average behavior" of the catalyst during the period it is used and regenerated, one would naturally use the estimated values from the many adsorption experiments. □

9.2.6 Design of Experiments

Experimental design is a large topic and we can only mention several of the important issues here. To keep this discussion focused on parameter estimation for reactor models, we must assume the reader has had exposure to a course in basic statistics [5]. We assume the reader understands the source of experimental error or noise, and knows the difference between correlation and causation. The process of estimating parameters in reactor models is part of the classic, iterative scientific method: hypothesize, collect experimental data, compare data and model predictions, modify hypothesis, and repeat. The goal of experimental design is to make this iterative learning process *efficient*.

Our goal at this point can be quite specific because we have already built the causal models of interest to us in Chapters 2–8, and require only the remaining small task of finding best parameter values and quantifying the parameter uncertainty. We are not concerned at this point with embedding this step in the larger task described previously: deciding if our model structure is well chosen and the model assumptions are appropriate given the data.

The main goal of our experimental design then is to choose experiments that make the uncertainty in the parameter estimates small. We also must decide what small means when it comes to parameter uncertainty.

Optimal experimental design. *Optimal* experimental design is an experimental design with the express purpose of making the parameter uncertainty *as small as possible.* Before developing precise, quantitative methods of experimental design, let's build up some physical intuition about the problem we are addressing. Imagine we wish to use a straightedge to draw a line on a piece of paper — a task most of us face early in childhood. We mark two points on the paper, lay the straightedge next to the two points, and draw the line. To make the problem more interesting and realistic, consider that no matter how carefully

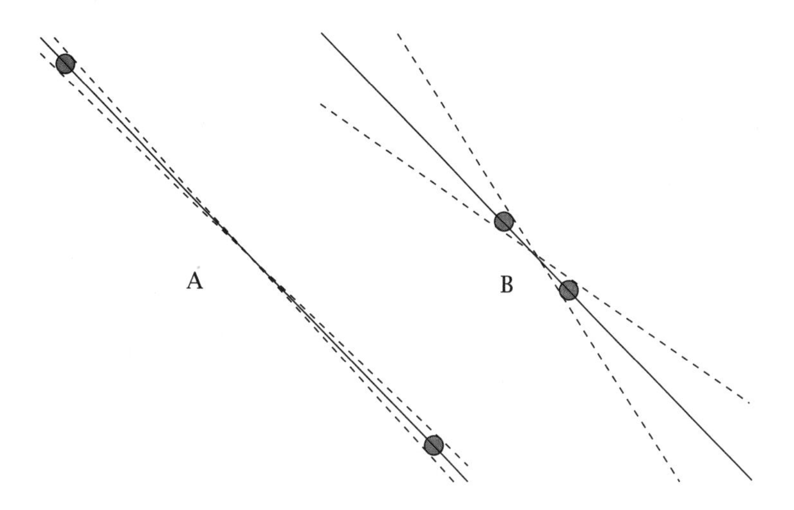

Figure 9.17: Drawing a line through two points under measurement error: points far apart (A); points close together (B).

we measure, our placement of the points is subject to small errors. As shown in Figure 9.17, the small circle represents the regions in which we might actually mark a point, with the center of the circle representing the ideal location where we are supposed to mark the point. The size of these circles tells us how much uncontrolled error exists in our experiment. An experienced draftsman or carpenter with steady hands may be able to measure carefully and place points within fairly small circles; a young child may require rather large circles to bound the expected errors. In all cases we face error in point placement and that error produces error in the final line. If we wish to minimize the error in the constructed line, one option is to make the measurement error circles small. Improvements in measurement technology are always an attractive option. A reasonably high-resolution printer, such as the one used to print this page, for example, may be able to locate a point to within 1/1200 inch. But regardless of the current state of technology, the measurement noise persists at some nonzero level.

Experimental design provides another option. Now we ask the question: where on the line should we *locate* the two points to give us the smallest error in the line placement? Our experience and intuition probably lead us to guess rather quickly that we want to spread the two points apart as far as possible. As shown in Figure 9.17, if the points are far apart, the spread in the slopes of the lines is small. If

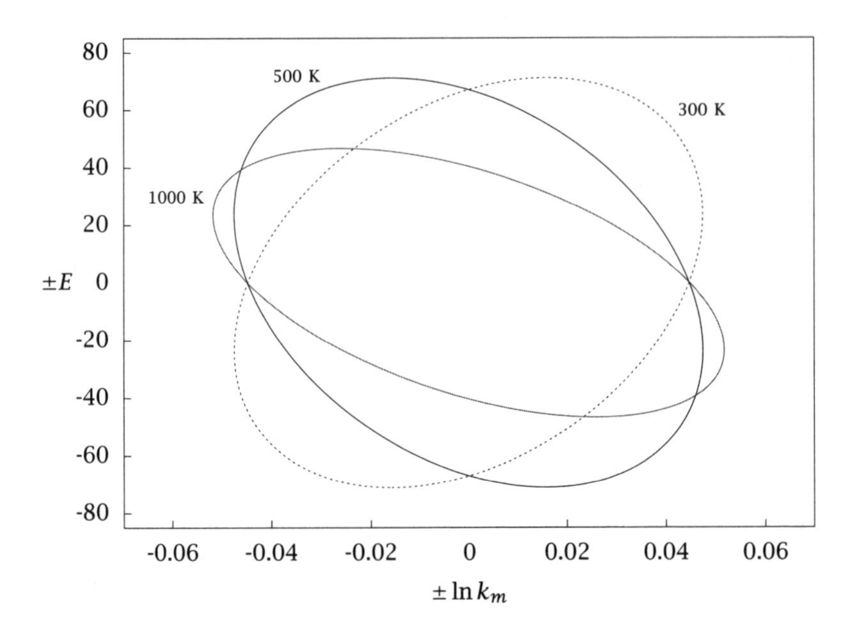

Figure 9.18: Effect of next measurement temperature on parameter confidence intervals.

the points are close together, the spread in slopes is large. If the points are close enough that the circles overlap, then a line with *any* slope is possible.

Optimal experimental design generalizes this intuitive notion of how to construct the experiment to minimize the effect of the experimental errors on the estimated parameters. It also can provide experimental designs in situations in which our intuition does not provide ready answers. We explore one such nonobvious design issue in an industrial case study in Section 9.3. But for illustrative purposes, consider again the least-squares estimation of the mean rate constant and activation energy from measuring the rate constant at several temperatures. Assume we have temperature constraints in our experimental system, and can only operate the experiment and collect rate data in the temperature range

$$300 \text{ K} \leq T \leq 500 \text{ K}$$

We cannot estimate two parameters unless we have at least two measurements, so, motivated by Figure 9.17, we choose the first two measurements at the extreme limits of the temperatures, $T = 300$ K, $T = 500$ K. We next explore what happens to our confidence intervals as we make different choices for the next experiment. Figure 9.18 shows the confidence intervals for three different choices of the next experiment. The choices of 300 K and 500 K produce two 95% confidence regions of the same size, but with different orientations. This result is consistent with our physical intuition of drawing straight lines. If we add points at the *right* end of the line shown in Figure 9.13 ($T = 300$ K), then that anchors somewhat the right end of the line, and to increase the value of $\ln k_m$ we must also *increase* the slope. So the 300 K ellipse is stretched in the direction with positive correlation between $\ln k_m$ and E. Alternatively, if we add points at 500 K, then we have anchored somewhat the *left* end of the line in Figure 9.13, and to increase the value of $\ln k_m$, we must *decrease* the slope. The 500 K ellipse is therefore stretched in the direction with negative correlation between $\ln k_m$ and E. If we were able to violate our temperature limits and perform the next experiment at 1000 K, we see that we can achieve more certainty in the activation energy, at the cost of slightly more uncertainty in the intercept. This result is exactly what we observed in Figure 9.17 — to obtain less uncertainty in the slope, spread the points as far apart as is practical.

With this background, we now ask very specific questions and design experiments with precise goals. If experiments are either time consuming or expensive, we ask the natural question: *where should we place the next experiment to obtain as much information about the parameters as possible?* To answer this question we have to decide what information we seek. For example, we may want to know only the activation energy, we may want to know only the intercept, we may want to know both of them, but one is somewhat more important than the other. Figure 9.19 displays the size of the sides of the box tangent to the 95% confidence ellipse as we vary the temperature of the next experiment. If we are interested in the activation energy, we see that the next experiment should be placed at either $T = 300$ K or $T = 500$ K. Both values are equally good. In this problem we chose $T_m = 375$ K, i.e., the mean of inverse temperature,

$$T_m = \frac{2}{1/300 + 1/500} = 375 \text{ K}$$

If we are interested in the value of the rate constant at the mean temperature, Figure 9.19 shows we should place the next experiment at the mean temperature $T = T_m$.

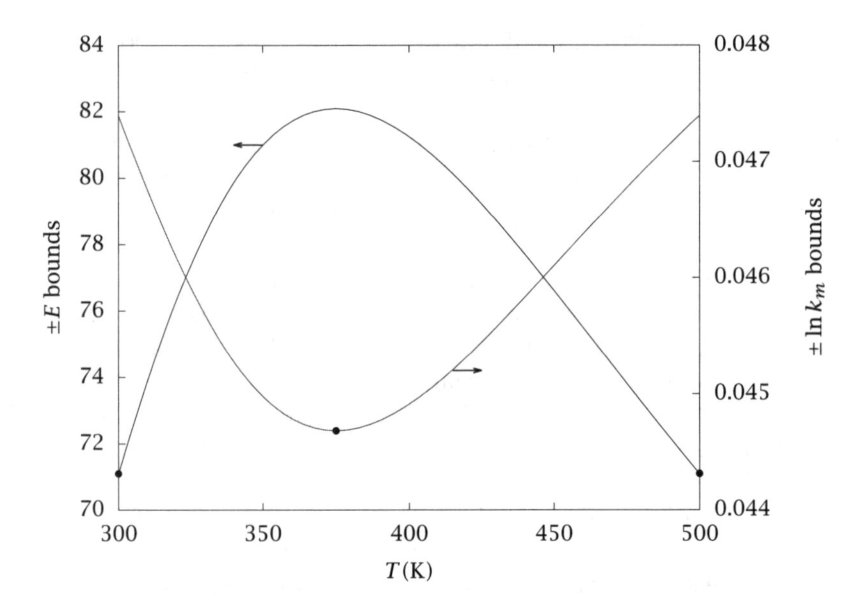

Figure 9.19: Uncertainty in activation energy E and rate constant $\ln k_m$ versus next measurement temperature.

The results in this case are not too surprising; but that is the main reason we studied this problem, to build some intuition about what to ask and expect of optimal experimental design. One big advantage of the approach is that it is quantitative. In this problem, we can guess immediately to put the next experiment at the upper or lower temperature limit if we want to know the activation energy, but would we know that the uncertainty would change from 82 to 72 when we did so? If that improvement is not large enough for our purposes, we would know further experiments are required. We can answer the question of how many further experiments are required to meet a given bound on uncertainty. The second big advantage of this approach is that it is general and can be applied routinely in even complex experimental situations in which our intuition is not highly developed. As we mentioned previously, experimental design is a large topic, and the interested reader can find a wealth of literature discussing its many aspects [1].

Replicate experiments. Finally, we examine the impact of replicating experiments. The main reason one replicates an experiment is to obtain a direct measure of the experimental reproducibility. By replica-

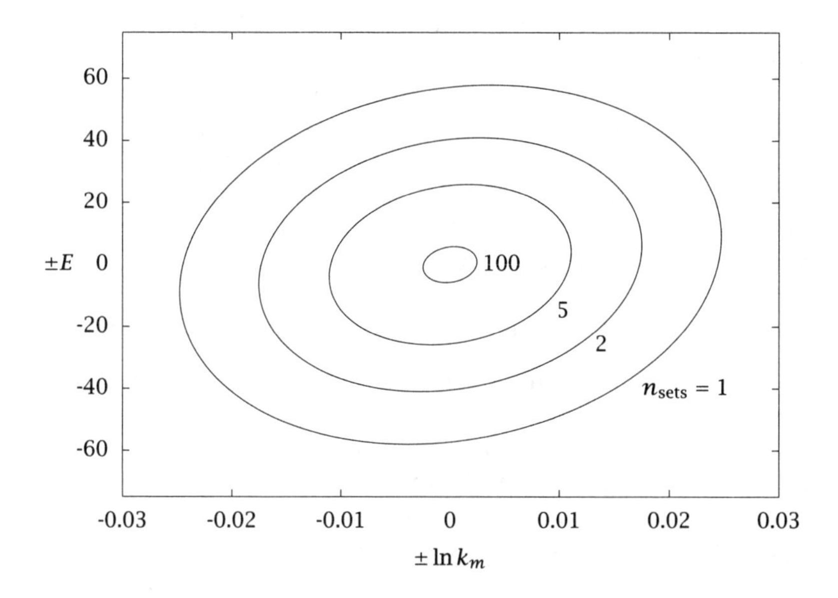

Figure 9.20: Uncertainty in activation energy E and rate constant $\ln k_m$ versus number of replicated experiments; 10 equally spaced temperatures per experiment.

tion we are confronted with the simple experimental fact that no matter how much care we take and how much effort we exert, the results of two experiments are never exactly the same. The differences among replicated experiments give us a direct quantitative measure of the experimental error or noise. These errors are partly due to our inability to measure accurately the outputs of interest, but they are also due to our inability to control completely the experimental environment. The irreproducibility of the experiment is one of the key motivations for the random variable description of $\boldsymbol{e}$ in Equation 9.7.

But replicating an experiment has another interesting and expected benefit. It increases our confidence in our conclusions. To see this benefit, we examine again the least-squares problem for estimating k_m, E. Assume now the experiments are *inexpensive* and we are not concerned about performing optimal experiments. We simply make measurements at 10 evenly spaced temperatures as shown in Figure 9.13. If we perform one set of 10 measurements we achieve the confidence intervals we have shown previously in Figure 9.14. What happens if we simply repeat the experiment and perform two sets of 10 measurements? The confidence region becomes smaller, as shown in Figure 9.20. We also show the result of using 5 and 100 replicates of the experiment. The confidence intervals continue to shrink as we replicate the experiments. So replicating experiments provides another general avenue for increasing the confidence in the parameter estimates. Be aware, however, as discussed further in Exercise 9.10, that this "brute force" approach may be time consuming and expensive.

9.2.7 Parameter Estimation with Differential-Equation Models

Now we turn to the single most important parameter estimation problem in chemical reactor modeling: determining reaction-rate constants given dynamic concentration measurements. We devote the rest of the chapter to developing methods for this problem.

To get started, we consider a simple reactor model consisting of a single differential equation with a single experimentally measured quantity

$$\frac{dx}{dt} = f(x;\boldsymbol{\theta}) \tag{9.21}$$

$$x(0) = g(x_0;\boldsymbol{\theta}) \tag{9.22}$$

$$y = h(x) \tag{9.23}$$

in which x is the single material balance of interest, $\boldsymbol{\theta}$ are the unknown model parameters, x_0 is the initial condition, and y is the experimentally measurable quantity. For simplicity let us assume here that x itself is measured, in which case $h(x) = x$. As we see in several of the examples, it may be necessary to include some of the initial conditions also as unknown parameters. Often t is time, but in steady-state tubular PFRs, reactor volume or length can take the place of time without changing the structure of the parameter-estimation problem.

As before, we define a least-squares objective to measure our fit to the data

$$\Phi(\boldsymbol{\theta}) = \sum_i (\tilde{x}_i - x_i)^2 \tag{9.24}$$

in which $\tilde{x}_i$ is the experimental measurement at time t_i, and x_i is the solution to the model at time t_i, $x_i = x(t_i;\boldsymbol{\theta})$. Note x_i is the only part of the objective function that depends on the model parameters. Again, we minimize this objective function to obtain our parameter estimates

$$\min_{\boldsymbol{\theta}} \Phi(\boldsymbol{\theta}) \tag{9.25}$$

subject to Equations 9.21–9.23.

The major change is that the model constraint consists of nonlinear differential equations rather than linear or nonlinear algebraic equations as in the previous sections. The differential equations make it much more expensive to evaluate the constraints while solving the optimization problem. We can increase the efficiency of the optimizer if we provide an accurate gradient of the objective function. Recall the gradient of the objective function is the vector of first derivatives of Φ with respect to the model parameters. Differentiating Equation 9.24 gives

$$\frac{\partial \Phi}{\partial \theta_j} = -2 \sum_i (\tilde{x}_i - x_i) \frac{\partial x_i}{\partial \theta_j} \tag{9.26}$$

Differentiating a second time gives the Hessian of the objective function, which we again use to construct approximate confidence intervals

$$\frac{\partial^2 \Phi}{\partial \theta_k \partial \theta_j} = 2 \sum_i \left[\frac{\partial x_i}{\partial \theta_k} \frac{\partial x_i}{\partial \theta_j} - (\tilde{x}_i - x_i) \frac{\partial^2 x_i}{\partial \theta_k \theta_j} \right] \tag{9.27}$$

Gauss-Newton approximation and sensitivities. In the Gauss-Newton approximation of the Hessian, we neglect the second term in Equation 9.27 to yield,

$$\frac{\partial^2 \Phi}{\partial \theta_k \partial \theta_j} \approx 2 \sum_i \frac{\partial x_i}{\partial \theta_k} \frac{\partial x_i}{\partial \theta_j} \tag{9.28}$$

Two arguments support the Gauss-Newton approximation. If the model fits the data well at the optimal value of parameters, the residuals are small in magnitude and of different signs. The sum in the second term is then small. Alternatively, the second derivative of the model may be small compared to the first derivative. If the model is linear in the parameters, for example, the second derivatives are identically zero and the Gauss-Newton approximation is exact. The Gauss-Newton approximation is not valid if the model solution is a highly nonlinear function of the parameters, or if the residuals are large and not randomly distributed about zero at the optimal value of parameters. Of course, in the latter case one should question the model structure because the model does not well represent the data. For the highly nonlinear case, one may try numerical finite difference formulas to compute the Hessian. Computing a reliable finite difference approximation for a *second* derivative is not a trivial matter either, however, and the step size should be carefully chosen to avoid amplifying the errors introduced by the finite numerical precision.

The first derivatives of the model solution with respect to the model parameters are known as the model sensitivities,

$$S_{ij} = \frac{\partial x_i}{\partial \theta_j}$$

The sensitivities also can be described as the solution to a set of differential equations; these equations are derived in Appendix A. This fact allows us to solve the model and sensitivity equations simultaneously with an ODE solver, rather than use finite difference formulas to obtain the sensitivities. Bard provides a readable account for further study on these issues [2]. Caracotsios, Stewart and Sørensen developed this approach and produced an influential software code (GREG) for parameter-estimation problems in chemical reaction engineering [6, 29, 7].

From the sensitivities and model solution, we can then calculate the gradient of the objective function and the Gauss-Newton approximation of the Hessian matrix. Reliable and robust numerical optimization programs are available to find the optimal values of the parameters. These programs are generally more efficient if we provide the gradient in addition to the objective function. The Hessian is normally needed only to calculate the confidence intervals after the optimal parameters are determined. If we define $\boldsymbol{e}$ to be the residual vector

$$e_i = \tilde{x}_i - x_i$$

We can express Equation 9.26 in matrix notation as

$$\nabla\Phi = -2\boldsymbol{S}^T\boldsymbol{e} \tag{9.29}$$

In terms of the sensitivities, we can express Equation 9.28 as

$$H_{kj} = 2\sum_i S_{ik}S_{ij} = 2\sum_i S_{ki}^T S_{ij} \tag{9.30}$$

If we write the last sum as a matrix multiplication, we can summarize this relationship in matrix notation

$$\boldsymbol{H} = 2\boldsymbol{S}^T\boldsymbol{S}$$

Given these expressions for the gradient and Hessian, we can construct a fairly efficient parameter-estimation method for differential equation models using standard software for solving nonlinear optimization and solving differential equations with sensitivities.

Parameter estimation algorithm:

① Guess initial parameter values.

② Using an appropriate ODE solver, solve the model and sensitivity equations simultaneously given the current parameter values. Compute x_i and S_{ij}.

③ Evaluate Φ and $\nabla\Phi$ using Equations 9.24 and 9.29.

④ Update parameter values to minimize Φ. This step and the next are usually controlled by an optimization package.

⑤ Check convergence criteria. If not converged, go to ②.

⑥ On convergence, set $\hat{\boldsymbol{\theta}}$ to current parameter values. Calculate $\boldsymbol{H}$ using Equation 9.30. Calculate confidence intervals using Equation 9.19.

Example 9.4: Fitting reaction-rate constant and order

We illustrate these methods on a classic reactor modeling problem: finding the rate constant and reaction order from isothermal concentration measurements in a batch reactor.

Consider an irreversible, nth order reaction

$$\mathrm{A} + \mathrm{B} \longrightarrow \text{products}, \qquad r = kc_A^n$$

taking place in a liquid-phase batch reactor containing a large excess of reactant B. Given the measured concentration of component A shown in Figure 9.21, determine the best values of the model parameters.

Solution

The material balance for species A is

$$\frac{dc_A}{dt} = -kc_A^n \tag{9.31}$$

$$c_A(0) = c_{A0} \tag{9.32}$$

Given the data shown in Figure 9.21, it does not seem reasonable to assume we know c_{A0} any more accurately than the other measurements, so we include it as a parameter to be estimated. The model therefore contains three unknown parameters

$$\boldsymbol{\theta}^T = \begin{bmatrix} k & c_{A0} & n \end{bmatrix}$$

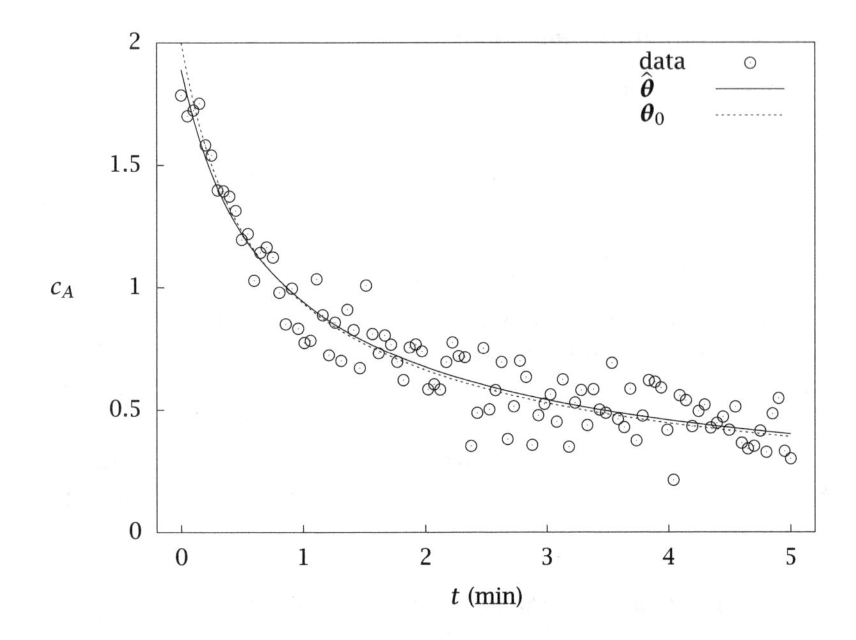

Figure 9.21: Experimental measurement and best parameter fit for nth-order kinetic model, $r = kc_A^n$.

We can generate a reasonable initial parameter set by guessing values and solving the model until the model simulation is at least on the same scale as the measurements. We provide this as the starting point, and then solve the nonlinear optimization problem in Equation 9.25 using the least-squares objective as shown in Equation 9.24. We then compute the approximate confidence intervals using Equation 9.19 with Equation 9.30 for $\boldsymbol{H}$. The solution to the optimization problem and the approximate confidence intervals are given in Equation 9.33.

$$\boldsymbol{\theta}_0 = \begin{bmatrix} k \\ c_{A0} \\ n \end{bmatrix} = \begin{bmatrix} 0.5 \\ 2.0 \\ 2.5 \end{bmatrix} \qquad \hat{\boldsymbol{\theta}} = \begin{bmatrix} 0.47 \\ 1.89 \\ 2.50 \end{bmatrix} \pm \begin{bmatrix} 0.052 \\ 0.18 \\ 0.42 \end{bmatrix} \tag{9.33}$$

The parameters that we used to generate the data in Figure 9.21 also are given in Equation 9.33. Notice the estimates are close to the correct values, and we have fairly tight approximate confidence intervals.

Next we examine the quality of these approximate confidence intervals for this problem. Figure 9.22 shows the results of a Monte Carlo

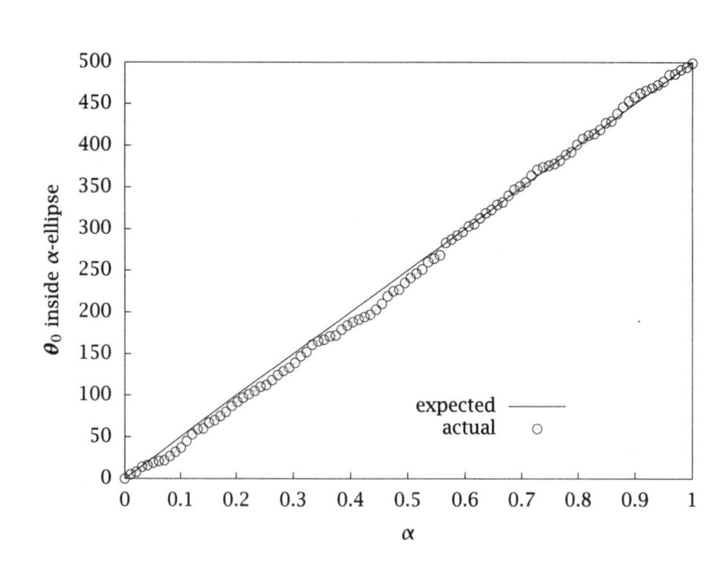

Figure 9.22: Monte Carlo evaluation of confidence intervals; the points indicate the actual number of times that the true parameter falls within the α-level ellipse of the estimate for 500 replicated datasets.

simulation study. In this study we generate 500 datasets by adding zero-mean measurement noise with variance $\sigma^2 = 0.01$ to the model solution with the correct parameters. For each of these 500 datasets, we solve the optimization problem to obtain the parameter estimates. We also produce a value for the Hessian, sample variance for each dataset, and the α-level contour distance to the actual $\boldsymbol{\theta}_0$. Finally we calculate how frequently the true parameter falls within α-level ellipse of the estimates for the 500 replicated datasets as we vary α. This result is plotted as the points in Figure 9.22. Notice for all α values, the true value of the parameter is within the corresponding α-level ellipse centered at the individual parameter estimates, approximately the correct number of times, indicating that the approximate confidence intervals given in Equation 9.19 are fairly reliable for this problem. Be aware that this conclusion may not be true for other nonlinear problems, and should be checked. This computational check is fairly expensive; note that we had to solve 500 optimization problems to produce Figure 9.22, for example. But given the dramatic increase in computational speed, a few Monte Carlo simulations in the final stages of a modeling study are well justified. □

Differential-equation models with multiple measurements. Most reactor models of interest contain balances for several species and therefore consist of several differential equations. To determine the parameters in such models, usually the concentrations of several species are measured as well. To treat this case, we consider the differential-equation model

$$\frac{d\boldsymbol{x}}{dt} = \boldsymbol{f}(\boldsymbol{x};\boldsymbol{\theta}) \tag{9.34}$$

$$\boldsymbol{x}(0) = \boldsymbol{g}(\boldsymbol{x}_0;\boldsymbol{\theta}) \tag{9.35}$$

$$\boldsymbol{y} = \boldsymbol{h}(\boldsymbol{x}) \tag{9.36}$$

in which $\boldsymbol{x}$ is the vector of states that defines the reactor model, $\boldsymbol{\theta}$ are the unknown model parameters, $\boldsymbol{x}_0$ are the initial conditions, and $\boldsymbol{y}$ are the experimentally measurable quantities.

In this case, we again define a scalar objective function that measures our fit to the data. When we have different measured quantities, however, it often does not make sense to sum the squares of the residuals. The measured variables may differ in size from each other by orders of magnitude. The influence a measurement has on the objective also would be influenced by the arbitrary choice of the units of the measurement, which is obviously undesirable. The simplest way to address this issue is to employ *weighted* least squares. The reader should be aware that more general procedures are available for the multiple measurement case, including the maximum likelihood method [29, 30, 31]. For simplicity of presentation, we focus here on weighted least squares.

In weighted least squares, we combine the different measurements by forming the *weighted* sum of the residuals. Let $\boldsymbol{e}_i$ be the residual vector at the ith sample time

$$\boldsymbol{e}_i = \tilde{\boldsymbol{y}}_i - \boldsymbol{h}(\boldsymbol{x}(t_i;\boldsymbol{\theta}))$$

The objective function is then

$$\Phi(\boldsymbol{\theta}) = \sum_i \boldsymbol{e}_i^T \boldsymbol{W} \boldsymbol{e}_i \tag{9.37}$$

in which $\boldsymbol{W}$ is a symmetric, positive-definite weighting matrix. Usually $\boldsymbol{W}$ is chosen to be a diagonal matrix. The elements on the diagonal are the weights assigned to each measurement type. To estimate the parameters we now solve

$$\min_{\boldsymbol{\theta}} \Phi(\boldsymbol{\theta}) \tag{9.38}$$

subject to Equations 9.34–9.36. The sensitivities are now a time-varying matrix,

$$S_{jk}(t_i) = \frac{\partial x_j(t_i;\boldsymbol{\theta})}{\partial \theta_k}$$

We can compute the gradient as before

$$\nabla\Phi = -2\sum_i \boldsymbol{S}_i^T \frac{\partial \boldsymbol{h}_i^T}{\partial \boldsymbol{x}_i} \boldsymbol{W}\boldsymbol{e}_i$$

and the Gauss-Newton approximation of the Hessian is

$$\boldsymbol{H} = 2\sum_i \boldsymbol{S}_i^T \frac{\partial \boldsymbol{h}_i^T}{\partial \boldsymbol{x}_i} \boldsymbol{W} \frac{\partial \boldsymbol{h}_i}{\partial \boldsymbol{x}_i^T} \boldsymbol{S}_i$$

in which $\boldsymbol{S}_i = \boldsymbol{S}(t_i)$, $\boldsymbol{x}_i = \boldsymbol{x}(t_i)$, and $\boldsymbol{h}_i = \boldsymbol{h}(\boldsymbol{x}(t_i))$.

Example 9.5: Fitting rate constants in hepatitis B virus model

We illustrate using multiple measurements by revisiting the hepatitis B model introduced in Chapter 1

$$\text{nucleotides} \xrightarrow{\text{cccDNA}} \text{rcDNA} \qquad (9.39)$$

$$\text{nucleotides} + \text{rcDNA} \longrightarrow \text{cccDNA} \qquad (9.40)$$

$$\text{amino acids} \xrightarrow{\text{cccDNA}} \text{envelope} \qquad (9.41)$$

$$\text{cccDNA} \longrightarrow \text{degraded} \qquad (9.42)$$

$$\text{envelope} \longrightarrow \text{secreted or degraded} \qquad (9.43)$$

$$\text{rcDNA} + \text{envelope} \longrightarrow \text{secreted virus} \qquad (9.44)$$

Find rate constants k_1–k_6 from the cccDNA, rcDNA and envelope protein measurements given in Figures 9.23–9.25. As before we assume the nucleotides and amino acids are in large excess, and cccDNA catalyzes Reactions 9.39 and 9.41.

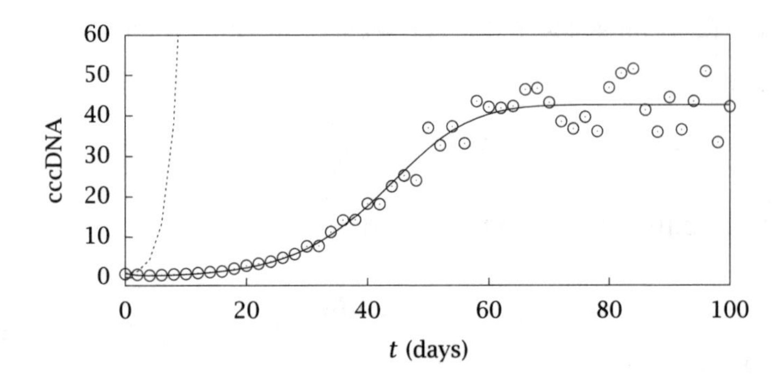

Figure 9.23: Species cccDNA versus time for hepatitis B virus model; initial guess and estimated parameters fit to data.

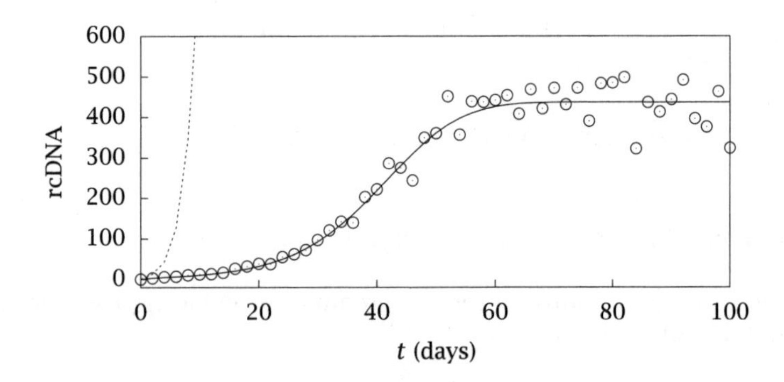

Figure 9.24: Species rcDNA versus time for hepatitis B virus model; initial guess and estimated parameters fit to data.

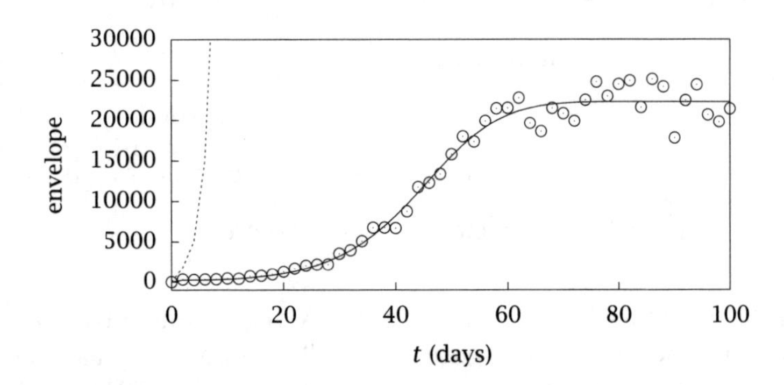

Figure 9.25: Envelope versus time for hepatitis B virus model; initial guess and estimated parameters fit to data.

Solution

The reaction rates and production rates for Reactions 9.39–9.44 are given by

$$\begin{bmatrix} r_1 \\ r_2 \\ r_3 \\ r_4 \\ r_5 \\ r_6 \end{bmatrix} = \begin{bmatrix} k_1 c_A \\ k_2 c_B \\ k_3 c_A \\ k_4 c_A \\ k_5 c_C \\ k_6 c_B c_C \end{bmatrix} \qquad \begin{bmatrix} R_A \\ R_B \\ R_C \end{bmatrix} = \begin{bmatrix} r_2 - r_4 \\ r_1 - r_2 - r_6 \\ r_3 - r_5 - r_6 \end{bmatrix} \tag{9.45}$$

in which A is cccDNA, B is rcDNA, and C is envelope. We write mass balances for these components to produce three differential equations. The initial condition for the experiment is

$$\begin{bmatrix} c_A & c_B & c_C \end{bmatrix}^T = \begin{bmatrix} 1 & 0 & 0 \end{bmatrix}^T$$

and is assumed to be known precisely. The data are generated by assuming values for the rate constants and adding noise to the model solution. The measurement noise is assumed proportional to the size of the measured value. We choose

$$\boldsymbol{W} = \text{diag}\left[1, 10^{-2}, 10^{-4}\right]$$

to reflect the relative sizes of the cccDNA, rcDNA and envelope species. This weight matrix gives each measurement roughly the same relative weight in the objective function.

When estimating several rate constants that may range in values by several orders of magnitude, *it often is useful to define the model parameters to be the logarithms of the rate constants.*

$$\theta_i = \log_{10}(k_i) \tag{9.46}$$

This transformation often makes it easier for optimization software to find accurate solutions, and automatically enforces the natural constraints that the rate constants are positive. The data in Figures 9.23–9.25 were generated using the following rate constants

$$\boldsymbol{k}_0 = \begin{bmatrix} 2.0 \\ 0.025 \\ 1000 \\ 0.25 \\ 2.0 \\ 7.5 \times 10^{-6} \end{bmatrix} \qquad \boldsymbol{\theta}_0 = \begin{bmatrix} 0.30 \\ -1.60 \\ 3.00 \\ -0.60 \\ 0.30 \\ -5.12 \end{bmatrix} \tag{9.47}$$

We generated a random initial guess for the parameters and produced the following optimal set of parameters and approximate confidence interval

$$\boldsymbol{\theta}_{\text{init}} = \begin{bmatrix} 0.80 \\ -1.13 \\ 3.15 \\ -0.77 \\ -0.16 \\ -5.46 \end{bmatrix} \qquad \hat{\boldsymbol{\theta}} = \begin{bmatrix} 0.32 \\ -1.43 \\ 2.52 \\ -0.42 \\ -0.20 \\ -5.13 \end{bmatrix} \pm \begin{bmatrix} 0.20 \\ 1.57 \\ 0.99 \\ 1.61 \\ 1.01 \\ 0.22 \end{bmatrix} \tag{9.48}$$

The dashed lines in Figures 9.23–9.25 show the solution of the model with the initial parameter values. As you can see, we are not giving the optimizer a very good initial guess. In spite of the poor initial guess, the optimizer is able to find a good solution. The model solution using the optimal parameter estimates is shown by the solid lines in the figures. Notice we have an excellent fit to the data. But notice also in Equation 9.48 that the confidence intervals are quite large, which means *we have little confidence in the estimated parameters in spite of the good fit to the data.* This outcome is not unusual when estimating many rate constants.

We may be satisfied with fitting the data in this way, but we may wish to reduce the model so that the estimated parameters have tighter confidence intervals. We pursue model reduction next. The Hessian is a 6×6 matrix for this problem because we have six parameters. We cannot graphically represent the six-dimensional confidence hyperellipsoid as we did in Figure 9.6 in two dimensions; but we can do the next best thing, which is to examine the eigenvalues and eigenvectors. The eigenvalues are given in Equation 9.49

$$\boldsymbol{\lambda} = \begin{bmatrix} 1.26 \times 10^8 \\ 4.7 \times 10^6 \\ 1.5 \times 10^6 \\ 2.1 \times 10^5 \\ \boxed{3.7 \times 10^3} \\ \boxed{1.3 \times 10^3} \end{bmatrix} \qquad \boldsymbol{v}_5 = \begin{bmatrix} -0.02 \\ -0.14 \\ \boxed{0.68} \\ -0.14 \\ \boxed{0.70} \\ 0.01 \end{bmatrix} \qquad \boldsymbol{v}_6 = \begin{bmatrix} 0.07 \\ \boxed{0.68} \\ 0.14 \\ \boxed{0.70} \\ 0.14 \\ -0.09 \end{bmatrix} \tag{9.49}$$

The two smallest eigenvalues, λ_5 and λ_6, are two orders of magnitude smaller than the largest four, which indicates we should remove about two degrees of freedom from the parameter-estimation problem. The eigenvectors associated with these two eigenvalues, v_5 and v_6, also are given in Equation 9.49. Notice that v_6 is aligned almost entirely in the (θ_2, θ_4) coordinate directions; the other elements of this eigenvector are near zero. Similarly, v_5 is aligned almost entirely in the (θ_3, θ_5) coordinate directions. These eigenvectors tell us that we can move the parameters a significant distance in the v_5, v_6 directions without greatly changing the value of the least-squares objective.

To demonstrate this fact, consider the model solutions shown in Figures 9.26–9.28. The solid lines correspond to the estimated parameters in Equation 9.48, and the dashed lines correspond to the following parameters

$$\boldsymbol{\theta} = \begin{bmatrix} 0.28 & -1.77 & 2.45 & -0.77 & -0.27 & -5.08 \end{bmatrix}^T \tag{9.50}$$

We added $(1/2)v_6$ to $\widehat{\boldsymbol{\theta}}$ to obtain Equation 9.50. Notice that even though we have changed the parameters by a large amount, we have not changed the simulations much at all. In other words, because of the logarithmic transformation, if we *multiply* parameters k_3 and k_5 by the same constant, we do not change the fit to the data. Similarly, if we multiply parameters k_2 and k_4 by some other constant, we do not change the fit to the data. We guess immediately that we should find the *ratios* of the rate constants k_3/k_5 and k_2/k_4.

We also can gain insight by examining the steady-state solution to the model. If we set the production rates given in Equation 9.45 to zero and solve for the steady-state concentrations we obtain[2]

$$c_{As} = \frac{k_1 - k_4}{k_6\left(\frac{k_4}{k_2}\right)\left(\frac{-k_1+k_4+k_3}{k_5}\right)}$$

$$c_{Bs} = \left(\frac{k_4}{k_2}\right) c_{As} \qquad c_{Cs} = \left(\frac{-k_1 + k_4 + k_3}{k_5}\right) c_{As}$$

Notice in Equation 9.47 that $k_3 \gg k_1, k_4$, and, therefore, to an excellent approximation

$$c_{As} = \frac{k_1 - k_4}{k_6\left(\frac{k_4}{k_2}\right)\left(\frac{k_3}{k_5}\right)} \qquad c_{Bs} = \left(\frac{k_4}{k_2}\right) c_{As} \qquad c_{Cs} = \left(\frac{k_3}{k_5}\right) c_{As} \tag{9.51}$$

We see again that the ratios of the rate constants k_3/k_5 and k_2/k_4 appear. Also the difference $k_1 - k_4$ is the important parameter, not the individual values. We therefore choose the model parameters to be

$$\boldsymbol{\phi} = \log_{10}\begin{bmatrix} k_1 - k_4 & k_2/k_4 & k_3/k_5 & k_6 \end{bmatrix}^T$$

[2]Note that the zero solution also is a steady state.

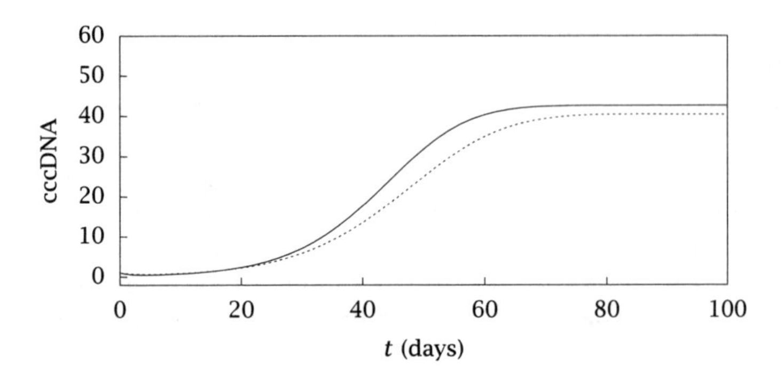

Figure 9.26: Species cccDNA versus time for hepatitis B virus model; estimated (solid) and perturbed (dashed) parameter values.

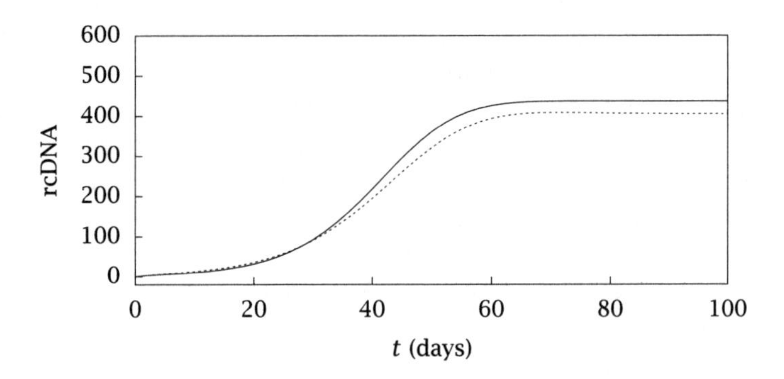

Figure 9.27: Species rcDNA versus time for hepatitis B virus model; estimated (solid) and perturbed (dashed) parameter values.

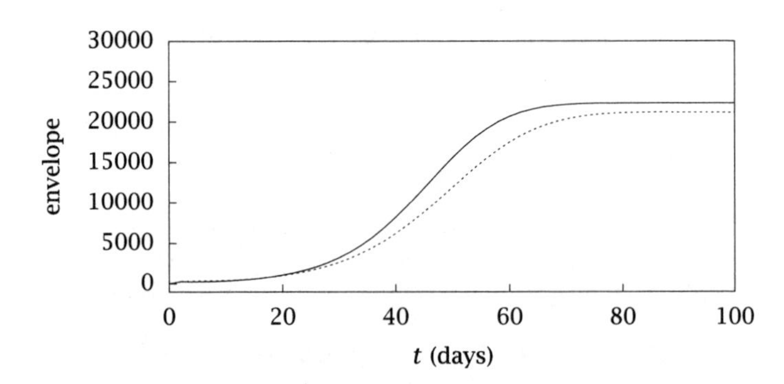

Figure 9.28: Envelope versus time for hepatitis B virus model; estimated (solid) and perturbed (dashed) parameter values.

and set k_3 and k_4 (or k_3 and k_1) to arbitrary values. Note they cannot be completely arbitrary because we require $k_3 \gg k_1, k_4$ for this analysis to be valid. If we then perform parameter estimation with the reduced set of four parameters we obtain the tight confidence intervals shown in Equation 9.52.

$$\boldsymbol{\phi}_0 = \begin{bmatrix} 0.24 \\ -1.00 \\ 2.70 \\ -5.12 \end{bmatrix} \qquad \hat{\boldsymbol{\phi}} = \begin{bmatrix} 0.25 \\ -1.00 \\ 2.71 \\ -5.10 \end{bmatrix} \pm \begin{bmatrix} 0.12 \\ 0.14 \\ 0.10 \\ 0.05 \end{bmatrix} \tag{9.52}$$

We do *not* need to use the correct values of k_3 and k_4 to estimate $\boldsymbol{\phi}$; even if we use values of k_3 and k_4 that are off by an order of magnitude from the values used to generate the data, the estimate of $\boldsymbol{\phi}$ is the same as in Equation 9.52. In fact, we must design a new experiment to determine the values k_3 and k_4 if they are of interest, because they cannot be found using data shown in Figures 9.23–9.25. Please note that these particular conclusions depend on the particular values of the rate constants we used to generate the data. If very different parameter values are chosen, the method of analysis can be used again but the model reduction may change. □

The complexity of this model probably precludes us from guessing quickly and intuitively that parameter pair k_3 and k_4 (or k_3 and k_1) cannot be determined from the experimental data, and we should estimate certain ratios and differences of rate constants instead of the rate constants themselves. By estimating all parameters, and analyzing the eigenvalues and eigenvectors of the Hessian, we have a systematic approach to develop such understanding fairly quickly when presented with a new problem of interest.

9.3 An Industrial Case Study

Reactor modeling in the industrial environment is a challenging task. When successful models are constructed, however, the payoff can be large, resulting in more efficient reactor operation, more consistent and higher quality product, and improved reactor designs. Given the difficulty of the modeling challenge in the best industrial circumstances, the least we should expect is that the computational and numerical procedures being employed are efficient and reliable, and are not compounding the difficulty of extracting models from the available data. To conclude this chapter we examine a small prototypical case study that illustrates some of the challenges we can expect when we model data from industrial facilities. This study was conducted in collaboration with colleagues at Kodak and more details are available [26].

End-point problems. In many chemical reactions, two main reactants are combined to yield a primary desired product. A small quantity of one of the reactants often remains at the end of the reaction due to batch-to-batch variability in the purity and reactivity of the starting materials, and variability in the rate of side reactions. In some cases, an excess of one reactant at the final time does not present a problem and the opportunity for improvement with better control is proportional to the quantity remaining, which is usually small. In these cases, an excess of one reactant is commonly added to ensure complete consumption of the other reactant.

In some cases, however, a small amount of neither starting material is tolerable. These materials may be difficult (impossible for practical purposes) to separate from the products, and their presence even in small quantities may prevent further processing steps from occurring. The excess addition of one reactant could also result in undesired reactions, usually after the limiting reagent is exhausted. In these cases, the cost of having the unreacted materials present in some amount greater than a small threshold is the cost of the entire batch of chemicals. This general problem in which neither reagent can exist at the end of the reaction is called the "end-point control" problem.

The reaction of interest is the dehalogenation of a dihalogenated starting material to form the divinyl product, which is used in photographic film production. It is assumed that the halide groups are removed from the starting material in two consecutive reactions:

$$\mathrm{A + B \xrightarrow{k_1} C + B{\cdot}HX}$$

$$\mathrm{C + B \xrightarrow{k_2} D + B{\cdot}HX}$$

A	$XH_2CCH_2RCH_2CH_2X$
B	Organic base
C	$H_2C{=}CHRCH_2CH_2X$
D	$H_2C{=}CHRCH{=}CH_2$

The dihalogenated starting material (A) loses HX to the base (B) to form the mono-halogenated intermediate (C), which subsequently loses HX

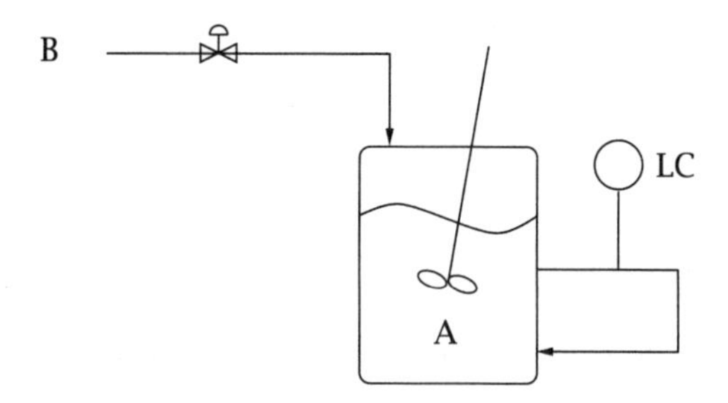

Figure 9.29: Semi-batch reactor addition of component B to starting material A.

to the base to produce the desired final product (D). Over-addition of base causes polymerization and loss of the batch.

Modeling. The reaction is carried out in a well-stirred semi-batch reactor as shown in Figure 9.29. The reactor is initially charged with a weighed amount of component A and component B is added. The material balances for the semi-batch reactor are

$$\begin{aligned}
\frac{dV_R}{dt} &= Q_f(t) \\
\frac{d(c_A V_R)}{dt} &= -k_1 c_A c_B V_R \\
\frac{d(c_B V_R)}{dt} &= Q_f c_{Bf} - (k_1 c_A c_B + k_2 c_C c_B) V_R \\
\frac{d(c_C V_R)}{dt} &= (k_1 c_A c_B - k_2 c_C c_B) V_R \\
\frac{d(c_D V_R)}{dt} &= k_2 c_C c_B V_R
\end{aligned}$$

in which Q_f is the volumetric flowrate of base and c_{Bf} is the feed concentration of B. The primary assumptions in this model are: isothermal operation, negligible volume change upon reaction or liquid chromatograph (LC) sampling, perfect mixing, negligible side reactions such as polymerization of the C and D, and reaction with impurities and inhibitors in the starting material. The initial conditions for the ODEs are:

$$\begin{aligned}
V_R(0) &= V_{R0} \\
n_A(0) &= n_{A0} \\
n_B(0) = n_C(0) &= n_D(0) = 0
\end{aligned}$$

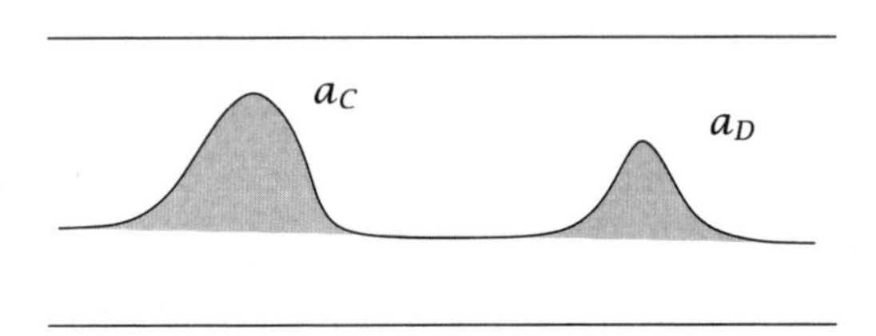

Figure 9.30: Depiction of an LC curve for determining the concentration of intermediate C and product D.

It is assumed that c_{Bf} and V_{R0} are known precisely, and therefore the unknown model parameters are given by $\boldsymbol{\theta} = [k_1 \; k_2 \; n_{A0}]^T$. Although the initial A charged to the reactor is weighed, the parameter n_{A0} is chosen to be adjustable to account for the unknown level of impurities and the neglected side reactions. Notice that making good decisions of this type depend on experience and judgment, and reflect more the art rather than the science of reactor modeling.

The feed flowrate of base is measured by a mass flow meter. For composition analysis there is an on-line LC that draws a sample every 8 to 10 minutes and, after a 7-minute delay, reports the relative amounts of C and D. The unknown parameters, $\boldsymbol{\theta}$, are estimated with the dynamic LC data. The LC-detector wavelength is set to detect the R-vinyl bond. Therefore only the C and D species show peaks in the LC output as depicted in Figure 9.30. The areas of the two peaks are related to the molar concentrations by

$$a_D = 2k_{\rm lc}c_D$$

$$a_C = k_{\rm lc}c_C$$

in which $k_{\rm lc}$ is the proportionality constant for the LC. By calculating normalized areas this constant can be removed,

$$y(t) = \frac{a_C}{a_C + a_D} = \frac{c_C}{c_C + 2c_D}$$

This normalized C peak area, y, is used for the parameter estimation. The parameter estimation is performed by minimizing the following relative least-squares objective function,

$$\Phi = \sum_i \left(\frac{\tilde{y}_i - y_i}{y_i} \right)^2$$

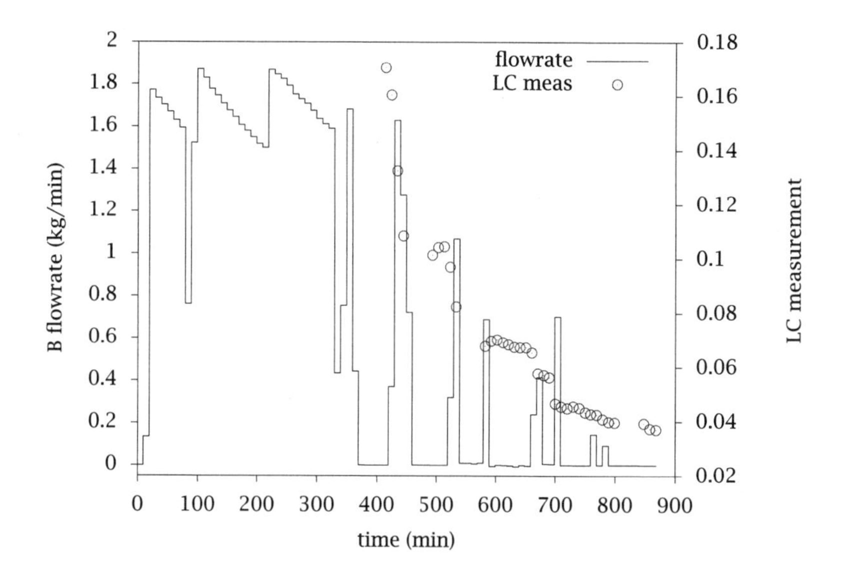

Figure 9.31: Base addition rate and LC measurement versus time.

in which $\tilde{y}_i$ is the measured normalized C peak and y_i is the model prediction at the ith sampling time.

Conventional manufacturing procedure. Normal operating procedure is to charge the reactor with some (only approximately known) amount of A, and add an initial amount of B that is sure not to overshoot the end point. An example of a typical B addition profile is given in Figure 9.31. After the initial B is added, the LC is switched on and the operator waits until the readings stabilize. The operator then sees how much C is remaining, and — based on experience — adds more B. The objective is to add enough B to consume all but 3% of the A. Ideally the operator would like to consume all the A, but the target is set at 3% to allow a margin for error. The penalty for overshoot is so high that only conservative addition steps are generally ever taken. After making the addition, the LC readings are again allowed to stabilize, and the operator again checks to see how close he is to the target before making another addition. This cycle repeats until the operator is satisfied that he is close enough (between 2-4%) to the target. As can be seen in Figure 9.31, the operator required 7 additions and about 500 minutes after first turning on the LC before he had determined how much B was required to reach end point.

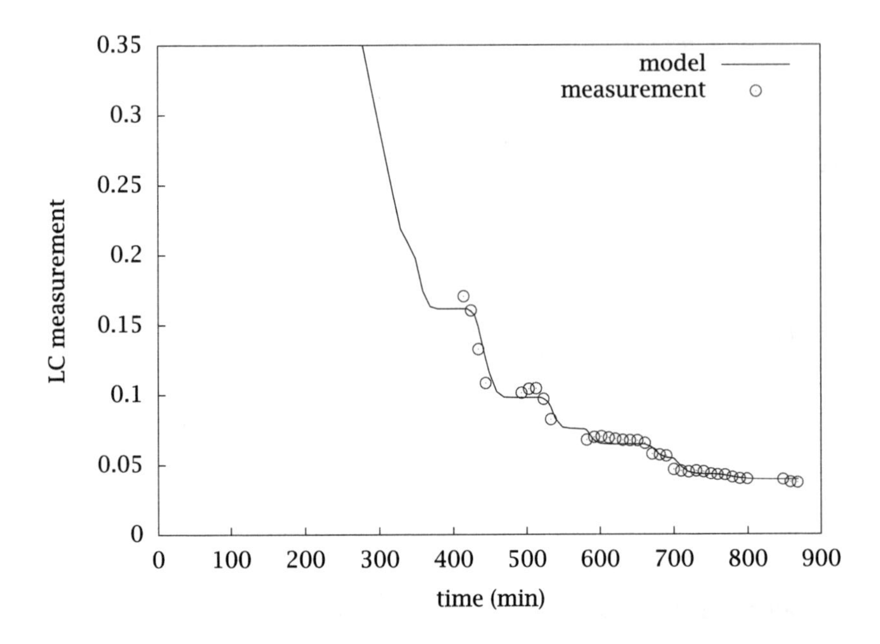

Figure 9.32: Comparison of data to model with optimal parameters.

Parameter estimation. If we perform parameter estimation with these data we achieve the model fit shown in Figure 9.32 The model predictions of all the concentrations are shown in Figures 9.33 and 9.34. Note that the concentration of B provides the most information on the rate constants. Unfortunately this concentration is not measured.

The optimal values of the parameters and their approximate 95% confidence intervals are given by

$$\begin{array}{|lcrcr|} n_{A0} & = & 2.35 & \pm & 0.0073 \\ k_1 & = & 3628 & \pm & 4590 \\ k_2 & = & 1687 & \pm & 2044 \end{array}$$

Notice that the initial amount of A is determined to within 0.6%, but the rate constants have 200% uncertainty. That uncertainty is a direct result of the problem structure. The relative amounts of C and D pin down accurately the amount of starting material. The rate constants determine the speed at which we arrive at these values. After each base addition, we have only rough information about the reaction rates by

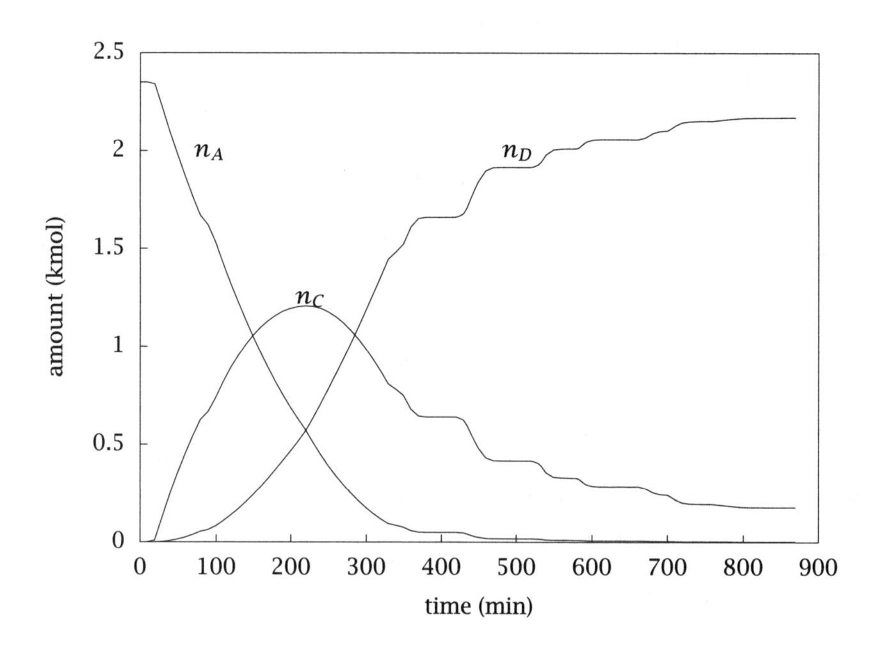

Figure 9.33: Total amount of species A, C and D versus time.

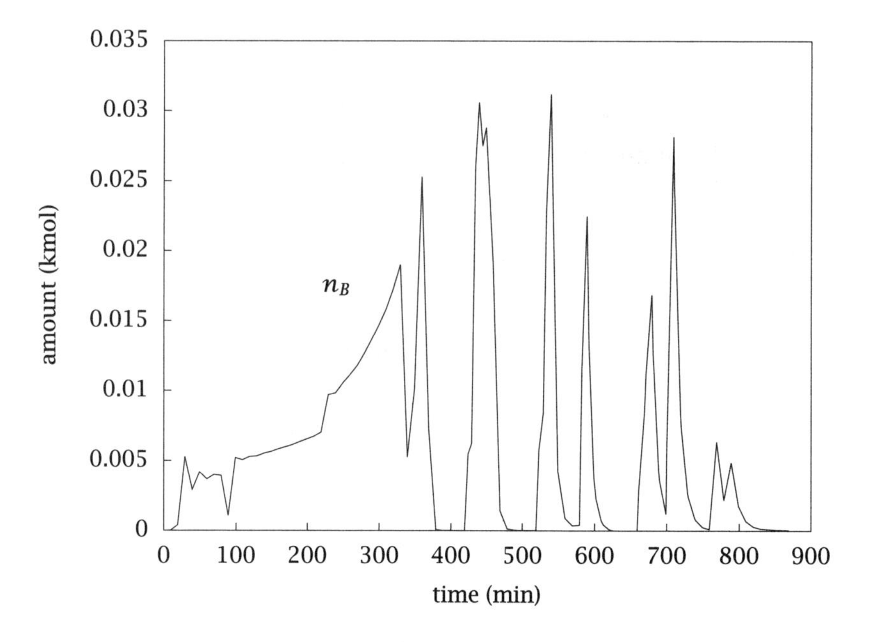

Figure 9.34: Total amount of species B versus time.

observing the relative amounts of C and D. If we could measure the B concentration, we could achieve narrow confidence intervals for the rate constants as well. For the purposes of end-point control, however, the rate constants are irrelevant. We seek to know how much total base to add, not how quickly it will be consumed. In this application, we regard k_1 and k_2 as nuisance parameters. We must estimate them to determine the parameter we care about, but their values are not useful to us.

Model reduction. The large parameter uncertainty tells us that the experimental data do not contain sufficient information to determine the rate constants. That diagnosis is essential because it motivates our next step: model reduction.

Although it may not be immediately apparent, the difficulty we face is caused by the presence of large rate constants. We wish to make the equilibrium assumption as described in Chapter 5 to reduce the model. This analysis is clearest if we first rewrite the material balances in terms of extents of the two reactions as in Chapter 5

$$\begin{aligned}\frac{dV_R}{dt} &= Q_f \\ \frac{d\varepsilon_1}{dt} &= r_1 = k_1 c_A c_B V_R = k_1 n_A n_B / V_R \\ \frac{d\varepsilon_2}{dt} &= r_2 = k_2 c_C c_B V_R = k_2 n_C n_B / V_R\end{aligned} \tag{9.53}$$

The initial conditions for this model are

$$\begin{aligned}V_R(0) &= V_{R0} \\ \varepsilon_1(0) &= 0 \\ \varepsilon_2(0) &= 0\end{aligned}$$

We can easily translate from the two reaction extents to total moles of species via

$$\begin{aligned}n_A &= n_{A0} - \varepsilon_1 \\ n_B &= n_{B\text{add}} - (\varepsilon_1 + \varepsilon_2) \\ n_C &= \varepsilon_1 - \varepsilon_2 \\ n_D &= \varepsilon_2\end{aligned} \tag{9.54}$$

in which

$$n_{B\text{add}}(t) = \int_0^t Q_f(t') c_{Bf} dt'$$

From the two reaction-rate expressions in Equations 9.53, if we assume equilibrium is established with these irreversible reactions, either the concentration of B is zero or the concentrations of A and C are zero. Due to the reactor-addition policy, we know B is the limiting reagent, and conclude that under large k_1, k_2, the B concentration is zero. The material balance for B in Equations 9.54 then provides one algebraic equation for the two extents

$$\varepsilon_1 + \varepsilon_2 = n_{B\text{add}} \tag{9.55}$$

The additional equation comes from examining the two extents' differential equations and noticing

$$\frac{d\varepsilon_1}{dt} = \left(k\frac{n_A}{n_C}\right)\frac{d\varepsilon_2}{dt} \qquad k = \frac{k_1}{k_2} \tag{9.56}$$

We see that the concentration of B disappears from this slow time-scale model and the ratio $k = k_1/k_2$ appears instead of the individual rate constants. If we wish to cast the reduced model in differential-equation form, we can differentiate Equation 9.55 and substitute Equation 9.56 to eliminate ε_1 to obtain

$$\frac{d\varepsilon_2}{dt} = \frac{Q_f c_{Bf}}{1 + k n_A/n_C}$$

Substituting Equations 9.54 for the moles of A and C, and using Equation 9.55 again to eliminate ε_1 produces a differential equation for the second extent

$$\frac{d\varepsilon_2}{dt} = Q_f c_{Bf}\left(1 + k\frac{n_{A0} - n_{B\text{add}} + \varepsilon_2}{n_{B\text{add}} - 2\varepsilon_2}\right)^{-1}$$

Solving this reduced model for various values of k produces the results shown in Figure 9.35. If we perform parameter estimation with this model and these data we obtain the following results.

$$\boxed{\begin{array}{rcl} n_{A0} & = & 2.35 \pm 0.0048 \\ k = k_1/k_2 & = & 2.25 \pm 0.43 \end{array}} \tag{9.57}$$

We have improved the situation compared to the uncertainty in the full model. But we still have more than 10% uncertainty in the ratio of rate constants k. These results may be adequate, but if we wish to further improve the confidence in $k = k_1/k_2$ and n_{A0}, we proceed by reexamining the experimental design.

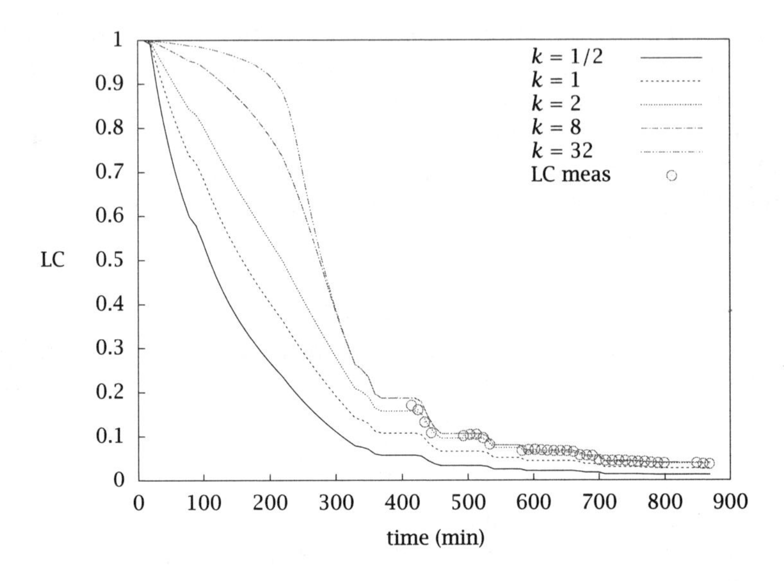

Figure 9.35: Predictions of LC measurement for reduced model.

Experimental design. We have removed the rate constants from the estimation problem by assuming their values are arbitrarily large and making the equilibrium assumption. But we still have a reasonably large uncertainty in the ratio of rate constants. Given what we know at this point, we can easily remedy this final, remaining problem. Consider again the results shown in Figure 9.35. We see that the ratio of the rate constants has its primary effect at early times. By the end of the semi-batch addition, there is little information left. For this reason, we have the relatively large uncertainty in k shown in Equation 9.57. But at early times, when significant amounts of A and C remain in the reactor, the data are much more informative. Although under the original reactor-addition policy, the operator had no reason to turn on the LC until later times, we see that these early measurements are actually the informative ones. We do not have industrial operating data with early LC measurements, but we can simulate the effect with our model. Notice the advantage of modeling. We have the ability to query the model instead of performing expensive experiments to evaluate the impact of a proposed change. Consider Figure 9.36 in which we have simulated early LC measurements by solving the model with

$$k = k_1/k_2 = 2.0, \qquad n_{A0} = 2.35$$

and adding measurement noise. Notice this noise is greater than the measurement noise shown in the actual operating data. We next perform parameter estimation on the data shown in Figure 9.36 and obtain the following parameters and approximate confidence intervals

$$\begin{array}{rcl} n_{A0} & = & 2.35 \pm 0.0025 \\ k = k_1/k_2 & = & 2.05 \pm 0.072 \end{array}$$

Figure 9.37 shows the confidence-interval ellipses and boxes when using the original dataset and using the dataset augmented with early-time data. As summarized in Figure 9.37 the primary benefit in adding the early time data is a more precise estimate of k. The uncertainty in k is reduced by more than a factor of six when adding the early time LC measurements. The initial number of moles of A is relatively accurately determined by both datasets.

Improved reactor operation. Given these modeling results, we can shorten the batch time significantly. First we switch the LC on at time $t = 0$. Then as the LC measurements become available, we estimate the initial number of moles n_{A0} and monitor its confidence interval. As soon as the uncertainty in n_{A0} reaches a sufficiently low threshold, we are confident how much B is required and can add the remainder in one shot. Testing this approach with many datasets at Kodak allowed us to conclude that by the time the first large addition was completed, we obtained sufficient confidence on n_{A0} that the rest of the B could be added immediately. Such a procedure reduces the batch time from about 900 minutes with the conventional approach to about half that time with the model-based operation. The new operation essentially doubles the production rate without constructing new reactor facilities, which is significant for this capacity-limited chemical.

9.4 Summary

In this chapter we first summarized some of the analytical methods and experimental reactors used to collect reactor data, focusing the discussion on:

- infrared spectroscopy
- gas chromatography
- mass spectrometry

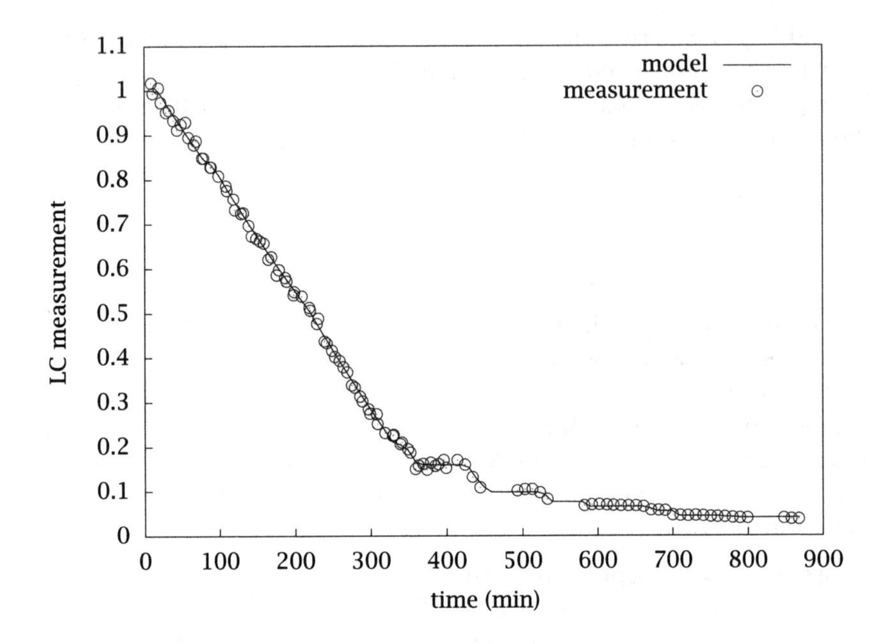

Figure 9.36: Fit of LC measurement versus time for reduced model; early time measurements have been added.

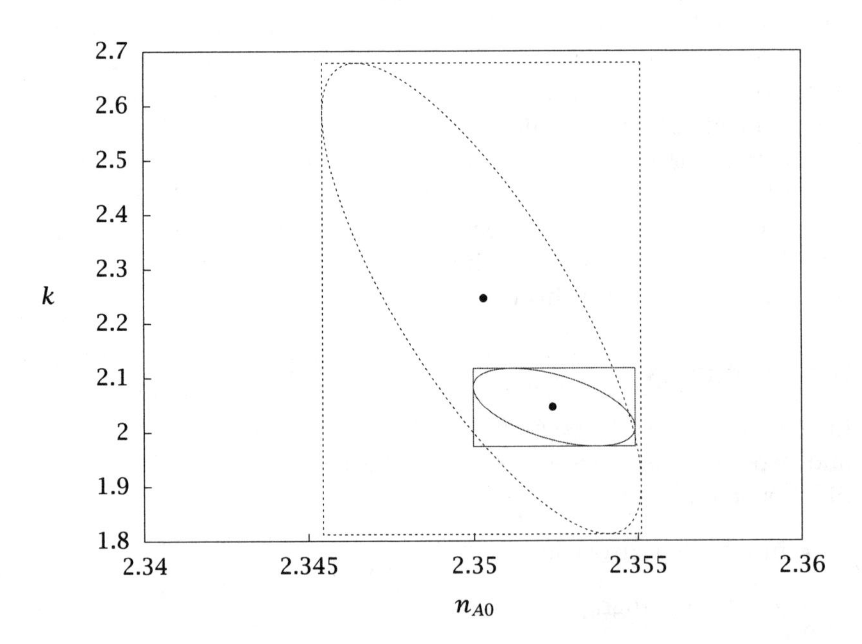

Figure 9.37: Confidence intervals for reduced model without (dashed) and with (solid) redesigned experiment.

We also discussed how to model differential reactors, which are convenient reactors for measuring reaction rates as a function of species concentrations.

Next we covered analysis of data. We used probability and random variables to model the irreproducible part of the experiment. For models that are linear in the parameters, we can perform parameter estimation and construct exact confidence intervals analytically. For models that are nonlinear in the parameters, we compute parameter estimates and construct approximate confidence intervals using nonlinear optimization methods.

We then covered the important topic of estimating parameters for differential-equation models. We employed computational methods for solving differential equations and sensitivities, and solving nonlinear optimization problems in order to tackle this challenging problem.

The chapter concluded with an industrial case study. In this case study we illustrated the following steps:

- Proposing an initial model given a rough idea of the reaction chemistry.
- Estimating the parameters for the full model given composition measurements.
- Reducing the model based on the parameter confidence intervals.
- Designing new experiments based on the reduced model's confidence intervals.
- Using the final model to find new operating policies to double the production rate.

We hope the examples and methods in this chapter serve to inspire students and practicing engineers to build models as part of understanding new processes and chemistries of interest. The modeling experience often leads to deeper process understanding and produces a compact summary of current knowledge that is easily and efficiently communicated to other colleagues and team members. Process understanding coupled with creativity often leads to process improvement and new discovery.

Notation

A	absorbance
b	path length for the absorbance measurement
c	concentration of absorbing species
c_j	concentration of component j
$\hat{c}_j(t_i)$	measured concentration of component j at sampling time t_i
D_p	catalyst particle diameter
D_t	fixed-bed reactor tube diameter
$\boldsymbol{e}$	residual vector
$\boldsymbol{e}$	measurement error vector
H_{kj}	Hessian matrix, $H_{kj} = \partial^2\Phi/\partial\theta_k\partial\theta_j$
L	fixed-bed reactor tube length
m	mass velocity
n_d	number of data points
n_p	number of model parameters
N_j	molar flow of component j
$p(\boldsymbol{x})$	probability density function of random variable $\boldsymbol{x}$
Q	volumetric flowrate
Q_f	feed volumetric flowrate
r	reaction rate of (single) reaction
r_i	reaction rate for ith reaction
$\boldsymbol{r}$	reaction-rate vector
R_j	production rate for jth species
$\boldsymbol{R}$	production-rate vector
S	open tube area for flow
S_{jk}	sensitivity of state x_j with respect to parameter θ_k
t_i	sampling time
V_R	reactor volume
$\boldsymbol{x}$	state vector
$\boldsymbol{y}$	vector of measured responses
ϵ	molar absorptivity
ϵ_B	bed porosity
$\boldsymbol{\theta}$	parameter vector
μ	fluid viscosity
ν_{ij}	stoichiometric number for the jth species in the ith reaction
$\boldsymbol{\phi}$	transformed parameter vector
Φ	objective function

9.5 Exercises

Exercise 9.1: Bringing it all back home

Given the new tools in this chapter, return to the data displayed in Chapter 2, Figures 2.2 and 2.3. Reaction rates and production rates are related by

$$\boldsymbol{R} = \boldsymbol{\nu}^T \boldsymbol{r}$$

If we measure production rates and wish to estimate reaction rates, we can model the measurement process with

$$\boldsymbol{R} = \boldsymbol{\nu}^T \boldsymbol{r} + \boldsymbol{e}$$

in which $\boldsymbol{e}$ is assumed to have a normal distribution.

(a) What is the formula for the 95% confidence interval on estimated reaction rates $\hat{\boldsymbol{r}}$?

(b) Place your 95% confidence interval on the set of points displayed in Figure 2.3. These points are available at `www.engineering.ucsb.edu/~jbraw/chemreacfun`. Given the Octave code displayed in Chapter 2 that generated $\boldsymbol{R}_{\text{meas}}$, what is the value of the measurement error variance. Of the 500 estimated reaction rates, how many of the values fall outside your 95% confidence interval ellipse?

Notice that if you know (or have some means of estimating) the measurement error variance beforehand, you do not need the experimental measurements to know the confidence intervals. They are computable from the model before you do any experiments.

(c) Now assume the measured production rates are only the first column in $\boldsymbol{R}_{\text{meas}}$ in Chapter 2. Compute the least-squares estimate and 95% confidence interval in which you assume you do not know the variance in the measurement error and must estimate it from these data also. How do your two ellipses compare? Do you obtain a reasonable idea of the parameter uncertainty when you have only one measurement of the six species production rates?

Exercise 9.2: Estimating activation energy and preexponential factor

(a) Calculate the parameter estimates ($\ln k_m$, E) for the data given in Figure 9.7. Assume the measurement error variance is unknown and you must estimate it from the data. The numerical values for the data points are available at the website `www.engineering.ucsb.edu/~jbraw/chemreacfun`.

(b) What are the coordinates of the corners of the box corresponding to the 95% confidence interval?

(c) What are the coordinates of the semi-major axes of the ellipse corresponding to the 95% confidence interval?

(d) Sketch your result by plotting the parameter estimate, ellipse and box. Sketch the ellipse by hand if it is difficult to make your favorite plotting package draw what you want. Are the parameter estimates highly correlated? Why or why not?

(e) Finally, plot the fit of the model to the data with both (T, k) and ($1/T$, $\ln k$) as the (x, y) axes.

t (min)	0	4	8	12	16	20	24	28	32	36	40
c_A (mol/L)	0.83	0.68	0.59	0.51	0.43	0.38	0.33	0.30	0.26	0.24	0.21

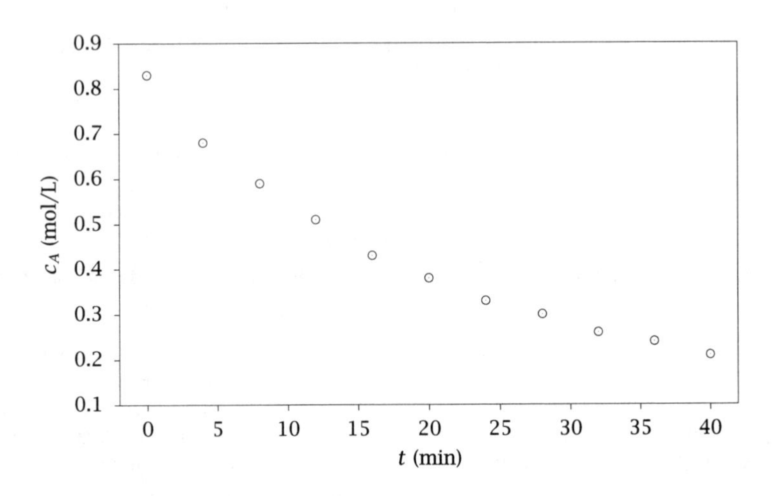

Figure 9.38: Batch-reactor data for Exercise 9.3.

Exercise 9.3: Batch-reactor parameter estimation

A well-mixed, isothermal, constant-volume batch reactor is initially charged with reactant A and the following reaction takes place

$$A \longrightarrow 2B$$

The reactor volume is 10 L and the initial charge of A is 8.3 mol. The concentration of A versus time is measured and shown in Figure 9.38. You wish to fit these data using the following kinetic rate-law expression,

$$r = kc_A^n$$

(a) What is the solution to the model, $c_A(t)$, for this rate law?

(b) Calculate from the data rough estimates of the parameters k and n. What are their units?

Hint: consider approximating the time derivative dc_A/dt by $\Delta c_A/\Delta t$ and compare those approximate values of the rate using the data to kc_A^n. A nonlinear transformation of the rate expression can make the estimation problem linear.

(c) Estimate the parameters by solving a least-squares problem. Plot the fit of both models to the data and compare. Are the model fits comparable?

Exercise 9.4: The perils of differentiating data

Consider the nth-order, irreversible reaction

$$A \longrightarrow \text{products}, \qquad r = kc_A^n$$

t (min)	0.0	4.0	8.0	12.0	16.0	20.0	24.0	28.0	32.0	36.0	40.0
c_A	0.830	0.696	0.591	0.507	0.440	0.384	0.339	0.301	0.268	0.241	0.217
(mol/L)	0.833	0.683	0.593	0.515	0.446	0.380	0.324	0.303	0.256	0.225	0.205
	0.862	0.699	0.561	0.511	0.412	0.407	0.346	0.276	0.322	0.254	0.215

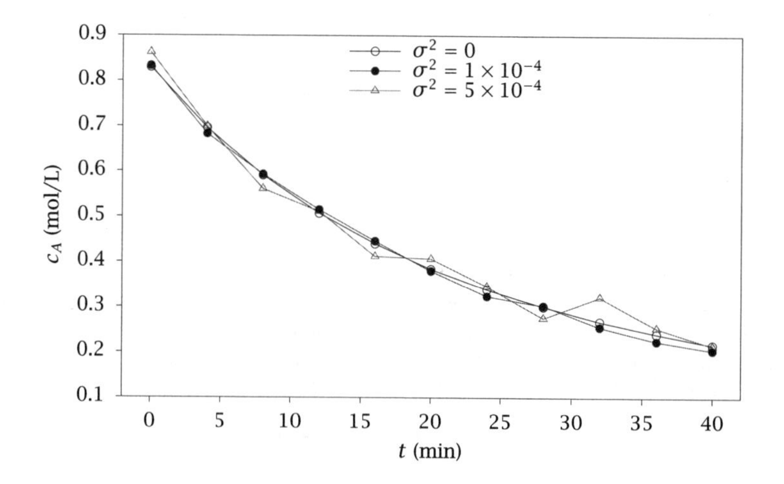

Figure 9.39: Batch-reactor data for Exercise 9.4; 3 runs with different measurement error variance.

taking place in an isothermal, liquid-phase batch reactor. Measurements of c_A versus time for different measurement accuracies are displayed in Figure 9.39. The following method is sometimes recommended for finding the rate constant and order of this reaction.

1. Compute an approximation to the reaction rate by "differentiating" the data

$$r_l = -\frac{c_{A,l+1} - c_{Al}}{t_{l+1} - t_l}, \qquad l = 1, \ldots, n_d - 1$$

   in which c_{Al} is the measured concentration at sample time t_l and r_l is the "measured" rate at this sample time. Notice the rate is not actually measured; the concentration is measured.

2. Take the logarithm of the rate expression to obtain a model linear in the parameters

$$\ln(r) = \ln(k) + n \ln(c_A)$$

3. Use linear least squares with this model and the "measured" rate r_l versus c_{Al} to find k and n.

Let's examine how well this method performs on the three datasets shown in Figure 9.39.

(a) Estimate the rate constant and order for each of the three datasets. Delete any data points that produce negative reaction rates before taking the logarithm in step 2.

(b) Plot the model fit and data for $c_A(t)$ versus t for each dataset. Plot the model fit (straight line) and "data" for $\ln(r)$ versus $\ln(c_A)$ for each dataset.

What causes the lack of fit in c_A versus t for the first two datasets? What causes the lack of fit in c_A versus t for the third dataset?

Exercise 9.5: Estimation without differentiation

Resolve Exercise 9.4 as a nonlinear optimization over the parameters (k, n, c_{A0}) using an ODE/sensitivity solver and an optimizer. Discuss the advantages and disadvantages of the two approaches regarding accuracy of the estimates and sensitivity to noise in the measurements.

Exercise 9.6: Fitting the RTD step response

Consider the RTD test data given in Exercise 8.11. Use an ODE/sensitivity solver and optimizer to find the best value of parameter ρ, given in Equations 8.46, that fits these data. Solve the model with this value of ρ and show the fit to the data. Resolve Exercise 8.11 and predict the reactor yield and conversion.

Exercise 9.7: The differential reactor and volume change

Your colleague has an idea that, since not much conversion occurs in the differential reactor, we should be able to assume the volumetric flowrate is constant to simplify the model.

Consider the gas-phase reaction $A \longrightarrow 2B$ using a pure A stream and compare analyzing the data from a differential reactor in two ways. First, use Equation 9.5, which accounts for the changing volumetric flowrate. Second, neglect the change in volumetric flowrate, assume $Q = Q_f$, and show that

$$\tilde{r} = \frac{Q_f}{V_R} \frac{(c_{Af} - c_A)}{-1} \qquad \tilde{r} = \frac{Q_f}{V_R} \frac{(c_{Bf} - c_B)}{2}$$

Calculate the relative error $e = (r - \tilde{r})/r$ committed when neglecting the volumetric flowrate change in a differential reactor assuming you are measuring the concentration of component A. Is this a large or small error? If it is a large error, can you explain what went wrong with your colleague's logic?

Exercise 9.8: Estimating rate constant from catalyzed CSTR reaction data

The heterogeneously catalyzed reaction

$$A \xrightarrow{k} B \qquad r = kc_A$$

was studied in a small, research-scale CSTR. The CSTR design eliminates external mass-transfer limitations (i.e., you may assume the bulk fluid concentration equals the surface concentration). The feed consisted of pure A. The reactor pressure was 1.0 atm and the temperature was 573 K. The residence time for the CSTR is 2.0 s. The following table presents steady-state effluent concentration data for different catalyst sizes. The catalyst particles can be assumed spherical. The effective diffusivity of A in the pellet is $D_A = 0.0008$ cm^2/s. The pellet density equals the bed density, $\rho_p = \rho_B$.

Radius (cm)	c_A (mol/cm^3)
0.016	2.13×10^{-6}
0.071	3.89×10^{-6}
0.121	5.84×10^{-6}
0.191	8.20×10^{-6}
0.246	9.07×10^{-6}

Estimate the value of the rate constant at 573 K. Plot the fit of the model with the best estimate of the rate constant to the measurements provided in the table.

Exercise 9.9: Oxygen adsorption, revisited

Consider again the adsorption data in Exercise 5.12. The data are listed in Table 5.8 and plotted in Figure 5.25. The adsorption is dissociative

$$\mathrm{O_2 + 2X \underset{k_{-1}}{\overset{k_1}{\rightleftharpoons}} 2O \cdot X}$$

(a) Write out again the expression for the Langmuir isotherm to model the concentration of adsorbed oxygen in terms of the concentration of gas-phase oxygen. How many unknown parameters does your model contain?

(b) Estimate these parameters using the nonlinear least-squares technique described in this chapter.

(c) Consider again the transformation of the model into the linear form, i.e., consider $1/\bar{c}_\mathrm{O}$ as a function of $1/\sqrt{c_{\mathrm{O_2}}}$. Estimate the parameters again using linear least squares.

(d) Compare the parameter estimates obtained with these two approaches by plotting the fit to the original data. Why are the fits so different? Which technique do you recommend and why?

Exercise 9.10: Replication and confidence

Imagine you have performed an experiment and set up the least-squares problem

$$\boldsymbol{y} = \boldsymbol{X\theta} + \boldsymbol{e}$$

in which $\boldsymbol{y}$ contains the vector of measurements, $\boldsymbol{X}$ contains the matrix of independent variables, $\boldsymbol{\theta}$ is the parameter vector, and $\boldsymbol{e}$ is the measurement error, assumed to have known variance σ^2. We developed the following formula for the parameter estimates and α-level confidence intervals

$$\hat{\boldsymbol{\theta}} = (\boldsymbol{X}^T\boldsymbol{X})^{-1}\boldsymbol{X}^T\boldsymbol{y}$$
$$(\boldsymbol{\theta} - \hat{\boldsymbol{\theta}})^T\boldsymbol{X}^T\boldsymbol{X}(\boldsymbol{\theta} - \hat{\boldsymbol{\theta}}) \le \sigma^2\chi^2(n_p, \alpha) \tag{9.58}$$

If you replicate this experiment n_s times, the new data matrix is

$$\boldsymbol{X}_s = \left.\begin{bmatrix} \boldsymbol{X} \\ \boldsymbol{X} \\ \vdots \\ \boldsymbol{X} \end{bmatrix}\right\} n_s \text{ times}$$

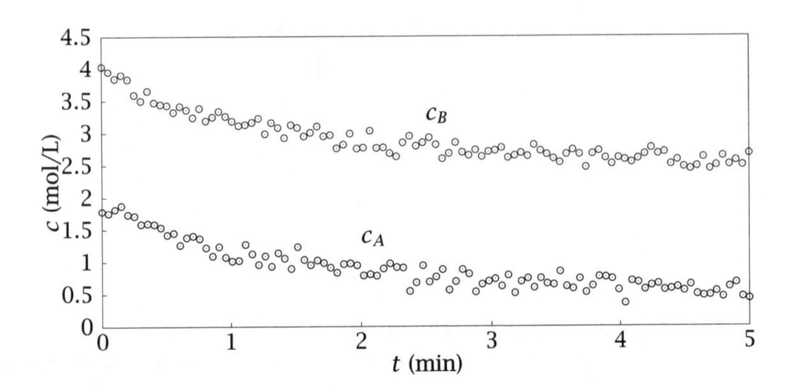

Figure 9.40: Batch-reactor data for Exercise 9.11.

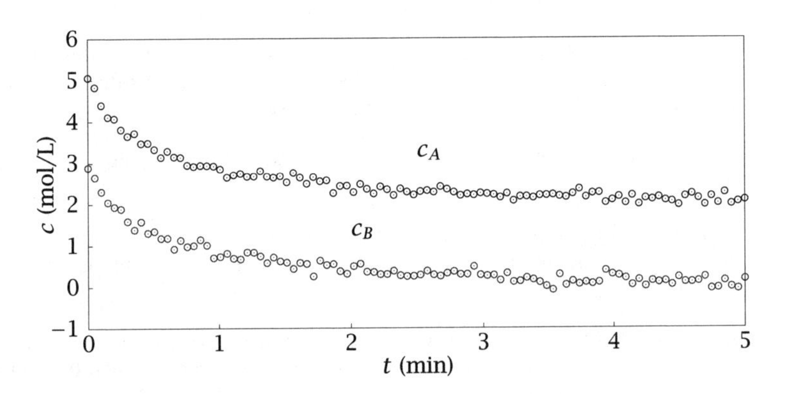

Figure 9.41: A second experiment for Exercise 9.11.

(a) Calculate $X_s^T X_s$. From Equation 9.58, how much smaller is the confidence interval ellipse after 2, 5 and 100 replicate experiments?

(b) Does the confidence ellipse change shape as well as size upon replication? Does your result agree with Figure 9.20?

(c) If you want to reduce uncertainty by a factor of β, how many replicate experiments are required? Explain why straight replication is a potentially expensive way to reduce uncertainty.

Exercise 9.11: Estimating a rate expression involving two reactants

Consider the irreversible reaction

$$\mathrm{A} + \mathrm{B} \longrightarrow \text{products}, \qquad r = kc_A^n c_B^m$$

taking place in an isothermal, liquid-phase batch reactor. Measurements of c_A and c_B versus time are displayed in Figure 9.40. We wish to determine from the data the rate constant and the order of the reaction with respect to both A and B.

(a) Estimate rate parameters, k, n and m, and initial concentrations, c_{A0} and c_{B0}, from the data in Figure 9.40 *without* differentiating the concentration data. The data are available at `www.engineering.ucsb.edu/~jbraw/chemreacfun`. Notice your confidence intervals for k, n and m are rather large. Why is this model poorly identified from these data?

(b) Consider a second experiment shown in Figure 9.41. Estimate the parameters using only the second experiment. Are your confidence intervals for k, n and m any better using the second experiment?

(c) Estimate the parameters using both experiments simultaneously. Are your confidence intervals any better using both experiments compared to using either experiment alone? Explain why or why not.

(d) Based on these results, to design experiments to determine the order of the rate expression with respect to two different species, how should you choose the initial conditions of the two species?

Exercise 9.12: Living and dying with linear least squares

Consider differentiating the data of Exercise 9.11 as described in Exercise 9.4. Notice differentiating the c_A data and the c_B data provide *two* independent estimates of the same reaction rate. You may wish to average these to obtain a good single estimate of the rate. Taking the logarithm of the rate expression again produces a model linear in the parameters

$$\ln(r) = \ln(k) + n\ln(c_A) + m\ln(c_B)$$

(a) To ensure that your method is coded correctly, consider first the "error free" data depicted in Figures 9.42 and 9.43. Estimate k, n and m using linear least squares and three different datasets: only the data in Figure 9.42, only the data in Figure 9.43, and the data in both figures. Do you get similar estimates with all three datasets? Are the confidence intervals similar? Explain why or why not.

(b) Now apply the linear least-squares approach to the data in Figures 9.40 and 9.41. Are your estimates and their uncertainties comparable to the result of Exercise 9.11? Which method of estimating parameters do you recommend and why?

Exercise 9.13: Least-squares estimate formula

Consider the least-squares objective

$$\Phi = \sum_{i=1}^{n_d} \left(y_i - \sum_{j=1}^{n_p} X_{ij}\theta_j \right)^2$$

and the parameter estimation problem

$$\min_{\boldsymbol{\theta}} \Phi(\boldsymbol{\theta})$$

(a) Differentiate Φ with respect to θ_k, $k = 1, \ldots, n_p$, set the result to zero, and derive Equation 9.9

$$\hat{\boldsymbol{\theta}} = (\boldsymbol{X}^T\boldsymbol{X})^{-1}\boldsymbol{X}^T\boldsymbol{y}$$

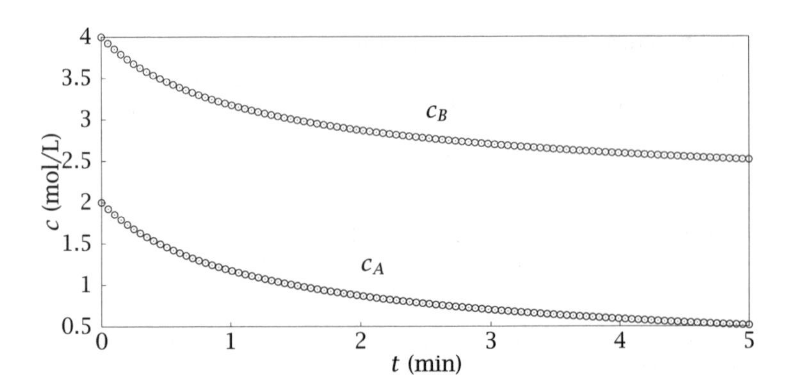

Figure 9.42: Batch-reactor data for Exercise 9.12.

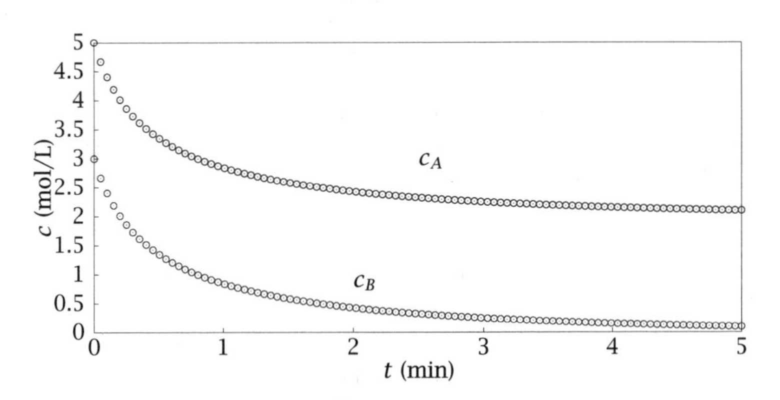

Figure 9.43: A second experiment for Exercise 9.12.

(b) Repeat for the weighted least-squares problem

$$\Phi = \sum_{i=1}^{n_d} W_i \left(y_i - \sum_{j=1}^{n_p} X_{ij}\theta_j \right)^2$$

and show

$$\hat{\boldsymbol{\theta}} = \left(\boldsymbol{X}^T \boldsymbol{W} \boldsymbol{X} \right)^{-1} \boldsymbol{X}^T \boldsymbol{W} \boldsymbol{y}$$

in which

$$\boldsymbol{W} = \begin{bmatrix} W_1 & 0 & \cdots & 0 \\ 0 & W_2 & \cdots & 0 \\ \vdots & \vdots & \ddots & \vdots \\ 0 & 0 & \cdots & W_{n_d} \end{bmatrix}$$

Exercise 9.14: Linear transformation of a normal is normal

Let $\boldsymbol{x}$ be a normally distributed random variable with mean $\boldsymbol{m}_x$ and covariance $\boldsymbol{P}_x$

$$\boldsymbol{x} \sim N(\boldsymbol{m}_x, \boldsymbol{P}_x)$$

Let a new random variable $\boldsymbol{z}$ be defined as a linear (affine) transformation of $\boldsymbol{x}$

$$\boldsymbol{z} = \boldsymbol{A}\boldsymbol{x} + \boldsymbol{b}$$

in which $\boldsymbol{A}$ is a constant matrix and $\boldsymbol{b}$ is a constant vector. Then one can show that $\boldsymbol{z}$ is also normally distributed with mean $\boldsymbol{m}_z$ and covariance $\boldsymbol{P}_z$ given by

$$\boldsymbol{z} \sim N(\boldsymbol{m}_z, \boldsymbol{P}_z) \qquad \boldsymbol{m}_z = \boldsymbol{A}\boldsymbol{m}_x + \boldsymbol{b} \qquad \boldsymbol{P}_z = \boldsymbol{A}\boldsymbol{P}_x\boldsymbol{A}^T$$

Consider again the linear model and least-squares estimate

$$\boldsymbol{y} = \boldsymbol{X}\boldsymbol{\theta} + \boldsymbol{e} \qquad \widehat{\boldsymbol{\theta}} = (\boldsymbol{X}^T\boldsymbol{X})^{-1}\boldsymbol{X}^T\boldsymbol{y}$$

in which

$$\boldsymbol{e} \sim N(\boldsymbol{0}, \sigma^2\boldsymbol{I})$$

Use the above result on linear transformation of normals to show that the parameter estimates are distributed as given in Section 9.2.3

$$\widehat{\boldsymbol{\theta}} \sim N(\boldsymbol{\theta}, \boldsymbol{P}) \qquad \boldsymbol{P} = \sigma^2(\boldsymbol{X}^T\boldsymbol{X})^{-1}$$

Exercise 9.15: Parameters of Michaelis-Menten kinetics

Consider the enzyme kinetics

$$\mathrm{E} + \mathrm{S} \underset{k_{-1}}{\overset{k_1}{\rightleftharpoons}} \mathrm{ES}$$

$$\mathrm{ES} \xrightarrow{k_2} \mathrm{P} + \mathrm{E}$$

in which the free enzyme E binds with substrate S to form bound substrate ES in the first reaction, and the bound substrate is converted to product P and releases free enzyme in the second reaction.

If the rates of these two reactions are such that either the E or ES is present in small concentration, the mechanism can be reduced by making the QSSA leading to an irreversible decomposition of S to P

$$\mathrm{S} \longrightarrow \mathrm{P} \qquad r = \frac{kc_S}{1 + Kc_S}$$

This rate expression is known as Michaelis-Menten kinetics.

The following measurements of c_S versus time were taken in your laboratory.

t (min)	c_S (cmol/L)
0.0	0.978
3.0	0.671
6.0	0.423
9.0	0.241
12.0	0.110
15.0	0.049
18.0	0.014

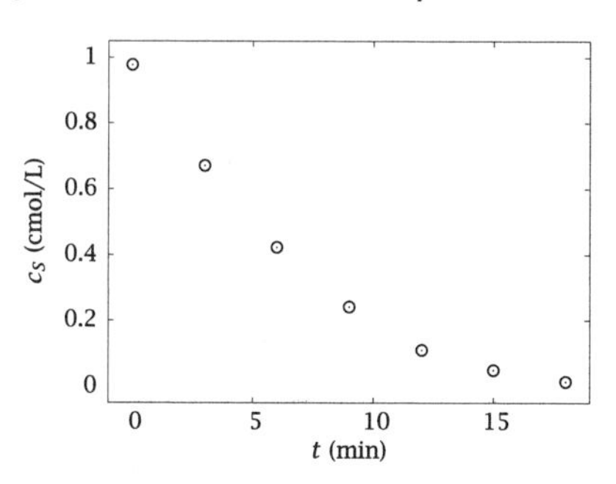

Estimate the parameters k and K appearing in the Michaelis-Menten rate expression (be sure to give the parameters' units). Hint: consider the transformation $1/r$ and formulate a linear least-squares problem that you can solve on a calculator.

Bibliography

[1] A. C. Atkinson and A. N. Donev. *Optimum Experimental Designs.* Oxford University Press, New York, 1992.

[2] Y. Bard. *Nonlinear Parameter Estimation.* Academic Press, New York, 1974.

[3] L. J. Bellamy. *The Infra-red Spectra of Complex Molecules*, volume one. Chapman and Hall, London, third edition, 1975.

[4] G. E. P. Box and G. C. Tiao. *Bayesian Inference in Statistical Analysis.* John Wiley & Sons, New York, 1973.

[5] G. E. P. Box, W. G. Hunter, and J. S. Hunter. *Statistics for Experimenters.* John Wiley & Sons, New York, 1978.

[6] M. Caracotsios and W. E. Stewart. Sensitivity analysis of initial value problems with mixed ODEs and algebraic equations. *Comput. Chem. Eng.*, 9 (4):359–365, 1985.

[7] M. Caracotsios and W. E. Stewart. Sensitivity analysis of initial-boundary-value problems with mixed PDEs and algebraic equations. *Comput. Chem. Eng.*, 19(9):1019–1030, 1995.

[8] J. J. Carberry. Designing laboratory catalytic reactors. *Ind. Eng. Chem.*, 56:39–46, 1964.

[9] L. G. Christophorou, editor. *Electron-Molecule Interactions and Their Applications*, volume one. Academic Press, Orlando, FL, 1984.

[10] L. E. Davis, N. C. MacDonald, P. W. Palmberg, G. E. Riach, and R. E. Weber, editors. *Handbook of Auger Electron Spectroscopy.* Physical Electronics Industries, Eden Prairie, Minnesota, second edition, 1976.

[11] A. A. Davydov. *Infrared Spectroscopy of Adsorbed Species on the Surface of Transition Metal Oxides.* John Wiley & Sons, Chichester, England, 1990.

[12] C. B. Duke, editor. *SURFACE SCIENCE The First Thirty Years.* Elsevier Science Publishers, Amsterdam, 1994.

[13] F. H. Field and J. L. Franklin. *Electron Impact Phenomena and the Properties of Gaseous Ions.* Academic, New York, 1957.

[14] F. W. Fifield and D. Kealey. *Principles and Practice of Analytical Chemistry.* Blackie Academic and Professional, Glasgow, UK, third edition, 1990.

[15] C. F. Gauss. *Theory of the combination of observations least subject to errors: part one, part two, supplement.* SIAM, Philadelphia, 1995. Translated by G. W. Stewart.

[16] R. L. Grob, editor. *Modern Practice of Gas Chromatography.* John Wiley & Sons, New York, 1977.

[17] M. L. Hair. Transmission infrared spectroscopy for high surface area oxides. In A. T. Bell and M. L. Hair, editors, *Vibrational Spectroscopies for Adsorbed Species, ACS Symposium Series 137,* pages 1–11. American Chemical Society, Washington, D.C., 1980.

[18] E. Heftmann, editor. *Chromatography, Part A Fundamentals and Techniques.* Elsevier, Amsterdam, fifth edition, 1992.

[19] G. Herzberg. *Molecular Spectra and Molecular Structure II. Infrared and Raman Spectra of Polyatomic Molecules.* Van Nostrand Reinhold, New York, 1945.

[20] J. M. Lafferty, editor. *Foundations of Vacuum Science and Technology.* John Wiley & Sons, New York, 1998.

[21] M. L. Lee, F. J. Yang, and K. D. Bartle. *Open Tubular Column Gas Chromatography: Theory and Practice.* John Wiley & Sons, New York, 1984.

[22] J. F. Moulder, W. F. Stickle, P. E. Sobol, and K. D. Bomben. *Handbook of X-ray Photoelectron Spectroscopy.* Perken-Elmer Corporation, Eden Prairie, Minnesota, 1992.

[23] K. Nakamoto. *Infrared and Raman Spectra of Inorganic and Coordination Compounds.* John Wiley & Sons, New York, third edition, 1978.

[24] M. A. Natal-Santiago, S. G. Podkolzin, R. D. Cortright, and J. A. Dumesic. Microcalorimetric studies of interactions of ethene, isobutene, and isobutane with silica-supported Pd, Pt, and PtSn. *Catal. Lett.*, 45:155–163, 1997.

[25] S. G. Podkolzin and J. A. Dumesic. Adsorption data for H_2 on Pd. Personal communication, 2001.

[26] J. B. Rawlings, N. F. Jerome, J. W. Hamer, and T. M. Bruemmer. Endpoint control in semi-batch chemical reactors. In *Proceedings of the IFAC Symposium on Dynamics and Control of Chemical Reactors, Distillation Columns, and Batch Processes,* pages 323–328, 1989.

[27] G. A. Somorjai. *Chemistry in Two Dimensions, Surfaces.* Cornell University Press, Ithaca, New York, 1981.

[28] J. L. Stephenson and W. E. Stewart. Optical measurements of porosity and fluid motion in packed beds. *Chem. Eng. Sci.*, 41(8):2161 - 2170, 1986.

[29] W. E. Stewart, M. Caracotsios, and J. P. Sørensen. Parameter estimation from multiresponse data. *AIChE J.*, 38(5):641–650, 1992.

[30] W. E. Stewart, T. L. Henson, and G. E. P. Box. Model discrimination and criticism with single-response data. *AIChE J.*, 42(11):3055–3062, 1996.

[31] W. E. Stewart, Y. Shon, and G. E. P. Box. Discrimination and goodness of fit of multiresponse mechanistic models. *AIChE J.*, 44(6):1404–1412, 1998.

[32] J. M. Thomas and W. J. Thomas. *Principles and Practice of Heterogeneous Catalysis.* VCH, Weinheim, 1997.

[33] I. E. Wachs, editor. *Characterization of Catalytic Materials.* Butterworth-Heinemann, Stoneham, MA, 1992.

10
Particulate Reactors

10.1 Particle Size Distributions

In this chapter we would like to treat more complex situations in which the reactor of interest contains one phase of matter dispersed in a second phase. This dispersion is often a solid phase, sometimes called a particle phase, dispersed in a liquid. The crystallization and purification of solid crystalline products, such as pharmaceuticals, from a solvent mixture is an important example of this kind of dispersion. But the dispersion may also be one liquid phase in small domains that are encapsulated by a separating membrane or stabilization layer and dispersed in a second, continuous liquid phase. Biological cells and emulsion polymers are examples of this type of dispersion. The models developed in previous chapters do not have the required structure to describe this kind of system, and the goal of this chapter is to provide this required extra structure.

When a collection of (solid or liquid) particles is dispersed in a second continuous (usually liquid) phase, one of the characteristics of the particle phase that is often of interest is the particle size distribution (PSD). A suitably generalized particle "size" distribution is the main new quantity of interest, and we would like to develop evolution equations for this quantity so that we can predict and control its properties with the same facility that we established in Chapter 4 to predict species concentrations in single-phase reacting systems. This kind of new evolution equation is called a population balance. A population balance essentially describes how the population of particles changes with time due to birth, growth, and death processes that are taking place in the reactor of interest.

To make matters explicit let $f(L,t)$ denote a particle size distribution in which L is some (single) characteristic length of the particles,

t is time, and $f(L,t)dL$ represents the number of particles in the size range L to $L + dL$. Note that f is often normalized by the volume of the reactor as well, so that $f(L,t)dL$ is the number of particles per volume of reactor. If we integrate over the particle size distribution we can obtain total particle properties of interest such as the following:

$\int_0^\infty f(L,t)dL$	total number of particles (per reactor volume)
$\int_0^\infty f(L,t)a_p L^2 dL$	total area of particles
$\int_0^\infty f(L,t)v_p L^3 dL$	total volume of particles

Note that a_p is an area "shape factor" such that $a_p L^2$ is the surface area of a particle of size L. Similarly, v_p is the volume shape factor such that $v_p L^3$ is the volume of a particle of size L. For example, if the particles are well approximated as spherical, and $L = r$ is the characteristic size, the shape factors are $a_p = 4\pi$ and $v_p = (4/3)\pi$. Total particle area is often important when particles are adsorbing reactants from a continuous phase, because the mass transfer rates are related to area. Total particle volume is often important in calculating total rates of reaction in the particle phase, or calculating total mass of particulate phase produced.

10.2 Applications

Before launching into the modeling of the particle size distribution, it may be helpful to provide some context and present a physical picture of some important applications. The chosen examples are obviously not intended to be exhaustive, but merely representative of the kinds of important industrial processes for which this kind of modeling is useful.

10.2.1 Crystallization

Crystallization is an ideal starting place because it often is the simplest possible physical situation in which population balance modeling is essential. In many crystallizations, the continuous phase is a single solvent with a single dissolved solute, and the particulate phase is (essentially) pure crystalline solute. Because the crystal lattices of the solid phase are almost always of extremely high purity (or the lattice cannot form), crystallization is an ideal separation method for pure

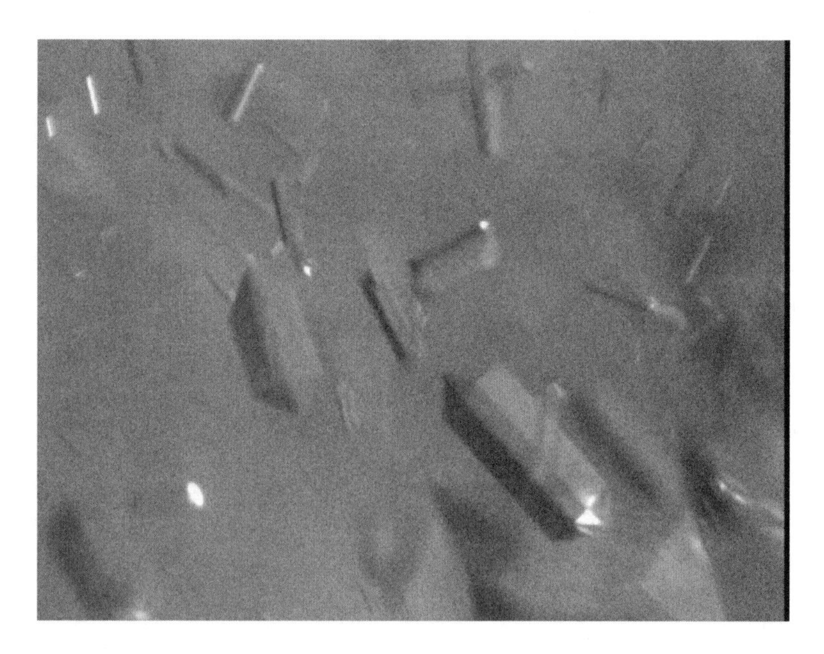

Figure 10.1: Growing glycine crystals displaying a distribution of sizes and shapes. Images taken in a laboratory reactor vessel.

solid products. The phase equilibrium of the solid particle phase and the solution phase is a classic case covered in all chemical engineering thermodynamic textbooks. At equilibrium the chemical potential of the solute in the pure particle phase is equal to the chemical potential of the solute in solution. When this condition is met, the solution is termed "saturated." Of course we are mainly interested in describing the nonequilibrium situation in which the solution is supersaturated, and mass transfer from solution to solid phase, i.e., crystallization, is taking place. In this situation the particle phase is growing. Sometimes we are interested in describing the rate of dissolution of solid phase when the solution phase is undersaturated.

In the crystallization of solid products, spherical particles are rarely observed. Faceted crystals of a wide variety of shapes are the norm. For example, Figure 10.1 shows typical sizes and shapes obtained when crystallizing glycine from solution. These images were obtained in the laboratory reactor depicted in Figure 10.2. The light source strobe is used to freeze the particle motion in the well-stirred, transparent, glass

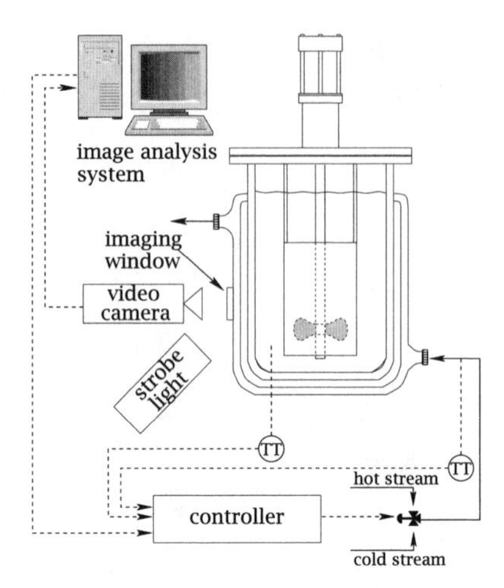

Figure 10.2: Laboratory crystallizer equipped with real-time imaging of growing crystals.

vessel, and the images are captured by a video camera.

Other common particle shapes are high aspect ratio "plates" or "needles." Many pharmaceutical products crystallize with this particle morphology, which can lead to significant challenges in washing, filtering, drying, and handling the solid product. Figure 10.3 shows the results of a typical pharmaceutical crystallization in a laboratory crystallizer. The particles are highly elongated with aspect ratios of 10:1 or more. In this case, the particles are first settled on a microscope stage so that they are all lying flat before analyzing the video image taken from the microscope. If these high aspect ratio particles had been frozen in place with a strobe light in random orientations caused by the stirring in the well-stirred reactor, a large variety of aspect ratios would be presented in the image. Such an image would be misleading because, as shown in Figure 10.3, all of the particles are in fact of similar shape and only differ in size. Recent reviews of the research into modeling and controlling crystal size and shape are provided in [10, 16]. Significant recent research into modeling the growth rates of different crystal faces, and hence crystal shape is presented in [8].

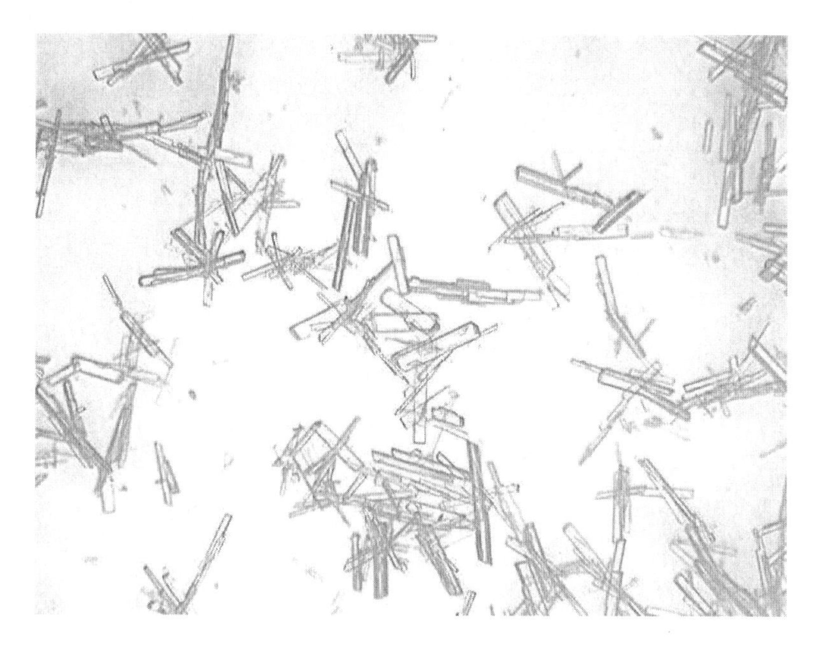

Figure 10.3: Growing pharmaceutical drug crystals with a needle-like particle shape. Images taken after particles have settled on a microscope stage.

10.2.2 Emulsion Polymerization

Emulsion polymerization is used to manufacture a wide variety of industrial polymers including adhesives, rubbers, latex paints, ink pigments, and wall and floor coverings. This process accounts for many billion pounds of polymer per year. The physical picture for emulsion polymerization is depicted in Figure 10.4. Surfactant, which has both a hydrophilic head group, and a hydrophobic long chain tail is used to stabilize polymer particles and monomer droplets dispersed in a continuous aqueous phase. When the surfactant concentration in the aqueous phase exceeds a critical concentration, known as the critical micelle concentration (CMC), the surfactant forms a second phase, known as the micellar phase or micelle. The micelle is thermodynamically favorable because small collections of surfactant molecules are able to arrange themselves such that their hydrophobic tails are aggregated with each other and separated from the aqueous phase by the outwardly pointing hydrophilic head groups. Monomer also swells the

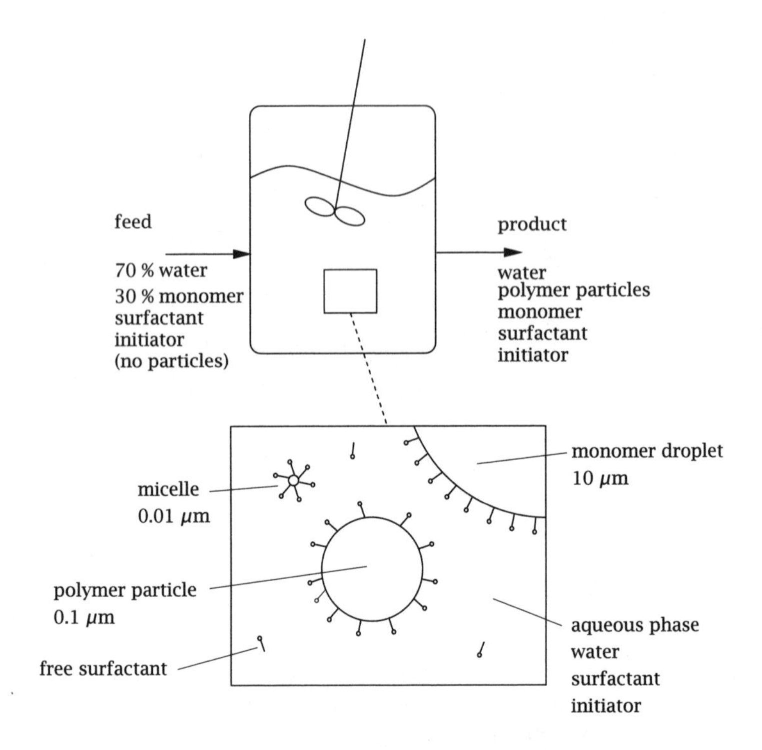

Figure 10.4: Well-stirred emulsion polymerization reactor. Surfactant forms micelles, which serve as the nucleation sites for the growing polymer particles.

interior of the micelles because of the attractive interactions between the hydrophobic monomer and the surfactant tail groups.

The micelles play an important role in the polymerization process. Normally a water-soluble free-radical initiator is fed to the reactor. The initiator decomposes into free radicals as discussed in the free-radical polymerization kinetics in Example 5.4 of Chapter 5. These free radicals enter the monomer-swollen micelles and initiate polymerization. As the initiated polymer particles grow and consume the monomer, monomer is continually resupplied by mass transfer from the large monomer droplets through the aqueous phase to establish the equality of chemical potential for monomer in the polymer particle, aqueous phase, and monomer droplets. Radicals may of course also enter the large monomer droplets and initiate polymerization, but the total surface area of the micelle phase is usually orders of magnitude larger than

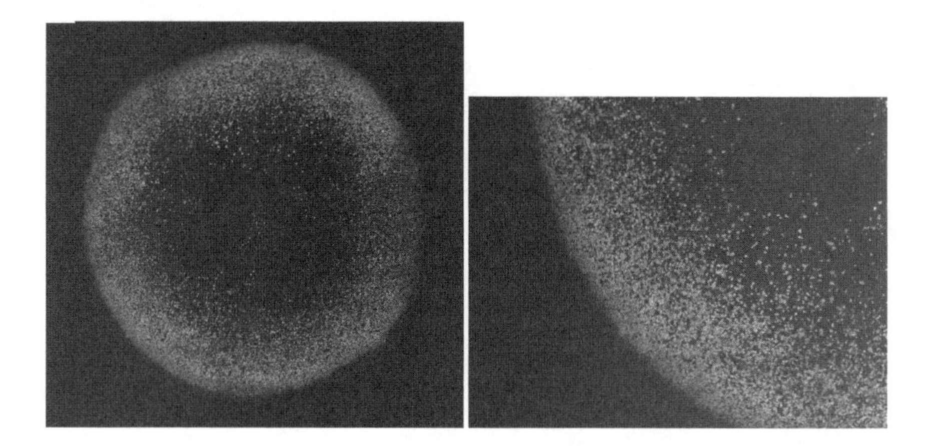

Figure 10.5: Cell infection by a virus. The cells are uniformly distributed, and the virus is placed initially in the center and diffuses outward. The cells fluoresce after they become infected. The dark inner core shows dead cells.

the surface area of the large monomer droplets and the polymerization in the droplets can be neglected.[1]

10.2.3 Biological Cells

Reactions involving living cells are ubiquitous in nature and, today, also in the bioprocess industries. Improving our understanding of and control over these reactions has large implications for human health. Consider the interactions of cells and viruses, for example. Viral infection of cells is responsible for maladies such as the common cold, influenza, chickenpox, cold sores, Ebola hemorrhagic fever, AIDS, avian influenza, and SARS. Figure 10.5 shows images of cell infection by a virus. The cells are uniformly distributed, and the virus is placed initially in the center and diffuses outward. The cells fluoresce after they become infected. The dark inner core shows the cells killed by the virus. Modeling the distribution of cells as a function of time since infection and interaction with the signaling molecules released into their

[1] Alternatively, in *dispersion* polymerization, the monomer droplets are made much smaller, usually by exposing the monomer phase to high shear rates, and the small monomer droplets serve as the locus of polymerization without the presence of any micellar phase.

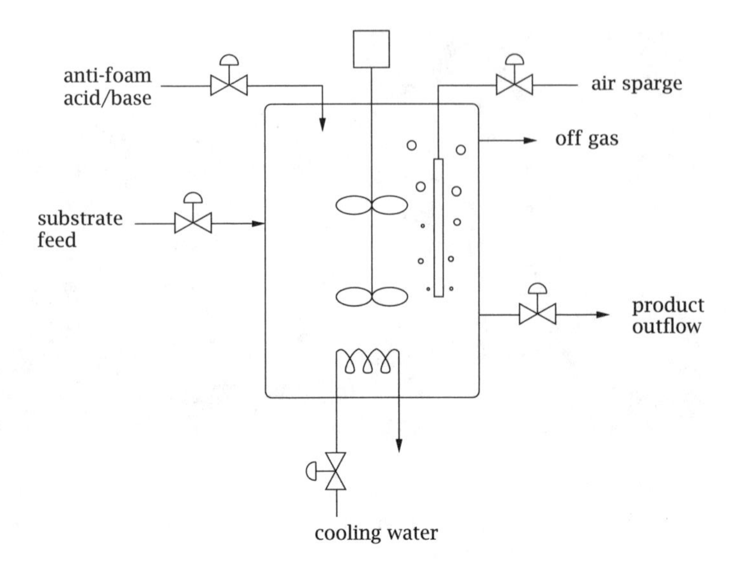

Figure 10.6: Basic fermentation system.

environment enables quantitative predictions that can be tested and verified with experimental measurement [5].

Living cells are also used to manufacture many important antibiotics and other pharmaceuticals. Figure 10.6 displays a simple schematic of a fermentation system used for antibiotic manufacture. The cells can grow exponentially quickly so that although there are no cells in the feed, they can establish a nonzero steady state in the CSTR, commonly known as a chemostat in the bioprocess industries. The classic paper [3] was one of the first to lay out the fundamentals for modeling the dynamic behavior of these kinds of cell populations. We further develop the modeling of chemostats later in the chapter.

10.3 Population Balance

Given this brief overview of applications, we next develop the evolution equation for the particle size distribution, known as the population balance. We first treat deterministic models with a single size coordinate and single source of nucleation of new particles at a single (zero) size. Next we extend the model to include multiple sources of nucleation at multiple sizes. Then we extend the model to handle the case when multiple internal coordinates are required to specify the state of the particle. Size and shape, or mass and age are common examples of

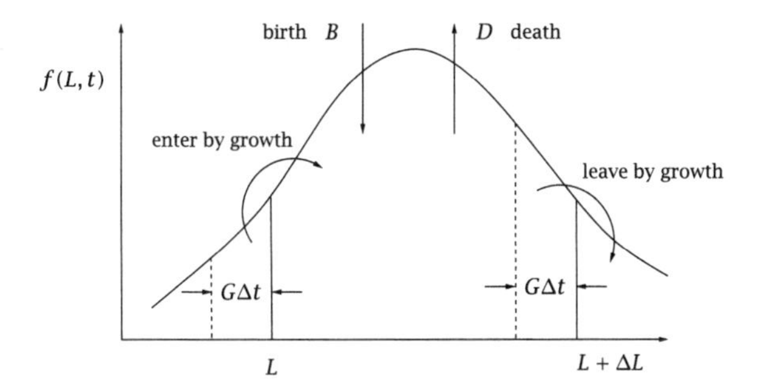

Figure 10.7: Particle size distribution showing particles entering and leaving size class L to $L + \Delta L$ due to particle growth, birth, and death.

required pairs of internal coordinates. Finally, we treat stochastic population models and develop the connections between the deterministic population balance and the stochastic population models.

10.3.1 Single Size Coordinate

Consider a population of particles well described by a single size coordinate. A collection of similarly-shaped particles, such as spheres, cylinders, or platelets, which differ by only the characteristic length, would be an example of when this description suffices. For illustrative purposes, all of the reactors considered in this chapter are assumed to be well stirred, so we do not need to consider the spatial location of the particles in the list of coordinates.[2]

As depicted in Figure 10.7, we wish to write a balance for the number of particles, $f(L,t)dL$, in size class L to $L+\Delta L$ for small ΔL. We consider the change in the particle number in this size class from time t to time $t+\Delta t$ due to particle growth and any mechanisms that create or destroy particles. In words, the balance for the particles is

$$\left\{\begin{matrix}\text{change}\\ \text{in particle}\\ \text{number}\end{matrix}\right\} = \left\{\begin{matrix}\text{particles}\\ \text{that enter}\\ \text{by growth}\end{matrix}\right\} - \left\{\begin{matrix}\text{particles}\\ \text{that leave}\\ \text{by growth}\end{matrix}\right\}$$
$$+ \left\{\begin{matrix}\text{particles}\\ \text{that are born}\end{matrix}\right\} - \left\{\begin{matrix}\text{particles}\\ \text{that die}\end{matrix}\right\} \qquad (10.1)$$

[2] It is not difficult to add spatial location to the coordinate list when considering spatially distributed (or nonideally mixed) particulate reactors.

Substituting in the defined particle size distribution $f(L,t)$ gives

$$\int_L^{L+\Delta L} f(L,t+\Delta t)dL - \int_L^{L+\Delta L} f(L,t)dL = \int_{L-G(L,t)\Delta t}^{L} f(L,t)dL - \int_{L+\Delta-G(L+\Delta L,t)\Delta t}^{L+\Delta L} f(L,t)dL + \int_L^{L+\Delta L} B(L,t)dL\Delta t - \int_L^{L+\Delta L} D(L,t)dL\Delta t \quad (10.2)$$

in which $G(L,t)$, $B(L,t)$, and $D(L,t)$ are the growth, creation, and destruction rates of particles of size L at time t. Since ΔL is small, we may approximate the integrals with the trapezoid rule giving

$$[f(L,t+\Delta t) - f(L,t)]\,\Delta L = f(L,t)G(L,t)\Delta L\Delta t - f(L+\Delta L,t)G(L+\Delta L,t)\Delta L\Delta t + [B(L,t) - D(L,t)]\,\Delta L\Delta t \quad (10.3)$$

Dividing by ΔL and Δt gives

$$\frac{f(L,t+\Delta t) - f(L,t)}{\Delta t} = -\left[\frac{f(L+\Delta L,t)G(L+\Delta L,t) - f(L,t)G(L,t)}{\Delta L}\right] + B(L,t) - D(L,t) \quad (10.4)$$

and taking the limit as $\Delta L, \Delta t \to 0$ gives finally

$$\frac{\partial f}{\partial t} = -\frac{\partial (fG)}{\partial L} + B - D \quad (10.5)$$

which is our first example of a population balance. This equation is analogous to the continuity equation of fluid mechanics. Exercise 10.1 makes this analogy precise. The population balance holds for all positive sizes and times, $0 \le L < \infty$, $0 \le t < \infty$. Next we consider the necessary boundary conditions.

10.3.2 Boundary Conditions

We see that the population balance, Equation 10.5, is a first-order (hyperbolic) partial differential equation in t and L, so we require a boundary condition (initial condition) at some time and a boundary condition at some L. The initial condition specifies the entire PSD at $t = 0$

$$f(L,t) = f_0(L) \qquad t = 0 \quad 0 < L < \infty \quad (10.6)$$

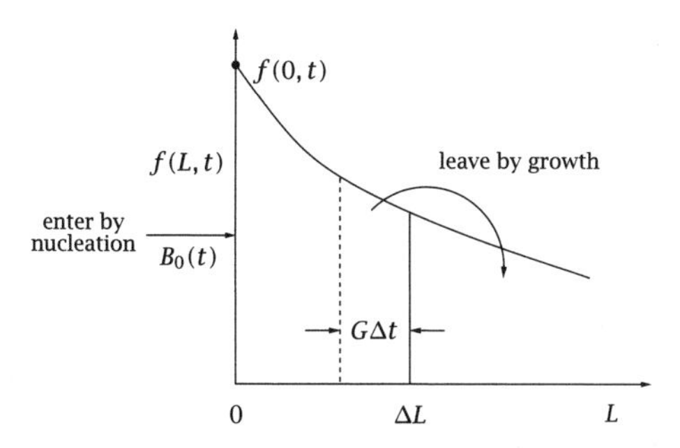

Figure 10.8: Particle size distribution near $L = 0$. Particles are nucleated at $L = 0$ and grow out of the interval 0 to ΔL.

The initial condition simply describes the condition of the reactor when it is started.

One nucleation term at zero size. For the boundary condition, we consider first the case in which particles are nucleated at small size, assumed to be $L = 0$, at some nucleation rate $B(L,t) = B^0(t)\delta(L)$. Consider the PSD in a region near $L = 0$ depicted in Figure 10.8. The change in the number of particles in the interval 0 to ΔL during time t to $t + \Delta t$ is given by

$$\underbrace{\int_0^{\Delta L} f(L, t+\Delta t)dL - \int_0^{\Delta L} f(L,t)dL}_{\text{change in particle number}} = \underbrace{B^0(t)\Delta t}_{\substack{\text{enter} \\ \text{by nucleation}}} - \underbrace{\int_{\Delta L - G\Delta t}^{\Delta L} f(L,t)dL}_{\text{leave by growth}}$$

Because we are taking ΔL small, we can approximate the integrals to obtain

$$\Delta L\,[f(\Delta L, t+\Delta t) - f(\Delta L, t)] = B^0(t)\Delta t - G(\Delta L, t)\Delta t f(\Delta L, t)$$

Dividing both sides by Δt and taking the limit as both ΔL and Δt go to zero gives

$$f(0,t) = \frac{B^0(t)}{G(0,t)}$$

which is the $L = 0$ boundary condition for the population balance. This boundary condition basically tells us that the number of particles at zero size is a competition between the rate at which they are being nucleated and the rate at which they are moving out into larger sizes due to particle growth.

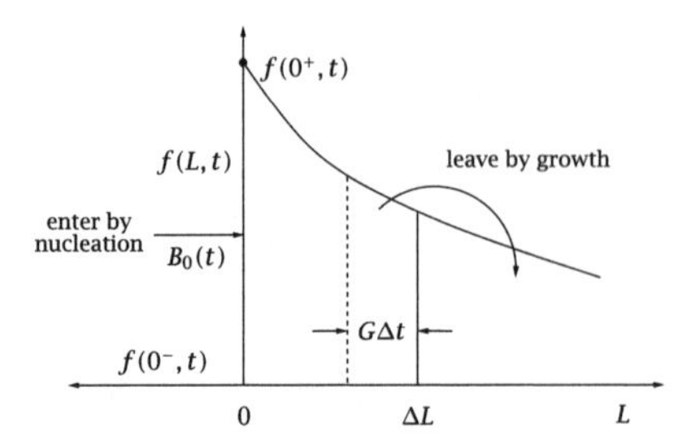

Figure 10.9: Integrating over a small volume element near $L = 0$ to obtain a boundary condition at $L = 0$.

Alternative derivation. When we consider reactors in which particles are created at more than a single size, we find it convenient to use delta functions to describe the particle source terms. We introduce the ideas here in the simplest setting. Consider again particle nucleation at size $L = 0$; we may also consider this event as part of the particle birth term B as follows

$$B(L,t) = B^0(t)\delta(L)$$

This equation implies that particles are nucleated in a vanishingly small size range at $L = 0$ at some rate given by $B^0(t)$. For this assumed nucleation term, the population balance is then

$$\frac{\partial f}{\partial t} = -\frac{\partial (fG)}{\partial L} + B^0(t)\delta(L) - D$$

If we wish to derive the boundary condition at $L = 0$, we integrate this equation across a narrow region containing $L = 0$.

As illustrated in Figure 10.9, we have

$$\int_{0^-}^{0^+} \frac{\partial f}{\partial t} dL = -\int_{0^-}^{0^+} \frac{\partial (fG)}{\partial L} dL + \int_{0^-}^{0^+} B^0(t)\delta(L)dL - \int_{0^-}^{0^+} D dL$$

The first term and the last term go to zero because they are standard functions integrated across a vanishingly small interval. But the second term we integrate by parts and the third term picks up the action of the delta function giving

$$\begin{aligned} 0 &= -\left. fG \right|_{0^-}^{0^+} + B^0(t) \\ 0 &= f(0^-,t)G(0^-,t) - f(0^+,t)G(0^+,t) + B^0(t) \end{aligned}$$

As we see in Figure 10.9, $f(0^-,t) = 0$ and we obtain

$$0 = -f(0^+,t)G(0^+,t) + B^0(t)$$

which we can solve for $f(0,t)$ to obtain

$$f(0,t) = \frac{B^0(t)}{G(0,t)}$$

Notice that this result for the boundary condition agrees with the result obtained previously.

We summarize the population balance, initial condition, and boundary condition for this one-dimensional case with a single nucleation term at zero size

$$\boxed{\begin{aligned} \frac{\partial f}{\partial t} &= -\frac{\partial (fG)}{\partial L} - D \\ f(L,t) &= f_0(L) && L > 0 \quad t = 0 \\ f(L,t) &= \frac{B^0(t)}{G(0,t)} && L = 0 \quad t > 0 \end{aligned}} \tag{10.7}$$

Multiple source terms. The second derivation is ideal for treating systems with several nucleation mechanisms that produce particles at different sizes. The total nucleation rate is then the sum of these several nucleation rates

$$B(L,t) = \sum_i B^i(t)\delta(L - L_i)$$

Substituting this nucleation rate into the population balance and integrating across a small region containing L_i, the size at which particles are nucleated by the ith mechanism gives

$$\int_{L_i^-}^{L_i^+} \frac{\partial f}{\partial t} dL = -\int_{L_i^-}^{L_i^+} \frac{\partial (fG)}{\partial L} dL + \int_{L_i^-}^{L_i^+} \sum_j B^j(t)\delta(L - L_j) dL - \int_{L_i^-}^{L_i^+} D dL$$

Evaluating the integrals as before gives

$$\begin{aligned} 0 &= -fG\big|_{L_i^-}^{L_i^+} + B^i(t) \\ 0 &= f(L_i^-,t)G(L_i^-,t) - f(L_i^+,t)G(L_i^+,t) + B^i(t) \\ 0 &= -\left(f(L_i^+,t) - f(L_i^-,t)\right) G(L_i,t) + B^i(t) \end{aligned}$$

The last equality follows because $G(L,t)$ is continuous at $L = L_i$; only the particle size distribution $f(L,t)$ jumps at $L = L_i$ because of the particle nucleation term. Each nucleation mechanism then creates a jump in the particle size distribution at size L_i where the nucleation takes place. We then have the following jump boundary conditions:

$$f(L_i^+,t) - f(L_i^-,t) = \frac{B^i(t)}{G(L_i,t)} \qquad \text{all } i$$

10.3.3 Multiple Internal Coordinates

We now consider the case in which the particle is characterized by a *vector* of coordinates, $\boldsymbol{x}$. In the previous discussion we had assumed that $\boldsymbol{x}$ was the scalar, L. Now we admit more complex particles that may require further descriptors, such as size plus an aspect ratio, if the particles exhibit a variety of shapes as well as sizes. In the case of biological cells, we may require size or mass plus time since last cell division (cell age) to model the rates of cell growth and division. The many different application areas where population balances are useful produce a wide variety of coordinates that characterize the particles of interest. Our only requirement at this point is that the list of coordinates is finite. The generalized particle size distribution is denoted $f(\boldsymbol{x},t)$. We continue to refer to f as the particle size distribution or particle density, even though the $\boldsymbol{x}$ coordinates may not even contain a particle size coordinate.

Consider an arbitrary volume element in the $\boldsymbol{x}$ coordinate space depicted in Figure 10.10. The element has volume V that may change with time, bounding surface S that changes with velocity $\boldsymbol{v}_s$ with respect to some fixed coordinate system, and outwardly pointing normal vector $\boldsymbol{n}$. We next write a macroscopic particle balance over this volume element.

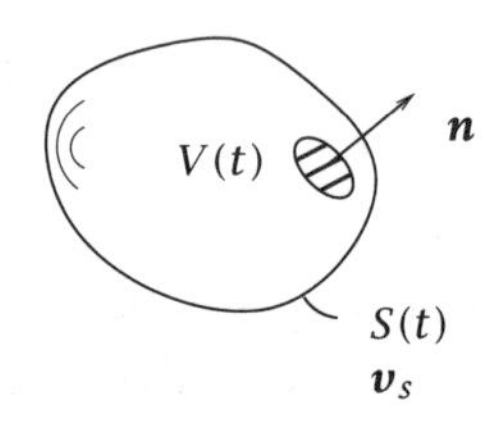

Figure 10.10: Volume element.

Let $\boldsymbol{v}_x$ be the rate of change of the particle's coordinates in some fixed coordinate system. The population balance for the particles is then

$$\frac{d}{dt}\int_{V(t)} f(\boldsymbol{x},t)d\Omega = -\int_{S(t)} f(\boldsymbol{x},t)(\boldsymbol{v}_x - \boldsymbol{v}_s)\cdot \boldsymbol{n}d\sigma + \int_{V(t)} (B-D)d\Omega \qquad (10.8)$$

in which the difference $(\boldsymbol{v}_x - \boldsymbol{v}_s) \cdot \boldsymbol{n}$ accounts for the net rate at which particles enter the volume element due to both change in their coordinates (the $\boldsymbol{v}_x$ term) and motion of V (the $\boldsymbol{v}_s$ term). We require two fundamental results to move from the macroscopic particle balance to the microscopic population balance. These are the Leibniz rule for differentiating an integral, and the divergence theorem for converting a surface integral into a volume integral. We state these results next and then apply them.

The Leibniz rule for differentiating an integral is [2, p.824]

$$\frac{d}{dt}\int_{V(t)} f(\boldsymbol{x},t)d\Omega = \int_{V(t)} \frac{\partial f}{\partial t}d\Omega + \int_{S(t)} f(\boldsymbol{v}_s \cdot \boldsymbol{n})d\sigma \qquad (10.9)$$

in which, as depicted in Figure 10.10, $V(t)$ is an arbitrary closed volume, which may be moving with arbitrary velocity, $S(t)$ is V's bounding surface, and f is an arbitrary function that is continuous in V. The velocity of the bounding surface is $\boldsymbol{v}_s$ and $\boldsymbol{n}$ is the outwardly pointing unit normal to S. The divergence theorem is [2, p.824]

$$\int_V (\nabla \cdot \boldsymbol{v})d\Omega = \int_S (\boldsymbol{v} \cdot \boldsymbol{n})d\sigma \qquad (10.10)$$

in which V is an arbitrary volume element with bounding surface S and $\boldsymbol{v}$ is any vector field that has continuous partial derivatives (with respect to $\boldsymbol{x}$) in V. With these two results one can move easily between microscopic and macroscopic balances.

Applying the Leibniz rule to the first term in Equation 10.8 and the divergence theorem to the second term yields

$$\int_{V(t)} \left[\frac{\partial f(\boldsymbol{x},t)}{\partial t} + \nabla \cdot (f(\boldsymbol{x},t)\boldsymbol{v}_x) - (B - D)\right] d\Omega = 0 \qquad (10.11)$$

In general, the vanishing of an integral does not allow one to conclude that the integrand is zero. In this case, however, we have placed almost no restrictions on the volume element V. If the integrand is continuous in the $\boldsymbol{x}$ coordinates, and it is nonzero at some value of $\boldsymbol{x}$, we can choose V to be a very small element at this same $\boldsymbol{x}$ value and violate Equation 10.11 unless the integrand vanishes everywhere. Therefore, we conclude that the integrand is zero yielding

$$\boxed{\frac{\partial f(\boldsymbol{x},t)}{\partial t} + \nabla \cdot (f(\boldsymbol{x},t)\boldsymbol{v}_x) - (B - D) = 0}$$

which is the population balance for the particles. As before, this population balance requires an initial condition and a boundary condition

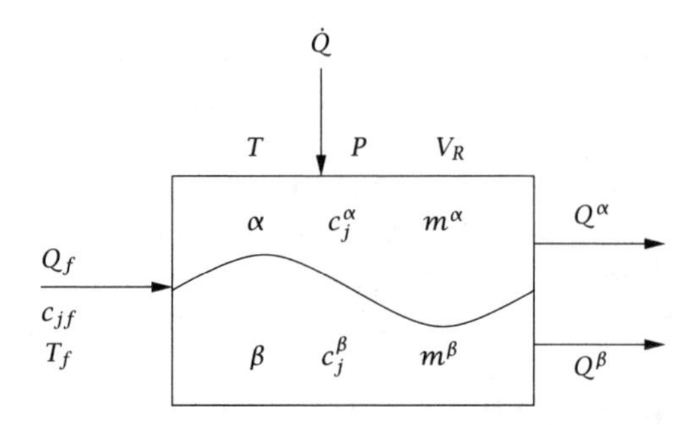

Figure 10.11: Reactor containing two well-mixed phases of matter.

for each component of the $\boldsymbol{x}$ coordinates. For those interested in reading further on the development of population balances, the classic paper [6] is still highly recommended reading.

10.4 Multiphase Mass and Energy Balances

Given the presence of the particle phase, the single-phase mass and energy balances of the previous chapters must be generalized to handle the multiple phases that are present.

Example 10.1: Mass and energy balances with multiple phases

Consider the process depicted in Figure 10.11 in which the reactor contents and the streams entering and leaving the reactor consist of multiple, well-mixed phases. Without loss of generality, we shall consider two phases: α and β. Consider the state of the reactor to be described by the following set of $2(n_s + 2)$ intensive variables and 2 extensive variables

$$T^\alpha, P^\alpha, m^\alpha, c_j^\alpha \quad j = 1, \dots n_s$$
$$T^\beta, P^\beta, m^\beta, c_j^\beta \quad j = 1, \dots n_s$$

in which m^α, m^β are the masses of the α and β phases, respectively.

Let us assume the phases equilibrate with each other even though the reactor is not assumed to be at complete chemical equilibrium. As discussed in Chapter 3, the conditions of phase equilibrium, Equations 3.50, imply that both phases are at the same temperature, T, and both phases are at the same pressure, P, which allows us to describe the state of the reactor with $2n_s + 2$ intensive variables and 2 extensive variables

$$T, P, m^\alpha, m^\beta, c_j^\alpha, c_j^\beta \quad j = 1, \dots n_s$$

Since the process in Figure 10.11 is an open system, we also have the two effluent flowrates and the rate of heat transfer to the system

$$Q^\alpha, Q^\beta, \dot{Q}$$

that need to be determined, bringing the total number of unknown variables to $2n_s + 7$. We assume we have specified the feed conditions, Q_f, c_{jf}, T_f. Let us also assume that the system pressure and volume are known constants and the rate of heat transferred to the system is specified. So we have specified two of our unknown variables

$$P, \dot{Q}$$

which brings us back to $2n_s + 5$ unknowns, and we have one constraint on the total system volume.

(a) Write the mass and energy balances for this multiphase reactor.

(b) Is the process now fully specified? If so, write the $2n_s + 5$ equations that fully determine the system. If underdetermined, what natural additional constraint(s) would you apply to fully specify this system?

Solution

(a) Let us consider the balances that can be written. First we have n_s component balances of the following form

$$\frac{dn_j}{dt} = Q_f c_{jf} - Q^\alpha c_j^\alpha - Q^\beta c_j^\beta + \sum_i \nu_{ij} \left(r_i^\alpha V^\alpha + r_i^\beta V^\beta \right) \quad j = 1, \dots, n_s \tag{10.12}$$

in which r_i^α and r_i^β are the rates of the ith chemical reaction in the α and β phases, respectively, and V^α and V^β are the volumes of the α and β phases, respectively. Note that some of these rates may be zero in one phase but nonzero in the other phase. The rates are functions of the temperature and concentrations. We can evaluate the phase volumes from the other variables by

$$V^\alpha = \frac{m^\alpha}{\rho^\alpha} \qquad V^\beta = \frac{m^\beta}{\rho^\beta}$$

We also state that the total system volume, which is known and fixed, is the sum of the two phase volumes

$$V_R = V^\alpha + V^\beta \tag{10.13}$$

The concentrations and total moles are related by

$$n_j^\alpha = c_j^\alpha V^\alpha \quad n_j^\beta = c_j^\beta V^\beta \qquad j = 1, \ldots n_s$$
$$n_j = n_j^\alpha + n_j^\beta$$

so we have not introduced any new unknowns by using n_j in the material balances. The total mass of each phase is expressible as the weighted sum of the moles of each species

$$m^\alpha = \sum_j n_j^\alpha M_j \qquad m^\beta = \sum_j n_j^\beta M_j \tag{10.14}$$

Because the phases are in equilibrium with each other, we have the following n_s conditions of phase equilibrium

$$\hat{\mu}_j^\alpha = \hat{\mu}_j^\beta \quad j = 1, \ldots, n_s \tag{10.15}$$

We next write a total energy balance. Using the same assumptions that lead to Equation 6.27, we have

$$\frac{dU}{dt} = Q_f \rho_f \hat{H}_f - Q^\alpha \rho^\alpha \hat{H}^\alpha - Q^\beta \rho^\beta \hat{H}^\beta + \dot{Q} \tag{10.16}$$

in which we set the boundary work term to zero because the system has constant volume. The total internal energy is related to the masses and specific volumes of the two phases by

$$U = m^\alpha \hat{U}^\alpha + m^\beta \hat{U}^\beta$$

so we have not introduced a new unknown by using U in the energy balance. Counting up the equations in Equations 10.12, 10.13, 10.14, 10.15 and 10.16, we have written $2n_s + 4$ equations and the system is underspecified by one equation. We require one additional constraint to specify the system.

(b) A natural additional constraint in many reactors is to fix the mass of the condensed phase, phase β, say. If we are heating the reactor, for example, we may measure the height of the condensed liquid phase and use feedback control to determine Q^β that maintains constant m^β, so we do not accidentally run the reactor dry. Notice that we do not require the heat of reaction nor the heat of phase change for the energy balance if we have the thermochemical data to compute the enthalpies of the streams.

□

Example 10.2: Crystallization model

We start with the population balance and continuous phase solute balance required to model isothermal crystallization of a single solute in a CSTR. Assume a simple power law empirical expression for the nucleation and growth rates, and assume that the growth rate is independent of particle size[3]

$$B^0 = k_b(c - c_{\text{sat}})^b \qquad G = k_g(c - c_{\text{sat}})^g$$

Solution

Substituting the growth and nucleation rates into Equation 10.7 gives

$$\begin{aligned}
\frac{\partial f}{\partial t} &= -k_g(c - c_{\text{sat}})^g \left(\frac{\partial f}{\partial L}\right) - \frac{1}{\tau} f \\
f(L,t) &= f_0(L) && L > 0 \quad t = 0 \\
f(L,t) &= \frac{k_b}{k_g}(c - c_{\text{sat}})^{b-g} && L = 0 \quad t > 0
\end{aligned}$$

Given a particle of characteristic size L with volume $v_p L^3$, if the particle experiences growth rate G, i.e., $dL/dt = G$, then the corresponding volume rate of change is $dV/dt = 3v_p L^2 G$. The rate of mass transfer to the particle is then the crystal density ρ_c times this rate of particle volume change, integrated over the population of particles. The solute mass balance then becomes

$$\frac{dc}{dt} = \frac{1}{\tau}(c_f - c) - 3\rho_c v_p k_g (c - c_{\text{sat}})^g \int_0^\infty f(L,t) L^2 dL$$

The coupling of the solute balance and the population balance can give rise to interesting dynamic behavior such as long-period oscillations, which sometimes plague operation of industrial crystallizers [9]. □

[3]The assumption that linear growth rate is independent of particle size is often referred to as McCabe's ΔL law [11]. Many systems are known to violate this assumption, but it remains popular mainly because it simplifies solving the population balance.

Example 10.3: Emulsion polymerization model

Next we write down the population balance and continuous phase balances required to model emulsion polymerization in a CSTR.

Solution

We use the particle volume, V, for the size coordinate and the population balance is

$$\frac{\partial f}{\partial t} = -\frac{\partial (f G_V)}{\partial V} - \frac{1}{\tau} f$$

The polymerization rate inside the particles is given by $r_p = k_p i \phi$ in which i is the number of free radicals in the particle and ϕ is the volume fraction of monomer. For simplicity of exposition, we assume both i and ϕ are constant and independent of particle size. The polymerization rate then determines the particle growth rate as monomer is continually added to maintain ϕ constant. The rate of volume change due to polymerization can be shown to be [15]

$$G_V = \frac{k_p i \phi}{\rho_p (1 - \phi)}$$

in which ρ_p is the density of the polymer. The nucleation rate is determined by the number of micelles and the free radical concentration in the aqueous phase. The total surfactant can be taken as a constant, and the amount of free surfactant available to form micelles is taken as the total surfactant in the reactor less what is required to saturate the aqueous phase and stabilize the particles' surface,

$$m'(t) = (s - s_{wc}) a_{em} / a_m - a_p / a_m \int_0^\infty f(L, t) L^2 dL$$

$$m(t) = m' H(m')$$

The constant a_{em} is the surface area occupied by a single surfactant molecule, and a_m is the total surface area of a micelle. The integral computes the amount of surfactant required to stabilize the particles. The Heaviside function is used to set the number of micelles to zero when the aqueous phase surfactant concentration drops below the critical micelle concentration, s_{wc}. For simplicity we assume the free radical concentration, R, is constant giving

$$B^0(t) = k_{mm} R m(t)$$

in which k_{mm} is the mass transfer coefficient for the radical entry into micelles. Finally, the monomer balance is given by

$$\frac{dM}{dt} = \frac{1}{\tau}(M_f - M) - v_p \int_0^\infty k_p i\phi f(L,t) L^3 dL$$

which accounts for the flow streams and the consumption of monomer due to the polymerization taking place inside the particles.

Here we see already the fairly complicated set of equations required to track the particle composition and size distribution, as well as the continuous phase species. Again, the interactions between the micelle balance and the population balance lead to complex dynamic behavior such as sustained oscillations. See [15] for details on how to relax the many simplifying assumptions made here. □

Example 10.4: Fermentation model

Write down a population balance for the cells and the continuous phase balances for substrate and product to model fermentation in a CSTR.

Solution

In the bioprocess literature, a *segregated* model is one that explicitly models the population of cells, i.e., includes a population balance. A *structured* model is one that requires more than one chemical species to describe the state of the cell. We choose cell mass to be the internal coordinate describing the cell population.[4] The rate of change of a given cell's mass, $\dot{m}$, due to the metabolic reactions is called the cell growth rate, denoted μ. Cell growth rate is usually normalized by the cell mass, so it has units of inverse time, and we have $\dot{m} = \mu m$. The population balance is then

$$\frac{\partial f(m,t)}{\partial t} = -\frac{\partial(\mu m f(m,t))}{\partial m} + B - D - \frac{1}{\tau} f \qquad (10.17)$$

in which B accounts for production of new cells, usually by cell division, which produces two new cells of roughly half the mass of the mother cell. The death term D accounts for losses of cells of size m by, for example, cell death and cell division. Note that, for convenience, the outflow term is treated separately from the other death terms. The continuous phase balances consist of the substrate(s), S, that the cells

[4]Cell mass alone may be inadequate to predict cell division; cell age may also be used if that is a better predictor of cell division.

consume for cell growth, and the product(s), P, secreted by the cells as side products of their metabolic processes. The cell growth yield, y, is the ratio of cell mass increase to substrate mass consumed. In the segregated models, the cell growth rate and growth yield may be functions of any of the continuous phase concentrations and the cell's mass. A typical balance would be [17]

$$\frac{dS}{dt} = \frac{1}{\tau}(S_f - S) - \int_0^\infty f(m,t)\frac{\mu(S,m)m}{y(S,m)}dm$$

The formation rate of some products is often well correlated with the cell growth rate. These are the so-called growth-associated products such as enzymes and proteins. Many secondary metabolites, such as antibiotics, are nongrowth-associated products, and they form at a relatively constant rate, even if the cell growth rate is zero. A typical product formation rate expression accounting for both forms is

$$q_p(S,m) = \alpha\mu(S,m) + \beta$$

and the product balance would be

$$\frac{dP}{dt} = -\frac{1}{\tau}P + \int_0^\infty f(m,t)q_p(S,m)m\,dm$$

□

10.5 Nonsegregated Fermentation Model

Given the complexity of determining B and D, and the metabolic functions μ, y, and q_p for cell lines and products of interest, fermentor models are often simplified further to make them more tractable. The first common simplification is to ignore the distribution of cells and lump all cells together in a single species, biomass. Although this simplification does violence to the known biology, we shall see that these models still provide insight and useful predictions of aspects of fermentor behavior. Given our previous starting point, we can define total biomass as

$$X := \int_0^\infty f(m,t)m\,dm$$

If we assume that μ is only a weak function of cell mass (otherwise this model simplification is not accurate), we can integrate the population balance as follows. Multiply the population balance, Equation 10.17, by m and integrate over all cell mass. The left-hand side becomes

$$\int_0^\infty \frac{\partial f(m,t)}{\partial t} m \, dm = \frac{d}{dt} \int_0^\infty f(m,t) m \, dm = \frac{dX}{dt}$$

Using integration by parts, the first term on the right-hand side of the population balance becomes

$$-\int_0^\infty \frac{\partial(\mu m f(m,t))}{\partial m} m \, dm = - \left. m^2 \mu f \right|_0^\infty + \int_0^\infty f(m,t) \mu m \, dm$$

The integrand vanishes at the two limits, and μ can be taken outside the integral giving

$$-\int_0^\infty \frac{\partial(\mu m f(m,t))}{\partial m} m \, dm = \mu X$$

The cell division terms cancel in the integration over B and D because cell mass is conserved on cell division. If cell death is negligible, then these terms disappear completely. The final result, neglecting cell death, is that the population balance reduces to the following total biomass balance

$$\frac{dX}{dt} = \mu(S) X - \frac{1}{\tau} X$$

The substrate and product balances can be simplified if we assume that the yield and product formation rate do not vary appreciably over the cell population. Taking these terms outside the integrals gives

$$\begin{aligned} \frac{dS}{dt} &= \frac{1}{\tau}(S_f - S) - \frac{\mu(S)}{y(S)} X \\ \frac{dP}{dt} &= -\frac{1}{\tau} P + q_p(S) X \end{aligned}$$

We next discuss the functional form of the cell growth rate and its dependence on the substrate.

Substrate limited growth. Many different cell growth expressions have been found useful [1, 17]

$$\text{Monod equation:} \qquad \mu = \frac{\mu_m S}{K_s + S}$$

$$\text{Blackman equation:} \qquad \mu = \begin{cases} \mu = \mu_m & S \geq 2K_s \\ \mu = \frac{\mu_m S}{2K_s} & S < 2K_s \end{cases}$$

$$\text{Tessier equation:} \qquad \mu = \mu_m(1 - e^{-K_s S})$$

$$\text{Moser equation:} \qquad \mu = \frac{\mu_m S^n}{K_s + S^n}$$

$$\text{Contois equation:} \qquad \mu = \frac{\mu_m S}{K_{sx} X + S}$$

We recognize the Monod equation [12] for cell growth rate as the simplest form of the Langmuir adsorption isotherm and the resulting Hougen-Watson kinetics for reaction rates on catalyst surfaces discussed in Chapter 5. When multiple substrates, $S_1, S_2, \ldots$, affect cell growth, a simple model for overall growth rate is to take the smallest rate as the limiting growth rate

$$\mu = \min_j \mu(S_j)$$

Growth inhibitors. At high substrate or product concentrations, cell growth is inhibited. Common functional forms for this inhibition are the following:

$$\text{substrate inhibition:} \qquad \mu = \frac{\mu_m S}{K_s + S + K_1 S^2}$$

$$\text{product inhibition:} \qquad \mu = \frac{\mu_m S}{K_s\left(1 + \frac{P}{K_p}\right) + S}$$

Reactor behavior. Assuming Monod kinetics for cell growth, the combined biomass and substrate mass balances are

$$\begin{aligned} \frac{dX}{dt} &= \left(-D + \frac{\mu_m S}{K_s + S}\right)X \\ \frac{dS}{dt} &= D(S_f - S) - \frac{1}{y}\left(\frac{\mu_m S}{K_s + S}\right)X \end{aligned} \tag{10.18}$$

in which we have used dilution rate, $D = 1/\tau$, instead of residence time for the outflow terms.[5] Since there is no product inhibition in

[5]There should be no confusion with the D in the population balance death term in this section.

this model, the product balance is not required to solve the biomass and substrate balances. We first analyze the steady state of this model. Setting the time derivatives to zero, we notice first from the biomass balance that $X_s = 0$ is a steady state, and substituting this result into the substrate balance gives $S_s = S_f$. We can find a second steady state by setting the bracketed term to zero in the biomass equation and solving for S_s, which gives $S_s = DK_s/(\mu_m - D)$. Substituting this result into the substrate balance and solving yields $X_s = y(S_f - S_s) = y(S_f - DK_s/(\mu_m - D))$. So we see that there are two steady states for all values of parameters:

$$X_{s1} = 0 \qquad S_{s1} = S_f$$
$$X_{s2} = y(S_f - S_{s2}) \qquad S_{s2} = \frac{DK_s}{\mu_m - D} \tag{10.19}$$

Consider the dilution rate to be the parameter of interest, and notice that the second steady state makes physical sense only if $D < D_c$. Otherwise X_s is negative and $S_s > S_f$. We can solve $S_{s2} = S_f$ to find this critical value of dilution rate and obtain

$$D_c = \frac{\mu_m S_f}{K_s + S_f}$$

For high dilution rate, $D > D_c$, we have only one physically meaningful steady state in which $X_s = 0$ and $S_s = S_f$. In this parameter regime, the dilution rate is too large for the system to support any biomass and any initial biomass simply washes out of the reactor. This steady state is known as the "washout" steady state. For low dilution rate $D < D_c$, there are two possible steady states, the washout steady state, and a steady state with positive biomass production and substrate consumption. Here we have another classic case of steady-state multiplicity as studied in Chapter 6. Exercise 10.7 asks you to show that the washout steady state is stable for $D > D_c$ and unstable for $D < D_c$. The second, nontrivial steady state has the opposite stability; it is unstable for $D > D_c$ and stable for $D < D_c$. These results are shown in Figure 10.12 for a range of dilution rates. The other parameter values used to prepare the figure are

$$\mu_m = 1 \qquad K_s = 1 \qquad S_f = 5 \qquad y = 1$$

Finally, increasing dilution rate further, we notice that there is a singularity in S_{s2} at $D = \mu_m$, and the substrate steady state changes sign and

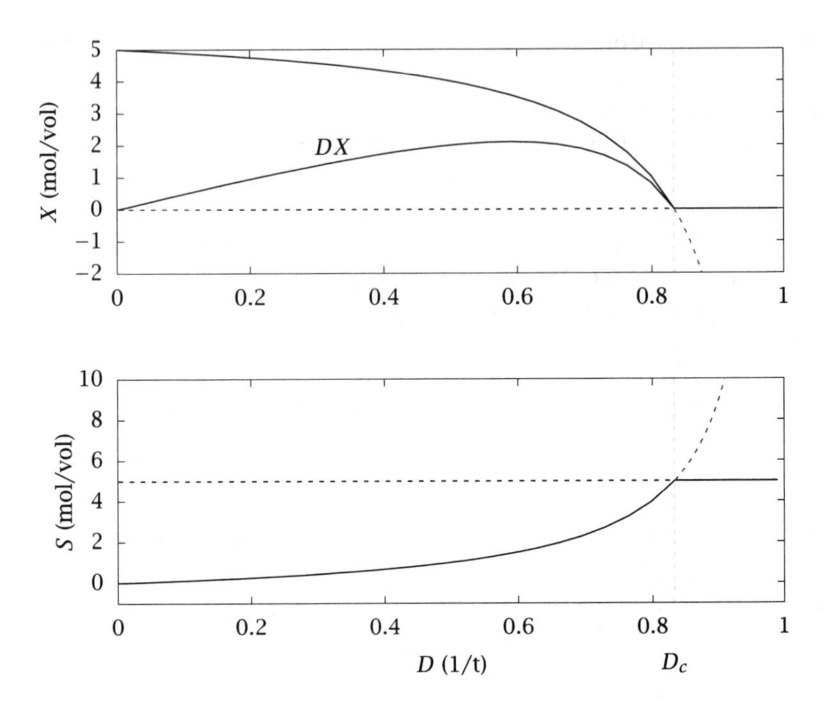

Figure 10.12: The two steady-state biomass and substrate concentrations versus dilution rate; stable (solid), unstable (dashed). Stability changes at $D = D_c$. Also shown is total biomass production rate, DX, for the stable steady state.

becomes negative and the biomass becomes positive. Although this steady state is then also stable, it is not physically meaningful because of the negative substrate concentration (see also Exercise 10.7).

Notice that the steady-state production rate of biomass is given by the product DX_s, which has units of mass per volume of reactor per time. This quantity is also plotted in Figure 10.7. Notice that it has an optimum, which can be found by differentiating DX_{s2} with respect to D and setting to zero. The result is

$$D^0 = \mu_m \left(1 - \sqrt{\frac{K_s}{K_s + S_f}} \right) \tag{10.20}$$

A low dilution rate gives a high reactor biomass *concentration* but little biomass *outflow* from the reactor.[6] Operating at a high dilution rate leads to washout and zero biomass production. An optimum naturally exists between these operational extremes.

10.6 Stochastic Models of Nucleation and Growth

Turning attention back to the general topic of modeling particulate reactors, we can also consider the phenomena of particle nucleation and growth entirely from a stochastic perspective. As we saw in the discussion of stochastic kinetics in Chapter 4, the stochastic perspective provides valuable understanding of certain experimental observations, such as dispersion (spreading) in the particle size distribution with time. We also can derive the population balance of the previous sections starting with the stochastic equations and taking the limit of large numbers and small sizes of the solute molecules compared to the small numbers and large sizes of the growing crystals.

10.6.1 Modeling Particle Growth and Dissolution

We start with a simple experiment to make the discussion concrete. Imagine we have a single, pure-solid particle of some initial size in a well-stirred supersaturated solution of solute molecules with some initial supersaturation. This experiment is easy to conceptualize and also easy to perform in the laboratory. We assume that the particle's crystal structure and geometric shape are not important variables needed to describe the growth rate of the particle. This assumption is valid for certain kinds of particles. Because we have a single particle experiencing only growth, the particle size distribution is arbitrarily narrow at this single size. The supersaturation in the solution phase is the driving force for particle growth. If the particle were to grow large enough that it removes enough solute from the solution phase, then the solution phase approaches saturation and the driving force for further growth drops to zero. The equilibrium state for this simple system is a single particle of a size larger than the initial size, coexisting with a saturated solution phase. Similarly, if the solution phase is initially undersaturated, then the particle dissolves, releasing solute back to the solution phase until either the particle dissolves completely, or the solution phase reaches saturation.

[6]Recall that one sells the mass leaving the reactor, not the reactor concentration.

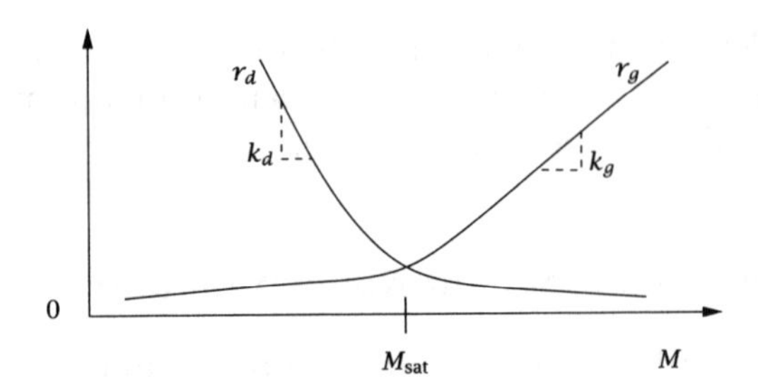

Figure 10.13: Typical growth rate and dissolution rate as a function of solute concentration in the solution phase.

We now build a simple stochastic kinetic model corresponding to this growth/dissolution picture. Let P_n denote a particle with integer n denoting the particle size coordinate, and L_n denote the size of the particle in the nth size class. The number of particles in size class n is denoted f_n. We let M denote the integer number of solute or monomer molecules in the solution phase. We denote the saturation number of solute molecules by M_{sat}. Particle growth corresponds to the addition of a solute molecule to the particle, which is modeled by the reaction

$$\mathrm{P_n} + \mathrm{M} \xrightarrow{k_g} \mathrm{P_{n+1}}$$

Similarly, particle dissolution is the transfer of a solute molecule from the solid phase to the solution phase, which corresponds to the reaction

$$\mathrm{P_n} \xrightarrow{k_d} \mathrm{P_{n-1}} + \mathrm{M}$$

We expect the rates of these events to look like the curves depicted in Figure 10.13. The simplest model we can use that is consistent with this picture is the following

$$r_{gn} = \begin{cases} k_{gn} f_n (M - M_{\text{sat}}) & M > M_{\text{sat}} \\ 0 & M \leq M_{\text{sat}} \end{cases}$$

$$r_{dn} = \begin{cases} 0 & M \geq M_{\text{sat}} \\ k_{dn} f_n (M_{\text{sat}} - M) & M < M_{\text{sat}} \end{cases}$$

in which r_{gn} is the probability (rate) of growth of particle P_n and r_{dn} is the probability (rate) of dissolution of particle P_n. If we further assume

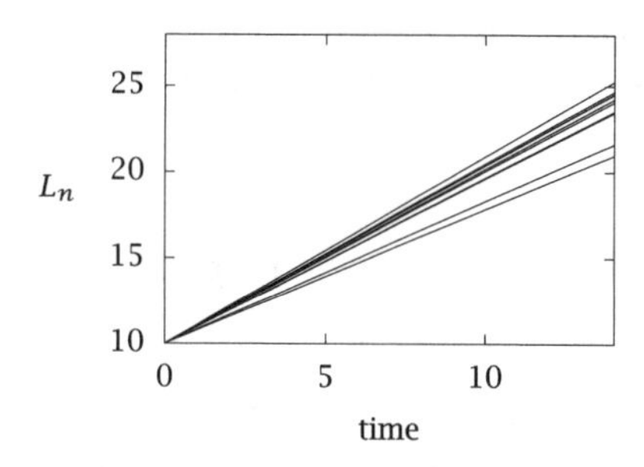

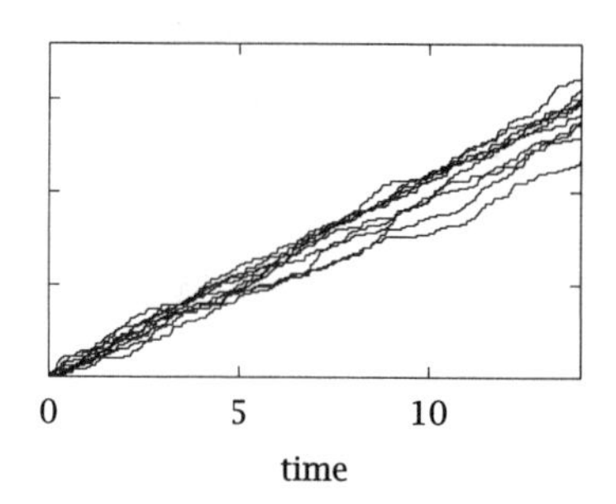

Figure 10.14: Tracking 10 particles undergoing different growth-rate dispersion mechanisms; intrinsic growth-rate dispersion (left), and growth-dependent dispersion (right).

that the rate constants do not depend on the size of the growing or dissolving particle, we have

$$r_g = k_g f_n (M - M_{\text{sat}})$$
$$r_d = k_d f_n (M_{\text{sat}} - M)$$

and we drop the reminder that the rate of growth (dissolution) of a particle P_n is zero if the solution is undersaturated (supersaturated).

Models of Growth-Rate Dispersion. Growth rate dispersion refers to the general experimental observation that different particles experience different growth rates. The dispersion in the growth rates is normally observed by watching the spread in the particle size distribution increase with time. There are many possible mechanisms explaining growth rate dispersion. For example, particles may be nucleated with different intrinsic growth characteristics, which they maintain during the entire growth period. Alternatively, all nucleated particles may be identical, but the growth process is itself an inherently random process. In Figure 10.14 we simulate the behavior of 10 particles undergoing these two different growth mechanisms. Notice that the total spread in particle sizes is about equal in the two mechanisms, even though the mechanisms causing the spread in size are quite different. Under the intrinsic growth-rate dispersion mechanism, the ten particles grow deterministically, but with different rates. Under the growth-dependent dispersion mechanism, the 10 particles undergo independent random walks, and the dispersion is caused by their random motions. It can even be a challenge to infer from experimental measurements of particle size which mechanism is the better model of the observed dispersion in particle size.

10.7 Stochastic Simulation and Deterministic Population Balances

Because the population balance describes a particle size "distribution," there always seems to be some initial confusion about the probability "distribution" of stochastic models when used to describe a system with a population balance. Hopefully we can clear up this confusion, especially since we start with a discrete set of size classes. Given a stochastic model of particle growth, each particle size class has its own inherent probability distribution, $p(f_n, t)$, the probability that we have f_n particles of size L_n at time t. Monte Carlo methods provide samples from these distributions, and the samples can be averaged to yield the mean value. Tabulating the mean values yields the mean of the stochastic model.

That the deterministic population balance is equal to the mean of the stochastic model seems to be widely believed, but it is rarely true. There are only two general situations in which the mean of the stochastic model is equal to the deterministic model:

1. When all rate processes are linear in the species numbers.

2. In the limit as the numbers of molecules/particles goes to infinity.

When the first condition holds, it does not matter if we have a large or a small number of molecules/particles. When the second condition holds, it does not matter what form the rate expressions take. But in this second case, we can say more. It is not just that the mean of the stochastic model converges to the deterministic model. In the limit of large numbers of molecules, the randomness of the stochastic model disappears entirely, and the probability density of the stochastic model becomes arbitrarily sharp at the solution to the deterministic model. This result was established in a classic paper by Kurtz [7]. This second case therefore has a much stronger link between the stochastic and deterministic models. This case is commonly referred to as the "thermodynamic limit," because this convergence was first established between statistical (molecular) thermodynamics and classical (deterministic, macroscopic) thermodynamics. We have already illustrated the large species number thermodynamic limit when introducing stochastic kinetics in Chapter 4. See Figures 4.29–4.31 and the accompanying discussion. So here we would like to illustrate the first case. We shall do this by considering a pure random growth process.

10.7.1 Linear rate processes.

We consider a large bath of monomer, whose concentration remains constant, and a smaller number of particles exhibiting random growth by the reactions

$$\mathrm{P_n} + \mathrm{M} \xrightarrow{k_g} \mathrm{P_{n+1}} \qquad r_n = k_g c_M f_n \qquad n = 1, 2, \ldots, n_T \quad (10.21)$$

We consider the monomer concentration, c_M to be constant, and let f_n denote the number of particles in size class $n = 1, 2, \ldots, n_T$. We consider the reaction (growth) rate constant k_g to be independent of particle size, but this assumption is merely for convenience and is not important to the development. We denote the probability density

$$p(f_1, f_2, \ldots, f_{n_T}, t)$$

as the probability that at time t, the system has f_1 particles in size class 1, f_2 particles in size class 2, and so on up to the largest size class n_T. The deterministic model for this growth process follows immediately from the production rates of Reactions 10.21.

Deterministic.

$$\begin{aligned}
\frac{d}{dt} f_1 &= -\lambda f_1 \\
\frac{d}{dt} f_n &= \lambda f_{n-1} - \lambda f_n \qquad n = 2, \ldots n_T - 1 \\
\frac{d}{dt} f_{n_T} &= \lambda f_{n_T - 1}
\end{aligned}$$

in which we use λ to denote the constant $\lambda = k_g c_M$. For the random model, the equation governing the probability density is

$$\begin{aligned}
\frac{d}{dt} p(f_1, f_2, \ldots, f_{n_T}, t) = \\
& - \lambda (f_1 + f_2 + \cdots f_{n_T - 1})\, p(f_1, f_2, \ldots, f_{n_T}, t) \\
& + \lambda (f_1 + 1)\, p(f_1 + 1, f_2 - 1, \ldots, f_{n_T}, t) \\
& + \lambda (f_2 + 1)\, p(f_1, f_2 + 1, f_3 - 1, \ldots, f_{n_T}, t) \\
& + \cdots + \lambda (f_{n_T - 1} + 1)\, p(f_1, f_2, \ldots, f_{n_T - 1} + 1, f_{n_T} - 1, t) \quad (10.22)
\end{aligned}$$

This balance accounts for loss of probability from state $(f_1, f_2, ..., f_{n_T})$ due to the firing of any reaction. These are the $n_T - 1$ negative terms in the probability balance. The probability of the state increases due to the firing of reaction n when in state $(f_1, \ldots, f_n + 1, f_{n+1} - 1, \ldots, f_{n_T})$. These are the $n_T - 1$ positive terms in the probability balance. To compute the evolution equations for the *average particle numbers* in each size class n, we multiply the probability density by f_n and sum

$$\langle f_n \rangle = \sum_{f_{n_T}=0}^{\infty} \cdots \sum_{f_2=0}^{\infty} \sum_{f_1=0}^{\infty} f_n p(f_1, f_2, \ldots, f_{n_T}, t)$$

Performing this operation on the evolution equation of the probability density gives the following result. See Exercise 10.10 for more details and some hints for performing this step.

Mean Stochastic.

$$\begin{aligned} \frac{d}{dt}\langle f_1 \rangle &= -\lambda \langle f_1 \rangle \\ \frac{d}{dt}\langle f_n \rangle &= \lambda \langle f_{n-1} \rangle - \lambda \langle f_n \rangle \qquad n = 2, \ldots n_T - 1 \\ \frac{d}{dt}\langle f_{n_T} \rangle &= \lambda \langle f_{n_T - 1} \rangle \end{aligned} \tag{10.23}$$

We can see by inspection that these are identical to the deterministic case. Note that this last step hinges entirely on the rate expressions being linear in the species numbers f_n. If these expressions were nonlinear, the equivalence would not hold. As we see here, the equivalence of mean stochastic and deterministic cases at low particle number requires a very special situation to hold for the reaction rates. Exercise 10.11 presents two other representative processes, *particle agglomeration* and *particle breakage*, and asks you to decide whether the equivalence holds for these two processes. Exercise 10.12 explores the situation in which the supersaturation is low so that the monomer species also must be treated as a random variable in the model. Exercise 10.16 shows that the equivalence does not hold also in this case.

Limit as monomer size goes to zero. The size classes in the discrete model reflect the change in particle size when adding one monomer unit by Reaction 10.21. Consider now the case of a *single* particle starting in the first size class in a bath of monomer. For a single particle, the evolution of the probability density that the particle is in class n is given by

$$\frac{d}{dt}p(n,t) = \lambda p(n-1,t) - \lambda p(n,t) \qquad p(n,0) = \delta_{n1} \tag{10.24}$$

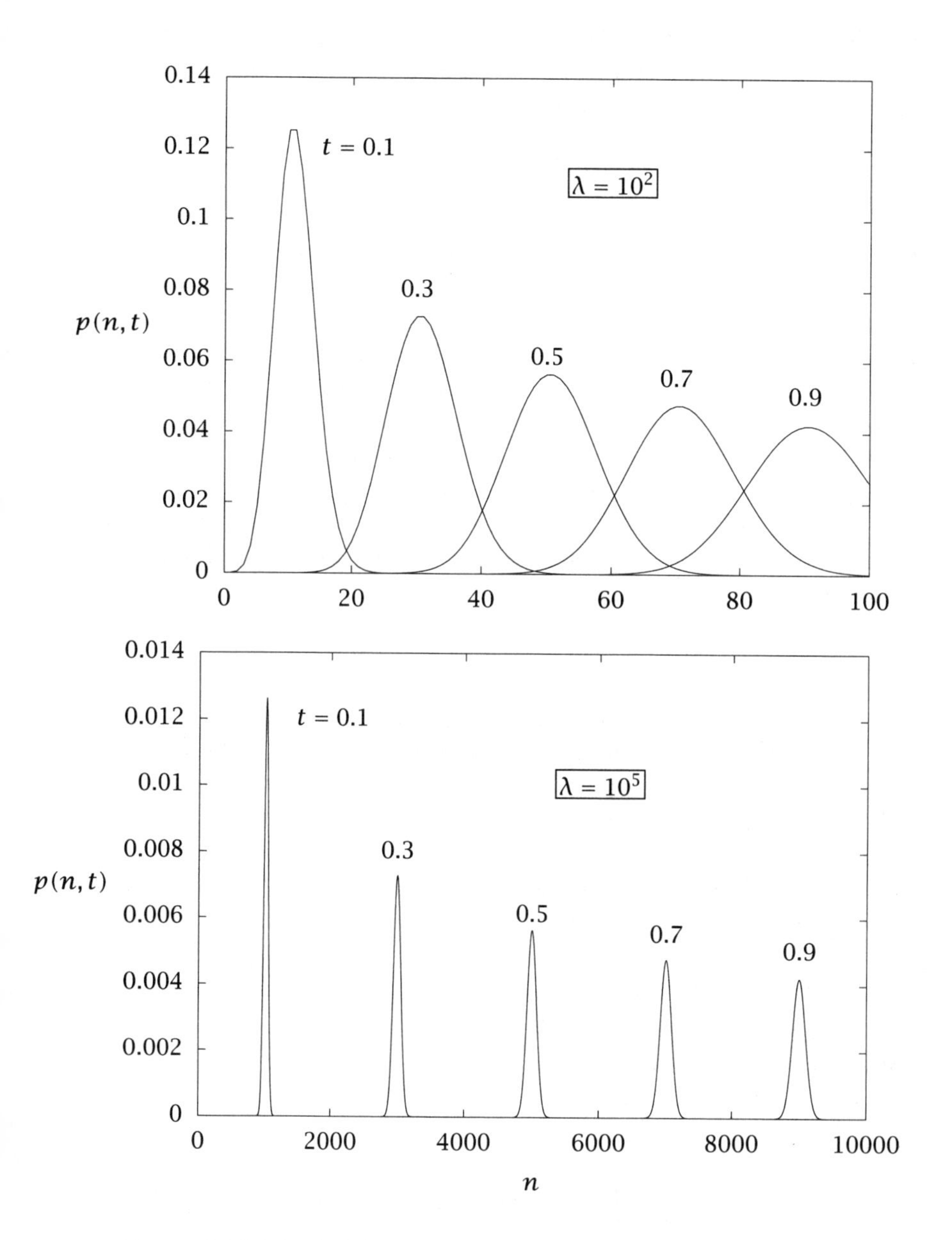

Figure 10.15: Probability density for single particle stochastic growth model at different times. The stochastic model converges to deterministic as $\lambda \to \infty$.

which we see is the same equation as the deterministic model and also the mean of the stochastic model for many particles. We can solve this equation analytically to obtain

$$p(n,t) = \frac{e^{-\lambda t}(\lambda t)^{n-1}}{\Gamma(n)} \qquad n = 1, 2, \ldots, n_T - 1$$

If this equation looks familiar, it may be because we have seen a close relative before. This is also the residence-time distribution of a chain of n CSTRs, given in Equation 8.16 with parameter λ playing the role of n/τ and time t being residence time θ. If we wish to treat the monomer as molecular size, we take the limit of this equation as $\lambda \to \infty$, and also as the number of size classes $n_T \to \infty$. The results are displayed in Figure 10.15. Some growth rate dispersion is apparent, and the probability density widens as time increases. But decreasing the monomer size and increasing the number of size classes sharpens the probability density. In the limit $\lambda \to \infty$, the density will be a delta function moving at constant speed to larger sizes.

10.7.2 Thermodynamic limit.

We conclude the chapter by taking the limit as the number of solute molecules and the number of particles both go to infinity. We show that the discrete stochastic model approaches the continuous population balance in this limit [4]. In this discussion it is convenient to use the volume of the particle, V, as the characteristic size, and we derive the population balance in that particle coordinate system. It is not difficult to convert to the length coordinate system if desired (see Exercise 10.3 for the details).

The discrete stochastic kinetic system converges to the *discrete* deterministic system in the thermodynamic limit, and the kinetic equations are

$$\frac{d}{dt}f_1 = (1/2)k_n M^2 - k_{g1} M f_1 \tag{10.25}$$

$$\frac{d}{dt}f_n = -k_{gn} M f_n + k_{gn-1} M f_{n-1} \quad n = 2, 3, \ldots \tag{10.26}$$

$$\frac{d}{dt}M = -k_n M^2 - \sum_{n=1}^{\infty} k_{gn} M f_n \tag{10.27}$$

We now show that the discrete deterministic equations converge to the continuous population balance. First we equate the numbers of particles in the discrete size classes with the appropriate integrals of the continuous distribution by

$$f_n(t) = \int_{V_n-\Delta/2}^{V_n+\Delta/2} f(V,t)dV \qquad n = 2,3,\ldots$$

$$f_n(t)G_n(t) = \int_{V_n-\Delta/2}^{V_n+\Delta/2} f(V,t)G(V,t)dV \qquad n = 2,3,\ldots$$

in which $V_n = (n+1)\Delta$ and $G_n = k_{gn}M$. Differentiating this equation with respect to time gives

$$\frac{d}{dt}f_n = \int_{V_n-\Delta/2}^{V_n+\Delta/2} \frac{\partial f(V,t)}{\partial t} dV \qquad n = 2,3,\ldots$$

Substituting Equation 10.26 gives

$$\begin{aligned}\int_{V_n-\Delta/2}^{V_n+\Delta/2} \frac{\partial f(V,t)}{\partial t} dV &= -G_n f_n + G_{n-1} f_{n-1} \\ &= -\int_{V_n-\Delta/2}^{V_n+\Delta/2} \left(f(V,t)G(V,t) - f(V-\Delta,t)G(V-\Delta,t)\right) dV\end{aligned}$$

in which we again convert the discrete particle number to an integral over the continuous distribution. Setting the integrand to zero gives

$$\frac{\partial f(V,t)}{\partial t} = -f(V,t)G(V,t) + f(V-\Delta,t)G(V-\Delta,t)$$

Next we use a Taylor series to approximate the change in argument on the right-hand side

$$\begin{aligned}&f(V-\Delta,t)G(V-\Delta,t) = \\ &\qquad f(V,t)G(V,t) - \Delta\frac{\partial(fG)}{\partial V} + \frac{\Delta^2}{2}\frac{\partial^2(fG)}{\partial V^2} + \cdots\end{aligned} \tag{10.28}$$

Since Δ is small, truncating at the first-order term in the Taylor series and substituting into the previous equation gives

$$\frac{\partial f}{\partial t} = -\Delta\frac{\partial(fG)}{\partial V}$$

Finally, we define $G_V(V,t) := \Delta G(V,t) = \Delta k_g(V,t)M(t)$ as the continuous particle growth rate and obtain

$$\boxed{\frac{\partial f}{\partial t} = -\frac{\partial(fG_V)}{\partial V}}$$

Note the correspondence with Equation 10.7. We also obtain a kinetic relationship for the particle growth rate, although we are still free to specify the size (and time) dependence of the rate constant $k_g(V,t)$, so this is not completely specified.

The boundary condition at $V = 0$ can be derived by considering the total particle number. Integrating the population balance over all sizes gives

$$\frac{d}{dt}\int_0^\infty f(V,t)dV = f(0,t)G_V(0,t)$$

Summing Equations 10.25 and 10.26 gives the rate of change of total particle number to be $(1/2)k_n M^2 := B^0$, and we obtain a kinetic relationship for the nucleation rate as well. Equating these two expressions for the change in total particle number and solving for $f(0,t)$ gives the boundary condition

$$\boxed{f(0,t) = \frac{B^0(t)}{G_V(0,t)} = \frac{k_n M(t)}{2\Delta k_g(0,t)}}$$

in agreement with Equation 10.7. Finally we obtain the solute balance by converting Equation 10.27 to the continuous distribution

$$\frac{d}{dt}M = -k_n M^2 - \int_0^\infty G(V,t)f(V,t)dV$$

Normally the amount of monomer consumed in nucleation is negligible compared to that consumed by growth of all the particles, and this result simplifies to

$$\boxed{\frac{d}{dt}M = -\int_0^\infty G(V,t)f(V,t)dV}$$

10.8 Summary

In this chapter, we introduced multiphase, particulate systems consisting of a particulate phase distributed in a continuous solution phase. Several examples were chosen to illustrate such systems including: crystallization from solution, emulsion polymerization, and fermentation using living cells. A suitably general particle "size" distribution was introduced to aid in modeling the particulate phase.

An evolution equation for the particle size distribution, known as the population balance, was derived next. The population balance is the major new modeling concept in the chapter. It is a first-order, hyperbolic partial differential equation. The associated boundary conditions describing nucleation of new particles and initial condition were presented. We then generalized the coordinates describing the state of a particle from a single "size" coordinate, to any finite dimensional vector of coordinates. Next the mass and energy balances were generalized to treat multiphase systems. With these new tools, models describing crystallization, emulsion polymerization, and fermentation were presented.

Because cell processes are so complex, scientists and engineers often wish to simplify the models and lump all the cells together into a single biomass species, and ignore the segregated nature of the cell population. The next section derived these simplified overall chemostat balances. Several popular forms for the overall cell growth rate were presented including the Monod, Blackman, Tessier, Moser, and Contois equations. The simplified chemostat balances proved useful in understanding the phenomena of multiple steady states, balanced growth, and washout.

The chapter concluded with a discussion of *stochastic* models for particle nucleation and growth. The two general situations in which the mean of a stochastic model is equal to the deterministic model were given: (i) when all rate processes are linear, and (ii) in the large number limit. In the large number or thermodynamic limit, the randomness disappears entirely, and the stochastic system becomes arbitrarily narrowly distributed at the deterministic solution. Both of these cases were illustrated and the chapter concluded with a derivation of the continuous deterministic population balance starting from a discrete, stochastic kinetic model of the nucleation and growth reactions.

Notation

a_p	area shape factor
b	nucleation order
B	birth rate of particles
B^0	birth rate at zero particle size
B^i	birth rate for ith nucleation mechanism
c	mass density of solute in solution
c_j	molar concentration of species j
c_M	molar concentration of monomer
c_{sat}	solute saturation concentration

D	death rate of particles
D	dilution rate, $D = 1/\tau$
D_c	critical dilution rate above which washout occurs
$f(L,t)$	particle size distribution
$f(\boldsymbol{x},t)$	generalized particle size distribution
$\langle f \rangle$	mean or expectation of random variable f
f_0	initial condition of particle size distribution
f_n	number of particles in size class n
g	growth order
G	growth rate in particle length coordinate
G_V	growth rate in particle volume coordinate
$\hat{H}$	enthalpy per mass
k_b	nucleation rate constant
k_g	growth kinetic rate constant
K_s	cell growth constant
L	characteristic length of particle
L_n	characteristic length of particle in nth size class
m	mass of phase
M	monomer or solute species
n_j	molar number of species j
n_T	number of size classes
$\boldsymbol{n}$	outwardly pointing unit normal
$p(n,t)$	probability of particle in size class n at time t
P	pressure
P	product concentration
P_n	particle of size class n
Q	volumetric flowrate
$\dot{Q}$	rate of heat transfer
r_i	rate of ith reaction
r_{dn}	dissolution rate for particle of size class n
r_{gn}	growth rate for particle of size class n
S	substrate concentration
$S(t)$	bounding surface of time-varying volume element
t	time
T	temperature
U	internal energy
v_p	volume shape factor
$\boldsymbol{v}_s$	velocity of bounding surface
$\boldsymbol{v}_x$	velocity of particles with $\boldsymbol{x}$ coordinates

V_R	reactor volume
$V(t)$	time-varying volume element
$\boldsymbol{x}$	vector of coordinates specifying state of a particle
X	biomass concentration
Γ	complete gamma function
δ	Dirac delta function
Δ	volume of solute molecule
λ	growth rate parameter, $\lambda = k_g c_M$
μ	overall cell growth rate
μ_j	chemical potential of species j
μ_m	cell growth constant constant
ν_{ij}	stoichiometric number for the jth species in the ith reaction
ρ	mass density
ρ_c	density of solid crystal
τ	residence time

10.9 Exercises

Exercise 10.1: From population balance to continuity equation

Starting with the population balance, Equation 10.5

$$\frac{\partial f}{\partial t} = -\frac{\partial (fG)}{\partial L} + B - D$$

consider an even simpler situation in which spatial location x rather than L is the only required coordinate. Let the mass density $\rho(x,t)$ take the place of $f(L,t)$, the particle density; let the fluid velocity $v = dx/dt$ take the place of the particle growth rate $G = dL/dt$; and set $B = D = 0$ because mass is not created nor destroyed. What does the population balance reduce to in this simple situation?

Exercise 10.2: Solving population balance with method of characteristics

Consider a constant growth rate process in a well-mixed batch reactor, and a population balance with no birth and death mechanisms

$$\frac{\partial f}{\partial t} = -G\frac{\partial f}{\partial L}$$

The boundary conditions are

$$f(L,t) = \begin{cases} f_0(L) & L > 0 \quad t = 0 \\ B^0(t) & L = 0 \quad t > 0 \end{cases}$$

in which we have absorbed the constant G into the defined nucleation rate $B^0(t)$. In the (L,t) plane in which we solve the population balance, lines of constant $L - Gt$ are known as characteristic lines for this hyperbolic partial differential equation. Note that the characteristic lines are straight lines with slope equal to G in the (L,t) plane.

(a) Show that the solution to the population balance is given by

$$f(L,t) = \begin{cases} f_0(L - Gt) & L > Gt \\ B^0(t - L/G) & L < Gt \end{cases}$$

Notice that for $L > Gt$, the solution is determined by the initial condition, and for $L < Gt$, the solution is determined by the nucleation rate. The characteristic line $L - Gt = 0$ separates these two regimes.

(b) Next consider the CSTR with no particles in the feed so the population balance becomes

$$\frac{\partial f}{\partial t} = -G\frac{\partial f}{\partial L} - \frac{1}{\tau}f$$

Given the same boundary conditions as in the last part, what is the solution to this population balance?

Exercise 10.3: Changing particle coordinates

Population balance models can appear quite different in different coordinate systems. Consider changing the characteristic size coordinate from particle length, L, to particle volume, V. The coordinates are related by

$$V = v_p L^3$$

in which v_p is again the particle shape factor for volume. The particle size distributions in these two coordinate systems are related by

$$f(L,t) = f(L(V),t)\frac{\partial V}{\partial L} = \tilde{f}(V,t)\frac{\partial V}{\partial L}$$

The corresponding population balances for the well-stirred batch reactors in these two coordinate systems are

$$\frac{\partial f}{\partial t} = -\frac{\partial (fG)}{\partial L} \qquad \frac{\partial \tilde{f}}{\partial t} = -\frac{\partial (\tilde{f}G_V)}{\partial V}$$

Say we wish to express the rate of change of the total particle volume, $\dot{V}_p$

$$\dot{V}_p := \frac{d}{dt}\int_0^\infty f(L,t)(v_p L^3)dL$$

$$\dot{V}_p := \frac{d}{dt}\int_0^\infty \tilde{f}(V,t)VdV$$

(a) Show that the results for the two coordinate systems are

$$\dot{V}_p = 3v_p\int_0^\infty GL^2 f(L,t)dL \qquad \dot{V}_p = \int_0^\infty G_V\tilde{f}(V,t)dV$$

Hint: differentiate the defining integrals for $\dot{V}_p$ and substitute the corresponding population balances.

Note that we used the first form to write the solute balance in Example 10.2.

(b) Although these two expressions look quite different, show that they are in fact equivalent.
Hint: First differentiate the coordinate relationship to show that the two growth rates are related by $G_V = 3v_pL^2G$

Exercise 10.4: Moving volume elements around

The macroscopic statement of conservation of particles is provided in Equation 10.8,

$$\frac{d}{dt}\int_{V(t)} f(\boldsymbol{x},t)d\Omega = -\int_{S(t)} f(\boldsymbol{x},t)(\boldsymbol{v}_x - \boldsymbol{v}_s)\cdot\boldsymbol{n}d\sigma + \int_{V(t)}(B-D)d\Omega$$

in which $\boldsymbol{v}_x$ is the rate of change of a given particle's coordinate (size, mass, etc.) and $\boldsymbol{v}_s$ is the velocity of the outer boundary of the volume element V. There are two popular choices for volume elements in transport discussions: (i) the stationary volume element, in which case $\boldsymbol{v}_s = 0$, and (ii) the volume element moving with the particle, in which case $\boldsymbol{v}_s = \boldsymbol{v}_x$.

(a) Show that both popular choices of volume element lead to the same Equation 10.11 and therefore the same population balance.

(b) What is not emphasized in most transport discussions is that the motion of the volume element is *completely irrelevant* to the resulting microscopic equation. One is not restricted to the two popular choices of volume element to derive the microscopic continuity equation. Explain why $\boldsymbol{v}_s$ does not appear in Equation 10.11 and hence the population balance regardless of how it is chosen.

Exercise 10.5: Pedagogy and Newton's law of universal gravitation

The following approach to the introduction of cell growth rate is commonplace [17, p. 155]. First the cell growth rate μ is *defined* to be the rate of change of total reactor biomass, X

$$\mu := \frac{1}{X}\frac{dX}{dt} \tag{10.29}$$

in which we use the symbol := to denote not just equality, but equal by definition. Although equation 10.29 is meaningful in the case of a well-stirred bioreactor without any flow streams, its use as a *defining relation* for cell growth rate is problematic. This statement is rather a mass balance for biomass in a well-stirred, constant volume reactor, i.e., reactor biomass changes with time because cells are growing.

The justification for *defining* a reaction rate (cell growth rate) as the time rate of change of a species (biomass) concentration is usually pedagogical. Some argue that new students find this approach more intuitive and easier to grasp. If Isaac Newton were to have thought this way, he would have said,

> Let's see, people understand and can measure mass and position, x. If we watch an apple falling, then we can approximately measure velocity, dx/dt, by calculating differences in position; then we can approximately measure acceleration, d^2x/dt^2, by calculating differences in velocity. So, let's use as the *definition* of gravitational force
>
> $$F_g := m\frac{d^2x}{dt^2}$$
>
> This is pedagogically advantageous because now the beginner can get up and running based on tangible things that he knows are measurable, and he will be off to a great start in understanding gravity and the new laws of motion.

Of course, Newton did not do this. What Newton did instead was to say

> Look, I think the Earth and the apple are pulling on each other with a force. What is the nature of that force? My proposal is that the magnitude of the gravitational attractive force between *any* two point masses is
>
> $$F_g := k\frac{m_1 m_2}{r_{12}^2} \tag{10.30}$$
>
> and the vector points along the line joining the points. The next key step is how to relate an applied force to a change in momentum. My proposal for the conservation law for momentum is
>
> $$\frac{d}{dt}(mv) = F \tag{10.31}$$
>
> There, those two statements should get the beginner up and running and able to solve all kinds of problems involving gravity and the laws of motion.

(a) From the conservation of momentum and the definition of the force of gravitational attraction, find out how long it takes Newton's apple to hit the ground when released from rest at height h above the ground. What further assumptions did you make? Explain physically why the apple's mass is irrelevant in this time it takes to hit the ground.

(b) Which statement do you think contains more hard thinking and is less obvious to a beginner: Equation 10.30 or Equation 10.31? Give your reasons. Also, which one of these statements is more clearly attributable to Newton himself and not his contemporaries.

Exercise 10.6: Optimal dilution rate

Plot the steady-state biomass production rate, DX_s, versus D over the range of dilution rates given in Figure 10.12. Verify that the optimal dilution rate D^0 is given by Equation 10.20.

Increase feed substrate concentration to $S_f = 10$. Describe the effect on the production rate curve and the location of D^0. What happens in the limit of large substrate feed concentration? Do you foresee any difficulty in operating at this optimal steady-state biomass production rate for large S_f?

Exercise 10.7: Stability of the chemostat steady states

Compute the Jacobian matrix of the chemostat model, Equations 10.18, and verify that it is

$$J = \begin{bmatrix} \dfrac{\mu_m S}{K_s + S} - D & X\left(\dfrac{\mu_m K_s}{(K_s + S)^2}\right) \\[2ex] \dfrac{1}{y}\dfrac{\mu_m S}{K_s + S} & -D - \dfrac{X}{y}\left(\dfrac{\mu_m K_s}{(K_s + S)^2}\right) \end{bmatrix}$$

Plot the eigenvalues of the Jacobian evaluated at the two steady states given in Equation 10.19 over the range of dilution rates given in Figure 10.12.

Discuss the stability of the two steady states in Figure 10.12 as a function of dilution rate. Give the range of dilution rates for which the washout steady state is stable. Give the range of dilution rates for which the positive biomass steady state is stable.

Exercise 10.8: Stability for larger dilution rate

Remake Figure 10.12 for the extended range of dilution rates $0 \le D \le 2\mu_m$. We saw in Figure 10.12 that with $D_c < D < \mu_m$, the second steady state has $X_{s2} < 0$ and $S_{s2} > S_f$, and noted that negative biomass concentration is not physically realistic. We also saw in the previous exercise that this steady state is *unstable*.

Now consider what happens in the range $\mu_m < D < 2\mu_m$. Show that both the washout steady state and the other steady state are *stable* for $\mu_m < D < 2\mu_m$. If the second steady state is mathematically stable, why is it not observed in nature? Hint: think about its region of attraction.

Exercise 10.9: Substrate inhibition

Consider the substrate inhibition growth rate model examined in [18]

$$\mu(S) = \frac{\mu_m S}{K_s + S + K_1 S^2}$$

(a) Sketch the growth rate function $\mu(S)$ as a function of substrate concentration S.

(b) First find the steady-state solutions of the chemostat mass balances, Equations 10.18, using this form of the growth rate. Note that to obtain a nonwashout steady state, the biomass balance must satisfy $\mu = D$. Consider varying D from high values to zero. In how many places can the straight line $\mu = D$ intersect with your sketch of the growth rate curve $\mu(S)$ as you vary D?

(c) In the range of D values for which there are *two* nonwashout steady states, which of these two steady states are stable? Compare your results to [18].

Hint: first show that the Jacobian of Equations 10.18 evaluated at the non-washout steady state is

$$J = \begin{bmatrix} 0 & X\frac{d\mu}{dS} \\ -\frac{D}{y} & -D - \frac{X}{y}\frac{d\mu}{dS} \end{bmatrix}$$

Exercise 10.10: Evolution of the stochastic mean for linear growth model

We wish to derive the mean stochastic model for constant monomer concentration, given in Equations 10.23. We require a few preliminaries.

(a) By changing variables in the summations show that for all $n = 1, 2, \ldots, n_T - 1$

$$\sum_{f_{n+1}=0}^{\infty} \sum_{f_n=0}^{\infty} f_n p - (f_n + 1) p(f_1, f_2, \ldots, f_n + 1, f_{n+1} - 1, \ldots f_{n_T}, t) = 0$$

where we condense the notation using $p = p(f_1, f_2, \ldots, f_n, \ldots f_{n_T}, t)$.

(b) Next show that

$$\sum_{f_{n+1}=0}^{\infty} \sum_{f_n=0}^{\infty} f_n^2 p - f_n (f_n + 1) p(f_1, f_2, \ldots, f_n + 1, f_{n+1} - 1, \ldots f_{n_T}, t) = \sum_{f_{n+1}=0}^{\infty} \sum_{f_n=0}^{\infty} f_n p$$

(c) Finally show that

$$\sum_{f_n=0}^{\infty} \sum_{f_{n-1}=0}^{\infty} f_n f_{n-1} p - f_n (f_{n-1} + 1) p(f_1, f_2, \ldots, f_{n-1} + 1, f_n - 1, \ldots f_{n_T}, t) = - \sum_{f_n=0}^{\infty} \sum_{f_{n-1}=0}^{\infty} f_{n-1} p$$

(d) Multiply Equation 10.22 by f_n, sum over all particle numbers, and use the results of the previous three parts to derive Equation 10.23.

Exercise 10.11: Particle agglomeration and breakage

Particle agglomeration between particles of sizes n and m is modeled by the event

$$\mathrm{P_n} + \mathrm{P_m} \xrightarrow{k_{nm}} \mathrm{P_{n+m}} \qquad r_{nm} = k_{nm} f_n f_m$$

Particle breakage is modeled by the event

$$\mathrm{P_n} \xrightarrow{k'_{nm}} \mathrm{P_m} + \mathrm{P_{n-m}} \qquad r_{nm} = k'_{nm} f_n$$

in which a particle of size n breaks into two particles of sizes m and $n - m$. Here we assume that size reflects a quantity, such as particle mass, that is conserved during the agglomeration or breakage process.

For which of these processes is the mean of the stochastic model at low particle number equal to the deterministic population balance? Justify your answer. Compare your answer to that in [13]; see the last paragraph before the conclusions. Compare also to the later version [14, pp. 305,311].

Exercise 10.12: Monomer balance included in random growth model

To treat the case of low supersaturation, include the monomer number, M, in the species list of random variables so the probability density is

$$p(M, f_1, f_2, \ldots, f_{n_T}, t)$$

for the growth reactions

$$\mathrm{M} + \mathrm{P_n} \xrightarrow{k_g} \mathrm{P_{n+1}} \qquad n = 1, 2, \ldots, n_T - 1$$

(a) Develop the analogous probability balance to Equation 10.22 for this case.

(b) Write the deterministic model for this case including the monomer material balance.

(c) At low particle number, is the mean of the stochastic model equal to the deterministic model for this case?

Exercise 10.13: Dispersion in the deterministic population balance

Consider the case in which the monomer addition size, Δ, is not negligible compared to the particle size, V. In this case, we should consider at least the next higher-order term in the Taylor expansion given in Equation 10.28. Add this term to the derivation and show that the population balance becomes

$$\frac{\partial f}{\partial t} = -\frac{\partial (f G_V)}{\partial V} + \frac{\Delta}{2} \frac{\partial^2 (f G_V)}{\partial V^2}$$

Notice the appearance of the diffusion term in the model. The "diffusivity" in this model is proportional to Δ. See also [4] for some illustrative solutions of this model.

Exercise 10.14: Solving the deterministic growth model

Solve the set of differential equations leading to Equation 10.24 listed in the text. Since these equations are linear, you may find the Laplace transform useful for this step. Note that this result for $p(n, t)$ holds up to only $n = n_T - 1$. In the last size class, n_T, there is no reaction to move the particle to a next larger size class. So this last size class is a unique absorbing state and the probability of the particle being in this state must go to unity as time goes to infinity.

(a) Show that for the last size class

$$p(n_T, t) = \frac{\gamma(n_T - 1, \lambda t)}{\Gamma(n_T - 1)}$$

What is the limit of this equation as $t \to \infty$?

(b) Show that the sum over all size classes satisfies the axiom of probability that the particle must be in some size class for all time

$$\sum_{n=1}^{n_T} p(n,t) = 1, \qquad t \geq 0$$

Note that in many applications, the number of size classes is quite large. Consider, for example, how many molecular-sized solute molecules must be incorporated into a macroscopic crystal before we can detect a noticeable size difference under an optical microscope. Because n_T is large, $p(n_T, t)$ given above remains close to zero until long times. For this reason, we often neglect the last size class.

Exercise 10.15: Particle size instead of particle number

Instead of tracking the random variable P_n, the *number* of particles of size n, to model particle growth, consider using the random variable X_i, which is defined to be the *size* of the ith particle, for $i = 1, 2, \ldots, X$.[7] To obtain the particle number in each size class from the size of each particle, we use the following transformation

$$P_n = \sum_{i=1}^{X} \delta(X_i, n)$$

in which

$$\delta(X_i, n) := \begin{cases} 1, & X_i = n \\ 0, & X_i \neq n \end{cases} \tag{10.32}$$

for all $n = 1, 2, \ldots, N_T$.

It is perhaps not too surprising that if we track the sizes of all the particles, we have enough information to evaluate the numbers of particles in all size classes. But why is this alternate approach useful?

(a) It may be simpler to compute probabilities that particles change size. For example, compute the probability that particle i is the one that grows from state $X_i = n$ to $X_i = n + 1$ when the next reaction fires. Note that this probability is independent of the particle, i, the size of the particle, X_i, the size class of interest, n, and the current solute concentration, M.

(b) To compute the expectation or mean of the particle size distribution, show that

$$\mathcal{E}(P_n) = \sum_{i=1}^{X} \Pr(X_i = n), \quad n = 1, 2, \ldots, N_T$$

This result proves useful in the next exercise.

Exercise 10.16: The final nail in the coffin

Although we have shown that the expectation of the stochastic model is equal to the deterministic model when the rate processes are linear in the species numbers, that does not *prove* that equality does *not* hold in all cases of nonlinear rate laws. So we would like to establish the lack of equality for the particle growth model given in Reactions 10.21.

---

[7] The authors would like to thank D. F. Anderson of the UW Department of Mathematics for suggesting this and the next exercise.

(a) Derive the deterministic model and show that

$$\frac{d}{dt}m = -kXm, \quad m(0) = M_0$$
$$\frac{d}{dt}p_1 = -kmp_1, \quad p_1(0) = X$$
$$\frac{d}{dt}p_n = kmp_{n-1} - kmp_n, \quad p_n(0) = 0, \quad n = 2, 3, \ldots, n_T - 1$$

Show that the following holds for the number of particles in the first size class at steady state

$$p_1 = Xe^{-M_0/X}$$

(b) Next we establish that the mean of P_1 of the stochastic model at steady state is not equal to this value. We do this in several steps and take advantage of the X_i random variable description of the previous exercise. First show that the probability that particle i remains in the first size class after all M_0 solute molecules are consumed is

$$\Pr(X_i = 1) = \left(1 - \frac{1}{X}\right)^{M_0}, \quad i = 1, 2, \ldots, X$$

Note that the stochastic system is at steady state after all solute molecules are consumed because no further reactions are possible, i.e., all reaction rates are zero.

(c) Next use the result of the previous exercise to show that at steady state

$$\mathcal{E}(P_1) = X\left(1 - \frac{1}{X}\right)^{M_0}$$

Notice that this result does not agree with the steady state of the deterministic model for any finite values of X and M_0.

(d) Finally show that in the limit $M_0, X \to \infty$ the two results do agree. Equality in this limit is a consequence of the thermodynamic limit discussed in the text. Notice that in taking this limit we keep the ratio M_0/X constant and nonzero to avoid the uninteresting cases in which there are zero or an unbounded number of particles in the first size class at steady state.

Bibliography

[1] J. E. Bailey and D. F. Ollis. *Biochemical Engineering Fundamentals.* McGraw-Hill, New York, second edition, 1986.

[2] R. B. Bird, W. E. Stewart, and E. N. Lightfoot. *Transport Phenomena.* John Wiley & Sons, New York, second edition, 2002.

[3] A. G. Fredrickson, D. Ramkrishna, and H. M. Tsuchiya. Statistics and dynamics of procaryotic cell populations. *Math. Biosci.*, 1:327–374, 1967.

[4] E. L. Haseltine, D. B. Patience, and J. B. Rawlings. On the stochastic simulation of particulate systems. *Chem. Eng. Sci.*, 60(10):2627–2641, 2005.

[5] E. L. Haseltine, J. B. Rawlings, and J. Yin. Dynamics of viral infections: Incorporating both the intracellular and extracellular levels. *Comput. Chem. Eng.*, 29:675–686, 2005.

[6] H. M. Hulburt and S. Katz. Some problems in particle technology: A statistical mechanical formulation. *Chem. Eng. Sci.*, 19:555–574, 1964.

[7] T. G. Kurtz. The relationship between stochastic and deterministic models for chemical reactions. *J. Chem. Phys.*, 57(7):2976–2978, 1972.

[8] Z. B. Kuvadia and M. F. Doherty. Spiral growth model for faceted crystals of non-centrosymmetric organic molecules grown from solution. *Cryst. Growth Des.*, 11:2780–2802, 2011.

[9] P. A. Larsen, D. B. Patience, and J. B. Rawlings. Industrial crystallization process control. *IEEE Ctl. Sys. Mag.*, 26(4):70–80, August 2006.

[10] M. A. Lovette, A. R. Browning, D. W. Griffin, J. P. Sizemore, R. C. Snyder, and M. F. Doherty. Crystal shape engineering. *Ind. Eng. Chem. Res.*, 47: 9812–9833, 2008.

[11] W. L. McCabe. Crystal growth in aqueous solutions. *Ind. Eng. Chem.*, 21 (1):30–33, 1929.

[12] J. Monod. The growth of bacterial cultures. *Ann. Rev. Microbiol.*, 3:371–394, 1949.

[13] D. Ramkrishna. Analysis of population balance—IV: The precise connection between Monte Carlo simulation and population balances. *Chem. Eng. Sci.*, 36:1203–1209, 1981.

[14] D. Ramkrishna. *Population Balances.* Academic Press, San Deigo, 2000.

[15] J. B. Rawlings and W. H. Ray. The modelling of batch and continuous emulsion polymerization reactors. Part I: Model formulation and sensitivity to parameters. *Polymer Eng. & Sci.*, 28(5):237–256, March 1988.

[16] J. B. Rawlings, S. M. Miller, and W. R. Witkowski. Model identification and control of solution crystallization processes: a review. *Ind. Eng. Chem. Res.*, 32(7):1275–1296, July 1993.

[17] M. L. Shuler and F. Kargi. *Bioprocess Engineering.* Prentice-Hall, Upper Saddle River, NJ, second edition, 2002.

[18] T. Yano and S. Koga. Dynamic behavior of the chemostat subject to substrate inhibition. *Biotech. Bioeng.*, 11(2):139–153, 1969.

Index

Abbott, M.M., 63
Abraham, T.K., 474
Acetone, 130
  thermal decomposition, 191, 226
Activation energy, 156, 204, 279, 515
Activity, 65
Activity coefficient, 82, 200
Adamson, A.W., 238
Adiabatic operation, 279
Adiabatic temperature rise, 281
Adler, R.J., 451, 455, 458, 465
Adomaitis, R.A., 324
Adsorption, 193, 239, 242, 525, 565
  chemical, 239
  dissociative, 246, 249
  heat of, 240
  physical, 239, 354
  rate of, 241
Affine transformation, 569
Aftalion, F., 325
Alchemists, 59
Alumina, 354
Ammonia synthesis, 325, 345
Applied Materials, 11
Aris, R., 27, 358, 379, 390, 391
Arrhenius, 204, 279, 387
Atkinson, A.C., 532
Attainable region, 473
Auger electron spectroscopy, 509
Avogadro's number, 42, 163

Baccaro, G.P., 314
Bailey, J.E., 314, 595, 596
Bakker, A., 484
Banerjee, S., 11
Bansleben, D.A., 17
Bard, Y., 536
Bartholomew, C.H., 13
Bartle, K.D., 499
Batch reactor, *see* Reactor, 110
Beach, D., 11
Beckenbach, E., 72
Beenackers, A.A.C.M., 490
Beer's law, 498
Bell, A.T., 193, 238, 241, 252, 499
Bell, J.P., 465
Bellamy, L.J., 498
Bellman, R., 72
Benson, S.W., 55
Benzene pyrolysis, 150, 153
Bessel function, 371
BET isotherm, 354
Bezzo, F., 484
Biegler, L.T., 474
Bimolecular, 33
Biot number, 382, 401
Bird, R.B., 80, 145, 274, 277, 360, 379, 425, 442, 488, 587
Bischoff, K.B., 321, 379, 443
Blackman equation, 596
Bodenstein number, 446
Bodenstein, M., 218
Bomben, K.D., 509
Borwanker, J.D., 165
Boudart, M., 238, 246
Boundary-value problem, *see* BVP
Box, G.E.P., 514, 515, 520, 528
Boyce, W.E., 115
Bramblett, T., 11
Branch
  lower, 296
  middle, 296
  upper, 296
Breiland, W., 11
Brenan, K.E., 183, 645
Brenner, H., 490

Brown, P.N., 646
Browning, A.B., 576
Bruemmer, T,M., 548
Brunauer, S., 354
Buffham, B.A., 430, 465, 493
Burri, J.F., 474
Bush, S.F., 314
Butene, 66
Butt, J.B., 359
BVP, 327, 391, 652

Callen, H.B., 63
Campbell, S.L., 183, 645
Caracotsios, M., 536
Carberry, J.J., 506
Carbon monoxide
  adsorption, 244
  oxidation, 239, 250, 393
Carslaw, H.S., 490
Catalysis, 190, 354
Catalyst, 238, 353, 506
  pellet, 351
Catalyst particle
  characteristic length, 371
  cylinder, 370
  diffusion limited, 386
  external mass transfer, 381
  first-order, 362
  general balances, 359
  Hougen-Watson kinetics, 376
  nonisothermal, 388
  nth-order reaction, 373
  reaction limited, 386
  slab, 370
  spherical, 362
  steady-state multiplicity, 391
Catalytic converter, 393, 408
Catalytic process, 354
Cauchy inequality, 72, 106
Cavendish, J.C., 393, 405
Chang, M., 314
Chapman, D.L., 218
Chapman-Enskog relation, 357
Characteristic lines, 612
Chauhan, S.P., 465
Chemical
  kinetics, 189
  equilibrium, 85
  potential, 64, 79, 200
Chemical vapor deposition, *see* CVD
Chemisorption, 239, 507
Chemostat, 580
Chianelli, R., 15
Christofides, P.D., 216
Christophorou, L.G., 502
Cinar, A., 324
Clark, A., 238
Clausius-Clapeyron equation, 101
CMC
  see micelle, 577
Coefficient of expansion, 276
Collinear collision, 197
Coltrin, M.E., 33
Complete segregation, 451
Complex mechanisms, 226
Compressibility factor, 102
Concentration gradient, 356
Confidence interval, 515, 520, 534
Configurational diffusion, 356
Connor, E.F., 17
Conservation of mass, *see* Mass
Conservation of sites, 242
Constant density, 134
Continuous-stirred-tank reactor, *see* CSTR
Contois equation, 596
Conversion, 127, 149
Convolution integral, 432
Corrigan, T.E., 261, 262
Cortright, R.D., 525, 527
Crystallization, 574
CSTR, 2, 5, 6, 127, 156, 284, 435
CSTR equivalence principle, 161, 474
CSTRs in series, 5, 440, 457
Cumene hydroperoxide, 130
CVD, 10, 11, 28, 32
Czarnowski, J., 259

DAE, 23, 135, 172, 182, 213, 645
  index, 183
Danckwerts boundary conditions, 443
Danckwerts, P.V., 443, 451
Dantzig, G.B., 92, 105
Daoutidis, P., 216
Davis, L.E., 509
Davydov, A.A., 499
de Azevedo, E.G., 82
de Pablo, J.J., 63
DeAcetis, J., 393
Deal, M., 11
Decomposition
  thermal, 191
  unimolecular, 191
Delta function, 431, 433
Denbigh, K.G., 63
Denn, M.M., 178, 332
Density
  constant, 134
  nonconstant, 134
Desorb, 193
Deviation variables, 487
Differential equation
  analytical solution, 111
  linear, 115
  nonhomogeneous, 115
  nonlinear, 117
  sensitivities, *see* Sensitivities
Differential-algebraic equation, *see* DAE
Diffusion
  bulk, 356
  coefficient, effective, 353, 358
  combined, 357
  Knudsen, 356
Diffusional flux, 110
Dihora, J.O., 480
Dilution rate, 596
DiPrima, R.C., 115
Dirac delta function, *see* Delta function
Disilane, 11
Dispersed flow, 442
Djega-Mariadassou, G., 238, 246
Doherty, M.F., 576
Donev, A.N., 532
Douglas, J.M., 314
Dryer, F.L., 205
Duke, C.B., 506
Dumesic, J.A., 525, 527

Eaton, J.W., xii, 634
Effective diffusivity, 356
Effectiveness factor, 367, 390, 400
Eigenvalue, 512, 544
Eigenvector, 511, 544
Ekerdt, J.G., 11, 193, 252
Element conservation, 39, 58
Elementary reaction kinetics, 195
Elliott, J.R., 63, 82
Ellison, P., 161, 474
Emmett, P.H., 354
End-point control, 548
Endothermic reaction, 101, 290
Energy balance, 273, 280, 351
  batch reactor, 275
  CSTR, 284
  multiphase, 588
  PFR, 315
  semi-batch reactor, 314
Energy states, 200
Engel, T., 239
Enquist, L.W., 18
Enthalpy, 275, 276
Equation of state, 133
Equilibrium
  approximation, 209
  assumption, 189, 209
  chemical, 63, 85
  composition, 68
  constant, 66, 95
  ideal-gas, 66
  multiple reaction, 88, 89, 92
  reaction, 639
Equivalence principle
  CSTR, 161, 474
Ergun equation, 397
Ergun, S, 397

Error function, 444
Ertl, G., 239
Ethane pyrolysis, 154, 232
Exothermic reaction, 101, 290
Experimental design, 497, 528
Experimental reproducibility, 532
Exponential of a matrix, 183
External transport control, 353
Extinction point, 290

Farrauto, R.J., 13
Fast time scale, 212
Feinberg, M., 104, 105, 161, 473, 474
Feller, W., 492
Fermentation, 580
Field, F.H., 502
Fifield, F.W., 499, 502
First-order
  irreversible reaction, 112
  reversible reaction, 114
Fixed-bed reactor, 351
  pressure drop, 397
Flame ionization detector, 500
Flint, S.J., 18
Floudas, C., 474
Flow controller, 135
Flow streams, 110, 274
Flowrate
  volumetric, 109, 145, 146
Franklin, J.L., 502
Fredrickson, A.G., 580
Friedrich, S.K., 17
Froment, G.F., 321
Fuel cells, 28
Fugacity, 65
  coefficient, 82
Furusawa, T., 177

Gaedtke, H., 192
Gaitonde, N.Y., 314
Gamma function, 440
Gandhi, K., 217
Garver, J.C., 261, 262
Gas chromatography, 499
Gas-phase reactor, 146
Gas-solid systems, 238
Gates, B.C., 353
Gates, S., 11
Gauss, C.F., 510
Gauss-Newton approximation, 535
Germane, 11
Gibbs energy, 64, 66, 200, 639
  difference, 195
  of formation, 75
  temperature dependence, 76
Gibbs, J.W, 104
Gibbs-Duhem relations, 81, 104, 316
Gillespie algorithm, 164
Gillespie, D.T., 164, 165
Glandt, E.D., 264
Glasser, D., 469, 473, 474
Glycine, 575
Godorr, S., 469, 473
Graziani, K.R., 314
Greenlief, C., 11
Griffin, D.W., 576
Griffin, P.B., 11
Grob, R.L., 499
Grossmann, I.E., 474
Growth Inhibition, 596
Grubbs, R.H., 17
Guckenheimer, J., 300

Hagen-Poiseuille flow, 488
Haidari, A.H., 484
Hair, M.L., 499
Hamer, J.W., 548
Hancock, M.D., 314
Harker, A.B., 192
Haseltine, E.L., 580, 606, 617
Hayward, D.O., 238, 246
HDPE, 15
HDS, 13
Heat capacity, 276
  partial molar, 78
Heat of reaction, *see* Reaction
Heat-transfer coefficient, 279
Heaviside function, 438

Heftmann, E., 499
Hegedus, L.L., 393, 405
Helmholtz energy, 200
Henderson, J.I., 17
Henderson, L.S., 276
Henry's law, 81
Henson, T.L., 514
Hepatitis B virus, 18, 541
Herzberg, G., 498
Herzfeld, K.F., 56, 191, 226, 260
Hessian, 524, 535
Heterogeneous, 351
  catalyzed reactions, 351
  reaction, 82
Hicks, J.S., 389
Hildebrandt, D., 469, 473, 474
Hill, C.G., 194
Hill, T.L., 246
Hindmarsh, A.C., 642, 646
Hochgreb, S., 205
Holmes, P., 300
Homogeneous, 351
  equation, 115
Horn, F.J.M., 473
Horwich, A.L., 18
Hougen, O.A., 150
Hougen-Watson kinetics, 252, 376, 405, 424
Huang, M., 18
Hudson, J.L., 314
Hulburt, H.M., 588
Hunter, J.S., 515, 520, 528
Hunter, W.G., 515, 520, 528
Hydrodesulfurization, *see* HDS

Ideal mixing, 2
Ideal mixture, 80
Ideal-gas
  equation of state, 146
  equilibrium, 66
  mixture, 86, 132
Ideal-liquid mixture, 132
Ignition point, 290
Impulse function, 431
Impulse response, 433
Incompressible fluid, 277, 317
Index, *see* DAE
Induction time, 223
Industrial case study, Kodak, 548
Infrared spectroscopy, 498, 499
Initial-value problem, *see* IVP
Intermediate species, 218
Internal combustion engine, 28
Internal energy, *see* Energy
Interstitial region, 351
Intraparticle transport control, 351, 353
Intrinsic parameters, 386
Isobutane, 66
Isoenergetic, 197
Isola, 348
Isothermal compressibility, 278
IVP, 393

Jacobian matrix, 183, 299
Jaeger, J.C., 490
Jensen's inequality, 492
Jerome, N.F., 548
Johnson, S.M., 92, 105
Johnston, H.S., 192

Kargi, F., 594, 595, 614
Katz, S., 588
Kauchali, S., 474
Kealey, D., 499, 502
Kee, R.J., 33
Kenney, C.N., 314
Kinetic control, 351, 353
Kirk, R.S., 261
Kisliuk, P., 246
Kissen, Y.V., 15, 17
Knudsen flow, 356
Koga, S., 615
Kokossis, A.C., 474
Krug, R.M., 18
Kumar, A., 216
Kumar, R., 217
Kunigita, E., 490

Kurtz, T.G., 602
Kuvadia, Z.B., 576
Kyle, B.G., 63

Lafferty, J.M., 502
Laidler, K.J., 154, 194, 195, 232
Lakshmanan, A., 474
Laminar flow, 488
Langer, R., 359
Langmuir adsorption, 241
Langmuir isotherm, 238, 243
Langmuir, I., 443, 446
Langmuir-Hinshelwood kinetics, *see* Hougen-Watson kinetics
Laplace transform, 220
Larsen, P.A., 591
LDPE, 15
Le Chatelier's principle, 74
Leach, B.E., 12, 353
Least squares, 27, 50, 515, 567
  nonlinear, 523
  weighted, 540
Least-squares estimation, 514
  unknown variance, 520
Lee, H.H., 11
Lee, M.L., 499
Leibniz rule, 425
Levenspiel, O., 159, 443, 444, 469
Lichtenhaler, R.N., 82
Lightfoot, E.N., 80, 145, 277, 360, 379, 425, 442, 488, 587
Limit cycles, 303
Lin, C.C., 216
Line broadening, 508
Linear algebra
  fundamental theorem, 58, 104
Linear independence, 27, 35–39
Linear transformation, 569
Liquid chromatography, 549
Lira, C.T., 63, 82
LLDPE, 15
Loeb, D., 18
Lombardo, S.J., 238, 241
Lovette, M.A., 576
Low-index crystal planes, 238
Lu, Q., 11
Lubben, D., 11

Macchietto, S., 484
MacDonald, N.C., 509
Macromixing, 450, 461
Manousiouthakis, V., 474
Marek, M., 314
Margules equation, 82
Marshall, E.M., 484
Mason, W.S., 18
Mass
  basis, 276
  conservation of, 27, 34, 37, 109
  transfer, 79
Mass balance
  multiphase, 588
  steady-state, 393
  total, 132, 134
Mass spectroscopy, 501, 508
Mass-transfer coefficient
  external, 382
Material balance, 110, 111, 150, 225, 279, 351, 588
MATLAB, 47, 634–663
Matranga, K.R., 264
Matrix
  eigenvalue, 512
  eigenvector, 511
  exponential, 183
  positive definite, 105
  positive semidefinite, 105
  rank, 37
  sparse, 33
  stoichiometric, 27, 29, 33, 37, 45, 53
  transpose, 35
Maximum mixedness, 454, 459, 488
McCabe's ΔL law, 591
McCabe, W.L., 591
McCulloch, D.C., 12
Mean free path, 356
Mean residence time, 128, 435

Mechanism, 193
Metal oxides, 354
Methane synthesis, 193, 252
Micelle, 577, 578
  critical micelle concentration, 577
Michelsen, M.L., 653
Microelectronic materials, 32
Micromixing, 450, 462
Microscopic reversibility, 194, 199
Middlebrooks, S.A., 12
Miller, J.A., 33
Miyauchi, T., 177
Modell, M., 82
Molar flowrate, 233
Molecular partition function, 200
Monod equation, 596
Monod, J., 596
Monolayer, 240
Monte Carlo simulation, 539
Moore, J.W., 195, 197, 219
Morse function, 196
Moser equation, 596
Motooka, T., 11
Moulder, J.F., 509
Multiphase mass and energy balances, 588
Multiple steady states, *see* Steady-state multiplicity
Multiple-reaction system, 150
Myers, A.L., 264

Nakamoto, K., 498
Natal-Santiago, M.A., 525, 527
Nauman, E.B., 430, 465, 493
Niemantsverdriet, J.W., 268
Nishimura, H., 177
Noll, W., 104
Nonconstant density, 134
Nonhomogeneous equation, 115
Nonideal mixtures, 82
Nonisothermal pellet, 389
Nonlinear differential equation, 117
nth-order
  irreversible reaction, 120
Null space, 58, 104
Numerical methods, 634–663
Numerical solution, 111

O'Connell, J.P., 82, 337, 343
O'Malley, R.E.J., 216
O-xylene, 321
Occelli, M.L., 15
Octave, xii, 47, 634–663
ODE, 23, 135, 171, 213, 453, 642
  implicit, 645
Ogg, R.G., 258
Oh, S.H., 393, 405
Ohm's law, 500
Olefin polymerization, 15
Ollis, D.F., 595, 596
Optimization, 538
Ordinary differential equation, *see* ODE
Orthogonal collocation, 446, 462, 652
Oscillations, sustained, 303
Otake, T., 490
Ottino, J.M., 460
Overall selectivity, 127
Oxidation
  carbon monoxide, 250

Palmberg, P.W., 509
Pantelides, C.C., 484
Parallel reactions, 124
Parameter estimation, 528, 534, 552
  differential-equation models, 534
  multiple measurements, 540
Partial differential equation, *see* PDE
Partial molar properties, 77
Particle
  shape factor, 574, 613
  agglomeration, 604, 616
  breakage, 604, 616
  growth, 599, 603
  imaging, 576
  nucleation, 599
Particle density, 354
Partition function, 200
Parulekar, S.J., 443

Pathway, 193
Patience, D.B., 591, 606, 617
PDE, 446
  hyperbolic, 582, 612
Pearson, R.G., 195, 197, 219
Pebsworth, L.W., 15, 17
Péclet number, 444, 485
Pedersen, L., 197
Pellet density, 354, 399
Periodic table, 59
Perturbation parameter, 216
Petersen, E.E., 358, 377, 379, 425
Petzold, L.R., 183, 645, 646
PFR, 2, 144, 150, 156, 225, 315, 438
PFR-CSTR arrangement, 449
PFR-CSTR comparisons, 156, 449
Phase equilibrium, 79, 80, 96
Phase rule, 103–104
Phenol, 130
Photochemical smog, 192
Phthalic anhydride, 321
Physisorption, 240, 354
Pilling, M., 218, 223
Plausibility argument, 293, 295
Plug flow, *see* PFR
Plug-flow reactor, *see* PFR
Plummer, J.D., 11
Podkolzin, S.G., 525, 527
Point selectivity, 126
Polanyi, J.C., 197
Polànyi, M., 56
Poling, B.E., 82, 337, 343
Polyethylene
  high-density, *see* HDPE
  linear-low-density, *see* LLDPE
  low-density, *see* LDPE
Polymerization
  dispersion, 579
  emulsion, 578
  free-radical, 230
  gas-phase, 17
  kinetics, 230
Poore, A.B., 288, 348
Pore structure, 354, 358
Pore volume, 354
Porosity, 354, 399
Porter, R.N., 197
Potential-energy surface, 195, 197
Poynting correction factor, 81
Prausnitz, J.M., 82, 337, 343
Pressure drop correlation, 397
Probability, 510
  density, 431
  function, 431
Probability distribution
  chi-squared, 515
  F, 520
  Gaussian, 510
  normal, 510
Production rate, 27, 43, 47, 53, 110
Propylene oxidation, 393
Pujado, P.R., 130
Pulse response, 433
Pyzhev, V., 325

QSSA, 189, 218, 226
QSSA species, 218, 226
Quasi-steady-state assumption, *see* QSSA

Racaniello, V.R., 18
Raff, L.M., 197
Ramkrishna, D., 165, 217, 443, 580, 617
Rank, *see* Matrix
Rase, H.F., 261
Rate
  constant, 203, 362, 386
  expression, 226
  laws, 111
  production, 41–46, 110
  reaction, 41–46, 110
Rate constant
  effective, 228
  first-order, 228
  units, 112, 117, 120, 121
Rate expression
  concave, 463
  convex, 463, 489

Rate-limiting step, 209
Rawlings, J.B., 548, 580, 591–593, 606, 617, 634
Ray, W.H., 288, 314, 348, 480, 592, 593
Razon, L.F., 314
Reactants, 29
Reaction
  bimolecular surface, 250
  chemical, 28
  chemical vapor deposition(CVD), 32
  coordinate, 195, 198
  coordinate diagram, 199, 201
  elementary, 43, 189, 190, 193, 226
  equilibrium, 85
  extent, 42, 64, 151
  first-order, 112, 362, 367
  first-order, irreversible, 112, 139, 367
  first-order, reversible, 114
  gas-phase, 146, 201, 504
  heat of, 277
  heterogeneously catalyzed, 237, 238, 505, 564
  independent, 36
  inhibition, 121
  intermediates, 218, 226
  kinetics, 239, 388
  linear combination, 36
  liquid-phase, 146, 147, 279, 281, 285, 503
  network, 28, 29, 37
  nitric oxide, 34
  nth-order, irreversible, 120, 373, 537, 562
  parallel, 124
  photochemical, 192
  polymerization, 139, 230
  probability, 163–165, 184
  rate, 27, 28, 42, 47, 53, 110, 636
  rate expression, 226
  reversible, 116
  second-order, heterogeneous, 402
  second-order, irreversible, 117
  statements, 190
  surface, 238
  volume change of, 278
  water gas shift, 28, 35, 44, 661
Reaction-rate constant, 537
Reaction-rate expression, 189, 555
Reactions
  in series, 123
  independent, 35–39
  isomerization, 48
  linearly independent, 39, 53
Reactor
  adiabatic, 288
  autothermal, 345
  autothermal plug-flow, 324
  batch, 2, 4, 110, 219, 224, 275, 438, 503
  chemical vapor deposition (CVD), 11
  combinatorial, 7
  constant-mass, 135
  constant-pressure, 277
  constant-volume, 135, 278
  continuous-stirred-tank, *see* CSTR
  differential, 504
  experimental, 502
  fixed-bed, 6, 14, 351, 399
  gas-phase, batch, 279
  imperfectly mixed, 476
  isothermal, 111, 280
  isothermal, fixed-bed, 402
  nonisothermal, 273
  plug-flow, *see* PFR
  segregated, 458
  semi-batch, 4, 131, 314, 549
  single-wafer, 11
  stability, 298, 328
  tubular, 446
  volume element, 109, 110
  well-stirred, 127
Reichelt, M.W., 644
Residence-time distribution, *see* RTD
Reynolds number, 3, 506
Riach, G.E., 509

Rice, F.O., 191, 226, 260
Rooney, W.C., 474
RTD, 431, 437
Russell, N.M., 11
Russell, T.W.F., 178

Saddle point, 199
Saltzman, W.M., 359
Sandler, S.I., 82
Schieber, J.D., 63
Schmitz, R.A., 314
Schork, F.J., 314
Schreiber, I., 314
Schreiber, J.L., 197
Schuit, G.C.A., 268
Schumacher, H.J., 259
Schwarz inequality, 72
Schwarz, A., 195
Second-order
  irreversible reaction, 117, 118
Seeger, C., 18
Segel, L.A., 216
Segregated model, 593
Selectivity
  overall, 127
  point, 126
Semi-batch
  polymerization, 139
  reactor, 131
Sensitivities, 23, 535, 536, 541, 648–651, 661
Seubold, J.F.K., 130
Shah, B.H., 165
Shampine, L.F., 644
Shapiro, N.Z., 105
Shapley, L.S., 105
Sherman, M., 11
Shinnar, R., 434
Shon, Y., 514
Shuler, M.L., 594, 595, 614
Sifniades, S., 130
Silicon-germanium alloys, 10
Silver, B.L., 59
Sims, L.B., 197
Single crystal, 238
Single-phase system, 276
Sinke, G.C., 67, 98, 99
Sinniah, K., 11
Site balance, 249
Sizemore, J.P., 576
Skalka, A.M., 18
Slab, 370, 376
Slemrod, M., 216
Slow time scale, 212
Smith, J.M., 63, 337, 338, 343, 344
Smith, P.M., 18
Snyder, R.C., 576
Sobol, P.E., 509
Solution
  least-squares, 50, 635
  lower-branch, 293
  middle-branch, 293
  unstable steady-state, 293
  upper-branch, 293
Somorjai, G.A., 238, 506
Sørensen, J.P., 536
Spherical pellet, 362, 367
Srinivasan, R., 55
Srivastava, R., 18, 171
Stability analysis, 298
Standard state, 65
Steady-state
  concentration, 116
  multiplicity, 288, 290, 328, 391
  operation, 145
  solution, 288
Stefan-Maxwell relations, 357
Step response, 432
Stephenson, J.L., 506
Stevens, M.P., 17
Stewart, W.E., 80, 145, 277, 360, 379, 425, 442, 462, 488, 506, 514, 536, 587
Sticking coefficient, 241
Stickle, W.F., 509
Stivers, L., 197
Stochastic simulation, 162
Stoichiometric coefficient, 30

Stoichiometric matrix, *see* Matrix
Stoichiometry, 27
overall, 190
Strang, G., 37, 58
Streetman, B.G., 11
Structure model, 593
Stull, D.R., 67, 98, 99
Suda, Y., 11
Summers, J., 18, 171
Surface diffusion, 356
Surface order, 238
Surface-site balance, 241
Surfaces, 237
Symyx Technologies, Inc., 7

Taylor dispersion, 442
Teller, E., 354
Temkin, M., 325
Tessier equation, 596
Tester, J.W., 82
Teymour, F., 314
Thermal conductivity, 500
Thermodynamics, 63
Thermoneutral reaction, 290
Thiele modulus, 363, 367, 390, 400
Thiele, E.W., 363
Thodos, G., 393
Thomas, J.M., 506
Thomas, W.J., 506
Thompson, D.L., 197
Tiao, G.C., 515, 520
Titania, 354
Tolman, R.C., 194
Tomlin, A.S., 218, 223
Tortuosity factor, 359
Total mass balance, 134
Tracer molecule, 431, 439
Transition-state complex, 199, 201
Transition-state theory, *see* TST
Transmission electron microscopy, 508
Transpose, *see* Matrix
Trapnell, B.M.W., 238, 246
Troe, J., 192
TST, 194, 195, 204
Tsuchiya, H.M., 580
Turànyi, T., 218, 223
Turbulent flow, 3, 144

Underhill, L.K., 218
Unimolecular, 33
UOP LLC, 15
Uppal, A., 288, 348

Vacant site, 240
Vacant surface sites, 239
van Heerden diagrams, 298
van Heerden, C., 293, 325, 328
Van Ness, H.C., 63
van Reijen, L.L., 268
van Santen, R.A., 268
Van Swaaij, W.P.M., 490
Van Welsenaere, R.J., 321
van 't Hoff equation, 78
Vanadia, 354
Vaughan, W.E., 130
Vibration, 202
Vibrational partition function, 202
Villa, C.M., 480
Villadsen, J., 462, 653
Virus model, 541
Void fraction, 354, 399
Volume balance, 134
Volume change of reaction, *see* Reaction
Volumetric flowrate, 145, 146

Wachs, I.E., 506, 508
Washout, 597
Water gas shift reaction, 28, 661
Watson, K.M., 150
Weast, R.C., 177
Weber, R.E., 509
Weinstein, H., 451, 455, 458
Weisz, P.B., 354, 389
Weisz-Hicks problem, 389
Weller, S.W., 446
Westerberg, A.W., 474
Westerterp, K.R., 490

Weston, R.E., 195
Westrum, E.F., Jr., 67, 98, 99
White, W.B., 92, 105
Wilson, S.D., 474
Wojciechowski, B.W., 154, 232
Work, 274

X-ray diffraction, 508
X-ray photoelectron spectroscopy, 509

Yang, F.J., 499
Yano, T., 615
Yield, 127
Yin, J., 18, 171, 580
You, L., 18, 171
Younkin, T.R., 17
Yu, M., 18

Zeolite, 506
Ziegler catalyst, 17
Zwietering, T.N., 451, 454, 457, 488

Note: Appendix A can be found at `www.engineering.ucsb.edu/~jbraw/chemreacfun`